STRIPLINE CIRCULATORS • *Joseph Helszajn*

THE STRIPLINE CIRCULATOR: THEORY AND PRACTICE • *Joseph Helszajn*

LOCALIZED WAVES • *Hugo E. Hernández-Figueroa, Michel Zamboni-Rached, and Erasmo Recami (eds.)*

MICROSTRIP FILTERS FOR RF/MICROWAVE APPLICATIONS • *Jia-Sheng Hong and M. J. Lancaster*

MICROWAVE APPROACH TO HIGHLY IRREGULAR FIBER OPTICS • *Huang Hung-Chia*

NONLINEAR OPTICAL COMMUNICATION NETWORKS • *Eugenio Iannone, Francesco Matera, Antonio Mecozzi, and Marina Settembre*

FINITE ELEMENT SOFTWARE FOR MICROWAVE ENGINEERING • *Tatsuo Itoh, Giuseppe Pelosi, and Peter P. Silvester (eds.)*

INFRARED TECHNOLOGY: APPLICATIONS TO ELECTROOPTICS, PHOTONIC DEVICES, AND SENSORS • *A. R. Jha*

SUPERCONDUCTOR TECHNOLOGY: APPLICATIONS TO MICROWAVE, ELECTRO-OPTICS, ELECTRICAL MACHINES, AND PROPULSION SYSTEMS • *A. R. Jha*

OPTICAL COMPUTING: AN INTRODUCTION • *M. A. Karim and A. S. S. Awwal*

INTRODUCTION TO ELECTROMAGNETIC AND MICROWAVE ENGINEERING • *Paul R. Karmel, Gabriel D. Colef, and Raymond L. Camisa*

MILLIMETER WAVE OPTICAL DIELECTRIC INTEGRATED GUIDES AND CIRCUITS • *Shiban K. Koul*

MICROWAVE DEVICES, CIRCUITS AND THEIR INTERACTION • *Charles A. Lee and G. Conrad Dalman*

ADVANCES IN MICROSTRIP AND PRINTED ANTENNAS • *Kai-Fong Lee and Wei Chen (eds.)*

SPHEROIDAL WAVE FUNCTIONS IN ELECTROMAGNETIC THEORY • *Le-Wei Li, Xiao-Kang Kang, and Mook-Seng Leong*

ARITHMETIC AND LOGIC IN COMPUTER SYSTEMS • *Mi Lu*

OPTICAL FILTER DESIGN AND ANALYSIS: A SIGNAL PROCESSING APPROACH • *Christi K. Madsen and Jian H. Zhao*

THEORY AND PRACTICE OF INFRARED TECHNOLOGY FOR NONDESTRUCTIVE TESTING • *Xavier P. V. Maldague*

METAMATERIALS WITH NEGATIVE PARAMETERS: THEORY, DESIGN, AND MICROWAVE APPLICATIONS • Ricardo Marqués, Ferran Martín, and Mario Sorolla

OPTOELECTRONIC PACKAGING • *A. R. Mickelson, N. R. Basavanhally, and Y. C. Lee (eds.)*

OPTICAL CHARACTER RECOGNITION • *Shunji Mori, Hirobumi Nishida, and Hiromitsu Yamada*

ANTENNAS FOR RADAR AND COMMUNICATIONS: A POLARIMETRIC APPROACH • *Harold Mott*

INTEGRATED ACTIVE ANTENNAS AND SPATIAL POWER COMBINING • *Julio A. Navarro and Kai Chang*

ANALYSIS METHODS FOR RF, MICROWAVE, AND MILLIMETER-WAVE PLANAR TRANSMISSION LINE STRUCTURES • *Cam Nguyen*

FREQUENCY CONTROL OF SEMICONDUCTOR LASERS • *Motoichi Ohtsu (ed.)*

WAVELETS IN ELECTROMAGNETICS AND DEVICE MODELING • *George W. Pan*

OPTICAL SWITCHING • Georgios Papadimitriou, Chrisoula Papazoglou, and Andreas S. Pomportsis

SOLAR CELLS AND THEIR APPLICATIONS • *Larry D. Partain (ed.)*

ANALYSIS OF MULTICONDUCTOR TRANSMISSION LINES • *Clayton R. Paul*

INTRODUCTION TO ELECTROMAGNETIC COMPATIBILITY, Second Edition • *Clayton R. Paul*

ADAPTIVE OPTICS FOR VISION SCIENCE: PRINCIPLES, PRACTICES, DESIGN AND APPLICATIONS • *Jason Porter, Hope Queener, Julianna Lin, Karen Thorn, and Abdul Awwal (eds.)*

ELECTROMAGNETIC OPTIMIZATION BY GENETIC ALGORITHMS • *Yahya Rahmat-Samii and Eric Michielssen (eds.)*

INTRODUCTION TO HIGH-SPEED ELECTRONICS AND OPTOELECTRONICS • *Leonard M. Riaziat*

NEW FRONTIERS IN MEDICAL DEVICE TECHNOLOGY • *Arye Rosen and Harel Rosen (eds.)*

ELECTROMAGNETIC PROPAGATION IN MULTI-MODE RANDOM MEDIA • *Harrison E. Rowe*

ELECTROMAGNETIC PROPAGATION IN ONE-DIMENSIONAL RANDOM MEDIA • *Harrison E. Rowe*

HISTORY OF WIRELESS • *Tapan K. Sarkar, Robert J. Mailloux, Arthur A. Oliner, Magdalena Salazar-Palma, and Dipak L. Sengupta*

PHYSICS OF MULTIANTENNA SYSTEMS AND BROADBAND PROCESSING • *Tapan K. Sarkar, Magdalena Salazar-Palma, and Eric L. Mokole*

SMART ANTENNAS • *Tapan K. Sarkar, Michael C. Wicks, Magdalena Salazar-Palma, and Robert J. Bonneau*

NONLINEAR OPTICS • *E. G. Sauter*

APPLIED ELECTROMAGNETICS AND ELECTROMAGNETIC COMPATIBILITY • *Dipak L. Sengupta and Valdis V. Liepa*

COPLANAR WAVEGUIDE CIRCUITS, COMPONENTS, AND SYSTEMS • *Rainee N. Simons*

ELECTROMAGNETIC FIELDS IN UNCONVENTIONAL MATERIALS AND STRUCTURES • *Onkar N. Singh and Akhlesh Lakhtakia (eds.)*

ELECTRON BEAMS AND MICROWAVE VACUUM ELECTRONICS • *Shulim E. Tsimring*

FUNDAMENTALS OF GLOBAL POSITIONING SYSTEM RECEIVERS: A SOFTWARE APPROACH, Second Edition • *James Bao-yen Tsui*

RF/MICROWAVE INTERACTION WITH BIOLOGICAL TISSUES • *André Vander Vorst, Arye Rosen, and Youji Kotsuka*

InP-BASED MATERIALS AND DEVICES: PHYSICS AND TECHNOLOGY • *Osamu Wada and Hideki Hasegawa (eds.)*

COMPACT AND BROADBAND MICROSTRIP ANTENNAS • *Kin-Lu Wong*

DESIGN OF NONPLANAR MICROSTRIP ANTENNAS AND TRANSMISSION LINES • *Kin-Lu Wong*

PLANAR ANTENNAS FOR WIRELESS COMMUNICATIONS • *Kin-Lu Wong*

FREQUENCY SELECTIVE SURFACE AND GRID ARRAY • *T. K. Wu (ed.)*

ACTIVE AND QUASI-OPTICAL ARRAYS FOR SOLID-STATE POWER COMBINING • *Robert A. York and Zoya B. Popović (eds.)*

OPTICAL SIGNAL PROCESSING, COMPUTING AND NEURAL NETWORKS • *Francis T. S. Yu and Suganda Jutamulia*

SiGe, GaAs, AND InP HETEROJUNCTION BIPOLAR TRANSISTORS • *Jiann Yuan*

ELECTRODYNAMICS OF SOLIDS AND MICROWAVE SUPERCONDUCTIVITY • *Shu-Ang Zhou*

Physics of Multiantenna Systems and Broadband Processing

Physics of Multiantenna Systems and Broadband Processing

Tapan K. Sarkar
Magdalena Salazar-Palma
Eric L. Mokole

With Contributions from:

Santana Burintramart
Jeffrey T. Carlo
Wonsuk Choi
Arijit De
Debalina Ghosh
Seunghyeon Hwang
Jinhwan Koh
Raúl Fernández Recio
Mary Taylor
Nuri Yilmazer
Yu Zhang

WILEY

A JOHN WILEY & SONS, INC., PUBLICATION

Published by John Wiley & Sons, Inc., Hoboken, New Jersey.
Published simultaneously in Canada.

Library of Congress Cataloging-in-Publication Data:

Sarkar, Tapan (Tapan K.)
Physics of multiantenna systems and broadband processing / Tapan K. Sarkar, Magdalena Salazar-Palma, Eric L. Mokole ; with contributions from Santana Burintramart . . . [et al.].
p. cm. — (Wiley series in microwave and optical engineering)
Includes index.
ISBN 978-0-470-19040-1 (cloth)
1. Antenna arrays—Mathematical models. 2. MIMO systems—Mathematical models. 3. Broadband communication systems—Mathematical models. I. Salazar-Palma, Magdalena. II. Mokole, Eric L. III. Burintramart, Santana. IV. Title.
TK7871.6.S27 2008
621.384'135—dc22 2007050158

Printed in the United States of America.

10 9 8 7 6 5 4 3 2 1

Contents

Preface . . . . **xv**

Acknowledgments . . . . **xxi**

Chapter 1 What Is an Antenna and How Does It Work? 1

1.0 Summary 1
1.1 Historical Overview of Maxwell's Equations 2
1.2 Review of Maxwell-Heaviside-Hertz Equations 4
1.2.1 Faraday's Law 4
1.2.2 Generalized Ampère's Law 7
1.2.3 Generalized Gauss's Law of Electrostatics 8
1.2.4 Generalized Gauss's Law of Magnetostatics 9
1.2.5 Equation of Continuity 10
1.3 Solution of Maxwell's Equations 10
1.4 Radiation and Reception Properties of a Point Source Antenna in Frequency and in Time Domain 15
1.4.1 Radiation of Fields from Point Sources 15
1.4.1.1 Far Field in Frequency Domain of a Point Radiator 16
1.4.1.2 Far Field in Time Domain of a Point Radiator. 17
1.4.2 Reception Properties of a Point Receiver 18
1.5 Radiation and Reception Properties of Finite-Sized Dipole-Like Structures in Frequency and in Time 20
1.5.1 Radiation Fields from Wire-like Structures in the Frequency Domain 20
1.5.2 Radiation Fields from Wire-like Structures in the Time Domain 21
1.5.3 Induced Voltage on a Finite-Sized Receive Wire-like Structure Due to a Transient Incident Field 21
1.6 Conclusion 22
References 23

Chapter 2 Fundamentals of Antenna Theory in the Frequency Domain 25

2.0 Summary 25
2.1 Field Produced by a Hertzian Dipole 25
2.2 Concept of Near and Far Fields 28

2.3 Field Radiated by a Small Circular Loop 30
2.4 Field Produced by a Finite-Sized Dipole 32
2.5 Radiation Field from a Linear Antenna 34
2.6 Near- and Far-Field Properties of Antennas 36
2.6.1 What Is Beamforming Using Antennas 36
2.6.2 Use of Spatial Antenna Diversity 43
2.7 The Mathematics and Physics of an Antenna Array 46
2.8 Propagation Modeling in the Frequency Domain 49
2.9 Conclusion 57
References 57

Chapter 3 Fundamentals of an Antenna in the Time Domain 59

3.0 Summary 59
3.1 Introduction 59
3.2 UWB Input Pulse 61
3.3 Travelling-Wave Antenna 62
3.4 Reciprocity Relation Between Antennas 63
3.5 Antenna Simulations 65
3.6 Loaded Antennas 65
3.6.1 Dipole 65
3.6.2 Bicones 71
3.6.3 TEM Horn 74
3.6.4 Log-Periodic 78
3.6.5 Spiral 80
3.7 Conventional Wideband Antennas 83
3.7.1 Volcano Smoke 83
3.7.2 Diamond Dipole 85
3.7.3 Monofilar Helix 86
3.7.4 Conical Spiral 88
3.7.5 Monoloop 90
3.7.6 Quad-Ridged Circular Horn 91
3.7.7 Bi-Blade with Century Bandwidth 93
3.7.8 Cone-Blade 94
3.7.9 Vivaldi 96
3.7.10 Impulse Radiating Antenna (IRA) 97
3.7.11 Circular Disc Dipole 99
3.7.12 Bow-Tie 100
3.7.13 Planar Slot 101
3.8 Experimental Verification of the Wideband Responses from Antennas 102
3.9 Conclusion 108
References 109

Chapter 4 A Look at the Concept of Channel Capacity from a Maxwellian Viewpoint 113

4.0 Summary 113
4.1 Introduction 114
4.2 History of Entropy and Its Evolution 117
4.3 Different Formulations for the Channel Capacity 118
4.4 Information Content of a Waveform 124
4.5 Numerical Examples Illustrating the Relevance of the Maxwellian Physics in Characterizing the Channel Capacity 130
4.5.1 Matched Versus Unmatched Receiving Dipole Antenna with a Matched Transmitting Antenna Operating in Free Space 131
4.5.2 Use of Directive Versus Nondirective Matched Transmitting Antennas Located at Different Heights above the Earth for a Fixed Matched Receiver Height above Ground 133
4.5.2.1 Transmitting Horn Antenna at a Height of 20 m 135
4.5.2.2 Transmitting Dipole Antenna at a Height of 20 m 136
4.5.2.3 Orienting the Transmitting Horn or the Dipole Antenna Located at a Height of 20 m Towards the Receiving Antenna 137
4.5.2.4 The Transmitting Horn and Dipole Antenna Located at a Height of 2 m above Ground 137
4.5.2.5 Transmitting Horn and Dipole Antenna Located Close to the Ground but Tilted Towards the Sky ... 138
4.5.2.6 Channel Capacity as a Function of the Height of the Transmitting Dipole Antenna from the Earth 139
4.5.2.7 Presence of a Dielectric Wall Interrupting the Direct Line-of-sight Between Transmitting and Receiving Antennas 141
4.5.2.8 Increase in Channel Capacity when Matched Receiving Antenna Is Encapsulated by a Dielectric Box 143
4.6 Conclusion 146
4.7 Appendix: History of Entropy and Its Evolution 148
References 164

Chapter 5 Multiple-Input-Multiple-Output (MIMO) Antenna Systems 167

5.0 Summary ... 167
5.1 Introduction ... 168
5.2 Diversity in Wireless Communications ... 168
5.2.1 Time Diversity ... 169
5.2.2 Frequency Diversity ... 170
5.2.3 Space Diversity ... 170
5.3 Multiantenna Systems ... 172
5.4 Multiple-Input-Multiple-Output (MIMO) Systems ... 173
5.5 Channel Capacity of the MIMO Antenna Systems ... 176
5.6 Channel Known at the Transmitter ... 178
5.6.1 Water-filling Algorithm ... 179
5.7 Channel Unknown at the Transmitter ... 180
5.7.1 Alamouti Scheme ... 180
5.8 Diversity-Multiplexing Tradeoff ... 182
5.9 MIMO Under a Vector Electromagnetic Methodology ... 183
5.9.1 MIMO Versus SISO ... 184
5.10 More Appealing Results for a MIMO system ... 189
5.10.1 Case Study: 1 ... 189
5.10.2 Case Study: 2 ... 190
5.10.3 Case Study: 3 ... 191
5.10.4 Case Study: 4 ... 194
5.10.5 Case Study: 5 ... 197
5.11 Physics of MIMO in a Nutshell ... 199
5.11.1 Line-of-Sight (LOS) MIMO Systems with Parallel Antenna Elements Oriented Along the Broadside Direction. ... 200
5.11.2 Line-of-Sight MIMO Systems with Parallel Antenna Elements Oriented Along the Broadside Direction ... 202
5.11.3 Non-line-of-Sight MIMO Systems with Parallel Antenna Elements Oriented Along the Broadside Direction ... 204
5.12 Conclusion ... 206
References ... 207

Chapter 6 Use of the Output Energy Filter in Multiantenna Systems for Adaptive Estimation 209

6.0 Summary ... 209
6.1 Various Forms of the Optimum Filters ... 210
6.1.1 Matched Filter (Cross-correlation filter) ... 211
6.1.2 A Wiener Filter ... 212
6.1.3 An Output Energy Filter (Minimum Variance Filter) ... 213
6.1.4 Example of the Filters ... 214

6.2 Direct Data Domain Least Squares Approaches to Adaptive Processing Based on a Single Snapshot of Data ... 215
6.2.1 Eigenvalue Method ... 218
6.2.2 Forward Method ... 220
6.2.3 Backward Method ... 221
6.2.4 Forward-Backward Method ... 222
6.2.5 Real Time Implementation of the Adaptive Procedure ... 224
6.3 Direct Data Domain Least Squares Approach to Space-Time Adaptive Processing ... 226
6.3.1 Two-Dimensional Generalized Eigenvalue Processor ... 230
6.3.2 Least Squares Forward Processor ... 232
6.3.3 Least Squares Backward Processor ... 236
6.3.4 Least Squares Forward-Backward Processor ... 237
6.4 Application of the Direct Data Domain Least Squares Techniques to Airborne Radar for Space-Time Adaptive Processing ... 238
6.5 Conclusion ... 246
References ... 247

Chapter 7 Minimum Norm Property for the Sum of the Adaptive Weights in Adaptive or in Space-Time Processing 249

7.0 Summary ... 249
7.1 Introduction ... 250
7.2 Review of the Direct Data Domain Least Squares Approach ... 251
7.3 Review of Space-Time Adaptive Processing Based on the D3LS Method ... 253
7.4 Minimum Norm Property of the Adaptive Weights at the DOA of the SOI for the 1-D Case and at Doppler Frequency and DOA for STAP ... 255
7.5 Numerical Examples ... 258
7.6 Conclusion ... 273
References ... 274

Chapter 8 Using Real Weights in Adaptive and Space-Time Processing 275

8.0 Summary ... 275
8.1 Introduction ... 275
8.2 Formulation of a Direct Data Domain Least Squares Approach Using Real Weights ... 277
8.2.1 Forward Method ... 277
8.2.2 Backward Method ... 281

8.2.3. Forward-Backward Method 282
8.3 Simulation Results for Adaptive Processsing 283
8.4 Formulation of an Amplitude-only Direct Data Domain Least Squares Space-Time Adaptive Processing 289
8.4.1 Forward Method 289
8.4.2 Backward Method 291
8.4.3 Forward-Backward Method 292
8.5 Simulation Results 292
8.6 Conclusion 299
References 300

Chapter 9 Phase-Only Adaptive and Space-Time Processing 303

9.0 Summary 303
9.1 Introduction 303
9.2 Formulation of the Direct Data Domain Least Squares Solution for a Phase-Only Adaptive System 304
9.2.1 Forward Method 304
9.2.2 Backward Method 310
9.2.3 Forward-Backward Method 310
9.3 Simulation Results 311
9.4 Formulation of a Phase-Only Direct Data Domain Least Squares Space-Time Adaptive Processing 318
9.4.1 Forward Method 318
9.4.2 Backward Method 318
9.4.3 Forward-Backward Method 318
9.5 Simulation Results 319
9.6 Conclusion 322
References 322

Chapter 10 Simultaneous Multiple Adaptive Beamforming 323

10.0 Summary 323
10.1 Introduction 323
10.2 Formulation of a Direct Data Domain Approach for Multiple Beamforming 324
10.2.1 Forward Method 324
10.2.2 Backward Method 327
10.2.3 Forward-Backward Method 328
10.3 Simulation Results 328
10.4 Formulation of a Direct Data Domain Least Squares Approach for Multiple Beamforming in Space-Time Adaptive Processing 332
10.4.1 Forward Method 332
10.4.2 Backward Method 336
10.4.3 Forward-Backward Method 337

10.5 Simulation Results ... 338
10.6 Conclusion ... 345
References ... 345

Chapter 11 Performance Comparison Between Statistical-Based and Direct Data Domain Least Squares Space-Time Adaptive Processing Algorithms 347

11.0 Summary ... 347
11.1 Introduction ... 347
11.2 Description of the Various Signals of Interest ... 348
11.2.1 Modeling of the Signal-of-Interest ... 349
11.2.2 Modeling of the Clutter ... 349
11.2.3 Modeling of the Jammer ... 350
11.2.4 Modeling of the Discrete Interferers ... 350
11.3 Statistical-Based STAP Algorithms ... 351
11.3.1 Full-Rank Optimum STAP ... 351
11.3.2 Reduced-Rank STAP (Relative Importance of the Eigenbeam Method) ... 352
11.3.3 Reduced-Rank STAP (Based on the Generalized Sidelobe Canceller) ... 353
11.4 Direct Data Domain Least Squares STAP Algorithms ... 356
11.5 Channel Mismatch ... 356
11.6 Simulation Results ... 357
11.7 Conclusion ... 368
References ... 368

Chapter 12 Approximate Compensation for Mutual Coupling Using the In Situ Antenna Element Patterns 371

12.0 Summary ... 371
12.1 Introduction ... 371
12.2 Formulation of the New Direct Data Domain Least Squares Approach Approximately Compensating for the Effects of Mutual Coupling Using the In Situ Element Patterns ... 373
12.2.1 Forward Method ... 373
12.2.3 Backward Method ... 376
12.2.4 Forward-Backward Method ... 377
12.3 Simulation Results ... 378
12.4 Reason for a Decline in the Performance of the Algorithm When the Intensity of the Jammer Is Increased ... 386
12.5 Conclusion ... 386
References ... 386

Chapter 13 Signal Enhancement Through Polarization Adaptivity on Transmit in a Near-Field MIMO Environment 389

13.0 Summary....389
13.1 Introduction....389
13.2 Signal Enhancement Methodology Through Adaptivity on Transmit....391
13.3 Exploitation of the Polarization Properties in the Proposed Methodology....395
13.4 Numerical Simulations....395
13.4.1 Example 1....396
13.4.2 Example 2....402
13.4.3 Example 3....406
13.5 Conclusion....410
References....411

Chapter 14 Direction of Arrival Estimation by Exploiting Unitary Transform in the Matrix Pencil Method and Its Comparison with ESPRIT 413

14.0 Summary....413
14.1 Introduction....413
14.2 The Unitary Transform....415
14.3 1-D Unitary Matrix Pencil Method Revisited....416
14.4 Summary of the 1-D Unitary Matrix Pencil Method....419
14.5 The 2-D Unitary Matrix Pencil Method....419
14.5.1 Pole Pairing for the 2-D Unitary Matrix Pencil Method....425
14.5.2 Computational Complexity....426
14.5.3 Summary of the 2-D Unitary Matrix Pencil Method....426
14.6 Simulation Results Related to the 2-D Unitary Matrix Pencil Method....427
14.7 The ESPRIT Method....430
14.8 Multiple Snapshot-Based Matrix Pencil Method....432
14.9 Comparison of Accuracy and Efficiency Between ESPRIT and the Matrix Pencil Method....432
14.10 Conclusion....435
References....436

Chapter 15 DOA Estimation Using Electrically Small Matched Dipole Antennas and the Associated Cramer-Rao Bound 439

15.0 Summary....439
15.1 Introduction....440
15.2 DOA Estimation Using a Realistic Antenna Array....441
15.2.1 Transformation Matrix Technique....441

15.3 Cramer-Rao Bound for DOA Estimation 444
15.4 DOA Estimation Using 0.1 λ Long Antennas...................... 445
15.5 DOA Estimation Using Different Antenna Array Configurations 448
15.6 Conclusion 461
References 462

Chapter 16 Non-Conventional Least Squares Optimization for DOA Estimation Using Arbitrary-Shaped Antenna Arrays 463

16.0 Summary.. 463
16.1 Introduction 463
16.2 Signal Modeling.. 464
16.3 DFT-Based DOA Estimation... 465
16.4 Non-conventional Least Squares Optimization 466
16.5 Simulation Results 467
16.5.1 An Array of Linear Uniformly Spaced Dipoles...... 468
16.5.2 An Array of Linear Non-uniformly Spaced Dipoles 470
16.5.3 An Array Consisting of Mixed Antenna Elements 471
16.5.4. An Antenna Array Operating in the Presence of Near-Field Scatterers 472
16.5.5 Sensitivity of the Procedure Due to a Small Change in the Operating Environment 473
16.5.6 Sensitivity of the Procedure Due to a Large Change in the Operating Environment 474
16.5.7 An Array of Monopoles Mounted Underneath an Aircraft 476
16.5.8. A Non-uniformly Spaced Nonplanar Array of Monopoles Mounted Under an Aircraft 477
16.6 Conclusion 479
References 479

Chapter 17 Broadband Direction of Arrival Estimations Using the Matrix Pencil Method 481

17.0 Summary.. 481
17.1 Introduction 481
17.2 Brief Overview of the Matrix Pencil Method 482
17.3 Problem Formulation for Simultaneous Estimation of DOA and the Frequency of the Signal................................ 488
17.4 Cramer-Rao Bound for the Direction of Arrival and Frequency of the Signal 494
17.5 Example Using Isotropic Point Sources 505
17.6 Example Using Realistic Antenna Elements 512

17.7 Conclusion 521
References 521

Chapter 18 ADAPTIVE PROCESSING OF BROADBAND SIGNALS 523

18.0 Summary 523
18.1 Introduction 523
18.2 Formulation of a Direct Data Domain Least Squares Method for Adaptive Processing of Finite Bandwidth Signals Having Different Frequencies 524
18.2.1 Forward Method for Adaptive Processing of Broadband Signals 524
18.2.2 Backward Method 529
18.2.3 Forward-Backward Method 529
18.3 Numerical Simulation Results 530
18.4 Conclusion 535
References 535

Chapter 19 Effect of Random Antenna Position Errors on a Direct Data Domain Least Squares Approach for Space-Time Adaptive Processing 537

19.0 Summary 537
19.1 Introduction 537
19.2 EIRP Degradation of Array Antennas Due to Random Position Errors 540
19.3 Example of EIRP Degradation in Antenna Arrays 544
19.4 Simulation Results 547
19.5 Conclusion 551
References 551

Index 553

Preface

The objective of the book is to present a scientific methodology that can be used to analyze the physics of multiantenna systems. The multiantenna systems are becoming exceedingly popular because they promise a different dimension (spatial diversity) than what is currently available to the communication systems engineers. Simultaneously using multiple transmit and receive antennas provides a means to perform spatial diversity, at least from a theoretical standpoint. In this way, one can increase the capacities of existing systems that already exploit time and frequency diversity. The deployment of multiantenna systems is equivalent to using an overmoded waveguide, where information is simultaneously transmitted via not only the dominant mode but also through all the higher-order modes. We look into this interesting possibility and study why communication engineers advocate the use of such a system, whereas electromagnetic and microwave engineers have avoided such propagation mechanisms in their systems. Most importantly, we study the physical principles of multiantenna systems through Maxwell's equations and utilize them to perform various numerical simulations to observe how a typical system will behave in practice. The first five chapters of this book are devoted to this topic.

Specifically, Chapter 1 describes Maxwell's equations in the frequency and time domains and shows how to solve practical problems in both domains.

Chapter 2 presents the frequency domain properties of antennas, and specifically what is meant by near field and far field of antennas, which are relevant to our discussions as an antenna beam can only be defined in the far field. In particular, an antenna has no nulls in the near field, which is independent of distance, and is only a function of the azimuth and elevation angles. We also study how the presence of a ground plane, namely the earth, modifies our concepts and how it affects the electrical performance of a system.

Chapter 3 describes the properties of antennas in the time domain and illustrates how a broadband antenna should behave. Using the terminology broadband implies a finite width time domain pulse that can be either transmitted or received by an antenna without severe distortion. From this perspective, a spread spectrum system will not be considered broadband, since the instantaneous spectrum of its signals is still small. In dealing with wideband signals, one observes that the impulse response of the antenna in the transmit mode is the time derivative of the impulse response of the antenna in the receive mode. We also look at the impulse response of some of the conventionally used wideband antennas, including a century bandwidth antenna.

Chapter 4 looks at the concept of channel capacity from a Maxwellian viewpoint. The concept of channel capacity is intimately connected with the concept of entropy – hence related to physics. We present two forms of the channel capacity, the usual Shannon capacity which is based on power; and the seldomly used definition of Hartley which uses values of the voltage. These two definitions of capacities are shown to yield numerically very similar values if one is dealing with conjugately matched antennas. However, from an engineering standpoint, the voltage-based form of the channel capacity is more useful as it is related to the sensitivity of the receiver to an incoming electromagnetic wave. Furthermore, we illustrate through numerical simulations how to apply the channel capacity formulas in an electromagnetically proper way. To perform the calculations correctly, first in the simulations, the input power fed to the antennas need to remain constant in a comparison. Second, the expression of power often used by most communication engineers in the channel capacity is related to the radiated power and not to the input power, which is not correct. In a fair comparison, one should deal with the gain of antenna systems and not their directivities, which is an alternate way of referring to the input power fed to the antennas rather than to the radiated power. The problem is, the radiated power essentially deals with the directivity of an antenna and theoretically one can get any value for the directivity of an aperture. Hence, the distinction needs to be made between gain and directivity in a proper way to compare systems. Finally, one needs to use the Poynting's theorem to calculate the power in the near field and not using exclusively either the voltage or the current. This applies to the power form of the Shannon channel capacity theorem. For the voltage form of the capacity due to Hartley is applicable to both near and far fields. Use of realistic antenna models in place of representing antennas by point sources further illustrates the above points, as the point sources by definition generates only far field.

Chapter 5 presents the concept of a multi–input–multi–output (MIMO) antenna system and illustrates the strengths and the weaknesses of this multiantenna deployments in both the transmitters and the receivers. Sample simulations show that only the classical phased array mode out of the various spatial modes that characterize spatial diversity is useful and the other spatial modes are not efficient radiators. Hence, it is more useful to use the concept of adaptive beam forming using a phased array mode.

The next seven chapters address a new phased array methodology for accurate and efficient adaptive processing. In Chapter 6, three classes of optimum filters are presented to illustrate in what sense they are optimal. Of the three classes, one has the promise of performing estimation rather than the usual detection process carried out in conventional adaptive processing. We illustrate that it is possible to perform adaptive processing using a single snapshot of the data, which may be more useful for a highly dynamic environment or in the presence of blinking jammers. A single snapshot based adaptive procedure generates a least squares solution and does not require any statistical description of the signals. In fact, it has been illustrated in the literature and summarized in this book that processing a single snapshot of the data has essentially the same number of degrees of freedom for coherent interferers as a classical multiple-snapshot processing that is based on conventional sample matrix inversion techniques. In addition, this new method is at least an order of magnitude faster in computational speed than the sample matrix inversion techniques when using the same number of degrees of freedom. This new methodology is then extended to space-time adaptive processing, where a single snapshot is applied to a range cell and requires neither secondary data nor a statistical description of clutter. Recently, this methodology was applied to real airborne data and demonstrated to provide a better solution than conventional statistical methods.

In Chapter 7, we show that the minimum of the sum of the absolute value of the weights can be used for further or equivalently secondary processing for improving the estimation of the direction of arrival of the signal of interest in an adaptive processing methodology. In this way, one can further improve the estimates for both the direction of arrival and the Doppler frequency for the signal of interest in a space-time adaptive algorithm. In particular, the minimum value for the norm of the adaptive weights is obtained at the true value for the direction of arrival for adaptive processing or at the true value for direction of arrival and Doppler frequency in space-time adaptive processing (STAP).

Chapter 8 illustrates that the direct-data-domain least-squares (D3LS) adaptive methodology is quite flexible and it can easily be modified to deal with real values of the adaptive weights for both adaptive and space-time adaptive processing. How this adaptive processing approach can be achieved and implemented for phase-only weights is illustrated in Chapter 9. In Chapter 10, the D3LS method is used for simultaneously forming more than one main beam, which makes it possible to track multiple targets in the same adaptive process.

In Chapter 11, a performance comparison is made between four versions of the statistical-based STAP and D3LS STAP algorithms, when the number of training data is varied. The four statistical-based methods are: the full-rank statistical method; the relative importance of the eigenbeam (RIE) method; the principle component generalized sidelobe canceller (GSC) method; and the cross-spectral GSC method. In contrast to the D3LS approach utilizes only a single snapshot of data (space and time corresponding to one range cell only), one needs to know the rank of the interference covariance matrix for multiple-snapshots to make the statistically-based methods work. The D3LS performs better when the number of training data available for the statistical-based methods is less than the rank of the interference covariance matrix. The channel mismatch is also introduced to all methods to evaluate their performance.

Chapter 12 shows the effects of mutual coupling among the antenna elements in the array and illustrates how a nonplanar array with nonuniformly spaced elements can be used for adaptive processing. One method that can be used to compensate for the mutual coupling is using the embedded in-situ element patterns. This simple widely used method, however, breaks down when the intensity of the interferer increases. In those situations, implementing a more accurate compensation technique through the transformation matrix approach is necessary. When the strengths of the interferers are comparable to the signal of interest, using dummy antenna elements at the edges of an array can minimize the effects of mutual coupling.

Chapter 13 illustrates how reciprocity can be used in directing a signal to a preselected receiver when there is a two way communication between a transmitter and the receiver. This embarrassingly simple method is much simpler in computational complexity than a traditional MIMO and can even exploit the polarization properties for effectively decorrelating multiple receivers in a multi–input–single–output (MISO) system.

The next three chapters treat the estimation of the direction of arrival (DOA). Chapter 14 describes the Matrix Pencil method for DOA estimation, as knowledge of the DOA for the signal of interest is often necessary in many problems. A unitary transform is applied to illustrate how this method can be implemented in a real system using real arithmetic. The Matrix Pencil method is a direct data domain approach as opposed to ESPRIT, which uses a correlation matrix of the data. For situations, where few available snapshots of the data are available, we show that the Matrix Pencil method provides a more accurate estimate of the DOA than the ESPRIT method. In Chapter 15, DOA estimation is carried out using electrically small antennas and presents the associated Cramer-Rao bound to illustrate the accuracy of this estimation procedure. It is shown that conjugately matched electrically small antennas can be as effective, if not more effective, than their resonant versions. Chapter 16 presents a nonconventional least squares methodology for DOA estimation using arbitrary shaped nonplanar conformal arrays.

The next two chapters discuss broadband processing of signals operating at different frequencies or those having a finite bandwidth. Chapter 17 presents a broadband DOA estimation algorithm that uses the Matrix Pencil method, with the main objective of finding not only the azimuth and the elevation angles of arrival for the signals of interest but also their operating frequencies. Simulations illustrate how one can use realistic antennas to perform broadband DOA estimation. In Chapter 18, D3LS STAP of Chapter 6 is applied to show how broadband adaptive processing can be performed.

Finally, Chapter 19 analyzes how random position errors in the location of the antenna elements in an array can affect its STAP performance.

To recapitulate, the primary goal of this book is to develop a basic understanding of the physics of multiantenna and the concept of channel capacity by using Maxwell's theory. Since an antenna is a temporal filter as well as a spatial filter, any analysis dealing with antennas needs to merge both their spatial and temporal properties to obtain a physically meaningful solution. These two diverse properties are reflected in Maxwell's equations and throughly understanding these four century old equations, first articulated by Heinrich

Hertz in the scalar form and then by Oliver Heaviside in the vector form that we use nowadays, can address most of the problems dealing with space-time properties of antennas. Because, the classical phased array mode is dominant in multiantenna systems, we show how to do adaptive processing in a least squares fashion in an accurate and efficient way without requiring any statistical information as an a priori description of the signals. Demonstrating that this type of methodology is also amenable to broad band processing is a secondary goal of this book.

Every attempt has been made to guarantee the accuracy of the materials in the book. We would however appreciate readers bringing to our attention any errors that may have appeared in the final version. Errors and /or any comments may be emailed to any of the authors.

Acknowledgments

We gratefully acknowledge Carlos Hartmann (Syracuse University, Syracuse, New York), Michael C. Wicks, Darren M. Haddad, and Gerard J. Genello (Air Force Research Laboratory, Rome, New York), John S. Asvestas and Oliver E. Allen (NAVAIR, Patuxent River, Maryland), Miguel Lagunas (CTTC, Barcelona, Spain) and Steven R. Best (MITRE Corporation, Bedford, Massachusetts) for their continued support in this endeavor. We gratefully acknowledge Dipak L. Sengupta, Robert C. Hansen, and Deb Chatterjee for help and suggestions.

Thanks are also due to Ms. Christine Sauve, Ms. Brenda Flowers, and Ms. Maureen Marano, (Syracuse University) for their expert typing of the manuscript. We would also like to express sincere thanks to Seongman Jang, Mengtao Yuan, Hongsik Moon, LaToya Brown, Ying Huang, Xiaomin Lin and Weixin Zhao for their help with the book.

Tapan K. Sarkar (*tksarkar@syr.edu*)
Magdalena Salazar-Palma (*salazar@tsc.uc3m.es*)
Eric L. Mokole (*eric.mokole@nrl.navy.mil*)
Syracuse, New York
June 2008

1

WHAT IS AN ANTENNA AND HOW DOES IT WORK?

1.0 SUMMARY

An antenna is a structure that is made of material bodies that can be composed of either conducting or dielectric materials or may be a combination of both. Such a structure should be matched to the source of the electro-magnetic energy so that it can radiate or receive the electromagnetic fields in an efficient manner. The interesting phenomenon is that an antenna displays selectivity properties not only in frequency but also in space. In the frequency domain an antenna is capable of displaying a resonance phenomenon where at a particular frequency the current density induced on it can be sufficiently significant to cause radiation of electromagnetic fields from that structure. An antenna also possesses an impulse response that is a function of both the azimuth and elevation angles. Thus, an antenna displays spatial selectivity as it generates a radiation pattern that can selectively transmit or receive electromagnetic energy along certain spatial directions. As a receiver of electromagnetic fields, an antenna also acts as a spatial sampler of the electromagnetic fields propagating through space. The voltage induced in the antenna is related to the polarization and the strength of the incident electromagnetic fields. The objective of this chapter is to illustrate how the impulse response of an antenna can be determined. Another goal is to demonstrate that the impulse response of an antenna when it is transmitting is different from its response when the same structure operates in the receive mode. This is in direct contrast to antenna properties in the frequency domain as the transmit radiation pattern is the same as the receive antenna pattern. An antenna provides the matching necessary between the various electrical components associated with the transmitter and receiver and the free space where the electromagnetic wave is propagating. From a functional perspective an antenna is thus related to a loudspeaker, which matches the acoustic generation/receiving devices to the open space. However, in acoustics, loudspeakers and microphones are bandlimited devices and so their impulse responses are well behaved. On the other hand, an antenna is a high pass device and therefore the transmit and the receive impulse responses are not the same; in fact, the former is the time derivative of the latter. An antenna is like our lips, whose instantaneous change of shapes provides the necessary match between the vocal cord and the outside environment as the frequency of the voice changes. By proper shaping of the antenna structure one can focus the radiated energy along certain specific directions in space. This spatial directivity occurs only at certain specific frequencies, providing selectivity in frequency. The interesting point is that it is difficult to separate these two spatial and temporal properties of the antenna, even though in the literature they are treated separately. The tools that deal with the dual-coupled space-time analysis are *Maxwell's equations*. We first present the background of Maxwell's equations and illustrate how to solve for them analytically. Then we utilize them in the subsequent sections and chapters to illustrate how to obtain the impulse responses of antennas both as transmitting and receiving elements and illustrate their relevance in the saga of smart antennas.

1.1 HISTORICAL OVERVIEW OF MAXWELL'S EQUATIONS

In the year 1864, James Clerk Maxwell (1831–1879) read his "Dynamical Theory of the Electromagnetic Field" [1] at the Royal Society (London). He observed theoretically that electromagnetic disturbance travels in free space with the velocity of light [1–7]. He then conjectured that light is a transverse electromagnetic wave by using dimensional analysis [7]. In his original theory Maxwell introduced 20 equations involving 20 variables. These equations together expressed mathematically virtually all that was known about electricity and magnetism. Through these equations Maxwell essentially summarized the work of Hans C. Oersted (1777–1851), Karl F. Gauss (1777–1855), André M. Ampère (1775–1836), Michael Faraday (1791–1867), and others, and added his own radical concept of displacement *current* to complete the theory.

Maxwell assigned strong physical significance to the magnetic vector and electric scalar potentials $\boldsymbol{A}$ and ψ, respectively (bold variables denote vectors; italic denotes that they are function of both time and space, whereas roman variables are a function of space only), both of which played dominant roles in his formulation. He did not put any emphasis on the sources of these electromagnetic potentials, namely the currents and the charges. He also assumed a hypothetical mechanical medium called *ether* to justify the existence of displacement currents in free space. This assumption produced a strong opposition to Maxwell's theory from many scientists of his time. It is well known that Maxwell's equations, as we know them now, do not contain any potential variables; neither does his electromagnetic theory require any assumption of an artificial medium to sustain his displacement current in free space. The original interpretation given to the displacement current by Maxwell is no longer used; however, we retain the term in honor of Maxwell. Although modern Maxwell's equations appear in modified form, the equations introduced by Maxwell in 1864 formed the foundation of electromagnetic theory, which together is popularly referred to as *Maxwell's electromagnetic theory* [1–7].

Maxwell's original equations were modified and later expressed in the form we now know as Maxwell's equations independently by Heinrich Hertz (1857–1894) and Oliver Heaviside (1850–1925). Their work discarded the requirement of a medium for the existence of displacement current in free space, and they also eliminated the vector and scalar potentials from the fundamental equations. Their derivations were based on the impressed sources, namely the current and the charge. Thus, Hertz and Heaviside, independently, expressed Maxwell's equations involving only the four field vectors $\boldsymbol{E}$, $\boldsymbol{H}$, $\boldsymbol{B}$, and $\boldsymbol{D}$: the electric field intensity, the magnetic field intensity, the magnetic flux density, and the electric flux density or displacement, respectively. Although priority is given to Heaviside for the vector form of Maxwell's equations, it is important to note that Hertz's 1884 paper [2] provided the Cartesian form of Maxwell's equations, which also appeared in his later paper of 1890 [3]. Thus, the coordinate forms of the four equations that we use nowadays were first obtained by Hertz [2,7] in scalar form and then by Heaviside in 1888 in vector form [4,7].

It is appropriate to mention here that the importance of Hertz's theoretical work [2] and its significance appear not to have been fully recognized [5]. In this 1884 paper [2] Hertz started from the older action-at-a-distance theories of electromagnetism and proceeded to obtain Maxwell's equations in an alternative way that avoided the mechanical models that Maxwell used originally and formed the basis for all his future contributions to electromagnetism, both theoretical and experimental. In contrast to the 1884 paper, in his 1890 paper [3] Hertz postulated Maxwell's equations rather than deriving them alternatively. The equations, written in component forms rather than in vector form as done by Heaviside [4], brought unparalleled clarity to Maxwell's theory. The four equations in vector notation containing the four electromagnetic field vectors are now commonly known as Maxwell's equations. However, Einstein referred to them as *Maxwell–Heaviside–Hertz equations* [6,7].

Although the idea of electromagnetic waves was hidden in the set of 20 equations proposed by Maxwell, he had in fact said virtually nothing about electromagnetic waves other than light, nor did he propose any idea to generate such waves electromagnetically. It has been stated [6, Ch. 2, p. 24]: "*There is even some reason to think that he [Maxwell] regarded the electrical production of such waves as impossibility.*" There is no indication left behind by him that he believed such was even possible. Maxwell did not live to see his prediction confirmed experimentally and his electromagnetic theory fully accepted. The former was confirmed by Hertz's brilliant experiments, his theory received universal acceptance, and his original equations in a modified form became the language of electromagnetic waves and electromagnetics, due mainly to the efforts of Hertz and Heaviside [7].

Hertz discovered electromagnetic waves around the year 1888 [8]; the results of his epoch-making experiments and his related theoretical work (based on the sources of the electromagnetic waves rather than on the potentials) confirmed Maxwell's prediction and helped the general acceptance of Maxwell's electromagnetic theory. However, it is not commonly appreciated that

"*Maxwell's theory that Hertz's brilliant experiments confirmed was not quite the same as the one Maxwell left at his death in the year 1879*" [6]. It is interesting to note how the relevance of electromagnetic waves to Maxwell and his theory prior to Hertz's experiments and findings are described in [6]: "*Thus Maxwell missed what is now regarded as the most exciting implication of his theory, and one with enormous practical consequences. That relatively long electromagnetic waves or perhaps light itself, could be generated in the laboratory with ordinary electrical apparatus was unsuspected through most of the 1870's.*"

Maxwell's predictions and theory were thus confirmed by a set of brilliant experiments conceived and performed by Hertz, who generated, radiated (transmitted), and received (detected) electromagnetic waves of frequencies lower than light. His initial experiment started in 1887, and the decisive paper on the finite velocity of electromagnetic waves in air was published in 1888 [3]. After the 1888 results, Hertz continued his work at higher frequencies, and his later papers proved conclusively the optical properties (reflection, polarization, etc.) of electromagnetic waves and thereby provided unimpeachable confirmation of Maxwell's theory and predictions. English translation of Hertz's original publications [9] on experimental and theoretical investigation of electric waves is still a decisive source of the history of electromagnetic waves and Maxwell's theory. Hertz's experimental setup and his epoch-making findings are described in [10].

Maxwell's ideas and equations were expanded, modified, and made understandable after his death mainly by the efforts of Heinrich Hertz, George Francis Fitzgerald (1851–1901), Oliver Lodge (1851–1940), and Oliver Heaviside. The last three have been christened as "the Maxwellians" by Heaviside [2, 11].

Next we review the four equations that we use today due to Hertz and Heaviside, which resulted from the reformulation of Maxwell's original theory. Here in all the expressions we use SI units (Système International d'unités or International System of Units).

1.2 REVIEW OF MAXWELL–HEAVISIDE–HERTZ EQUATIONS

The four Maxwell's equations are among the oldest sets of equations in mathematical physics, having withstood the erosion and corrosion of time. Even with the advent of relativity, there was no change in their form. We briefly review the derivation of the four equations and illustrate how to solve them analytically [12]. The four equations consist of Faraday's law, generalized Ampère's law, generalized Gauss's law of electrostatics, and Gauss's law of magnetostatics, respectively.

1.2.1 Faraday's Law

Michael Faraday (1791–1867) observed that when a bar magnet was moved near a loop composed of a metallic wire, there appeared to be a voltage induced between the terminals of the wire loop. In this way, Faraday showed that a magnetic field produced by the bar magnet under some special circumstances can indeed generate an electric field to cause the induced voltage in the loop of wire and there is a connection between the electric and magnetic fields. This physical principle was then put in the following mathematical form:

$$V = -\oint_L \boldsymbol{E} \bullet \mathbf{d}\ell = -\frac{\partial \Phi_m}{\partial t} = -\frac{\partial}{\partial t} \iint_S \boldsymbol{B} \bullet \mathbf{d}s \tag{1.1}$$

where:

V	=	voltage induced in the wire loop of length L,
$\mathbf{d}\ell$	=	differential length vector along the axis of the wire loop,
$\boldsymbol{E}$	=	electric field along the wire loop,
Φ_m	=	magnetic flux linkage with the loop of surface area S,
$\boldsymbol{B}$	=	magnetic flux density,
S	=	surface over which the magnetic flux is integrated (this surface is bounded by the contour of the wire loop),
L	=	total length of the loop of wire,
$\bullet$	=	scalar dot product between two vectors,
$\mathbf{d}s$	=	differential surface vector normal to the surface.

This is the integral form of Faraday's law, which implies that this relationship is valid over a region. It states that the line integral of the electric field is equivalent to the rate of change of the magnetic flux passing through an open surface S, the contour of which is the path of the line integral. In this chapter, the variables in italic, for example ***B***, indicate that they are functions of four variables, *x*, *y*, *z*, *t*. This consists of three space variables (*x*, *y*, *z*) and a time variable, *t*. When the vector variable is written as **B**, it is a function of the three spatial variables (*x*, *y*, *z*) only. This nomenclature between the variables denoted by italic as opposed to roman is used to distinguish their functional dependence on spatial-temporal variables or spatial variables, respectively.

To extend this relationship to a point, we now establish the differential form of Faraday's law by invoking Stokes' theorem for the electric field. Stokes' theorem relates the line integral of a vector over a closed contour to a surface integral of the curl of the vector, which is defined as the rate of spatial change of the vector along a direction perpendicular to its orientation (which provides a rotary motion, and hence the term curl was first introduced by Maxwell), so that

$$\oint_L \boldsymbol{E} \bullet \mathbf{d}\ell = \iint_S (\nabla \times \boldsymbol{E}) \bullet \mathbf{d}s \tag{1.2}$$

where the curl of a vector in the Cartesian coordinates is defined by

$$\nabla \times \boldsymbol{E}\,(x,y,z,t) = \textit{determinant of} \begin{bmatrix} \hat{\boldsymbol{x}} & \hat{\boldsymbol{y}} & \hat{\boldsymbol{z}} \\ \dfrac{\partial}{\partial x} & \dfrac{\partial}{\partial y} & \dfrac{\partial}{\partial z} \\ E_x & E_y & E_z \end{bmatrix}$$

$$= \hat{\boldsymbol{x}} \left[\frac{\partial E_z}{\partial y} - \frac{\partial E_y}{\partial z}\right] + \hat{\boldsymbol{y}} \left[\frac{\partial E_x}{\partial z} - \frac{\partial E_z}{\partial x}\right] + \hat{\boldsymbol{z}} \left[\frac{\partial E_y}{\partial x} - \frac{\partial E_x}{\partial y}\right] \tag{1.3}$$

Here $\hat{\boldsymbol{x}}$, $\hat{\boldsymbol{y}}$, and $\hat{\boldsymbol{z}}$ represent the unit vectors along the respective coordinate axes, and E_x, E_y, and E_z represent the x, y, and z components of the electric field intensity along the respective coordinate directions. The surface S is limited by the contour L. ∇ stands for the operator $\left[\hat{\boldsymbol{x}}(\partial/\partial x) + \hat{\boldsymbol{y}}(\partial/\partial y) + \hat{\boldsymbol{z}}(\partial/\partial z)\right]$. Using (1.2), (1.1) can be expressed as

$$\oint_L \boldsymbol{E} \bullet \mathbf{d}\ell = \iint_S (\nabla \times \boldsymbol{E}) \bullet \mathbf{d}s = -\frac{\partial}{\partial t} \iint_S \boldsymbol{B} \bullet \mathbf{d}s \tag{1.4}$$

If we assume that the surface S does not change with time and in the limit making it shrink to a point, we get Faraday's law at a point in space and time as

$$\begin{aligned} \nabla \times \boldsymbol{E}(x, y, z, t) &= \frac{1}{\varepsilon} \nabla \times \boldsymbol{D}(x, y, z, t) \\ &= -\frac{\partial \boldsymbol{B}(x, y, z, t)}{\partial t} = -\mu \frac{\partial \boldsymbol{H}(x, y, z, t)}{\partial t} \end{aligned} \tag{1.5}$$

where the constitutive relationships between the flux densities and the field intensities are given by

$$\boldsymbol{B} = \mu \boldsymbol{H} = \mu_0 \mu_r \boldsymbol{H} \tag{1.6a}$$

$$\boldsymbol{D} = \varepsilon \boldsymbol{E} = \varepsilon_0 \varepsilon_r \boldsymbol{E} \tag{1.6b}$$

D is the electric flux density and ***H*** is the magnetic field intensity. Here, ε_0 and μ_0 are the permittivity and permeability of vacuum, respectively, and ε_r and μ_r are the relative permittivity and permeability of the medium through which the wave is propagating.

Equation (1.5) is the point form of Faraday's law or the first of the four Maxwell's equations. It states that at a point the negative rate of the temporal variation of the magnetic flux density is related to the spatial change of the electric field along a direction perpendicular to the orientation of the electric field (termed the curl of a vector) at that same point.

1.2.2 Generalized Ampère's Law

André M. Ampère observed that when a wire carrying current is brought near a magnetic needle, the magnetic needle is deflected in a very specific way determined by the direction of the flow of the current with respect to the magnetic needle. In this way Ampère established the complementary connection with the magnetic field generated by an electric current created by an electric field that is the result of applying a voltage difference between the two ends of the wire. Ampère first illustrated how to generate a magnetic field using the electric field or current. Ampère's law can be stated mathematically as

$$I = \oint_L \boldsymbol{H} \bullet \mathbf{d}\ell \tag{1.7}$$

where I is the total current encircled by the contour. We call this the *generalized Ampère's law* because we use the total current, which includes the displacement current due to Maxwell and the conduction current. In principle, Ampère's law is connected strictly with the conduction current. Since we use the term *total current*, we use the prefix *generalized* as it is a sum of both the conduction an displacement currents. Therefore, the line integral of $\boldsymbol{H}$, the magnetic field intensity along any closed contour L, is equal to the total current flowing through that contour.

To obtain a point form of Ampère's law, we employ Stokes' theorem to the magnetic field intensity and integrate the current density $\boldsymbol{J}$ over a surface to obtain

$$\begin{aligned} I &= \iint_S \boldsymbol{J} \bullet \mathbf{d}s = \oint_L \boldsymbol{H} \bullet \mathbf{d}\ell = \iint_S (\nabla \times \boldsymbol{H}) \bullet \mathbf{d}s \\ &= \frac{1}{\mu} \iint_S (\nabla \times \boldsymbol{B}) \bullet \mathbf{d}s \end{aligned} \tag{1.8}$$

This is the integral form of Ampère's law, and by shrinking S to a point, one obtains a relationship between the electric current density and the magnetic field intensity at the same point, resulting in

$$\boldsymbol{J}(x,y,z,t) = \nabla \times \boldsymbol{H}(x,y,z,t) \tag{1.9}$$

Physically, it states that the spatial derivative of the magnetic field intensity along a direction perpendicular to the orientation of the magnetic field intensity is related to the electric current density at that point. Now the electric current density $\boldsymbol{J}$ may consist of different components. This may include the conduction current (current flowing through a conductor) density $\boldsymbol{J}_c$ and displacement current density (current flowing through air, as from a transmitter to a receiver without any physical connection, or current flowing through the dielectric between the plates of a capacitor) $\boldsymbol{J}_d$, in addition to an externally applied impressed current density $\boldsymbol{J}_i$. So in this case we have

$$\boldsymbol{J} = \boldsymbol{J}_i + \boldsymbol{J}_c + \boldsymbol{J}_d = \boldsymbol{J}_i + \sigma \boldsymbol{E} + \frac{\partial \boldsymbol{D}}{\partial t} = \nabla \times \boldsymbol{H} \tag{1.10}$$

where $\boldsymbol{D}$ is the electric flux density or electric displacement and σ is the conductivity of the medium. The conduction current density is given by *Ohm's law*, which states that at a point the conduction current density is related to the electric field intensity by

$$\boldsymbol{J}_c = \sigma \boldsymbol{E} \tag{1.11}$$

The displacement current density introduced by Maxwell is defined by

$$\boldsymbol{J}_d = \frac{\partial \boldsymbol{D}}{\partial t} \tag{1.12}$$

We are neglecting the convection current density, which is due to the diffusion of the charge density at that point. We consider the impressed current density as the source of all the electromagnetic fields.

1.2.3 Generalized Gauss's Law of Electrostatics

Karl Friedrich Gauss established the following relation between the total charge enclosed by a surface and the electric flux density or displacement $\boldsymbol{D}$ passing through that surface through the following relationship:

$$\oiint_S \boldsymbol{D} \bullet \mathbf{d}s = Q \tag{1.13}$$

where integration of the electric displacement is carried over a closed surface and is equal to the total charge Q enclosed by that surface S.

We now employ the divergence theorem. This is a relation between the flux of a vector function through a closed surface S and the integral of the divergence of the same vector over the volume V enclosed by S. The divergence of a vector is the rate of change of the vector along its orientation. It is given by

$$\oiint_S \boldsymbol{D} \bullet \mathbf{d}s = \iiint_V \nabla \bullet \boldsymbol{D} \, dv \tag{1.14}$$

Here dv represents the differential volume. In Cartesian coordinates the divergence of a vector, which represents the rate of spatial variation of the vector along its orientation, is given by

$$\begin{aligned} \nabla \bullet \boldsymbol{D} &= \left[\hat{\boldsymbol{x}} \frac{\partial}{\partial x} + \hat{\boldsymbol{y}} \frac{\partial}{\partial y} + \hat{\boldsymbol{z}} \frac{\partial}{\partial z} \right] \bullet \left[\hat{\boldsymbol{x}} D_x + \hat{\boldsymbol{y}} D_y + \hat{\boldsymbol{z}} D_z \right] \\ &= \frac{\partial D_x(x,y,z,t)}{\partial x} + \frac{\partial D_y(x,y,z,t)}{\partial y} + \frac{\partial D_z(x,y,z,t)}{\partial z} \end{aligned} \tag{1.15}$$

So the divergence $(\nabla \bullet)$ of a vector represents the spatial rate of change of the vector along its direction, and hence it is a scalar quantity, whereas the curl $(\nabla \times)$ of a vector is related to the rate of spatial change of the vector perpendicular to its orientation, which is a vector quantity and so possesses both a magnitude and a direction. All of the three definitions of *grad*, *Div* and *curl* were first introduced by Maxwell.

By applying the divergence theorem to the vector $\boldsymbol{D}$, we get

$$\oiint_S \boldsymbol{D} \bullet \mathbf{d}s = \iiint_V \nabla \bullet \boldsymbol{D}\, dv = Q = \iiint_V q_v\, dv \qquad (1.16)$$

Here q_v is the volume charge density and V is the volume enclosed by the surface S. Therefore, if we shrink the volume in (1.16) to a point, we obtain

$$\begin{aligned}\nabla \bullet \boldsymbol{D} &= \frac{\partial D_x(x,y,z,t)}{\partial x} + \frac{\partial D_y(x,y,z,t)}{\partial y} + \frac{\partial D_z(x,y,z,t)}{\partial z} \\ &= q_v(x,y,z,t)\end{aligned} \qquad (1.17)$$

This implies that the rate change of the electric flux density along its orientation is influenced only by the presence of a free charge density at that point.

1.2.4 Generalized Gauss's Law of Magnetostatics

Gauss's law of magnetostatics is similar to the law of electrostatics defined in Section 1.2.3. If one uses the closed surface integral for the magnetic flux density $\boldsymbol{B}$, its integral over a closed surface is equal to zero, as no free magnetic charges occur in nature. Typically, magnetic charges appear as pole pairs. Therefore, we have

$$\oiint \boldsymbol{B} \bullet \mathbf{d}s = 0 \qquad (1.18)$$

From the application of the divergence theorem to (1.18), one obtains

$$\iiint_V \nabla \bullet \boldsymbol{B}\ \mathbf{d}v = 0 \qquad (1.19)$$

which results in

$$\nabla \bullet \boldsymbol{B} = 0 \qquad (1.20)$$

Equivalently in Cartesian coordinates, this becomes

$$\frac{\partial B_x(x,y,z,t)}{\partial x} + \frac{\partial B_y(x,y,z,t)}{\partial y} + \frac{\partial B_z(x,y,z,t)}{\partial z} = 0 \qquad (1.21)$$

This completes the presentation of the four equations, which are popularly referred to as Maxwell's equations, which really were developed by Hertz in scalar form and cast by Heaviside into the vector form that we use today. These four equations relate all the spatial-temporal relationships between the electric and magnetic fields.

1.2.5 Equation of Continuity

Often, the equation of continuity is used in addition to equations (1.18)–(1.21) to relate the impressed current density $\boldsymbol{J}_i$ to the free charge density q_v at that point. The equation of continuity states that the total current is related to the negative of the time derivative of the total charge by

$$I = \frac{-\partial Q}{\partial t} \tag{1.22}$$

By applying the divergence theorem to the current density, we obtain

$$I = \oiint_S \boldsymbol{J} \bullet \mathbf{d}s = \iiint_V (\nabla \bullet \boldsymbol{J})\, dv = \frac{-\partial}{\partial t} \iiint_V q_v\, dv \tag{1.23}$$

Now shrinking the volume V to a point results in

$$\nabla \bullet \boldsymbol{J} = \frac{-\partial q_v}{\partial t} \tag{1.24}$$

In Cartesian coordinates this becomes

$$\frac{\partial J_x(x,y,z,t)}{\partial x} + \frac{\partial J_y(x,y,z,t)}{\partial y} + \frac{\partial J_z(x,y,z,t)}{\partial z} = -\frac{\partial q_v(x,y,z,t)}{\partial t} \tag{1.25}$$

This states that there will be a spatial change of the current density along the direction of its flow if there is a temporal change in the charge density at that point. Next we obtain the solution of Maxwell's equations.

1.3 SOLUTION OF MAXWELL'S EQUATIONS

Instead of solving the four coupled differential Maxwell's equations directly dealing with the electric and magnetic fields, we introduce two additional variables $\boldsymbol{A}$ and ψ. Here $\boldsymbol{A}$ is the magnetic vector potential and ψ is the scalar electric potential. The introduction of these two auxiliary variables facilitates the solution of the four equations.

We start with the generalized Gauss's law of magnetostatics, which states that

$$\nabla \bullet \boldsymbol{B}(x, y, z, t) = 0 \tag{1.26}$$

Since the divergence of the curl of any vector $\boldsymbol{A}$ is always zero, that is,

$$\nabla \bullet \nabla \times \boldsymbol{A}(x,y,z,t) = 0 \tag{1.27}$$

one can always write

$$\boldsymbol{B}(x,y,z,t) = \nabla \times \boldsymbol{A}\,(x,y,z,t) \tag{1.28}$$

which states that the magnetic flux density can be obtained from the curl of the magnetic vector potential $\boldsymbol{A}$. So if we can solve for $\boldsymbol{A}$, we obtain $\boldsymbol{B}$ by a simple differentiation. It is important to note that at this point $\boldsymbol{A}$ is still an unknown quantity. In Cartesian coordinates this relationship becomes

$$\begin{aligned}\boldsymbol{B}(x,y,z,t) &= \hat{\boldsymbol{x}}\, B_x(x,y,z,t) + \hat{\boldsymbol{y}}\, B_y(x,y,z,t) + \hat{\boldsymbol{z}}\, B_z(x,y,z,t) \\ &= \textit{determinant} \text{ of } \begin{bmatrix} \hat{\boldsymbol{x}} & \hat{\boldsymbol{y}} & \hat{\boldsymbol{z}} \\ \dfrac{\partial}{\partial x} & \dfrac{\partial}{\partial y} & \dfrac{\partial}{\partial z} \\ A_x & A_y & A_z \end{bmatrix} \\ &= \hat{\boldsymbol{x}} \left[\frac{\partial A_z(x,y,z,t)}{\partial y} - \frac{\partial A_y(x,y,z,t)}{\partial z} \right] + \\ &\quad \hat{\boldsymbol{y}} \left[\frac{\partial A_x(x,y,z,t)}{\partial z} - \frac{\partial A_z(x,y,z,t)}{\partial x} \right] + \\ &\quad \hat{\boldsymbol{z}} \left[\frac{\partial A_y(x,y,z,t)}{\partial x} - \frac{\partial A_x(x,y,z,t)}{\partial y} \right] \end{aligned} \tag{1.29}$$

Note that if we substitute $\boldsymbol{B}$ from (1.28) into Faraday's law given by (1.5), we obtain

$$\nabla \times \boldsymbol{E} = -\frac{\partial \boldsymbol{B}}{\partial t} = -\frac{\partial}{\partial t}\left[\nabla \times \boldsymbol{A}\right] = -\nabla \times \frac{\partial \boldsymbol{A}}{\partial t} \tag{1.30}$$

or equivalently,

$$\nabla \times \left[\boldsymbol{E} + \frac{\partial \boldsymbol{A}}{\partial t} \right] = 0 \tag{1.31}$$

If the curl of a vector is zero, that vector can always be written in terms of the gradient of a scalar function ψ, since it is always true that the curl of the gradient of a scalar function ψ is always zero, that is,

$$\nabla \times \nabla\, \psi(x,y,z,t) = 0 \tag{1.32}$$

where the gradient of a vector is defined through

$$\nabla \psi = \left(\hat{\boldsymbol{x}} \frac{\partial}{\partial x} + \hat{\boldsymbol{y}} \frac{\partial}{\partial y} + \hat{\boldsymbol{z}} \frac{\partial}{\partial z} \right) \psi(x,y,z,t) \tag{1.33}$$

We call ψ the electric scalar potential. Therefore, we can write the following (we choose a negative sign in front of the term on the right-hand side of the equation for convenience):

$$\boldsymbol{E} + \frac{\partial \boldsymbol{A}}{\partial t} = -\nabla \psi \tag{1.34}$$

or

$$\boldsymbol{E}(x,y,z,t) = -\frac{\partial \boldsymbol{A}(x,y,z,t)}{\partial t} - \nabla \psi(x,y,z,t) \tag{1.35}$$

This states that the electric field at any point can be given by the time derivative of the magnetic vector potential and the gradient of the scalar electric potential. So we have the solution for both $\boldsymbol{B}$ from (1.28) and $\boldsymbol{E}$ from (1.35) in terms of $\boldsymbol{A}$ and ψ. The problem now is how we solve for $\boldsymbol{A}$ and ψ. Once $\boldsymbol{A}$ and ψ are known, $\boldsymbol{E}$ and $\boldsymbol{B}$ can be obtained through simple differentiation, as in (1.35) and (1.28), respectively.

Next we substitute the solution for both $\boldsymbol{E}$ [using (1.35)] and $\boldsymbol{B}$ [using (1.28)] into Ampère's law, which is given by (1.10), to obtain

$$\boldsymbol{J}_i(x,y,z,t) + \sigma \boldsymbol{E}(x,y,z,t) + \frac{\partial \boldsymbol{D}(x,y,z,t)}{\partial t} = \nabla \times \boldsymbol{H}(x,y,z,t) \tag{1.36}$$

Since the constitutive relationships are given by (1.6) (i.e., $\boldsymbol{D} = \varepsilon \boldsymbol{E}$ and $\boldsymbol{B} = \mu \boldsymbol{H}$), then

$$\boldsymbol{J}_i + \sigma \boldsymbol{E} + \varepsilon \frac{\partial \boldsymbol{E}}{\partial t} = \frac{1}{\mu} \nabla \times \nabla \times \boldsymbol{A} \tag{1.37}$$

Here we will set $\sigma = 0$, so that the medium in which the wave is propagating is assumed to be free space, and therefore conductivity is zero. So we are looking for the solution for an electromagnetic wave propagating in a non-conducting medium. In addition, we use the following vector identity:

$$\nabla \times \nabla \times \boldsymbol{A} = \nabla(\nabla \bullet \boldsymbol{A}) - (\nabla \bullet \nabla)\boldsymbol{A} \tag{1.38}$$

By using (1.38) in (1.37), one obtains

$$\begin{aligned} \nabla \times \nabla \times \boldsymbol{A} &= \nabla(\nabla \bullet \boldsymbol{A}) - (\nabla \bullet \nabla)\boldsymbol{A} \\ &= \mu \boldsymbol{J}_i + \mu\varepsilon \frac{\partial}{\partial t}\left[-\frac{\partial \boldsymbol{A}}{\partial t} - \nabla \psi\right] \\ &= \mu \boldsymbol{J}_i - \mu\varepsilon \frac{\partial^2 \boldsymbol{A}}{\partial t^2} - \mu\varepsilon \nabla \frac{\partial \psi}{\partial t} \end{aligned} \tag{1.39}$$

or equivalently,

$$(\nabla \bullet \nabla)\boldsymbol{A} - \mu\varepsilon \frac{\partial^2 \boldsymbol{A}}{\partial t^2} + \mu \boldsymbol{J}_i = \nabla\left[\nabla \bullet \boldsymbol{A} + \mu\varepsilon \frac{\partial \psi}{\partial t}\right] \tag{1.40}$$

Since we have introduced two additional new variables, $\boldsymbol{A}$ and ψ, we can without any problem impose a constraint between these two variables or these two potentials. This can be achieved by setting the right-hand side of the expression in (1.40) equal to zero. This results in

$$\nabla \bullet \boldsymbol{A} + \mu \varepsilon \frac{\partial \psi}{\partial t} = 0 \tag{1.41}$$

which is known as the *Lorenz gauge condition* [13]. It is important to note that this is not the only constraint that is possible between the two newly introduced variables $\boldsymbol{A}$ and ψ. This is only a particular assumption, and other choices will yield different forms of the solution of the Maxwell–Heaviside–Hertz equations. Interestingly, Maxwell in his treatise [1] chose the Coulomb gauge [7], which is generally used for the solution of static problems.

Next, we observe that by using (1.41) in (1.40), one obtains

$$(\nabla \bullet \nabla)\, \boldsymbol{A} - \mu \varepsilon \frac{\partial^2 \boldsymbol{A}}{\partial t^2} = -\mu\, \boldsymbol{J}_i \tag{1.42}$$

In summary, the solution of Maxwell's equations starts with the solution of equation (1.42) first, for $\boldsymbol{A}$ given the impressed current $\boldsymbol{J}_i$. Then the scalar potential ψ is solved for by using (1.41). Once $\boldsymbol{A}$ and ψ are obtained, the electric and magnetic field intensities are derived from

$$\boldsymbol{H} = \frac{1}{\mu} \boldsymbol{B} = \frac{1}{\mu} \nabla \times \boldsymbol{A} \tag{1.43}$$

$$\boldsymbol{E} = -\frac{\partial \boldsymbol{A}}{\partial t} - \nabla \psi \tag{1.44}$$

This completes the solution in the time domain, even though we have not yet provided an explicit form of the solution. We now derive the explicit form of the solution in the frequency domain and from that obtain the time domain representation. We assume the temporal variation of all the fields to be time harmonic in nature, so that

$$\boldsymbol{E}(x, y, z, t) = \mathbf{E}(x, y, z)\, e^{j\omega t} \tag{1.45}$$

$$\boldsymbol{B}(x, y, z, t) = \mathbf{B}(x, y, z)\, e^{j\omega t} \tag{1.46}$$

where $\omega = 2\,\pi f$ and f is the frequency (Hertz) of the electromagnetic fields. By assuming a time variation of the form $e^{j\omega t}$, we now have an explicit form for the time differentiations, resulting in

$$\frac{\partial}{\partial t}\left[\mathcal{A}(x,y,z,t)\right] = \frac{\partial}{\partial t}\left[\mathbf{A}(x,y,z)\, e^{j\omega t}\right] = j\omega\,\mathbf{A}(x,y,z)\, e^{j\omega t} \tag{1.47}$$

Therefore, (1.43) and (1.44) are simplified in the frequency domain after eliminating the common time variations of $e^{j\omega t}$ from both sides to form

$$\mathbf{H}(x,y,z) = \frac{1}{\mu}\,\mathbf{B}(x,y,z) = \frac{1}{\mu}\,\nabla\times\mathbf{A}(x,y,z) \tag{1.48}$$

$$\mathbf{E}(x,y,z) = -\, j\omega\,\mathbf{A}(x,y,z) - \nabla\psi(x,y,z) \tag{1.49}$$

Furthermore, in the frequency domain (1.41) transforms into

$$\nabla\bullet\mathbf{A} + j\omega\mu\varepsilon\,\psi = 0$$

or equivalently,

$$\psi = -\,\frac{\nabla\bullet\mathbf{A}}{j\omega\mu\varepsilon} \tag{1.50}$$

In the frequency domain, (1.42) transforms into

$$\nabla^2\mathbf{A} + \omega^2\mu\varepsilon\mathbf{A} = -\,\mu\mathbf{J}_i \tag{1.51}$$

The solution for **A** in (1.51) can now be written explicitly in an analytical form through [12]

$$\mathbf{A}(x,y,z) = \frac{\mu}{4\pi}\int_V \frac{\mathbf{J}_i(x',y',z')\, e^{-jkR}}{R}\, dv \tag{1.52}$$

where

$$\mathbf{r} = \hat{x}\,x + \hat{y}\,y + \hat{z}\,z \tag{1.53}$$

$$\mathbf{r}' = \hat{x}\,x' + \hat{y}\,y' + \hat{z}\,z' \tag{1.54}$$

$$R = \left|\mathbf{r} - \mathbf{r}'\right| = \sqrt{(x-x')^2 + (y-y')^2 + (z-z')^2} \tag{1.55}$$

$$k = \frac{2\pi}{\lambda} = \frac{2\pi f}{c} = \sqrt{\omega^2\mu\varepsilon} = \sqrt{(2\pi)^2 f^2\mu\varepsilon} \tag{1.56}$$

$$c = \text{velocity of light in the medium} = \frac{1}{\sqrt{\mu\varepsilon}} \tag{1.57}$$

$$\lambda = \text{wavelength in the medium} \tag{1.58}$$

In summary, first the magnetic vector potential **A** is solved for in the frequency domain given the impressed currents $\mathbf{J}_i(\mathbf{r})$ through

$$\mathbf{A}(\mathbf{r}) = \mathbf{A}(x,y,z) = \frac{\mu}{4\pi} \iiint_V \frac{\mathbf{J}_i(\mathbf{r}')\ e^{-jk\left|\mathbf{r}-\mathbf{r}'\right|}}{\left|\mathbf{r}-\mathbf{r}'\right|} \mathrm{d}\mathbf{r}' \tag{1.59}$$

then the scalar electric potential ψ is obtained from (1.50). Next, the electric field intensity $\mathbf{E}$ is computed from (1.49) and the magnetic field intensity $\mathbf{H}$ from (1.48).

In the time domain the equivalent solution for the magnetic vector potential $\boldsymbol{A}$ is then given by the time-retarded potentials:

$$\boldsymbol{A}(\mathbf{r},t) = \boldsymbol{A}(x,y,z,t) = \frac{\mu}{4\pi} \iiint_V \frac{\boldsymbol{J}_i\left(\mathbf{r}', t - \frac{\left|\mathbf{r}-\mathbf{r}'\right|}{c}\right)}{\left|\mathbf{r}-\mathbf{r}'\right|} \mathrm{d}\mathbf{r}' \tag{1.60}$$

It is interesting to note that the time and space variables are now coupled and they are not separable. That is why in the time domain the spatial and temporal responses of an antenna are intimately connected and one needs to look at the complete solution. From the magnetic vector potential we obtain the scalar potential ψ by using (1.41). From the two vector and scalar potentials the electric field intensity $\boldsymbol{E}$ is obtained through (1.44) and the magnetic field intensity $\boldsymbol{H}$ using (1.43).

We now use these expressions to calculate the impulse response of some typical antennas in both the transmit and receive modes of operations. The reason that impulse response of an antenna is different in the transmit mode than in the receive mode is because the reciprocity principle in the time domain contains an integral over time. The reciprocity theorem in the time domain is quite different from its counterpart in the frequency domain. For the former a time integral is involved, whereas for the latter no such relationship is involved. Because of the frequency domain reciprocity theorem, the antenna radiation pattern when in the transmit mode is equal to the antenna pattern in the receive mode. However, this is not true in the time domain, as we shall now see through examples.

1.4 RADIATION AND RECEPTION PROPERTIES OF A POINT SOURCE ANTENNA IN FREQUENCY AND IN TIME DOMAIN

1.4.1 Radiation of Fields from Point Sources

In this section we first define what is meant by the term *radiation* and then observe the nature of the fields radiated by point sources and the temporal nature of the voltages induced when electromagnetic fields are incident on them. In contrast to the acoustic case (where an isotropic source exists), in the electromagnetic case there are no isotropic point sources. Even for a point source, which in the electromagnetic case is called a *Hertzian dipole*, the radiation pattern is not isotropic, but it can be omnidirectional in certain planes.

We describe the solution in both the frequency and time domains for such classes of problems.

Any element of current or charge located in a medium will produce electric and magnetic fields. However, by the term *radiation* we imply the amount of finite energy transmitted to infinity from these currents. Hence, radiation is related to the far fields or the fields at infinity. This will be discussed in detail in Chapter 2. A static charge may generate near fields, but it does not produce radiation, as the field at infinity due to this charge is zero. Therefore, radiated fields or far fields are synonymous. We will also explore the sources of a radiating field.

1.4.1.1 Far Field in Frequency Domain of a Point Radiator. If we consider a delta element of current or a Hertzian dipole located at the origin represented by a constant J times a delta function δ (0, 0, 0), the magnetic vector potential from that current element is given by

$$\mathbf{A}(x,y,z) = \frac{\mu}{4\pi}\frac{e^{-jkR}}{R}\mathbf{J}_i \tag{1.61}$$

where

$$R = \sqrt{x^2+y^2+z^2} \tag{1.62}$$

Here we limit our attention to the electric field. The electric field at any point in space is then given by

$$\begin{aligned}\mathbf{E}(x,y,z) &= -j\omega\mathbf{A} - \nabla\psi = -j\omega\mathbf{A} + \frac{\nabla(\nabla \bullet \mathbf{A})}{j\omega\mu\varepsilon} \\ &= \frac{1}{j\omega\mu\varepsilon}\left[k^2\mathbf{A} + \nabla(\nabla \bullet \mathbf{A})\right]\end{aligned} \tag{1.63}$$

In rectangular coordinates, the fields at any point located in space will be

$$\mathbf{E}(x,y,z) = \frac{1}{j\omega\mu\varepsilon}\left[\begin{array}{l} k^2\,\mathbf{A}(x,y,z) + \left\{\hat{\boldsymbol{x}}\dfrac{\partial}{\partial x} + \hat{\boldsymbol{y}}\dfrac{\partial}{\partial y} + \hat{\boldsymbol{z}}\dfrac{\partial}{\partial z}\right\} \\ \times\left\{\dfrac{\partial \mathrm{A}_x(x,y,z)}{\partial x} + \dfrac{\partial \mathrm{A}_y(x,y,z)}{\partial y} + \dfrac{\partial \mathrm{A}_z(x,y,z)}{\partial z}\right\}\end{array}\right] \tag{1.64}$$

However, some simplifications are possible for the far field (i.e., if we are observing the fields radiated by a source of finite size at a distance of $2D^2/\lambda$ from it, where D is the largest physical dimension of the source and λ is the wavelength – the physical significance of this will be addressed in chapter 2.). For a point source, everything is in the far field. Therefore, for all practical purposes, observing the fields at a distance $2D^2/\lambda$ from a source is equivalent to observing the fields from the same source at infinity. In that case, the far fields can be obtained from the first term only in (1.63) or (1.64). This first term due to the magnetic vector potential is responsible for the far field and there is no contribution from the scalar electric potential ψ. Hence,

$$\mathbf{E}_{\text{far}}(x, y, z) = -j\,\omega\mathbf{A} = -j\,\frac{\omega\mu}{4\pi}\,\mathbf{J}_i\,\frac{e^{-jkR}}{R} \tag{1.65}$$

and one obtains a spherical wavefront in the far field for a point source. However, the power density radiated is proportional to E_θ and that is clearly zero along $\theta = 0°$ and is maximum in the azimuth plane where $\theta = 90°$. The characteristic feature is that the far field is polarized and the orientation of the field is along the direction of the current element. It is also clear that one obtains a spherical wavefront in the far field radiated by a point source.

The situation is quite different in the time domain, as the presence of the term ω in the front of the expression of the magnetic vector potential will illustrate.

1.4.1.2 Far Field in Time Domain of a Point Radiator. We consider a delta current source at the origin of the form

$$\boldsymbol{J}_i\,\delta(0,0,0,t) = \hat{z}\,\delta(0,0,0)\,f(t) \tag{1.66}$$

where $\hat{z}$ is the direction of the orientation of the elemental current element and $f(t)$ is the temporal variation for the current fed to the point source located at the origin. The magnetic vector potential in this case is given by

$$\boldsymbol{A}(\mathbf{r},t) = \frac{\mu}{4\pi}\,\frac{\hat{z}\,f(t - |R|/c)}{R} \tag{1.67}$$

There will be a time retardation factor due to the space-time connection of the electromagnetic wave that is propagating, where R is given by (1.62).

Now the transient far field due to this impulsive current will be given by

$$\boldsymbol{E}(\mathbf{r},t) = -\frac{\partial \boldsymbol{A}(\mathbf{r},t)}{\partial t} = -\frac{\mu\hat{z}}{4\pi R}\,\frac{\partial f(t - |R|/c)}{\partial t} \tag{1.68}$$

Hence, the time domain field radiated by a point source is given by the time derivative of the transient variation of the elemental current element. Therefore, a time-varying current element will always produce a far field and hence will cause radiation. However, if the current element is not changing with time, there will be no radiation from it. Equivalently, the current density $\boldsymbol{J}_i$ can be expressed in terms of the flow of charges; thus it is equivalent to $\rho\upsilon$, where ρ is the charge density and υ is its velocity. Therefore, radiation from a time-varying current element in (1.68) can occur if any of the following three scenarios occur:

1. The charge density ρ may change as a function of time.
2. The direction of the velocity vector $\boldsymbol{\upsilon}$ may change as a function of time.
3. The magnitude of the velocity vector $\boldsymbol{\upsilon}$ may change as a function of time, or equivalently, the charge is accelerated or decelerated.

Therefore, in theory any one of these three scenarios can cause radiation. For example, in a dipole the current goes to zero at the ends of the structure and hence the charges decelerate when they come to the end of a wire. That is why radiation seems to emanate from the ends of the wire and also from the feed point of a dipole where a current is injected or a voltage is applied and where the charges are induced and hence accelerated. Current flowing in a loop of wire can also radiate as the direction of the velocity is changing as a function of time even though its magnitude is constant. So a current flowing in a loop of wire may have a constant angular velocity, but the temporal change in the orientation of the velocity vector may cause radiation. To maintain the same current along a cross-section of the wire loop, the charges located along the inner circumference of the loop have to decelerate, whereas the charges on the outer boundary have to accelerate. This will cause radiation. In a klystron, by modulating the velocity of the electrons, one can have bunching of the electrons or a change of the electron density with time. This also causes radiation. In summary, if any one of the three conditions described above occurs, there will be radiation.

By observing (1.68), we see that a transmitting antenna acts as a differentiator of the transient waveform fed to its input. The important point to note is that an antenna impulse response on transmit is a differentiation of the excitation on transmit. Therefore in all baseband broadband simulations the differential nature of the point source must be taken into account. This implies that if the input to a point radiator is a pulse, it will radiate two impulses of opposite polarities – a derivative of the pulse. Therefore, when a baseband broadband signal is fed to an antenna, what comes out is the derivative of that pulse. It is rather unfortunate that very few simulations dealing with baseband broadband signals really take this property of an isotropic point source antenna into account in analyzing systems.

1.4.2 Reception Properties of a Point Receiver

On receive, an antenna behaves in a completely different way than on transmit. We observed that an isotropic point antenna acts as a differentiator on transmit. On receive, the voltage received at the terminals of the antenna is given by

$$V = \int \boldsymbol{E} \bullet \mathbf{d}\ell \tag{1.69}$$

where the path of the integral is along the length of the antenna. Equivalently, this voltage, which is called the *open-circuit voltage* V_{oc}, is equivalent to the dot product of the incident field vector and the effective height of the antenna and is given by [14, 15]

$$V_{oc} = \mathbf{E} \bullet \mathbf{H}_{eff}$$

The effective height of an antenna is defined by

$$H_{eff} = \int_0^H I(z)\, dz = H\, I_{av} \tag{1.70}$$

where H is the length of the antenna and it is assumed that the maximum value of the current along the length of the antenna $I(z)$ is unity. I_{av} then is the average value of the current on the antenna. This equation is valid at only a single frequency. Therefore, when an electric field E^{inc} is incident on a small dipole of total length L from a broadside direction, it induces approximately a triangular current on the structure [15]. Therefore, the effective height in this case is $L/2$ and the open-circuit voltage induced on the structure in the frequency domain is given by

$$V_{oc}(\omega) = -\frac{L\, E^{inc}(\omega)}{2} \tag{1.71}$$

and in the time domain as the effective height now becomes an impulse-like function, we have

$$V_{oc}(t) = -\frac{L\, E^{inc}(t)}{2} \tag{1.72}$$

Therefore, in an electrically small receiving antenna called a *voltage probe* the induced waveform will be a replica of the incident field provided that the frequency spectrum of the incident electric field lies mainly in the low-frequency region, so that the concept of an electrically small antenna is still applicable.

In summary, the impulse response of an antenna on transmit given by (1.65) is the time derivative of the impulse response of the same antenna when it is operating in the receive mode as given by (1.72). In the frequency domain as we observe in (1.65) the term $j\omega$ is benign as it merely introduces a purely imaginary scale factor at a particular value of ω. However, the same term when transferred to the time domain represents a time derivative operation. Hence, in frequency domain the transmit radiation antenna pattern is identical to the antenna pattern when it is operating in the receive mode. In time domain, the transmit impulse response of the same antenna is the time derivative of the impulse response in the receive mode for the same antenna. At this point, it may be too hasty to jump to the conclusion that something is really amiss as it does not relate quite the same way to the reciprocity theorem which in the frequency domain has shown that the two patterns in the transmit–receive modes are identical. This is because the mathematical form of the reciprocity theorem is quite different in the time and in the frequency domains. Since the reciprocity theorem manifests itself as a product of two quantities in the frequency domain, in the time domain then it becomes a convolution. It is this phenomenon that makes the impulse response of the transmit and the receive modes different. We use another example, namely a dipole, to illustrate this point further.

1.5 RADIATION AND RECEPTION PROPERTIES OF FINITE-SIZED DIPOLE-LIKE STRUCTURES IN FREQUENCY AND IN TIME

In this section we describe the impulse responses of transmitting and receiving dipole-like structures whose dimensions are comparable to a wavelength. Therefore, these structures are not electrically small. Detailed analysis of these structures will be done in Chapter 3. In this section, the main results are summarized. The reason for choosing finite-sized structures is that the impulse responses of these wire-like structures are quite different from the cases described in the preceding section. For a finite-sized antenna structure, which is comparable to the wavelength at the frequency of operation, the current distribution on the structure can no longer be taken to be independent of frequency. Hence the frequency term must explicitly be incorporated in the expression of the current.

1.5.1 Radiation Fields from Wire-like Structures in the Frequency Domain

For a finite-sized dipole, the current distribution that is induced on it can be represented mathematically to be of the form [14]

$$\mathrm{I}(z) = \sin\left[k\left(L/2-|z|\right)\right] \tag{1.73}$$

where L is the wire antenna length. We here assume that the current distribution is known. However, in a general situation we have to use a numerical technique to solve for the current distribution on the structure before we can solve for the far fields. This is particularly important when mutual coupling effects are present or there are other near-field scatterers. For a current distribution given by (1.73), the far fields can be obtained [14] as

$$\mathrm{E}_\theta = \frac{\mathrm{I}_0\,\eta\, e^{j\omega[t-r/c]}}{2\pi r}\left[\frac{\cos\left(\frac{kL}{2}\cos\theta\right) - \cos\frac{kL}{2}}{\sin\theta}\right] \tag{1.74}$$

where η is the characteristic impendence of free space and I_0 represents the maximum value of the current. Here L is the length of the antenna. k is the free-space wavenumber and is equal to $2\pi/\lambda = \omega/c$, where c is the velocity of light in that medium. It is important to note that only along the broadside direction and in the azimuth plane of $\theta = \pi/2 = 90°$ is the radiated electric field omni-directional in nature.

1.5.2 Radiation Fields from Wire-like Structures in the Time Domain

When the current induced on the dipole is a function of frequency, the far-zone time-dependent electric field at a spatial location r is given approximately by [15]

$$E_\theta(t) = \frac{\eta}{2\pi r \sin\theta} \left\{ \begin{aligned} & I\left(t - \frac{r}{c}\right) + I\left[t - \frac{r}{c} - \frac{L}{c}\right] \\ & - I\left[t - \frac{r}{c} - \frac{L}{2c}(1 + \cos\theta)\right] \\ & - I\left[t - \frac{r}{c} - \frac{L}{2c}(1 - \cos\theta)\right] \end{aligned} \right\} \tag{1.75}$$

where $I(t)$ is the transient current distribution on the structure. It is interesting to note that for L/c small compared to the pulse duration of the transient current distribution on the structure, then from [15] the approximate far field can be written as

$$E_\theta \simeq \frac{\eta}{2\pi r}\left(\frac{L}{2c}\right)^2 \frac{\partial^2 I\left(t - \frac{r}{c}\right)}{\partial t^2} \sin\theta \tag{1.76}$$

that is, the far-field now is proportional to the second temporal derivative of the transient current on the structure.

1.5.3 Induced Voltage on a Finite-Sized Receive Wire-like Structure Due to a Transient Incident Field

For a finite-sized antenna of total length L, the effective height will be a function of frequency and it is given by

$$\mathrm{H}_{\mathrm{eff}}(\omega) = \int_{-L/2}^{L/2} \sin\left[k\left(\frac{L}{2} - |z|\right)\right] dz = \frac{2c}{\omega}\left[1 - \cos\left(\frac{kL}{2}\right)\right] \tag{1.77}$$

Hence the induced voltage for a broadside incidence will be given approximately by

$$\mathrm{V}_{\mathrm{oc}}(\omega) = -\mathrm{H}_{\mathrm{eff}}(\omega)\,\mathrm{E}^{\mathrm{inc}}(\omega) \tag{1.78}$$

In the time domain, the effective height will be given by

$$H_{\mathrm{eff}}(t) = jc \begin{cases} +1 & 0 < t < \dfrac{L}{2c} \\[2ex] -1 & \dfrac{-L}{2c} < t < 0 \end{cases} \tag{1.79}$$

Hence the transient received voltage in the antenna due to an incident field will result in the following convolution (defined by the symbol ✪) between the incident electric field and the effective height, resulting in

$$V_{oc}(t) = -E^{inc}(t) \;✪\; H_{eff}(t) \tag{1.80}$$

This illustrates that when (1.79) is used in (1.80), the received open-circuit voltage will be approximately the derivative of the incident field when L/c is small compared to the duration of the initial duration of the transient incident field. In Chapter 2, we study the properties of arbitrary shaped antennas in the frequency domain using a general purpose computer code described in [16]. Furthermore, we focus on the implications of near and far fields. The near/far field concepts are really pertinent in the frequency domain as they characterize the radiation properties of antennas. However, in the time domain this distinction is really not applicable as everything is near field unless we have a strictly band limited signal! In Chapter 3 the properties of arbitrary shaped antennas embedded in different materials are studied in the time domain using the methodology of [17].

1.6 CONCLUSION

The objective of this chapter has been to present the necessary mathematical formulations, popularly known as Maxwell's equations, which dictate the space-time behavior of antennas. Additionally, some examples are presented to note that the impulse response of antennas is quite complicated and the waveshapes depend on both the observation and the incident angles in azimuth and elevation of the electric fields. Specifically, the transmit impulse response of an antenna is the time derivative of the impulse response of the same antenna in the receive mode. This is in contrast to the properties in the frequency domain where the transmit antenna pattern is the same as the receive antenna pattern. Any broadband processing must deal with factoring out the impulse response of both the transmitting and receiving antennas. The examples presented in this chapter do reveal that the waveshape of the impulse response is indeed different for both transmit and receive modes, which are again dependent on both the azimuth and elevation angles. For an electrically small antenna, the radiated fields produced by it along the broadside direction are simply the differentiation of the time domain waveshape that is fed to it. While on receive it samples the field incident on it. However, for a finite-sized antenna, the radiated fields are proportional to the temporal double derivative of the current induced on it, and on receive, the same antenna differentiates the transient electric field that is incident on it. Hence all baseband broadband applications should deal with the complex problem of determining the impulse responses of the transmitting and receiving antennas. This is in contrast to spread spectrum methodologies where one deals with an instantaneous narrowband signals even when frequency hopping. For the narrowband case, determination of the impulse response is not necessary. The goal of this chapter is to outline the methodology that will be necessary to determine the impulse response of the transmit/receive antennas. By thus combining the electromagnetic analysis with the signal-processing algorithms, it will be possible to design better systems.

REFERENCES

[1] J. C. Maxwell, "A Dynamical Theory of the Electromagnetic Field", *Philosophical Transactions*, Vol. 166, pp. 459–512, 1865 (reprinted in the *Scientific Papers of James Clerk Maxwell*, Vol. 1, pp. 528–597, Dover, New York, 1952).

[2] H. Hertz, "On the Relations between Maxwell's Fundamental Equations of the Opposing Electromagnetics", (in German), *Wiedemann's Annalen*, Vol. 23, pp. 84–103, 1884. (English translation in [9, pp. 127–145]).

[3] H. Hertz, "On the Fundamental Equations of Electromagnetics for Bodies at Rest," in [9, pp. 195–240].

[4] P. J. Nahin, *Oliver Heaviside: Sage in Solitude*, IEEE Press, New York, 1988.

[5] C-T Tai and J. H. Bryant, "New Insights into Hertz's Theory of Electromagnetism", *Radio Science*, Vol. 29, No. 4, pp. 685–690, July–Aug. 1994.

[6] B. J. Hunt, *The Maxwellians*, Chap. 2, p. 24, Cornell University Press, Ithaca, NY, 1991.

[7] T. K. Sarkar, R. J. Mailloux, A. A. Oliner, M. Salazar-Palma, and D. L. Sengupta, *History of Wireless*, John Wiley and Sons, 2006.

[8] H. Hertz, "On the Finite Velocity of Propagation of Electromagnetic Action", *Sitzungsber ichte der Berliner Academic der Wissenschaften*, Feb. 2, 1888; *Wiedemann's Annalen*, Vol. 24, pp. 551, reprinted in H. Hertz (translated by D. E. Jones), *Electric Waves*, Chap. 7, pp. 107–123, Dover, New York, 1962.

[9] H. Hertz, *Electric Waves* (authorized English translation by D. E. Jones), Dover, New York, 1962.

[10] J. H. Bryant, *Heinrich Hertz: The Beginning of Microwaves*, IEEE Service Center, Piscataway, NJ, 1988.

[11] J. G. O'Hara and W. Pritcha, *Hertz and Maxwellians,* Peter Peregrinus, London, 1987.

[12] J. D. Kraus, *Electromagnetics,* McGraw-Hill, New York, 1980.

[13] R. Nevels and C. Shin, "Lorenz, Lorentz, and the Gauge", *IEEE Antennas and Propagat. Magazine*, Vol. 43, No. 3, pp. 70–72, June 2001.

[14] J. D. Kraus, *Antennas,* McGraw-Hill, New York, 1988.

[15] D. L. Sengupta and C. T. Tai, "Radiation and Reception of Transients by Linear Antennas", Chap. 4, pp. 182–234 in L. B. Felsen (ed.), *Transient Electromagnetic Fields,* Springer-Verlag, New York, pp. 182–234, 1976.

[16] B. M. Kolundzija, J. S. Ognjanovic, and T. K. Sarkar, *WIPL-D: Electromagnetic Modeling of Composite Metallic and Dielectric Structures*, Artech House, Norwood, MA, 2000.

[17] T. K. Sarkar, W. Lee, and S. M. Rao, "Analysis of Transient Scattering from Composite Arbitrarily Shaped Complex Structures", *IEEE Transactions on Antennas and Propagation*, Vol. 48, No. 10, pp. 1625–1634, Oct. 2000.

2

FUNDAMENTALS OF ANTENNA THEORY IN THE FREQUENCY DOMAIN

2.0 SUMMARY

In this chapter we are going to look at the fields produced by some simple antennas and their current distributions in the frequency domain. Specifically, we will observe the fields produced by a Hertzian dipole, a finite sized dipole antenna, and a small loop antenna. We will define how to separate the near fields from the far fields and how that is tied to radiation. And finally, we will define what is meant by the term radiation and what does it signify and how does it relate to directivity and gain. Examples will be presented to illustrate what are valid near field/far field modeling including analysis of antennas over an imperfectly conducting earth as in a wireless communication environment.

2.1 FIELD PRODUCED BY A HERTZIAN DIPOLE

Consider an element of current oriented along the z-direction. The current element, a Hertzian dipole, is an infinitesimally small current element so that [1]

$$\iiint\limits_{\substack{\text{volume encompassing}\\ \text{the source}}} J_z^i \, dv = I_d \, l \tag{2.1}$$

where J_z^i is the current density (amp/m^3). This particular current distribution belongs to an antenna called an electric dipole. As shown in Figure 2.1, l is the length of the wire along which the current I_d flows. This element of current terminates and originates from two charges. The uniform current element I_d distributed along the entire length l gives rise to the charges $+Q$, and $-Q$, which in turn gives rise to a displacement current that flows through space. The current is related to the charge by $I_d = dQ/dt$ [1–4].

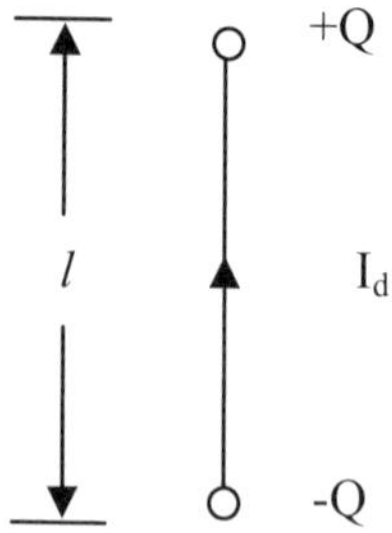

Figure 2.1. An equivalent circuit of a Hertzian dipole.

The magnetic vector potential produced by this elementary dipole located at $\delta(x', y', z')$ is given by

$$\mathbf{A}(x, y, z) = \hat{z}\,\frac{\mu_0\,\mathrm{I_d}\,l}{4\pi}\,\frac{\exp(-j\,k\,R)}{R} \tag{2.2}$$

where x, y, z are the coordinate points at which the potential is evaluated. In addition $k = 2\pi/\lambda = 2\pi\,f/c = 2\pi f\sqrt{\mu_0\,\varepsilon_0}$, where λ is the wavelength of the wave corresponding to a frequency f and c is the velocity of light. Since the dipole is located at the origin, we have

$$R = \sqrt{(x-x')^2 + (y-y')^2 + (z-z')^2} = \sqrt{x^2 + y^2 + z^2} = r \tag{2.3}$$

It is now convenient to obtain the potential in spherical coordinates (r, θ, ϕ). The components of the magnetic vector potential in spherical coordinates are:

$$\mathrm{A_r} = \mathrm{A_z}\cos\theta = \frac{\mu_0\,\mathrm{I_d}\,l\,e^{-jkr}\cos\theta}{4\pi r} \tag{2.4}$$

$$\mathrm{A_\theta} = -\mathrm{A_z}\sin\theta = \frac{-\mu_0\,\mathrm{I_d}\,l\,e^{-jkr}\sin\theta}{4\pi\,r} \tag{2.5}$$

$$\mathrm{A_\varphi} = 0 \tag{2.6}$$

Since, the magnetic field intensity is given by, $\mathbf{H} = (\nabla\times\mathbf{A})/\mu_0$, then we obtain

$$\mathrm{H_\varphi} = \frac{1}{\mu_0\,r}\left[\frac{\partial}{\partial r}\left(r\,\mathrm{A_\theta}\right) - \frac{\partial\,\mathrm{A_r}}{\partial\theta}\right] = \frac{\mathrm{I_d}\,l\,e^{-jkr}\sin\theta}{4\pi r^2} + \frac{jk\,\mathrm{I_d}\,l\,e^{-jkr}\sin\theta}{4\pi r} \tag{2.7}$$

with $\mathrm{H_r} = \mathrm{H_\theta} = 0$. The first term in (2.7) is due to the induction field of a current element and can be obtained from Ampere's law of (1.8). This is also the induction field with a time retardation or phase delay. It is in phase with the exciting current I_d and decreases as the inverse square law. The second term in (2.7) is in phase quadrature with the excitation current and remains so, even if we come in close proximity to the current element. This second component does not occur in the application of Ampere's law. We shall see later that this term is related to radiation.

Next, we find the electric field intensity associated with this magnetic field. Since $\mathbf{E} = (\nabla \times \mathbf{H})/(j\omega\varepsilon_0)$

$$E_r = \frac{-j I_d l e^{-jkr} \cos\theta}{2\pi\omega\varepsilon_0 r^3} + \frac{k I_d l e^{-jkr} \cos\theta}{2\pi\omega\varepsilon_0 r^2} \tag{2.8}$$

$$E_\theta = \frac{k I_d l e^{-jkr} \sin\theta}{4\pi\omega\varepsilon_0 r^2} - \frac{j I_d l e^{-jkr} \sin\theta}{4\pi\omega\varepsilon_0 r^3} + \frac{j k^2 I_d l e^{-jkr} \sin\theta}{4\pi\omega\varepsilon_0 r} \tag{2.9}$$

with $E_\varphi = 0$. There are two different variations of the field components with respect to the distance. The $1/r^3$ variation of the field intensity is due to the fields produced by a charge and it represents the static fields from a dipole and the $1/r^2$ variation of the fields is due to electromagnetic induction or is often referred to as transformer action. Therefore, very close to the current element, the **E** field reduces to that of a static charge dipole and the **H** reduces to that of a constant current element, and the fields are said to be quasi-static. At intermediate values of r the field is said to be induction field.

Next, we look at the direction and the amplitude of the power flow density (power per unit area) from the dipole. That is given by the Poynting vector $\mathbf{S}_{dipole}$ characterizing the complex power density and can be mathematically expressed as [1–4]

$$\begin{aligned} \mathbf{S}_{dipole} &= \frac{1}{2}\left(\mathbf{E}\times\mathbf{H}^*\right) = \frac{1}{2}(\hat{\boldsymbol{r}}\, E_\theta H_\varphi^* - \hat{\boldsymbol{\theta}}\, E_r H_\varphi^*) \\ &= \hat{\boldsymbol{r}}\, \frac{\eta k^2 |I_d\, l|^2 \sin^2\theta}{32\pi^2 r^2}\left[1 - \frac{j}{(kr)^3}\right] + \hat{\boldsymbol{\theta}}\, \frac{j\eta k |I_d\, l|^2 \cos\theta \sin\theta}{16\pi^2 r^3}\left[1 + \frac{1}{(kr)^2}\right] \end{aligned} \tag{2.10}$$

Here the superscript * denotes the complex conjugate. The factor ½ will not be there if we use the root mean square (rms) values rather than the peak values for the fields. In the later chapters we shall be using the rms values and therefore the factor of ½ will not be there. And $\eta = k/(\omega\varepsilon_0) = \sqrt{\mu_0/\varepsilon_0} = 120\pi$ defines the characteristic impedance of free space and is numerically equal to 377 ohms. $\hat{r}$ is the unit vector in the radial direction and $\hat{\theta}$ is the unit vector along the elevation angle, or θ direction. The outward directed power along the radial direction is obtained by integrating the Poynting vector over a sphere of radius r

$$P_{dipole} = \oiint \mathbf{S}_{dipole} \bullet \boldsymbol{dS} = \oiint \mathbf{S}_{dipole} \bullet (\hat{r}\, r^2 \sin\theta \; d\theta\, d\phi)$$

$$= \int_0^{2\pi} d\phi \int_0^{\pi} d\theta \,(r^2 \sin\theta \; \mathrm{E}_\theta \mathrm{H}_\varphi^*) = \frac{\pi\eta}{3}\left|\frac{\mathrm{I_d}\, l}{\lambda}\right|^2 \left[1 - \frac{j}{(kr)^3}\right] \tag{2.11}$$

The second term, or the θ-component of the Poynting vector of (2.10) does not contribute to the radiated power. Therefore the entire outward power flow comes from the first term. This real part of the power is independent of the distance r and represents the power crossing the surface of a sphere, even if the radius of the sphere is infinity. The reactive part of the power diminishes with the distance and vanishes when $r = \infty$. Since the reactive power is negative, it indicates that there is an excess of electric energy over the magnetic energy in the near field, at a finite distance r from the dipole.

2.2 CONCEPT OF NEAR AND FAR FIELDS

An alternating current in a circuit has a near field and a far field. In the near field, it is assumed that the fields are concentrated near the source (*http://www.majr.com/docs/Understanding_Electromagnetic_Fields_And_Antenna.pdf*). The radiating field is referred to as the far field as its effect extends beyond the source, particularly, the received power at infinity. To illustrate this point, we observe that the scalar power density S_{dipole} characterized by (2.10) from the dipole at a finite distance from a source can be represented by

$$\mathrm{S}_{dipole} = \frac{C_1}{r^2} + \frac{C_2}{r^3} + \frac{C_3}{r^4} + \cdots \tag{2.12}$$

Now consider a sphere with radius r, centered at the source. Then the total power passing the surface of the sphere will be given by

$$P = \left(4\pi r^2\right)\mathbf{S}_{dipole} = 4\pi\left(C_1 + \frac{C_1}{r^2} + \frac{C_2}{r^3} + \cdots\right) \tag{2.13}$$

It is seen that the first term is a constant. So for this term no matter what size of the sphere is chosen, the same amount of power flows through it and it demonstrates that power is flowing away from the source and is called the radiation field. The other terms become negligible as r gets large. Consequently at small distance r, the other terms become much larger compared to the radiative field. These other terms taken together represent the power in the near field or termed the reactive field.

Therefore in the far field there are only two components of the field for the dipole. They are

$$\mathrm{H}_\varphi \approx \frac{jk\,\mathrm{I_d}\, l\, e^{-jkr} \sin\theta}{4\pi r} = \frac{j\, e^{-jkr} \sin\theta}{2r} \frac{\mathrm{I_d}\, l}{\lambda} \tag{2.14}$$

$$\mathrm{E}_\theta \approx \frac{j k^2 \, \mathrm{I_d} \, l \, e^{-jkr} \sin\theta}{4\pi\omega\varepsilon_0 r} = \frac{j\eta \, e^{-jkr} \sin\theta}{2r} \frac{\mathrm{I_d} \, l}{\lambda} \tag{2.15}$$

It is now seen that their ratio is given by $\mathrm{E}_\theta/\mathrm{H}_\varphi = k/(\omega\varepsilon_0) = \sqrt{\mu_0/\varepsilon_0} =$ $= \sqrt{\mu_0/\varepsilon_0} = \eta = 120\pi$ where η is the characteristic impedance of free space which is 377 ohms. Therefore, in the far field the electric and the magnetic fields are orthogonal to each other in space, but coherent in time. In addition, they are related by the characteristic impedance of free space. The emanating wave generated from the source in the far field is a plane wave. The factor sin θ in the two expressions represents the radiation pattern. It is the field pattern at $r \to \infty$. This field pattern is independent of the distance r and is associated with the far field component. The Poynting vector and the power flow can then be approximated in the far field as

$$\mathbf{S}_{dipole} = \hat{r}\,\frac{\eta k^2 |\mathrm{I_d}\, l|^2 \sin^2\theta}{32\pi^2 r^2} = \hat{r}\,\frac{15\pi \sin^2\theta}{r^2}\left|\frac{\mathrm{I_d}\, l}{\lambda}\right|^2 ; \tag{2.16}$$

$$\mathrm{P}_{dipole} = \frac{2\pi\eta}{3}\left|\frac{\mathrm{I_d}\, l}{\lambda}\right|^2 = 80\pi^2\left|\frac{\mathrm{I_d}\, l}{\lambda}\right|^2 \tag{2.17}$$

In the far field the power flow is independent of the distance r.

So the differences between the properties of the reactive near field and the radiated far field can be summarized as follows:

1. In the far field there is always a transmission of energy through space whereas in the reactive near field the energy is stored in a fixed location.
2. In the far field, the energy is radiated outward from the source and is always a real quantity as the waves of **E** & **H** fields move through space, whereas, in the reactive near field, energy is recoverable as they oscillate in place.
3. In the far field, the **E** & **H** fields decrease as the inverse of the distance, whereas in the near fields they follow an inverse squared law or higher.
4. The electric and the magnetic far fields are in space quadrature as seen from (2.14) and (2.15) but they are coherent in time. The fields form a plane wave. Whereas in the reactive near field, the electric and the magnetic fields can have any spatial or temporal orientation with respect to each other.
5. The radiated far field is related to the real part of the antenna impedance, whereas the reactive part of the impedance is related to the reactive near field.
6. In the far field, the antenna field pattern is defined independent of the spatial distance from the antenna but is related only to the spatial angles as the r terms in the field expressions are generally not deemed relevant.

7. In the far field, the power radiated from the antenna is always real and is considered to exist even at infinity whereas the reactive power component becomes zero.

We will define a rule of thumb by placing some mathematical constraints on the distance r later on to demarcate the near and the far field regions.

At this point it might be useful to introduce the concept of a point source [2] instead of an infinitesimal current element of a finite length l. If we focus our attention only to the far field region of the dipole, we observe that the electric and the magnetic fields are transverse to each other and the power flow or the Poynting vector is oriented in the radial direction. It is convenient in many analyses to assume that the fields from the antenna are everywhere of this type. In fact, we may assume, by extrapolating inward along the radii of the circle along the direction r that the waves originate at a fictitious volume less emitter, or a point source at the center of the observation circle of radius r. The actual field variation near the antenna, or the near fields is ignored and we describe the source of the waves only in terms of the far field it produces. Provided that the observations are made at a sufficient distance, any antenna, regardless of its size or complexity, can be represented in this way by a point source [2]. A complete description of the far field of a source requires knowledge of the electric field as a function of both space and time. For many purposes, however, such a complete knowledge is not necessary. It may be sufficient to specify merely the variation with angle of the power density from the antenna. In this case the vector nature of the field is disregarded, and the radiation is treated as a scalar quantity. When polarization of the fields is of interest then the variation of the nature of the fields must be specified as a function of time. Hence, we lose all the vector nature of the problem when we approximate an antenna by a point source [2].

Next, we consider the fields produced by a small circular loop of current.

2.3 FIELD RADIATED BY A SMALL CIRCULAR LOOP

Consider a small loop antenna of radius a placed symmetrically on the x-y plane at $z = 0$ as shown in Figure 2.2. The wire loop is considered to be very thin and is assumed to have a constant current I_ℓ distribution.

To calculate the field radiated by the loop, we first need to find the magnetic vector potential **A** as [2-4]

$$\mathbf{A}(x,y,z) = \frac{\mu}{4\pi} \int_c \mathbf{J}(x',y',z') \frac{e^{-jkR}}{R} dl' \tag{2.18}$$

$$R = \sqrt{r^2 + a^2 - 2ar\sin\theta\cos(\phi - \phi')}, \quad \text{and} \tag{2.19}$$

$$dl' = a\, d\phi \tag{2.20}$$

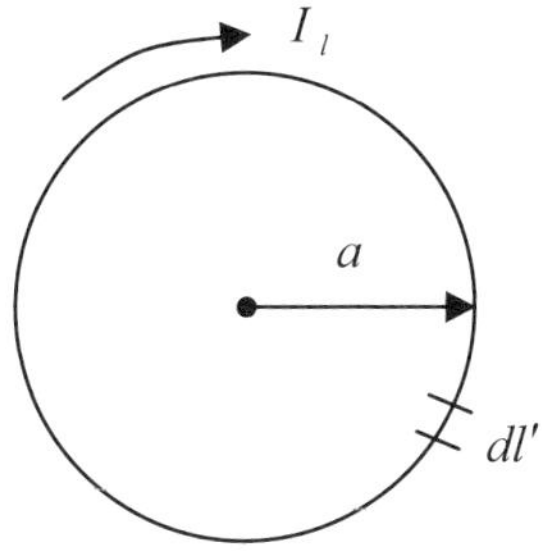

Figure 2.2. A loop made of very thin wire.

We have,

$$\mathbf{J} = \hat{\boldsymbol{r}}\ \mathrm{I}_\ell \sin\theta \sin(\phi - \phi') + \hat{\boldsymbol{\phi}}\ \mathrm{I}_\ell \cos\theta \sin(\phi - \phi') + \hat{\boldsymbol{\theta}}\ \mathrm{I}_\ell \cos(\phi - \phi') \tag{2.21}$$

$$\begin{aligned} x &= r \sin\theta \cos\theta & x' &= a\cos\phi' \\ y &= r\sin\theta \sin\phi & y' &= a\sin\phi' \\ z &= r\cos\theta & z' &= 0 \\ x^2 + y^2 + z^2 &= r^2 & x'^2 + y'^2 + z'^2 &= a^2 \end{aligned} \tag{2.22}$$

Therefore,

$$\mathrm{A}_\varphi \cong \frac{a\,\mu}{4\pi} \int_0^{2\pi} \mathrm{I}_\ell \cos(\phi - \phi')\, \frac{e^{-jkR}}{R}\, d\phi' \tag{2.23}$$

If we set $\phi = 0$ for simplicity, then the radiated field can be obtained as

$$\begin{aligned} \mathrm{A}_\varphi &\cong \frac{a\,\mu \mathrm{I}_\ell}{4\pi} \int_0^{2\pi} \cos\phi' \left[\frac{1}{r} + a\left(\frac{jk}{r} + \frac{1}{r^2}\right) \sin\theta \cos\phi'\right] e^{-jkr}\, d\phi \\ &\simeq \frac{a\,\mu \mathrm{I}_\ell}{4} e^{-jkr} \left(\frac{jk}{r} + \frac{1}{r^2}\right) \sin\theta \end{aligned} \tag{2.24}$$

The integration for A_r and A_θ becomes zero. Therefore

$$\mathrm{H}_r = j\frac{k\,a^2\, e^{-jkr} \mathrm{I}_\ell \cos\theta}{2r^2} \left[1 + \frac{1}{jkr}\right] \tag{2.25}$$

$$\mathrm{H}_\theta = -\frac{(k\,a)^2\ \mathrm{I}_\ell \sin\theta}{4r} \left[1 + \frac{1}{jkr}\right] e^{-jkr} \tag{2.26}$$

$$\mathrm{E}_\varphi = \frac{\eta (k\,a)^2\, e^{-jkr}\ \mathrm{I}_\ell \sin\theta}{4r} \left(1 + \frac{1}{jkr}\right) \tag{2.27}$$

$$\text{with } H_\varphi = E_r = E_\theta = 0 \tag{2.28}$$

It is interesting to note that the complex power density **S** radiated by the loop is given by

$$\mathbf{S}_{loop} = \frac{1}{2}\left(\mathbf{E}\times\mathbf{H}^*\right) \tag{2.29}$$

(Here we are using the peak values. For rms values the factor ½ will not be there) which yields

$$\mathbf{S}_{loop} = \hat{\boldsymbol{r}}\frac{(ka)^4}{32}\left|I_\ell\right|^2\frac{\sin^2\theta}{r^2}\left[1+j\frac{1}{(kr)^3}\right] \tag{2.30}$$

and the complex power P_r is given by

$$P_r = \oiint_S \mathbf{S}_{loop}\bullet d\boldsymbol{S} = \eta\left(\frac{\pi}{12}\right)(ka)^4\left|I_\ell\right|^2\left[1+\frac{j}{(kr)^3}\right] \tag{2.31}$$

Therefore, for a small loop of area $A_{loop} = \pi a^2$ carrying a constant current I_l, the far or the radiation fields are given by

$$H_\theta = -\frac{(ka)^2 e^{-jkr} I_\ell \sin\theta}{4r} = -\frac{\pi e^{-jkr} I_\ell \sin\theta}{r}\frac{A_{loop}}{\lambda^2} \tag{2.32}$$

$$E_\varphi = \frac{\eta(ka)^2 e^{-jkr} I_\ell \sin\theta}{4r} = \frac{120\pi^2 e^{-jkr} I_\ell \sin\theta}{r}\frac{A_{loop}}{\lambda^2} \tag{2.33}$$

It is now of interest to compare the far field expressions for a small loop with that of a short dipole. The presence of the factor j in the dipole expressions (2.14) and (2.15) and its absence in the field expressions for the loop indicate that the fields of the electric dipole and of the loop are in time-phase quadrature, the current I being assumed to be the same phase in both cases. The dipole is considered to be oriented parallel to the z-axis and the loop is located in the x-y plane. Therefore if a short electric dipole is mounted inside a small loop antenna and both the dipole and the loop are fed in phase with equal power, then the radiated fields are circularly polarized with the doughnut-shaped field pattern of the dipole.

2.4 FIELD PRODUCED BY A FINITE-SIZED DIPOLE

In order to calculate the fields from antennas, it is necessary to know the current distribution along the length of the antenna. For that we need to solve Maxwell's equations, subject to the appropriate boundary conditions along the antenna. In the absence of a known antenna current, it is possible to assume certain current distribution and from that calculate the approximate field distribution. The accuracy of the fields calculated will depend on how good an assumption was made for the current distribution. It turns out that for a thin linear wire antenna; the sinusoidal current distribution is a very good approximation. When greater accuracy is desired, and for the cases where the sinusoidal approximation breaks down (for thick or short dipole antennas, where the diameter of the wire is greater than one-tenth of its length) it is necessary to use a distribution closer to the true one. Our objective in this section is to bring out certain characteristic properties of antennas as they relate to the field distribution and hence applicable to beamforming.

Consider a symmetrical center-fed dipole antenna of length $L = 2H$. It will be assumed for convenience that the current distribution on the structure [2–4]

$$\begin{aligned} \mathbf{J}(z') &= \hat{\mathbf{z}}\, I_m \sin k\left(H - z'\right) \quad \text{for } z' > 0 \\ &= \hat{\mathbf{z}}\, I_m \sin k\left(H + z'\right) \quad \text{for } z' < 0 \end{aligned} \tag{2.34}$$

where I_m is the value of the current maximum. The vector potential at a point $P\,(x, y, z)$ due to the current element will be

$$\mathrm{A}_z = \frac{\mu_0}{4\pi}\left[\int_{-H}^{0} \frac{I_m \sin k\left(H + z'\right)e^{-jkR}}{R}\, dz' + \int_{0}^{H} \frac{I_m \sin \beta\left(H - z'\right)e^{-jkR}}{R}\, dz'\right] \tag{2.35}$$

where $R = \sqrt{(z - z')^2 + y^2}$, with $x' \equiv y' \equiv 0$. In cylindrical coordinates, the magnetic field in the Y-Z plane, H_ϕ at the point P can be obtained as

$$\mathrm{H}_\varphi = -\frac{I_m}{4\pi j}\left(\frac{e^{-jkR_1}}{y} + \frac{e^{-jkR_2}}{y} - \frac{2\cos kH\, e^{-jkr}}{y}\right) \tag{2.36}$$

The electric field in the $x = 0$ plane is given by

$$\mathrm{E}_z = -j30I_m\left(\frac{e^{-jkR_1}}{R_1} + \frac{e^{-jkR_2}}{R_2} - \frac{2\cos \beta H\, e^{-jkr}}{r}\right) \tag{2.37}$$

and

$$\mathrm{E}_y = j30J_m\left(\frac{z - H}{y}\frac{e^{-jkR_1}}{R_1} + \frac{z + H}{y}\frac{e^{-jkR_2}}{R_2} - \frac{2\cos kH}{y}\frac{e^{-jkr}}{r}\right) \tag{2.38}$$

The parameters R_1, R_2, and r are given by Figure 2.3. These equations provide the electric and the magnetic fields both near and far from an antenna carrying a sinusoidal current distribution. This is an exact analytic expression for the assumed current distribution.

It is interesting to observe that E_z represents a juxtaposition of three spherical waves. One each from the ends of the antenna and the last term is originating from the feed point. However, for a half wavelength long dipole

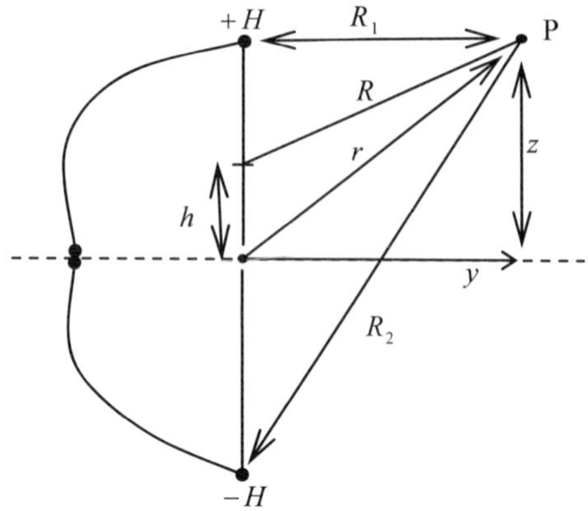

Figure 2.3. The geometry of a dipole antenna and the location of the field point.

$H = \lambda/4$, the third term in (2.38) disappears and the total field is a combination of two spherical waves originating from two isotropic sources located at the ends of one dipole. The other important point to note is that the field can never be zero at any finite distance from the dipole. The only place where the field can be zero is when the distance $y \to \infty$ and we talk about the radiation field. The electric field is a complex quantity and the real part of E_z is zero when the imaginary part is a maximum and vice-versa. The absolute value of the fields decays monotonically as we recede from the antenna.

Therefore, it is interesting to observe that the fields from a finite-sized dipole actually never become zero as a function of angle independent of the distance from the antenna. In addition, in the near field, an antenna beam pattern cannot be defined as the beam will be range dependent and has no nulls except at a few isolated points. We now look at the radiation fields from any finite arbitrary shaped antenna radiating in free space as in real life there are no elementary Hertzian dipoles.

2.5 RADIATION FIELD FROM A LINEAR ANTENNA

To calculate the radiated far field from a linear antenna certain approximations can be made. Even though these mathematical approximations are independent of the nature of the antenna and are universal in nature, here we apply it to a simple dipole antenna. The magnetic vector potential for a z-directed straight wire of length L and carrying a current $I(z)$ can be written as [2–4]

$$A_z = \frac{1}{4\pi} \int_{-L/2}^{L/2} I(z') \frac{e^{-jkR}}{R} dz' \tag{2.39}$$

$$R = |\boldsymbol{r} - \boldsymbol{r}'| = \sqrt{r^2 + z'^2 - 2rz'\cos\theta} \tag{2.40}$$

where $r^2 = x^2 + y^2 + z^2$ and $z = r\cos\theta$. If we consider that $r >> z$, i.e., we are quite far away from the antenna, then one can simplify the expression for R using a binomial expansion

$$R = r\left[1 + \frac{z'^2 - 2rz'\cos\theta}{r^2}\right] \approx r\left[1 + \frac{1}{2}\frac{z'^2 - 2rz'\cos\theta}{r^2}\right]^{1/2}$$
$$\approx r - z'\cos\theta + \frac{1}{2}\frac{z'^2}{r}\sin^2\theta + \frac{1}{2}\frac{z'^3}{r^2}\cos\theta\,\sin^2\theta + \text{other terms} \qquad (2.41)$$

while neglecting the remaining of the higher order terms. It is important to note that when the third term becomes maximum at $\theta = \pi/2$, the fourth term becomes zero.

Now "R" appears in both the denominator of (2.39) as an amplitude term and in the numerator in the form of a phase term. It is sufficient to approximate $R \approx r$ for all practical purposes for the denominator. For the phase term we need to consider the function kR, by neglecting the last term of (2.41)

$$kR = k\left[r - z'\cos\theta\right] + \frac{1}{2}\frac{kz'^2}{r} \qquad (2.42)$$

If we consider a three bit phase shifter, then it has eight phase states between 0° and 360° in steps of 45°. Therefore, the maximum phase error that can be tolerated across the aperture in this case is a maximum of 45°. Over half of the aperture the acceptable phase error is 22.5°. [It is not clear what the origin of this choice was. Here, we explained the choice by an example]. If we bound the error by this magical number, and find the value of r (it is a rule of thumb for far field approximation) at which the following equality is satisfied, i.e.,

$$\frac{1}{2}\frac{k\left(L^2/2\right)}{r} \leq \pi/8 \quad \text{or} \quad \frac{1}{2}\frac{kL^2}{4r} \leq \pi/8 \text{ ; therefore,} \quad r \geq \frac{2L^2}{\lambda} \qquad (2.43)$$

Here L is the entire length of the antenna. The rules of thumb for small ($L < 0.33\lambda$) and intermediate ($0.33\lambda < L < 2.5\lambda$) values of L, dictate that the far field commence at $r > 5D$ and $r > 1.6\lambda$, respectively [23] instead of (2.43).

The question that is now addressed is if this dipole is placed over a perfect ground plane, what will be the size of the effective aperture. In that case, we assert that this length L must be an equivalent length as it should include the image of the antenna below the ground plane. Equivalently, L is the diameter of the circle that encompasses the entire source antenna along with all the images produced by its operational environment, i.e., this circle must encompass both the original source and its image. For illustration purposes, consider a center-fed half wave dipole antenna operating in free space. From (2.36) to (2.38), it is observed that only two spherical waves emanate from the two ends of the dipole and nothing contributes from the middle section of the dipole. The field from a center-fed dipole is then equivalent to two point sources separated by $\lambda/2$. The far field from such a configuration starts at $2L^2/\lambda$. This is also equivalent to a $\lambda/4$ monopole radiating over a ground plane where the field is produced by a single spherical wave source over a perfect ground plane. One can then use this observation to predict the far field of center fed dipoles that are located on top of towers. For antennas located on top of high towers operating over earth, the value for L in (2.43) should encompass all the antenna sources along with their images.

Alternately, the Fresnel number, N_F, has been introduced in the context of diffraction theory to arbitrarily demarcate the separation between the regions of the near and the far fields. It is a dimensionless number and it is given in optics in the context of an electromagnetic wave propagation through an aperture of size V (e.g. radius) and then the wave propagates over a distance W to a screen. The Fresnel number is given by

$$N_F = \frac{V^2}{W\lambda} \tag{2.44}$$

where λ is the wavelength. For values of the Fresnel number well below 1, one has the case of Fraunhoffer diffraction where the screen essentially shows the far field diffraction pattern of the aperture, which is closely related to the spatial Fourier transform of the complex amplitude distribution of the fields after the aperture. In Fraunhoffer diffraction, $N_F << 1$; and the diffraction pattern is independent of the distance between the aperture and the screen, depending only on the angles to the screen from the aperture. For Fresnel number around 1 or larger, i.e., $N_F \geq 1$, one has the case of Fresnel diffraction (near-field diffraction) where mathematical description of the fields is somewhat complicated. If we assume the Fresnel number to be $N_F = 0.125$ which is $<< 1$, i.e., we are in the far field Fraunhoffer region, then (2.44) transforms to (2.43) as $W > 8V^2/\lambda$ and V is the radius and so we need to consider $L = 2V$ providing essentially the same approximation for the far field condition.

Next we look at some examples.

2.6 NEAR- AND FAR-FIELD PROPERTIES OF ANTENNAS

2.6.1 What Is Beamforming Using Antennas

Another interesting topic is beamforming using antenna arrays. An antenna array is a group of antennas in which the relative phases of the respective signals feeding the antennas is reinforced along a defined direction and suppressed along undesired directions. The first point one should think about is: what is meant by the terms *an antenna beam* or *an antenna pattern*? As we have observed it is the field at infinity and the pattern is independent of the distance r. The antenna pattern is only part of the fields produced by the source which decays linearly with the distance.

In summary, when we talk about the antenna beam pattern, we always imply the far field pattern. Conversely, in a Maxwellian context one cannot speak of a near field antenna pattern since the radiated power is a function of the distance from the antenna and is a complex quantity. Hence, pattern nulls, as functions of the angular sector, are not produced. These subtle issues do not come into the picture, if one uses an unrealistic characterization of an antenna by an isotropic point source as then every region is part of the far field!

As an example, let us consider a half-wave center-fed dipole (L = 15 cm) of radius 1 mm radiating in free space. The far field of the dipole operating, let us say, at 1 GHz (λ = 0.3 m) starts at a distance ($2D^2/\lambda = 2 \times 0.15 \times 0.15/0.3$) = 0.15 m from the antenna, where D, the effective span of the radiating element, is D = 0.15 m. At a distance greater than 0.15 m, the field pattern of the dipole is given by the radiation plot as shown in Fig. 2.4 [1–4] as a function of θ from 0 to 90° and is independent of the distance from the antenna. One fourth of the radiation pattern is shown. The far field pattern has a null along the zenith. This is well known and is available in any standard textbook on antennas [1–4]. However, the principles presented have very far-reaching implications in a wireless communications environment, as we shall illustrate next.

Now, let us consider the same half wave dipole antenna on top of a 20-m high tower as is conventionally done in a mobile communication environment. Typically the transmitting antennas are placed on top of a tower to serve a cell or a region, where the tower is physically located. Here, we consider the effects of the Earth, first as a perfectly conducting [2] and then as an imperfectly conducting ground plane [5,6].

First let us assume that the dipole is situated over a perfect ground plane. This is an idealistic situation, but is used as an example to illustrate the physics. For a half wave dipole situated on top of a tower and radiating in free space, the far field will start not at 0.15 m but at a distance of $2 \times 40 \times 40/0.3 = 10.6$ km. As the antenna is situated on top of a 20-m tower

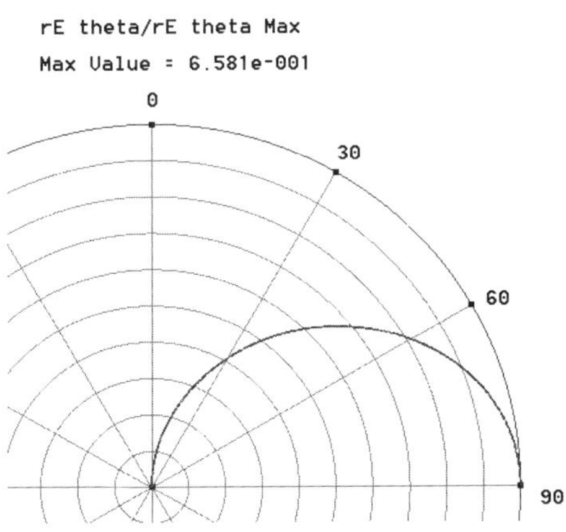

Figure 2.4. The radiation pattern of a half wave dipole in free space (only one fourth shown).

over a perfectly conducting ground, the effective transmission aperture D is 40 m and not 0.15 m. This is because the dipole antenna has an image which is 20 m below the ground, as the source dipole is located 20 m above the ground (Earth) [2,3]. It can be shown that, if we consider a half wave-length radiating dipole situated 20 m above a perfectly conducting Earth, its far field radiation pattern will be given by Fig. 2.5. The image produced by the perfectly conducting ground will be vectorially additive to the original free space field, producing the maximum value of the field strength along the horizon which is exactly doubled as seen from Figs 2.4 and 2.5. It is obvious from a Maxwellian point of view that this deployment of the antenna will produce a large number of maxima and minima and the antenna pattern will look awful in a vertical plane (as shown in the expanded version in Fig. 2.6) due to the vectorial interaction between the original field and the field produced by the image. This also affects the channel capacity as we will illustrate later on, as part of this radiation pattern (in the form of the complex vector **E**, the electric field) when integrated over the length of the receiving antenna induces a voltage $\left(V = \int \mathbf{E} \cdot d\mathbf{l}\right)$ at the feed point of the receiver. However, the radiation pattern is omni-directional along a horizontal plane. In any case, the far field pattern is irrelevant for this system, since a mobile user will invariably be in the near field for all practical purposes. Hence,

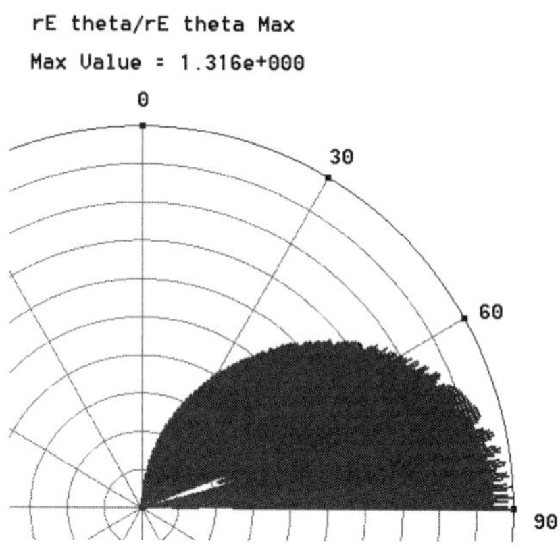

Figure 2.5. Radiation pattern of a half wave-length dipole situated on top of a 20-m tower located over a perfectly conducting ground (only one fourth shown).

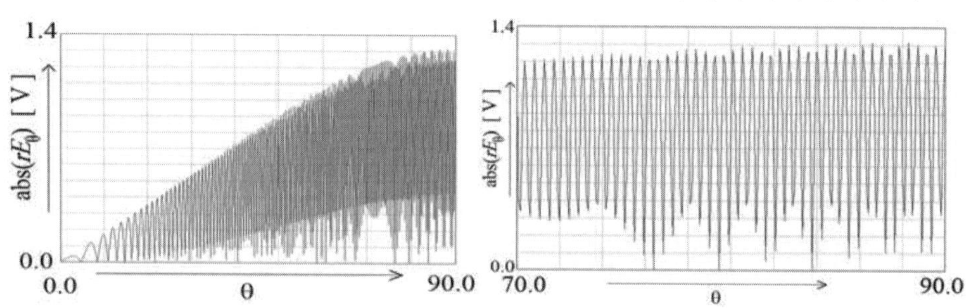

Figure 2.6. Expanded elevation radiation pattern of the half wave dipole situated on top of a 20-m tower located over a perfectly conducting ground.

one needs to carefully look at the near field of this antenna, which can be computed accurately using the Maxwell's equations and taking into account the effects of the environment [5,6].

For the sake of completeness, we consider the same half wave dipole antenna now placed on top of the tower situated over an imperfectly conducting ground, characterizing the real Earth with sandy soil, for an assumed relative permittivity of $\varepsilon_r = 10$, and a conductivity of $\sigma = 2\times10^{-3}$ mhos/m [p.893, 2]. The radiation pattern for this elevated dipole is shown in Fig. 2.7. (The field pattern is calculated using a commercial software package AWAS [6],which uses the Sommerfeld formulation). Observe that near the ground, the pattern is essentially the same as in Fig. 2.4, except that the fields are enhanced by 6 dB, or by a factor of 2 as the fields from the original source and its image add up at the ground. But near the zenith it is different as the strength of the field from the image is weaker due to the assumption of an imperfectly conducting ground. The expanded version of the pattern along the ground is shown in Fig. 2.8. The interference pattern in Fig. 2.7 is produced by the interaction of the free space field due to the dipole located in free space and the surface wave produced by the imperfectly conducting Earth [5]. The transmitting antenna launches the surface wave along the air-earth interface which attenuates exponentially as we go away from the antenna but otherwise decays slower than the space wave. This will be discussed at length later on.

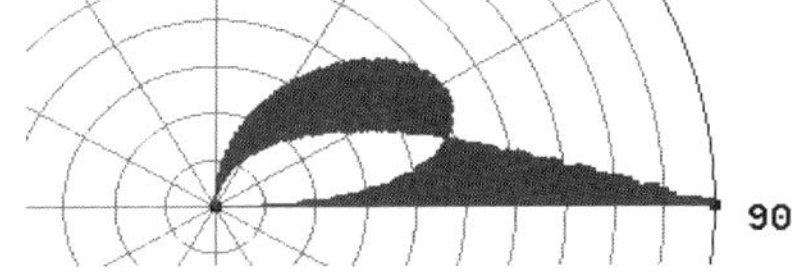

Figure 2.7. Radiation pattern of a dipole situated on top of a 20 m tower located over an imperfectly conducting ground (only one fourth shown).

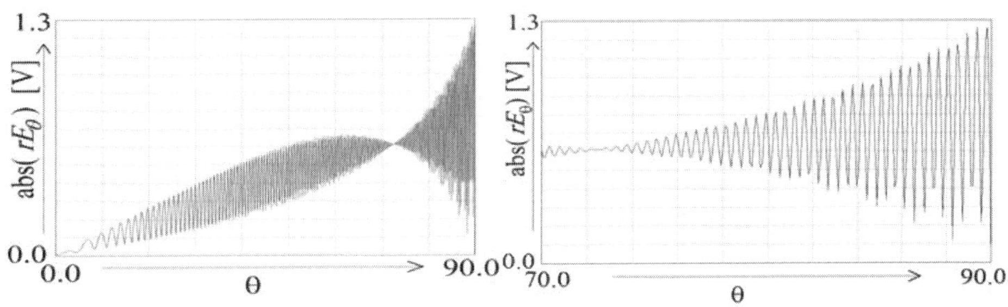

Figure 2.8. Expanded radiation pattern of a dipole situated on top of a 20 m tower located over an imperfectly conducting ground.

It is quite clear from Figs. 2.5 to 2.8 that placing an antenna on top of a tower produces a highly undesirable radiation pattern. Moreover, the signal strength goes through a series of maxima and minima, even though the antenna is located and radiating in free space. This type of radiation pattern for a transmitting dipole antenna is natural due to the interference between the fields from the original dipole with its image created by the presence of the ground. It is important to point out that if the antenna is located 100 m or higher above the Earth then the lobing effects in the pattern will be minimized as the strength of the image will be reduced due to the losses in the ground. The measurements carried by Okamura et al. [7] had the transmit antenna located in some cases 800 m above the ground plane. In that case, as the Sommerfeld theory predicts [5,6] the effect of the surface wave will be minimal as we go away from the antenna and the interference between the space wave and the surface wave will be less reducing the lobing effects in the far field patterns. However, what happens in the near field, can be calculated quite accurately but it is not possible to provide as clear a picture as shown for the far field patterns.

It is quite pertinent to ask at this point how with such a severely distorted radiation pattern one is going to perform beam steering or even beam forming! Beam forming or beam steering is out of the question in the near field as a beam or field pattern is not defined. In the far field pattern with so many nulls it is difficult to see what beam steering implies. Moreover, since the far field for a dipole situated 20 m on top of a ground plane starts at 10.6 km from the tower, then the cell, which this antenna is going to serve, will be definitely be in the near field. If that is the case, then what does beam forming mean in a near field environment? This point must be addressed from a Maxwellian point of view, based on actual physics, if we want to have an effective operating system! Assumption of a point source for the antenna bypasses all these subtleties as a point source has no near fields. But as shown before, such an assumption is not correct!

In view of these subtleties in the analysis of the vector antenna problem, if we consider a set of transmitting antennas directing energy to a receiving antenna in a communication system, then, what does it mean to direct a beam if the receiving antenna is in the near field of the transmitting antenna. For example, what do we mean by directing a beam formed by using an array of transmitting antennas to a designated receiving antenna in a cell or microcell? Another important issue, in the plots of Figures 2.5–2.8 is that the antenna is operating in a free space environment and there are no other scatterers. It is often assumed that if there are many scatterers present near the tower then these nulls in the pattern will fill up! However, for a vector electromagnetic scattering problem, it is not clear why that will be the case. In fact, the problem will get much more complicated if the presence of other near field scatterers is taken into account in the computation of the beam pattern. In the near field, the antenna beam can only be defined at a specific distance as it varies from point to point and moreover there cannot be any pattern nulls independent of the distance. Therefore, we need to know precisely the spatial locations of the transmitters and the receivers, if one is willing to direct the signal-of-interest (SOI) to the desired receiver. In addition, we must also know the near field environment, i.e., the location and shape of any possible scatterers in order to compute a distance dependent antenna beam. If such is the case then it is difficult to see how the simplistic approach to multiple-input multiple-output (MIMO) that many communication papers present will really work in practice, with such a severely distorted element pattern. A disbeliever of the Maxwellian principles may say, MIMO does not deal with beams but direct energies, or employs the principle of superposition to increase the probability of communication. However, these principles must be demonstrated from a physics-based Maxwellian point of view. In fact, the message we are trying to convey is that all these systems can be analyzed with great numerical accuracy using currently available numerical electromagnetics code, for example [8], and most physical principles can be verified.

Secondly, in a microcell and in a picocell and with a deployment of the antennas at a large distance from the ground, we are always operating in the near field. Hence under these conditions, this complicated element fields can be analyzed accurately using Maxwell's equations as is done in [8]. Many researchers often use ray tracing to characterize the multipaths in this propagation MIMO environment in order to characterize the channel. We need to initiate a discourse about how meaningful is near field ray tracing from a Maxwellian point of view, and even in the far field when the antenna element pattern has a large number of maxima and minima!

Finally, when dealing with radiation pattern of antennas, a distinction must be made between antenna directivity and antenna gain [2]. The antenna directivity is defined as the ratio between the maximum radiation intensity or watts per square radian from the antenna under consideration and the radiation intensity from an isotropic point source radiating the same power. Therefore the directivity $\mathcal{D}$ is defined by [2,9–11]

$$\mathcal{D} = \frac{\text{Maximum Radiated Power Density}}{\text{Average Radiated Power}} = \frac{4\pi\,(\text{Maximum Radiated Power Density})}{\text{Total Power Radiated}} \tag{2.45}$$

The definition of directivity is based entirely on the shape of the radiated antenna pattern. The power input and antenna efficiency is not involved. Ideally, the number of point sources an antenna can resolve is numerically equal to the directivity of the antenna [2]. Equivalently, directivity is equal to the number of beam areas into which the antenna pattern can subdivide the observation region of interest and provides the added significance that the directivity is equal to the number of point sources in the sky that the antenna can resolve under the assumed ideal conditions of a uniform source distribution. C. T. Tai [12], in his paper on optimum directivity of linear arrays, shows the onset of superdirectivity to occur when the element spacings in the antenna array are less than half a wavelength, i.e., $\lambda/2$. As the element spacings are further decreased and the optimum directivity sought, the degree of super-directivity is increased. As the degree of superdirectivity is increased, so is the difficulty in practically implementing the synthesis. Hansen quotes the data of Yaru [13] who studied a nine element array with $\lambda/32$ spacing between the elements. The required tolerance for maintaining the designed –26 dB sidelobes was one part in 10^{10}. Hence, in a supergain condition when one uses super resolution methodology, theoretically the directivity of an antenna can take any value. For example, in the limit the spacing between the elements approach zero (i.e., for a continuous source distribution) it can be shown that the end fire directivity of an array can increase as the square of the number of antenna elements [9–11]!

However, when we talk about the gain of an antenna, the antenna efficiency is involved. The gain $\mathcal{G}$ of an antenna is defined by [2]

$$\mathcal{G} = \frac{\text{Maximum Radiation Intensity}}{\begin{array}{c}\text{Maximum Radiation Intensity from a}\\ \text{Reference Antenna with the Same Power Input}\end{array}} \tag{2.46}$$

Any type of antenna may be taken as the reference. Often the reference is a half wavelength long dipole. Gain includes the effect of losses both in the antenna under consideration (subject antenna) and in the reference antenna. It is important to note that the IEEE Standard definition for gain does not include impedance or polarization mismatches of the antenna. However, a practical system needs to consider these two very important factors. Sometimes it is convenient to assume that the reference antenna is an isotropic source of 100% efficiency. The gain so defined for the subject antenna is called the gain with respect to an isotropic source and is designated $\mathcal{G}_o$. Thus [2]

$$\mathcal{G}_o = \frac{\text{Maximum Radiation Intensity from Subject Antenna}}{\begin{array}{c}\text{Radiation Intensity from a Lossless Isotropic}\\ \text{Source with the Same Power Input}\end{array}} \tag{2.47}$$

The gain $\mathcal{G}_o$ is then related to the directivity by [2]

$$\mathcal{G}_o = k\,\mathcal{D} \tag{2.48}$$

So the gain of an antenna over a lossless isotropic point radiator equals the directivity if the antenna is 100% efficient (i.e., $k = 1$) but is less than the directivity if any losses are present in the antenna ($k < 1$).

Except for very small arrays superdirective antenna arrays have proven to be impractical because they have large current elements and high loss, and require very precise excitation. In most cases, they are also very narrow band. Superdirectivity is produced using rapid phase variations across an array of closely spaced elements. Unfortunately, the higher directivity results from an interference process, and only the sidelobes are in real angular space θ, with the pattern main beam in or partly in "invisible space" ($\sin\theta > 1$). The resulting ratio of stored to radiated energy (Q) is extremely high and so the frequency bandwidth is very small. Furthermore, since the radiation resistance is very low, efficiency is poor and the antenna noise temperature is high in the presence of losses due to finite antenna and matching network conductivity. Since the high directivity depends on cancellation of the contribution from the array currents, superdirective array behavior is dependent on highly accurate current determination and small errors in array excitation can destroy the properties of superdirective arrays.

Therefore, as has clearly been pointed out by Hansen [p. 102, 9]: *A fixed aperture size can achieve* (*in theory*) *any desired directivity value*; delineating the fact that a superdirective array can have any value for the channel capacity unless the input power is restricted! However, the value for the gain cannot increase arbitrarily and there will be a limit, as the input power is limited. Hence, for practical usage it is the gain that is very useful. The value for directivity in a practical system does not enter into the picture as it is not tied to any of the system parameters like input power and therefore has little practical relevance/importance in antenna engineering. This subtle distinction between directivity and gain is often ignored by communication and even antenna engineers resulting in erroneous conclusions. It is important to note that almost all communication theory materials related to communication using antenna arrays including the Shannon Channel capacity theorem do not mention anything about the input power to the system. They only refer to the radiated power and therefore to the directivity. Hence, we come to the familiar result for antenna pattern synthesis: if one is operating in a supergain condition and is focused only on the results for the directivity of the antenna array without considering the radiation efficiency then one can obtain any result one wants! Therefore, all meaningful presentation of the channel capacity and comparison of performance should be made when the input power to the antenna is kept fixed.

2.6.2 Use of Spatial Antenna Diversity

To correct for signal cancellation in a complex environment, as illustrated even with a simple example in the previous section (an antenna located on top of a tower), it is often suggested that the use of more transmitting antennas by increasing spatial transmit diversity is going to mitigate the signal loss due to multipaths at the receiving antenna. Again, it is difficult to accept such a conjecture under the present scenario when we observe the radiation patterns of Figures 2.4–2.8. The electric field is a vector quantity and obviously adds according to the specific rules that apply for these types of vector quantities. In Figs 2.4–2.8, the interference of the signal produced by the antenna and its image due to the ground over which the antenna is placed occur in free space in the absence of any obstacles. The presence of obstacles will further complicate the scenario. Given the highly distorted element pattern of the half wave dipole located on top of a tower, it is impossible to see how adding more antenna elements in the transmit mode is going to eliminate some of the pattern nulls. To the contrary, addition of more antenna elements will deteriorate the scenario further as Maxwell's equations demonstrate. For example, we know from a simplified point source antenna array theory that the resultant radiation pattern is the product of the element radiation pattern and the array factor [2–4]. If we now consider an array of isotropic omni-directional point radiators (radiators as mentioned in section 2.2, by the way, do not exist in reality), then the array pattern will no longer be isotropic, as the array factor by itself produces pattern nulls. Thus, if an antenna element pattern has Q number of nulls, an array of such elements will definitely have more than Q nulls under all conditions! Therefore, one needs to initiate a dialogue to see how diversity with a highly distorted element pattern makes any scientific sense in improving the mode of communication.

However, we may reach some bizarre conclusions if we apply probability theory. For example, because the antenna element pattern has so many maxima and minima then the antenna pattern can be treated hypothetically as a random variable, and using the *central limit theorem* [14] one can reach the erroneous conclusion that the addition of more antenna elements is going to improve the scenario and the field strength at a spatial location will improve due to ensemble averaging. Often when dealing with the analysis of very large communication system problems, we introduce the concepts of probability theory either to introduce uncertainty into the model or supplement our incomplete knowledge about the system [14]. In addition when a problem is too large or too complex to solve by any other method we take recourse to using probability. One point to be made is that a probabilistic model cannot be used to supplement insufficient or inadequate knowledge about a system [14]. Through the introduction of a probabilistic model we cannot obtain *more* knowledge about a system. We merely make some assumptions about the underlying process (through the use of the ensemble) so that we can apply available analytical tools to solve a problem without giving further thought as to whether the assumptions made for the system are indeed relevant. A probabilistic description of a random process is not adequate since additional information about the system may be required, with such information generally unavailable. This is why assumption of a Gaussian or a Markov process is often made, as for such probability distributions all the higher order probability density functions can easily be determined. In short, the analysis is often simplified through the use of wide sense stationarity or the assumption of an ergodic system. Again, it should be pointed out, the scalar probability theory does not handle correctly the various vector interactions of the different components of the electric fields. Also, the introduction of the concept of a random process in the analysis of a system presupposes that one will obtain a solution, which will be a possible outcome, in an ensemble. However, the probabilistic solution may not be an appropriate one. Sometimes the application of probability theory may provide some very interesting counterintuitive result [15].

Taking a physics-based approach not only tells us which system does not perform as we would like, but also we could predict what can be a viable solution. Therefore, to understand basic properties of an antenna and its fundamental limits in performance, it is imperative to apply Maxwell's theory to obtain the design guidelines rather than using statistical modeling. Unless backed by the Maxwellian physics the statistical modeling will lead to erroneous conclusions. For example, if we use a different deployment of the transmitting antennas instead of using the concept of spatial diversity, then one might get a better solution to the signal transmission problem instead of applying some statistical procedures. For example, the signal cancellation problem can be largely diminished if the antenna is placed close to the Earth and not on top of a tower as the following example will illustrate. We have seen that the gain of a vertically oriented half wave dipole located in free space is given by the plot marked '1' of Fig. 2.9 and this gain is minimum only along the zenith. We now place the same antenna almost touching an infinitely conducting ground plane. The gain pattern of such an antenna system is given by the plot marked '6' in Fig. 2.9. For the same voltage excitation, the gain of an antenna very close to the ground along the horizontal plane is increased by 6 dB over the same antenna when it is located in free space. This is the positive aspect due to the image in the ground.

Next, we consider the same half wave dipole antenna on top of a 20-m tower. It is seen in the plot marked '2' of Fig. 2.9 that the gain pattern has too many nulls due to the ground and the interesting part is that it does not get any better whether we use 3 (plot marked '3' of Fig. 2.9) or 21 (plot marked '4' of Fig. 2.9) dipole elements spaced half a wavelength apart on top of the tower. The only difference is that the radiated fields get stronger. The y-axis provides the value of the gain in dB and the x-axis is the elevation angle. If the single antenna is now brought close to the ground and placed 15 cm on top of it, the gain is given by the plot marked '5' of Fig. 2.9. Observe that the gain along the horizontal plane of the plots '2' and '5' are very similar along the azimuth

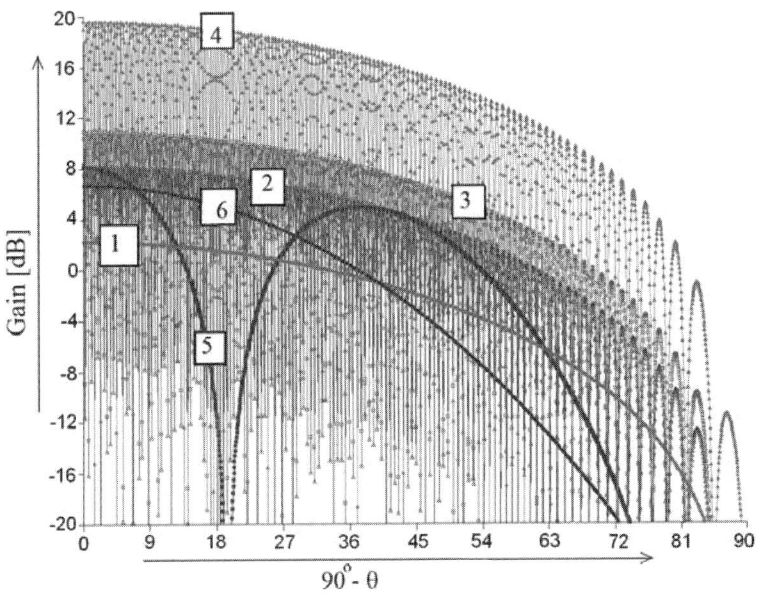

Figure 2.9. Radiation pattern of a half wave dipole elements located in various environments.

directions. Hence, it is quite illustrative that diversity for antennas located on top of a tower has very little real physics. However, by treating the plot marked '2' in Fig. 2.9 as a scalar random variable one can essentially play any theoretical game one likes but it is not going to change the reality that this antenna will have an awful pattern and putting in more antennas in reality will not improve the situation! The only way to make things work is to place the antenna close to the ground and generate similar gain pattern marked as '5' in Fig 2.9, which does not contain a large number of antenna pattern nulls.

Bringing the transmitting antenna closer to the ground may at first sound paradoxical because most broadcast antennas in TV or mobile communications are placed on top of a tower for wide area coverage. Even though these devices work at frequencies, where propagation is primarily line-of-sight (LOS), we all know that the TV and mobile communication signals are often received inside a house or a car. So first, we must realize that the propagation is not strictly LOS. Secondly, for TV, often we transmit or receive the signals with a high gain Yagi-Uda antenna [It is a directive antenna and for details refer to the encyclopedia in reference 16] and thus we discriminate between the source and its image, thereby eliminating the patterns with multiple nulls. In mobile communication, since we typically do not use such highly directive receiving antennas it makes sense to place the antenna closer to the ground and use more transmitting antennas, if desired, strategically located to increase the signal coverage rather than using the concept of spatial diversity by deploying more antennas on top of a tall tower as seen in the gain patterns marked by '5' of Fig. 2.9. Deploying transmitting antennas close to the ground will also increase the channel capacity as we will demonstrate later in Chapter 4. Addition of more antennas under the guise of diversity clearly indicates that there is no improvement in the transmit radiation pattern. What makes an antenna pattern more useful is if the antenna is located closer to the ground than on top of a tower, or very far away from the Earth as done by Okamura et al. [7] in his seminal work on measurements of field strength in an urban environment.

However, in a receiver one can use multiple antennas and then use a switch to couple to the signal of largest amplitude in any of the receiving antennas to improve the received signal to noise ratio. But then, we are not vectorially adding all of the electromagnetic field components incident on the receiving antenna simultaneously, but selecting an individual candidate.

A disbeliever of the Maxwellian principles may argue that if there are buildings or other near field scatterers then the pattern nulls shown in Fig. 2.9 may fill up due to the multiple scattering of the patterns and that perhaps the pattern may have no nulls! Yet no evidence is provided, to demonstrate that such a situation may even remotely be true for the vector electromagnetic problem.

2.7 THE MATHEMATICS AND PHYSICS OF AN ANTENNA ARRAY

When dealing with phased arrays [16–18], we want to establish one of the mathematical underpinnings of phased array theory, which has to deal with the computation of the direction of arrival (DOA) of the various signals impinging on the array. It is also connected with adaptively enhancing the signal in the presence of interferers and noise. When dealing with sensor elements in acoustics, we define them primarily as isotropic point radiators. This makes perfect sense as a loudspeaker does not distort the input electrical signal when transforming it to acoustics or a microphone does not distort the input acoustic signal when transforming it to electrical ones. A uniform linear array consisting of isotropic point radiators separated by a distance d, is shown in Fig. 2.10 (the figure is specialized to an one-dimensional array). The signal of interest (SOI) impinging on the array is denoted by S, the interferers by J and the clutter, which is diffused reflected transmitted energy is characterized by C. Each of these signals has a complex amplitude A_i that impinge on the array from an elevation angle of $(90°-\theta_i)$ and an azimuth angle of φ_i. Then the voltage $V_{n,m}$ induced at the $(n, m)^{\text{th}}$ element of a two-dimensional array oriented along the x and y axes is given by

$$V_{n,m} = \sum_{i=1}^{P} A_i \exp\left[\begin{array}{l} \dfrac{j2\pi(n-1)d}{\lambda} \cos\varphi_i \; \sin\theta_i \; + \\ \dfrac{j2\pi(m-1)d}{\lambda} \sin\varphi_i \; \sin\theta_i \end{array}\right] \tag{2.49}$$

where λ is the wavelength of transmission and P is the total number of signals impinging on the array. Here, it is assumed that P is much less than the total number of the antenna elements. Therefore, for Fig. 2.10, when we examine equation (2.49), it is seen that the DOA $\{\cos\varphi_i \sin\theta_i\, ; \sin\varphi_i \sin\theta_i\}$ of the various signals impinging on the array and the voltages $\{V_{n,m}\}$ induced in the antenna elements in an acoustic phased array form a two-dimensional discrete Fourier transform (DFT-implemented through the Fast Fourier Transform, FFT) pair. This is the mathematical basis for an acoustic phased array theory for performing estimation of DOA of the various signals impinging on the antenna array.

Now, let us assume that we are dealing not with the scalar acoustic problem, but with the vector electromagnetic problem. In an electromagnetic problem, isotropic point radiators do not exist. The smallest antenna element is a

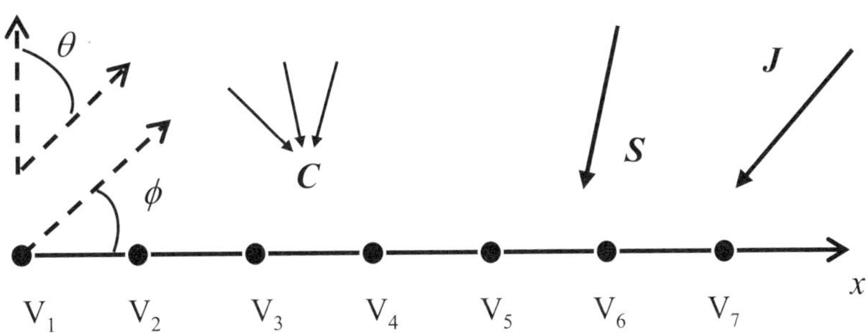

Figure 2.10. A phased array (one-dimensional) consisting of isotropic point radiators.

dipole and it has a radiation pattern which has a doughnut-shaped pattern, even though it may be omni-directional along certain planes. Also, in electromagnetics one has the added dimension of polarization, which is missing in acoustics. Finally, one has to be careful about the mutual coupling between the antenna elements and the presence of near field scatterers which may distort the basic assumption of an isotropic radiator. Now, if we assume that a plane electromagnetic wave with a linear polarization is impinging on an array consisting of z-directed Hertzian dipoles as shown in Fig. 2.11, then the voltages induced in the Hertzian dipoles will be a function of the polarization of the incident field. For the appropriate polarization, the voltage induced at the $(n, m)^{\text{th}}$ Hertzian dipole element could be expressed as

$$V_{n,m} = \sum_{i=1}^{P} A_i \sin\theta_i \exp\left[\begin{matrix} \dfrac{j2\pi(n-1)d}{\lambda} \cos\varphi_i \sin\theta_i + \\ \dfrac{j2\pi(m-1)d}{\lambda} \sin\varphi_i \sin\theta_i \end{matrix}\right] \tag{2.50}$$

Observe that we have now an extra factor of $\sin\theta_i$ in (2.50) when compared with (2.49). By comparing (2.50) with (2.49), it is seen that the DOA $\{\cos\varphi_i \sin\theta_i ; \sin\varphi_i \sin\theta_i\}$ and the induced voltages $\{V_{n,m}\}$ do not form a Fourier transform pair, even though they can be when $\theta_i = 90°$, or the incident field is polarized parallel to the Hertzian dipole elements. Moreover, in a real situation, when there are circularly polarized fields or when we have a finite dimensional antenna and not a small Hertzian dipole, it is necessary to modify (2.50) to obtain the DOA of the various signals impinging on an array. This modification is also necessitated due to the existence of mutual coupling effects between the antenna elements or other near field interactions between the antenna and the surroundings or the platform on which they are mounted [19].

In the presence of mutual coupling among the antenna elements and other near field effects, (2.50) actually transforms into a very complicated integro-differential equation and one has to take recourse to Maxwell's equations in addition to signal processing algorithms to solve the complete DOA estimation

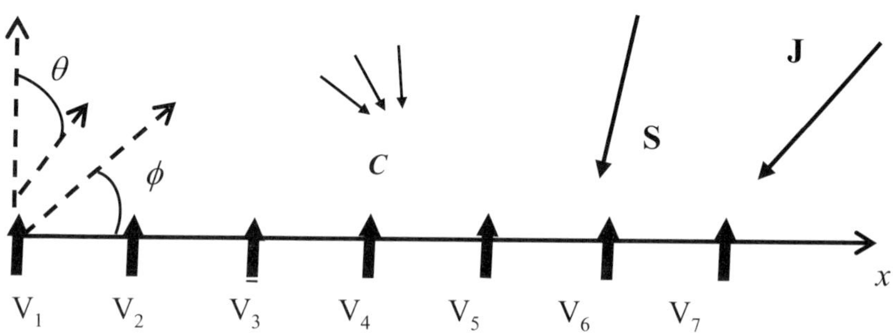

Figure 2.11. A phased array (one-dimensional) of Hertzian dipoles as antenna elements.

problem [3]. An example for an antenna, which is not a point source or a small Hertzian dipole, is a finite-sized dipole antenna. Even for a z-directed thin-linear straight wire antenna of finite length, the incident field E^{inc} and the current distribution $I(z')$ on the antenna are related by the following integro-differential equation [2–4]:

$$E_z^{inc} = -\mu_0 \int_{axes} I(z') \frac{e^{-jkR}}{4\pi R} dz' + \frac{1}{\varepsilon_0} \frac{\partial}{\partial z} \int_{axes} \frac{\partial I(z')}{\partial z'} \frac{e^{-jkR}}{4\pi R} dz' \tag{2.51}$$

where $R = \sqrt{(z - z')^2 + a^2}$, and a is the radius of the wire antenna. The integral is taken along the axis of the antenna. The important point is that one needs to solve the Maxwell's equations to compute the voltages induced in actual electromagnetic problems, due to the presence of mutual coupling and the polarization of the incident fields. Thus, we need to refine our conventional scalar acoustic phased array model and its connected theories to deal with the vector electromagnetic case. Other aspects of this issue will be illustrated later. Furthermore, the impulse responses of an antenna are different in transmit and receive modes for the vector electromagnetic problem as will be discussed in Chapter 3. Use of point sources in the model instead of using the realistic antenna sizes and shape often ignores the mathematical subtleties introduced by Maxwell's equations. To simulate realistic scenarios it is necessary to characterize antennas by the complex integro-differential equations which correctly encompasses the principles of the vector electromagnetic problem.

2.8 PROPAGATION MODELING IN THE FREQUENCY DOMAIN

Propagation modeling is intimately related to the electromagnetic analysis of antennas. The basic formulation is still the same. However, in this section we look at the justification of the specific electromagnetic methodology used in propagation modeling. The connection is established because the propagation model uses the far field response of antennas in the field calculations. Our objective here is to demonstrate that such practices are questionable from the concept of near field and far field analysis. We now illustrate the problem.

Propagation modeling is an active area of research in mobile wireless communication and a significant number of papers [20,21] have appeared in numerous journals. It is interesting to note that most propagation models are based on the seminal experimental work of Okamura et al. [7] and many models are being developed to fit this experimental data. Okamura et al. [7] carried out electromagnetic field measurements in the city of Tokyo by placing an antenna on top of a mountain and then measuring the electric fields at various distances from the mountain starting at a few kilometers and up to 100 km from the mountain. The effective antenna heights for his measurements varied from 30 to 1000 m above the ground. They carried out the measurements at four different frequencies. One of the interesting points of the experimental data of Okamura et al. is that even in an urban environment, where there may be parks, tall buildings, trees and so on – namely a highly diverse environment – the plots for the fields are relatively smooth [7] on the average. The measured data covered both the near field and the far field regions. It is also interesting to observe that Okamura et al. in their measurements used a directive reflector antenna and a horn antenna located on top of a mountain. As the transmitting antenna is not a simple vertical dipole situated over a flat earth, this will clearly minimize the effect of the image of the antenna below the ground as the beam is directed more in the forward direction due to the directivity of the transmitting antenna. In addition, when an antenna is placed over a real ground, the lobing of the antenna pattern shown in Fig. 2.9 is due to the interference between the surface wave which decays exponentially from the transmitting antenna as it propagates and the space wave [5]. However, if the antenna is situated far above the ground like 100 m or higher then the strength of the image will be weak due to the ground losses and the lobing of the transmit antenna pattern will be dramatically reduced and will approach the pattern of Fig. 2.12–2.13 [6]. The lobing of the antenna pattern will be further reduced due to the use of a high gain antenna instead of a dipole. This is the reason why, in Okamura et al.'s measured data, the measured field comes closer to the free space values as the antenna is located 820 m from the ground plane and the measured data have a wide variability when the transmit antenna is located about 50 m from the ground. These can be explained quite well by looking at the explicit mathematical expression for the fields as a function of the polarizations [5].

Various researchers have then attempted to approximate the various slopes in the measured curves using many different ad hoc formulas. For example, Hata [20] used some curve fitting to generate various rules of thumbs for the propagation model. To this day, these papers are treated as the major source of information even though various evolutionary modifications to that formula have been made by various researchers [20,21]. A more revolutionary concept in the context of propagation modeling in mobile communication would be to observe how the field strength varies as a function of distance, if we propose to carry out an essentially exact field analysis of a vertical dipole located over an imperfect ground plane. Here, we will present the result using Maxwell's equations. This is the same problem treated by Arnold Sommerfeld in 1908. For the imperfectly conducting ground, we consider a real Earth with an urban ground [p.893, 2] of relative permittivity $\varepsilon_r = 4$ and with a conductivity of $\sigma = 2\times10^{-4}$ mhos/m. We use the classical Sommerfeld formulation and calculate the fields using a receiving antenna at a height of 3 m from the ground, as was done by Okamura et al. using the commercially available wire antenna analysis code AWAS [6] for analysis of antennas over an imperfect ground plane. This is similar to the experimental setup of Okamura et al. The various plots of the fields are shown in Figs. 2.12 and 2.13 for different transmitting frequencies.

It comes as no surprise that the Sommerfeld analysis duplicates the measurement results of Okamura et al. for the near and the far fields with reasonable correlation. The field plots obtained by the computer program AWAS are shown as dots in Fig. 2.12 with an assumed $\varepsilon_r = 4$ and $\sigma = 2\times10^{-4}$ for the

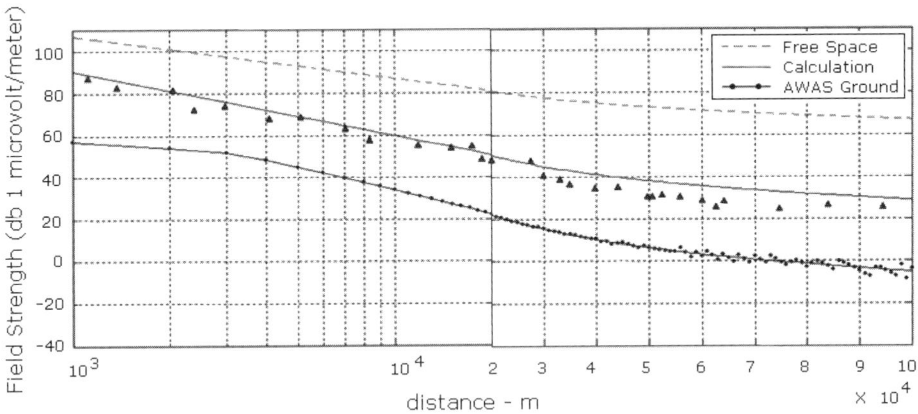

Figure 2.12. The field strength as a function of distance measured in an urban environment at 453 MHz.

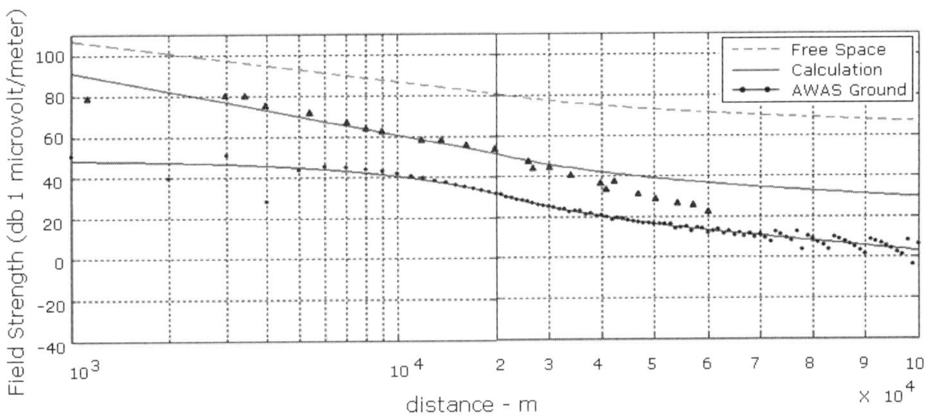

Figure 2.13. The field strength as a function of distance measured in an urban environment at 1430 MHz.

parameters of the ground in Tokyo. For this figure the operating frequency is 453 MHz. The height from the ground of the base station transmitting antenna is 140 m while the height of the receiving antenna is 3 m. The electric field variation of the antenna over a ground plane is shown by a dotted line and has the same slope as in the measurements [7]. The line (termed calculation) is the curve fitted model of Hata [20] which agrees with the experimental results of Okamura et al. [7], shown as triangles. The dashed is the free space field produced by the antenna. The difference in the absolute level of the electric field between the calculated (curve fitting of measurement data by Hata) and AWAS (numerical analysis) is due to the choice of different values of excitations for the transmitting antenna. For the measurement, the transmitting antenna had a high gain, whereas in numerical computation we considered a simple dipole, centrally excited by 1 V. However, the decay of the fields, both near and far show almost parallel curves. Fig. 2.13 shows another plot for the variation of the electric field from the transmitter at a frequency of 1430 MHz.

These two plots demonstrate that instead of curve fitting measured data to generate the characteristics of the signal strength in an urban environment it is possible to use computer codes based on an accurate analysis of the Sommerfeld theory derived from Maxwell's postulates to generate the predicted results which correlate with experiments. Second, if one knows the approximate electrical parameters of the Earth, then it is possible to predict how the field strength will vary with distance without carrying out expensive experimental measurements. Thirdly, even though the realistic urban environment is quite complex these results indicate that even under those circumstances it is possible to carry out accurate numerical electromagnetic simulations with commercially available computer codes in the frequency domain which approximates the natural environment. An interesting question that arises at this point is, how come Okamura's measured results and an accurate electromagnetic simulation do not show any fading?

Another problem that is often treated extensively in the Antennas and Propagation literature is the use of ray tracing for an indoor mobile communication environment [20,22]. Typically, to find the fields inside a room, a hallway, or for roof top propagation, one generally uses a ray tracing technique to compute the field distribution in the neighboring environment. Such a methodology should be highly questionable from a scientific view point because of the existence of the floors and walls in the model which will produce multiple images of complex amplitudes. Hence, it is not sufficient to use just the antenna element pattern in ray tracings but one should take into account the existence of the various images for the antenna element and moreover one will be dealing with a near field prediction problem which precludes the use of ray tracing as the latter is more suitable for far field predictions. As an example, consider a center fed half wave $\hat{z}$ -directed dipole located at the center (given by the origin of the coordinate system) of a $7\,\lambda \times 7\,\lambda \times 7\,\lambda$ room which has a wall of thickness 0.03 λ. The relative dielectric constant of the wall all around this cube is considered to be $\varepsilon_r = 2.5$. Next, we plot the various components of the fields inside this dielectric cube produced by the center fed half wave dipole as computed by the computer code of [8]. The fields inside the room are shown in Figs. 2.14–2.17 at various locations and compared with the field that would exist if the dipole were to be operating in free space.

As seen from Figs. 2.14–2.17, it is clear that the effects of the images of the initial dipole source generated by the dielectric walls, floor and the ceiling will clearly distort the fields in such a way that simple ray tracing using the antenna element pattern of free space is not sufficient to predict the correct fields. Because of the generations of the various images computation of the fields inside the dielectric cube is a near field problem. Hence the applicability of ray tracing itself may be questionable. This clearly demonstrates that in the computation of the fields inside a room, a hall way or for a roof top propagation, the effect of the

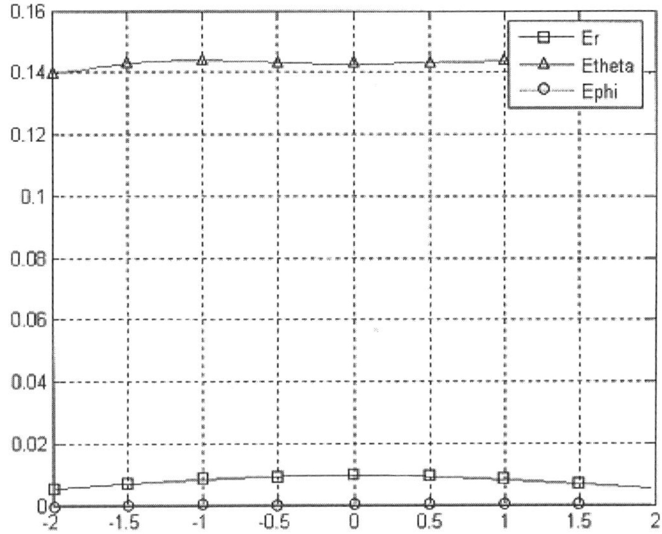

a) free space

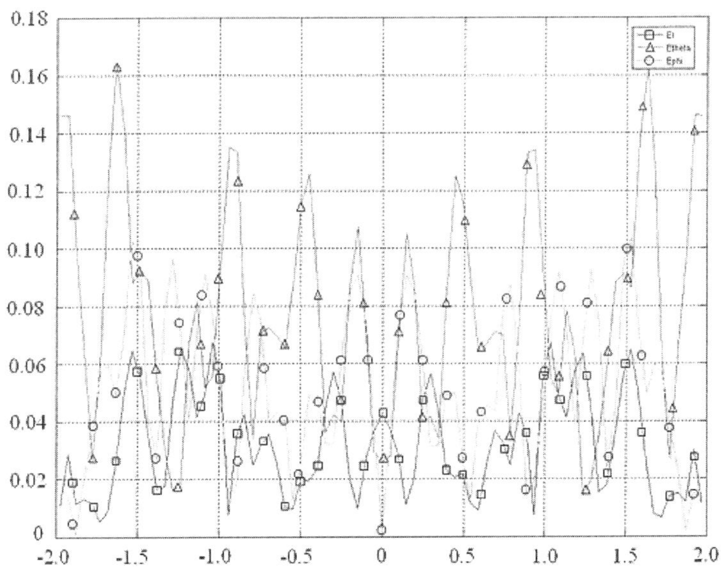

b) dielectric room

Figure 2.14. The three components of the fields E_r, E_θ, E_φ inside the room at $x = -3.75\lambda$, $z = -3.75\lambda$, as a function of y.

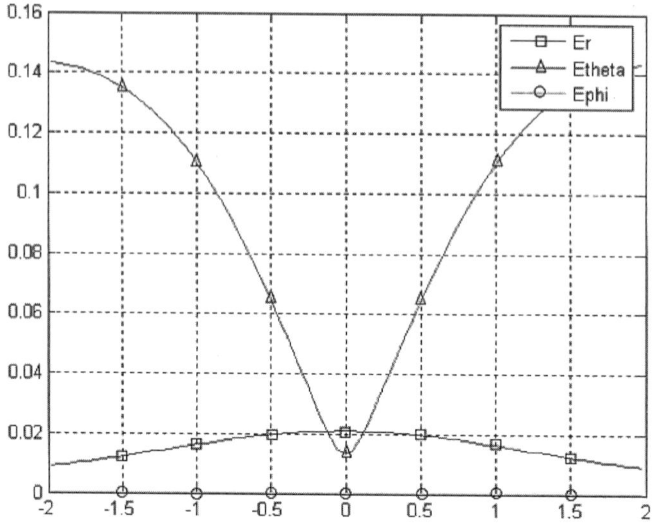

a) free space

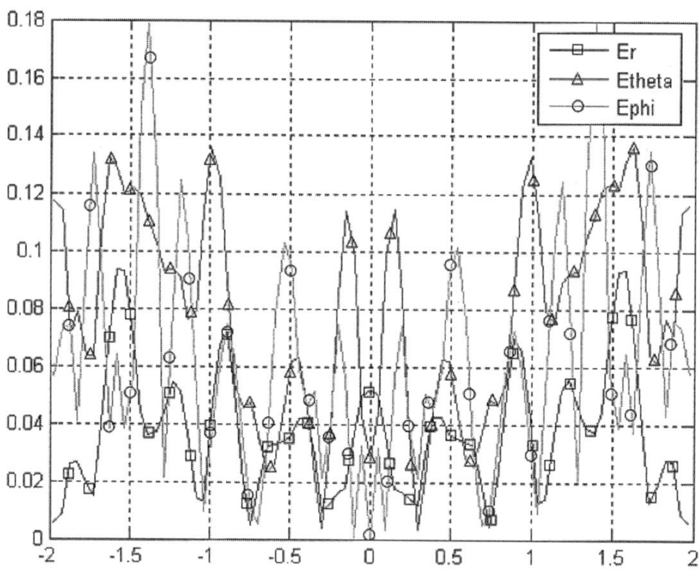

b) dielectric room

Figure 2.15. The three components of the fields E_r, E_θ, E_φ inside the room at $x = -0.0198\lambda$, $z = -3.75$, as a function of y.

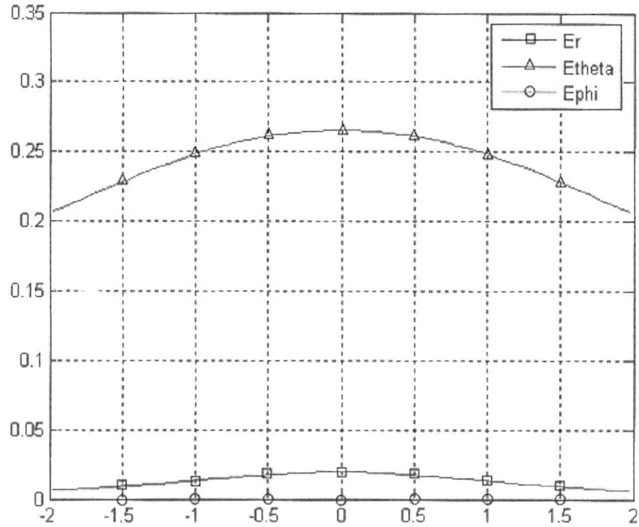

a) free space

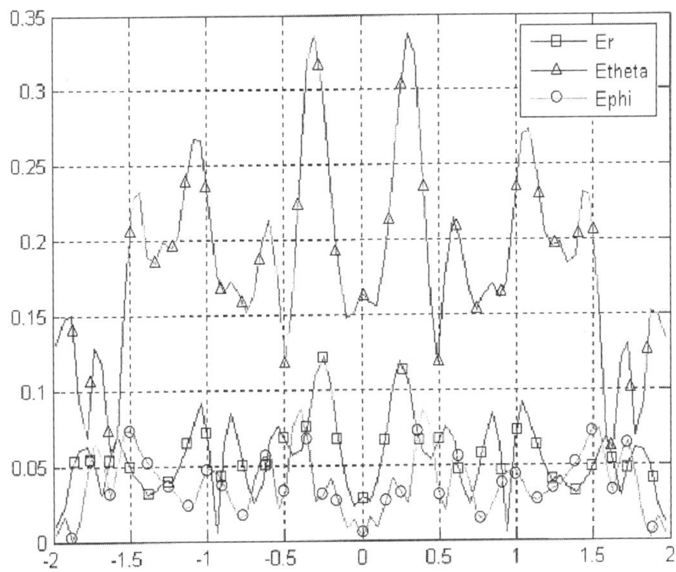

b) dielectric room

Figure 2.16. The three components of the fields E_r, E_θ, E_φ inside the room at $x = -3.75\lambda$, $z = -0.0198\lambda$, as a function of y.

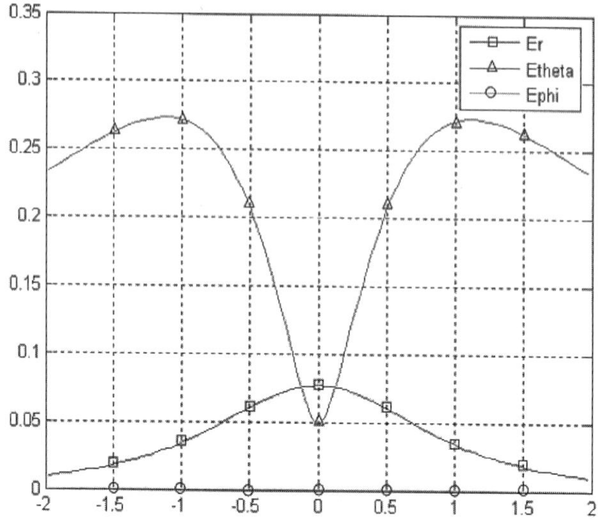

a) free space

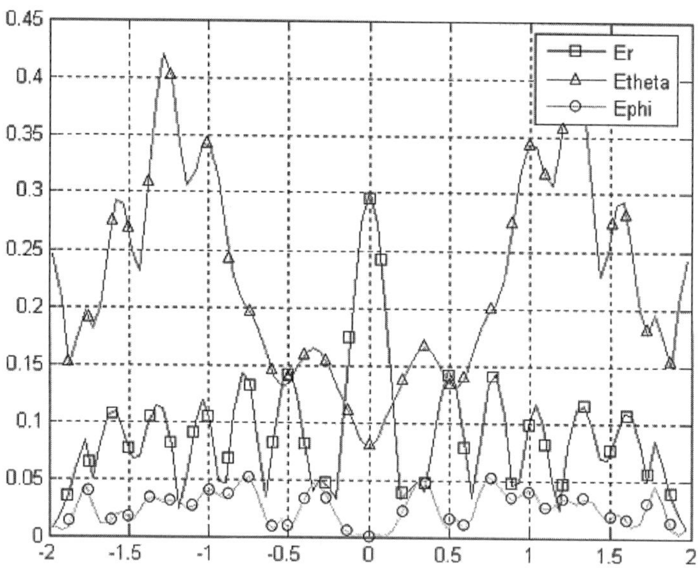

b) free space

Figure 2.17. The three components of the fields E_r, E_θ, E_φ inside the room at $x = -0.0198\lambda$, $z = -0.0198\lambda$, as a function of y.

images produced by the sources must be taken into account to produce any meaningful simulation results for the fields. Instead of using ray tracing, one can solve the entire problem accurately using a numerical electromagnetics code [8] providing stable and accurate results.

2.9 CONCLUSION

In this chapter, frequency domain analysis of some simple antennas is carried out. Concept of the near field and far field properties of the fields is extremely essential to carry out system design and modeling of the electromagnetic environment in which the antennas are deployed. Several examples have been presented to illustrate the subtleties involved in the near field and far field electromagnetic modeling and correct prediction of the fields emanating from the structures. Even though a realistic environment may be quite complex, numerical tools that solve Maxwell's equations can provide a very good approximation to the actual results even for a very realistic environment.

REFERENCES

[1] R. G. Brown, R A. Sharpe, W. L. Hughes and R. E. Post, *Lines, Waves, and Antennas*, John Wiley and Sons, New York, 1973, Second Edition.

[2] J. D. Kraus and R. J. Marhefka, *Antennas*, McGraw Hill, New York, Third Edition, 2002.

[3] R. F. Harrington, *Time-Harmonic Electromagnetic Fields*, McGraw-Hill, 1961.

[4] C. A. Balanis, *Antenna Theory: Analysis and Design*, Harper and Row Publishers, New York, 1982.

[5] T. K. Sarkar, "Analysis of Arbitrarily Oriented Thin Wire Antennas over a Plane Imperfect Ground", *AEU*, Band 31, Heft 11, pp. 449-457, 1977.

[6] A. R. Djordjevic, M. B. Bazdar, V. V. Petrovic, D. I. Olcan, T. K. Sarkar and R. F. Harrington, *AWAS for Windows Analysis of Wire Antennas and Scatterers*: V2.0, Artech House, Norwood, Mass, 2002.

[7] T. Okumura, E. Ohmori, and K. Fukuda, "Field Strength and Its Variability in VHF and UHF Land Mobile Service", *Review Electrical Communication Laboratory*, Vol. 16, No. 9–10, pp. 825–873, 1968.

[8] B. M. Kolundzija, J. S. Ognjanovic, and T. K. Sarkar, *WIPL-D: Electromagnetic Modeling of Composite Metallic and Dielectric Structures*, Norwood, MA: Artech House, 1995.

[9] R. C. Hansen, *Electrically Small, Superdirective, and Superconducting Antennas*, John Wiley and Sons, New York, 2006.

[10] R. C. Hansen, *Phased Array Antennas*, John Wiley & Sons, New York, 1998.

[11] R. J. Mailloux, *Phased Array Antenna Handbook*, Artech House, Boston, 1994.

[12] C. T. Tai, "The Optimum Directivity of Uniformly Spaced Broadside Array of Dipoles", *IEEE Transactions on Antennas and Propagation*, Vol. 12, 1964, pp. 447-454.

[13] N. Yaru, "A Note on Supergain Arrays", *Proc. IRE*, Vol. 39, pp. 1081-1085. Vol. 39, pp. 1081-1085.

[14] T. K. Sarkar, H. Schwarzlander, S. Choi, M. Salazar-Palma and M. C. Wicks, "Stochastic versus Deterministic Models in the Analysis of Communication

Systems", *IEEE Antennas and Propagation Mag.*, Vol. 44, No. 4, Aug. 2002, pp. 40-50.

[15] S. Kay, "Can Delectability Be Improved by Adding Noise?", *IEEE Signal Processing Letters*, Vol. 7, No. 1, Jan 2000, pp. 8-10.

[16] T. K. Sarkar, R. S. Adve, and M. Salazar-Palma, "Phased Array Antennas" in *Wiley Encyclopedia of Electrical and Electronic Engineering*, John G. Webster, editor, John Wiley & Sons, New York, 2001.

[17] T. K. Sarkar, K. Kim, and M. Salazar-Palma, "Adaptive Antennas", in *Telecommunication Handbook*, John Proakis, editor, John Wiley & Sons, 2003.

[18] T. K. Sarkar, S. Burintramart, N. Yilmazer, S. Hwang, Y. Zhang, A. De, M. Salazar-Palma, "A Discussion About Some of the Principles/Practices of Wireless Communication Under a Maxwellian Framework", *IEEE Transactions on Antennas and Propagation*, Vol. 54, No. 12, Dec. 2006, Page(s):3727-3745.

[19] T. K. Sarkar, M. C. Wicks, M. Salazar-Palma, and R. Bonneau, *Smart Antennas*, John Wiley & Sons, New York, 2003.

[20] M. Hata, "Empirical Formula for Propagation Loss in Land Mobile Radio Service", *IEEE Transactions on Vehicular Technology*, Vol. 29, No. 3, pp. 317–325, 1980.

[21] H. L. Bertoni, *Radio Propagation for Modern Wireless Systems*, Prentice Hall, Upper Saddle River, NJ, pp. 90–92, 2000.

[22] T. K. Sarkar, Z. Ji, K. Kim, A. Medouri, and M. Salazar Palma, "A Survey of Various Propagation Models for Mobile Communication", *IEEE Antennas and Propagation Magazine*, Vol. 45, Issue 3, June 2003, Pages 51-82.

[23] R. Bansal, "The Far-Field: How Far is Far Enough?," *Applied Microwave and Wireless*, Nov. 1999, pp. 58–60.

3

FUNDAMENTALS OF AN ANTENNA IN THE TIME DOMAIN

3.0 SUMMARY

Broadband antennas are very useful in many applications as they operate over a wide range of frequencies. The objective of this chapter is to study the transient responses of various well-known antennas over broad frequency ranges. As such, the phase responses of these antennas as a function of frequency are of great interest. In the ensuing analysis, each antenna is excited by a monocycle pulse. Many antennas show resonant properties, and numerous reflections exist in the antenna outputs. The first part of this chapter deals with ways of converting various resonating antennas to traveling-wave antennas by using resistive loading. Appropriate loading increases the bandwidth of operation of the antennas. But the drawback is the additional loss in the load applied to the antenna structure, leading to loss of efficiency to around fifty percent. However, some of the antennas are inherently broadband, up to a 100:1 bandwidth. Hence, the transient responses of these antennas can be used to determine their suitability for wideband applications with low cutoff frequency. The second part of the chapter illustrates the radiation and reception properties of various conventional ultrawideband (UWB) antennas in the time domain. An antenna transient response can be used to determine the suitability of the antenna in wideband applications. Finally, experimental results are provided to verify the various time domain properties of some selected antennas carried out by Dr. J. R. Andrews of Picosecond Pulse Laboratory.

3.1 INTRODUCTION

In many applications, antennas have to be used effectively over a wide range of frequencies. Such antennas are called ultrawideband (UWB) antennas. According to the definition of the Federal Communications Commission (http://www.fcc.gov/oet/info/rules/PARTS15 8-26-03.pdf), an UWB device has a fractional bandwidth which is greater than 0.2 or occupies 500 MHz or more of the frequency spectrum, regardless of the fractional bandwidth [1]. Often, the inputs to UWB antennas have very short durations. These antennas have practical applications in impulse radar, electromagnetic pulse (EMP) measurement, and various communication systems. Much work has been done on pulsed antennas and on the analysis and synthesis of the associated radiated fields and received signals. However, very little research has investigated the wave shapes of the radiated and received pulses [2-18].

UWB antennas are gaining widespread popularity because of their various superior qualities. They can be modeled to use short-pulse techniques to implement high range resolution while directing energy in a particular direction. Consequently, accurate position determination and tracking of objects are possible with UWB signals. Furthermore, the performances of UWB short-pulse systems are superior to that of narrowband systems in multipath environments, because an UWB short-pulse allows the returns from distinct scatterers to be distinguished by using time delay. For short-pulse radiation, it is desired that UWB antennas radiate most of the energy in the direction where the pulse is most similar to the exciting waveform or its derivative. To achieve this end, one must reduce the dispersion from the antenna. One purpose of this chapter is to elaborate on an approach for obtaining traveling-wave antennas by reducing the reflections from the antenna structures. To a certain degree, this can be accomplished by properly designing the antenna, which is dependent on its structural characteristics. Additionally, for some antennas, loading with a predetermined resistive profile reduces the dispersion from the antenna. An early example of a resistive loading profile can be found in the work of Wu and King in 1965 [19]. To date, this profile has been applied only to cylindrical antennas like the dipole. In this chapter, we demonstrate their supposition that the Wu-King loading profile can be modified for application to other antennas.

The properties of several UWB antennas under unloaded and loaded conditions are investigated. The antennas include the dipole, the bicone, the TEM horn, the log-periodic, and the Archimedean spiral. Further, several conventional UWB antennas have been simulated under unloaded conditions to determine their performances in terms of their transient responses. These antennas include the volcano smoke, the diamond dipole, the helical spiral, the conical spiral, the monoloop, the quad-ridged horn, the bi-blade, the cone-blade, the Vivaldi, and the impulse radiating antenna (IRA). The transmitting and receiving properties of each antenna are simulated in the frequency and time domains. The antennas have been modeled with a program that utilizes the Electric Field Integral Equation (EFIE) to evaluate the currents on the structures [20]. The time-domain data is obtained by taking the inverse Fourier transform of the frequency data via the Fast-Fourier-Transform (FFT) technique. The FFT process imposes certain limits on the time-domain data. In particular, the time response is periodic and the total time span of the response is controlled by the lowest frequency of operation. The proper choice of frequency bands and frequency spacing makes this transformation procedure very helpful for simulating an experimental setup, as will be shown throughout the chapter. The efficiency of the structure has also been included in many cases.

Finally, experimental results are provided to enforce the various properties related to the transmit and receive impulse responses of some antennas. In general, it is experimentally verified that the impulse response in the transmit mode of any antenna is related to the time derivative of the impulse response of the same antenna when it is operating in the receive mode. In addition, a special transmit-receive system is discussed which is nondispersive over a very wide bandwidth of approximately tens of gigahertz.

3.2 UWB INPUT PULSE

For radiating antennas in the microwave range, the input must have a very short-duration pulse transition, typically in the picosecond regime. Commonly used baseband pulses are the impulse and the monocycle, the input of choice for this work. The monocycle is basically a doublet formed by differentiating an impulse or by doubly differentiating the unit step. As the monocycle has both positive and negative subpulses, it is useful for driving both halves of symmetrical antennas. Additionally, the anti-symmetrical nature of the monocycle does not permit the generation of dc currents, so the output of an antenna excited by a monopulse has no dc component.

In this chapter, all simulations have been carried out using the monocycle pulse as input. The monocycle is obtained by taking the time derivative of a short-duration Gaussian pulse:

$$\boldsymbol{E}^{inc}(t) = \hat{\mathbf{u}}_{\mathrm{i}} \frac{E_0}{\sigma\sqrt{\pi}} \frac{d}{dt}\left\{ \exp\left(-\frac{(t - t_0 - \mathbf{r}\bullet\mathbf{k})^2}{\sigma^2} \right) \right\} \tag{3.1}$$

where $\hat{\mathbf{u}}_{\mathrm{i}}$ is the unit vector that defines the polarization of the incoming plane wave, E_0 is the amplitude of the incoming wave (chosen to be 377 V/m), σ controls the width of the pulse, t_0 is the delay that is used to ensure the pulse rises smoothly from 0 at the initial time to its value at time t, $\boldsymbol{r}$ is the position of an arbitrary point in space, and $\boldsymbol{k}$ is the unit wave vector defining the direction of arrival of the incident pulse. The frequency spectrum of (3.1) is given by

$$\mathbf{E}(j\omega) = \hat{\mathbf{u}}_{\mathrm{i}}\, E_0\, j\omega\, \exp\left(-\frac{\sigma^2\omega^2}{4} - j\omega(t_0 + \boldsymbol{r}\bullet\boldsymbol{k}) \right) \tag{3.2}$$

where f is the frequency of the signal and $\omega = 2\pi f$.

Figures 3.1a and 3.1b, respectively, show the shape of a typical monocycle pulse in the time domain and its associated spectrum. The duration of the pulse is chosen corresponding to the frequency range of operation of each UWB antenna.

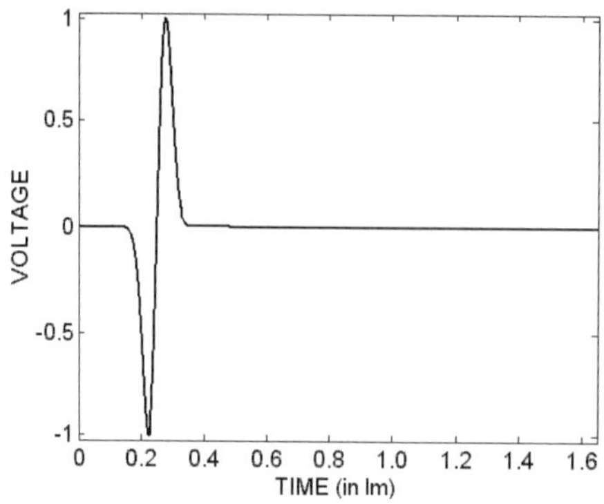

Figure 3.1a. A typical monocycle input.

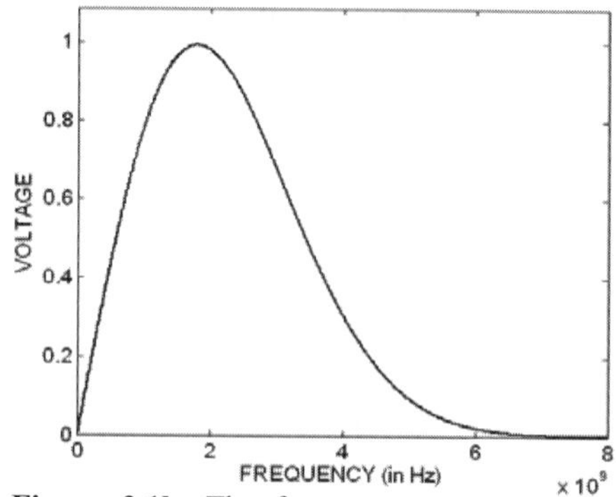

Figure 3.1b. The frequency spectrum of the input.

3.3 TRAVELLING-WAVE ANTENNA

Most antennas have resonant properties. In the time domain, the output waveform has components from numerous reflections on the antenna structure. The waves traveling outward from the feed point get reflected from the discontinuities of the structure such that a standing wave is formed. To obtain a traveling wave structure, one needs to focus the energy in a single pulse by preventing the reflections. This can be achieved by using matched loads at the ends of the structure. But, if the antenna is used over a broad band of frequencies, this approach is not useful because matched loads are effective only over a very narrow range of frequencies. The most effective way to reduce the reflections is to attenuate the outward traveling current towards the ends of the structure. This can be achieved by loading the antennas with a tapered resistive profile so that the resistance increases from the feed point to the ends of the structure. Various loading profiles have been analyzed in [21], where it is shown that a resistive loading is most effective in producing pulsed radiation.

The creation of an appropriate resistive profile was first proposed by Wu and King in 1965 [19]. Consider an antenna with height $2h$ and radius r. One can define a dimensionless parameter ψ for which the current on the antenna has a maximum, where

$$\psi = 2\left(\sinh^{-1}\frac{h}{r} - C(2kr, 2kh) - j\,S(2kr, 2kh) \right) + \frac{j}{kh}(1 - e^{-j2kh}) \tag{3.3}$$

and $C(a, x)$ and $S(a, x)$ are the generalized cosine and sine integrals defined by

$$C(a,x) = \int_0^x \frac{1-\cos W}{W}\,du \qquad S(a,x) = \int_0^x \frac{\sin W}{W}\,du \tag{3.4}$$

with

$$W = (u^2 + a^2)^{1/2} \tag{3.5}$$

Although Wu and King represent the input impedance of a linear dipole antenna by

$$Z_0 = R_0 - j\frac{1}{\omega C_0} = \frac{\psi \zeta_0}{2\pi} = 60\psi \tag{3.6}$$

with

$$\zeta_0 = \sqrt{\frac{\mu_0}{\varepsilon_0}} = 120\pi \tag{3.7}$$

Generally R_0 and C_0 are complex functions, and hence are not the resistance and capacitance except in special cases. However, in keeping with the notation of Wu and King, define the continuously varying loading profile along dimension κ by

$$z^i(\kappa) = r^i(\kappa) - j\frac{1}{\omega c^i(\kappa)} = \frac{\zeta_0 \psi}{2\pi}\frac{1}{h-|\kappa|} = \frac{60\psi}{h-|\kappa|} \tag{3.8}$$

The profile in (3.8) contains poles at $\kappa = \pm h$. However in practical applications, the antenna is divided into a finite number of sections, where the impedance is calculated at the midpoint of each section. Since κ actually never approaches h in (3.8), the presence of the poles does not limit practical applications. Moreover, the frequency dependence appears only in the form of a logarithm for small values of kh, so the antenna shows very broad frequency characteristics when compared to an antenna loaded with a lumped resistance as proposed earlier by Altshuler [22]. Because the reactive part of the impedance is very small compared to the resistive part, one does not need to implement the capacitive profile defined by (3.8) in this case. Consequently, we use only the resistive loading profile

$$r^i(\kappa) = \frac{60\,\Re\{\psi\}}{h-|\kappa|} \tag{3.9}$$

given by the real part of (3.8). $\Re\{\bullet\}$ represent the real part of the function.

To reduce the end reflections, we employ the resistive loading specified by (3.9). However, utilization of (3.9) to broaden the spectral content of the impulse response of the antenna is achieved at the expense of radiating efficiency and gain of the loaded antenna. The radiation efficiency is calculated by taking the ratio of the power lost in the resistive loading to the power radiated from the antenna. As the reader will observe in the ensuing analysis, the efficiency reduces by almost 50–60% in the case of a loaded antenna when compared to the unloaded case, where the efficiency is much higher, often close to 90–100%.

3.4 RECIPROCITY RELATION BETWEEN ANTENNAS

The UWB antennas discussed in this chapter deal with very short-duration pulses. If one can synthesize the input for a specified output, then the energy can be directed to a desired spatial position by using the synthesized input for that specific output. Therefore, two interesting major areas of study are the analysis of the output waveform for a specified input and the synthesis of the driving-point voltage for a specified output. Both analysis and synthesis can be achieved from a single set of simulated results if one applies the reciprocity relation for antennas, the Lorentz's Reciprocity theorem that is deduced from Maxwell's equations. Of the two derived types of reciprocity relations in the literature [23-25], reciprocity for circuits is used in this chapter. According to circuit theory for a two-port linear network, the reciprocity theorem states that, $V_1 I_1 = V_2 I_2$ if a voltage V_1 applied at port 1 produces a current I_2 at port 2 and if a voltage V_2 applied at port 2 produces a current I_1 at port 1.

The application of the principle of time reversal, which has been quite successful for the scalar acoustic problems in focusing the transmitted signals, is not very suitable for vector electromagnetic problems where the presence of distributed sources, frequency-dependent attenuation, polarization, and finite bandwidth are significant impediments [23, 24].

The reciprocity theorem for circuits can be modified to make it applicable to the case of antennas [25]. Reciprocity states that the source and the measurement points can be interchanged without changing the system response. If V_1 is the voltage applied to antenna 1, current I_2 is induced on antenna 2. Similarly, when voltage V_2 is applied to antenna 2, current I_1 is induced on antenna 1. Reciprocity implies that $V_1(\omega)\, I_1(\omega) = V_2(\omega)\, I_2(\omega)$. Consequently, the input voltage $v_2(t)$ required to produce the induced current $i_1(t)$ is obtained. If one knows $v_1(t)$ and the corresponding $i_2(t)$, then

$$v_2(t) = IFFT\left\{\frac{V_1(\omega)\; I_1(\omega)}{I_2(\omega)}\right\} \tag{3.10}$$

where *IFFT* denotes the inverse Fourier transform. Thus, by using two identical antennas, the input required for a specified induced current can be synthesized. Similarly, to synthesize the input for a specified radiation, one needs to have a single antenna and use the radiated fields instead of the induced currents in (3.10). If the radiation field due to input voltage V_1 is R_1 and if R_2 is the frequency domain characteristics of the required radiated pulse, the requisite time-domain input voltage is calculated from

$$v_2(t) = IFFT\left\{\frac{V_1(\omega)\, R_2(\omega)}{R_1(\omega)}\right\} \tag{3.11}$$

using the reciprocity principle. In a similar manner, this principle can be modified for application to the reception problem [23-25]. The goal here is to illustrate that by proper waveform shaping any pulse shape can be transmitted or received. The principle of reciprocity is vector in nature where power is scalar.

The far field radiated by a straight current element along the broadside direction $\mathbf{J}(z')$ is given by

$$\mathbf{E}_{\text{far}} = -\frac{j\,\omega\,\mu_0}{4\pi}\frac{e^{-jkr}}{r}\int_{z'} \mathbf{J}(z')e^{-jkz'}dz'\,. \tag{3.12}$$

Now if we consider a long straight wire irradiated by a plane wave of constant amplitude, then the open circuit voltage received by the wire will be

$$V_{rec} = \frac{1}{I(0)} \int I(z')\, e^{-jkz'} dz' , \tag{3.13}$$

where $I(z')$ is the current distribution on the wire when it is in the receive mode. The term $j\omega$ in $\mathbf{E}_{far}$ indicates that it is proportional to V_{rec} when multiplied by a linear function of ω. This represents a derivative operation in the time domain.

Therefore, the impulse response of an antenna in the transmit mode is the time derivative of the impulse response in the receive mode.

3.5 ANTENNA SIMULATIONS

Several broadband antennas are simulated in this chapter. In each case, the resulting radiated field is calculated from the antenna that is excited by a monocycle voltage pulse (Figs. 3.1a and 3.1b) at the feed point, which is short circuited after passing of the initial pulse. For reception, the same antenna is used as a scatterer that is illuminated by a monocycle pulsed wave, and the induced current at the feed point is observed. The simulated results are described in the following subsections. Throughout the chapter the unit of time is chosen as a light meter (lm), a measure of the time required by light to travel 1 m. So $1 \text{ lm} = (\text{speed of light})^{-1}$ and therefore $1 \text{ lm} = 3.3333 \times 10^{-9}$ sec .

3.6 LOADED ANTENNAS

First we will study the performance of various common antennas like the dipole, the bicone, the TEM horn, the log-periodic, and the spiral in the time domain. Although all of these antennas exhibit time (phase) dispersions, they can be made suitable for UWB applications by various design modifications and proper loading. Each antenna is separately simulated as a radiating element and as a receiving element. The resulting electric fields are normalized to an absolute scale in order to illustrate their wave shapes.

3.6.1 Dipole

Consider a center-fed dipole antenna of length $2h$ and radius r and define the transit time of an input signal from the feed point to each end as $\tau = h / c$. The effective illumination of the antenna is determined by the parameter σ / τ, where σ is the width of the input pulse. To observe the effects of these parameters, two thin dipoles are considered: $h = 1$ m, $r = 1$ cm, and $\sigma / \tau < 1$ for Antenna 1; and $h = 0.05$ m, $r = 0.05$ cm, and $\sigma / \tau > 1$ for Antenna 2. The simulations are done for a σ of 0.15 lm over the frequency range from 10 MHz to 8 GHz. The time support of the response determined by the lowest frequency of operation is 30 lm. This artifact comes from the application of the FFT as it computes a Fourier series and not a real transform.

The spectral magnitudes of the radiations in the broadside direction for both transmitting antennas are shown in Figs. 3.2a and 3.2b. Antenna 1 ($\sigma/\tau < 1$) has significantly more spectral peaks, which results in sharper peaks and less oscillation of the time-domain radiated and received fields (Figs. 3.3 and 3.4) for the unloaded case and in Figs 3.5 and 3.6 for a loaded dipole antenna. In Figs. 3.7 and 3.8 the radiated and received fields for Antenna 2 ($\sigma/\tau > 1$) are presented.

In the case of an infinitely long antenna, the transmit transfer function is almost flat with respect to frequency [26]. So in the case of Antenna 1, an approximation to an infinitely long antenna, the radiated field in Fig. 3.3 is nearly a replica of the driving-point voltage, as expected. Kanda has shown that the transient response of an antenna in the transmit mode is proportional to the time derivative of the impulse response of the same antenna when it is operating in the receive mode [27]. Thus, it follows that the induced current will be an integral of the incident field as shown in Fig. 3.4. On the other hand, for the shorter dipole, Antenna 2 ($\sigma/\tau > 1$), the radiated far field is proportional to the second temporal derivative of the input voltage on the structure [18], whereas the received open-circuit voltage will be an approximate time derivative of the incident field [28].

The above observations are valid in the case that an antenna carries a purely traveling wave of current. However, as a consequence of reflections from the finite size of the dipole, the output of the antenna has a number of reflected pulses in addition to the initial pulse. In the radiated far field, the initial observed pulse is a replica of the input voltage that is radiated directly from the feed point; whereas subsequent pulses are radiations and reflections of the initial pulse from the end and feed points of the dipole [27, 29]. Specifically, the second and third pulses are radiations from the end points that occur h/c seconds (the time it takes the signal to travel from the feed to each end) after the initial pulse. Moreover, these two signals arrive at the far-field observation point at different times because the corresponding distances to the observation point are distinct, unless the observation point is in the broadside direction. The fourth and fifth components of the time-domain field are second radiations from the endpoints that occur $3h/c$ seconds after the initial pulse as a result of reflections from the dipole's endpoints. A second pulse from the feed, presented in [27] and [29], does not appear in this analysis, because short circuiting the feed after the initial radiation effectively eliminates the feed as a subsequent source of radiation.

In the case of Antenna 1, the pulse emanates from the feed 1 lm (6.67 pulse durations) before it radiates from the ends, since $\tau = 6.67\sigma$. Consequently, the radiations from the endpoints do not overlap in time with the initial pulse from the feed. An observer in the far field in any direction sees a replica of the source, followed by an aggregate radiation from the endpoints, followed by a reduced aggregate radiation from the endpoints 3 lm (20 pulse durations) after the initial emanation from the feed. Depending on the observation direction, the replicas from endpoints could overlap, especially near broadside. As expected, the radiated field exhibits separately identifiable pulses with the anticipated

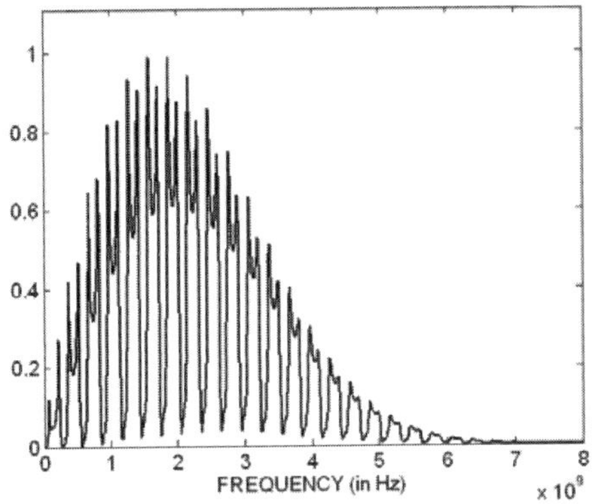

Figure 3.2a. Frequency spectrum with $\sigma/\tau < 1$.

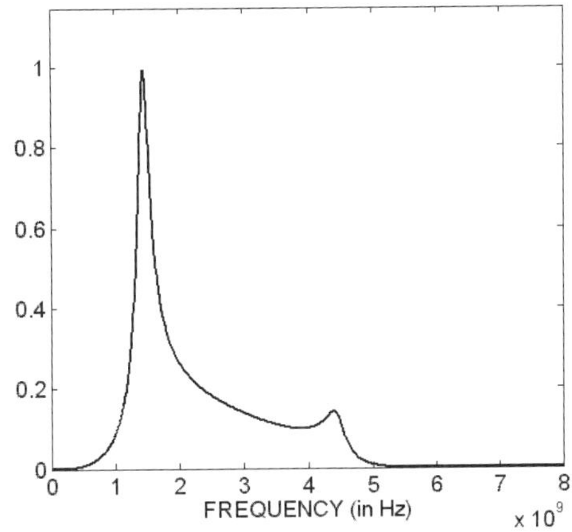

Figure 3.2b. Frequency spectrum with $\sigma/\tau > 1$.

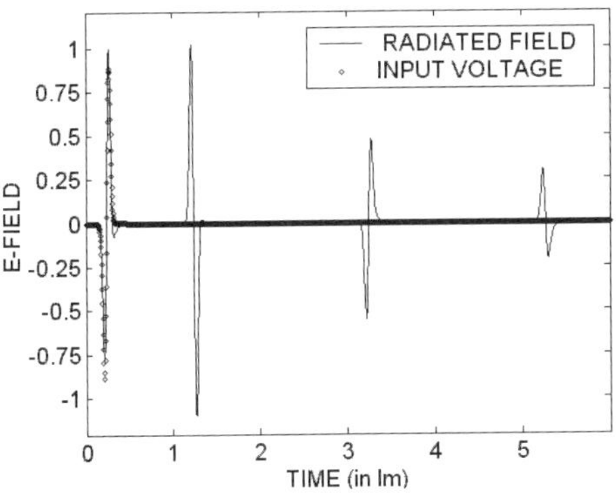

Figure 3.3. Radiation from the antenna with $\sigma/\tau < 1$.

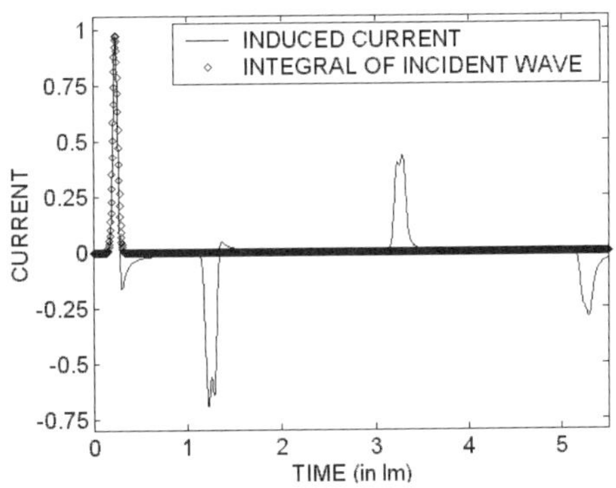

Figure 3.4. Reception of a monocycle pulse by the antenna with $\sigma/\tau < 1$.

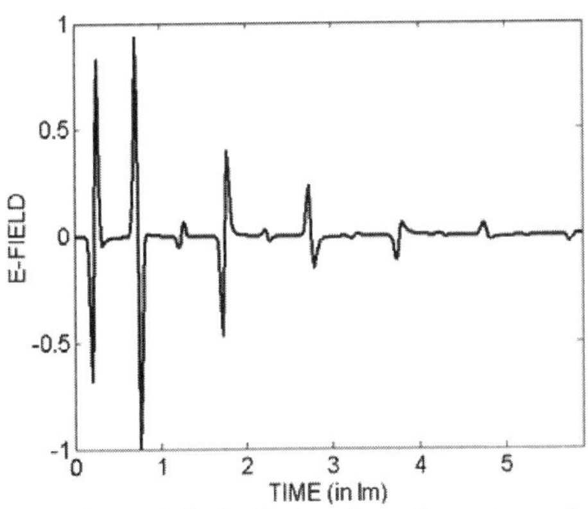

Figure 3.5. Radiation from the antenna 1 with a 50 ohms load at the feed.

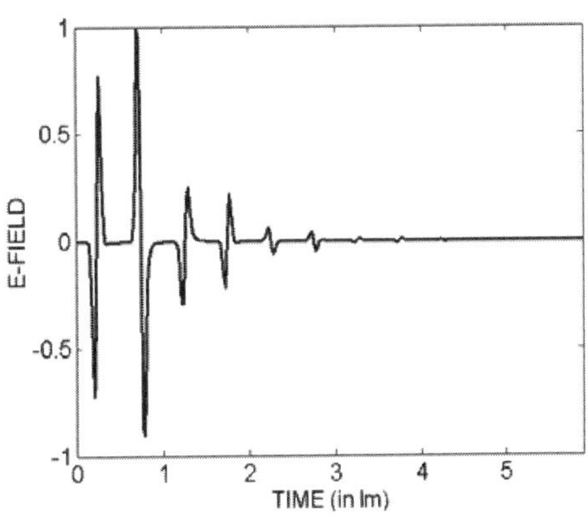

Figure 3.6. Radiation from the antenna 1 with a 500 ohms load at the feed.

temporal separations (Fig. 3.3). If you consider only the first pulse, then the properties of the infinitely long dipole antenna are verified; that is, the first radiated pulse is a replica of the input. When Antenna 1 is receiving the monocycle, the induced current at the feed shows a number of reflected waves (Fig. 3.4). To note is the fact that the first pulse of the received current is identical to the integration of the incident wave.

In the above analysis, the feed of the antenna is short circuited after the initial radiation effectively eliminates the feed as a subsequent radiation location. Thus a second pulse from the feed, as presented in [27] and [29], does not appear in this analysis. In the case when the feed is not short circuited after the initial radiation, instead a load is applied at the feed, reflections from the feed point are observed in addition to the reflections from the ends. The radiation from Antenna 1 when a load of 50 ohms is connected at its feed point is shown in Fig. 3.5. The radiated field shows a replica of the source followed by an aggregate radiation from the endpoints occurring 1 lm, an aggregate radiation from the feed point 2 lm, a reduced aggregate radiation from the endpoints 3 lm after initial emanation from the feed, and so on. When the 50 ohms load at the feed point is replaced by a load of 500 ohms, Fig. 3.6 shows the radiated field. As the loading at the feed point is increased, the reflection from the feed point also increases in magnitude as is evident from Figs. 3.5 and 3.6.

For Antenna 2 ($\sigma/\tau > 1$), the pulse emanates from the feed one third of a pulse duration (0.05 lm) and one pulse duration (0.15 lm) before the first and second endpoint radiations, respectively, since $\tau = \sigma/3$. To an observer in the far field, the temporal supports of radiation from the feed and the first radiations from the ends have a 67% overlap, and the supports of the first and second radiations from the ends overlap 33% of the time. Consequently, the first pulse cannot be distinguished from the reflected pulses as they are too closely spaced in time, which results in the longer interference signal of Fig. 3.7. The time-domain received signal behaves similarly (Fig. 3.8), when the monocycle field is incident on Antenna 2.

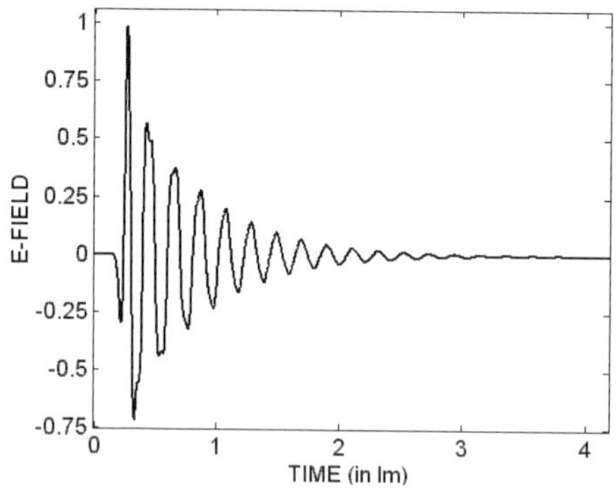

Figure 3.7. Radiation from the antenna with $\sigma/\tau > 1$.

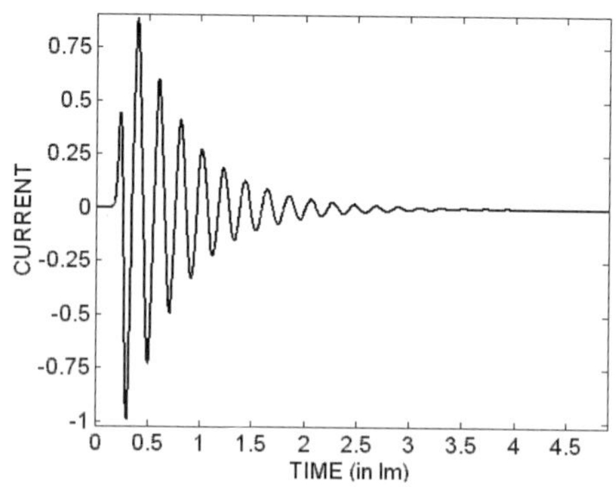

Figure 3.8. Reception of a monocycle pulse by the antenna with $\sigma/\tau > 1$.

To reduce the reflections, a tapered loading is applied along the length of the antenna as discussed in Section 3.3. According to (3.3), one can calculate the value of the parameter ψ at the frequency for which the dipole is a half-wavelength long. The surface resistance is calculated with (3.9), which depends on the length of the antenna and its radius, and is plotted as a function of location along the antenna for both cases of Figs. 3.9 and 3.10. To apply the loading, divide the antenna into a number of smaller sections, calculate the value of $r^i(\kappa)$ at the midpoint of each section, and apply the calculated resistance as a constant surface resistance on that section. This surface resistance can be implemented by coating the antenna with a resistive material. If the number of sections is sufficiently large, the step-function variation of $r^i(\kappa)$ will closely resemble the continuously varying resistive profile.

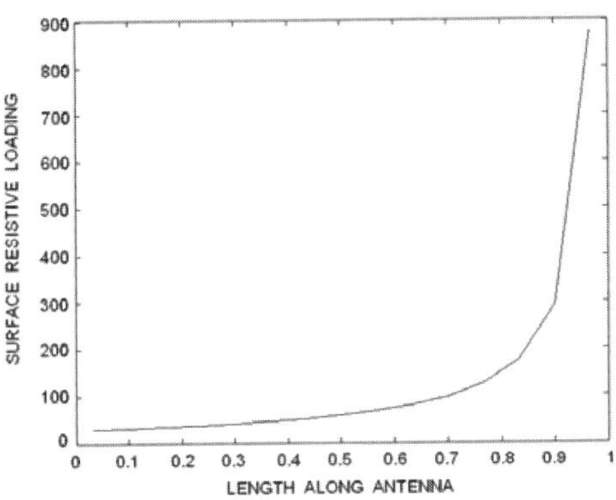

Figure 3.9. Resistive loading on the antenna with $\sigma/\tau < 1$.

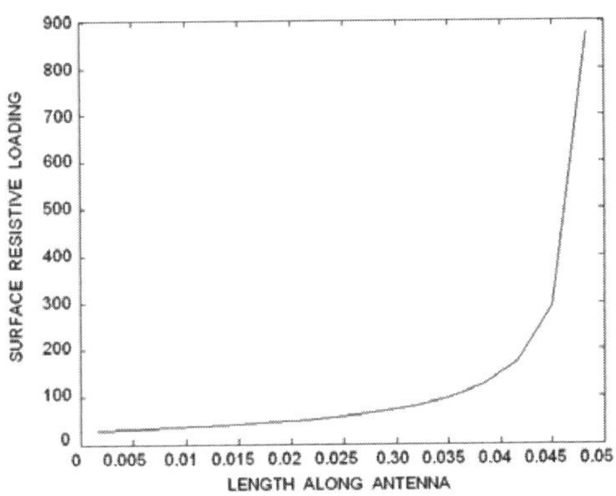

Figure 3.10. Resistive loading on the antenna with $\sigma/\tau > 1$.

The results after applying resistive loading to Antennas 1 and 2 for transmit and receive modes are depicted in Figs. 3.11–3.14. For Antenna 1, both responses show only a single pulse (Figs. 3.11–3.12), because the loading has converted this dipole to a traveling-wave antenna. These results further substantiate the property of a long dipole antenna which states that its far-field radiated signal is a replica of the input pulse. On the other hand, the form of the radiated field for $\sigma/\tau > 1$ (Antenna 2) in Fig. 3.13 concurs with an earlier statement that radiation from the antenna is the second temporal derivative of the input pulse [27–29]. The radiated and received (Fig. 3.13–3.14) pulsed signals associated with Antenna 2 indicate the effectiveness of the resistive loading in reducing the reflections from the end of the structure.

The efficiency of the loaded dipole antenna with $\sigma/\tau < 1$ is shown in Fig. 3.15, where a significant loss of the radiated energy from the loading is apparent at all frequencies (0.1–8.0 GHz). The efficiency monotonically increases with frequency, with a minimum of 16% at 0.1 GHz and a maximum of 50% at 8 GHz; whereas the efficiency of the unloaded antenna is expected to be 95% or better.

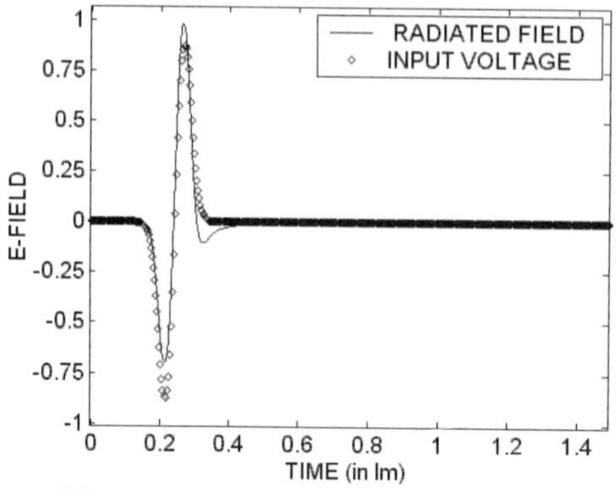

Figure 3.11. Radiation from Antenna 1 $(\sigma/\tau < 1)$ with loading.

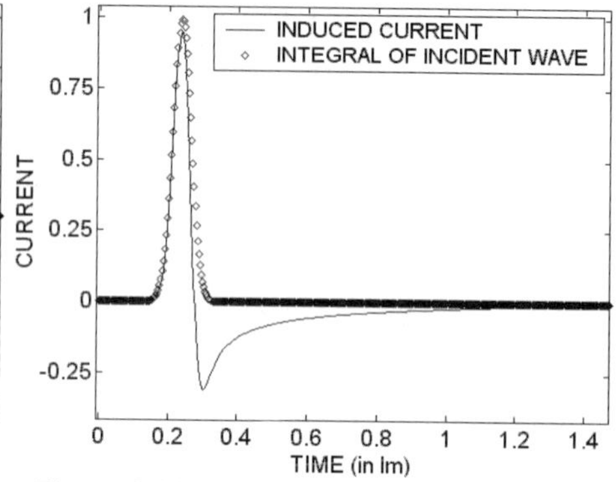

Figure 3.12. Reception of a monocycle pulse by Antenna 2 $(\sigma/\tau < 1)$ with loading.

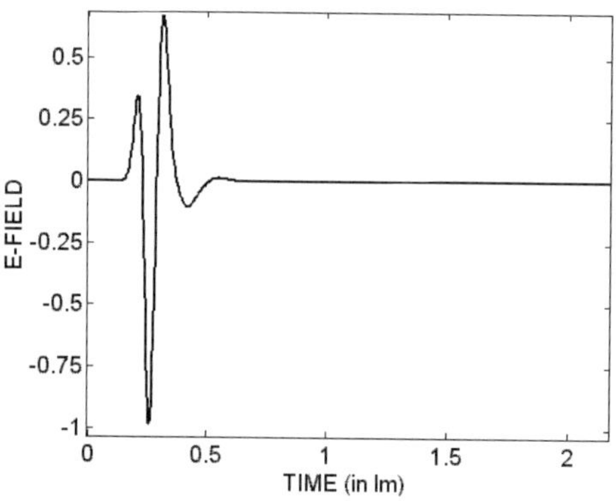

Figure 3.13. Radiation from Antenna 2 $(\theta/\tau > 1)$ with loading.

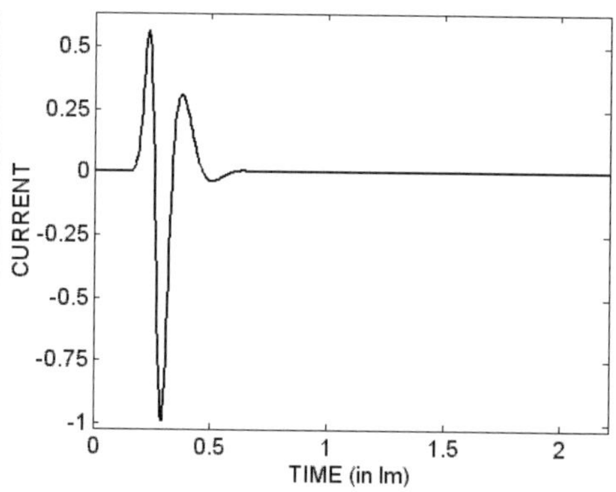

Figure 3.14 Reception of a monocycle pulse by Antenna 2 $(\sigma/\tau > 1)$ with loading.

Figure 3.15. Radiation efficiency of the loaded dipole Antenna 1.

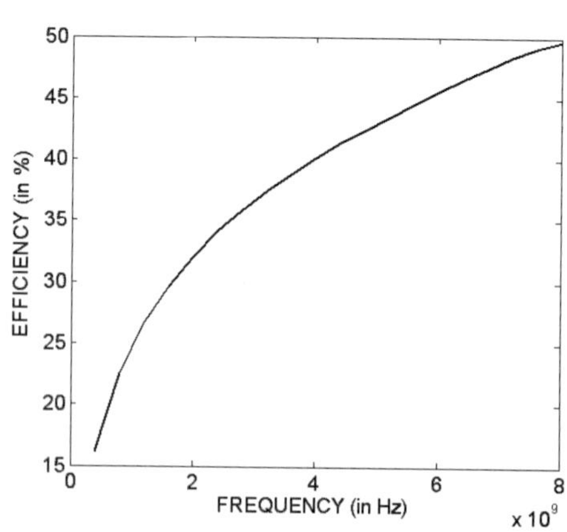

Finally, we discuss the excitation that needs to be applied to the unloaded dipole for $\sigma / \tau < 1$ so that it radiates a monocycle pulse. By using the principle of reciprocity of Section 3.4, the requisite exciting voltage is plotted in Fig. 3.16, along with the radiated field. In addition, Fig. 3.17 provides the incident waveform that is needed to induce a monocycle of current at the feed point of the dipole.

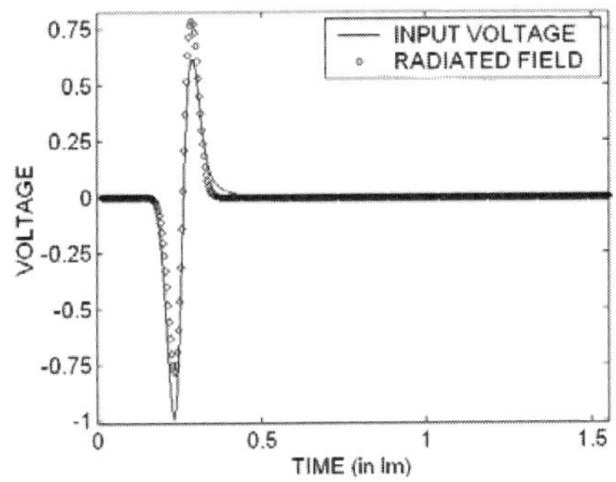

Figure 3.16. Input required for a monocycle transmission by the antenna.

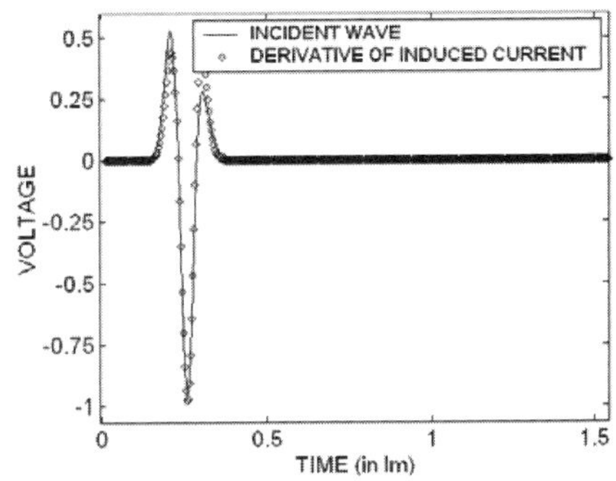

Figure 3.17. Incident pulse shape to generate a monocycle of current at reception.

3.6.2 Bicones

The radiation pattern of a biconical antenna is very similar to that of a long dipole antenna. As in the case of the dipole, the biconical antenna also suffers from the formation of standing waves due to the discontinuities on the structure. Reflections from the ends can be reduced by increasing the flare angle above 30° [30–31].

First consider a truncated bicone with open ends, 0.11m long along the lateral side, and a flare angle that varies from 24° to 90° (Fig. 3.18). The antenna has a feed wire of length 4 mm and radius 0.1 mm connecting the vertices of the two cones. The performance of the antenna is simulated between 150 MHz and 30 GHz. The time support of the response determined by the lowest frequency of operation is 2 lm, and the duration of the input pulse is 0.038 lm. The antenna is polarized along the vertical axis (the axis of symmetry through the vertices), and following Jasik [32], the antenna height is taken to be at least $\lambda /4$.

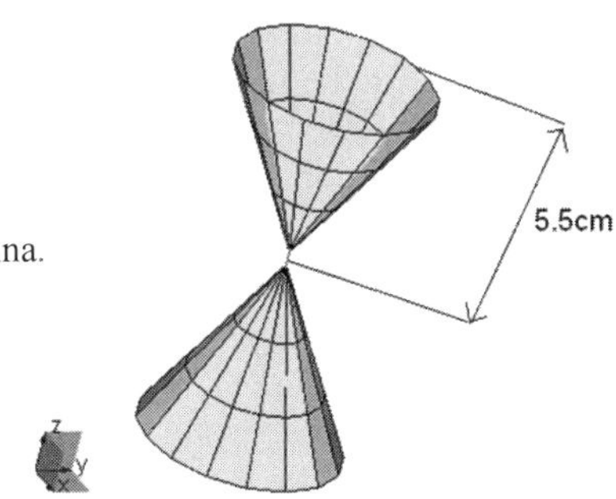

Figure 3.18. Structure of a biconical antenna.

The amplitudes of the frequency-domain radiated far fields for four flare angles (24°, 40°, 60°, 90°) illustrate that as the flare angle is increased, the responses have less oscillations, thereby resulting in less reflections in the time-domain fields (in Fig. 3.20). When the biconical antenna is used as a receiver, the induced currents (in Fig. 3.21) for the different flare angles are very similar for the first 0.1 s and are fairly closely banded for the remainder of the time.

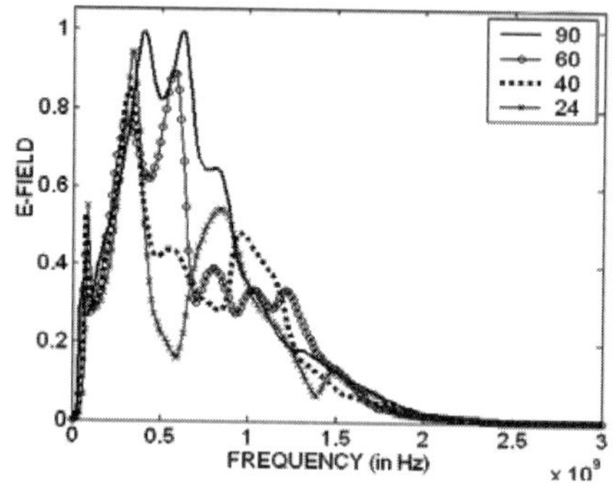

Figure 3.19. Frequency spectrum of the transmitted signal for different flare angles of the bicone.

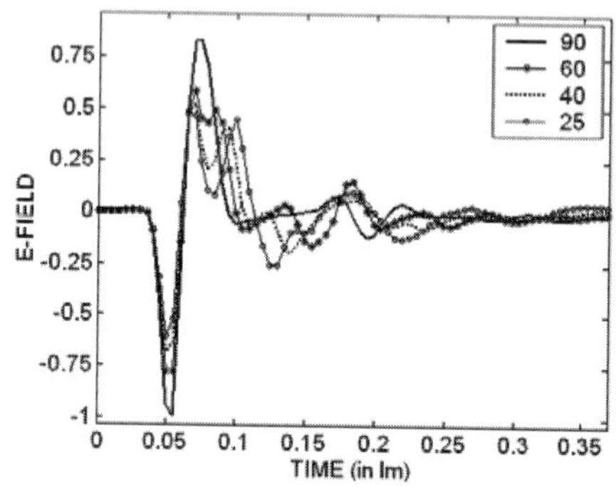

Figure 3.20. Radiated field from the biconical antenna for different flare angles.

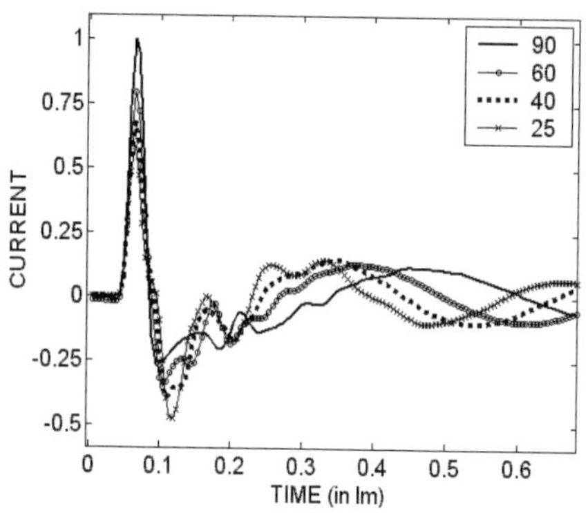

Figure 3.21. Reception of a monocycle by a biconical antenna for different flare angles.

To reduce the end reflections, the open truncated bicone can be made continuous (smoother) by closing the open ends with identical hemispherical caps (Fig. 3.22). Their inclusion prevents the sudden termination of the bicone. Each hemisphere has a radius of 0.055 m, and the origin of the coordinate system is placed at the midpoint of the feed wire that connects the apex of the two cones, with the z axis pointing upwards along the wire. Since this antenna is symmetric about the z axis, the computed radiated field at any angle in the xy plane is constant (Fig. 3.23) and is polarized along the vertical length of the cone (z axis). Figure 3.24 plots the received current when a monocycle field is incident to the bicone perpendicular to the z axis. Both figures show that the use of end caps reduces the reflections to a certain level but cannot negate the reflections totally.

To obtain a reflection-free structure, we use a flare angle of 90° and apply a tapered resistive loading outward from each apex along the lateral surface of the antenna. By using the principles stated in Section 3.3, we calculate the value of ψ at the frequency for which the bicone is a half-wavelength long by using (3.3). The effective radius for the load calculation in (3.3) is taken to be one-hundredth of the length of the bicone to meet the specification for a thin antenna. Thus one can obtain the resistive loading profile required to make the antenna a traveling wave structure. The initial model of the antenna has been divided into numerous plates along its length as shown in Fig. 3.18. Distributed loading is applied on each layer of plates by calculating the resistivity at the midpoint of that layer. If the number of sections is sufficiently large, then the step-functional variation of the resistance will bear a close resemblance to the continuously varying resistive profile plotted in Fig. 3.25.

The radiated field for this non-reflecting antenna nearly coincides with the input voltage (Fig. 3.26), which verifies that the reflectionless bicone behaves like

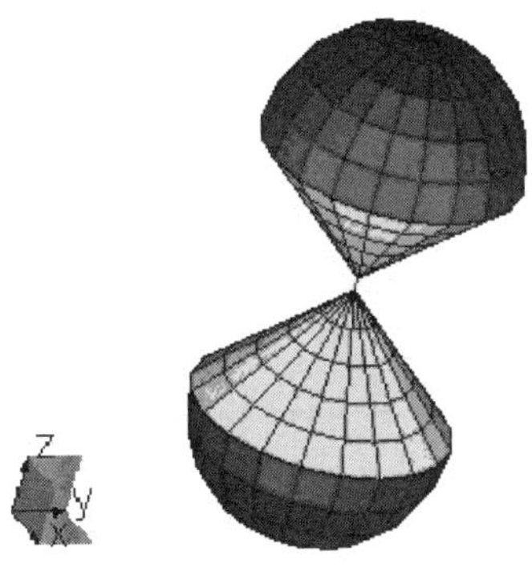

Figure 3.22. Structure of the biconical antenna with end caps.

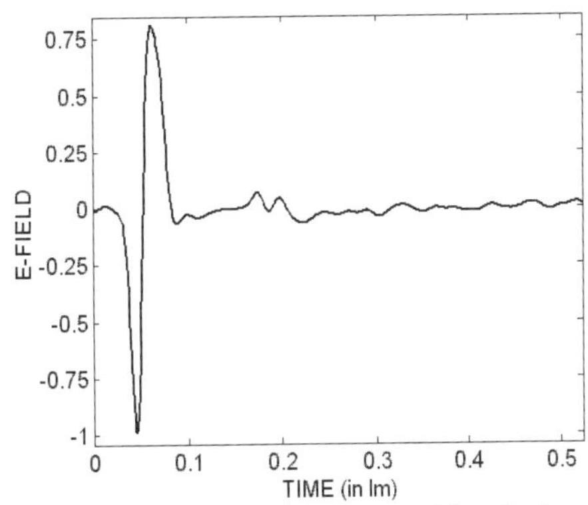

Figure 3.23. Radiation from a biconical antenna with end caps.

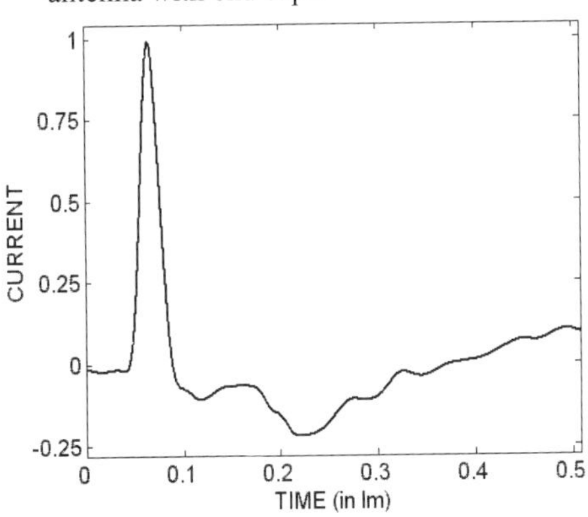

Figure 3.24. Reception of a monocycle pulse by a biconical antenna with end caps.

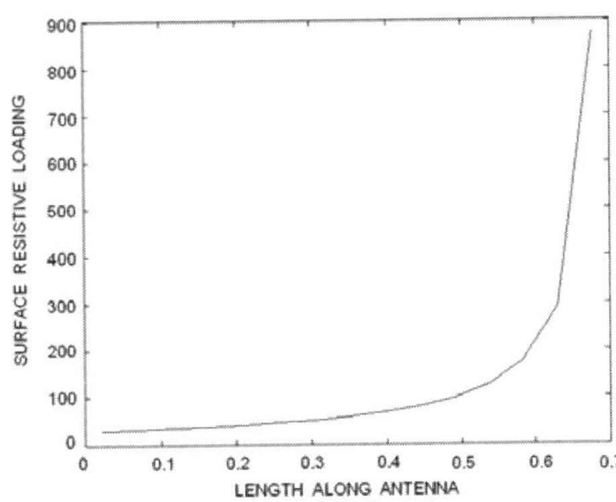

Figure 3.25. Resistive loading profile along the length of the biconical antenna.

a long dipole. If the antenna is used as a receiver, the received current at the feed point of the antenna deviates slightly from the integral of the input voltage (Fig. 3.27). This discrepancy is due to the absence of any dc current on the structure. As in the case of a dipole, we calculate the excitation that needs to be applied to the bicone so that it radiates a monocycle pulse (Fig. 3.28) and the required incident field for inducing a monocycle of current at the feed point of the bicone (Fig. 3.29).

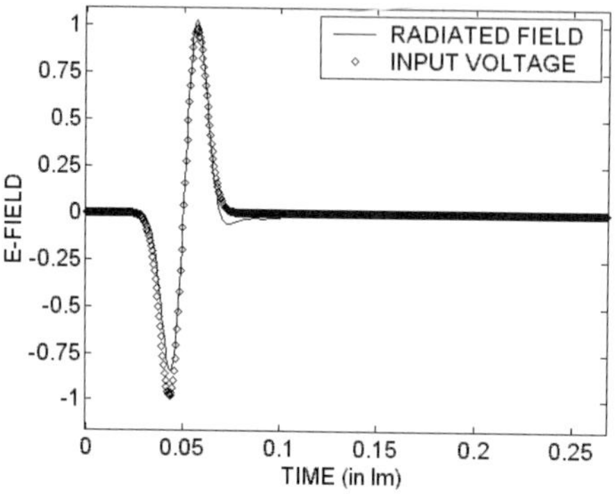

Figure 3.26. Radiation from the bicone antenna.

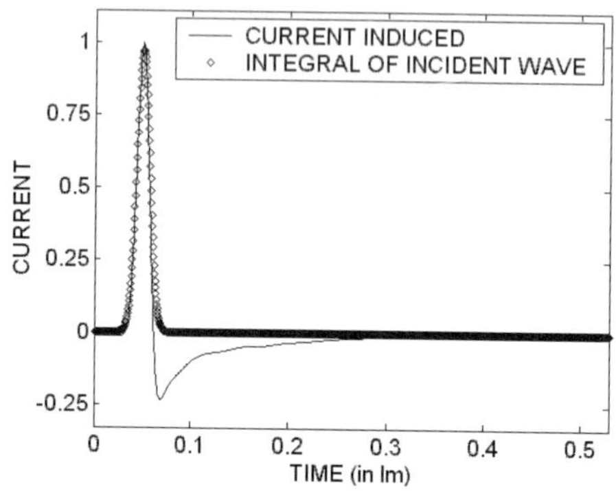

Figure 3.27. Reception of a monocycle by the bicone antenna.

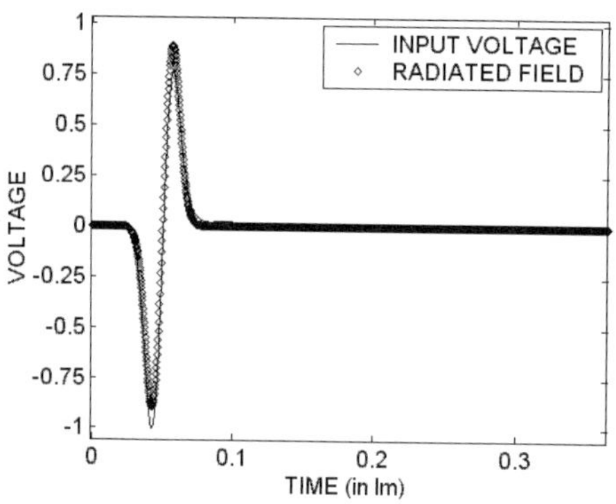

Figure 3.28. Input required for a monocycle transmission.

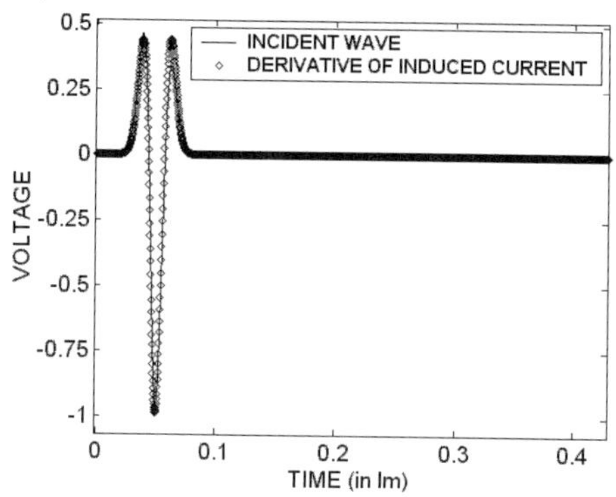

Figure 3.29. Incident field required for a monocycle of induced current.

3.6.3 TEM Horn

Theoretically, the TEM horn antenna gives a direct measurement of an incident wave when it is used as a receiving antenna. It is very popular in metrology. In our simulation, the overall length of the antenna is 0.15 m, the width at the mouth is 0.038 m, and the height at the mouth is 0.052 m. The feed wire of the antenna, which has a length of 4 mm and a radius of 0.05 mm, is situated at the narrower end of the antenna (Fig. 3.30). In this example, the origin of the coordinate system shown is the midpoint of the feed wire, and the z-axis runs along the feed wire. The top and bottom plates, which stretch along the x-axis, are not tapered; that is, the height-to-width ratio of the antenna is not maintained constant. The antenna is simulated from 200 MHz to 40 GHz, the time support of the response determined by the lowest frequency of operation is 1.5 lm, and the width of the input pulse is 0.04 lm.

The radiated time-domain field is polarized along the z-axis and has significant ripple when evaluated along the x-axis (Fig. 3.31), which represents a substantial departure from the monocycle. Similarly, if the antenna is receiving a z-polarized monocycle field traveling along the x-axis, the induced current has similar oscillatory behavior (Fig. 3.32). Neither temporal signals generate a clean pulse, which indicates that the TEM horn has some reflections. On modifying the structure by tapering the plates of the horn antenna to maintain a constant height-to-width ratio of 0.68 (Fig. 3.33) while keeping other parameters the same, the reflections are reduced (Figs. 3.34 and 3.35) but are not completely eliminated. For example, the radiated field in Fig. 3.34 is still not quite the derivative of the input pulse [33], because the structure resonates and still has some imperfections.

To convert this resonating structure to a traveling-wave structure, one must decrease the parallel-plate nature of the antenna near its feed point.

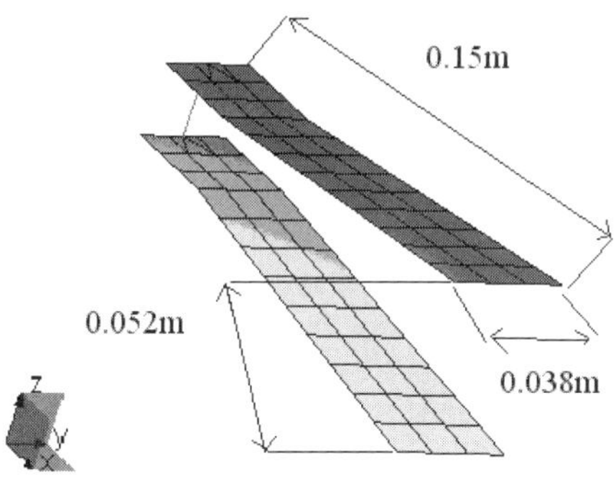

Figure 3.30. Structure of a TEM horn.

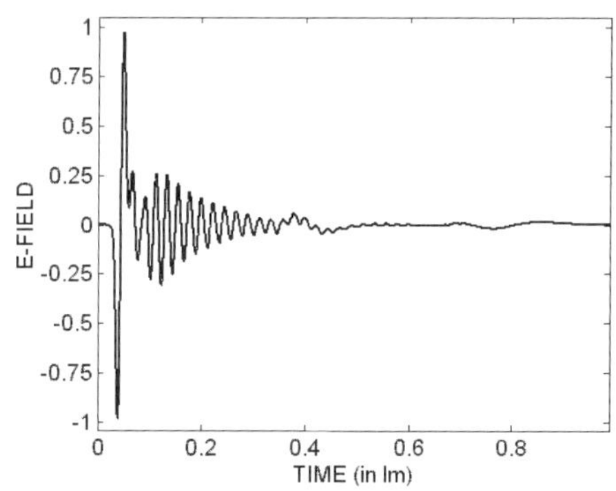

Figure 3.31. Radiation from a TEM horn antenna.

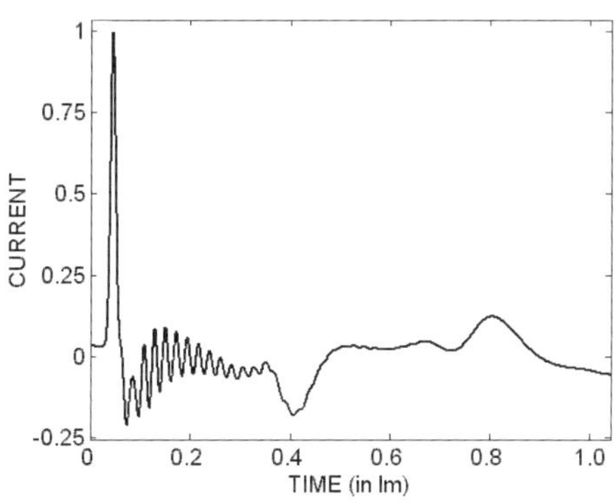

Figure 3.32. Reception of a monocycle by a TEM horn antenna.

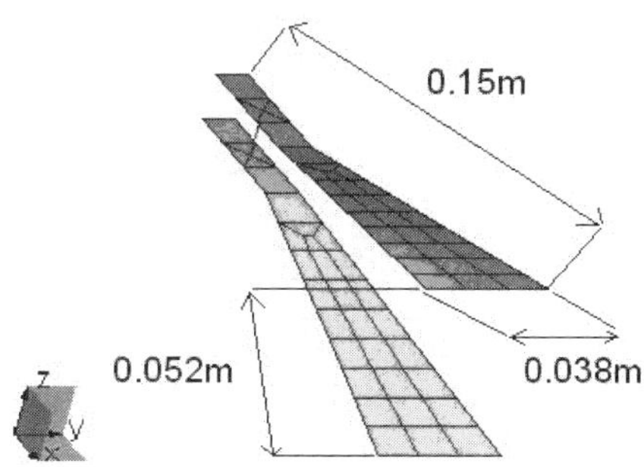

Figure 3.33. Structure of a tapered horn antenna.

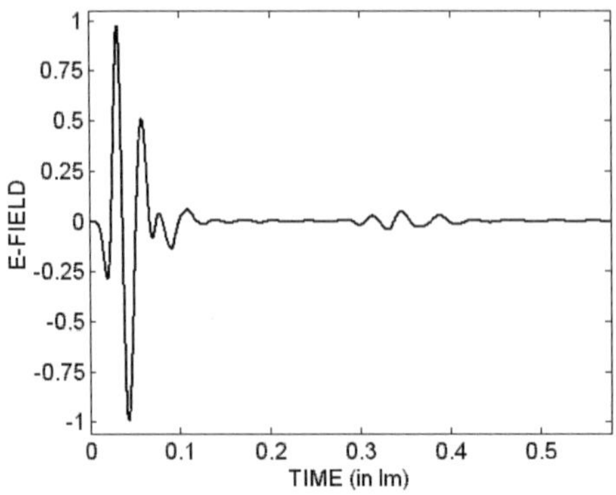

Figure 3.34. Radiation from a tapered horn antenna.

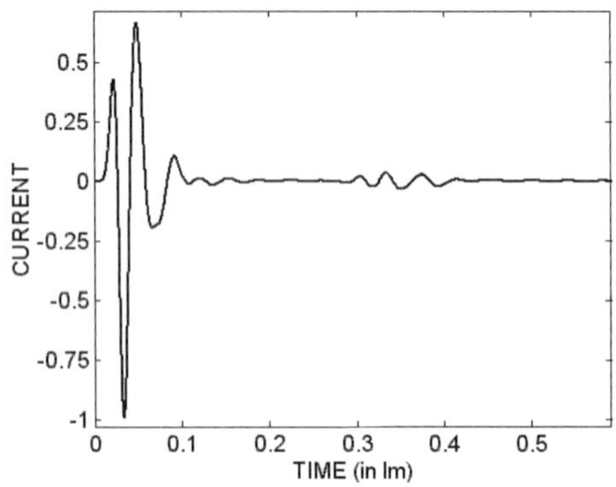

Figure 3.35. Reception of a monocycle by a tapered horn antenna.

Therefore, the lengths of the parallel plates are reduced to a minimum (Fig. 3.36). To reduce the dispersion further, a resistive loading is applied along the length of the antenna, with the profile specified by (3.9) and shown in Fig. 3.37. Originally, (3.9) was developed for cylindrical antennas only, but it is now modified for use with horn antennas. Instead of using the radius of the antenna in (3.3), we use an effective radius which is calculated as half of the average thickness of the antenna. The effective radius in this case is 0.045 m. We divide the antenna into separate zones along the length and calculate the surface resistance at the midpoint of each zone.

The radiated time-domain waveform and time-domain received current are displayed in Figs. 3.38 and 3.39. Both the transmission and the reception properties show that by properly loading the antenna, the unwanted reflections are reduced and the energy is focused in a single pulse. Previously, it was experimentally demonstrated that the radiated field is the first derivative of the driving point voltage, whereas the current induced during reception is identical to the incident wave [27,33]. Not only do the simulations contained herein verify the results of Ref. [33], they also indicate that the impulse response in the

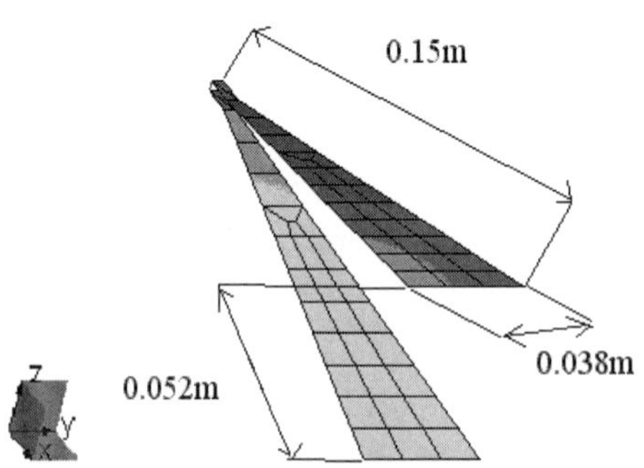

Figure 3.36. Structure of a non-resonating horn antenna.

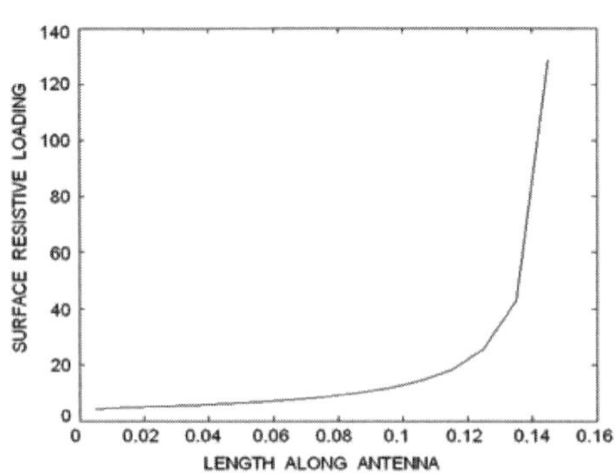

Figure 3.37. Resistive loading profile on the antenna.

transmit mode is proportional to the time derivative of the impulse response in the receive mode, as has been verified by Kanda [27].

Similar to the unloaded dipole, the efficiency of the unloaded TEM horn is approximately 96–100%. Over the range of simulated frequencies (200 MHz to 40 GHz), the efficiency of the loaded horn monotonically increases, with a maximum of 65% at 40 GHz and nearly linear behavior between 7 GHz and 27 GHz (Fig.3.40). The loaded horn is inefficient (< 10%) for frequencies < 7 GHz.

Finally, we calculate the excitation that needs to be applied to the TEM horn so that it radiates a monocycle pulse (Fig. 3.41) and the required incident field for inducing a monocycle of current at the feed point of the horn (Fig. 3.42).

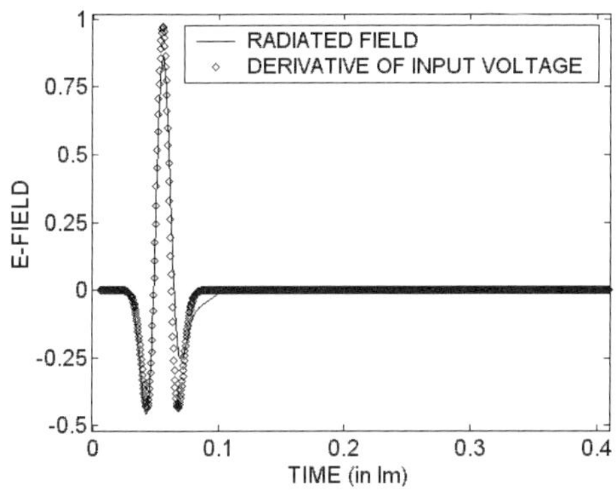

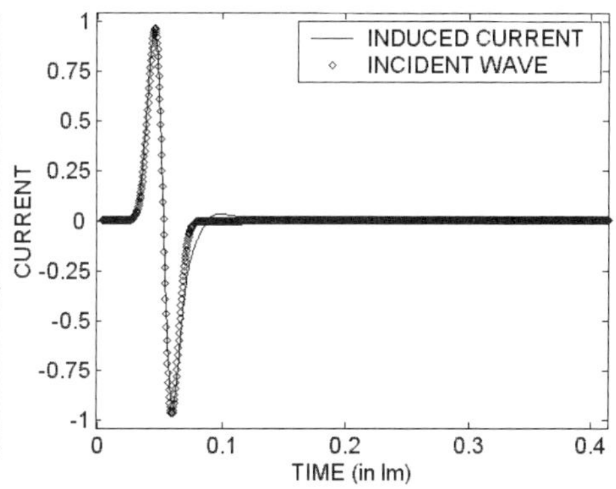

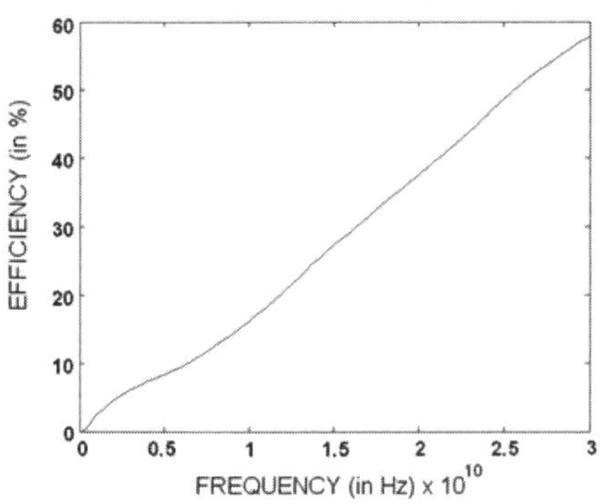

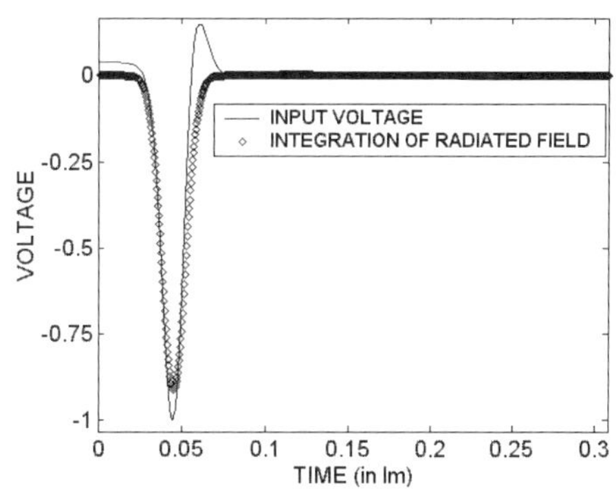

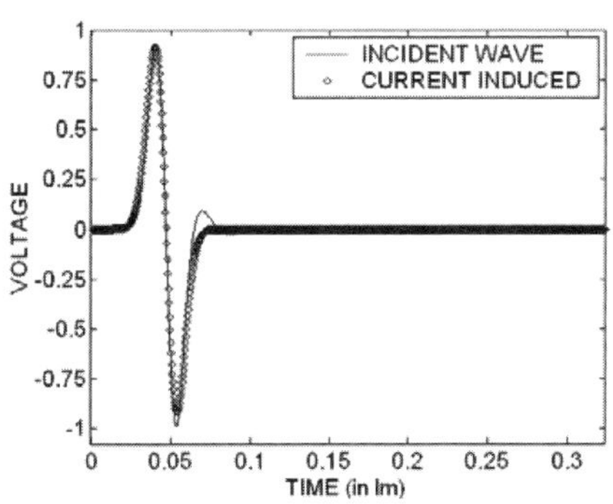

Figure 3.38. (top left) Radiation from the tapered horn antenna.

Figure 3.39. (top right) Reception of a monocycle by the tapered horn antenna.

Figure 3.40. (middle left) Radiation efficiency of the loaded tapered horn antenna.

Figure 3.41. (middle right) Input required for a monocycle transmission.

Figure 3.42. (left) Incident waveform shape to produce a monocycle of induced current.

3.6.4 Log-Periodic

The log-periodic dipole array (LPDA) antenna consists of parallel wire segments that form a structural geometry so that the impedance repeats periodically as a logarithmic function of frequency (Fig. 3.43). The dipoles are connected alternately to two central bars. The lengths of the antenna, the longest dipole, and the shortest dipole respectively are 0.6 m, 0.25 m, and 0.083 m. Consequently, the LPDA operates from 0.6 GHz to 1.8 GHz. The time support of the response determined by the lowest frequency of operation is 30 lm, and the width of the input pulse is 0.07 lm. The antenna has a feed wire connected between the two parallel bars along the length of the antenna near the shortest dipole. The feed wire has a length of 25 mm and a radius of 1 mm. The origin of the coordinate system of Fig. 3.43 is the midpoint of the feed wire.

Due to the presence of a number of dipoles along the antenna length, the antenna forms a resonating structure as the waves are reflected from the open ends of the dipoles. On observing the radiation along the end-fire direction (x-axis), it is seen that the transmitted waveform has more than one radiated pulse (Fig. 3.44). The second pulse arrives 0.6 lm after the first pulse, where 0.6 lm is the time required for light to travel the length of the antenna. The radiated field of the antenna is polarized along the z-axis of Fig. 3.43. Similarly, the received current for an incident monocycle field along the x-axis is shown in Fig. 3.45. Note the formation of standing waves.

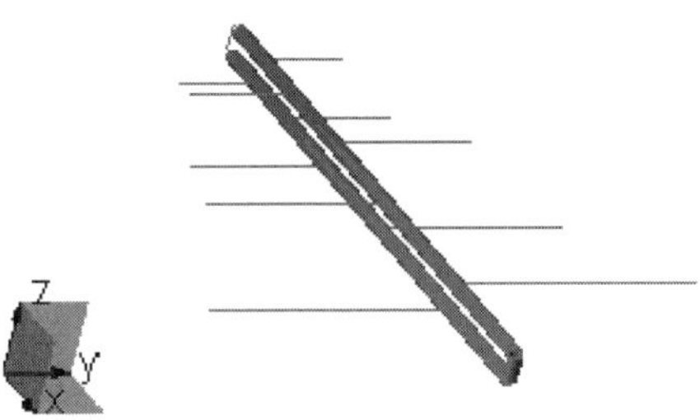

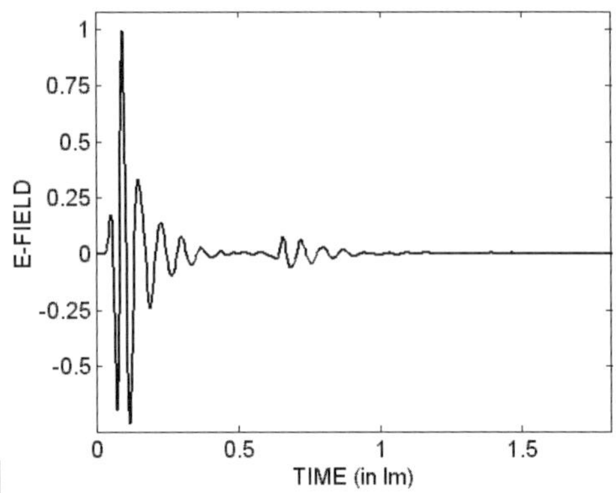

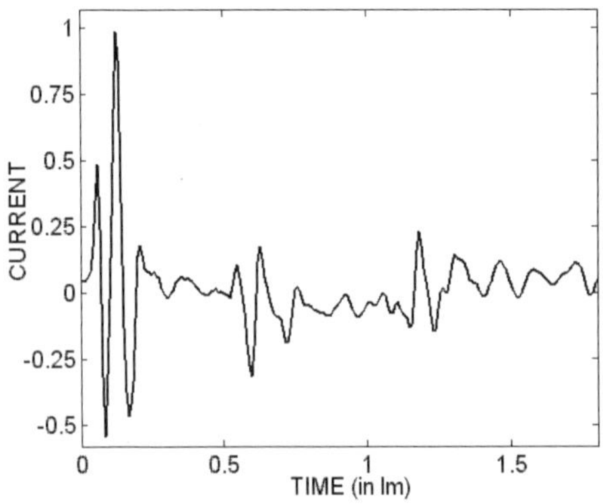

Figure 3.43. (top left) Structure of a log-periodic dipole array (LPDA).

Figure 3.44. (top right) Radiation from the LPDA.

Figure 3.45. (left) Reception of a monocycle by a LPDA.

To reduce the undesired reflections, a resistive loading is applied along the two bars located along the length of the antenna. The loading profile is calculated according to (3.9) by using an effective radius that is half of the average thickness of the antenna. The effective radius for this simulation is 0.333 m. The two parallel bars along the length of the antenna are divided into many segments, and surface resistivities are used along each segment according to the loading profile in Fig. 3.46. Although the radiated time-domain waveform for the loaded LPDA still has more than one clean pulse, the resistive loading minimizes the reflected groups of pulses so that the LPDA radiates only the first group (Fig. 3.47). If the same antenna is used as a receiver, then the induced received current at the feed point similarly consists of a single group of pulses and is devoid of unwanted dispersion (Fig. 3.48). Figures 3.49 and 3.50 display the calculated excitation and incident waveform that is required for radiation and reception of a monocycle, respectively. These figures reflect that for time domain applications this is not a good antenna to use.

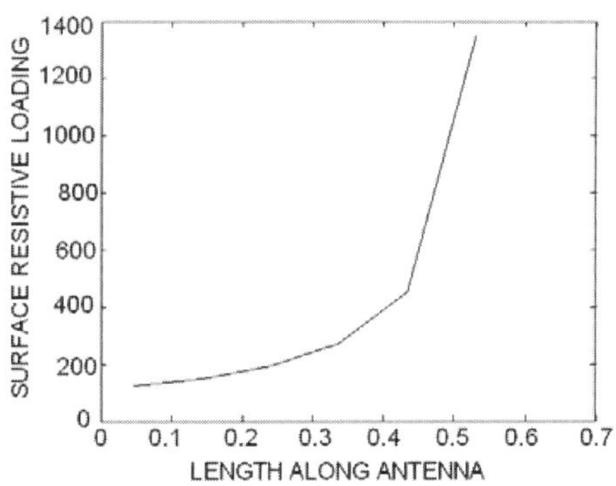

Figure 3.46. Resistive loading profile on LPDA.

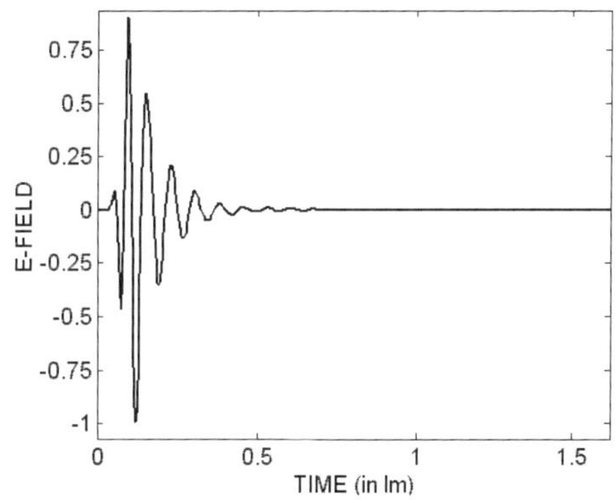

Figure 3.47. Radiation from a loaded LPDA.

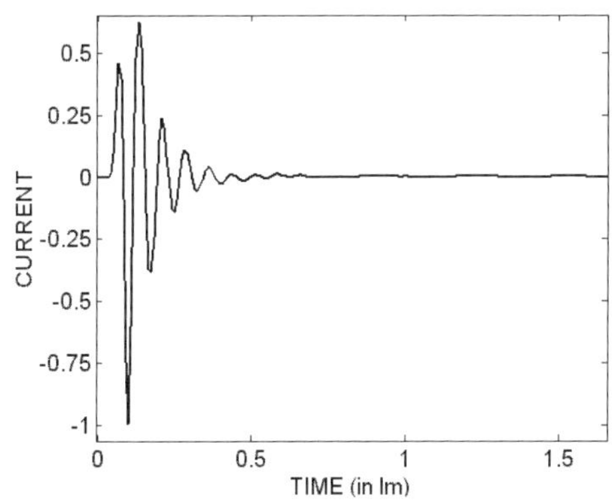

Figure 3.48. Reception of a monocycle pulse by a loaded LPDA.

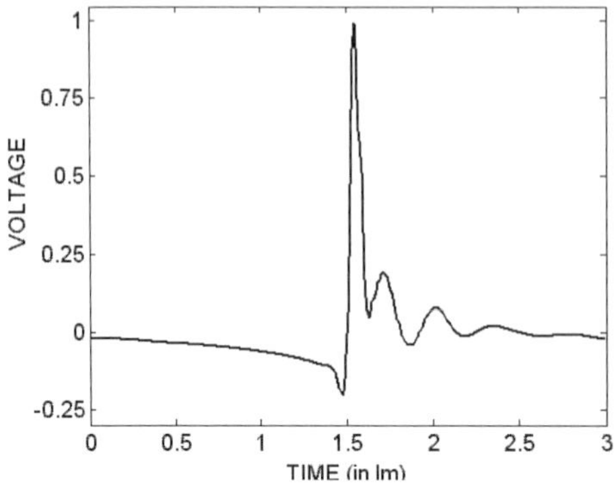

Figure 3.49. Input required for a monocycle transmission.

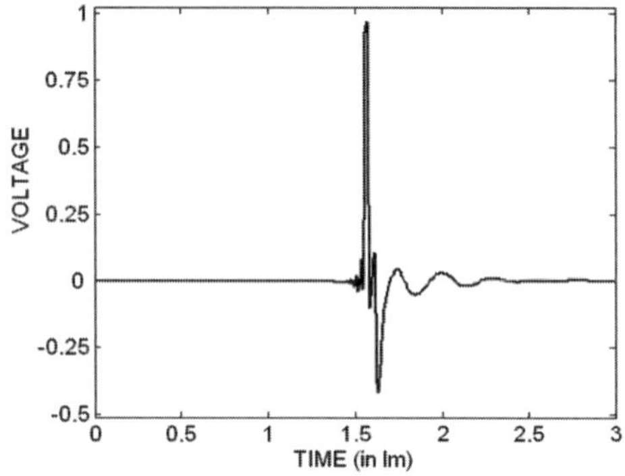

Figure 3.50. Incident wave shape to induce a monocycle of current.

3.6.5 Spiral

An antenna which can be completely specified by an angle is a frequency-independent antenna. A spiral antenna belongs to such a class of antennas (Fig. 3.51). It exhibits frequency independent behavior over a frequency range determined by its inner and outer radii. The pattern symmetry is best maintained by using a self-complementary spiral antenna, which is obtained if the width of each arm (w) and the spacing between each turn (s) are equal. In our design, we vary the width of each arm and the spacing between the limits of $0.01 \text{ m} < w < 0.1 \text{ m}$ and $0.01 \text{ m} < s < 0.1 \text{ m}$, respectively. The design parameters used for the spiral antenna are: inner radius = r_1 = 0.005 m; outer radius = r_2 = 0.12 m; low frequency = $f_{\text{low}} = c/(2\,\pi\, r_2)$ = 400 MHz; and high frequency = $f_{\text{high}} = c/(2\,\pi\, r_1)$ = 10 GHz. The feed wire of the antenna is connected between the two spiral arms at the center of the antenna and has a 10 mm length and a 0.1 mm radius.

The time support of the response determined by the lowest frequency of operation is 30 lm, and the width of the input pulse is 0.12 lm. The origin of the coordinate system is the midpoint of the feed wire with the antenna located in the xy-plane. The antenna is circularly polarized and radiates similar waveforms in both polarization planes. For the purpose of illustration, only the response in the polarization plane defined by the yz-plane (vertical plane) is calculated, because the response in the xy polarization plane is similar.

Figure 3.51. Structure of a spiral antenna.

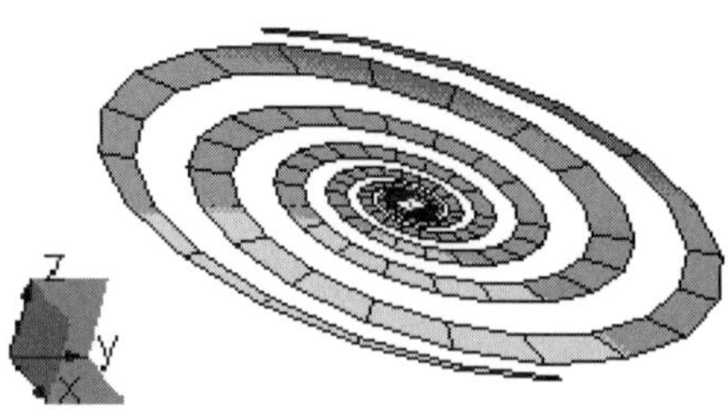

When the antenna radiates, the simulated far field along the z-axis (perpendicular to the plane of the antenna) is plotted in Fig. 3.52. Though the waveform does not show separately identifiable reflected waves, the radiated pulse is dispersed at the falling edge due to the resonant nature of the antenna. When the spiral is used as a receiving antenna and a circularly polarized monocycle pulse is incident along the z-axis, one draws a similar conclusion (Fig. 3.53).

To obtain a fairly constant input impedance for the whole range of frequencies, one can take the inner radius of the antenna equal to be the strip width or spacing between turns ($r_1 = w = s$). This has been experimentally verified in Ref. [34]. Thus another spiral antenna with $r_1 = 0.005$ m, $r_2 = 0.12$ m, and $w = s = 0.0025$ m was constructed. This modified spiral antenna is shown in Fig. 3.54. The radiated wave along the plane perpendicular to the antenna is shown in Fig. 3.55 for a monocycle excitation. When the antenna is receiving, the induced current from a circularly polarized monocycle field that is incident along the perpendicular to the plane of the antenna is shown in Fig. 3.56. On

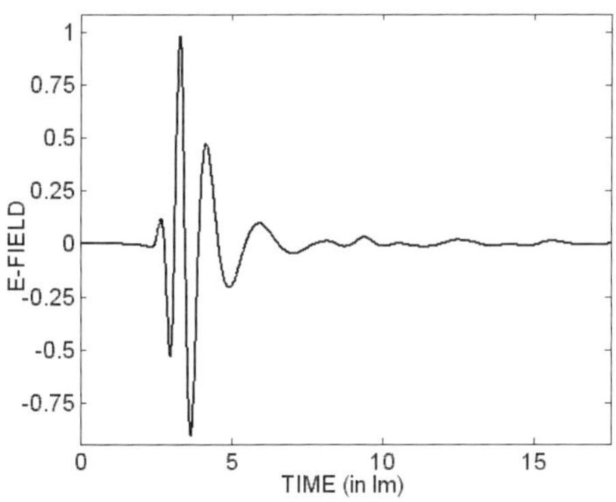

Figure 3.52. Radiation from a spiral antenna.

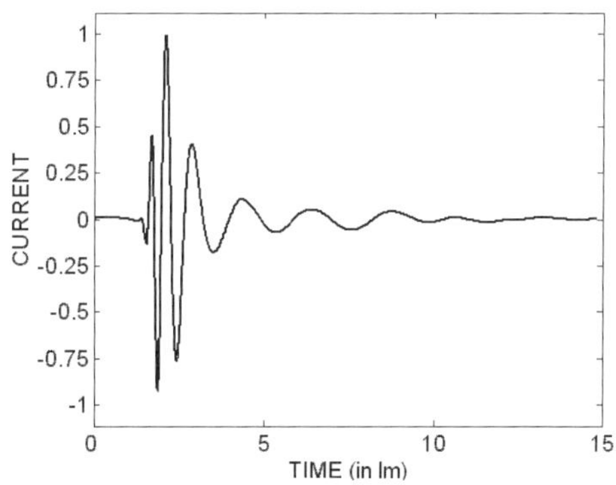

Figure 3.53. Reception of a monocycle by a spiral antenna.

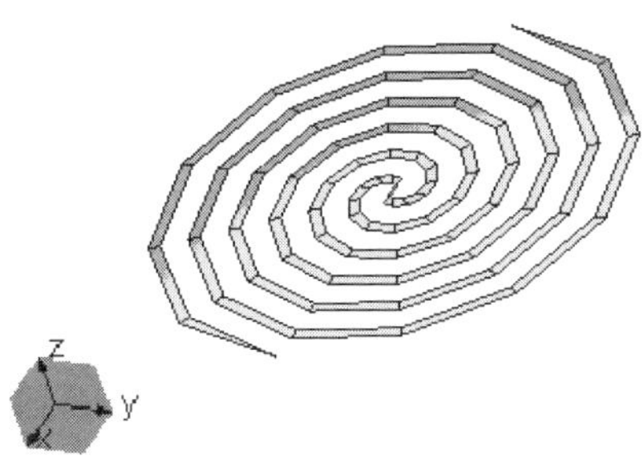

Figure 3.54. Structure of a modified spiral antenna.

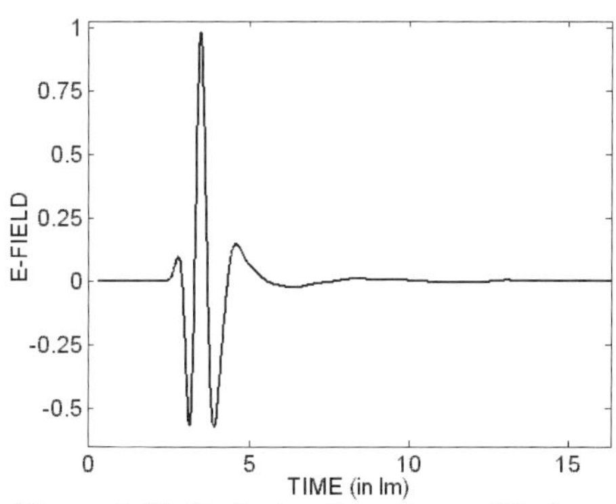

Figure 3.55. Radiation from the modified spiral antenna.

comparing the radiated fields from both spiral antennas (Figs. 3.52 and 3.55), some improvement in the pulse shape is apparent. Since the dispersion is less for the antenna with $r_1 = w = s$, one may conclude that the modification of the antenna structure subject to this constraint results in better frequency-independent behavior.

To enhance the antenna response further, apply a tapered resistive loading, with distributed loads applied to the arms of the antenna. The antenna is divided into the radial zones in Fig. 3.57. The plates within each radial zone are loaded with the same resistance. The loading profile of Fig. 3.58 is obtained from (3.3) by setting the length of the antenna to $2h = 0.1$ m and the effective radius to the width of each arm (0.0048 m). By this procedure, we can reduce the dispersion of the antenna to obtain a clean radiated pulse (Fig. 3.59). Note that the dispersion of the unloaded spiral in Fig. 3.55 has been eliminated by properly

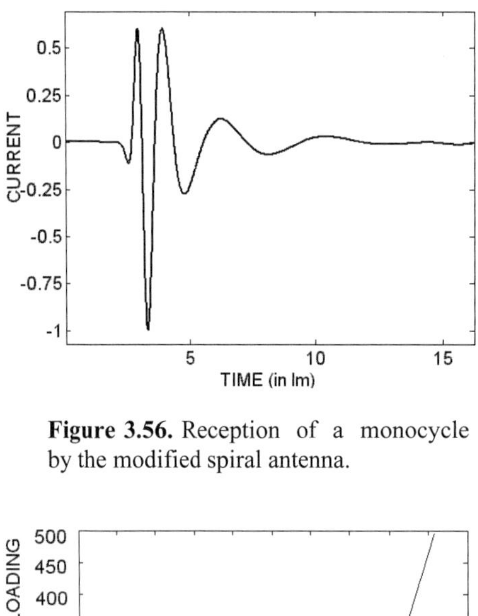

Figure 3.56. Reception of a monocycle by the modified spiral antenna.

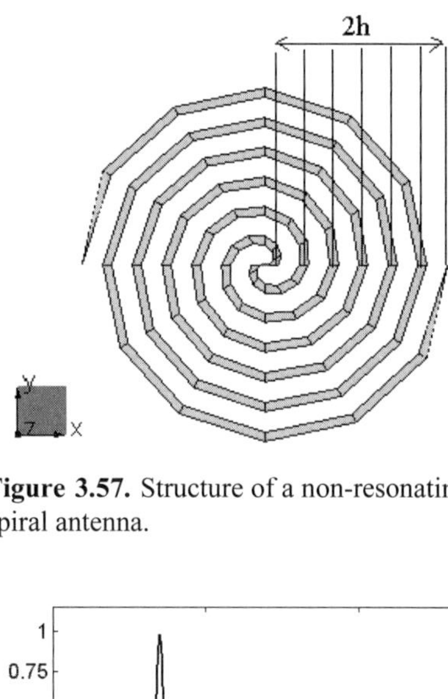

Figure 3.57. Structure of a non-resonating spiral antenna.

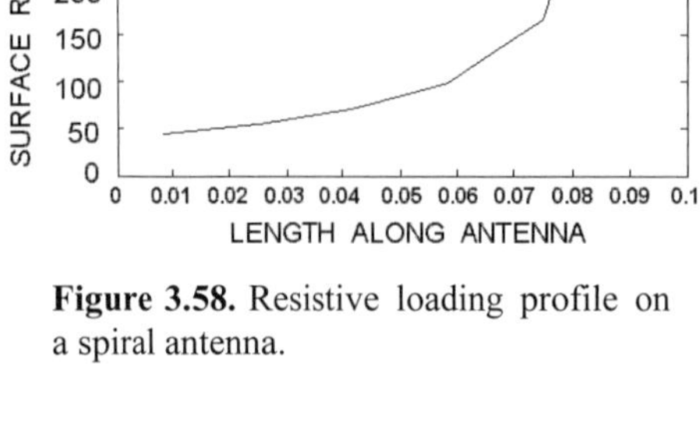

Figure 3.58. Resistive loading profile on a spiral antenna.

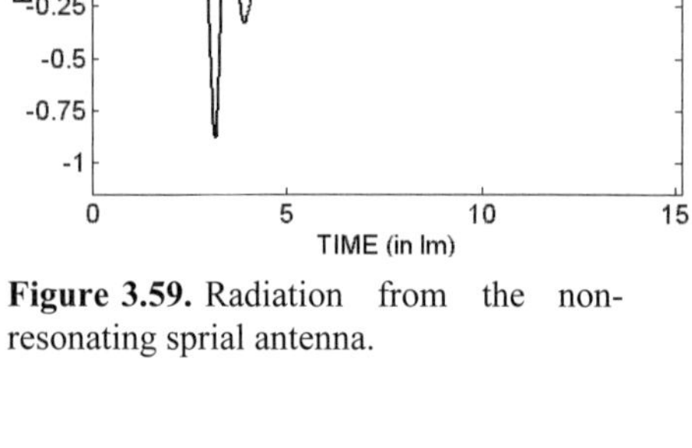

Figure 3.59. Radiation from the non-resonating sprial antenna.

loading the antenna. If the loaded antenna is used for reception, the induced current shows similar improvement (Fig. 3.60). The radiation efficiency of the loaded spiral monotonically increases with frequency, with a maximum efficiency of 55% at 10 GHz (Fig. 3.61). So the improvement in signal fidelity from loading is offset by a minimum loss in efficiency of 40–45%.

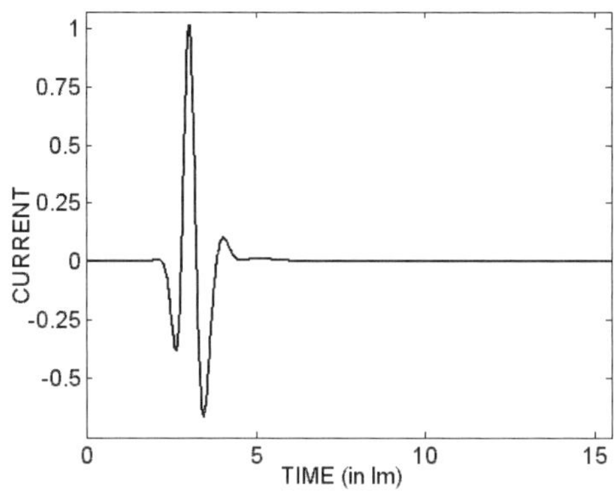

Figure 3.60. Reception of a monocycle by the non-resonating spiral antenna.

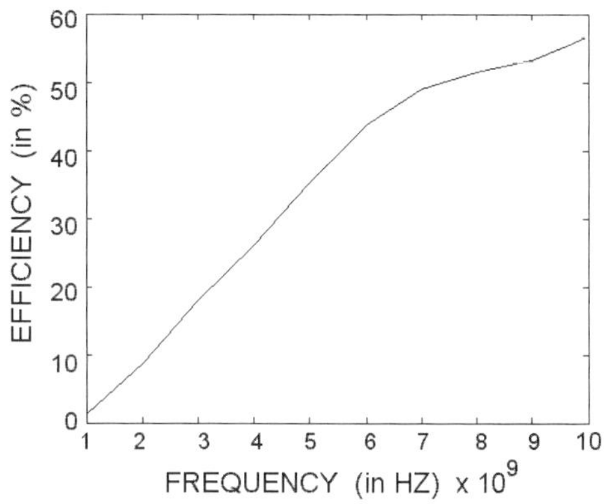

Figure 3.61. Radiation efficiency of the loaded spiral antenna.

3.7 CONVENTIONAL WIDEBAND ANTENNAS

In this section, we investigate the performances of various conventional UWB antennas. These antennas are simulated with the numerical electromagnetic code [20] from models and dimensions available in the published literature. Each antenna is simulated over wide frequency bands to study their radiation and scattering. The input to each of the antenna is a monocycle pulse of 1 V/m, so that the scales of their responses are comparable. The dynamic transient input resistance, the input resistance of the antenna due to a step excitation, is computed. The plot of the transient resistance illustrates the flow of the current along the antenna. In addition, to indicate how quickly the transient response of an antenna system stabilizes, the normalized dynamic transient resistance versus time in lm is plotted for each antenna. In some sense, this quantity provides a measure of the initial instantaneous match between the source and the antenna.

3.7.1 Volcano Smoke

The volcano smoke antenna was originally designed by Kraus [35]. It consists of a bulb and a ground plane, but no specific design equations and parameters are available. Thus we assume that the bulb has a teardrop shape, with its height approximately a quarter wavelength above a ground plane [36, 37]. Further, the teardrop consists of a finite cone, with a half-cone angle of 48°, and a sphere inscribed at the mouth of the inverted cone (Fig. 3.62). The height of the antenna is 0.1 m, and the height of the cone is 0.025 m. The input impedance of the antenna is governed by the half-cone angle, and the minimum height of the teardrop is considered to be a quarter wavelength at the lowest frequency of operation. The antenna is excited by a feed wire at the tip of the cone. The feed wire has a length of 3 mm and a radius of 0.1 mm. The origin of the coordinate system is the point where the feed wire is connected to the ground plane.

The antenna is simulated between 250 MHz and 25 GHz, time support of the response determined by the lowest frequency of operation is 1.75 lm, and the width of the exciting monocycle pulse is 0.05 lm for radiation and reception. The antenna radiates maximum energy in the horizontal *xy*-plane (Fig. 3.62), with nulls occurring at the zenith of the antenna, and the polarization of the far field is in the vertical direction along the *z*-axis. In the *xy*-plane, the radiated field closely resembles the monocycle input (Fig. 3.63), as should be the case for conical antennas. However, the field differs from the input between 0.18 and 0.25 lm as a consequence of a secondary pulse that corresponds to reflections from the ends of the antenna occurring 0.1 lm after the initial pulse. The induced current when this antenna receives a *z*-polarized monocycle field from any direction in the *xy*-plane is provided in Fig. 3.64. This antenna is reasonably suited for time domain applications if the top part of the antenna is shaped properly.

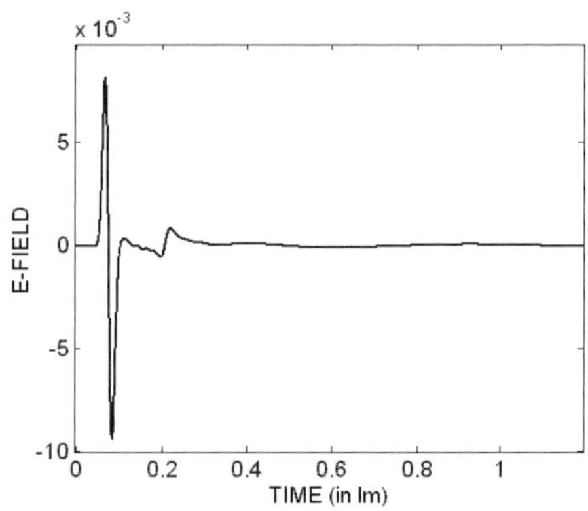

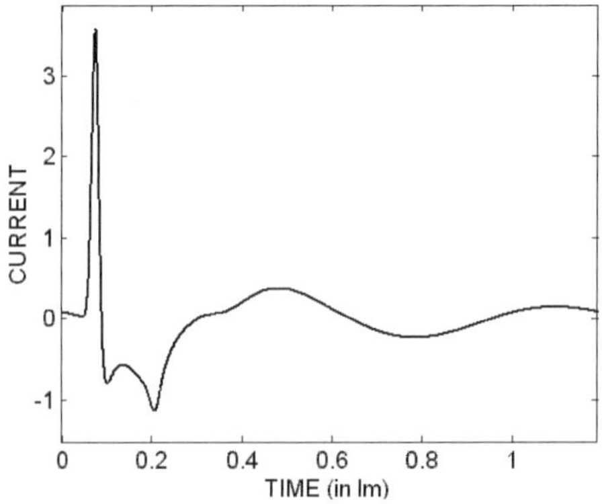

Figure 3.62. (top left) A volcano smoke antenna.

Figure 3.63. (top right) Radiation from a volcano smoke antenna.

Figure 3.64. (left) Reception of a monocycle by a volcano smoke antenna.

The normalized transient input resistance of the antenna due to a step excitation is shown in Fig. 3.65. The input resistance in this figure increases steadily until it stabilizes towards the end of the time span shown. The wave shape of the resistance plot illustrates the flow of the current along the antenna.

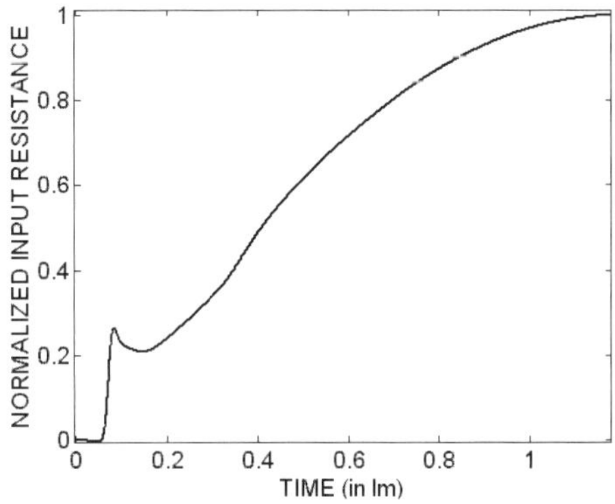

Figure 3.65. Transient input resistance of a volcano smoke antenna.

3.7.2 Diamond Dipole

The diamond dipole antenna [38] is an inverted bow-tie antenna. Unlike a bowtie, the diamond dipole has a response that is *not* temporally dispersive, leading to good UWB operation. The antenna consists of two isosceles triangular elements connected by a feed wire at their bases (Fig. 3.66). The height of each element is one-quarter wavelength at the operating frequency. In this simulation, the height of the antenna is chosen to be 3.8 cm. Thus the operating frequency of the antenna is 2 GHz. The origin of the coordinate system is the midpoint of the feed wire (1 mm long and radius 0.01 mm). At each frequency, this antenna has an azimuthally symmetric, z-polarized, donut-shaped radiation pattern about the z-axis similar to that of a dipole antenna, with maximum radiation in the xy-plane.

For this simulation, the frequencies vary between 40 MHz and 4 GHz, the time support of the response determined by the lowest frequency of operation is 75 lm, and the width of the monocycle is 0.312 lm. On viewing Fig. 3.67, one observes that the radiated field is essentially the second derivative of the input. In addition, the received current induced on the antenna from a monocycle incident field arriving along the x-axis (Fig. 3.68) looks like a smoother, less dispersive version of the induced current on the tapered horn (Fig. 3.35). From Figs. 3.67 to 3.68, it is evident that the diamond dipole antenna has little or no time (phase) dispersion. For this antenna, a typical plot of the transient input resistance is shown in Fig. 3.69.

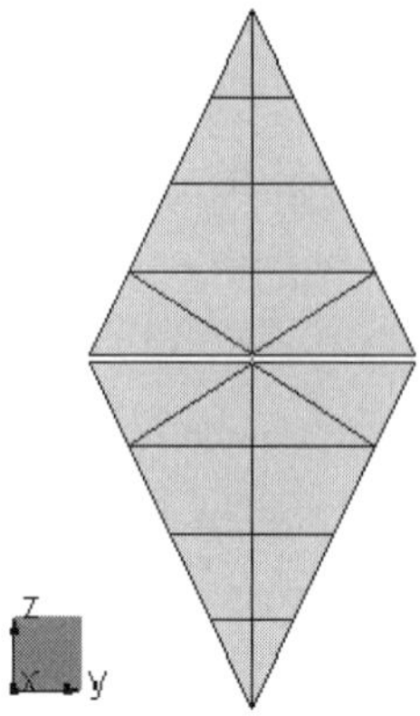

Figure 3.66. A diamond dipole antenna.

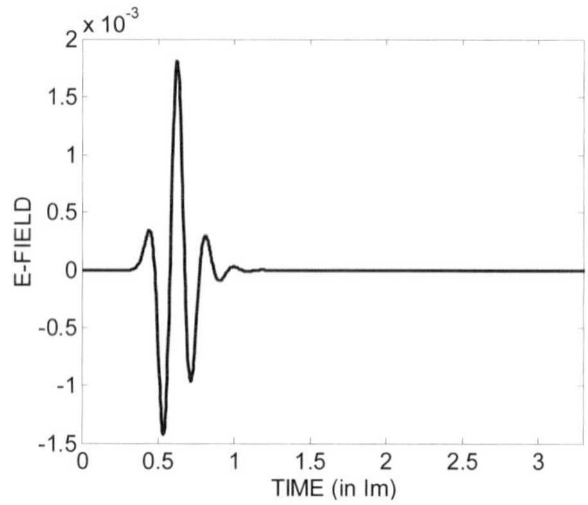

Figure 3.67. Radiation from a diamond dipole antenna.

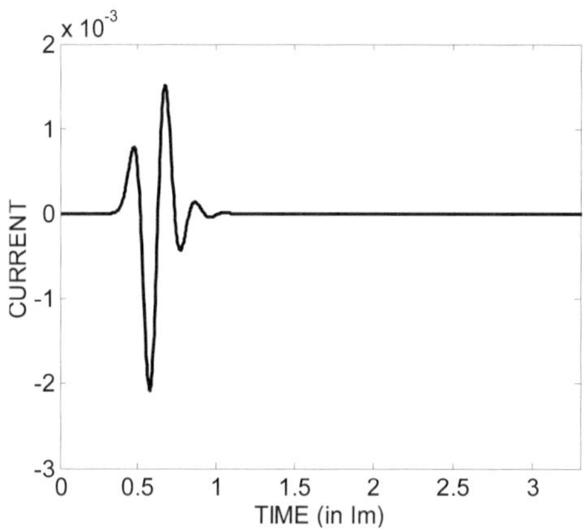

Figure 3.68. Reception of a monocycle by a diamond dipole antenna.

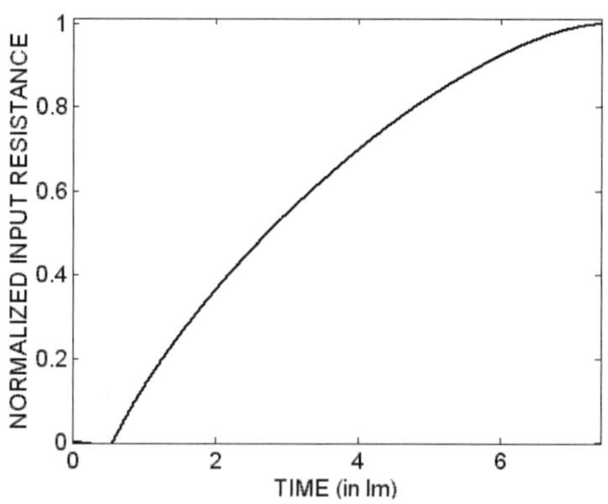

Figure 3.69. Transient input response of a diamond dipole antenna.

3.7.3 Monofilar Helix

The monofilar helical antenna has inherent broadband properties over a frequency range of 2 to 1. According to the experimental work of Kraus [35], the helical antenna can be optimized by maintaining its parameters within the following ranges: $3/4 < C/\lambda < 4/3$, $12° < \alpha < 14°$, and $N \geq 4$, where the circumference C of the helix is πD, S is the spacing between turns, and the pitch angle α is $\tan^{-1}(S/C)$. For these calculations, $\alpha = 13°$, $S = 0.16$ m, $N = 8$, and $C = 0.92\lambda = 0.7$ m. The helix is placed over a perfect electrically conducting (PEC) ground plane, with the feed wire attached to the ground plane (Fig. 3.70). The feed wire has a length of 20 mm and a radius of 0.5 mm. Furthermore, the

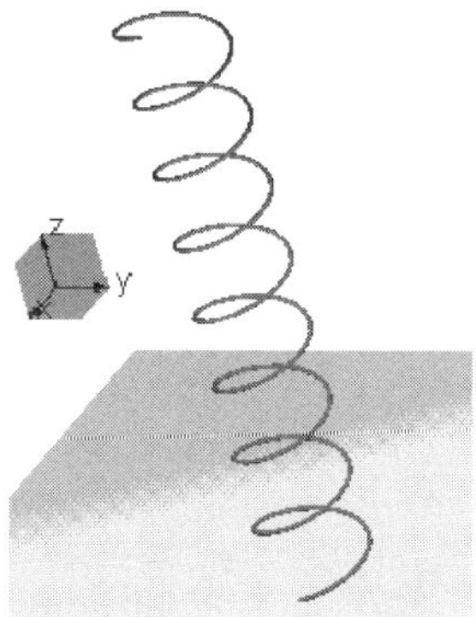

Figure 3.70. Structure of a monofilar helical antenna.

antenna is simulated between 100 MHz and 10 GHz, where the time support of the response determined by the lowest frequency of operation is 3 lm.

In the literature [35,39], helical antennas are typically designated as "normal mode" for radiation in the direction perpendicular to the axis or "axial mode" for beam-like radiation along the axis. If the circumferential length of one turn times the number of turns is much less than λ, normal-mode radiation occurs. If C is close to λ at the operating frequency, the helical antenna radiates the first-order axial mode only. So for lower frequencies, single-order mode radiation is observed. As the frequency is increased, the corresponding increase in C/λ results in the appearance of higher radiation modes. The first radiation mode has a directed beam with a maximum along the direction of the helical axis. In contrast, for the higher radiation modes, the pattern breaks down and maximum radiation does not occur along the axis of the helix [40]. Consequently, broadband users of this antenna must be cognizant that maximum radiation does not necessarily occur along the helical axis over the desired range of frequencies. By observing the radiation patterns for different frequencies along different spatial angles, it can be seen that for most regions, especially for the higher frequency region, there is significant radiation along the direction perpendicular to the axis of the antenna.

Since the radiated field is polarized along the helical axis, the field along a direction perpendicular to the axis of the helix is simulated. Clearly, the antenna has resonating characteristics, because the field displays more than one pulse (Fig. 3.71). The separation between adjacent pulses is around 0.7 lm, which corresponds to the time required by the current to travel the length of one turn of the helix. When this helix is used for reception, a similar conclusion can be drawn about the induced current at the feed point for an incident monocycle that is perpendicular to the axis and is polarized along the axis (Fig. 3.72). The transient input resistance of the helical antenna is shown in Fig. 3.73. From the temporal responses it appears this is not a good broadband antenna.

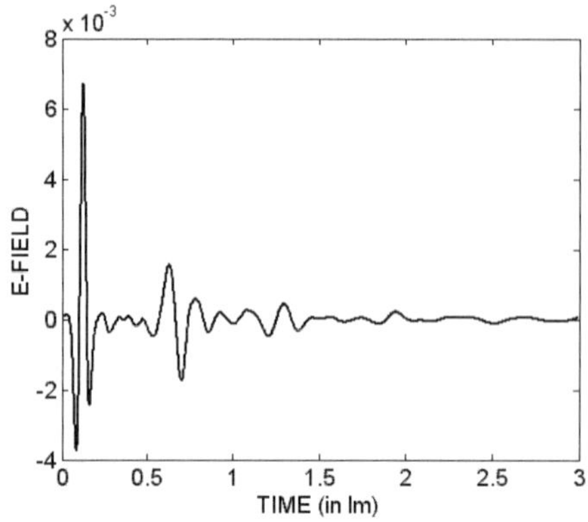

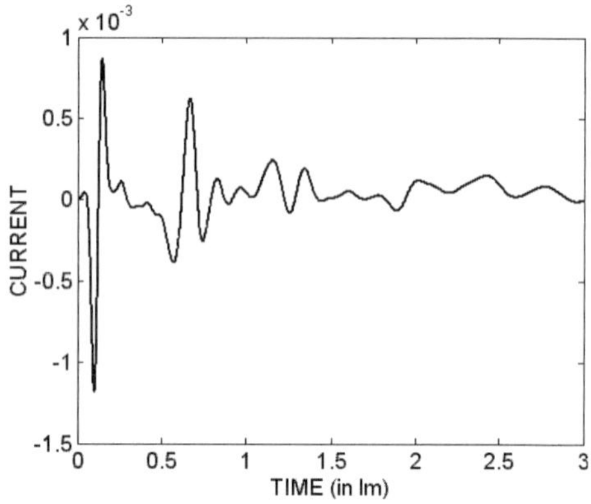

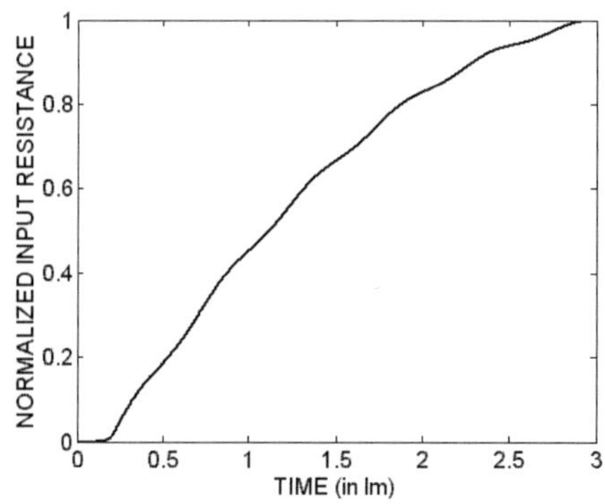

Figure 3.71. (above left)
Radiation from a helical antenna.

Figure 3.72. (above right)
Reception of a monocycle by a helical antenna.

Figure 3.73. (left)
Transient input resistance of a helical antenna.

3.7.4 Conical Spiral

Instead of using a helix of constant radius, one can use a helix whose radii and pitch angle vary with the distance from the feed, which produces a log-periodic appearance. The average circumference C is chosen to be 0.7 m, and the spacing S is maintained constant at 0.16 m. Several values of pitch angle are chosen ranging from $8° < \alpha < 32°$. The values of C and α for each turn are tabulated in Table 3.1. By using these parameters, an 8-turn conical spiral is constructed (Fig. 3.74). The conductor diameter is chosen to be 0.005 m. The antenna is placed

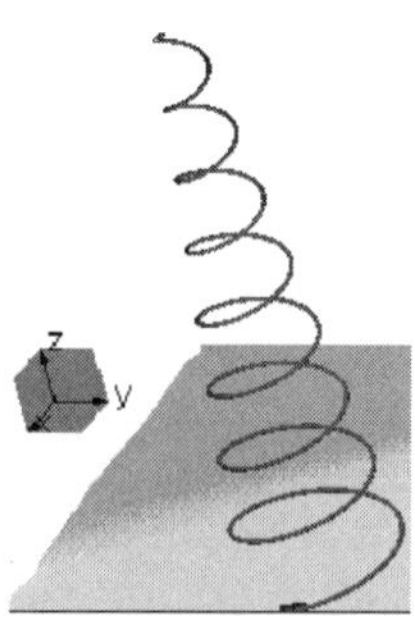

Figure 3.74. Structure of a conical spiral antenna.

Table 3.1. Circumferences and Pitch Angles for Conical-Spiral Simulations.

Turn #	C [m]	α [°]
0	1.1500	7.920
1	1.0375	8.767
2	0.9250	9.814
3	0.8125	11.140
4	0.7000	12.875
5	0.5875	15.235
6	0.4750	18.616
7	0.3625	23.816
8	0.2500	32.620

over a PEC ground plane, with the feed wire attached to the ground plane. The feed wire has a length of 20 mm and a radius of 0.5 mm. The antenna is simulated between 100 MHz and 10 GHz and the time support of the response determined by the lowest frequency of operation is 3 lm.

When the far-field observation point and the incident field are in the *xy*-plane, the corresponding radiated field and induced current are plotted in Figs. 3.75 and 3.76. The time-domain *z*-polarized radiated field consists of subpulses with decreasing amplitudes and increasing separations between consecutive pulses. The separation between the first and second pulses is seen to be around 1.1 lm, which corresponds to the time required by the induced current wave to

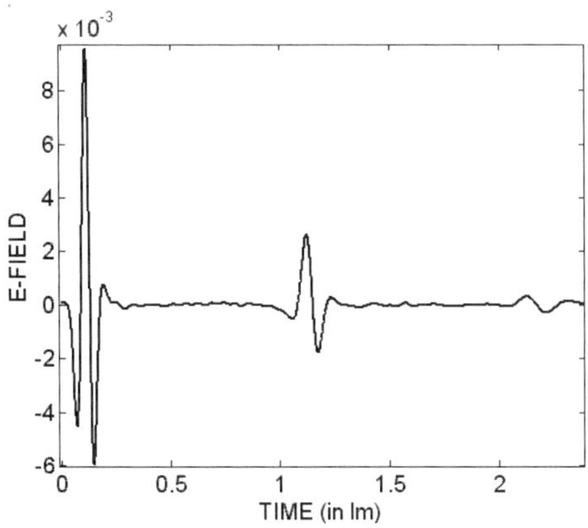

Figure 3.75. Radiation from a conical spiral antenna.

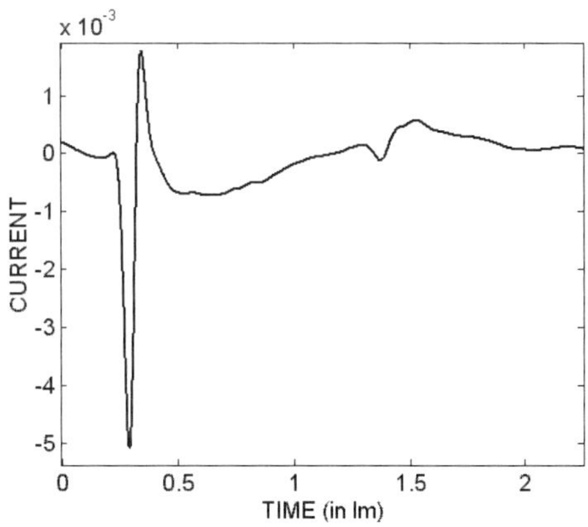

Figure 3.76. Reception of a monocycle by the conical spiral antenna.

travel the length of the first turn of the antenna. Similarly, the second and third pulses are separated by the distance associated with the length of the second turn of the spiral. The transient input resistance of the conical spiral (Fig. 3.77) looks like a smoother version of the resistance for monofilar helix. This has the potential of a broadband antenna if the reflections from the tip can be minimized.

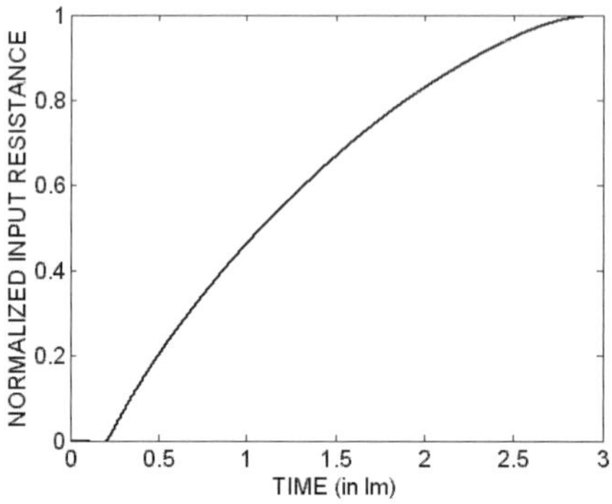

Figure 3.77. Transient input resistance of a conical spiral antenna.

3.7.5 Monoloop

The monoloop antenna [41,42] is a magnetic UWB antenna. Magnetic antennas have predominantly magnetic near fields. In embedded applications, magnetic antennas are more suitable because they do not cause undesired coupling with nearby objects. The monoloop antenna, like other small magnetic antennas, is a practical realization of the Hertzian magnetic dipole. It has a planar transceiving element that is affixed to a ground plane (Fig. 3.78). The transceiving or radiating element consists of a layer of a semicircular strip of metal arranged upon a dielectric substrate. The dielectric substrate has a relative dielectric constant of 4 and a conductivity of 0.01. The thickness of the dielectric substrate is 1 cm. The radiating element lies in the *xz*-plane and is perpendicular to the ground plane (*xy*-plane). The circumferential path along the middle of the radiating element is approximately equal to λ /4 with respect to the operating frequency. The bandwidth of the antenna is determined by the width of the radiating element. The antenna is fed symmetrically through a gap in the radiating element, where the feed wire is connected to the radiating element and has a length of 4 mm and a radius of 0.01 mm. The radiated field is observed in a plane that is perpendicular to the ground plane, and the polarization is in the plane of the antenna along the length of the feed.

The far field and induced current are simulated between 30 MHz and 3 GHz, and the time support of the response determined by the lowest frequency of operation is 10 lm. The radiated field is the derivative of the monocycle input (Fig. 3.79), and the induced received current is identical to the monocycle incident field (Fig. 3.80). These results clearly indicate that the monoloop has no time (phase) dispersion. Furthermore, its dynamic transient input resistance differs significantly from the resistances of previous antennas (Fig. 3.81).

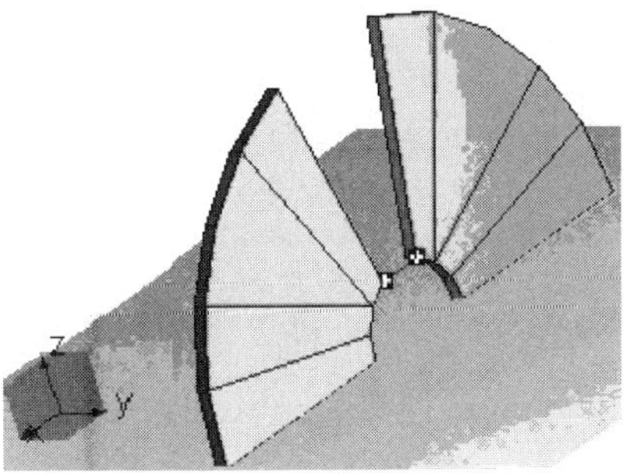

Figure 3.78. A monoloop antenna.

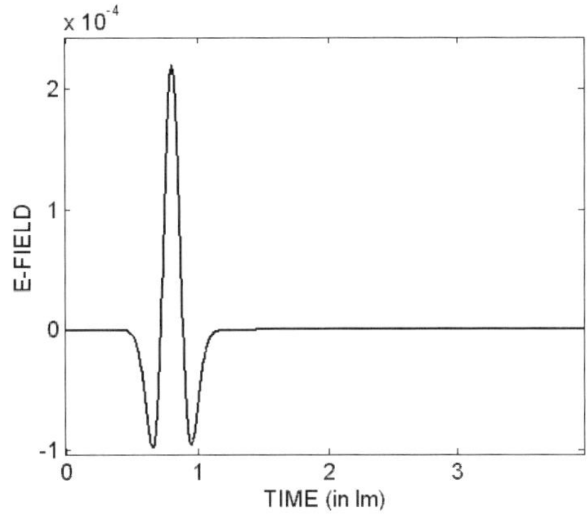

Figure 3.79. Radiation from a monoloop antenna.

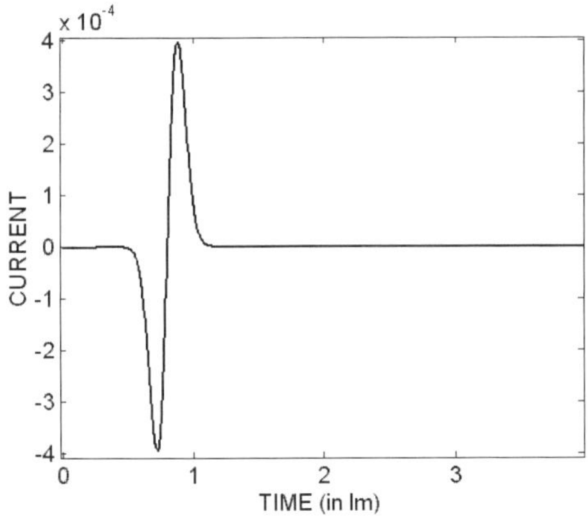

Figure 3.80. Reception of a monocycle by a monoloop antenna.

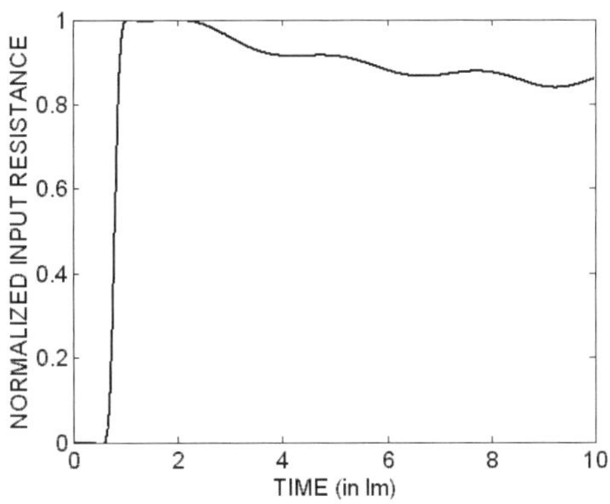

Figure 3.81. Transient input response of a monoloop antenna.

3.7.6 Quad-Ridged Circular Horn

The quad-ridged circular horn [43,44] is a conical horn with circular cross-sections and with four electrically conductive ridges placed orthogonal to each other inside the cone (Fig. 3.82). Due to its symmetric structure, this horn exhibits circular polarization. The ridges or blades are designed so that the edge of each ridge nearest the *z*-axis is tapered towards the open end of the antenna. The ridge taper is such that the spacing between the horn axis and the inner edge of the ridge appears as a truncated right hyperbolic triangle, giving rise to constant impedance along the device length. The top of each ridge is a circular arc that is designed to reduce reflections by making it a continuous wave structure. The length of the antenna is 0.35 m, and the diameter at the mouth of the antenna is 0.07 m. The antenna has two feed wires, each connecting the opposing ridges at the throat (narrowest part of the cylinder). Each feed wire has a length of 8 mm and a diameter of 0.1 mm. The antenna is excited by two generators connected to each of the two feed wires which are attached to the two diametrically opposite ridges. The generators are 90° out of phase with each other.

The antenna is simulated between 100 MHz and 15 GHz. The time support of the response determined by the lowest frequency of operation is 3 lm. The antenna radiates along the axis (z-axis) of the cone. Since the radiated field is circularly polarized, the radiation is calculated for one polarization. The radiated field is shown in Fig. 3.83. Due to the circularly polarized nature of the radiation, the field with the opposite polarization is identical. Figure 3.83 shows that the antenna suffers from time dispersion as a result of reflected pulses. The induced current between any two of the ridges on the antenna from a circularly polarized monocycle incident field arriving along the z-axis is shown in Fig. 3.84. As in the radiation case, multiple reflections cause temporal dispersion of the current. In this case, the antenna's transient input resistance has much shorter duration (Fig. 3.85).

Figure 3.82. A quad ridged horn antenna.

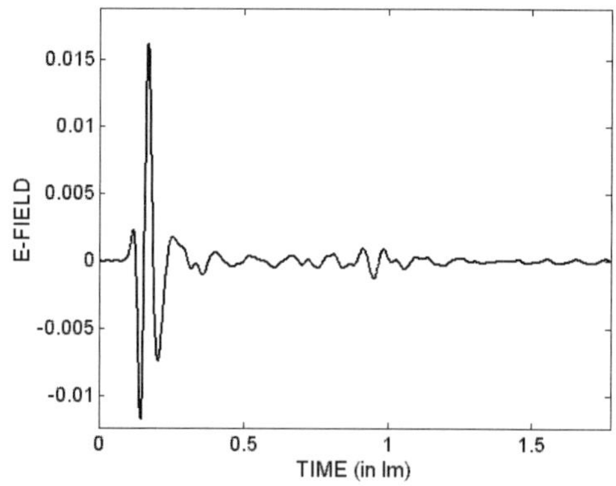

Figure 3.83. Radiation from a quad ridged horn antenna.

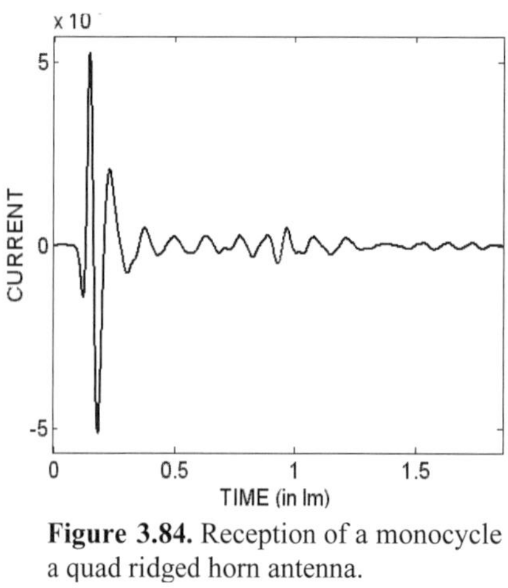

Figure 3.84. Reception of a monocycle by a quad ridged horn antenna.

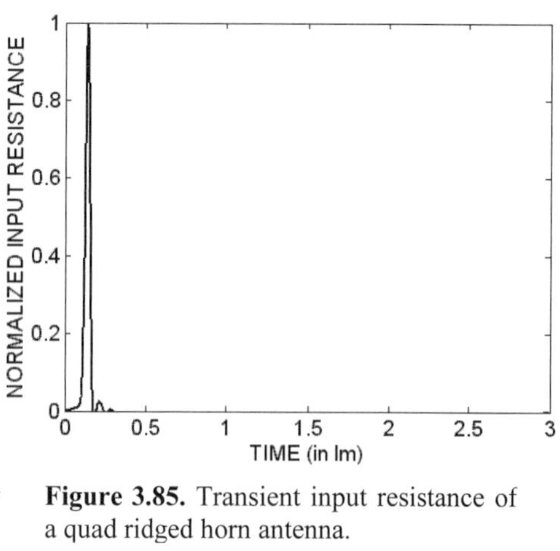

Figure 3.85. Transient input resistance of a quad ridged horn antenna.

3.7.7 Bi-Blade with Century Bandwidth

The bi-blade antenna is designed for multi-octave transmission with a century bandwidth, i.e., a ratio of 100:1. Due to its century bandwidth of operation, the bi-blade radiates and receives at UHF, L, C, S, and X bands [45]. Each of the two blades has a throat (narrowest part), a mouth (widest part), and a tip. The throat serves as the feed point. The tip is an arc of constant radius, thereby giving rise to a low voltage standing wave ratio of about 1.19 to 1 [45]. The radius of the arc determines the slope of the antenna's surge impedance. The blades are designed such that the slot width between the two blades increases logarithmically from the throat to the mouth of the antenna as shown in Fig. 3.86. The blade length is 0.56 m and the maximum slot width is 0.43 m. The blade is 0.11 m wide at its mouth. The antenna has a coplanar geometry and thus is easy to integrate into many systems. The antenna is fed by a generator placed between the two blades at the throat. The feed wire connecting the two blades of the antenna has a length of 4 mm and a radius of 0.01 mm. The origin of the coordinate system is the midpoint of the feed wire.

The antenna is simulated between 160 MHz and 16 GHz, the time support of the response determined by the lowest frequency of operation is 1.875 lm, and the width of the input pulse is 0.075 lm. The antenna radiates along the *z*-axis of Fig. 3.86, and the polarization of the radiated field is along the *y*-axis on the *yz*-plane. The radiated field has a strong return and a weak reflected pulse corresponding to the reflection from the tip of the antenna (Fig. 3.87), which implies that the antenna does not have significant phase dispersion. The first pulse is a good approximation of the input monocycle. When the antenna is illuminated by a monocycle pulse arriving along the *z*-axis, the induced current roughly has the same general shape as the incident field but with significant differences (Fig. 3.88). In particular, the current increases to a value that is commensurate with its maximum positive excursion. The transient input resistance of the antenna in response to a monocycle pulse is shown in Fig. 3.89.

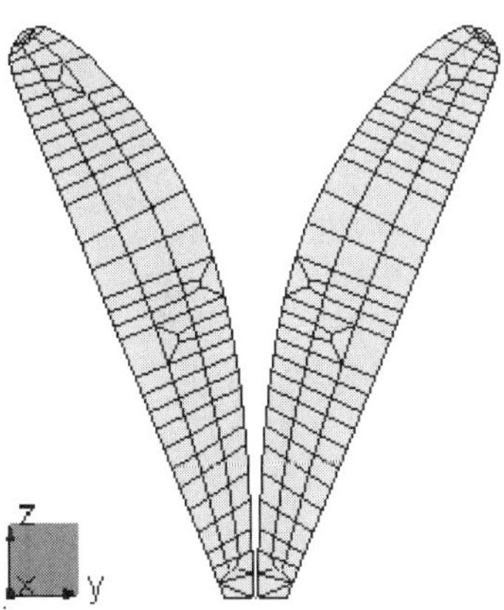

Figure 3.86. A bi-blade century bandwidth antenna.

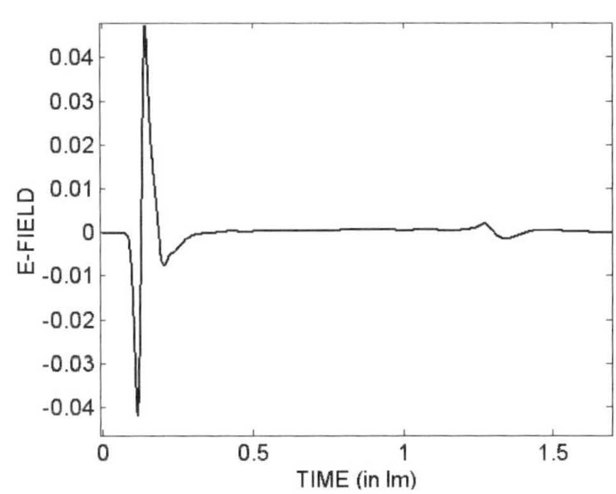

Figure 3.87. Radiation from a bi-blade antenna.

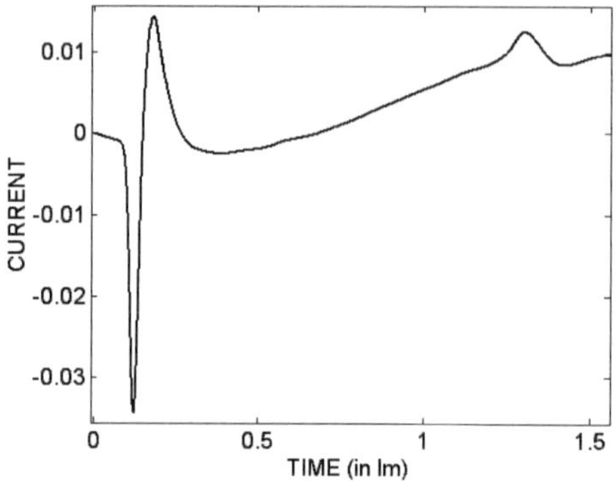

Figure 3.88. Reception of a monocycle by a bi-blade antenna.

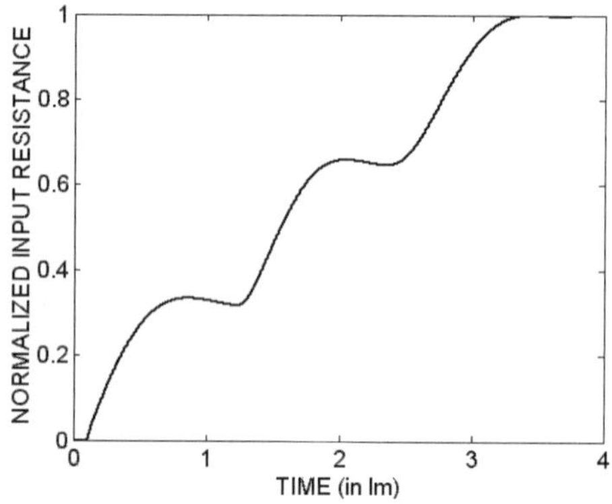

Figure 3.89. Transient input resistance of a bi-blade antenna.

This is an extremely wideband antenna and can also provide good temporal waveforms if the tips are properly shaped.

3.7.8 Cone-Blade

The cone-blade [46] is a circularly polarized antenna structure based on the conical antenna and is a modification of the bi-blade century bandwidth antenna. The cone-blade structure is a truncated closed half cone that is surrounded by four equally spaced blades that are separated by 90° (Fig. 3.90). The supporting base of the antenna is the flat truncated portion of the cone, and the blades are thin conducting plates that are placed near the base of the cone and extend upwards in the same direction as the vertex of the cone. The cone has a half angle of 11.53° and a height of 0.228 m. The antenna is fed via a quadri-phase monocycle excitation by four generators, one at each of the four wires connecting the blade and the cone. Each generator has a 90° phase shift from the generator on the adjacent feed arm. Since the main beam of the radiation pattern at each

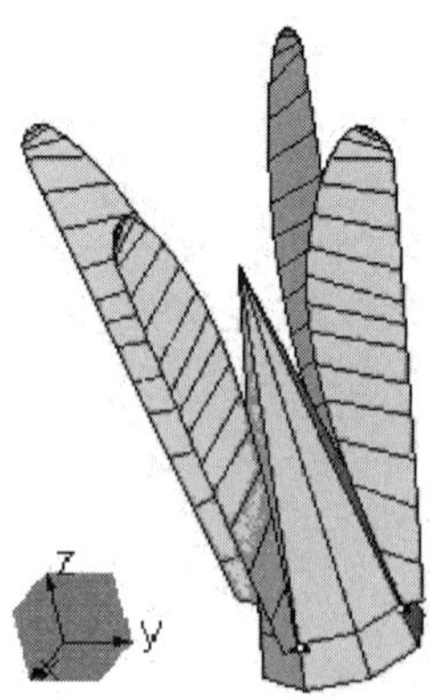

Figure 3.90. A cone-blade antenna structure.

frequency points along the axis of the cone (z-axis), the simulated far field is calculated at an observation point along the positive z-axis. Each of the feed wires has a length of 20 mm and a radius of 0.01 mm. The antenna is simulated between 200 MHz to 20 GHz, and the width of the input pulse is 0.06 lm.

The observed radiated field and the second derivative of the monocycle are shown in Fig. 3.91. Since no reflected pulses are present in the field and the curves almost coincide, this antenna is a traveling-wave antenna. For reception, the current induced in the antenna from a circularly polarized monocycle field that is incident along the z-axis is shown in Fig. 3.92. The induced current has no components associated with reflections along the antenna, thereby providing further verification that the antenna is a traveling-wave structure. Consequently, applying resistive loading to broaden the cone blade's transfer function is unnecessary, which means that reflectionless transmission and reception are achieved with no loss in efficiency. The dynamic transient input resistance of the antenna is shown in Fig. 3.93.

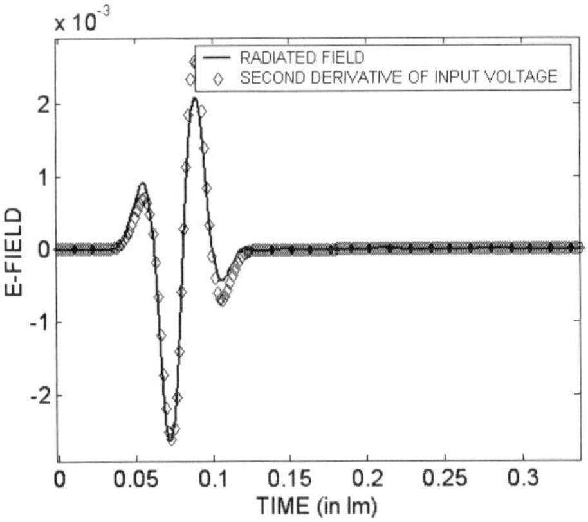

Figure 3.91. Radiation from a cone-blade antenna.

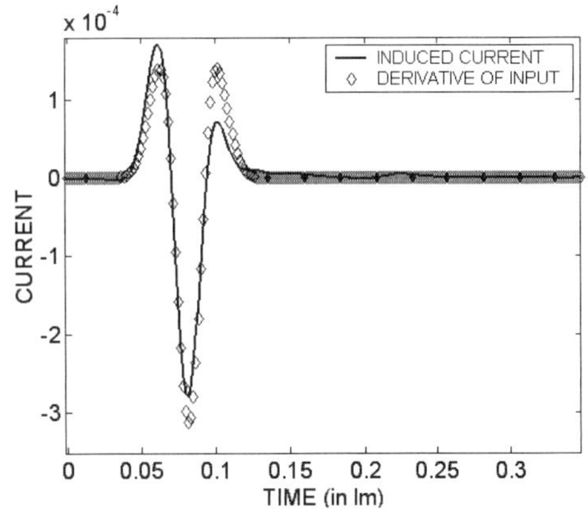

Figure 3.92. Reception of a monocycle by a cone-blade antenna.

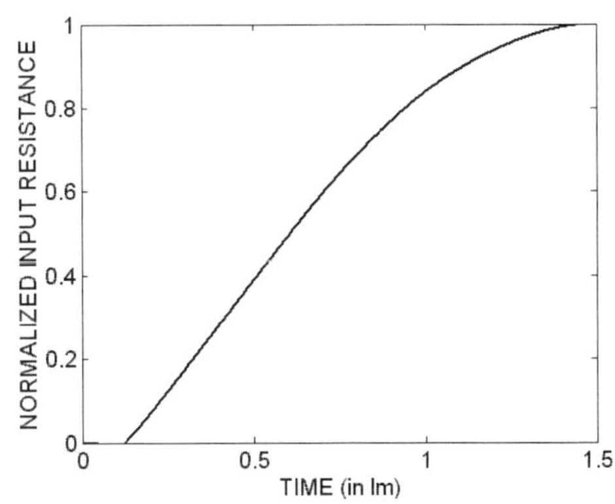

Figure 3.93. Transient input resistance of a cone-blade antenna.

Kanda has shown that the transmitting transient response is proportional to the time derivative of the receiving response [27]. In order to illustrate this property of the antenna, the radiated field of the transmitting antenna is compared to the derivative of the induced current on the receiving antenna (Fig. 3.94). By analyzing Fig. 3.94, one can conclude that this property of antennas stated by Kanda is applicable to any antenna, irrespective of the structure of the antenna.

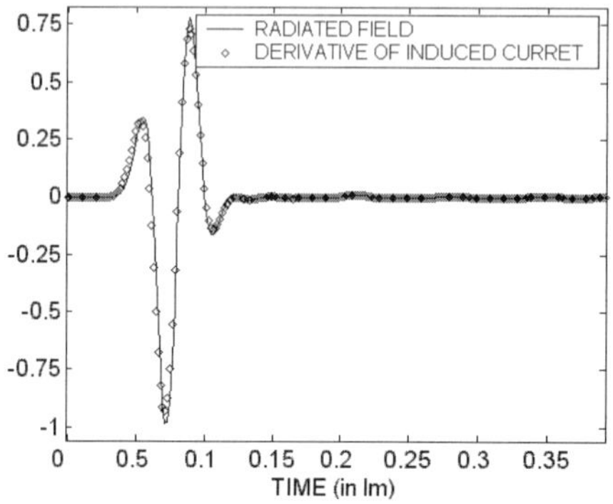

Figure 3.94. Comparison of the radiated field and the first derivative of the induced current when operating as a receiver.

3.7.9 Vivaldi

The Vivaldi antenna [47,48] under consideration here has an overall length of 8 cm and an aperture height of 2.2 cm (Fig. 3.95). It is simulated between 100 MHz to 10 GHz and the time support of the response determined by the lowest frequency of operation is 3 lm. The radiated field along the axis of the notch of the antenna is shown in Fig. 3.96. The polarization of the radiated field is perpendicular to the axis of the notch of the antenna. The induced current on the feed wire of the antenna due to a monocycle incident pulse arriving along the axis of the antenna notch is shown in Fig. 3.97. The polarization of the incident field is at a direction perpendicular to the axis of the notch of the antenna. The transient input resistance of the antenna is shown in Fig. 3.98. This antenna does not provide good temporal waveforms.

Figure 3.95. A Vivaldi antenna.

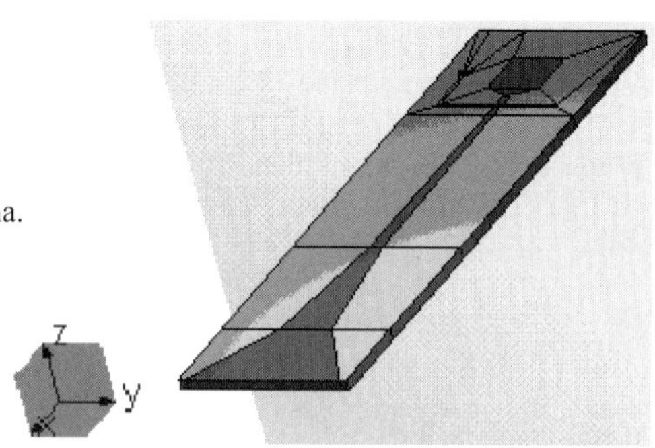

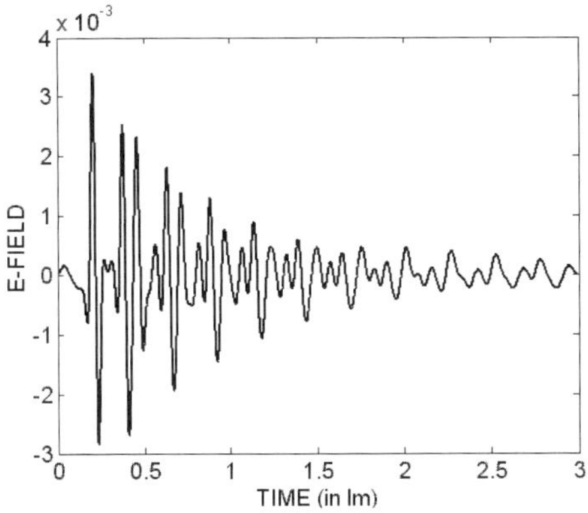

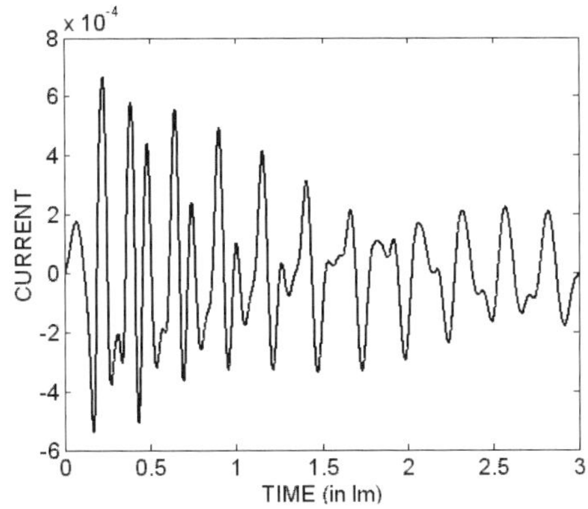

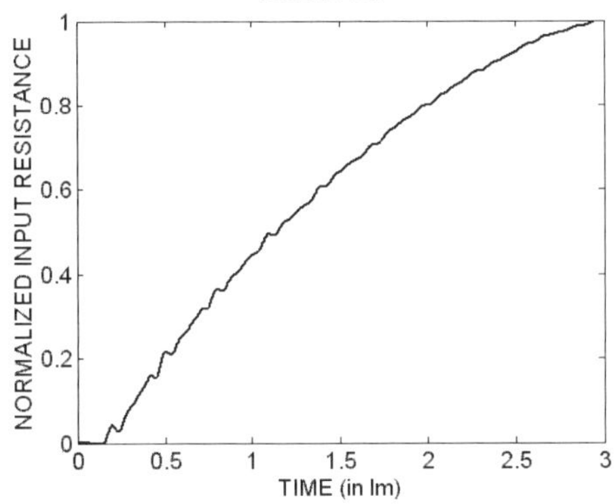

Figure 3.96. (above left) Radiation from a Vivaldi antenna.

Figure 3.97. (above right) Reception of a monocycle by a Vivaldi antenna.

Figure 3.98. (left) Transient input resistance of the Vivaldi antenna.

3.7.10 Impulse Radiating Antenna (IRA)

The structure is a 46-cm diameter IRA with 45° feed arms [49,50]. The ratio of the focal length to the aperture diameter is 0.4, and the ratio of the focal length to the feed arm of the parabolic reflector is 0.4. This antenna is simulated between 20 MHz to 20 GHz with a 20 MHz frequency step using a numerical electromagnetic analysis code, and the time support of the response determined by the lowest frequency of operation is 0.15 lm. The structure of the IRA is shown in Figs. 3.99 and 3.100. The Impulse Radiating Antenna (IRA) is a class of antennas designed to radiate a short pulse in a narrow beam for applications such as high-power pulse radiators and transient radars for mine detection. The IRA modeled here is a TEM-fed reflector. Baum et al. provide an excellent design procedure in Chapter 12 of Ref. [49]. This review summarizes the initial research, design, and development of the IRA and includes a comprehensive bibliography of references through 1999.

The time-domain radiation and reception of a monocycle pulse by the IRA are shown in Figs. 3.101 and 3.102, respectively. The transient input resistance of the antenna is shown in Fig. 3.103 and in a magnified form in Fig. 3.104. Hence, this antenna is truly capable of producing impulses of extremely high voltages and they have been deployed in practice for such applications.

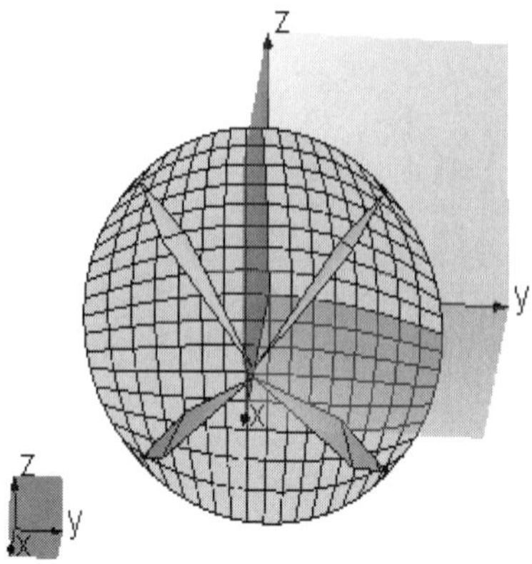

Figure 3.99. An Impulse Radiating Antenna (IRA).

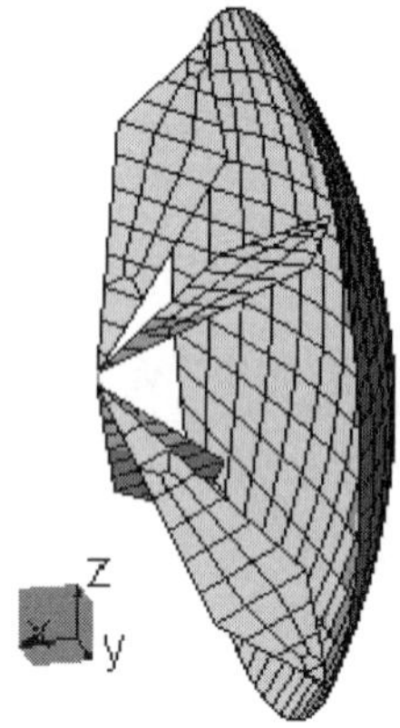

Figure 3.100. Sideview of the IRA.

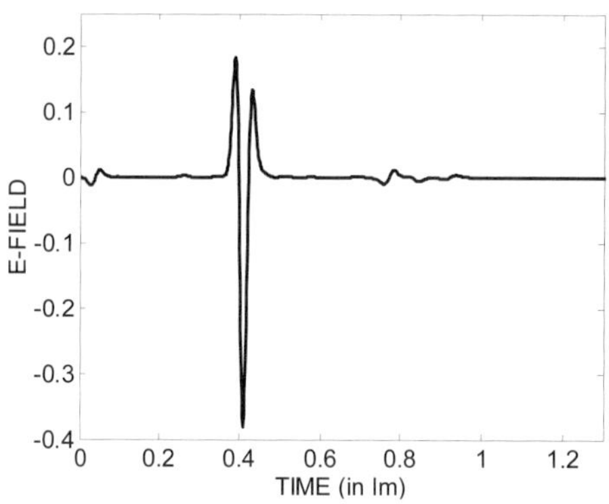

Figure 3.101. Radiation from an IRA.

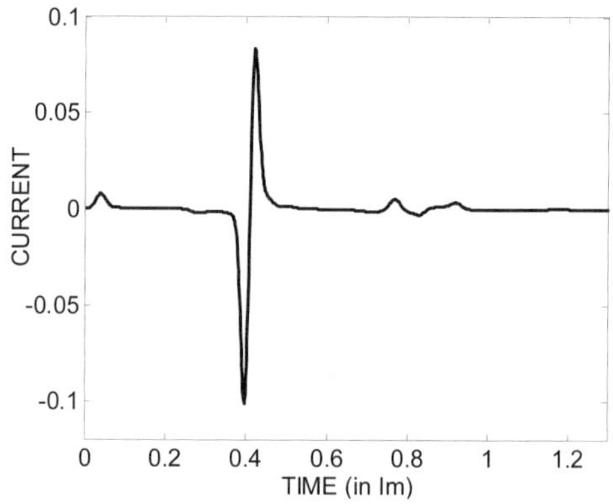

Figure 3.102. Reception of a monocycle pulse by the IRA.

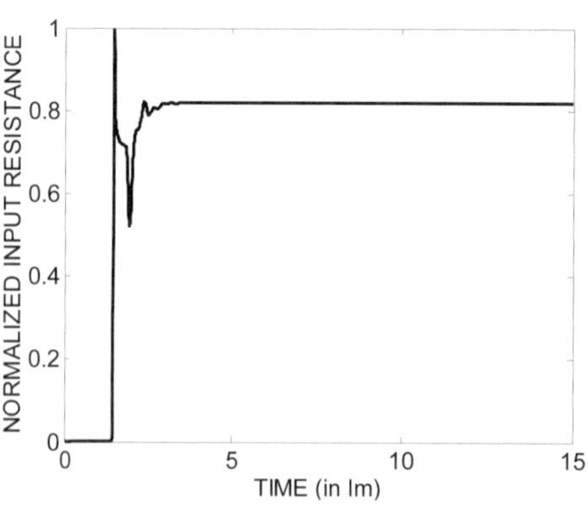

Figure 3.103. Transient input resistance of the IRA.

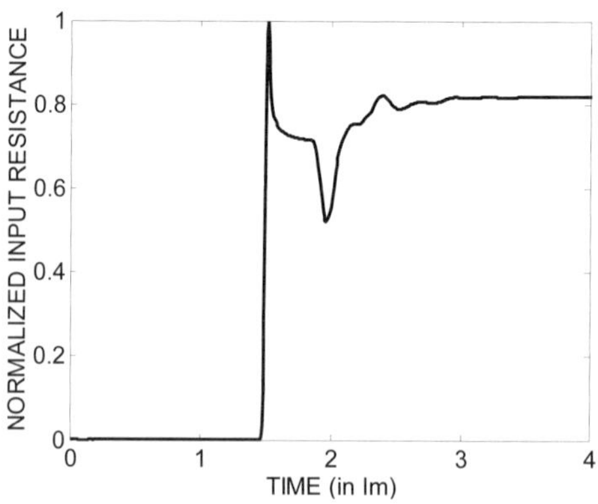

Figure 3.104. Magnified view of the transient input resistance of the IRA.

3.7.11 Circular Disc Dipole

The circular disc dipole antenna [51] falls in the special class of planar dipole antennas which are used for UWB arrays. This antenna exhibits wide bandwidth, small size, and conformability. The antenna consists of two circular discs of radius 1 cm, connected at their bases by a feed wire of length 1.1 mm and radius 0.01 mm as shown in Fig. 3.105. The origin of the coordinate system of Fig. 3.105 is the midpoint of the feed wire. The radiated field is predominant in the broadside direction (along the z-direction), and the radiated field is polarized along the y-axis.

The frequency range of operation is between 20 MHz and 10 GHz. The width of the radiated pulse is 0.03 lm. The radiated pulse along the predominant direction for the above polarization due to a monocycle input is shown in Fig. 3.106. The current induced in the receive antenna due to a monocycle incident field arriving from the positive *z*-axis shown in Fig. 3.107. The transient input resistance of the antenna is shown in Fig. 3.108. This has the potential to be a broadband antenna if appropriately loaded..

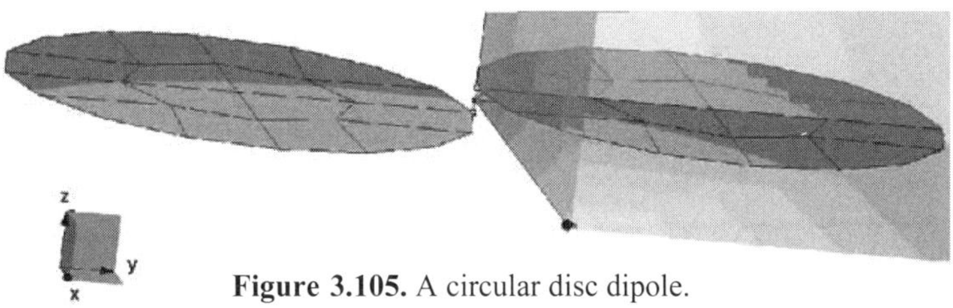

Figure 3.105. A circular disc dipole.

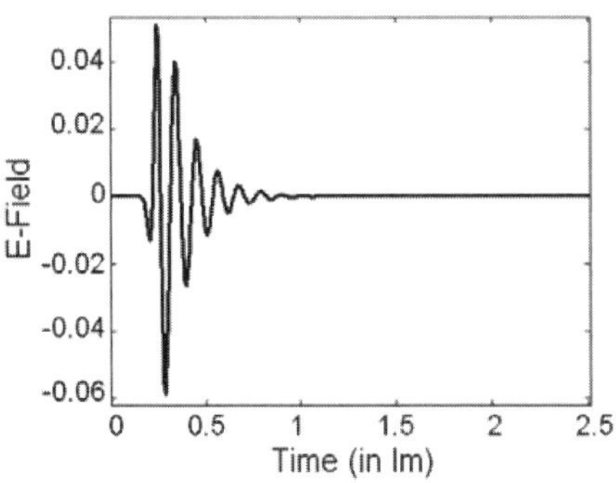

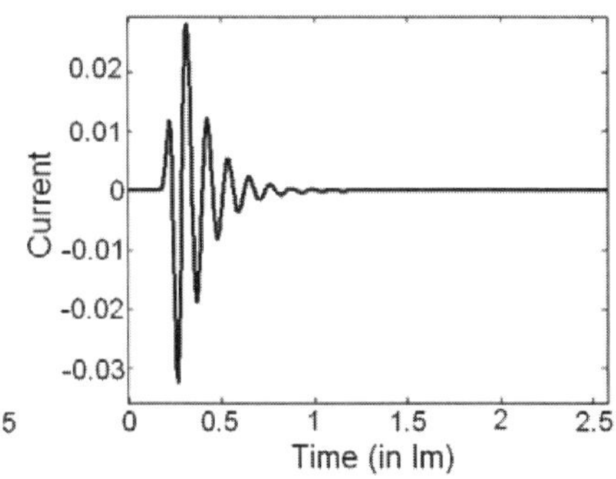

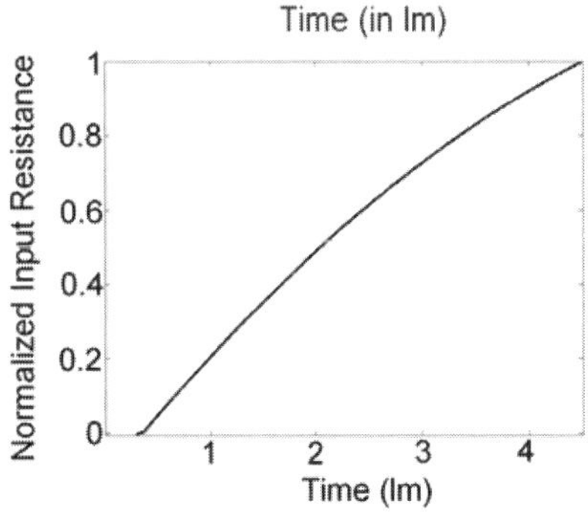

Figure 3.106. (above left) Radiation from a circular disc dipole.

Figure 3.107. (above right) Reception of a monocycle by the circular disc dipole.

Figure 3.108. (left) Transient input resistance of the circular disc dipole.

3.7.12 Bow-Tie

The bow-tie antenna [52] is made up of two isosceles triangular elements connected together by a feed wire at their vertices, which subtends an angle of 34°. The largest dimension of each element corresponds to a quarter wavelength at the first resonating frequency of the antenna. In this simulation, the height of each triangle is chosen to be 2.5 cm (Fig. 3.109). The feed wire is located in between the two triangles and has a length of 0.5 mm and a radius of 0.01 mm. The radiated field is predominant in the z-direction and is polarized along the y-axis (ϕ polarization). The radiated field, induced current, and transient input response are calculated for frequencies between 20 MHz and 10 GHz for a monocycle of 0.03-lm pulse duration. The radiated pulse along the predominant direction for the above polarization is shown in Fig. 3.110. The current induced for an incident field with ϕ polarization arriving along the z-axis is shown in Fig. 3.111. The normalized transient input resistance is shown in Fig. 3.112. This antenna also has the potential to be a broadband antenna if appropriately loaded.

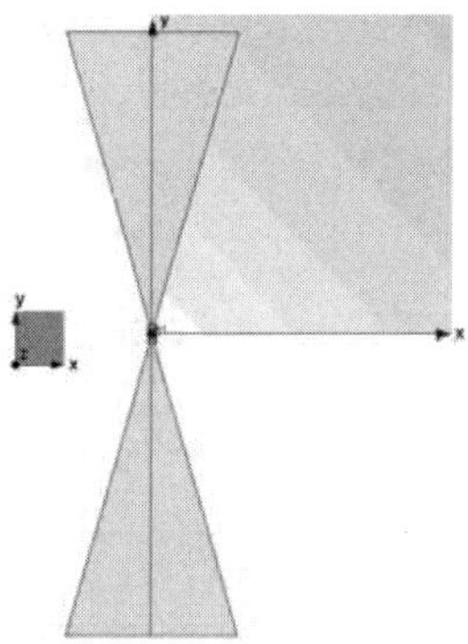

Figure 3.109. A bow-tie antenna.

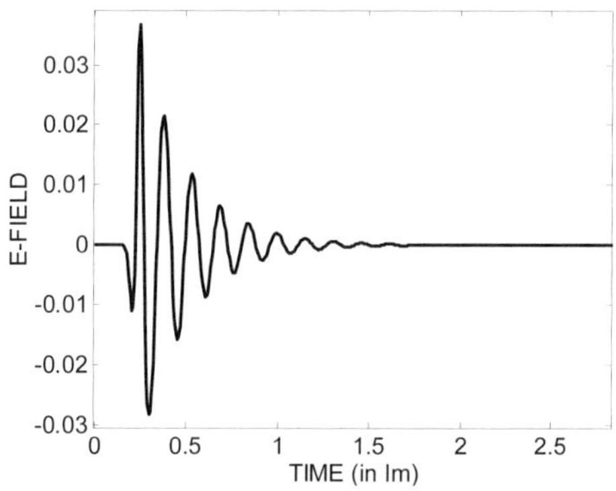

Figure 3.110. Radiation from a bow-tie antenna.

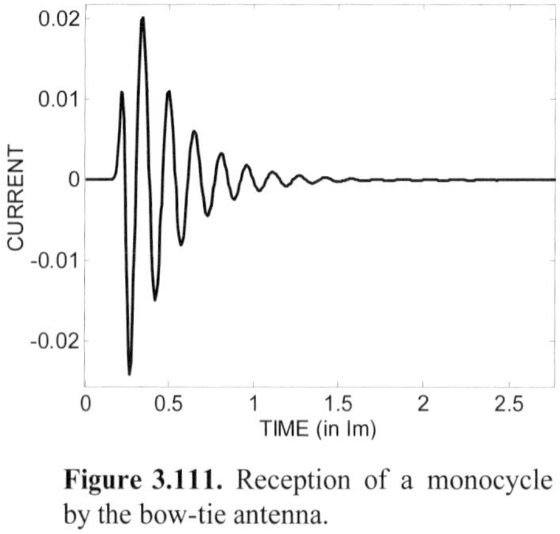

Figure 3.111. Reception of a monocycle by the bow-tie antenna.

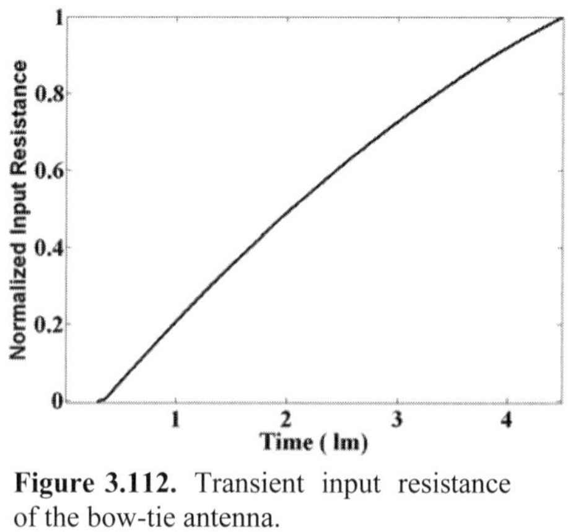

Figure 3.112. Transient input resistance of the bow-tie antenna.

3.7.13 Planar Slot

The structure of the planar slot antenna [53] is shown in Fig. 3.113. The antenna has two circular elements with a smaller disc of radius of 3.37 cm placed non-concentrically within the larger partly filled circular disc of radius 7.25 cm. The antenna is fed by a wire of length 0.1 mm and radius 0.01 mm and is placed within the smaller gap between the circular disc and the larger partly hollow disc. The slot gap length between the two discs in the x-direction is 3.8 cm. The first resonating frequency of the antenna is 0.75 GHz. The origin of the coordinate system is the center of the larger disc.

The frequency range of operation is between 20 MHz and 6 GHz, and the width of the monocycle is 0.05 lm. The ϕ-polarized radiated field of Fig. 3.114 is calculated for the far-field observation point at $\phi = 45° = \theta$, where ϕ is the angle from the x-axis in the xy-plane and θ is angle from the z-axis in spherical coordinates. The induced received current from an incident field arriving with the same direction and polarization is shown in Fig. 3.115. The transient input resistance of the antenna is shown in Fig. 3.116. This antenna has too much ringing.

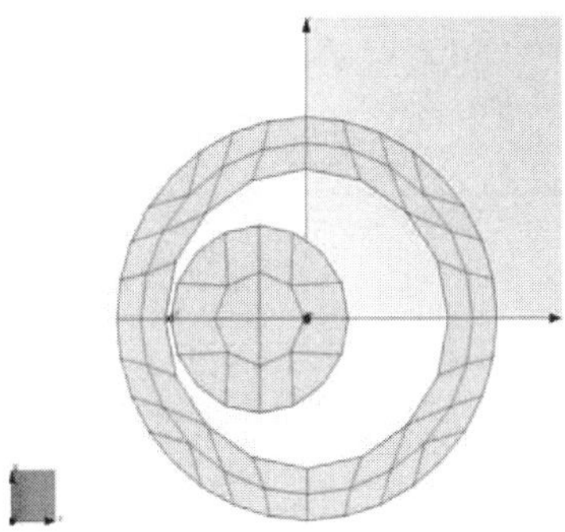

Figure 3.113. A planar slot antenna.

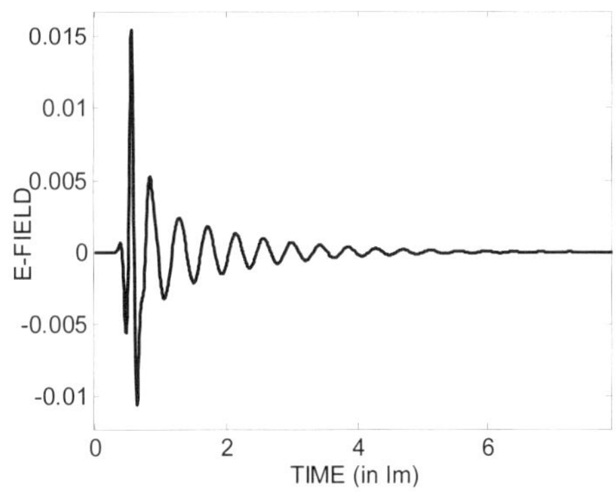

Figure 3.114. Radiation from a planar slot antenna.

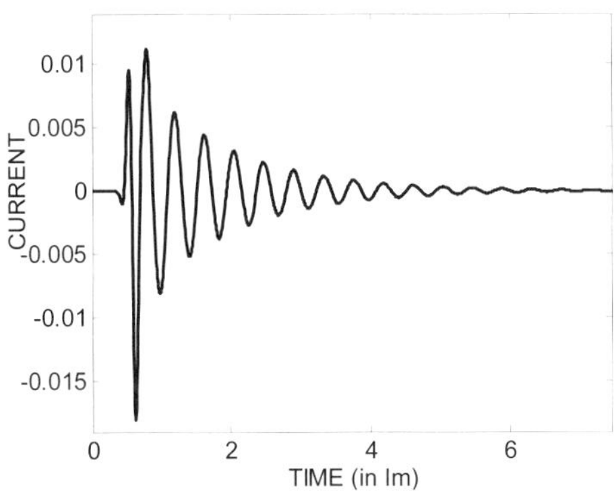

Figure 3.115. Reception of a monocycle by the planar slot antenna.

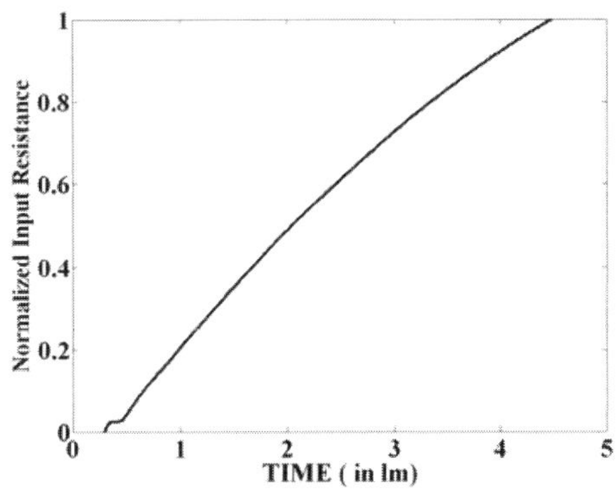

Figure 3.116. Transient input resistance of the planar slot antenna

3.8 EXPERIMENTAL VERIFICATION OF THE WIDEBAND RESPONSES FROM ANTENNAS

In order to experimentally verify the various theoretical simulations that were presented earlier, we include here the various experimental results obtained by Dr. James R. Andrews which are presented here with his permission [33,54]. A miniature UWB antenna range, Fig. 3.117, was designed to demonstrate the principles of UWB transmission, reception and propagation. The UWB metrology antennas demonstrated on this range include the conical, TEM horn and a D*dot probe. Monopole antennas and wave-guide horn antennas were also studied. Both differentiation and integration effects in the time domain will be demonstrated by these various antennas. The transmitting and the receiving antennas were placed on a 36 cm × 44 cm aluminum plate.

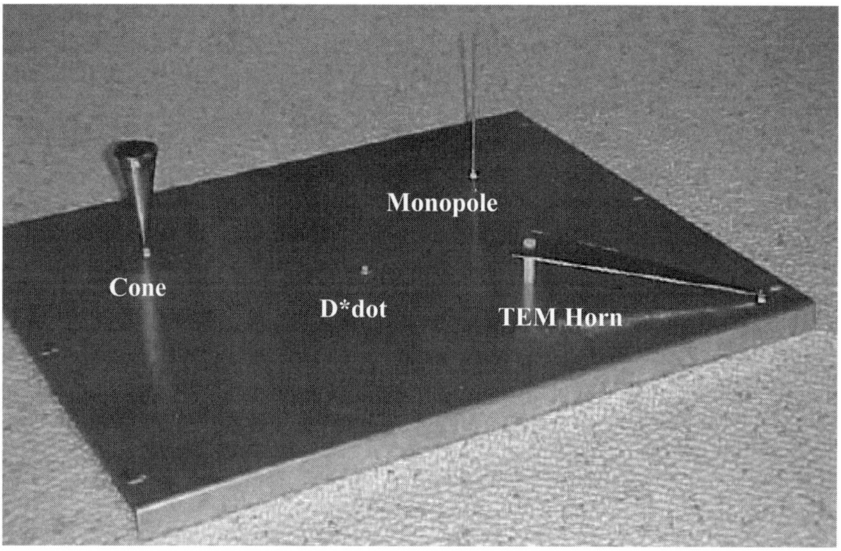

Figure 3.117. Miniature UWB Antenna Range. Antennas shown include a conical antenna, TEM Horn antenna, D*dot probe antenna, and Vertical Monopole Antenna.

The ultra-broadband test input signal used on this antenna range was a 4 V, 9 ps risetime step from a nonlinear transmission line (NLTL) pulse source of Picosecond Pulse Laboratory as shown in Fig. 3.118. All of the waveforms shown in this section were measured using a HP-54752B, 50 GHz, 9 ps risetime oscilloscope. A 35 cm, Gore, SMA coaxial cable was used to connect the antennas to the oscilloscope. The risetime of this cable was 9 ps. Thus the composite risetime of the pulse generator, coax cable and oscilloscope was 16 ps.

In the first example we use a conical antenna for transmission. The conical antenna suspended over a large metal ground plane is the preferred antenna for transmitting known transient electro-magnetic waves. This type of antenna is used by National Institute of Standards and Technology (NIST) as their reference standard transient transmitting antenna. This antenna radiates an E-M field that is a perfect replica of the driving point voltage waveform. However, when the pulse from the feed point reaches the top of the cone it is reflected back and so the perfect replica property of the transmitted pulse no longer holds unless resistive loading on the far end of the antenna is used to help suppress multiple reflections. The upper bandwidth of a conical antenna is mainly determined by the fidelity of the coax connector to conical antenna transition region. If a conical antenna is used as a receiving antenna, its output is the integral of the incident E field. The conical antenna was made of brass sheet with a half solid angle of 12° and having a height of 7 cm. This is shown in Figure 3.117.

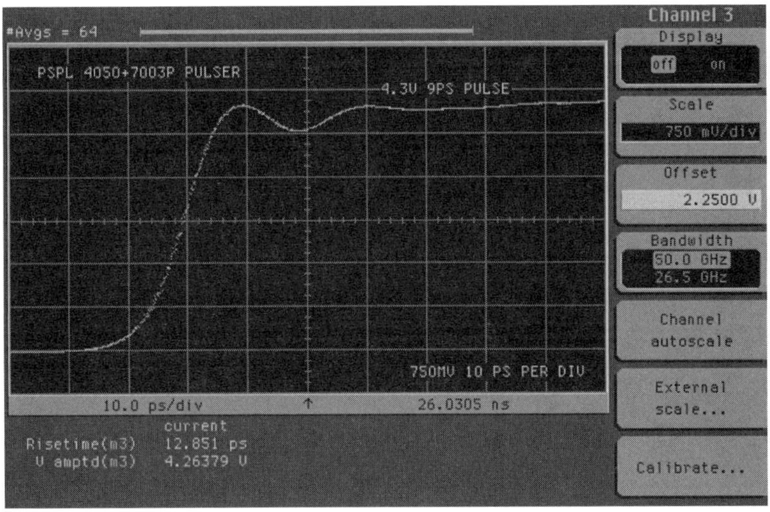

Figure 3.118. 4 Volt, 9 ps risetime, step pulse used for UWB antenna testing. Measured by an HP 50 GHz, 9 ps risetime, sampling oscilloscope. Scales are 750 mV/div & 10 ps/div.

TEM horns are the most preferred metrology receiving antenna for making a direct measurement of transient E-M fields. The TEM horn antenna is basically an open-ended parallel plate transmission line. It is typically built using a taper from a large aperture at the receiving input down to a small aperture at the coax connector output. The height to width ratio of the parallel plate is maintained constant along the length of the antenna to maintain an uniform characteristic impedance. However, to optimize sensitivity, most TEM antennas are designed with a 100 Ω antenna impedance. Practical UWB TEM horn antennas are usually designed with resistive loading near the mouth of the antenna to help suppress multiple reflections. The upper bandwidth of a TEM antenna is mainly determined by the size of its aperture and secondarily by the parallel plate to coax connector transition. When the aperture is too large relative to the wavelength of incident fields, the parallel plate line becomes a waveguide with higher order TE and TM modes present which in turn limit its bandwidth. If a TEM antenna is used as a transmit antenna, the radiated E field is the first derivative of the input driving point voltage.

To demonstrate faithful UWB radiation and propagation the conical antenna and TEM horn shown on the test range, Fig. 3.117, were used. The TEM horn was also fabricated from brass with a length of 15 cm. The height at the mouth was 2.6 cm and its width was 3.8 cm. At the mouth it was supported by a nylon spacer. The approximate incline angle was 9° as shown in Figure 3.117. The 4 V, 9 ps rise pulse, as shown in Fig. 3.118, was connected directly to the conical antenna. The output of the TEM horn antenna was connected through the 35 cm cable to the 50 GHz oscilloscope. The conical antenna was 7 cm high and had an impedance of 132 Ω. The TEM horn was 15 cm long and had an impedance of 106 Ω. For most of the experimental data shown in this chapter, the separation between the antennas was 25 cm. Fig. 3.119 is the output of the TEM horn antenna. This shows a 23 ps risetime step waveform that is in good agreement with the input waveform, Fig. 3.118. Fig. 3.120 shows the output from the TEM horn antenna when the two antennas were separated by 5 meters. This received signal is identical in wave shape to the 25 cm path signal, Fig. 3.119, except that it is weaker in amplitude to Fig. 3.119. It is clear that very little dispersion/distortion is introduced by the transmit receive system as earlier seen through numerical simulations.

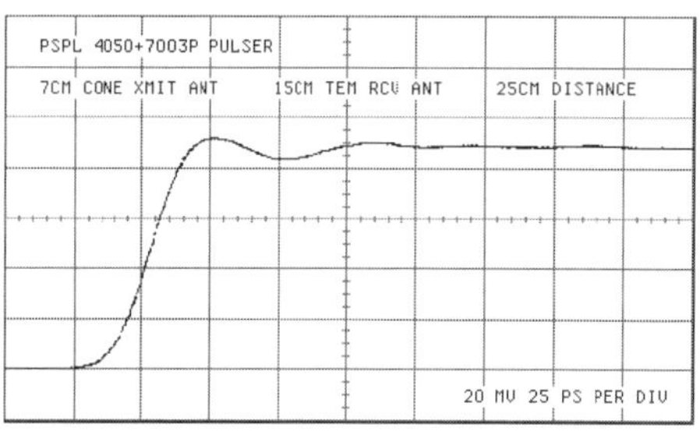

Figure 3.119. Transmission from a conical antenna to a TEM horn antenna over a 25 cm path.

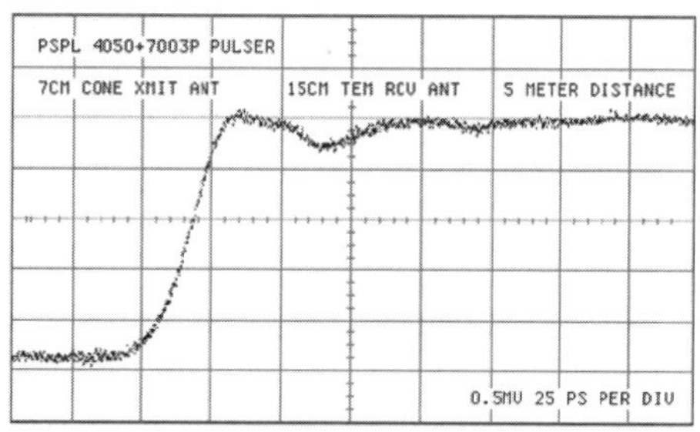

Figure 3.120. Transmission from a conical antenna to a TEM horn antenna over a 5 m path.

It is important to note that an antenna does not radiate dc. However, by observing Figs 3.118–3.120 one would get the impression that dc is being radiated. This is not true. These are waveforms that are used in time domain reflectometry (TDR), where trapezoidal pulses are used with positive and negative polarities. On the figures we focus on only the rise time of a single pulse to illustrate the fidelity with which the waveshapes have been radiated and received by the different antennas.

The above figures have tremendous implications when dealing with broadband wireless systems. Many researchers are used to routinely performing channel modeling of free space for a broadband wireless system. In perspective of the Figs 3.118, 3.119, and 3.120 it appears that modeling of a broadband wireless system in free space is equivalent to channel modeling of free space which is dispersionless! Hence, channel modeling of free space sounds very surprising at the first look because in Maxwell's theory air is typically assumed to be dispersionless, and so, why model a dispersionless channel? However, many researchers are used to only looking at the channel and not considering the effect of the transmitting and the receiving antennas which are an integral part of a wireless system. Alternately antennas are often modeled as point sources thereby excluding the fundamental physics from the analysis. Now, if one looks at Frii's transmission formula one observes that any transmitted signal strength will decay as $1/\lambda^2$ in free space due to the propagation of a spherical wave from the source. If such were the case, it will really be a highly dispersive channel. However, particularly for a broadband wireless system, one cannot only think about the isolated channel without the associated transmitting and receiving antennas, as they form an integral part of a wireless system. With the transmitting and the receiving antennas included in the broadband model, things behave in quite a different way. The gain of any transmitting antenna increases with $1/\lambda$ and so does the gain of any receiving antenna, which increases also with $1/\lambda$. Therefore, with the transmitting and the receiving antennas in a broadband wireless transmission system, Frii's transmission loss is compensated for by the two antennas making the channel completely dispersionless over a large band. The experimental data provided by Dr. James R. Andrews of the Picosecond Pulse Lab [33, 54] verifies the physical principles that we have discussed before.

This is where the vector electromagnetic problems differ from the scalar acoustic problems. One can observe that an acoustic signal is not distorted when it is converted to electrical energy through any microphone or when the electrical energy is converted to acoustic energy through any loudspeaker. However, the problem is quite different for the vector electromagnetic problem. In the time domain, the impulse response of the antenna when it is operating in the transmit mode is the time derivative of the impulse response when the same antenna is operating in the receive mode. Hence, even a point source, which is often used for modeling the antennas, differentiates the input waveform on transmit [3]. Therefore, the modeling of any wireless system and moreover that of a broadband system must include the antenna effects to obtain a physically meaningful solution [33].

Fig. 3.121 shows the received waveform when transmitting a step electromagnetic field between a pair of identical conical antennas. The resultant received output, rising ramp, waveform is the integral of the step pulse from the generator. This illustrates that a conical receive antenna integrates the incident electric field. Fig. 3.122 shows the received waveform when using a pair of identical TEM horn antennas for transmit and receive. The resultant output waveform, an impulse, is the first derivative of the pulse generator's step pulse. This illustrates that a TEM horn transmit antenna radiates an electric field which is the first derivative of the signal generator's waveform.

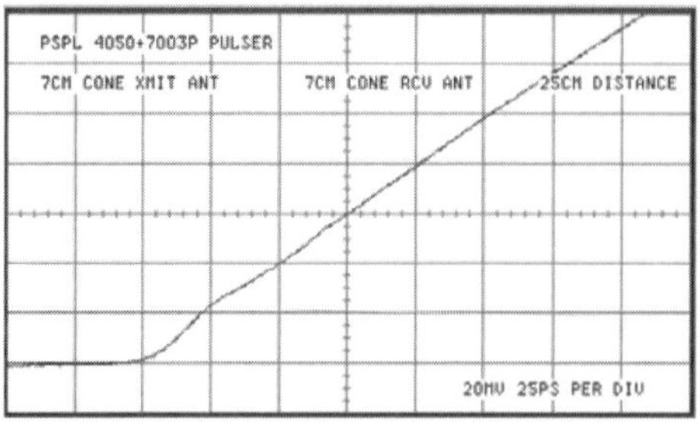

Figure 3.121. Transmission between a pair of identical conical antennas.

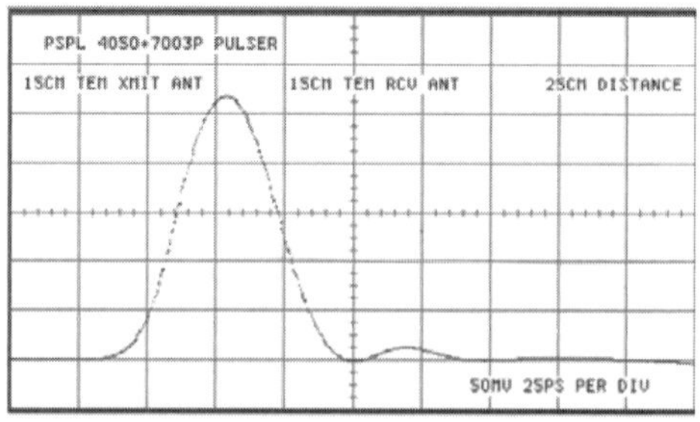

Figure 3.122. Transmission between a pair of identical TEM horn antennas.

Next, we consider a D*dot antenna which is another popular UWB metrology antenna. The D*dot antenna is basically an extremely short, monopole antenna. The equivalent antenna circuit consists of a series capacitance and a voltage generator. For a very short monopole, the antenna capacitance is very small and the capacitor thus acts like a differentiator to transient electromagnetic fields. Therefore the output from a D*dot probe antenna is the first derivative of the incident electric field. To determine the actual wave shape of the incident electric field, one must integrate the output from the D*dot probe. When the frequencies become too high, the D*dot probe loses its derivative properties and it becomes a monopole antenna. This happens when the length of the D*dot probe approaches a quarter wavelength of the incident wave. Fig. 3.123 is the output from a D*dot receive antenna when it was illuminated with a step electromagnetic field radiated from a conical transmit antenna. The lower trace is the actual receive antenna output. It is a 19 ps wide impulse which is consistent with the antenna's output being the first derivative of the incident field. The upper trace is the computed integral of the lower trace. It has a 19 ps risetime step and is a good representation of the incident electric field at the antenna. Fig. 3.124 shows the radiated field from a D*dot transmit antenna. It was received by

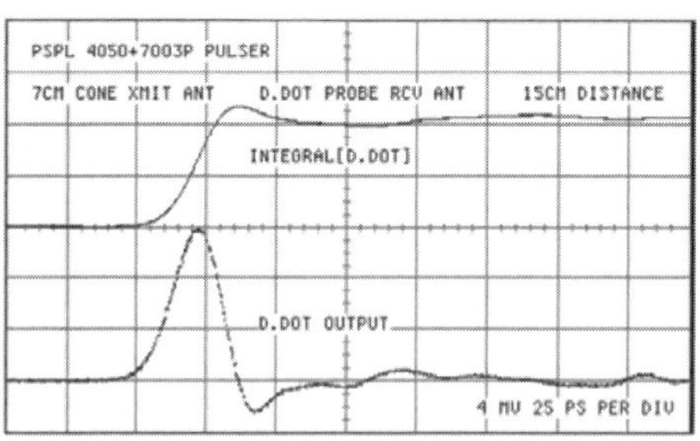

Figure 3.123. Transmission from conical antenna to D*dot antenna. The lower trace is the output from the receive antenna. The upper trace is the integral of the lower trace.

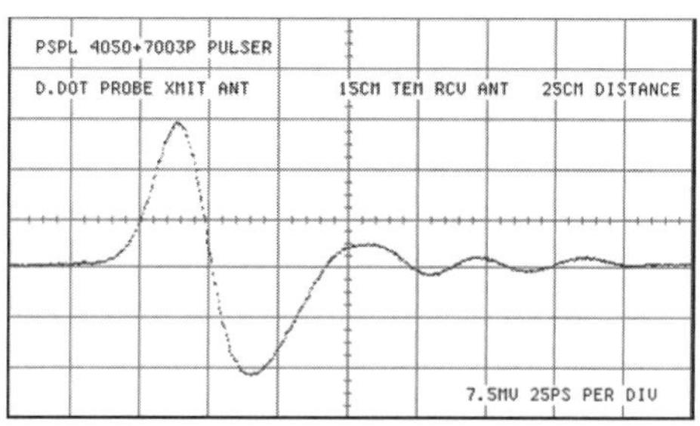

Figure 3.124. Transmission from a D*dot antenna to a TEM horn antenna.

a TEM horn antenna. This shows that the transmitting transient response of the D*dot antenna is the second derivative of the driving generator voltage. Fig. 3.125 is the received output signal using a pair of D*dot antennas for both transmit and receive. This signal is the third derivative of the excitation.

Finally, a monopole antenna is included in this UWB study, because it is the most fundamental building block for most antenna designs. It is the quarter wave whip antenna above an infinite ground plane. The monopole antenna is sometimes used as a simpler version of the conical antenna for transmitting UWB signals which are similar in wave shape to the driving point voltage. However, its radiated fields are not as uniform as those for the conical antenna. Its driving point impedance is not constant, but rises as a function of time. This leads to distortion of the radiated electromagnetic fields. Time domain reflectometer (TDR) studies of a monopole confirm this statement. Fig. 3.126 shows the radiated electric field of a 10 cm monopole. It resembles the step electric field from the conical antenna, Fig. 3.119. However its top line is not flat, but sags with increasing time. This is due to the non-uniform TDR impedance of this antenna. When a monopole is used for receiving transient electromagnetic fields, its output is the integral of the incident electric field. This is shown experimentally in Fig. 3.127. The lower trace is the antenna output. It shows an almost monotonically rising ramp which is consistent with the integral of the step incident electric field. The upper trace is the calculated first derivative of the lower trace and is somewhat representative of the step incident electric field.

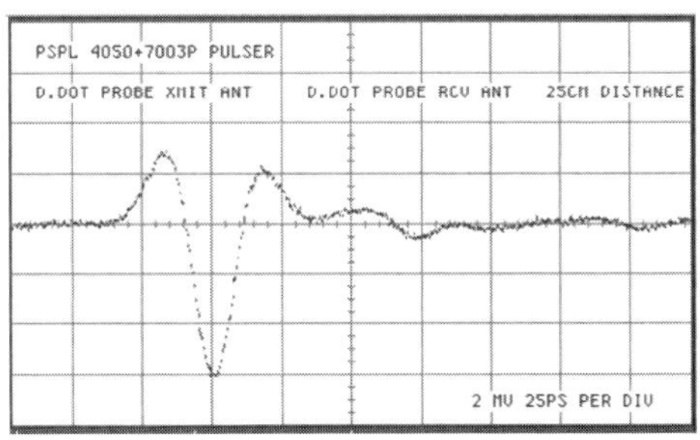

Figure 3.125. Transmission between a pair of identical D*dot antennas.

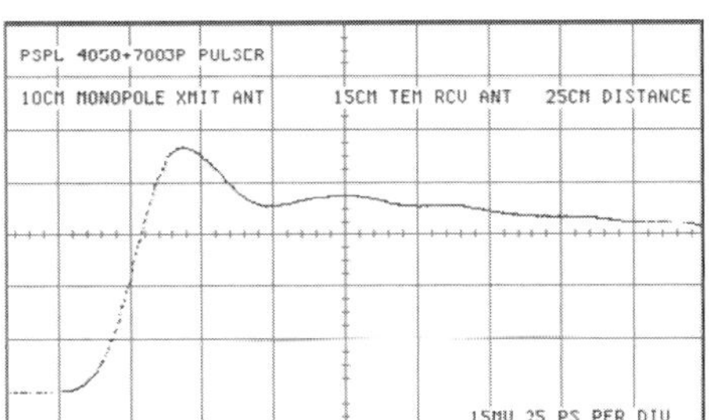

Figure 3.126. Transmission from monopole to the TEM horn.

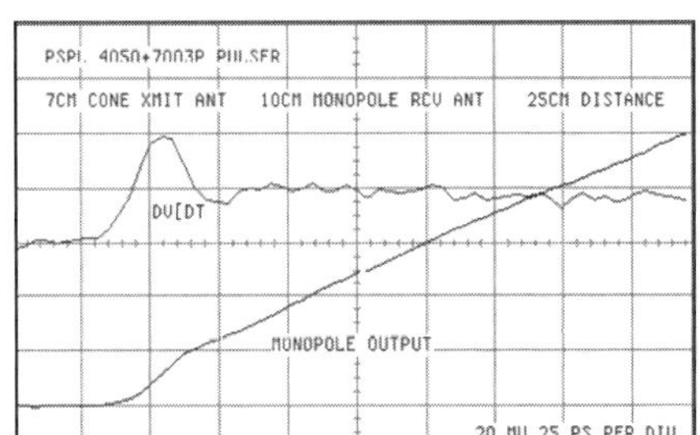

Figure 3.127. Transmission from the conical to the monopole. Lower trace receiving antenna output. Upper trace is dV/dt.

Fig. 3.128 shows the output using a pair of 10 cm monopoles for both transmit and receive antennas. This is essentially the integral of the generator's step pulse. The integration effect can be explained simply. Assume an impulsive electric field is incident upon the monopole at a 90° angle to its axis. Thus a current is induced simultaneously in each differential element, *dx*, of the antenna. These current elements, *dI*, thus start to flow towards the output connector of the antenna. They do not arrive at the output simultaneously, but in sequence. Thus the output appears as a step function, which is the integral of the incident impulse. The integrating effect of this antenna only lasts for $t < 1 / c$, where l is the length of the antenna. This effect is demonstrated in Figs. 3.129 and 3.130 for a transmitting/receiving pair of shorter 2 cm monopole antennas. Fig. 3.129 shows the output starts off like a rising linear ramp, as in Fig. 3.128, but due to the short length of the antenna, this stops after $1 / c = 67$ ps and then the multiple reflections on the antennas predominate. This is shown vividly in Fig. 3.130 on a slower speed of 100 ps/div in which the resonance frequency of the 2 cm antennas is obvious as a damped, ringing, sinusoid.

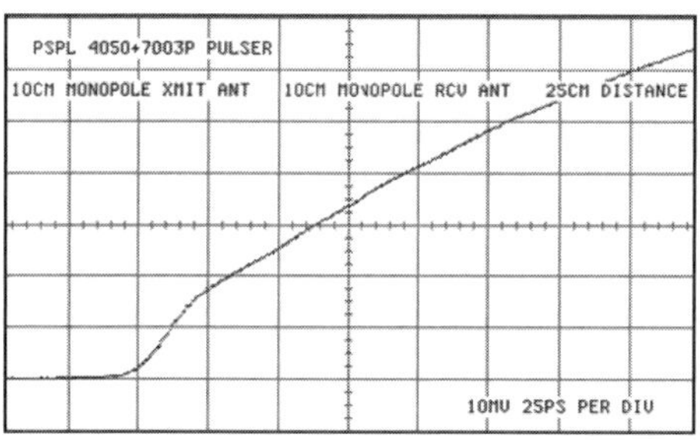

Figure 3.128. Transmission between a pair of identical 10 cm monopole antennas (25 ps/div).

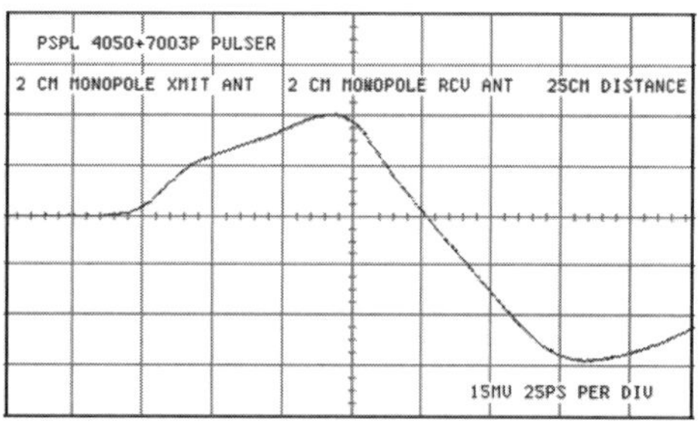

Figure 3.129. Transmission between a pair of identical 2 cm monopole antennas (25 ps/div).

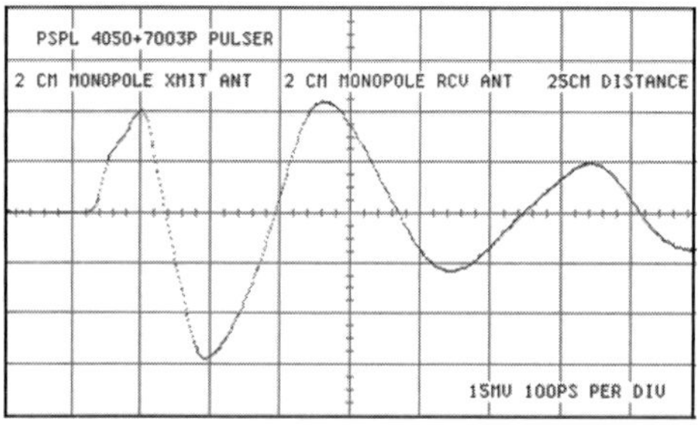

Figure 3.130. Transmission between a pair of identical 2 cm monopole antennas (100 ps/div).

3.9 CONCLUSION

This chapter presents a very broad survey of many diverse antennas. In particular, the transmitting and receiving properties of each antenna are discussed particularly in the time domain. Although many antenna structures exhibit resonating properties when used in the ultrawideband range, in many cases, proper design and modification of such antennas can minimize the reflections due to structural discontinuities, whereby the antenna can be converted to a guiding-wave structure. Additionally, the application of a tapered resistive loading along the length of the antenna helps reduce the outward traveling wave on the antenna so that reflections from the ends of the antenna are eliminated. This surface loading is designed according to the fundamental dimensions (length, average thickness, effective radius) of the antenna. Proper design of the loading profile minimizes the dispersion from the antenna, thus broadening its frequency response and focusing the energy in a single pulse. However, by loading the antenna, its efficiency and gain are reduced by a significant amount.

Many widely used UWB antennas have been studied in this chapter. Some of them have an excellent wideband response with little or no time (phase) dispersion, while others exhibit resonating properties which are not suitable for UWB applications. This chapter also verifies that the impulse response of an antenna structure in the transmit mode is proportional to the time derivative of the impulse response of the same antenna when it is operating in the receive mode, irrespective of the antenna type. Moreover, observations of the output wave shapes (far field and received current) from the antennas provide important information about their transmitting and receiving properties, and certain relationships are obtained between the input and output wave shapes. Consequently, one can conclude, for example, that if a bicone transmits to a receiving TEM horn, then the induced current in the horn will be exactly identical to the driving point voltage of the bicone. Such observations can have very important ramifications in broadband high-speed information transmission.

REFERENCES

[1] F. Sabath, E. L. Mokole, and S. N. Samaddar, "Definition and Classification of Ultra-wideband Signals and Devices", *Radio Science Bulletin*, URSI no. 313, pp. 12–26, June 2005.

[2] S. N. Samaddar and E. L. Mokole, "Biconical Antennas with Unequal Cone Angles", *IEEE Trans. Antennas Propagat.*, Vol. 46, no. 2, pp. 181–193, 1998.

[3] S. N. Samaddar and E. L. Mokole, "Transient Behavior of Radiated and Received Fields Associated with a Resistively Loaded Dipole", in *Ultra-Wideband Short-Pulse Electromagnetics 4* (edited by J. Shiloh, B. Mandelbaum, and E. Heyman), pp. 165–179. New York, NY: Kluwer Academic/Plenum Publishers, 1999.

[4] A. K. Choudhury, S. N. Samaddar, and E. Mokole, "Transient Field of Non-reflecting Resistively Loaded Dipole Excited by Ultrawideband Signals", *IEEE Antennas and Propagation Society International Symposium,* Vol. 4, pp. 2314–2317, June 1998.

[5] E. L. Mokole, "Behavior of Ultrawideband-radar Array Antennas", IEEE International Symposium on Phased Array Systems and Technology, pp. 113–118, Oct. 1996.

[6] V. H. Rumsey, "Frequency-independent Antennas", *IRE National Convention Record,* pp. 114–118, 1957.

[7] M. Kanda, "Time Domain Sensors for Radiated Impulsive Measurements", *IEEE Trans. Antennas and Propagat.*, Vol. AP-31, pp. 438–444, 1983.

[8] D. Lamensdorf and L. Susman, "Baseband-pulse-antenna Techniques", *IEEE Antennas Propagat. Mag.*, Vol. 36, No. 1, pp. 20–30, Feb. 1994.

[9] A. Shlivinski, E. Heyman, and R. Kastner, "Antenna Characterization in the Time Domain", *IEEE Trans. Antennas Propagat.*, Vol. 45, No. 7, pp. 1140–1149, July 1997.

[10] H. G. Schantz, "Dispersion and UWB Antennas," *Ultra Wideband Systems, Joint with Conference on Ultrawideband Systems and Technologies*, Kyoto, Japan, pp. 161–165, 18-21 May 2004.

[11] H. Schantz, *The Art and Science of Ultrawideband Antennas*. Norwood, MA: Artech House, 2005.

[12] M. A. Peyrot-Solis, G. M. Galvan-Tejada1, and H. Jardon-Aguilar, "State of the Art in Ultra-wideband Antennas," 2nd *International Conference on Electrical and Electronics Engineering* (ICEEE) and XI *Conference on Electrical Engineering* (CIE 2005), Mexico City, Mexico, pp. 101–105, September 7-9, 2005.

[13] L-C. Shen and T. T. Wu, "Cylindrical Antenna with Tapered Resistive Loading", *Radio Sci.*, Vol. 2, pp. 191–201, 1967.

[14] L-C. Shen, "An Experimental Study of the Antenna with Non-reflecting Resistive Loading", *IEEE Trans. Antennas Propagat.*, Vol. AP-15, No. 5, pp. 606–611, 1967.

[15] J. Lally and D. T. Rouch, "Experimental Investigation of the Broad-band Properties of a Continuously Loaded Resistive Monopole", *IEEE Trans. Antennas Propagat.*, Vol. AP-18, No. 6, pp. 764–768, Nov. 1970.

[16] D. L. Sengupta and Y-P. Liu, "Analytic Investigation of Waveforms Radiated by a Resistively Loaded Linear Antenna Excited by a Gaussian Pulse," *Radio Sci.*, Vol. 9, No. 6, pp. 621–630, June 1974.

[17] K. P. Esselle and S. S. Stuchly, "Pulse Receiving Characteristics of Resistively Loaded Dipole Antennas", *IEEE Trans. Antennas Propagat.*, Vol. AP-38, pp. 1677–1683, 1990.

[18] S. N. Samaddar and E. L. Mokole, "Some Basic Properties of Antennas Associated with Ultrawideband Radiation," in *Ultra-Wideband Short-Pulse Electromagnetics 3* (edited by C. E. Baum, L. Carin, and A. P. Stone), pp. 147–164. New York: Plenum Press, 1997.

[19] T. T. Wu and R. W. P. King, "The Cylindrical Antenna with Non-reflecting Resistive Loading," *IEEE Trans. Antennas Propagat.*, Vol. AP-13, No. 3, pp. 369–373, 1965.

[20] B. M. Kolundzija, J. S. Ognjanovic, and T. K. Sarkar, *WIPL-D: Software for Electromagnetic Analysis of Composite Wire, Plate and Dielectric Structures*. Norwood, MA: Artech House, 2000.

[21] P. T. Montoya and G. S. Smith, "A Study of Pulse Radiation from Several Broad-band Loaded Monopoles", *IEEE Trans. Antennas Propagat.*, Vol. 44, No. 8, pp. 1172–1182, 1996.

[22] E. E. Altshuler, "The Traveling Wave Linear Antenna", *IEEE Trans. Antennas Propagat.*, Vol. AP-9, No. 4, pp. 324–329, 1961.

[23] Z. Ji, T. K. Sarkar, and B. H. Jung, "Transmitting and Receiving Wideband Signals Using Reciprocity", *Microwave & Optical Technology Letters*, Vol. 38, No. 5, pp. 359–362, Sep. 2003.

[24] A. Medouri and T. K. Sarkar, "Signal Enhancement in Multiuser Communication through Adaptivity on Transmit", *Microwave and Optical Technology Letters*, Vol. 38, No. 4, pp. 265–269, Aug. 2003.

[25] S. Hwang, A. Medouri, and T. K. Sarkar, "Signal Enhancement in a Near Field MIMO Environment through Adaptivity on Transmit", *IEEE Trans. Antennas Propagat.*, Vol. 53, No. 2, pp. 685–693, 2005.

[26] C. W. Harrison and R. W. P. King, "On the Transient Response of an Infinite Cylindrical Antenna", *IEEE Trans. Antennas Propagat.*, Vol. 15, pp. 301–302, 1967.

[27] M. Kanda, "Time Domain Sensors and Radiators", in *Time Domain Measurements in Electromagnetics*, edited by E. K. Miller, Ch. 5. New York: Van Nostrand Reinhold, 1986.

[28] T. K. Sarkar, M. C. Wicks, M. Salazar-Palma, and R. J. Bonneau, *Smart Antennas.* Hoboken, NJ: Wiley-IEEE Press, 2003.

[29] S.N. Samaddar, "Transient Radiation of a Single-cycle Sinusoidal Pulse from a Thin Dipole", *J. Franklin Inst.*, Vol. 10, pp. 259–271, 1992.

[30] W. L. Stutzman and G. A. Thiele, *Antenna Theory and Design*, 2nd Edition, New York: Wiley, 1997.

[31] C. P. Papas and R. W. P. King, "Radiation from Wide-angle Conical Antennas Fed by a Coaxial Line", *Proc. IRE*, Vol. 39, pp. 49–50, 1951.

[32] H. Jasik, Ed., *Antenna Engineering Handbook*, Ch. 6. New York: McGraw Hill, 1961.

[33] J. R. Andrews, "UWB Signal Sources, Antennas and Propagation," *IEEE Tropical Conference on Wireless Communication Technology*, Hawaii, pp. 439–440, Oct. 2003.

[34] E. D. Caswell, *Design and Analysis of Star Spiral with Application to Wideband Arrays with Variable Element Sizes*, Virginia Polytechnic and State University, MS Thesis, Dec. 2001.

[35] J. D. Kraus, *Antennas*, 2nd Edition, Ch. 7, New York: McGraw Hill, 1988.

[36] T. Taniguchi and T. Kobayashi, "An Omnidirectional and Low-VSWR Antenna for the FCC-approved UWB Frequency Band", *IEEE Antennas and Propagation Society International Symposium*, Vol. 3, Columbus, OH, pp. 460–463, 22-27 June 2003.

[37] L. Paulsen, J. B. West, W. F. Perger, and J. Kraus, "Recent Investigations on the Volcano Smoke Antenna", *IEEE Antennas and Propagation Society International Symposium*, Vol. 3, Columbus, OH, pp. 845–848, 22-27 June 2003.

[38] H. G. Schantz and L. Fullerton, "The Diamond Dipole: a Gaussian Impulse Antenna", *IEEE Antennas and Propagation Society International Symposium*, Vol. 4, Boston, MA, pp. 100–103, 8-13 July 2001.

[39] T. S. Maclean, *Principles of Wire Antennas: Wire and Aperture*. Cambridge: Cambridge University Press, 1986.

[40] S. N. Samaddar, "Preliminary Study of Some Antenna Elements Which Have Potential Usefulness as Ultrawideband Radiators", *Memorandum Report* NRL/MR/5341-95-7794, Naval Research Laboratory, Washington, DC, Nov. 24, 1995.

[41] H. G. Schantz, "Single Element Antenna Apparatus", *US Patent* 6,437,756 BI, Aug. 20, 2002.

[42] H. G. Schantz, "UWB Magnetic Antennas", *IEEE Antennas and Propagation Society International Symposium*, Vol. 3, Columbus, OH, pp. 604–607, 22-27 June 2003.

[43] K. L. Walton and V. C. Sundberg, "Broadband Ridged Horn Design", *Microwave Journal*, Vol. 7, pp. 96–101, Mar. 1964.

[44] G. A. Rief, D. E. Heckman, and R. J. Schrimpf, "Antenna Horn and Associated Methods", *US Patent* 6,271,799 BI, Feb. 15, 2000.

[45] P. VanEtten and M. C. Wicks, "Bi-blade Century Bandwidth Antenna", *US Statuary Invention Registration* H1,913, Nov. 7, 2000.

[46] M. C. Wicks and P. VanEtten, "Orthogonally Polarized Quadraphase Electromagnetic Radiator," *US Patent* 5,068,671, Nov. 26, 1991.

[47] J. Shin and D. H. Schaubert, "A Parameter Study of Stripline-fed Vivaldi Notch-antenna Arrays", *IEEE Trans. Antennas Propagat.*, Vol. 47, No. 5, pp. 879–886, May 1999.

[48] M. Yuan, T. K. Sarkar, and B. Kolundzija, "Solution of Large Complex Problems in Computational Electromagnetics Using Higher Order Basis in MOM with Out-of-core Solvers", *IEEE Antennas Propagation Magazine*, Volume 48, Issue 2, Apr. 2006 Page(s):55-62.December 2005.

[49] C. E. Baum, E. G. Farr, and D. V. Giri, "Review of Impulse-radiating Antennas", in *Review of Radio Science 1996-1999*, edited by W. S. Stone, Ch. 12, New York: Wiley-IEEE Press, 1999.

[50] L. H. Bowen, E. G. Farr, C. E. Baum, T. C. Tran, and W. D. Prather, "Results of Optimization Experiments on a Solid Reflector IRA", *Sensor and Simulation Note 463*, Jan. 2002.

[51] M. Thomas and R. I. Wolfson, "Wideband Arrayable Planar Radiator", *US Patent* 5,319,377A, June 7, 1994.

[52] G. H. Brown and O. M. Woodward, Jr., "Experimentally Determined Radiation Characteristics of Conical and Triangular Antennas", *RCA Rev.,* Vol. 13, No. 4, pp. 425-452, December 1952.

[53] J. D. Powell and A. Chandrakasan, "Differential and Single Ended Elliptical Antennas", *US Patent Application* 20,050,280,582A1, December 22, 2005.

[54] James R. Andrews, UWB Signal Sources, Antennas and Propagation, Picosecond Pulse Lab, *Application Note AN-14a*, Aug. 2003, Boulder, CO. (Also presented at the 2003 Honolulu *IEEE Wireless Conference.) http://www.picosecond.com/objects/AN-14a.pdf/*. (Excerpts presented here with permission from Dr. James R. Andrews.)

4

A LOOK AT THE CONCEPT OF CHANNEL CAPACITY FROM A MAXWELLIAN VIEWPOINT

4.0 SUMMARY

Wireless communication is an active area of current research in communication technology. To assess the performance of a wireless system, one needs to quantify its ability to handle information. Typically, the performance of such systems is characterized in terms of the channel capacity. In this chapter, we look at the various mathematical representations of the channel capacity and trace how they have evolved from the initial concept of entropy. Two popular mathematical representations of channel capacity involve the power and the voltage related to the incident field at the receiver. If one uses similar values for the background noise power in the two formalisms and ensures that the transmit and receive antennas are matched, then the two formulas may yield similar results, even though they are functionally different. The essential point to be made here is that the channel capacity, like entropy, is an abstract mathematical number that has little connection to the electromagnetic properties of the system. However, introducing Maxwellian physics can help one interpret the channel capacity formulas in a physically realistic way by using the vector electromagnetic equations. In electromagnetics, power is carried by the fields and that is why the fields are fundamental in nature. In this case, a Maxwellian approach to wireless technology is not only relevant but also vitally important. Such a formalism will correct the variety of deficiencies in the current wireless communication literature.

The primary objective of this chapter is to apply the various formulas for channel capacity in a physically proper way. First, the channel capacity of any system needs to be characterized under the same input power constraints, while simultaneously accounting for the radiation efficiency of the transmitting and receiving antennas. Second, the voltage form of the channel capacity is more useful for wireless systems than the power form, since the sensitivities of the receivers are generally characterized in terms of the received electric fields. In addition, the received power is at least two orders of magnitudes larger than the background noise. Third, in a near-field scenario, it is not clear how to evaluate the power in a simple way. Consequently, we will show that the voltage form of the channel capacity, which depends only on the electric field, is always applicable to both the near and far fields; whereas, the power form depends on the electric and magnetic fields (unless the antennas are conjugately matched).

The solution of the vector electromagnetic problem also illustrates that deploying antennas near the ground yields a higher capacity than placing them on top of a high tower away from the earth. In addition, electrical tuning of the antennas further increases the capacity. These subtle points are generally missed by a statistical formulation. In addition, the practice of unrealistically representing antennas by point sources is devoid of any near-field effects, which require both the electric and the magnetic fields to compute the power. Examples are presented for single-input-single-output situations to illustrate the subtleties of the vector nature of the problem, which is missing in current formulations. In particular, the results of simulations of a dielectric box surrounding a receiving antenna suggest that the box enhances signals in some cases instead of impeding line-of-sight propagation.

4.1 INTRODUCTION

As Gabor [1] observed, *wireless communication systems are due to the generation, reception and transmission of electromagnetic signals. Therefore all wireless systems are subject to the general laws of radiation. Communication theory has up to now been developed mainly along mathematical lines, taking for granted the physical significance of the quantities which are fundamental in its formalism. But communication is the transmission of physical effects from one system to another. Hence communication theory should be considered as a branch of physics. Thus it is necessary to embody in its foundation such physical data. Hence we can apply to our problem the well known results of the theory of radiation by the Maxwell-Poynting theory.* Following this fifty-five-year-old suggestion of Gabor, we look at the quality of transmission of information in a wireless system by using the concept of channel capacity under very general circumstances, including a deterministic one.

Inherent in the idea of channel capacity is the concept of information content. If information is to be conveyed, signals must change unpredictably with time [2]. If information transmission is related to changing signals unpredictably with time, then why not change the signal as rapidly as we like and additionally vary the amplitude over all real numbers. Each change would imply increasing the information indefinitely [2]. Intuitively, more rarely occurring events thus carry more information than frequently occurring events. In practice, we deal with physical systems, which have two qualitative limitations on the amount of information per unit time (system capacity) a system can transmit [2]:

1. Inability of the system to respond instantaneously to signal changes due to presence of energy storage devices. In all networks, inherent

capacitance and inductance limit the time response. These limitations are related to the useful bandwidth of the operating system.

2. Inability of the system to distinguish infinitesimally small changes in signal levels (due to inherent voltage fluctuations or noise), which are related to the signal-to-noise ratio (SNR).

These two qualitative limitations determine system capacity, which is the maximum amount of information per second in bits (binary digits) that a system can transmit.

Currently, two forms of the channel capacity are in vogue. The first form does not consider background statistical noise to be important in impeding the reception of wireless communication signals. The quantification in this case is done in terms of signal amplitudes [2]. The second form, derived by Shannon, uses a stochastic model for the signal and additive noise [3,4]. The objective of Shannon's model was to introduce enough redundancy in the source, which he termed the transmitter, so that the redundancy in the form of coding will combat this induced noise very effectively when the corrupted signal is received after propagating through the noisy channel. Through the use of redundant codes it is possible to reduce signal power. Here, the quantization of the channel capacity is expressed in terms of the background SNR. However, real multipath interference appears as a convolution and hence should be multiplicative noise [2,5]. In that situation, the first form of channel capacity is more useful [2].

We now look at both methodologies and demonstrate that either form will provide similar answers if one properly accounts for the Maxwellian physics and the transmit and receive antenna systems are matched, even when the signal power is at least two orders of magnitude larger than the noise floor, which is typically the case for mobile wireless communication. This similarity stems from the fact that at resonance the power is real and the reactive component of the power does not enter into the picture. How to introduce the engineering aspects relating to the computation of power in the near field is described through the application of the Maxwellian physics. Another reason for introducing the physics related to radiation of energy is: the highly mathematical concept of channel capacity has its roots in physics via the concept of entropy [6].

One has to recognize that the channel capacity, like entropy, is a mathematical number and that this number has to be related to the parameters of an operating system. The key to establishing this relationship lies in properly interpreting this mathematical number and making the appropriate connection to real physics. Hence, it is instructive to observe how the concept of information content or capacity evolved from the thermodynamic concepts of entropy. We analyze the various equations of the channel capacity and determine which are suitable for handling realistic vector electromagnetic problems for wireless communication. The essential points are: the channel capacity is a mathematical entity defined by the governing equations; and *one can interpret those equations with some intelligent explanations, but one should always remember that the prose and the equations have to match up as the equations by themselves, do not contain any physics* [6]. For example, if we were to quantify the channel capacity in terms of entropy (a function of statistical probability) and the SNR (a function of the received power), then the pertinent question is: how should one apply these concepts in a near-field electromagnetic environment in which many cellular wireless communication systems operate? In addition, the concept of bandwidth is determined by the group velocity of the device and the phase response of the total transmit-receive system. Since the group velocity is related to the derivative of the phase response with respect to frequency, the deviation of the actual phase response from a linear phase response introduces undesired dispersion in the system. Thus the phase response essentially determines the useful bandwidth of a system. Therefore, we need to link the mathematical equations evolving from the concepts of entropy and information content to the fundamental physics that delineates the realistic parameters of a wireless system. In this chapter, we discuss the historical evolution of the concept of entropy, illustrate how the physics can be inserted into such a highly abstract mathematical concept, and relate entropy and physics to practical situations.

A historical perspective on understanding these concepts is important. In the words of Marcus T. Cicero (106–43 BC), the Roman statesman, orator, and philosopher [7]: *To be ignorant of what occurred before you were born is to remain always a child. For what is the worth of human life, unless it is woven into the life of our ancestors by the records of history? The causes of events are ever more interesting than the events themselves. History is the witness that testifies to the passing of time; it illuminates reality, vitalizes memory, provides guidance in daily life, and brings us tidings of antiquity.* Also, as the Greek philosopher Aristotle (384–322 BC) writes [7]: *If you would understand anything, observe its beginning and its development.*

The history of entropy, which is intimately connected to Information Theory, is essentially the development of ideas set forth several centuries ago to understand theoretically why a certain amount of available energy released from combustion reactions is always lost to dissipation or friction, that is, is unusable. Entropy is a concept that occurs in many areas of physics, including in thermodynamics, statistical mechanics, and information theory. The concepts of information and entropy have deep links with one another, although it took many years for the development of the theories of statistical mechanics and information theory to make this connection apparent. *This article is about information entropy, the information-theoretic formulation of entropy* [6]. *The popular literature is littered with articles, papers, books, and various and sundry other sources, filled to overflowing with prosaic explanations of entropy. But it should be remembered that entropy, an idea born from classical thermodynamics, is a quantitative entity, and not a qualitative one* [6]. *That means that entropy is not something that is fundamentally intuitive, but something that is fundamentally defined via an equation, via mathematics applied to physics. Remember in your various travails, that entropy is what the equations define it to be. There is no such thing as "entropy," without an equation that defines it. Entropy was born as a state variable in classical thermodynamics. But the advent of statistical mechanics in the late 1800's created a new look for entropy* [6]. Our goals are to extend this formalism to the evolution of channel capacity in a multipath near-field environment and to ascertain how to obtain meaningful, consistent, accurate results for real environments.

Section 4.2 provides a succinct overview of the concept of entropy, with a detailed exposition in the appendix that links entropy to the concepts of information content and channel capacity. The different forms of channel capacity are provided in section 4.3. In particular, it is important to use the gains of transmit and receive antennas instead of their directivities to obtain physically meaningful answers, when considering the various formulas for channel capacity. Three forms of the channel capacity by Shannon, Gabor, and Hartley are described. The Hartley version is pertinent in quantifying the transmission of information for wireless systems, as the information is carried by the electromagnetic fields. Since Hartley's representation deals with the amplitudes of the received signals, it is suitable for application in both near-field and far-field environments. Shannon's form of capacity was developed from a stochastic model for a system and deals with the signal and noise powers. Gabor's form is similar to Shannon's in the limit. In section 4.4, we argue from an engineering standpoint that it is more practical to deal with signal amplitudes in formulas for channel capacity. Also, in a near-field environment, using the power form of channel capacity requires specification of both the electric and the magnetic fields at the receiver, which may be practically untenable. However, this problem can be avoided by conjugately matching both the transmit and receive antennas. A few examples are presented in section 4.5 to illustrate how existing wireless systems need to be characterized by applying Maxwellian physics. These examples show how the channel capacity in current deployments of wireless systems can be improved. Finally, in the appendix, we trace the evolution of the concept of information content from the initial ideas of entropy. We also discuss how Shannon's singular contribution changed the communication world, even though many of his contemporaries had conceived and presented similar ideas.

4.2 HISTORY OF ENTROPY AND ITS EVOLUTION

A detailed history of information theory is provided and its link to entropy is delineated. Because this discussion is rather lengthy, it has been relegated to the appendix, as the historical connection between entropy and channel capacity is a secondary theme of this chapter.

In summary, the appendix discusses the question, *What is entropy and its related concept, channel capacity*?, and its answer, *Entropy is what the equations define it to be* [6]. Recall that entropy, an idea born from classical thermodynamics [8], is a quantitative entity and not a qualitative one. Consequently, entropy is not a fundamentally intuitive notion – rather it is defined by an equation via mathematics applied to physics. Essentially, entropy does not exist without a defining equation. One can interpret the defining equations to provide imaginative explanations, but remember that prose and equations must be consistent, because the equations give a firm, mathematical definition for entropy that cannot be circumvented [8]. In classical thermodynamics, the entropy of a system is the ratio of heat content to temperature, and the change in entropy represents the amount of energy input to the system which does not participate in the mechanical work done by the system. In statistical mechanics, the interpretation is perhaps more general in that the entropy is a function of statistical probability. In that case the entropy is a measure of the probability for a given macrostate, so that a large value for entropy indicates a high probability state, and a low entropy value indicates a low probability state. Entropy is also sometimes confused with *complexity*, the idea being that a more complex system must have higher entropy. In all likelihood the opposite is true. A system in a highly complex state is probably far from equilibrium and in a low entropy (improbable) state; whereas the equilibrium state would be simpler, less complex and have a higher value of entropy.

The connection between entropy and information was first presented by Leo Szilárd in 1929. His pioneering work was not well understood at that time [9–11] and remained obscure until it was rediscovered by Shannon in 1949. In his works, Shannon defined entropy with a negative sign, opposite to that of the standard thermodynamical definition [10]. Hence, what Shannon calls entropy of information actually represents *negantropy*. As outlined in the appendix, Szilard explained the phenomenon of Maxwell's demon [9–11] based on the principles of information content. Although most differences between these two definitions of entropy are minor, one very important difference distinguishes them: the information entropy H can be calculated for *any* probability distribution (if the "message" is taken to be that the event i which had probability of occurrence p_i out of the space of the possible events); but the thermodynamic entropy S refers specifically to thermodynamic probabilities p_i. Furthermore, S is dominated by the different arrangements of the system and, in particular, by the energies of the system that are possible at the molecular scale. In comparison, the information entropy of any macroscopic event is so small as to be completely irrelevant. However, a connection can be made between the two entropies: if the probabilities in question are the thermodynamic probabilities p_i, the (reduced) Gibbs entropy can then be seen as simply the amount of Shannon information needed to define the detailed microscopic state of the system, given its macroscopic description. In the inimitable words of physical chemist Gilbert Newton Lewis [11], when writing about chemical entropy in 1930, "Gain in entropy always means loss of information, and nothing more."

In summary, how the concept of information content is related to entropy has been described, with details of the various forms and applications of entropy described in the appendix. The two basic formulations of information content in terms of entropy, which can be either deterministic or statistical, are described next.

4.3 DIFFERENT FORMULATIONS FOR THE CHANNEL CAPACITY

The maximum capacity of transmission between a transmit/receive system in the communications literature is often characterized by the Shannon Channel

Capacity Theorem. According to Shannon, error-free transmission can be achieved over a channel, where the expression for the channel capacity is given by [3,4]

$$C_S = B \log_2\left(1 + \frac{P_S}{P_N}\right) \tag{4.1}$$

where B is the one-sided bandwidth of the channcl, P_S is the average received signal power, and P_N is the average noise power over the bandwidth of the channel. This expression does not use group velocity or the nature of dispersion in propagation that is generally used in the literature on electromagnetic theory. It would be useful perhaps to highlight some of the conditions that are associated with the various methodologies.

Per IEEE Standard 100, channel capacity is defined as "The Maximum possible information through a channel subject to the constraints of the channel (note that channel capacity can be defined by either per second or per symbol)." Shannon derived this formula using the concept of entropy, wherein the phenomenon of noise is purely random and neither multipath nor other interference sources are involved. Since the power is real in (4.1), it is also important to note that the magnitude must be understood in terms of far-field quantities in a wireless communication system. In a near-field scenario, power is complex and care must be taken in interpreting the variables of (4.1) in that case. In addition, Shannon did not develop this formula for a wireless system but for a wired system, where the voltages and currents are defined only at the terminals of the circuit. Now if the same formula is to be used for wireless systems, then it is necessary to define the nature of P_S. Generally, P_S is related to the received power, as all the power fed to the transmitter does not arrive at the receiver for various reasons. Hence, for a fair comparison between different systems, we need to supplement this formula by further stipulating that the input power to a transmitting system remains fixed, say at 1 W. The reason for this will subsequently be made clear. Before proceeding, we clarify an important point: when (4.1) is applied to wireless systems, the expression deals with the average received antenna power and not with the input power to the transmitting antennas. In other words, system losses are not considered, which means that the radiation efficiencies of the transmitting and receiving antennas do not enter the picture. Moreover, other near-field scatterers and mutual coupling between the antennas and the platform on which they are mounted are not considered, since the antennas are approximated by point sources. This omission of the efficiency and the inadequacy of the point-source model are serious impediments to system design. Therefore, (4.1) is not useful without an explicit specification of the input power.

In addition, to build a working wireless system, it is necessary to make a distinction between the directivity and the gain of an antenna. The antenna directivity $\mathfrak{D}$ has been defined in (2.45). The definition of directivity is based entirely on the shape of the radiated antenna pattern, and the power input and antenna efficiency are not involved. Ideally, the number of point sources an antenna can resolve is numerically equal to the directivity of the antenna [12]. Equivalently, directivity is equal to the number of beam areas into which the antenna pattern can subdivide the observation region of interest. Thus the directivity is equal to the number of point sources that an antenna can spatially resolve under the assumed ideal conditions of a uniform source distribution [p.27,12].

So, if one only uses the concept of directivity, it is natural to consider superdirectivity, because an aperture can assume any value for the directivity [p.102, 13]. This is also been discussed in section 2.6.1 and in [16,17]. Therefore, a superdirective condition will further increase the channel capacity! Hence, in a superdirective condition when the spacing between the antenna elements becomes less than half a wavelength at the operating frequency, the directivity of an antenna can theoretically take any large value. For example, in the limit when the spacing between the antenna elements approaches zero (that is, for a continuous source distribution), it can be shown that the end-fire directivity of an array can increase as the square of the number of antenna elements [13–15]!

However, when we talk about the gain of an antenna, the antenna efficiency is involved. The gain G of an antenna is defined in (2.46) and (2.47). To be physically relevant from a system perspective, antenna gain should include the effect of losses in the antenna under consideration (subject antenna) and in the reference antenna. Unfortunately, gain as defined in the IEEE Standard does not take these important system parameters into account. However, in the various examples of this chapter, the computed antenna gain includes all mismatch losses in a transmit-receive system. The gain G_0 is then related to the directivity by $\mathcal{G}_0 = \zeta\ \mathcal{D}$ as seen in (2.48), where ζ is the radiation efficiency of the antenna, including all the impedance and polarization mismatches. So the gain of an antenna over a lossless isotropic point radiator equals the directivity if the antenna is 100% efficient ($\zeta = 1$) but is less than the directivity if any losses are present in the antenna ($\zeta < 1$). The value for the input power does not enter into the expression for directivity but is included in the expression for the gain.

In summary [13–15], in the quest for systems with high capacity, it is natural for researchers unconsciously to delve into the region of superdirectivity – thereby unwittingly producing a non-workable system. Except for very small arrays, superdirective antenna arrays have proven to be impractical, because they have large current elements and high loss and require very precise excitation. In most cases, superdirective arrays are also very narrow band. Since the high directivity depends on cancellation of the contribution from the array currents, the behavior of a superdirective array is dependent on highly accurate current determination, and small errors in array excitation can destroy the properties of superdirective arrays.

So now we interpret Shannon's formula for wireless systems under constant input power to the transmitter. Otherwise, his channel capacity formula has no real utility. Therefore, as Hansen clearly observed [p. 102, 13], the moral of this story is: *A fixed aperture size can achieve (in theory) any desired directivity value*; delineating the fact that a superdirective array can have any value for the channel capacity, unless the input power is restricted! Subsequently in section 4.5, we compute the channel capacity for various antenna systems under a constant input power.

It is interesting to note that Gabor postulated a somewhat different but more sophisticated channel capacity theorem two years before Shannon [10,18]. In his 1946 paper, Gabor accounted for the interaction between the signal and thermal noise, which is not present in Shannon's formula. The expression for the Gabor capacity is given by [10,18]

$$C_G = B \log_2 \left(1 + \frac{\sqrt{1 + \dfrac{2P_M}{P_N}}}{2} \right) \tag{4.2}$$

where P_M represents the maximum power of the received signal over the bandwidth. However, if the system is observed over a very long time, the interactions between the signal and the noise average out so that C_G approaches C_S in the limit. Therefore, since we will be dealing with the average steady-state power in the frequency domain, we will be using (4.1) to compute the capacity instead of (4.2).

In the electromagnetics and signal processing literature, most authors use the power form of the Shannon Channel Capacity Theorem as represented by (4.1) and relate it to the SNR by transforming the levels in the amplitude of the signal to the square root of the power! This is invariably done by assuming the antenna is a point resistive source, which is an idealized artifice that does not exist in reality. So, the direct application of Shannon's formula in the near-field region of the source is doubtful. This point is very important, particularly in antenna problems as well as in near-field scenarios where the power may have a reactive component. In an arbitrary circuit, power is NOT proportional to the square magnitude of the voltage, scaled by a real number. These unresolved issues mandate the need to understand and characterize the real underlying physics of the form of the channel capacity theorem that is currently being used by most practitioners. This problem becomes more quixotic when one approaches this issue from the point of view of random variables, where the Fourier transform of the autocorrelation of a variable is termed as the power spectral density of that variable! In the original version of the Khinchine-Wiener theorem, the authors referred to the Fourier transform of the autocorrelation function as the Fourier spectrum and not the power spectral density [19]. This misuse of the terminology probably started after Nyquist, who was studying the thermal noise power in resistors. Of course, in a resistive circuit, the power is directly proportional to the square of the voltage, and no information about the phase of the current is necessary. Therefore, when we evaluate power in our computations, we will be following the correct procedure of evaluating (4.1) by using voltages and currents or equivalently by using the vector product of the electric and the magnetic fields, as is done in a near-field scenario. This procedure for obtaining the power is appropriate to apply in Shannon's formula as described in [22–24].

A more practical form of the channel capacity, which is more applicable to real systems, was first developed by Hartley [20,21] in 1928, where he generalized the earlier results. Among his conclusions, he stated, *The total amount of information which may be transmitted over a system whose transmission is limited to frequencies lying in a restricted range is proportional to the product of the frequency range which it transmits by the time during which it is available for transmission. Hartley's treatment represented a first step in the direction of measuring a message and the message transmitting capacity of a system* [21]. Both Nyquist [25,26] and Hartley [20,21] showed what must indeed have been common knowledge from the earliest days of telegraphy, namely, how the number of different signals available to a single receiver is limited by the relative magnitude of interference [21]. Also, as the signals traverse the transmission medium (or channel as it is commonly called), they are distorted, noise and interfering signals are added, and it becomes a major task to interpret the signals correctly at the desired destination [2]. We deal with physical systems, however, and these systems do not allow us to increase the rate of signal change indefinitely and to distinguish indefinitely many voltage amplitudes or levels [2]:

1. All of our systems have energy storage devices present, like inductances and capacitances, and changing the signal implies changing the energy content. Limits on making these changes are determined by bandwidth of a particular system.

2. Every system provides inherent (even if small) variations or fluctuations in voltage, or whatever parameter is used to measure the signal amplitude. One cannot subdivide amplitudes indefinitely. These unwanted fluctuations of a parameter to be varied are called noise.

The above two limitations are tied together by a simple expression developed for system capacity [2]: as $C = 1/\tau \log_2 n$, where τ is the minimum time required for the system to respond to signal changes and n is the number of distinguishable signal levels. The minimum response time is proportional to the reciprocal of the system bandwidth [2]. Thus system capacity can be written as $C = 2\ B \log_2 n$, where B is the one-sided system bandwidth in Hz. The information about the number of different amplitude levels can be sent by fewer numbers simply by binary encoding the different levels. For example, if the signal has 16 different levels then we require 4 bits. If the signal has 32 different levels, we require 5 bits. More generally, $\log_2 n$ bits are required for n levels.

The Hartley capacity is related to Shannon's capacity for matched antenna systems. The basic form of Shannon's theorem states [3,4] that *to distinguish between M different signal functions of duration T on a channel, we can say that the channel can transmit* $\log_2 M$ *bits in time T. The rate of transmission is then* $\log_2 M/T$*. More precisely the channel capacity, C′, may be defined as*

$$C' = \lim_{T\to\infty} \frac{\log_2 M}{T} \tag{4.3}$$

By following Hartley, one can obtain a simplified expression for the capacity without delving into the concept of power, if one requires the received signal to be separated into 2^Q distinct levels, so that the number of amplitudes which can be reasonably well distinguished is 2^Q. Furthermore, as Shannon points out: *Since in time T there are 2TB independent amplitudes then the total number of distinct signals is* $M = (2^Q)^{2BT}$*, where B is the one-sided bandwidth of the signal.* Hence, the number of bits that can be sent in this time is

$$log_2\,[M/T] = 2BQ \tag{4.4}$$

Equation (4.4) is more pertinent in characterizing near-field environments and is similar to the Hartley-Nyquist-Tuller form of the channel capacity theorem, which is [23]

$$C_{HNT} = 2B\log_2\left(1 + \frac{A_S}{\Delta V}\right) \tag{4.5}$$

where A_S is the received root-mean square (rms) amplitude voltage of the signal and ΔV is the rms level of the quantization noise voltage related to the discretization of the received signal.

Using equation (4.5) for the channel capacity is simpler and may provide a different value for the channel capacity than that obtained when it is computed with the power relationship. In equation (4.5), we need only the voltage. This form is ideally suited for wireless systems, since the capacity is directly related to the source, namely the fields, and is valid both for near-field and far-field characterizations. It is important to note, Nyquist or Hartley [21] did not neglect noise in deriving (4.5). Hartley postulated a particular detector system and assumed that the signal elements were of adequate power and were spaced sufficiently far apart so as to override noise and intersymbol interference, thus preventing errors in transmission. Recall that Shannon assumed a finite probability of error and obtained a statistical result. In comparing the formulas of Hartley (4.5) and Shannon (4.1), it is necessary to remember that these formulas do not represent the same transmission performance: one is essentially error-free; the other is error-free only in the statistical sense that the average rate of errors can be made arbitrarily small. Therefore, if one assumes the background noise characteristics are very similar in the two cases, then the two expressions might yield similar results for conjugately matched antennas as we demonstrate later. In (4.5), the limiting factor is the lowest discretization level to which the voltages are quantized. This formula was initially derived using a deterministic model and not assuming any statistical model. For those wireless systems where the signal levels are much larger than the background noise, the discretization ΔV may be related to quantization level of the signals and is more relevant for application to wireless systems. In such situations, (4.5) is a more appropriate formula to use than (4.1). Also, from a practical engineering point of view, (4.5) is more appropriate because the sensitivity of all wireless systems is expressed in terms of the sensitivity of the antenna relative to a particular electric field strength. This Hartley capacity uses only the voltages at the receiving antennas to define the capacity. Hence, (4.5) is useful from a system design point of view, as the sensitivity of a receiver is dictated primarily by the incident electric field strength and not by the power density! Additionally, the next section illustrates why this form may be more useful from a practical standpoint.

In summary, when using (4.1), care must be taken to account for the electromagnetic properties. Shannon's expression in (4.1) is often applied to wireless systems under the assumption that the average radiated power is fixed, which is equivalent to using the principle of directivity instead of the gain for the antennas. It is important to note that if the input power increases then the voltages inside the *log* function in (4.5) will approximately increase by the square root of that power, while in (4.1) the increase inside the *log* function will be approximately linear. Consequently, in evaluating capacity from different systems, it is necessary to fix the value for the input power. Because this requirement on the input power is necessary, the distinction between gain and directivity is very important. Recall that the gain of an antenna takes into account the amount of input power, while the directivity does not since it is related only to the radiated power. So in super-directive antennas, one could obtain any value of the directivity one wants, as the input power does not enter the picture. In contrast, the gain calculation always has an upper limit!

4.4 INFORMATION CONTENT OF A WAVEFORM

When one introduces a stochastic model into a system analysis, one needs to deal with a different mathematical formalism, as what is considered thermal background noise may not be Fourier integrable. This is because noise signals have finite power but infinite energy. Although a precise mathematical definition of the existence of a Fourier transform can be complicated, it is sufficient for a function to be absolutely integrable to ensure the existence of its transform. The stochastic model is the only method to use when the signal and the background noise have similar magnitudes, like in a global positioning system (GPS) or for deep-space communication networks. A deterministic formalism is not applicable for two reasons. First, the phase of a signal is noisy and may be hard to quantify to a specified degree of accuracy. Second, the background noise is not Fourier integrable; hence one must apply a different methodology that accommodates infinite-energy signals. Finally, we discuss whether it is more convenient to specify the design parameters of a mobile wireless system in terms of power or sensitivity of a receiver to the incoming electric field strength, as is done in various microwave engineering applications operating in near-field environments.

As mentioned earlier, there are two different ways (Shannon vs Hartley and others) of quantifying the channel capacity through entropy. Shannon's method is based on a stochastic model for the signal, where background noise plays an important role as the signal amplitude is comparable or possibly less than the background noise. For Hartley's model, noise is not deemed pertinent, and the capacity is determined by the discernable levels of the voltages of the signal. In this section, we illustrate these two procedures.

According to its definition as the logarithm of the number of possible, equally probable selections, entropy is a measure of the quantity of information. In the discussion to this point, the concepts of information and entropy are very similar, even identical [10]. But a somewhat new feature was introduced by Shannon, who defined the mean entropy per symbol for transmissions in which the symbols have probabilities p_i [10], which are relative frequencies. Consequently, the entropy becomes $H = - k\, p_i \log (p_i)$, where k is a positive constant. Because they are determined by the source, the probabilities p_i have no direct relation to the structure of a particular signal. A symbol can be represented by any configuration or group of configurations in a basic group of cells, provided that the basic group allows at least as many distinguishable configurations as symbols. Even this fairly general definition does not exhaust the almost limitless possible coding of signals. Hence the interesting properties implied by Shannon's representation must be attributed to its mathematical form rather than to its intrinsic relation with the physical concept of the same name, which led to modeling a communication system by a stochastic model. An intuitive understanding of information entropy thus relates to the amount of *uncertainty* associated with an event for a given probability distribution. For example, consider a box containing many colored balls. If the balls have different colors (hence no color predominates), the uncertainty of the color of a randomly drawn ball is maximal. On the other hand, if the box contains more red balls than any other color, then the result is slightly less uncertain: a ball drawn from the box has more chances of being red – so one would bet on drawing a red ball if forced to place a wager. Telling someone the color of every new drawn ball provides them with more information in the first case than it does in the second case, because more uncertainty exists about what might happen in the first case. Intuitively, if the outcome is completely certain, one would learn nothing by drawing the next ball, and the information content would be zero. As a result, the entropy of a "signal" (the sequence of drawn balls calculated from the probability distribution) is higher in the first case than in the second. Shannon, in fact, defined entropy as a measure of the average information content associated with a random outcome. His definition of information entropy makes this intuitive distinction mathematically precise [2,10,22,23]. Moreover, Shannon's characterization of channel capacity is based on this reasoning.

The Nyquist-Hartley-Tuller-Shannon [2,22,24] theory states that the time-bandwidth product $2BT$ of a waveform, with one-sided bandwidth B and time duration T, provides the number of degrees of freedom necessary to characterize any waveform in a finite-dimensional space. However, if the phases associated with waveforms are inconsequential, it should be possible to reduce the number of degrees of freedom by half, as only half the information is available. Halving the information content is true in optics, where the phase information is difficult to measure even though the signals may not be random and, in fact, led to the Wiener-Khinchine theory [19]. In the Wiener-Khinchine formula, instead of relating the random signal to its Fourier transform, we use an alternate formulation and relate the autocorrelation function of a stochastic signal to its Fourier spectrum, called the power spectral density as the signals are noise like and have finite energy. The original signal can have at most $2BT$ degrees of freedom; whereas the new expansion has approximately BT degrees of freedom, since the phase information is lost or not obtainable. This methodology has been applied with great success to any problems where signal phases cannot be obtained, either because a signal is random (noise problems) or because the experimental procedure used in the measuring device cannot detect the phase angles. Basically, the degrees of freedom dictate the channel capacity. The above discussions are related to the form of the channel capacity given by (4.1).

Next, we explore the practical characteristic features that are needed to define the sensitivity of a device in a real operating environment. In practice, the sensitivity of a receiver is often described in terms of the electric field strength at the receiving antenna rather than in terms of the signal power in form of the SNR. To characterize power in an arbitrary electromagnetic scenario, which is not necessarily the far field, one needs both the electric and the magnetic fields to define the Poynting vector or equivalently the power density. However, only one of the electric and magnetic fields is sufficient when concerned only with the far field. Since measuring the electric and magnetic fields simultaneously is difficult, characterizing the channel capacity in a near-field environment is hard to achieve. It is much simpler to characterize the sensitivity of the receiver in terms of the incident electric field. This discussion relates to the form of the channel capacity given by (4.5).

At this point, it is relevant to discuss the differences [24] between the works of two groups: (A) Nyquist [25,26], Hartley [20], Küpfmuller [27], and Gabor [18]; and (B) Tuller [23], Shannon [3,4], and Clavier [28]. The concept of a statistical model is absent in all presentations of the first group, which deal explicitly with a deterministic model of the channel contaminated with noise. In addition, they express the information content in terms of the amplitude of a signal, as the sensitivity of the receivers from a practical stand point is determined by the electric field strength. The reasoning and analysis of the second group are based on the special case of a stochastic model of the transmission channel. The genius of Shannon was to observe that the role of noise (or any other disturbances being additive or affecting the signal in any other way) was to introduce an element of *uncertainty* in the transmission of symbols from the source (defined as the transmitter) to destination (which is the receiver). The redundancy modeled by an uncertainty using a stochastic model for the transmitter is generally modeled by a probability distribution. The goal is to introduce enough redundancy in signals emanating from the transmitter so that the noise in the channel will not entirely corrupt the information. Norbert Wiener also shared Shannon's view of noise and uncertainty, but his attention was focused in different directions [24]. When dealing with noise, it makes more sense to deal with the power spectral density than with the amplitudes of the signals themselves. As the sensitivity of any receiver is characterized by the voltage induced at its input terminals due to the receiving antenna, it is necessary to specify the minimum strength of the incident electric field at the antenna for the system to operate. Hence application of (4.5) may be more appropriate for evaluating capacity in both near- and far-field conditions. In systems where the received signal power is approximately two orders of magnitude larger than background random noise, (4.5) may also be more useful than (4.1).

Furthermore, from an engineering standpoint in telecommunications and particularly in radio, the strength of a transmitted signal that is received, measured, or predicted at some significantly distant reference point from the transmitting antenna is generally given by the values of the electric field strength (not power). This parameter, also called the received signal level or field strength, typically is measured at the receiving antenna in volts per unit length. For example, transmission systems use higher powers, such as broadcasting units [29,30], of dB-millivolts per meter (dBmV/m). Very low-power transmission in devices like mobile phones have field strengths in dB-microvolts per meter (dBμV/m). Some examples of the minimum signal strength levels at the receiving antenna are:

- ★ 100 mV/m: at which blanketing interference occurs. Blanketing is interference caused by strong radio signals. Although the spectral mask of a radio station's transmitter suppresses spurious emissions on other frequencies in the band, being extremely close to a station may still allow them to be strong enough to cause significant interference. The strong station will appear on nearly every blank or weak channel in the band, especially in the FM broadcast band. This problem is greatly reduced by moderate-quality receivers, which have better selectivity than inexpensive disposable ones.
- ★ 1 mV/m: the edge of a radio station's protected area.
- ★ 100 μV/m: the minimum strength at which a station can be received.

Being engineers, these reasons are why Nyquist, Hartley, and Tuller expressed the channel capacity in terms of the number of different signal voltage levels that one can distinguish at the receiver rather than in terms of the received power. However, these signal field strengths can be transformed to a signal-power-level description by assuming that the field strength is dissipated in a 600-ohm resistor [30]. Using the units dBm implies a power ratio in dB of the measured power referenced to 1 mW. A typical transmission from an FM broadcast radio station is about 100 kW (80 dBm). Typical cellular phones transmit signals of about 27 dBm or 500 mW. The Bluetooth radio systems transmit 20 dBm (100 mW), 4 dBm (2.5 mW), or 0 dBm (1 mW) depending on whether they are a Class I, Class II, or Class III system with a range of 100 m, 10 m, or 1 m, respectively. In contrast, the thermal noise floor with 1 Hz bandwidth will be approximately −174 dBm (0.000004 fW = 10^{-15} W). So with a 500 MHz bandwidth as in an ultrawideband system, the noise floor will be 2 pW. By comparison, wireless signals over a network are in the range of 100 pW. Under these circumstances, it appears that the effect of thermal noise is negligible for mobile wireless systems, which raises an important issue on the rationale of modeling a wireless channel by additive Gaussian white noise, as discussed earlier. On the other hand, typical received signals from a GPS satellite are approximately −127.5 dBm (0.178 fW). The thermal noise floor for a commercial GPS signal with a bandwidth of 2 MHz is −111 dBm (8 fW). Therefore, under these conditions, it makes sense to apply the Shannon channel coding and capacity theorems to design an appropriate code that will introduce the necessary redundancy at the transmitter for GPS systems to combat the noisy characteristics of the channel, so that the received signal provides a reliable transmission of information. The launch of the first artificial satellite generated a need for conserving transmitted power over a real channel that is approximated by this model. Shannon's theorems thus served as a powerful stimulus to further theoretical work, particularly toward the goal of minimum-complexity decoding. By the late 1960s, it was feasible to code this additive white Gaussian noise (AWGN) channel with an error probability of 10^{-5}, so that the channel could conserve 5–6 dB of power relative to the uncoded operation at data rates in megabits per second. To gauge the economic advantage, consider that an alternative method of gaining 5 dB of power requires more than triple the area of the receiving antenna [31]. When the signal powers are at least two orders of magnitude stronger than the noise levels for mobile wireless communication, the relevance of considering the background noise as the limiting factor for transmission is unclear, and multipath and mutual coupling between antennas and the environment are more significant.

In short, it is important to note that if one uses a stochastic model one must use the definition of the sensitivity of the receiving system in terms of the SNR. However, in mobile wireless communication systems where the signal is much higher than the noise by at least two orders of magnitude, perhaps it may be more relevant for engineers to use the Hartley-Nyquist model and employ units which refer to the smallest recognizable electric field strength, which may be why Hartley defined the channel capacity in terms of a detectable voltage ratio. Even Tuller [21,2] expressed the capacity in terms of the *rms* amplitudes of the signal and the noise, even though he developed his formula via a stochastic model.

Furthermore, to have a communication system with a high level of performance, it is necessary to have high bit rate, low latency, and low bit error probability. The level of performance is often qualitatively described by the eye pattern or constellation diagram associated with the various signals [32,33]. In telecommunication, an eye pattern (also known as an eye diagram [32]) is an oscilloscope display in which a digital data signal from a receiver is repetitively sampled and applied to the vertical input, while the data rate is used to trigger the horizontal sweep. The eye diagram, used to assess impairments in the radio channel, is generally displayed in terms of the signal amplitudes. It derives its name from its appearance as a series of eyes between a pair of rails for several types of coding. Several system performance measures can be derived by analyzing the display. For example, if signals are too long, too short, poorly synchronized with the system clock, too high, too low, too noisy, and too slow to change or have too much undershoot or overshoot, these variations can be observed from the eye diagram. An opcn eye pattern corresponds to minimal signal distortion, while distortions of the signal waveform from inter-symbol interference and noise appear as closure of the eye pattern.

A constellation diagram [33] on the other hand is a representation of a signal modulated by a digital scheme such as quadrature amplitude modulation or phase-shift keying. It displays the signal as a two-dimensional scatter diagram in the complex plane at symbol sampling instants. In a more abstract sense, it represents the possible symbols that may be selected by a given modulation scheme as points in the complex plane. Measured constellation diagrams can be used to recognize the type of interference and distortion in a signal. By representing a transmitted symbol as a complex number and modulating a cosine and sine carrier signal with the real and imaginary parts (respectively), the symbol can be sent with two carriers on the same frequency. They are often referred to as *quadrature carriers*. A coherent detector is able to demodulate these carriers independently. This principle of using two independently modulated carriers is the foundation of quadrature modulation. In pure phase modulation, the phase of the modulating symbol is the phase of the carrier itself. Since the symbols are represented as complex numbers, they can be visualized as points on the complex plane. The real and imaginary axes are often called *in phase* (I-axis) and *quadrature* (Q-axis). Plotting several symbols in a scatter diagram produces the constellation diagram. The points on a constellation diagram are called *constellation points*, sets of *modulation symbols* which comprise the *modulation alphabet*.

From the preceding discussion, one can easily infer that it is necessary to quantify the signal amplitudes, because they play an important role in defining the quality of transmission and the SNR is a secondary factor. Perhaps this statement provides insight on why engineers like Nyquist, Hartley, and Tuller defined the capacity in terms of signal amplitudes while the mathematician Shannon expressed capacity in terms of power. When modeling realistic antennas, it is necessary to describe their performance in terms of the applied voltages and induced currents on these structures. In near fields, the power density is complex. Hence conventional use of point sources does not model the vector nature of an electromagnetic system – a serious shortcoming of current modeling methodologies. Consequently, for practical reasons in a near-field situation, one needs to quantify the capacity in terms of the signal levels, as they simplify the expressions. Based on the discussion in [2], this formulation was used in [34] to calculate the channel capacity of a parallel computer bus in the presence of cross-talk noise.

To illustrate applications of the two mathematical forms of the channel capacity, a two-dipole configuration is used for simulations involving (4.1) and (4.5). In a near-field scenario, often the case for most wireless communication systems, the voltage and the current are required for computing the power, unless the transmit and receive systems are conjugately matched for maximum power transfer, in which case the power will be a real quantity. The points of departure from previous methodologies are: realistic antennas are used in the simulations; both transmit and receive antennas are conjugately matched and the input power to the system is constrained.

4.5 NUMERICAL EXAMPLES ILLUSTRATING THE RELEVANCE OF THE MAXWELLIAN PHYSICS IN CHARACTERIZING THE CHANNEL CAPACITY

In all numerical examples, rms values for the voltages and currents are used in the computations instead of peak values. Therefore, the factor ½ will not be present in power computations using the Poynting vector. Results of these numerical examples are generated with an electromagnetic analysis code that accurately solves Maxwell's equations by analyzing the transmitted and received fields of various antennas. The electromagnetic techniques used in the code account for the vector nature of the problem and are known to provide physically meaningful results for all electrical parameters of interest like field distributions, induced currents, voltages, etc. Using numerical methods to obtain a stable numerical solution of Maxwell's equations by itself is a vast area of research. Unless one is intimately involved with computational electromagnetics, it is difficult to articulate the complexities and difficulties of this field in a few sentences. Having said that, computational electromagnetics is mature, and results typically obtained with ***any*** commercial numerical electromagnetic code for the simple problems discussed in this section are quite stable, accurate, and very reliable. Therefore, a non-electromagnetic practitioner can unquestionably rely on the numerical results generated by these commercial computer codes.

To illustrate that the vector principles of electromagnetics in wireless communication systems are needed for a correct and meaningfully relevant interpretation of equations (4.1) and (4.5), several simulations are now discussed. In these simulations, the total input power fed to the transmitting antenna of the various systems is limited to 1 W, which is equivalent to using the antenna gain in (4.1) and (4.5) instead of the directivity. The bandwidth is chosen as unity to calculate the channel capacity in Bits/Hz. In addition, the electromagnetic properties of the transmitting and the receiving antennas and the surrounding media have been included in computations of the channel capacity. The transmit and receive antennas are conjugately matched, resulting in real values for the received power even in a near field environment, which is more realistic than using point sources to represent the antennas. Not only does an appropriate electromagnetic model render a more accurate accounting of the mutual coupling between all the antenna elements and proximate structures, it also more realistically represents the effects of the ground, since the deployment of real antenna systems above the earth produces an image of the original source.

In this analysis, the earth is modeled by a perfectly conducting ground plane. Unless the transmitting antennas are far from the ground, the assumption of a perfectly conducting ground plane over the choice of an imperfectly conducting ground plane representing the earth suffices for our discussions. The interference pattern between the transmitting and receiving antennas and their images due to the presence of the ground plane are important factors that limit the received power. To ensure that the input power to the transmitter remains the same in all comparisons of the channel capacity for the different examples, the voltages and the currents at the transmitting antenna are calculated. In addition, the voltages and the currents are also calculated at the receiving antenna to evaluate the received power and to make it unnecessary to know the operating environment and whether one is located in the near-field or in the far-field of the transmitter.

4.5.1 Matched Versus Unmatched Receiving Dipole Antenna with a Matched Transmitting Antenna Operating in Free Space

As a first example, consider a pair of parallel, center-fed dipoles (one for transmission and one for reception) that are vertically oriented in free space and separated by 100 m. Both dipoles are a half-wavelength long at 1 GHz and have a radius of 1 mm. The transmitting dipole is conjugately matched at the input with a complex load impedance of $\{97.58 - j45.40\}\ \Omega$, so that maximum power can be radiated from it. The value for this complex load impedance is calculated by using an electromagnetic simulation code and principles of maximum power transfer. After exciting the transmitting antenna with an rms value of 1 V, the rms value of the current I_t in the transmit antenna and the corresponding transmitted power are calculated.

Consider two cases for the receiving antenna: (A) when it is terminated in a 50-Ω load; and (B) when it is conjugately matched with $\{97.58 - j45.40\}\ \Omega$ to receive maximum power. The induced currents on the receiving antenna and the corresponding received powers are computed by using $50 \times |I_r'|^2$ for case (A) and $97.58 \times |I_r''|^2$ for case (B), where I_r' and I_r'' respectively are the currents flowing through the loads of the receiving antennas. Then the voltages and currents are scaled such that the input power to the transmitter is always 1 W, under the matched conditions. This normalization is enforced to compare the various results for all simulations to a benchmark input power of 1 W.

After this normalization, the calculated received power when the receiver is terminated with a nominal impedance of 50 Ω is 0.0641 μW, while the received power is 0.078 μW when the receiving antenna is conjugately matched. Therefore, conjugately matching the receiving antenna yields an increase in the received power of approximately 22% [$100 \times (0.078 - 0.0641) \div 0.0641 = 22\%$] under the constraint of the same transmitted power. Under normal circumstances, this result would be considered good engineering since it follows

Gabor's approach [1] of using the Maxwell-Poynting theory to improve the electromagnetic performance of the system by almost 22%. However, in the channel capacity metric, such a *significant* improvement in performance is generally in the realm of the round-off error of the channel capacity formula. To demonstrate, assume that the background noise is 2.0 pW in (4.1). For computing the channel capacity using the signal voltages in (4.5), discretize the received voltage into 16 levels for 1 V. The reason for choosing this particular minimum quantization voltage level is because when it is considered as a noise source applied to the resistive part of the matched load, which is 97.6 Ω, the quantization noise power is approximately 2.4 pW. This quantization noise power is approximately equivalent to the white background noise power of 2 pW having a bandwidth of approximately 500 MHz. Through such a choice, the noise power related to the quantization noise of (4.5) is made approximately equal to the background thermal noise in (4.1). Consequently, Shannon's channel capacity for the 50 Ω terminated receive antenna is

$$C_S^{50} = B\log_2(1 + 0.0641/0.000002) = 14.968B,$$

while, for the conjugately matched case, it is

$$C_S^{Matched} = B\log_2(1 + 0.078/0.000002) = 15.25B.$$

As the difference between the values of the two capacities is only 0.282 B (a difference of only 1.85%), this raises some interesting questions. Does this imply that engineering physics based on electromagnetics is irrelevant in communication system design, since matching an antenna to receive maximum power (22% more power) does not seem to have much significance? An alternate question that may be posed is: Does this small difference between the two capacities between using a matched or an unmatched receiving antenna imply that the use of the channel capacity formula without paying attention to basic physics is simply a statistical aberration? A system designer must address these questions intelligently to deploy a meaningful practical working system! It is also important to note that for the 50 Ω terminated receive antenna, the background noise is kept the same as the matched-load case.

Next, for this problem, we use Hartley's channel capacity formula (4.5) to evaluate the performance of the same system. For a 1 W input power, the voltage induced in the receiving antenna with the 50 Ω load is 1.8 mV. Note that a 16-bit quantization yields approximately 4 pW (instead of 2 pW) of noise power when applied to a 50 Ω load. For a 16-bit quantization of the input received voltage, the Hartley-Nyquist-Tuller channel capacity is

$$C_{HNT}^{50} = 2B\log_2(1 + 0.0018/2^{-16}) = 13.79B.$$

When the receiving antenna is terminated by a matched load of {97.58 – *j*45.40} Ω, the voltage induced at the load of the receiving antenna is 3.05 mV. The channel capacity is

$$C_{HNT}^{Matched} = 2B \log_2 (1 + 0.003 / 2^{-16}) = 15.3B.$$

For the matched case, C_S computed by (4.1) and C_{HNT} computed from (4.5) are approximately equal, because the antennas are conjugately matched so that the received power is related to the square magnitude of the voltage, scaled by a real number. However, when the receiving antennas are not matched, to calculate the power one also needs to know the current, as the phase difference between the voltage and current is needed in (4.1). In contrast, only the voltage is required for (4.5). So in the unmatched case, C_S and C_{HNT} cannot be equal. Thus, it is anybody's guess as to which equation should be applied. This thrusts the user onto the horns of a philosophical dilemma – deciding which formula accurately characterizes reality given that the channel capacity is a mathematical formula with no associated physics. In the end, it is left to the user on how to interpret the results of the capacity formulas and then to decide which formula is more applicable to the real system under consideration!

When the two capacities numerically yield similar values, then of course one has some confidence in the results and can depend on using them to compare system performances. But that begs the question of what to do when the two capacities yield different numerical values, like in the present example where the difference in capacities (13.79 B and 14.97 B) is 8.6% when the receiving antenna is terminated by 50 Ω. Which result is correct? The answer to this question is stated in the introduction: *to interpret these numbers in the proper light, it is necessary to relate the problem setup to the actual Maxwellian physics.*

4.5.2 Use of Directive Versus Nondirective Matched Transmitting Antennas Located at Different Heights above the Earth for a Fixed Matched Receiver Height above Ground

As a second example, we illustrate how the capacity of a transmit-receive system is influenced by the deployment of directive and non-directive antennas over a perfect ground plane representing earth. In this case, the presence of the image due to the ground plane will significantly affect the results. Another goal is to illustrate how to employ the Hartley capacity in practice. In all the simulations, the input power is constrained to be the same. Therefore, if the formulation for the channel capacity is correct, based on relevant scientific and mathematical principles, then one should obtain approximately the same result for the capacity whether one uses Hartley's law through (4.5) or Shannon's formula through (4.1). Hartley's law is seldom used in the current literature, even though it may be more useful from an engineering viewpoint. We expect both formulations to predict similar values for the channel capacity, as the input power has been constrained, the received power in the near-field scenario has been computed using the Maxwell-Poynting theory, and all the physics have been properly incorporated! In addition, the limiting noise is essentially equivalent for both capacity expressions. Finally, the transmit and receive antenna systems are matched by applying conjugately matched loads for maximum power transfer.

For the initial simulation, we study the variation of the capacity as a function of the height of the transmitting antenna above the earth for a directive horn, as illustrated by Fig. 4.1 and a vertically oriented dipole which is sketched in Fig. 4.2. The receiving antenna is a vertically oriented dipole whose center point is 2 m above a perfectly conducting earth. Consider a horn antenna with 20 dB of gain operating at 1 GHz that produces vertically polarized fields. This unrealistic horn antenna has been obtained by scaling a commercial 20 dB gain horn antenna operating at X-band. Although this antenna is overly large for deployment, we use it to calculate the results for the capacity in deploying a high-gain antenna. The input impedance, $\{123 + j539\}$ Ω, of this probe-fed horn is computed by electromagnetically analyzing the horn antenna using Maxwell's equations. The internal impedance of the voltage source applied to the horn antenna needs to be $\{123 - j539\}$ Ω so that the horn is conjugately matched and can thus radiate the maximum power. The gain of the horn with this conjugately matched load is calculated to be 20.5 dB when it is operating in free space.

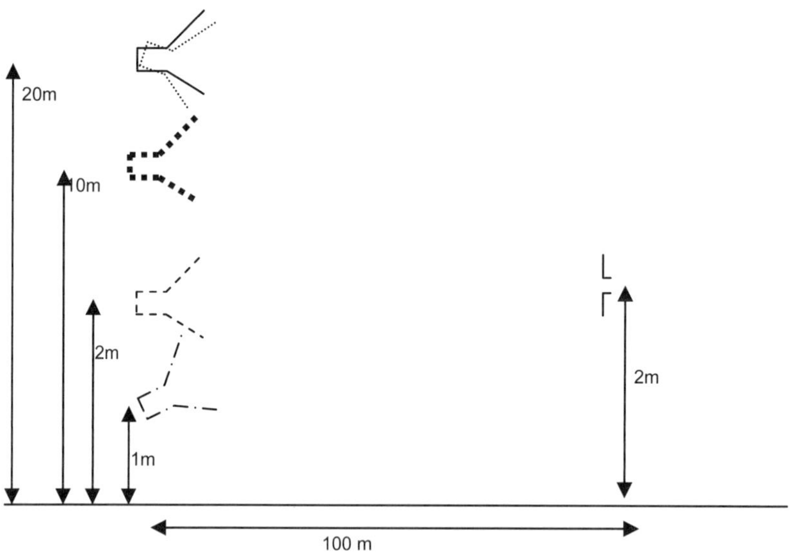

Figure 4.1. Different transmitting horn configurations at a height of 20 m, 20 m tilted downwards at an angle of 11°, 10 m, 2 m, and 1 m tilted upwards by an angle of 1° above the ground plane. The receiving dipole is located at 2 m above the ground plane and at a horizontal distance of 100 m from the transmitter.

4.5.2.1 Transmitting Horn Antenna at a Height of 20 m. The horizontally oriented transmitting horn is located on top of a tower so that the center point of the horn is situated at a height of 20 m above a flat perfectly conducting earth as shown in Fig. 4.1. A perfectly conducting ground plane is used to simulate moist earth. The transmitted fields from the matched horn are received by a half-wave center-fed dipole of 1-mm radius that operates at 1 GHz and is located at a horizontal distance of 100 m from the horn as seen in Fig. 4.1. The input impedance of the receiving dipole when it is operating in free space is $\{97.6 + j45.4\}$ Ω. The dipole is similarly conjugately matched with the impedance of $\{97.6 - j45.4\}$ Ω to receive the maximum power from the transmitting horn. For a 1-V excitation of the probe feed point of the horn antenna, the input current to the horn is computed to be $\{4.06 - j0.523\}$ mA. The computed current induced at the load located at the feed point of the receiving dipole is $\{3.41 - j6.40\}$ μA. Given all the voltages and the currents, we can scale them so that the input transmit power is limited to 1 W. This produces a received power of 1.26 μW for an input transmit power of 1 W to the horn. The received voltage across the load of the receiving dipole antenna is 0.012 V. All are considered to be *rms* values. The thermal noise floor is assumed to be 2 pW. The calculated Shannon capacity for a unit bandwidth is $\log_2(1+1.26/0.000002) = 19.26$ bits/Hz, and the Hartley capacity is $2\log_2(1+0.012/2^{-16}) = 19.24$ bits/Hz, assuming 16 bits of quantization levels for the voltage. Clearly, nearly equal values for the capacity are obtained for both formulations, because the input power to the transmitter is kept constant and the transmit/receive antennas are conjugately matched.

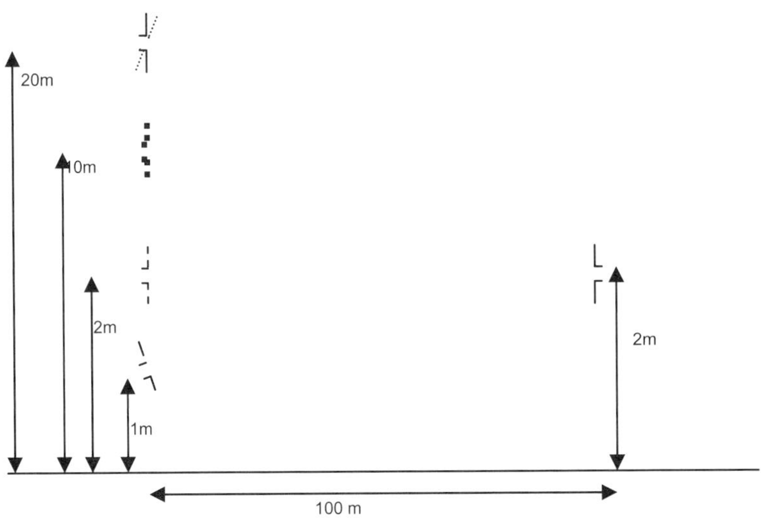

Figure 4.2. Different transmitting dipole configurations at a height of 20 m, 20 m tilted downwards at an angle of 11°, 10 m, 2 m, and 1 m tilted upwards by an angle of 1° above the ground plane. The receiving dipole is located at 2 m above the ground plane and at a horizontal distance of 100 m from the transmitter.

This example illustrates a near-field simulation where the conventional Shannon form of the channel capacity when applied to unmatched antennas will yield a complex value for the power and hence will not provide any sensible results. But, by fixing the input transmit power and making the transmit and the receive antennas conjugately matched the conventional Shannon form of the capacity can be used to yield meaningful results. However, it is quite clear that Hartley's law is much simpler to apply and it provides valid results if properly implemented. In this way, the shortcomings of the existing formulas for near-field applications are overcome. Note that a multipath scenario cannot be scientifically described in a near-field condition; therefore the conventional Rayleigh and Rician statistical fading models developed for a random medium are not applicable in near field environments. The limiting noise characteristics in (4.1) and (4.5) are also made equivalent by enforcing equality between the quantization noise power and the thermal background noise. If one assumes point sources for the antennas, then the basic physics are not included, because the antennas' radiation efficiencies and mutual coupling which affect the radiated power are never really characterized. Unless these elaborate precautions in carrying out experiments are made, incomplete experimental data that often appears in the literature do not provide meaningful conclusions.

4.5.2.2 Transmitting Dipole Antenna at a Height of 20 m. For the next simulation, replace the transmitting horn antenna with a center-fed, vertically oriented, half-wave dipole of length 15 cm and radius 1 mm, located at a height of 20 m above a perfectly conducting ground plane as shown in Fig. 4.2. The input impedance of the transmit dipole is again conjugately matched with a load of $\{97.6 - j45.4\}$ Ω so that it can radiate maximum power. The transmitted fields at 1 GHz are received by the same half-wave dipole receiving antenna, located at a horizontal distance of 100 m from the transmitting dipole. The receiving half-wave dipole is similarly conjugately matched as before and therefore can receive the maximum power at the load. Now, the transmitting dipole antenna is placed vertically at a height of 20 m above the ground plane, and the receiving dipole is situated at a height of 2 m above the ground plane, the same location as in the previous case. For a 1 V excitation of the transmit antenna, the input current at the transmit dipole is $\{5.12 - j0.002\}$ mA. The current induced at the load located at the feed point of the receiving dipole is $\{1.29 - j0.35\}$ μA. Given all the voltages and the currents, they are scaled so that the input transmit power is limited to 1 W. This results in a received power of 0.034 μW for a transmit input power of 1 W. For the same thermal noise floor as before, the Shannon channel capacity in this case is 14.05 bits/Hz. The corresponding voltage induced at the load of the receiving dipole for an input power of 1 W is 1.98 mV. For 16 levels of discretization of the received voltages, the capacity in terms of the voltages given by Hartley is 14.06 bits/Hz. In this antenna deployment scenario, the channel capacity is lower than when using a directive horn antenna, because the received power is smaller when the transmitter is a half-wave dipole versus a high-gain horn. However, the results for the capacity given by the power (Shannon) and voltage (Hartley) forms are the same for either case, whether one uses signal amplitudes in (4.5) or the classical Shannon channel capacity formula defined in (4.1).

4.5.2.3 Orienting the Transmitting Horn or the Dipole Antenna Located at a Height of 20 m Towards the Receiving Antenna. It is interesting to observe that the capacities for the horn and the dipole transmit antenna systems can be further increased, if one orients the vertically polarized transmitting horn towards the receiving half-wave dipole. This can be achieved by tilting the transmit horn antenna 11° towards the ground as seen in Fig. 4.1 so that the horn is directing its energy to the receiving dipole. The standard 20 dB gain horn, now directed to the receiving dipole, is operating at 1 GHz. The center of the horn is placed at a height of 20 m above a ground plane, and the center of the receiving dipole is situated 2 m above the same flat perfectly conducting ground plane. For 1 W of transmit power fed to the horn, the received power at the dipole is 5.27 μW. For the same thermal noise floor of the preceding examples, the Shannon's capacity is 21.33 bits/Hz, an expected significant increase when the directive antenna is pointed towards the receiver. Per Hartley' law, the capacity for a received voltage of 25 mV is 21.34 bits/Hz for a 16-level discretization. If one now replaces the transmit horn with the half-wave dipole that is tilted by 11° towards the ground as shown in Fig. 4.2, then the power at the receiving dipole is 36.1 nW, yielding a Shannon capacity of 14.14 bits/Hz for the same parameters. The calculated Hartley capacity for the 2.1 mV of received voltage is 14.23 bits/Hz. It is thus evident that if an antenna is pointed towards the receiving antenna system, the capacity of the system is increased proportional to the gain of the transmitting antenna.

4.5.2.4 The Transmitting Horn and Dipole Antenna Located at a Height of 2 m above Ground. Next, we bring the horn antenna closer to the ground by placing the center point at a height of 2 m above the ground plane as seen in Fig. 4.1. The transmitting horn and the receiving half-wave dipole are still separated horizontally by the same 100 m. Basic electromagnetic theory tells us that the severity of the interference pattern from the horn and its image will be much less [12,35], since the image of the transmitting horn formed by the ground will not produce an effective aperture of 40 m as before, but the effective aperture will approximately be 4 m. Because the interference between the source and its image will be reduced, one expects a higher value of capacity than before, since the radiation pattern will be smooth and the induced voltage, the integral of the radiation pattern over the dipole, will be higher [35]. Indeed, when the physical centers of the transmitting horn and the receiving dipole are both located 2 m from the ground, the received power across the load in the receiving dipole is 16.7 μW for a 1 W of power fed to the transmitting horn. For the same thermal noise floor of 2 pW as before, the Shannon channel capacity in this case is 22.99 bits/Hz. Furthermore, we can also determine the Hartley capacity in terms of the received voltage at the dipole, which is approximately 44.4 mV for an input power of 1 W to the horn. In this case, the Hartley capacity is 23.01 bits/Hz. It is also interesting to note that if the transmitting directive horn antenna is replaced by a conjugately matched half-wave dipole now located 2 m above the ground as seen in Fig.4.2, the received power at the dipole is 0.14 μW. For the same thermal noise floor as before, the Shannon Channel Capacity is 16.09 bits/Hz. There is also a significant increase in the channel capacity even when using the transmit half-wave dipole, since the interference pattern produced by the image dipole is much less as has been illustrated in [35]. Now if we let the received signal be 4.1 mV, the capacity in terms of the voltages given by Hartley is 16.1 bits/Hz.

It is thus clear that the channel capacity can be significantly increased if the transmitting antennas are located close to the ground. This will also make the systems much cheaper to deploy, since probably the cost of installing a tower will be typically more than the transmitting equipments. Also, using a directive horn antenna as a transmitter and moving it closer to the ground require a smaller input power. In this example, the needed input power is reduced by a factor of 13.25 (= 16.7/1.26), which is equivalent to pumping 75.4 mW instead of 1 W to achieve the same channel capacity as deploying the transmit antenna on top of a tower. For the dipole antenna systems that are brought closer to the ground, the input power to this transmit-receive system can be reduced by a factor of 4.12 (= 0.14/0.034), which is equivalent to pumping 242.9 mW instead of 1 W to achieve the same channel capacity when the transmitting antenna was deployed on top of a tower. The situation is more dramatic for directive antennas than for dipoles which are not so directive. The point to be made here is that communication theory provides an approximate analysis of the system without paying attention to the realizibility, whereas electromagnetic theory provides a recipe for system design.

4.5.2.5 Transmitting Horn and Dipole Antenna Located Close to the Ground but Tilted Towards the Sky. Next, we place the transmitting horn closer to the earth at a 1-m height above the ground, but tilted slightly upward. Specifically, the horn's axis is tilted 0.58° towards the sky as shown in Fig. 4.1. The rationale for this example is the practical elimination of the interference between the undesired reflections of the electromagnetic fields from the ground and the direct fields from the horn. As a result of the vector nature of the problem, this interference pattern between the direct and the reflected fields, when integrated over the receiving dipole, will produce a smaller induced voltage and hence a smaller received power. These undesired reflected fields, typically called ground clutter, can be described exactly by the image of the source antenna due to the ground. Because the antenna is slightly tilted towards the sky, the fields from the image will be directed into the ground, thereby reducing the effect on the interference pattern between the source and its image. This reduced interference and associated increase in the induced voltage should result in slightly higher capacities than computed in previous simulations. Indeed, with the transmitting horn tilted 0.58° towards the sky, instead of being directed along the line-of-sight towards the receiving dipole as before, the received power across the load in the receiving dipole will be 32.3 μW for a 1 W of power fed to the transmitting horn.

For the same thermal noise floor of 2 pW as before, the Shannon channel capacity is 23.95 bits/Hz. For the same input power to the horn, the received voltage at the dipole is approximately 62 mV, and the Hartley capacity is 23.98 bits/Hz.

Now substituting a conjugately matched half-wave dipole for the transmitting horn with the same 0.58° skyward tilt as shown in Fig. 4.2, the power at the receiving dipole is 0.26 μW. For the same thermal noise floor as before, Shannon's capacity is 16.99 bits/Hz. The induced voltage is 5.5 mV, which implies a Hartley capacity of 17.01 bits/Hz. Thus, decreasing a transmitting antenna's height and slightly tilting it skyward, to minimize the interference between transmitted signal from the direct path and its image from the ground plane, increase the channel capacity. These results illustrate that the mathematical form of the channel capacity is irrelevant, if the model is properly described by Maxwellian physics and the input power in the simulation is constant. When using the power or signal amplitudes in a deterministic model, the final results will be very similar. In addition, the strictly mathematical form of the channel capacity will have a physical basis, which can be exploited to design a good transmit-receive system.

In summary, placing transmitting antennas in current communication systems closer to the ground not only reduces system cost, since an expensive tower need not be built, but also requires less transmitter input power to achieve a higher capacity. One can further increase the capacity by bringing the transmitting antenna closer to the ground and by tilting it slightly upward so as to minimize further the interference between the direct fields from the antenna and the fields produced by the image due to the presence of the ground plane. Physics-based explanations by vector Maxwellian principles are straightforward and simply illustrate that the antenna interference pattern decreases as the effective aperture produced by the source and the image decreases. Because the receiving antenna integrates the signal incident on the structure to produce a voltage at its feed point, a waveform that goes through many nulls will induce a smaller voltage than one which has fewer nulls in the pattern [12,35]. This simple picture, however, does not hold when the transmit and receive antennas are operating in a near-field environment, since no radiation pattern can be defined in that case. These examples illustrate how the performance of existing communication systems can be significantly improved, based on the amount of received power, by conducting analyses using Maxwell's equations. In contrast, computed results based solely on statistical analysis do not capture the entire physics, and so may provide numbers that are hardly influenced by the critical parameters that actually control the transfer of power.

4.5.2.6 Channel Capacity as a Function of the Height of the Transmitting Dipole Antenna from the Earth. Since the previous example indicates that the antenna height can significantly affect the capacity, we compute the channel capacity as a function of the height of the transmitting antenna above the ground. Let the transmitting and receiving antennas be two parallel, center-fed, vertically oriented, half-wave dipoles with 100 m separation and assume they are appropriately matched, so that each dipole has high radiation efficiency. Since (4.1) and (4.5) yield similar values for the capacity for matched systems, these equations may be interchangeably used in the ensuing computations. The height of the receiving dipole's feed is 2 m above a perfectly conducting earth, while the height of the transmitting antenna's feed is varied. In addition, the total input power to the transmitting antenna is 1 W, which implies that the power captured by the receiving antenna and hence the channel capacity vary as a function of the height of the transmitting antenna. The channel capacity, which appears to vary almost cyclically with height, achieves an absolute maximum value in excess of 17 bits/Hz when the antenna is closest to the earth as shown in Fig. 4.3.

This cyclical pattern shown in Fig. 4.3 can be explained in the following way. The presence of the earth creates an image of the transmit antenna. As the height of the transmit antenna changes, so does the interference between the electric fields from the original source and its image produced by the earth – hence the cyclical pattern. Similar results have been observed in [35]. The received power is obtained by integrating the total incident field on the receiving

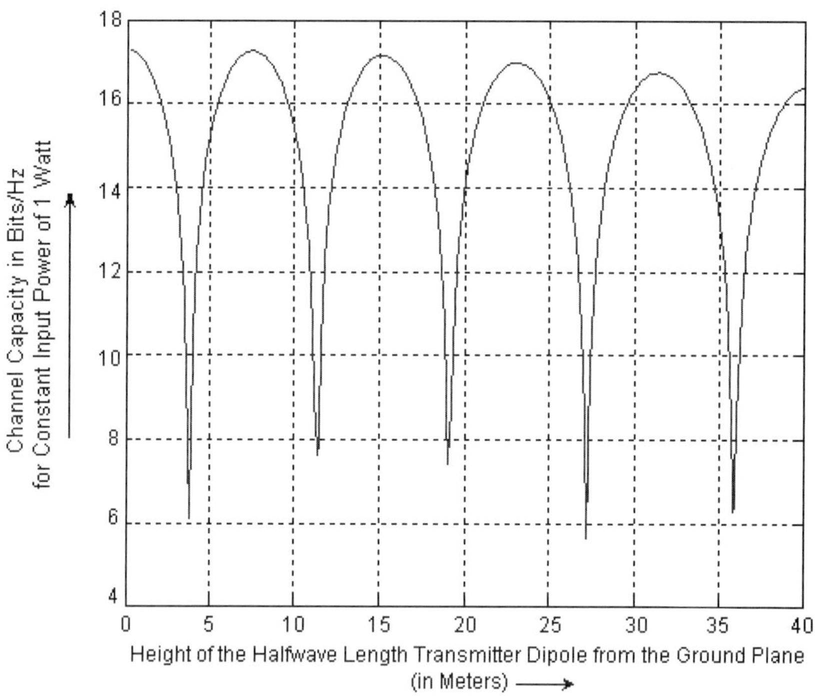

Figure 4.3. Plot of the variation of the channel capacity as a function of the height of the transmitting antenna above a perfectly conducting earth, for a fixed height of 2 m for the receiving antenna.

antenna. Although not depicted, similar plots have been generated for other antennas systems. This clear, accurate characterization of the channel capacity provides yet another demonstration of the immense value of incorporating the vector nature of electromagnetic wave propagation via Maxwell-Poynting theory!

So the best way to deploy transmitting antennas in a real environment for wireless communications is to locate them close to the ground plane, which is quite contrary to the popular belief that deployment of antennas on high towers provides better line-of-sight (LOS) coverage. This type of simplistic, electromagnetically naïve modeling that is not grounded in Maxwellian physics does NOT provide good solutions to real-life problems. One of the reasons why television transmitting antennas and microwave relay links are put on tall towers is to compensate for the spherical nature of the earth. As further demonstrated by the next numerical simulation, wireless communication does not obey strict LOS propagation characteristics.

4.5.2.7 Presence of a Dielectric Wall Interrupting the Direct Line-of-Sight Between Transmitting and Receiving Antennas. Assume the configurations, antennas, and specifications of section 4.5.2.6, but with the direct LOS between the transmitting and receiving antennas interrupted by a finite dielectric wall that is 85 m from the transmitter. The wall has a dielectric constant of $\varepsilon_r = 2.0$ and is 5 m high, 3 m wide, and 6 cm thick. A sequence of four subexamples illustrates the effects of varying the transmitter height and the radiating antenna.

When the transmitting antenna is a dipole that is placed 20 m above the ground plane and the receiving dipole is located at a height of 2 m as shown in Fig. 4.4, the power and the magnitude of the voltage at the receiving dipole are 34.06 nW and 2.01 mV, respectively. Consequently, Shannon's channel capacity is 14.08 bit/Hz, essentially the same value without the dielectric wall and having a direct LOS between transmitter and receiver. When the transmitting dipole is tilted 11° below horizontal towards the receiving dipole, the power and the magnitude of the voltage at the receiving dipole are 36.09 nW and 2.07 mV, with an associated Shannon channel capacity of 14.17 bit/Hz. Clearly, tilting the transmitting dipole and inserting the dielectric wall have little impact for a transmitter height of 20 m.

The situation is very different if the transmitting dipole is lowered to 2 m above the ground as shown in Fig. 4.4. The power at the receiving dipole in the presence of the dielectric wall significantly increases to 169.63 nW from 139.72 nW when the dielectric wall is not present. In addition, the magnitude of the received voltage increases from 4.07 mV to 4.49 mV, and the channel capacity increases from 16.12 to 16.4 bit/Hz. If the transmit dipole is lowered to 1 m above the ground and is tilted skyward by 0.58°, the received power increases from 260.64 nW to 300.17 nW in the presence of the dielectric wall, the magnitude of the received voltage increases from 5.56 mV to 5.97 mV, and the channel capacity increases from 17.02 to 17.22 bit/Hz. This clearly shows that bringing the antenna closer to the ground may still increase the channel capacity even in the absence of LOS propagation. A word of caution: in some

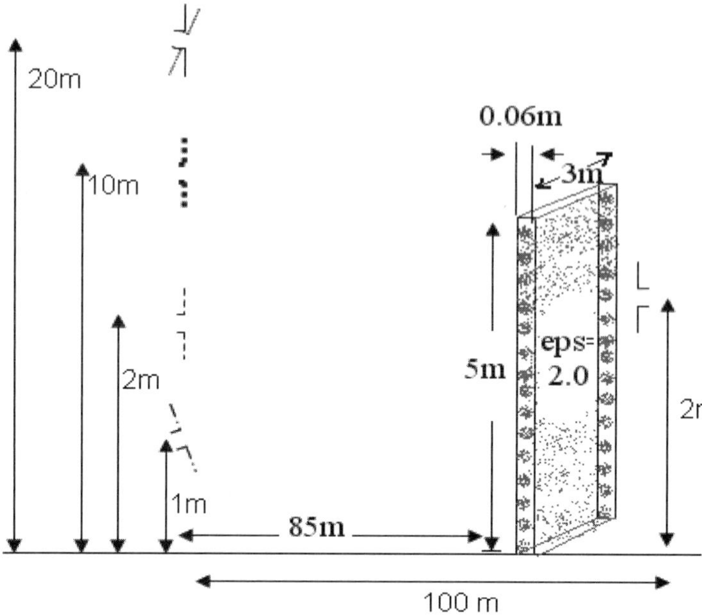

Figure 4.4. Different transmitting dipole configurations at a height of 20 m, 20 m tilted downwards at an angle of 11°, 10 m, 2 m, and 1 m tilted upwards by an angle of 1° above the ground plane. The receiving dipole is located at 2 m above the ground plane and at a horizontal distance of 100 m from the transmitter. The LOS component between the transmitter and the receiver is blocked by a dielectric wall.

instances the power at the receiving dipole antenna may be less as a consequence the vector nature of the problem!

The impact is even more dramatic when the transmitting dipole is replaced by a standard 20 dB gain horn. For example, even when the transmitting horn is 20 m above the ground and the LOS is blocked by the dielectric wall as seen in Fig. 4.5, the received power more than doubles from 1.26 μW to 2.88 μW, the magnitude of the received voltage increases by 52% from 12.24 mV to 18.5 mV, and the Shannon capacity increases from 19.3 to 20.5 bit/Hz. Furthermore, when the horn is tilted 11° down toward the receiving dipole with 1 W of transmit power, the received power increases dramatically from 5.27 μW to 9.127 μW, the magnitude of the received voltage increases from 25.02 mV to 32.9 mV, and Shannon's capacity increases slightly from 21.36 to 22.15 bit/Hz.

Next, when the horn is lowered to 2 m above the ground with 1 W of transmit power, the received power increases from 16.69 μW to 21.67 μW, the magnitude of the received voltage increases from 44.5 mV to 50.72 mV, and the Shannon channel capacity increases from 23.02 bit/Hz to 23.4 bit/Hz. Finally, when the horn is just 1 m from the ground and tilted upwards at 0.58° as seen in Fig. 4.5, with 1 W of transmit power, the received power increases from 32.35

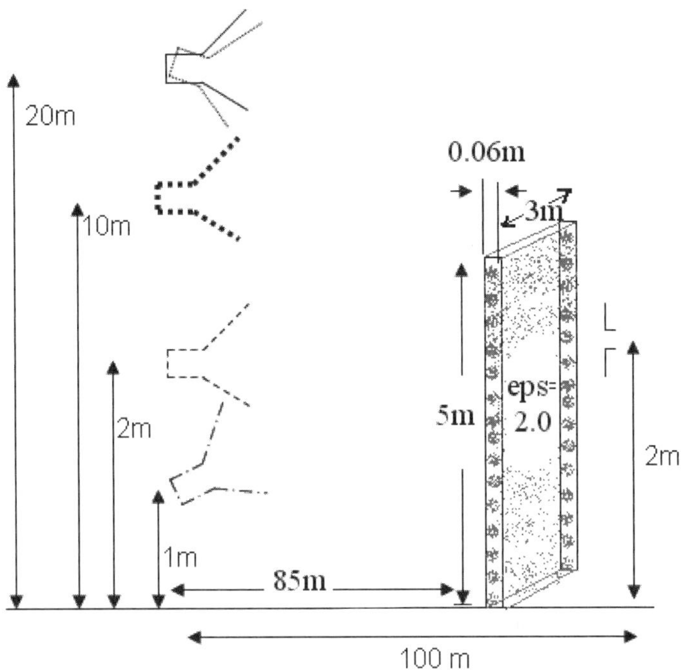

Figure 4.5. Different transmitting horn configurations at a height of 20 m, 20 m tilted downwards at an angle of 11°, 10 m, 2 m, and 1 m tilted upwards by an angle of 1° above the ground plane. The receiving dipole is located at 2 m above the ground plane and at a horizontal distance of 100 m from the transmitter. The LOS component between the transmitter and the receiver is blocked by a dielectric wall.

μW to 38.69 μW relative to the free-space value, the magnitude of the received voltage increases from 61.97 mV to 67.77 mV, and the capacity increases from 23.97 bit/Hz to 24.23 bit/Hz. So even the presence of an obstruction to the LOS mode of propagation does not mean that the received signal strength will be less. This non-intuitive result arises from the vector nature of electromagnetic propagation, where the diffracted and refracted fields may in fact add to provide a higher signal strength in some cases. Of course, this isolated example does not imply that such signal enhancements will occur all the time. For the particular problems under consideration, the nature of the signal strength is determined by the physical dimensions and electrical parameters of the system, and the results will be dictated solely by Maxwell's equations.

4.5.2.8 Increase in Channel Capacity when Matched Receiving Antenna Is Encapsulated by a Dielectric Box. For the next example, consider two parallel, half-wave, center-fed, vertically oriented, transmit-receive dipoles operating in free space over a perfectly conducting ground plane. The dipoles have a 1-mm radius and operate at 1 GHz. Initially, the bottoms of the transmitting and receiving dipoles are located at heights of 20 m and 2 m, respectively, above the ground. The basic differences between this example and the previous ones dealing with dipole transmitting antenna are: the feed of the transmit and receive dipoles are displaced vertically upward by 15 cm, the equivalent of a half wavelength; and the receive antenna is encapsulated by a dielectric box instead of having a wall interrupt the transmitter-receiver LOS. The outer dimensions of the dielectric box are: 1.5 m high and 2 m × 2 m in the cross-sectional plane. The thickness of the dielectric is 5 cm, and the bottom of the box is located 1.2 m above the ground as seen in Fig. 4.6. The dielectric constant of the box is assumed to be $\varepsilon_r = 2.5$ and has no loss tangent for simplicity.

A primary objective of this group of simulations is to gain insight into the impact of placing the receive antenna in a completely enclosed structure (building, container, etc.) vice on the other side of an obstruction (wall). In particular, the simulations illustrate that one can get very different values for the power induced in the receiving antenna, if correct vector Maxwellian physics is properly incorporated. Consequently, the basic question is: Which receiving dipole will have a larger induced voltage, hence larger received power from the transmitting antenna? – the receiving dipole in free space or the dipole encapsulated by the dielectric box? Equivalently, which receiving system will have a higher channel capacity? Since researchers in communication theory are more involved with space-time coding and intricate statistical details, they have neither addressed nor solved this system-related problem. Electromagnetic practitioners, who generally lack the computational resources to solve generic problems like the one presented above, may predict that the received power in the dielectric encapsulated dipole will be much less than when the receiving dipole is operating in free space along the direct LOS of the transmitting antenna. The possibility that the received power in the encapsulated dipole can be larger is often ignored as a possible solution in the contemporary wireless propagation literature. However, if one looks at a standard graduate text book on basic electromagnetic theory, it is quite clear that some situations do exist [p.370, 36] where the transmission coefficient of a system can actually exceed unity. This phenomenon generally occurs when the characteristic impedance of a medium is effectively greater than that of free space, which may occur near a parallel resonance. Under this condition, the radiated fields coming from the currents induced on the other side of the plate due to the field penetration through the aperture may add with the incident field going through the aperture. This may result in a higher value of the transmitted field, which does not imply that the transmitted power will be greater!

The following calculations address the issues raised by the preceding discussion. For example, the power induced at the receiving dipole operating in free space is 0.11 μW for a transmit input power of 1 W. When the receiving dipole is encapsulated by the dielectric box, the received power at the dipole is 0.128 μW for the same 1 W transmitter input power. The received power is considerably larger than before. It is also noteworthy that a shift of a half wavelength in the positions of the transmit and receive antennas also has an impact on the received power, even when the transmit powers are constrained to be the same. This clearly indicates the possibility that the presence of buildings may actually enhance the signals inside them in some cases. Since the voltage induced in the resistive load is 3.9 mV when the dipole is operating inside a dielectric box, as opposed to an induced voltage of 3.64 mV when the dipole is operating in free space with a direct LOS, the case of a larger induced power is substantiated. Moreover, Shannon's channel capacity increases from 15.8 bit/Hz in free space to 16.0 bit/Hz for the dielectric encapsulated dipole. It is important to note that these examples by no means provide a universal solution, but merely illustrate some of the possibilities when one employs the correct vectorial principles dictated by Maxwell's equations.

Next let the receiving antennas and their environments remain the same and lower the bottom of the transmitting dipole to 10 m above the ground as shown in Fig. 4.6. As discussed earlier [35], the induced power will be less than when the end of the transmitting antenna is located 20 m above the ground. Specifically, the induced powers at the receiving dipole in free space and when encapsulated by the dielectric box respectively are is 36.11 nW and 102.76 nW. As this simulation clearly indicates, situations exist for which the non-LOS power induced on a receiving antenna exceeds (significantly in this case) the received power achieved for a free-space transmit-receive configuration. For completeness, the induced voltage and the Shannon capacity for the free-space and dielectric-encapsulated cases increase from 2.07 mV to 3.49 mV and 14.17 bit/Hz to 15.68 bit/Hz, respectively.

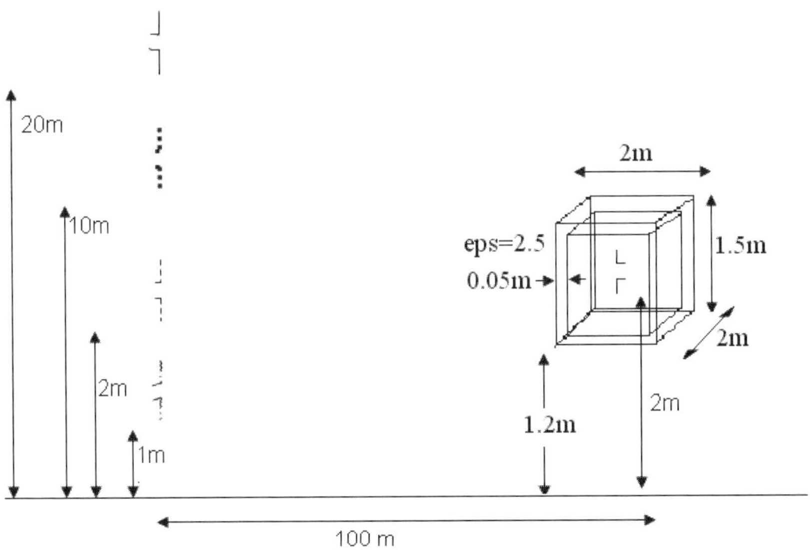

Figure 4.6. Different transmitting dipole configurations at a height of 20 m, 20 m tilted downwards at an angle of 11°, 10 m, 2 m, and 1 m tilted upwards by an angle of 1° above the ground plane. The receiving dipole is located at 2 m above the ground plane and at a horizontal distance of 100 m from the transmitter. The receiving dipole is enclosed by a dielectric shell.

Finally, the transmitting dipole is lowered so that its bottom is 2 m above the earth, while keeping the receiving dipoles at the same locations as seen in Fig. 4.6. For an input power of 1 W to the transmitting antenna, the received power (0.12 μW) in the free-space LOS mode is less than half the received power (0.281 μW) of the encased dipole. Similarly, the induced voltage increases from 3.77 mV to 5.78 mV, and the Shannon channel capacity increases from 15.9 bit/Hz to 17.13 bit/Hz. The last result is summarized in Fig. 4.7. As expected, the interference between the real source and its ground-plane image again causes cyclical behavior in the channel capacity as the height of the transmit dipole increases while keeping the receiving dipole in a dielectric box. However, what is most intriguing is that if one overlays a plot of the capacity for the transmit-receive configuration in the free-space LOS mode, it is always lower than the capacity for the dielectric-encapsulated receiving dipole! This phenomenon is not observed by a standard application of probability theory. Thus, it is clear that simple application of probability theory, without including the Maxwellian principles of basic electromagnetics, does not provide a complete and accurate solution to the problem.

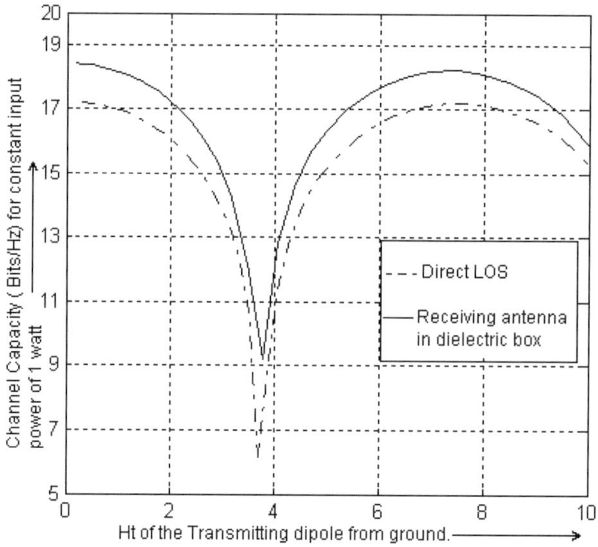

Figure 4.7. Comparison of the plots of the variation of the channel capacity as a function of the height of the transmitting antenna above a perfectly conducting earth, for a fixed height of 2 m for the receiving antenna. LOS stands for line-of-sight.

4.6 CONCLUSION

Current design and performance analyses for wireless communication systems usually assume the desired signals to be scalar quantities. On the basis of electromagnetic nature of the desired signals, the present chapter describes how the current design and analysis of communication system theory should be modified and improved by utilizing Maxwell's equations and appropriate results obtained therefrom for radiation and reception of electromagnetic signals. To this end, a historical overview of entropy has been presented to illustrate how the concept of information content of the signals evolved from the second law of thermodynamics. It is important to note that entropy is intimately related to the channel capacity as defined; however, they are not sufficiently useful for the design of a system.

For reliable and efficient design of a wireless communication system two forms of the channel capacity are described: the first based on the received voltages used in Hartley's law and the second on the received power used in Shannon's channel capacity formula. It is shown that both forms yield similar results if the Maxwellian theory is properly used to account for certain parameters involved. However, form an engineering point of view, Hartley's Law is more useful as it is applicable in both near-field and far-field environments. As explained in the text, Shannon's formula is not appropriate to use in the near field environment. Currently, Hartley's law is seldom used!

The limited numerical simulations discussed in the chapter demonstrate that a modern wireless communication system performance can be significantly improved by developing and deploying systems based on considerations of Maxwell's equations. It appears that deploying transmitting antennas closer to the ground would eliminate the use of expensive towers and also provide same channel capacities with reduced power input to the transmitter. It is important to appreciate that with cellular and wireless communication one often encounters the near field environment. In such cases, Maxwell's equations can be used to numerically account for the specific effects.

These subtle points are generally missed by the scalar statistical formalism used in communication theory, as it cannot handle the real nature of the antennas dictated by the vector Maxwellian equations, but rather treats them as unrealistic point sources which are devoid of any near-field concepts.

In summary, the goal of the present chapter is to establish the following principles:

1. Wireless Communication utilizes electromagnetic waves. Hence any analysis of system performance should use the vector nature of the signals. Therefore, Maxwell's equations are the logical choice to the analysis.
2. Shannon's expression for channel capacity was envisioned for wired systems. Hence, its application to wireless systems must be made with extreme care.
3. For wireless systems, the authors strongly suggest that Hartley's form of the channel capacity be used instead of Shannon's, because Hartley's law utilizes the fields in its expressions and most receiving wireless systems are sensitive to the electric field.
4. In comparing different communication systems in terms of the quality of service, the gain of the antenna must be used instead of their

directivity; this is because the total input power to the system must be kept the same and not the radiated power.

5. Hartley's form of the channel capacity can be applied very simply to near-field and far-field situations.
6. Care must be taken in using the Shannon's form of the channel capacity in the calculation of power when the receiver is located in the near field of the transmitter.
7. Both Hartley's law and Shannon's formula provide similar results for the capacity when the antennas are conjugately matched. If the antennas are not matched then in a near field scenario Shannon's formula should not be used.

4.7 APPENDIX: HISTORY OF ENTROPY AND ITS EVOLUTION

In this appendix, a brief history of information theory and its link to entropy and therefore to the theory of heat quantified mathematically through thermodynamics is provided. This appendix has been developed primarily from references in *Wikipedia* and various historical articles like [24], [31], [37]–[40]. In many instances, we have directly quoted the references without changing any words so as not to distort the intents of the original authors and consequently the historical perspective.

Basic physical notions of heat and temperature were established in the 1600s, and scientists of the time appear to have thought correctly that heat is associated with the motion of microscopic constituents of matter [6]. On pursuing this line of reasoning in 1698, engineer Thomas Savery built the first steam engine. Soon others followed, as Thomas Newcomen developed his steam engine in 1712 and Nicholas Joseph Cugnot developed the steam tricycle in 1769 [6]. These early engines, however, were inefficient because they converted less than 2% of the input energy into useful work output. In other words, large quantities of coal and wood had to be burned to yield a small fraction of work output. Hence the need for a new science of engine dynamics was born [6]. In the 1700s, it became widely believed that heat was instead a separate fluid-like substance, and in 1783 Antoine Lavoisier proposed the *caloric* theory. In the caloric theory, it was assumed that heat consists of a fluid called "caloric" that flows from hotter to colder bodies.

In 1803, physicist and mathematician Lazare Nicholas Marguerite Carnot [41], nicknamed "*the organizer of the victory*" for services rendered during the French revolution and whose protégé was Napoléon Bonaparte, published a work entitled *Fundamental Principles of Equilibrium and Movement.* This work includes a discussion on the efficiency of fundamental machines (pulleys and inclined planes). In his work, Carnot developed a general discussion on the conservation of mechanical energy [41]. Over the next three decades, Carnot's theorem was taken as a statement that in any machine the accelerations and shocks of the moving parts all represented losses of *moment of activity*, or the useful work done. From this Lazare drew the inference that perpetual motion was impossible. This *loss of moment of activity* was the first-ever rudimentary statement of the Second Law of Thermodynamics and the concept of "transformation energy" or entropy (energy lost to dissipation and friction).

After Carnot's death in exile in 1824, his son Nicolas Leonard Sadi Carnot [42], having graduated from the Ecole Polytechnique training school for engineers but then living on half pay with his brother Hippolyte in a small apartment in Paris, wrote *Reflections on the Motive Power of Fire*, wherein he defined "motive power" to be the expression of the *useful effect* that a motor is capable of producing. In addition, Sadi Carnot introduced the first modern-day definition of "work": *weight lifted through a height* [6,42]. He gave the first successful theoretical account of heat engines, now known as the Carnot cycle, thereby laying the foundations for the Second Law of Thermodynamics. In his paper, Sadi Carnot visualized an ideal engine in which the caloric converted into work could be reinstated by reversing the motion of the cycle, a concept subsequently known as thermodynamic reversibility. He thus showed that there was a fundamental limit to how much energy could be extracted from a heat engine. He further postulated, however, that some calorics are lost by not being converted to mechanical work. Hence no real heat engine could realize the Carnot cycle's reversibility and thus was condemned to be less efficient. The lost caloric content was a precursory form of entropy loss as we now know it. Later this result was generalized to the Second Law of Thermodynamics [6,24,41,42].

The name *thermodynamics*, however, did not exist until some twenty-five years later when, in 1849, the British mathematician and physicist William Thomson (Lord Kelvin) coined the term *thermodynamics* in a paper on the efficiency of steam engines. In 1850, Swiss physicist Rudolf Clausius [43] gave the first statements of the First and Second Laws of Thermodynamics, abandoning the caloric theory, but preserving Carnot's principles – the concept of reversibility and the consequence that mechanical work cannot be obtained from heat in the absence of the existence of a temperature difference. The Second Law of Thermodynamics states that *the total entropy of any isolated thermodynamic system tends to increase over time, approaching a maximum value*. The classical definition of entropy is given by dQ/T, where dQ is the quantity of heat to be given up or absorbed by the body and T is the absolute temperature of the body at that moment. In Clausius' first published memoir, he presented a verbal argument as to why Carnot's theorem, proposing the equivalence of heat and work ($Q = W$), was not perfectly correct and as such would need amendment. In 1854, Clausius stated [6,43]: "*In my memoir* On the Moving Force of Heat, and so on, *I have shown that the theorem of the equivalence of heat and work, and Carnot's theorem, are not mutually exclusive, by that, by a small modification of the latter, which does not affect its principle, they can be brought into accordance*." This small modification of Carnot's theorem is what developed into the Second Law of Thermodynamics. The idea that gases consist of molecules in motion had been discussed in some detail by Daniel Bernoulli in 1738, but had fallen out of favor, and was revived by Clausius in 1857. Perfect irreversibility was not possible as there was always a residual heat loss. In 1865, Clausius gave this heat loss a name [43]: *I propose to name the quantity S the entropy of the system, after the Greek word (trope), the transformation. I have deliberately chosen the word entropy to be as similar as possible to the word energy: the two quantities to be named by these words are so closely related in physical significance that a certain similarity in their names appears to be appropriate.* Clausius derived the concept of "entropy" as one of the characteristics specifying the state of a substance, as do its temperature, pressure, volume, and intrinsic energy. Entropy may be defined as the ratio of the heat energy intake or output in reversible changes of the condition of a body to the absolute temperature at which the change takes place. Another definition is possibly more illuminating. Since no actual engine can attain the efficiency of the ideal reversible engine, it follows that energy is wasted in all actual conversions of heat into mechanical energy. The change of entropy in such conversions may be taken as the measure of this unavoidable waste or equivalently as that entity which multiplied by the lowest available temperature gives the magnitude of the necessarily lost energy.

A finite universe may be considered an isolated system. As such, it may be subject to the Second Law of Thermodynamics, so that its total entropy is constantly increasing. It was speculated by Helmholtz in 1854 that the universe is fated to a heat death in which all the energy ends up as a homogeneous distribution of thermal energy, so that no more work can be extracted from any source [44]. Unlike almost all other laws of physics, Helmholtz' supposition associates thermodynamics with a definite arrow of time [45]. However, for a universe of infinite size, which cannot be regarded as an isolated system, the Second Law does not apply. Later, scientists like Ludwig Boltzmann, Willard Gibbs, and James Clerk Maxwell gave entropy a statistical framework based on the presence of the huge numbers of particles in macroscopic systems. For example, it is not impossible, in principle, for all 10^{23} atoms in a gas molecule to behave spontaneously in the same way, even though it may be highly improbable. In fact, this occurrence is so unlikely that no macroscopic violation of the Second Law has ever been observed.

Simultaneous to Clausius's research was the work of a Scottish mathematician and physicist James Clerk Maxwell [10,46], who formulated a new branch of thermodynamics called *Statistical Thermodynamics*, which proposed analyzing large numbers of particles at equilibrium; that is no changes occur in systems so that only the average properties of its defining quantities (temperature $\boldsymbol{T}$, pressure $\boldsymbol{P}$, volume $\boldsymbol{V}$) are important. Maxwell's discovery of the law of the distribution of molecular velocities in 1859 established the molecular theory for gases. From the mechanics of individual molecular collisions, he derived the expected distribution of molecular speeds in a gas. Over the next several years, the kinetic theory of gases developed rapidly, and many macroscopic properties of gases in equilibrium were computed. Maxwell was thus the first scientist to introduce a statistical law into the deterministic world of physics in 1861 through his paper *On Kinetic Theory of Gases* [6]. He also was the first to introduce the concept of *ensemble averaging* in 1879 through his paper *On the Average Distribution of Energy* [6]. Instead of characterizing the individual quantities of a system, his statistical theory evaluated a system's behavior through distributions or ensembles of each quantity. Thus the mathematical operation to which a function is subjected is judged not by its effect in a particular case, but by its average effect.

In 1872, the Austrian physicist Ludwig Boltzmann [47] stated his equation for the temporal development of distribution functions in phase space (a space in which all possible states of a system are represented). In 1877, he stated the relationship between entropy and probability through

$$S = -k_B \sum \left[P_i \log(P_i) \right] \tag{4.6}$$

In this equation, P_i is the probability that particle i will be in a given microstate. Each P_i is evaluated for the same macrostate of the system, and $\sum$ is a mathematical instruction to sum everything to the right of it. In this case, it means add the product of P_i and *log* (P_i) for all i particles of a system. The log function can have a base of e, 2, or even 10. The parameter k_B is an arbitrary constant which determines the units of measure of entropy. In thermodynamics, k_B is *Boltzmann's constant* (1.380658×10^{-23} Joules/Kelvin), but its value could just as easily be arbitrarily set to 1 without affecting the generality of the arguments presented here. Since $0 < P_i < 1$, the negative sign ensures the positivity of S by cancelling the negativity of each logarithm in the summand [47].

Contrary to the classical definition of entropy, which is heat energy divided by temperature, neither the temperature nor the heat energy explicitly appears in this equation. However, the restriction that all microstate probabilities must be calculated for the same macrostate assures that (as in the earlier case) the system must be in a state of thermal equilibrium. Equation (4.6) treats the microstate probabilities P_i individually. However, if all of the probabilities are the same, then one can simplify equation (4.6) to

$$S = k_B \log(M) \tag{4.7}$$

So what Boltzmann [6,47] accomplished was to visualize a probabilistic way to measure the entropy of an ensemble of ideal gas particles, in which he defined entropy to be proportional to the logarithm of the number of microstates M such a gas could occupy. Henceforth, the essential problem in statistical thermodynamics, according to Erwin Schrödinger, has been to determine the distribution of a given amount of energy E over M identical systems [6]. In this simplified form in the probability equation, the essential parameter of interest is M, the total number of microstates available to the system. It is important to note that M is *not* the total number of particles, but rather is the total number of microstates that the particles could occupy, with the constraint that all such microstate collections show the same macrostate. The basis of this explanation is found in the atomic structure of matter and in the quantum laws that provide for the stability of atoms, molecules, and crystals. A gram-atom of matter contains about 6×10^{23} actual atoms. Accordingly, one is completely unable to follow the motion of the individual atoms. In practice, the only observable properties are their average values obtained from statistical considerations. The Boltzmann formula shows the relationship between entropy and the number of ways the atoms or molecules of a thermodynamic system can be arranged.

The Second Law can also be described at the macroscopic level of everyday objects or at the microscopic level of tiny particles (like atoms and molecules), which is the most important level for most of our questions about origins. At the microscopic level, the entropy of a system is a property that depends on the number of ways that energy can be dispersed among the particles in the system. Therefore entropy is a measure of probability, since a particular state is more probable if energy can be dispersed in more ways in that state. For example, the chemicals in a system eventually tend to end up in the particular state (equilibrium) that can occur in the greatest number of ways when energy is dispersed among the multitude of molecules. Consequently, the Second Law is a probabilistic description which states that the entropy of an isolated system will increase during any reaction. More generally, the Second Law states that the entropy of *the universe* will increase during any reaction. An isolated system (one which cannot exchange energy or matter with its surroundings) is thermodynamically equivalent to a miniature universe, so during a reaction its entropy must increase. But the entropy of an open system (one which can exchange energy with its surroundings) can increase, decrease, or remain constant.

In 1876, a chemical engineer named Josiah Willard Gibbs [48,49], the first person in America to be awarded a PhD in engineering from Yale University, published an obscure 300-page paper entitled, *On the Equilibrium of Heterogeneous Substances*, wherein he formulated one grand equality, the Gibbs free energy equation, which gives a measure of the amount of "useful work" attainable in reacting systems. In this first paper of his trilogy on thermodynamics, Gibbs examined the possibilities of representing the properties of bodies and the work and heat relations of various processes, as expressed analytically by means of diagrams in a plane. Until the appearance of his paper, only a diagram using pressure and volume for the coordinates had been used. Gibbs introduced new diagrams with temperature and entropy as coordinates which had wide use in engineering. Through the new "molecular theory" of gases, Gibbs also extended the ideas of classical thermodynamics into the domain now called *statistical mechanics*. In classical thermodynamics, one deals with extensive individual systems; whereas one recognizes the role of the tiny constituents of the system in statistical mechanics. The temperature, for instance, of a system defines a *macrostate*, while the kinetic energy of each molecule in the system defines a *microstate*. Macrostate variables like temperature are recognized as expressions of the average of the microstate variables, an average kinetic energy for the system. Hence, if the molecules of a gas move faster, they have more kinetic energy, and the temperature naturally increases. In statistical mechanics, the Gibbs entropy formula [48,49] is the standard way to calculate the statistical mechanical entropy

$$S = -k_B \sum P_i \log(P_i) \tag{4.8}$$

of a thermodynamic system, where the summation is taken over the possible states p_i of the system as a whole (typically a $6N$-dimensional phase space, if the system contains N separate particles) and k_B is Boltzmann's constant. This calculation assumes no coherent quantum correlations exist between the probabilities. Even though the variables in Equations (4.6)–(4.8) define slightly different physical quantities, the equations are essentially the same!

Statistical mechanics [50,51] thus evolved as an application of probability theory, where one is dealing with large populations, into the field of mechanics, which is concerned with the motion of particles or objects when subjected to a force. Statistical mechanics provides a framework for relating the microscopic properties of individual atoms and molecules to the macroscopic or bulk properties of materials that can be observed in everyday life, by explaining thermodynamics as a natural result of statistics and mechanics (classical and quantum) at the microscopic level. In particular, statistical mechanics can be used to calculate the thermodynamic properties of bulk materials from the spectroscopic data of individual molecules. This ability to make macroscopic predictions based on microscopic properties is the salient feature that distinguishes statistical mechanics from thermodynamics. Both theories are governed by the Second Law of Thermodynamics through the concept of entropy. However, entropy in thermodynamics can only be known empirically; whereas in statistical mechanics, it is a function of the distribution of the system on its microstates. In statistical mechanics, Maxwell-Boltzmann statistics [10] describe the distribution of material particles over various energy states in thermal equilibrium, when the temperature is high enough and the density is low enough to render quantum effects negligible. Therefore, Maxwell-Boltzmann statistics are applicable to almost any terrestrial phenomena for which the temperature is above a few tens of degrees Kelvin. Statistical mechanics describes entropy as the amount of uncertainty (or "mixedupness" in the phrasing of Gibbs) associated with a system, after its observable macroscopic properties have been taken into account. For a given set of macroscopic quantities, like temperature and volume, the entropy measures the degree to which the probability of the system is spread over different possible quantum states. The more states with higher probability that are available to the system will result in greater entropy. In essence, the most general interpretation of entropy is that it is a measure of our ignorance about a system. The equilibrium state of a system maximizes the entropy because all information about the initial conditions is lost, except for the conserved quantities. Consequently, maximizing the entropy maximizes our ignorance about the details of the system [6]. In summary, the statistical interpretation of entropy was given by J. C. Maxwell, L. Boltzmann, and J. W. Gibbs, and was later completed by M. Planck [10].

According to Planck, a closed isolated system may have been created artificially with a very improbable structure. When left to itself, it will progressively follow a normal evolution toward a more probable structure. The probability of settling down to a stable structure has a natural tendency to increase and so does the entropy. The exact relation between entropy and probability states is given by the Boltzmann-Planck formula, the extension of Boltzmann's equation to quantum levels [50,51],

$$S = k \log\left(P_c\right) \tag{4.9}$$

where $k = 1.38\times10^{-16}$ ergs/°C (Boltzmann's constant), S is the entropy of the system under discussion, and P_c is the number of elementary complexions. In Quantum theory [50], an atom or a molecule can only exist in a finite number of distinct configurations of quantized physical states, each of which was called a complexion by Planck. Since only discrete stable (or metastable) structures exist at the quantum level, an atomic system suddenly jumps from one structure to another when absorbing or emitting energy.

In 1909, Carathéodory [51] linked entropy with a mathematical definition of irreversibility, in terms of trajectories and integrability. Until his work, proofs of the Second Law were based on thoughts and physical experiments associated with a perfect Carnot engine, or on statistical mechanics. Although rigorous, the scope of these prior theoretical approaches was limited. Carathéodory's generalization focused on the irreversibility of thermodynamic processes, rather than on entropy increases per se. By doing so, he circumvented the difficult problem of measuring entropy or entropy production in non-equilibrium situations. Since entropy production can be calculated only in systems that are at or will achieve equilibrium, Carathéodory's theoretical observations enabled subsequent researchers to rid themselves of the tight constraints imposed by previous measurements of entropy and provided greater theoretical clarity for the further exploration of thermodynamical systems.

In his 1927 generalization of Boltzmann-Planck formula [52], von Neumann introduced the concept of entropy in quantum statistical mechanics, the study of statistical ensembles of quantum mechanical systems. Although Quantum Mechanics is different from Statistical Mechanics, both are characterized in terms of statistics and probabilities. Therefore, entropy is handled by the two disciplines in much the same way. The quantum mechanical definition of entropy is identical to statistical mechanics definition in (4.6), with the only real difference being how the various probabilities are calculated. Even though Quantum Mechanics has its own peculiar rules for calculating probabilities, these rules are not relevant to the fundamental definition of entropy. As in the previous discussions, P_i are the microstate probabilities and must be calculated for the same microstates.

Qualitatively, entropy is often associated with the amount of *seeming* disorder in a system. For example, solids (which are typically ordered on the molecular scale) usually have smaller entropy than liquids [6], and liquids similarly have smaller entropy than gases. This phenomenon is attributable to the fact that the number of different microscopic states available to an ordered system is usually much smaller than the number of states available to a system that appears to be disordered. The Second Law of Thermodynamics forbids two bodies of equal temperature, brought in contact with each other and isolated from the rest of the Universe, from evolving to a state in which one of the two has a significantly higher temperature than the other. The Second Law is also interpreted as the assertion that entropy never decreases in an isolated system.

In opposition to the Second Law of Thermodynamics, Maxwell posed a thought experiment in 1867, wherein the Second Law is violated by the activities of a creature termed *Maxwell's demon* by Lord Kelvin [46]. Maxwell imagined two containers *A* and *B*, filled with gas at the same temperatures and placed next to each other. A little "demon" guards a trapdoor between the two containers and observes the molecules on both sides. When a faster-than-average molecule from *A* flies towards the trapdoor, the demon opens it and the molecule flies from *A* to *B*. Thus, the average speed of the molecules in *B* increases, while the molecules in *A* slow down on average. However, since average molecular speed corresponds to temperature, the temperature in *A* and *B* will decrease and increase, respectively, which is contrary to the Second Law of Thermodynamics. Maxwell's *thought experiment* has troubled physicists ever since he first published it. The question is: *Is Maxwell correct? Or equivalently, could such a demon, as he described it, actually violate the Second Law of Thermodynamics*?

Several physicists have presented calculations which show that the Second Law of Thermodynamics will not actually be violated, if a more complete analysis is made of the whole system, including the demon. The essence of these physical arguments is to show by calculation that any demon must "generate" more entropy segregating the molecules than it could ever eliminate by the method described. One of the most famous responses to this question was provided by Leó Szilárd [10,53], the inventor of nuclear fission, and later by Léon Brillouin [10]. During 1929 when he filed a patent for the cyclotron and was working with Einstein on the development of a home refrigerator without any moving parts, Szilárd noted that a real-life Maxwell's demon would need to have some means of measuring molecular speed and that the act of acquiring this information would require an expenditure of energy [10,53]. The Second Law states that the total entropy of an isolated system must increase. Since the demon and the gas are interacting, one must consider the total combined entropy of the gas and the demon. The expenditure of energy by the demon will cause an increase in the entropy of the system, which will be larger than the lowering of the entropy of the gas. For example, if the demon is checking molecular positions using a flashlight, the flashlight battery is a low-entropy device, a chemical reaction waiting to happen. As the flashlight's energy is depleted by emitting photons (whose entropy must now be counted as well!), the battery's chemical reaction will proceed and its entropy will increase, more than offsetting the decrease in the entropy of the gas. Put simply, no matter how it is done, both the act of the demon watching molecules and the act of opening and closing the trapdoor are by definition *work* and require the expenditure of energy. This observation by Szilárd is probably the earliest link between the concept of entropy and the information content of a system. Moreover, he showed that the entropy of a unit of information was equal to $k \ln 2$ (= $k \log_e 2$), where k is a constant [53]. In addition, relating entropy to information content, Szilárd provided the first mathematically quantifiable experiment of the relationship. It is interesting to note that this was the *only* paper written by Leó Szilárd on this topic!

In parallel to Szilárd's work, electrical engineers were also approaching the problem from the point of view of communication systems [3,24,37-40]. By 1920, one can safely say that telegraphy as a practical technological discipline had reached a mature level. Basic problems related to sending and receiving apparatuses, transmission lines, and cables were well understood. Even wireless transmission had been routine for several years. At this stage of development, when only small increases in efficiency were gained by technological improvements, it was natural to ask whether one was close to fundamental limits and to try to understand these limits. To this end, Harry Nyquist [25,26] seems to be the first person to touch upon, if not fully identify, some of the issues that were clarified by Shannon some twenty years later [24]. First, it was obvious to Nyquist that the "speed of transmission of intelligence," which he terms W, is limited by the bandwidth of the communication channel. Without much formal mathematical argument, Nyquist derived the following approximate formula for W

$$W = K \log m \tag{4.10}$$

where m is the "number of current values," which in modern terms is called "the size of the signaling alphabet," and K is a constant.

Whereas Nyquist's paper is mostly concerned with practical issues, such as choice of pulse waveform and different variants of the Morse code, a paper presented three years later by Hartley [20,24] on "Transmission of Information" is more fundamental in its approach to the problem. In the first paragraph, Hartley stated, "*What I hope to accomplish is to set up a quantitative measure whereby the capacities of various systems to transmit information may be compared.*" He also stated that the rate of transmission of information is limited by the distortion resulting from the storage of energy [20,24]. When the storage of energy is used to restrict the steady-state transmission to a limited range of frequencies, the amount of information that can be transmitted is proportional to the product of the width of the waveform's spectral occupancy (bandwidth) and its temporal duration (pulse duration). Hartley's observation is the first time the importance of the time-bandwidth product of a waveform was considered [24]. He further concluded that *the point of view developed in that it provides a ready means of checking whether or not claims made for the transmission possibilities of a complicated system lie within the range of physical possibility* [24]. Moreover, he formulated what would later be known as *Hartley's law*: information content is proportional to the product of time and bandwidth, and that two quantities provided a trade-off space for designing waveforms. It should also be mentioned that Hartley argued that the theory for telegraph signals (digital signals in modern terminology) could be generalized to continuous-time signals like speech and television [24]. Hartley's law can be written as

$$\textit{Amount of information} = \textit{const} \times B \times T \times \log m \tag{4.11}$$

where B is bandwidth, T is the time duration of the waveform, and m is number of distinguishable amplitude levels. The concept of entropy in information theory describes how much randomness (or, alternatively, "uncertainty") is present in a signal or random event. An alternative way to view this is to talk about how much information is carried by the signal. A relation between bandwidth B and time duration T, similar to the one found by Nyquist, was discovered simultaneously by Karl Küpfmüller in Germany [24,27].

In 1946, Gabor [18,23] presented an analysis which was more quantitative in nature than Hartley's purely qualitative reasoning. As noted by William G. Tuller, a graduate student from Massachusetts Institute of Technology who wrote his dissertation on this topic [23], a deficiency of the theories of Nyquist, Hartley, Küpfmuller, and Gabor was that their formulas did not include the effect of background random noise [24]. They considered the limiting features of information content to be the discretization noise of the signal. However, background random noise plays an essential role by setting a fundamental limit to the number m of levels that can be reliably distinguished by a receiver. Both Nyquist and Hartley [24] were aware of the fact that the amount of information depends on the number of distinguishable signal levels (or symbols). However, they merely included m in their formulas instead of deriving it from a more fundamental quantity, such as the signal-to-noise level [24]. This oversight may be attributed to the fact that Hartley and Nyquist were engineers, and in practical wireless communication problems, the sensitivity of the receiving antenna is quantified by the electrical field intensity at the receiver, as opposed to using the signal-to-noise ratio (SNR) at the receiver as done by Shannon who was a mathematician. In a short discussion, Nyquist mentions "interference" as one of the limiting factors of the signal alphabet size. In addition, Hartley also notes that inter-symbol interference from channel distortion is the most important limiting factor. Both are fundamentally wrong, as Tuller remarks [23], in that random background noise often plays an important role. The background noise is not the quantization noise of the signals under discussion, rather the signals of interest may possibly be embedded in thermal noise as in communications performed by the deep-space satellites to Jupiter or Saturn. For mobile wireless communication, the background noise is two orders of magnitude lower than the signal. In that case, the limiting factor for reliable communication is not the thermal noise but multipath and other undesired delayed reflections of the signal of interest. Hence, Tuller's remarks may be questionable for wireless communication systems.

Most of the work described thus far did not account for the effect of background thermal noise, where the signal amplitudes are on the same order of magnitude as the noise [24]. In addition to bandwidth, the SNR in a system plays a significant role for the capacity of information transmission. Noise is not a problem as long as it is sufficiently small compared to the signal amplitude. Attenuation along the transmission line is also not a problem as long as the signal is visible. However, what does one do when the signal attenuates below the noise amplitude [24]? Using an amplifier could rectify the situation, but an amplifier increases both the signal and the noise, which makes noise a limiting factor in transmission systems. According to Tuller [23], information transmission is in many ways identical to the problem of analyzing stationary time series. Stationary time series analysis was proposed by Norbert Weiner [54], who formulated communication theory as a statistical problem. In Wiener's formulation, an ensemble of functions is the appropriate mathematical representation of the messages produced by a continuous source of transmitted signals in the presence of perturbing noise. Thus in his view, communication theory is concerned not with operations on particular functions, but with operations on ensembles of functions [4]. In [54,55] Wiener addressed three important topics dealing with a theory of information, a theory for information content in a signal, and the transmission of this information through a channel, among other things. Wiener was, however, not a master of communicating his ideas to the technical community, and even though the relation to Shannon's formula is stated in [24,54,55], the notation is cumbersome and the relevance to practical communication systems is far from obvious.

By using the properties of sampling and quantization of a band-limited signal and by arguing that the inter-symbol interference introduced by a band-limited channel can in principle be eliminated, Tuller [23,24] correctly articulated in 1948 that under noise-free conditions an unlimited amount of information can be transmitted over such a channel. By using pulse code modulation (PCM), Tuller showed that the information H transmitted over a transmission link of bandwidth B during a time interval T with the *rms* (root-mean-square) amplitudes of the carrier-to-noise ratio C/N is limited by the inequality

$$H \leq 2BT \log\left(1 + C/N\right). \tag{4.12}$$

He arrived at the result, partly based on intuitive reasoning and partly on formal mathematics. It is important to note that, like Hartley and Nyquist, Tuller's results used *rms* values for the signal and noise rather than following Shannon's use of the concept of power [3]. This interesting and significant difference has been addressed through examples in sections 4.3 and 4.5.

Concurrently, in 1948, while working at Bell Telephone Laboratories, Claude Shannon set out to quantify mathematically the statistical nature of "lost information" in telephone-line signals [24]. Shannon developed the very general concept of information entropy in his classical 1948 paper [3], where it is defined by

$$S = -k\sum\left[P_i \log\left(P_i\right)\right] \tag{4.13}$$

Equation (4.13) is the same as the definition of entropy in statistical mechanics [30]. Through his Theorem 2 in Ref [3], Shannon proves that *this Boltzmann entropy is the only function which satisfies the requirements for a function to measure the uncertainty in a message* (*where a "message" is a string of binary bits*). In this case, the constant k is recognized as only setting the units; it is arbitrary, and can be set equal to unity without loss of generality. The parameter P_i is the probability for the value of a given bit (usually a binary bit, but not necessarily). In Shannon information theory, the entropy is a measure of the uncertainty over the true content of a message, but the task is complicated by the fact that successive bits in a string are not random, and therefore are not mutually independent in a real message. Shannon showed that the average number of bits per source symbol should be greater than or equal to the entropy of the source. Also note that "information" is not a subjective quantity here, but rather an objective quantity, measured in bits [24]. Roughly speaking, Shannon entropy is proportional to the minimum number of yes/no questions one has to ask to get the answer to some question. The statistical mechanical entropy is then proportional to the minimum number of yes/no questions one has to ask to determine the microstate, given that one knows the macrostate. Shannon shows that by sufficiently complicated coding systems it is possible to transmit binary digits at a rate

$$C = W \log_2 \left(1 + P/N\right) \tag{4.14}$$

where W is the bandwidth, P is the average transmitter power, N is the white thermal noise power, and C is the channel capacity in bits. This expression deals with mathematical modeling and analysis of a communication system rather than with physical sources and physical channels. Given an information source and a noisy channel, Shannon showed for the first time that the maximum number of bits required per symbol to represent fully the source and the maximum rate at which reliable communication can take place over the noisy channel [24]. In (4.14), the average output power is considered to be limited, but there is no constraint on the input power. Shannon's basic idea was to introduce redundancy in the transmitter, which he called the encoder. This redundancy will be exploited in the receiver, defined as the decoder, so that reliable transmission of information can be achieved in the presence of background noise. Shannon's work was both philosophically and mathematically different from any previous work and essentially laid the foundation for Information Theory [24]. However, for wireless transmit-receive systems, the model used by Nyquist, Hartley, and Tuller may be more appropriate in evaluating the capacity than that of Shannon's model!

In a similar fashion to Tuller, Clavier [24,28] began with Hartley's work and used PCM coding to obtain a formula that is essentially equivalent to (4.12) and Shannon's formula (4.14) for capacity using the power relationship. Another independent discovery of the capacity was made by Laplume in 1948 [24,56].

As the reader will hopefully have observed, research toward a general theory of communication had two major breakthroughs where several investigators made similar but independent discoveries. The first breakthrough came in the 1920s with the discovery of the relation between bandwidth, time, and information rate. The second breakthrough came twenty years later on how to extract the signal from the background noise which is comparable in amplitude to that of the signal. An important difference between the theories published during these two stages is that in the 1920s the signals were much larger in amplitude than the background noise. As explained by Lundheim [24]: Why did it take twenty years to fill the gap between Hartley's law and Shannon's formula?

The only missing step was to substitute m by $1 + C/N$ in equation (4.10). Why, all of a sudden, did three or more people independently and nearly simultaneously "see the light?" Why did neither Nyquist, Hartley, nor Küpfmüller realize that noise, or more precisely the SNR, plays as significant a role for the information transfer capacity of a system as does the bandwidth when the signals are deeply embedded in noise [24]?

One answer might be that they lacked the necessary mathematical tools for an adequate description of noise [24]. At that time, Wiener had just completed a series of papers on Brownian motion (1920–24) that would become a major contribution to what was later known as stochastic processes, the standard models for description of noise and other unpredictable signals, and one of Shannon's favorite tools of analysis. These ideas, based on probabilistic concepts, were however far from mature to be used by even the most sophisticated electrical engineers of the time [24]. Against this explanation, it may be argued that when Shannon's formula was discovered, only two of the independent researchers used a formal probabilistically based argument. The others based their reasoning on more common-sense reasoning, not resorting to mathematical techniques other than the popular tools of the trade in the 1920s [24]. However, one may offer another explanation for this twenty-year delay! The reason could be that Shannon was a mathematician and was using a stochastic model, because the only meaningful way to quantify any random signal is by its average power and not by its amplitude, which is what Shannon did. However, because the other researchers of that time were primarily addressing an engineering problem, where the sensitivity of a receiver is characterized by the minimum value of the electric field strength that an antenna is capable of detecting to produce an identifying signal which is larger than the background noise at the input of the first stage of the radio-frequency amplifier of a communication system, they had no need to pursue an alternate line of thought.

To complete the history of entropy, we now discuss the Maximum entropy school of thermodynamics (or more colloquially, the *MaxEnt* school of thermodynamics) that was initiated with two papers published in the Physical Review by Edwin T. Jaynes [6,11,57,58] in 1957. This group of physicists viewed statistical mechanics as an inference process: a specific application of inference techniques rooted in information theory, which relate not just to equilibrium thermodynamics, but also are general to all problems requiring prediction from incomplete or insufficient data (like image reconstruction, spectral analysis, and inverse problems). Central to the *MaxEnt* thesis is the principle of maximum entropy [6,11], which states that given certain *testable information* about a probability distribution (for example particular expectation values) that is not in itself sufficient to determine the distribution uniquely, one should prefer the distribution which maximizes the Shannon information entropy $S_I = -\sum p_i \ln(p_i)$. This principle is also known as the Gibbs algorithm to set up statistical ensembles to predict the properties of thermodynamic systems at equilibrium. A direct connection is thus made between the equilibrium thermodynamic entropy S_{Th}, a state function of pressure, volume, temperature, etc., and the information entropy for the predicted distribution with maximum uncertainty, conditioned only on the expectation values of those variables: $S_{Th}(P, V, T, \ldots) = k_B\, S_I(P,V,T,\ldots)$. The presence of Boltzmann's constant k_B has no fundamental physical significance here, but is necessary to retain consistency with the previous historical definition of entropy by Clausius.

In 1995, Tim Palmer [11,59] stated two unwritten assumptions about Shannon's definition of information that may make it inapplicable to quantum mechanics in its current form:

- the supposition that an observable state (for instance the upper face of a die or a coin) exists *before* the observation begins;
- the fact that knowing this state does not depend on the order in which observations are made (commutativity).

The article, "Conceptual inadequacy of the Shannon information in quantum measurement," [6,11] published in 2001 by Anton Zeilinger and Caslav Brukner [60], synthesized and further developed these remarks. The so-called Zeilinger's principle suggests that the quantization observed in Quantum Mechanics could be bound to *information* quantization – essentially, one cannot observe less than one bit, and what is not observed is by definition "random." These claims remain highly controversial as pointed out by Timpson [61].

In 1943, Erwin Schrödinger [62] used the concept of "negative entropy" in his popular-science book *What is life?* [62] Later, Léon Brillouin [10] shortened this expression to the single word, *negentropy*. Schrödinger introduced the concept when explaining that a living system exports entropy to maintain its own entropy at a low level [63,64]. By using the term *negentropy,* Brillouin expressed this fact in a more "positive" way: a living system imports negentropy and stores it. In a note to *What is Life?*, Schrödinger explains his usage of this term: *Let me say first, that if I had been catering for them* [physicists] *alone I should have let the discussion turn on free energy instead* [10,63]. Negentropy is the more familiar notion in this context, but this highly technical term seemed linguistically too near to *energy* for making the average reader aware of the contrast between these two physical quantities.

In 1974, Albert Szent-Györgyi proposed replacing the term *negentropy* with *syntropy,* a term which may have originated in the 1940s with the Italian mathematician Luigi Fantappiè, who attempted to construct a unified theory of the biological and physical worlds. His attempt has neither gained momentum nor borne great fruit since its postulation [63]. Buckminster Fuller attempted to popularize syntropy, though *negentropy* still remains common [63].

In 1964, James Lovelock [64] was among a group of scientists who were requested by NASA to create a theoretical detection system to look for life on Mars during an upcoming space mission. When contemplating this problem, Lovelock wondered *how can we be sure that Martian life, if any, will reveal itself to tests based on Earth's lifestyle?* To Lovelock, the basic question was "What is life, and how should it be recognized?" When speaking about this puzzling issue with some of his colleagues at the Jet Propulsion Laboratory, he was asked, *well what would you do to look for life on Mars?* To this Lovelock replied, *I'd look for an entropy reduction, since this must be a general characteristic of life.* Thus, according to Lovelock, to find signs of life, one must look *for a reduction or a reversal of entropy.*

Rolf Landauer [65] was an IBM physicist who in 1961 demonstrated that when information is lost in an irreversible circuit, the information becomes entropy and an associated amount of energy is dissipated as heat. Landauer realized that certain measurements need not increase thermodynamic entropy as long as they were thermodynamically reversible.

Szilárd's insight [9] into the working of Maxwell's demon was expanded by Charles H. Bennett in 1982 [66]. Bennett attempted to illustrate that computers may be thought of as engines for transforming free energy into waste heat and mathematical *work.* Due to the connection between thermodynamic entropy and information entropy, his thesis mandated that recorded measurements must not be erased. In other words, to determine what side of the gate a molecule must reside, the demon must store information about the state of the molecule. Bennett showed that, however well prepared, eventually the demon will run out of information storage space and must begin to erase the information it has previously gathered. Erasing information is a thermodynamically irreversible process that increases the entropy of a system. This discussion is relevant to reversible computing, quantum information, and quantum computing.

Thus far, we have considered entropy in its most common, well-known forms. Although most ordinary applications use one of these standard entropies, other forms of entropy exist [6,11]. For instance, Brazilian mathematician Constantino Tsallis [67] has derived a generalized form for entropy, which not only reduces to the Boltzmann-Gibbs entropy in equation (4.8) as a special case, but also can be used to describe the entropy of a system for which equation (4.8) would not work. Hungarian mathematician Alfréd Rényi [68] was able to construct a consistent definition of entropy for fractal geometries. Despite the other notions of entropy, the definitions of Tsallis and Rényi seem to be the most actively used. These relatively recent, generalized forms of entropy serve to show that "entropy" is not just an old friend that we know quite well, as in classical thermodynamics, but is also a concept that is rich in new ideas and scientific directions [6,11].

If the universe can be considered to have generally increasing entropy, then – as Roger Penrose has pointed out – an important role in the increase is played by gravity, which causes dispersed matter to accumulate into stars, which collapse eventually into black holes. Jacob Bekenstein and Stephen Hawking [69] have shown that black holes have the maximum possible entropy of any object of equal size. This makes them likely end points of all entropy-increasing processes, if they are totally effective matter and energy traps. Hawking has, however, recently changed his stance on this subject. Clearly, the role of entropy in cosmology remains a controversial subject. In 1975 Hawking and Bekenstein made a remarkable connection between thermodynamics, quantum mechanics, and black holes which predicted that black holes will slowly radiate, eventually losing all their energy. If black holes carried no entropy, it would be possible to violate the Second Law of Thermodynamics by throwing mass into the black hole. The only way to satisfy the Second Law is to admit that a black hole has entropy whose increase more than compensates for the decrease of the entropy carried by the object that was absorbed.

Recent work has cast considerable doubt on the Heat Death Hypothesis and the applicability of any simple thermodynamic model to the universe in general. Although entropy does increase in the model of an expanding universe, the maximum possible entropy rises much more rapidly and leads to an "entropy gap," thus pushing the system further from equilibrium with each time increment. Other complicating factors, such as the energy density of the vacuum and macroscopic quantum effects, are difficult to reconcile with thermodynamical models, making any predictions of large-scale thermodynamics extremely difficult. The relation between information entropy and thermodynamic entropy has become common currency in physics. Thus Stephen Hawking often speaks of the thermodynamic entropy of black holes in terms of their information content. Hence it is not surprising that computers must obey the same physical laws that steam engines do, even though they are radically different devices. Until 1995, no one was able to make a controlled calculation of black hole entropy based on statistical mechanics, that is, on counting the number of actual microstates.

The situation changed in 1995 when Andrew Strominger and Cumrun Vafa [70] calculated the right Bekenstein-Hawking entropy of a supersymmetric black hole in string theory, using methods based on D-branes. Their calculation was followed by many similar computations of entropy of large classes of other extremal and near-extremal black holes, and the result always agreed with the Bekenstein-Hawking formula.

To recapitulate, asking the question, *What is entropy?*, prompts an answer based on something that we said in the introduction [6]: *Entropy is what the equations define it to be.* Recall that entropy, an idea born from classical thermodynamics [8], is a quantitative entity, and not a qualitative one. Hence entropy is not something that is fundamentally intuitive, but something that is fundamentally defined via an equation, via mathematics applied to physics. Entropy does not have meaning without an equation to define it. Although one can postulate interesting interpretations for entropy-defining equations, the prose must be consistent with the equations [8]. In classical thermodynamics, the entropy of a system is the ratio of heat content to temperature, and the change in entropy represents the amount of energy input to the system which does not participate in mechanical work done by the system. In statistical mechanics where the entropy becomes a function of statistical probability, the interpretation is perhaps more general. In this case, the entropy is a measure of the probability for a given macrostate, so that a large value for entropy indicates a high probability state, and a low entropy value indicates a low probability state. Entropy is also sometimes confused with *complexity*, the idea being that a more complex system must have a higher entropy. In all likelihood, the opposite is probably true. A system in a highly complex state is probably far from equilibrium and in a low entropy (improbable) state, where the equilibrium state would be simpler, less complex and have higher entropy.

In summary, we have shown how the concept of information content is related to entropy and how different interpretations of entropy historically evolved from the principles of thermodynamics. These discoveries took place in two periods, 1924–1929 and 1948–1949. In the first period, Hartley [20], Nyquist [25,26], and Küpfmuller [27] presented a deterministic interpretation of information content of a signal based on the number of measurable discretization levels. During the same period, the connection between entropy and information was first presented by Szilárd in 1929 [9] when he explained the phenomenon of Maxwell's demon based on the principles of information content. Unfortunately, Szilard's pioneering work was not well understood at that time [9]. Twenty years later, the seminal research efforts of the first period were refined by Clavier [28], Shannon [3,4], Tuller [23], Wiener [54], Goldman [22], Laplume [56], and others. In particular, Shannon defined entropy with a negative sign, the opposite of the standard thermodynamical definition [10]. Hence, what Shannon calls entropy of information actually represents negentropy. The fundamental principle that determines channel capacity is intimately connected with entropy as was explained by Shannon and Von Neumann [52].

REFERENCES

[1] D. Gabor, "Communication Theory and Physics," *IEEE Trans. on Information Theory*, Vol. 1, No. 1, Feb. 1953, pp. 48-59.

[2] M. Schwartz, *Information, Transmission, Modulation, and Noise*, McGraw Hill Book Company, New York, 1959.

[3] C. E. Shannon, "A Mathematical Theory of Communication," *Bell System Technical Journal*, Vol. 27, July, 1948, pp. 379-423, and October, 1948, pp. 623-656.

[4] C. E. Shannon, "Communication in the Presence of Noise," *Proc. IRE*, Vo. 37 1949, pp. 10-21.

[5] R. Price and P. E. Green, "A Communication Technique for Multipath Channels," *Proc. Of the IRE*, Vol. 46, No. 3, March 1958, pp. 555-570.

[6] *http://en.wikipedia.org/wiki/Entropy*.

[7] T. K. Sarkar, R. J. Mailloux, A. A. Oliner, M. Salazar-Palma and D. Sengupta, *History of Wireless*, John Wiley & Sons, Hoboken, NJ, 2006.

[8] *http://www.tim-thompson.com/entropy1.html*

[9] L. Szilard, "On the Decrease of Entropy in a Thermodynamic System by the intervention of Intelligent Beings," *Behavioral Science*, Vol. 9, pp. 301-310, 1964, (Translated from *Zeits. Physik.*, Vol. 53, pp.84-856, 1929).

[10] L. Brillouin, *Science and Information Theory*, Academic Press, New York, 1956.

[11] *http://folk.uio.no/feder/Fys3130/Entropy/Entropy_in_thermodynamics_and_info.pdf*

[12] J. D. Kraus and R. J. Marhefka, *Antennas*, McGraw Hill, New York, Third Edition, 2002.

[13] R. C. Hansen, *Electrically Small, Superdirective, and Superconducting Antennas*, John Wiley and Sons, New York 2006.

[14] R. C. Hansen, *Phased Array Antennas*, John Wiley & Sons, New York, 1998.

[15] R. J. Mailloux, *Phased Array Antenna Handbook*, Artech House, Boston, 1994.
[16] C. T. Tai, "The Optimum Directivity of Uniformly Spaced Broadside Array of Dipoles," *IEEE Trans on Antennas and Propag.*, Vol. 12, 1964, pp. 447-454.
[17] N. Yaru, "A Note on Supergain Arrays," *Proc. IRE*, Vol. 39, pp. 1081-1085.
[18] D. Gabor, "Theory of Communication," *J. IEE*, Vol. 93, No. 3, 1946, pp. 429-457.
[19] A. Khinchin, *Mathematical Foundations of Information Theory*, Dover, New York, 1957.
[20] R. V. L. Hartley, "Transmission of Information," *Bell Syst. Techn. J.* Vol. 7, pp. 535 563, July 1928.
[21] H. S. Black, *Modulation Theory*, D. Van Nostrand Company Inc., New York, 1953.
[22] S. Goldman, *Information Theory*, Prentice Hall, Englewood Cliffs, NJ, 1953.
[23] W. G. Tuller, "Theoretical Limitations on the Rate of Transmission of Information," *Proc. of the IRE*, Vol. 37, No. 5, pp. 468-478, 1949.
[24] L. Lundheim, "On Shannon and 'Shannon's Formula'," *Telektronikk* (special issue on *Information Theory and Its Applications*), Vol. 98, No. 1-2002, 2002, pp. 20-29, ISSN 0085-7130, published by Telenor.
[25] H. Nyquist, "Certain Factors Affecting Telegraph Speed," *Bell Syst. Tech. J.*, Vol. 3, pp. 324-352, April 1924.
[26] H. Nyquist, "Certain Topics in Telegraph Transmission Theory," *AIEE Trans.*, Vol. 47, pp. 617-644, April 1928.
[27] K. Küpfmuller: "Über Einschwingvorgange in Wellen Filtern," *Elektrische Nachrichten- Technik*, Vol. 1, pp. 141-152, November 1924.
[28] A. G. Clavier, "Evaluation of Transmission Efficiency According to Hartley's Expression of Information Content," *Elec. Commun.: ITT Tech. J.*, Vol. 25, pp. 414-420, June 1948.
[29] *http://en.wikipedia.org/wiki/Signal_strength.*
[30] *http://en.wikipedia.org/wiki/DBm.*
[31] A. J. Viterbi, "Information Theory in the Sixties," *IEEE Trans. on Information Theory*, Vol. IT-19, No. 3, May 1973, pp. 257-262.
[32] *http://en.wikipedia.org/wiki/Eye_pattern.*
[33] *http://en.wikipedia.org/wiki/Constellation_diagram.*
[34] O. Ziv and J. H. Constable, "Interconnection Channel Capacity under Crosstalk Noise," *IEEE Trans. on Electromagnetic Compatibility*, Vol. 41, Nos. 4, 1999, pp. 361-365.
[35] T. K. Sarkar, S. Burintramart, N. Yilmazer, S. Hwang, A. De and M. Salazar-Palma, "A Discussion about some of the Principles/Practices of Wireless Communication under a Maxwellian Framework," *IEEE Trans. on Antennas and Propagation*, Vol. 54, Nos. 12, 2006, pp. 3727-3745.
[36] R. F. Harrington, *Time-Harmonic Electromagnetic Fields*, McGraw-Hill, 1961.
[37] E. C. Cherry, "A History of Information Theory," *Proc. Inst. Elec. Eng.* (London) , Vol. 98, Pt. 3, pp.383-393, Sept. 1951.
[38] J. R. Pierce, "The Early Days of Information Theory," *IEEE Trans. on Information Theory*, Vol. IT-19, No. 1, January 1973, pp. 3-8.
[39] S. Verdu, "Fifty Years of Shannon Theory," *IEEE Trans. on Information Theory*, Vol. 44, Nos. 6, October 1998, pp. 2057-2078.
[40] L. Varshney, "Engineering Theory and Mathematics in the Early Development of Information Theory," in *Proc. of the 2004 IEEE Conference on the History of Electronics*, Bletchley Park, England, June 2004.
[41] *http://en.wikipedia.org/wiki/Lazare_Carnot.*
[42] *http://en.wikipedia.org/wiki/Nicolas_L%C3%A9onard_Sadi_Carnot.*

[43] R. Clausius, *On the Motive Power of Heat, and on the Laws which can be deduced from it for the Theory of Heat.* Poggendorff's *Annalen der Physick*, LXXIX (Dover Reprint), 1850.
[44] *http://en.wikipedia.org/wiki/Heat_death.*
[45] *http://en.wikipedia.org/wiki/Entropy_%28arrow_of_time%29.*
[46] J. C. Maxwell, *Theory of Heat*, 1871, reprinted by Dover, New York, 2001.
[47] *http://en.wikipedia.org/wiki/Boltzmann's_entropy_formula.*
[48] *http://en.wikipedia.org/wiki/Willard_Gibbs.*
[49] J. W. Gibbs, *Elementary Principles in Statistical Mechanics*, Dover, New York, 1960.
[50] *http://en.wikipedia.org/wiki/Statistical_mechanics.*
[51] *http://omega.math.albany.edu:8008/cdocs/summer99/history-stat-mech/stat_m~1.htm.*
[52] *http://en.wikipedia.org/wiki/Quantum_statistical_mechanics.*
[53] *http://en.wikipedia.org/wiki/Le%C3%B3_Szil%C3%A1rd*
[54] N. Wiener, Extrapolation, Interpolation and Smoothing of Stationary Time Series with Engineering Applications, The MIT Press, Cambridge (MA), Wiley and Sons, New York, Chapman & Hall, London, 1949.
[55] N. Wiener, Cybernetics or Control and Communication in the Animal and the Machine, The MIT Press, Cambridge (MA), Wiley and Sons, New York, 1948. Second edition, revised, with two more chapters, The MIT Press, Cambridge (MA), Wiley and Sons, New York, 1961.
[56] J. Laplume, "Sur le nombre de signaux discernables en présence du bruit erratique dans un système de transmission à bande passante limitée," *Comp. Rend. Adac. Sci. Paris*, 226, pp 1348-1349, 1948.
[57] E. T. Jaynes, "Information Theory and Statistical Mechanics I," *Phys. Rev.,* Vol. 106, No. 4, pp. 620-630, 1957.
[58] E. T. Jaynes, "Information Theory and Statistical Mechanics II," *Phys. Rev.,* Vol. 108, No. 2, pp. 171-190, 1957.
[59] *http://www.answers.com/topic/entropy-in-thermodynamics-and-information-theory.*
[60] C. Brukner and A. Zeilinger, "Conceptual inadequacy of the Shannon information in quantum measurements," *Phys. Rev.* A, Vol. 63, 022113 (2001), 10 pages.
[61] C. Timpson, "On a Supposed Conceptual Inadequacy of the Shannon Information in Quantum Mechanics," *Studies in the History and Philosophy of Modern Physics*, Vol. 33, 2003, pp. 441-468.
[62] E. Schrödinger, *What is Life – the Physical Aspect of the Living Cell.* Cambridge University Press, 1944.
[63] *http://en.wikipedia.org/wiki/Negentropy.*
[64] *http://en.wikipedia.org/wiki/Entropy_and_life.*
[65] R. Landauer, "Irreversibility and heat generation in the computing process," *IBM Journal of Research and Development*, Vol. 5, pp. 183-191, 1961.
[66] C. H. Bennett, "The Thermodynamics of Computation – a Review," *Internat. J. Theoret. Phys.*, Vol. 21, pp. 905-940, 1982.
[67] C. Tsallis, "Possible generalization of Boltzmann-Gibbs statistics," *Journal of Statistical Physics*, Vol. 52, 1988, p. 479-487.
[68] A. Rényi, "On measures of information and entropy," *Proceedings of the 4th Berkeley Symposium on Mathematics, Statistics and Probability 1960*: 547-561.
[69] *http://en.wikipedia.org/wiki/Black_hole_thermodynamics.*
[70] A. Strominger and C. Vafa, "Microscopic origin of Bekenstein-Hawking entropy," *Phys. Lett. B*, pp. 379-399, 1996.

5

MULTIPLE-INPUT-MULTIPLE-OUTPUT (MIMO) ANTENNA SYSTEMS

5.0 SUMMARY

In this chapter, the principles of multiple-input-multiple-output (MIMO) antenna systems are presented. First, the classical development of a MIMO antenna system is described from a statistical viewpoint. It is seen through some numerical examples that the statistical methodology only treats a MIMO antenna system as a scalar problem and ignores the vector nature of the antennas. When the true vector nature of the problem is included in the MIMO antenna methodology it becomes difficult to justify the various claims related to multiplexing gain and the diversity gain, as the phasor nature of the basic principles of electrical engineering do not fully support such a theory from a Maxwellian point of view. The only mode in a MIMO system that is really effective is the dominant phased array mode as simultaneously the various other orthogonal modes from the multiple antennas cannot be combined effectively. It is often claimed that through space-time coding separation of the various spatial modes can be achieved but then similar performance in terms of grater radiation efficiency can be achieved with multiple single-input-single-output (SISO) antenna systems. The basic point of departure in this presentation over the classical statistical based methodology is that firstly, we use the antenna gain rather than the directivity which incorporates the radiation efficiency of the system. Secondly, the various electromagnetic effects and the mutual coupling between the antennas are taken into account in evaluating the system performance. The combination of the signals from the various antennas for the different modes poses a significant challenge due to the vector nature of the problem. Thus, MIMO can only be realized in the form of an adaptive phased array system. Thirdly, the total power input to the antennas in all the simulations is kept constant rather than the radiated power. Numerical examples have been presented to illustrate the vector nature of the problem.

5.1 INTRODUCTION

Multiple-Input-Multiple-Output (MIMO) wireless communication has become an active research area for sometimes after Foschini [1] developed the Bell Laboratories layered space-time system (BLAST). Since then the potential of using multiple antennas to improve the performance of communication systems, especially the capacity, has become more promising. The MIMO systems offer another domain, i.e., spatial domain, in a system design consideration. With this additional domain, it is possible to increase the system capacity or data transfer rate without increasing the bandwidth of the systems. This improvement in bandwidth efficiency of wireless systems is one of the ultimate goals in the Beyond 3G and the 4G era. The MIMO systems may provide a number of advantages over single-antenna systems. Such an advantage includes the reduction of sensitivity for multipath fading through diversity. However, what is not clear is whether a MIMO antenna system performs better than multiple disjoint single-input-single-output (SISO) antenna systems which are quite easy to deploy, for the same total input power.

Research areas in MIMO antenna systems range from information theory, communication system, and signal processing to antenna design. While signal processing and coding are key elements in the development of a MIMO antenna system, antennas and propagations also have a great impact to MIMO system performance. Recently, many researchers have investigated MIMO systems from an electromagnetic perspective [2-4]. It will be shown in this chapter that the various electromagnetic effects can deteriorate the performance of MIMO systems. Without considering the electromagnetic effects in the system design, one might not achieve the best performance or the performance may even get worse over a SISO system.

In the first part of this section, we start with the concept of diversity in general, and then this principle will be applied to the MIMO wireless systems. Some algorithms will be provided to give the reader insight on how the MIMO system works. Then, we will discuss the performance metric, known as *channel capacity*, for the MIMO systems along with assumptions on the channel knowledge and their impacts on the performance. The tradeoff between data rate gained and an increase in the signal-to-noise ratio (SNR) will be briefly reviewed. Then the discussion will move toward electromagnetic perspective on the MIMO systems based on numerical simulations, followed by conclusions.

Finally, in all the numerical computations we will be using the root mean square (rms) values and not the peak values for the voltages and the currents, so the factor ½ will not be there in the computation of the power.

5.2 DIVERSITY IN WIRELESS COMMUNICATIONS

It is generally assumed that wireless channels experience multipath fading, a signal transmitted over these channels is usually distorted, resulting in the reduction of overall transmission rate since the signal needs to be retransmitted. In this chapter we will not go into the cause of fading nor about the deployment of antenna systems which may eliminate them as they have been addressed in chapter 4. However, we will investigate whether existence of independent signal paths through the principles of diversity will likely increase the channel capacity for the problem. Assuming that there are deep fades that occur simultaneously, then the question is: Is it possible to transmit the same signal over these paths to mitigate the effect of multipath fading? This is the idea behind diversity. In general, diversity can be obtained over time, frequency, and space.

5.2.1 Time Diversity

When the wireless channel is time-varying, time diversity can be utilized. This is generally the case for wireless communications since the transmitter and/or the receiver is moving. The information symbols, bearing the same information, are repeatedly transmitted over a period of time so that the transmitted signals experience independent fading. When combining the received signals together, deep fades will be averaged out over time. This can be achieved if two repeated symbols are transmitted farther apart in time than a coherent time, T_{coh}, which is the amount of time over which the propagation remains correlated. This technique is called *interleaving*. We assume a signal symbol, $s(t)$, is transmitted repeatedly L times. The low-pass received signal corresponding to the l-th transmission can be described as follows:

$$r_l(t) = \sqrt{E_S}\,\alpha_l e^{-j\phi_l} s_l(t) + n_l(t), \qquad l = 1, \ldots, L \tag{5.1}$$

where $\alpha_l e^{-j\phi_l}$ represents the random complex gain for the l-th transmission, and $n_l(t)$ denotes the additive white Gaussian noise associated with the l-th transmission. E_S is the radiated signal power. At the receiver, a combiner combines these L signals together so that the received SNR per bit, γ_b, increases compared to when only one symbol is transmitted. This improvement in the received SNR is referred to as a *diversity gain*. The optimum combiner, *maximal ratio combiner* (MRC), can be achieved when the complex gain $\alpha_l e^{-j\phi_l}$ of each transmission is perfectly known [5]. The SNR per bit for this combiner is given by

$$\gamma_b = \frac{E_s}{N_0}\sum_{l=1}^{L}\alpha_l^2 \tag{5.2}$$

where $\left(E_s\alpha_l^2 / N_0\right)$ is the received SNR for the l-th transmission. The effects of errors in the estimates of α_i and ϕ_i for this technique are illustrated in [6, Appendix C].

Even though the output SNR can be improved by the repetition of L symbols in time, this obviously reduces the over all transmission rate of the system. Similar form of the diversity can be employed in other dimensions (frequency and space) to improve the rate of transmission as well.

In summary, the weighted average of the various returns improves the SNR, if the various components of the signal can be aligned in phase. This is the basic operational principle of a phased array and this topic will be discussed in details in chapter 6.

5.2.2 Frequency Diversity

For a frequency-selective fading channel, where its response is not flat in the bandwidth of interests, the constant complex gain for each channel as in (5.1) cannot be applied. This happens when the transmission bandwidth is much larger than the channel coherent bandwidth, B_{coh}, which is the frequency bandwidth over which the propagation channel remains correlated. For single-carrier systems, using equalization techniques can mitigate this effect, which can be implemented in either time or in frequency domain. For multicarrier systems, we represent the symbol into small frequency bands such that the frequency-selective fading channel is converted to narrowband flat fading channels. Then, the coded symbols are transmitted through multiple subcarriers, each of which experiences narrowband flat fading. When the subcarriers are orthogonal to each other, this technique is called *orthogonal frequency division multiplexing* or OFDM.

In the OFDM, the information symbol is first encoded to the coefficients of the discrete Fourier transform (DFT). We consider a burst of encoded N symbols to be transmitted at time t, $\mathbf{X}_t = \left[X_t(1),\ X_t(2),\ \ldots\ X_t(N)\right]$. The transmitter performs the inverse discrete Fourier transform to generate the transmitted signal $\tilde{x}_t$ as

$$\tilde{x}(t) = \frac{1}{\sqrt{N}} \sum_{k=1}^{N} X_k e^{j2\pi k t / N}, \quad 0 \le t \le T \tag{5.3}$$

where T is the transmitted signal interval. To avoid an interburst interference, the cyclic prefix can be appended to the burst of N symbols with a reduction in transmission rate [7]. On the receive channel, the receiver performs the DFT of the signal to extract the original transmitted symbols. Since all of the information is in the orthogonal frequencies, the OFDM methodology is a good candidate for frequency-selective fading channels. By a weighted combination of the various components of the signal the vector nature of the problem is retained and this enables one to increase the output signal-to-noise ratio.

5.2.3 Space Diversity

Similar to time and frequency diversities, the performance of a wireless system can also be enhanced over the spatial dimension as long as the two transmitted signals are spatially separated to ensure that the transmitted signals experience different fading. The distance over which the two signals have uncorrelated fading statistic is called the coherent distance, S_{coh} [8]. By using space diversity, one can improve the received signal quality or even increase the transmission rate without changing the bandwidth of the systems. However, the price one pays is the increased costs of deploying multiple transmitters and receivers. The space diversity can be implemented into diversity either at the receiver and/or at the transmitter.

At the receiver, many combining techniques can be used to obtain the diversity gain. One of the simplest ways to achieve space diversity is that the receiver with multiple antennas selects the received signal from the antenna that has the best reception according to some criteria such as received signal power, SNR, and so on. This technique is called *selection combining*. Another combining technique that is similar to beamforming is called *gain combining*, where the combiner produces weighted sums of the received signal from each antenna to improve the signal quality. The gain combining thus addresses correctly the vector nature of the combining various signal components. Examples of techniques based on the gain combiner are the equal gain combining (EGC), or maximal ratio combining (MRC), etc. Some of these techniques will be discussed in the following sections. However, such a clear physical picture is unavailable when the signal combining occurs in a near field scenario, where the concept of a beam does not exist.

Diversity techniques can also be implemented at the transmitters to prevent an increase in cost of using multiple receivers and receiving antennas. This method is more practicable for mobile phone systems when the mobile terminals are designed to keep their prices as low as possible. However, the diversity at the transmitter usually requires knowledge of the channels or existence of a feedback mechanism from the receivers to the transmitter. In a time division duplex (TDD) system, channel information can be fed back to the transmitter so that it can generate a weighted average to enhance the transmitted signal quality. For other systems, some kinds of feedback from the receiver to the transmitter need to be implemented; otherwise, the diversity cannot be applied at the transmitter side. When talking about diversity, in fact, we are actually trying to send signals through multiple independent paths using either time, frequency, or space. The effect of multipath fading can be mitigated due to independent use of the paths. The signals received through different paths can be combined to further enhance signal quality or transmission rate. MIMO-OFDM is one such technique where both space and frequency are used to increase the diversity. It is important to note that those paths need to be statistically independent or uncorrelated in order to utilize the concept of diversity. Many coding schemes known as *space-time coding* have been proposed in the literature to enhance the independency between various channels. Some of these coding schemes will be discussed in the next sections which are related to fading.

Finally, a word of caution. So far we have used the word "paths of propagation." This terminology is generally reserved to characterize far fields.

What happens in a near field scenario cannot be described by the "paths." However, this shortcoming can easily be removed if in the above discussions we replace the channel characteristic matrix by the voltages and the currents induced at the transmitting and receiving antennas. In addition, it is not clear that the fading often referred to in wireless communication is always due to the various multipaths of the transmitted signal. It is possible fading my be due to the interference between the antenna and its image produced by the earth. For example, we have seen in chapter 2 that an antenna does not produce any null in the near field nor it has any clearly defined path in an indoor propagation system. If this multipath is a big problem why does not it show up in TV transmission? So, it is difficult to see how spatial diversity will improve the scenario for the vector antenna problem. We will address this issue later in this chapter.

5.3 MULTIANTENNA SYSTEMS

We have seen previously that the combining techniques can be utilized in both time and spatial domains. For multiantenna systems, one is mainly concerned with the space diversity. This diversity can be implemented either at the transmitter or on the receiver side, leading to multiple-input-single-output (MISO) or single-input-multiple-output (SIMO) systems depending on where the diversity is applied. We first start with a SIMO which in principle has the same advantage as a MISO except that for a MISO system, the channel knowledge at the transmitter is more difficult to obtain than at the receiver. Later on we will extend this idea to multiple-input-multiple-output (MIMO) systems.

Recall the gain combining technique mentioned earlier, generates a system model at the receiver for the M-antenna systems which is shown in Figure 5.1. The weighted and summed output, y, entering the detector is given by

$$y = \sum_{i=1}^{M} h_i\, r_i = \sum_{i=1}^{M} \sqrt{E_S}\, h_i \alpha_i e^{-j\phi_i} s + \tilde{n}_i \tag{5.4}$$

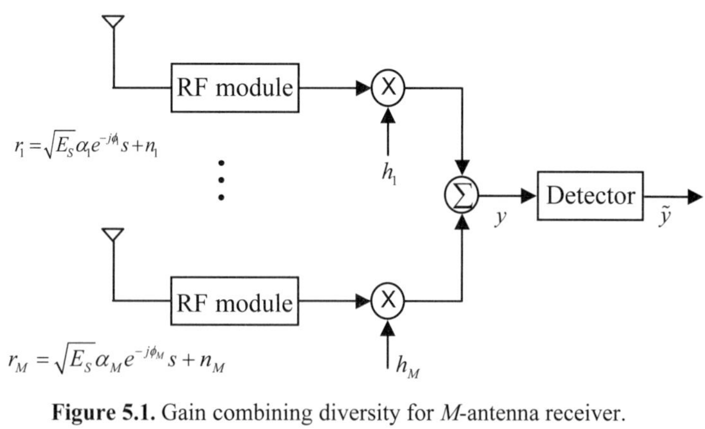

Figure 5.1. Gain combining diversity for M-antenna receiver.

where $\alpha_i e^{-j\phi_i}$ is the complex channel gain. The second term containing noise, is modeled by a zero-mean complex Gaussian random variable with unit variance at the ith receiver chain. h_i denotes a complex weight for the ith receiver. s is the transmitted signal from the transmitter and $\tilde{n}_i = h_i n_i$ is a zero-mean additive white Gaussian noise with variance N_0 at the receiver.

The weighting coefficient h_i can be chosen in several ways. In the equal gain combining (EGC) technique, these weights are chosen such that the received signals are cophased with each other or $h_i = e^{j\phi_i}$. The average output SNR of this combiner for, Rayleigh fading model, is given by [5]:

$$\overline{\gamma}_{EGC} = \frac{E_s}{N_0}\left(1+(M-1)\frac{\pi}{4}\right). \tag{5.5}$$

The MRC mentioned earlier can also be applied when the weighting coefficients are chosen such that $h_i = e^{j\phi_i}$. The average output SNR of the MRC is given by

$$\overline{\gamma}_{MRC} = \frac{E_s}{N_0}M \tag{5.6}$$

However, it is important to reemphasize that the receiver needs to have knowledge about the channel characteristics to obtain the optimum output SNR. These combining techniques can also be implemented at the transmitter in a similar fashion as long as the channel properties are known at the transmitter; otherwise, other techniques need to be implemented. Delay-diversity and phase-sweeping schemes [7] are examples of these techniques where the transmitter diversity is used to minimize multipath fading. In a real system, the up-link and the down-link operate at different frequencies. Therefore, the question is: how an assumed channel model developed at one frequency characterizes the real channel when the system is operated at a different frequency.

5.4 MULTIPLE-INPUT-MULTIPLE-OUTPUT (MIMO) SYSTEMS

Wireless systems with multiple antennas at the transmitter and receiver are considered as MIMO systems. The multiple antennas at the both ends of the systems provide independent paths in the multipath fading environment. This picture is valid if the receiver is located in the far field of the transmitter. As seen with multiple antennas, the signals can be enhanced through diversity gain. Thus, it is possible to utilize this diversity gain in MIMO systems as well. In MIMO systems, another achievable gain that increases the system performance is called the *multiplexing gain* – an increase in the transmission rate. The multiplexing gain comes from the fact that a MIMO channel can be decomposed into a number of independent channels. By having independent channels, multiple symbols can be transmitted simultaneously, which results in increasing the data transfer rate in comparison to a single antenna system.

A simple input-output relationship for a narrowband flat fading MIMO system with N_T transmit and N_R receive antennas can be expressed as

$$\mathbf{y} = \mathbf{Hx} + \mathbf{n} \tag{5.7}$$

wherein $\mathbf{y} = [y_1, y_2, \ldots, y_{N_R}]^T$ is the vector of received signals and $\mathbf{x} = [x_1, x_2, \ldots, x_{N_T}]^T$ is the vector of transmitted signals, $\mathbf{n} = [n_1, n_2, \ldots, n_{N_R}]^T$ is the noise vector where its components are assumed to be additive white Gaussian noise, and $\mathbf{H}$ denotes the $N_R \times N_T$ channel matrix, whose h_{ij} component represents the channel gain from the jth transmit antenna to the ith receive antenna. Figure 5.2 shows the MIMO system model. With the knowledge of $\mathbf{H}$, the MIMO channel can be decomposed into K parallel channels, where $K \le \min(N_R, N_T)$. A common approach in decomposing the MIMO channel is via the singular value decomposition (SVD) [9]. The SVD of the channel matrix $\mathbf{H}$ is given by

$$\mathbf{H} = \mathbf{U\Sigma V}^H \tag{5.8}$$

where $\mathbf{U} = [\mathbf{u}_1, \mathbf{u}_2, \ldots, \mathbf{u}_{N_R}] \in C^{N_R \times N_R}$, $\mathbf{V} = [\mathbf{v}_1, \mathbf{v}_2, \ldots, \mathbf{v}_{N_T}] \in C^{N_T \times N_T}$, and $\mathbf{\Sigma}$ is a diagonal matrix of the singular values, σ_i, of $\mathbf{H}$. $\{\cdot\}^H$ denotes a complex conjugate transpose. By using $\mathbf{U}$ and $\mathbf{V}$ in encoding the input signal $\mathbf{x}$, it is equivalent to transforming a transmission through MIMO channels into multiple transmissions using weighted parallel single-input-single-output (SISO) channels. This is shown mathematically as follows

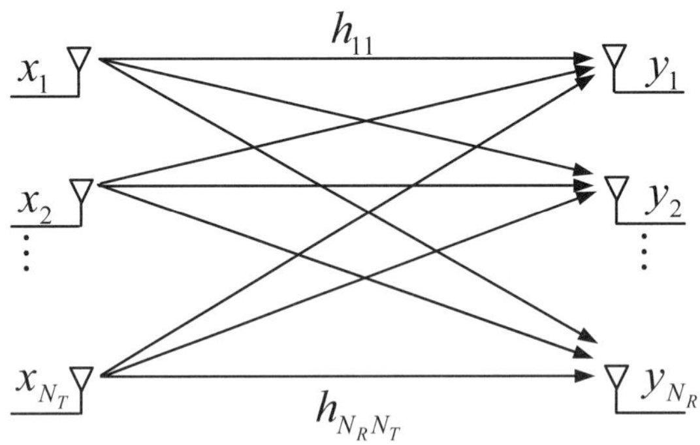

Figure 5.2. MIMO system model.

$$\tilde{\mathbf{y}} = \mathbf{U}^H \mathbf{y} = \mathbf{U}^H (\mathbf{U}\boldsymbol{\Sigma}\mathbf{V}^H)\mathbf{x} + \mathbf{U}^H \mathbf{n}$$

$$\tilde{\mathbf{y}} = \boldsymbol{\Sigma}\mathbf{V}^H \mathbf{x} + \mathbf{U}^H \mathbf{n}$$

$$\tilde{\mathbf{y}} = \boldsymbol{\Sigma}\mathbf{V}^H \mathbf{x} + \tilde{\mathbf{n}}$$

where $\tilde{\mathbf{y}}$ is the output of the encoded systems and $\tilde{\mathbf{n}} = \mathbf{U}^H \mathbf{n}$ is the transformed noise. Note that since $\mathbf{U}$ is a unitary matrix, $\mathbf{U}^H \mathbf{U} = \mathbf{I}$, where $\mathbf{I}$ is an identity matrix, and $\mathbf{U}^H \mathbf{n}$ does not change the distribution of the noise. Thus, $\tilde{\mathbf{n}}$ and $\mathbf{n}$ have the same distribution. However, $\mathbf{V}^H \mathbf{x}$ represents a linear transformation of the input signal $\mathbf{x}$. We, therefore, further encode the input signal as $\tilde{\mathbf{x}} = \mathbf{V}\mathbf{x}$, so that the input signal $\mathbf{x}$ will be transmitted through the uncoupled channels. This is sometimes referred to as *MIMO beamforming* in the literature, which should not be confused with the idea of beamforming from an antenna perspective, in which beam patterns are only defined in the far field. The overall channel decomposition can be illustrated through the following formulation and is intuitively explained by Figure 5.3. Thus,

$$\tilde{\mathbf{y}} = \mathbf{U}^H \mathbf{y} = \left(\mathbf{U}^H \mathbf{U}\right)\boldsymbol{\Sigma}\mathbf{V}^H \tilde{\mathbf{x}} + \mathbf{U}^H \mathbf{n}$$

$$\tilde{\mathbf{y}} = \mathbf{U}^H \mathbf{y} = \left(\mathbf{U}^H \mathbf{U}\right)\boldsymbol{\Sigma}\left(\mathbf{V}^H \mathbf{V}\right)\mathbf{x} + \mathbf{U}^H \mathbf{n}$$

$$\tilde{\mathbf{y}} = \boldsymbol{\Sigma}\mathbf{x} + \tilde{\mathbf{n}} \tag{5.9}$$

Equation (5.9) clearly demonstrates that a M-transmit and M-receive antennas is equivalent to M-SISO channels weighted by the appropriate singular values for each of the spatial modes. In a near field environment, where antenna patterns are not defined and antennas in general do not have nulls independent of the

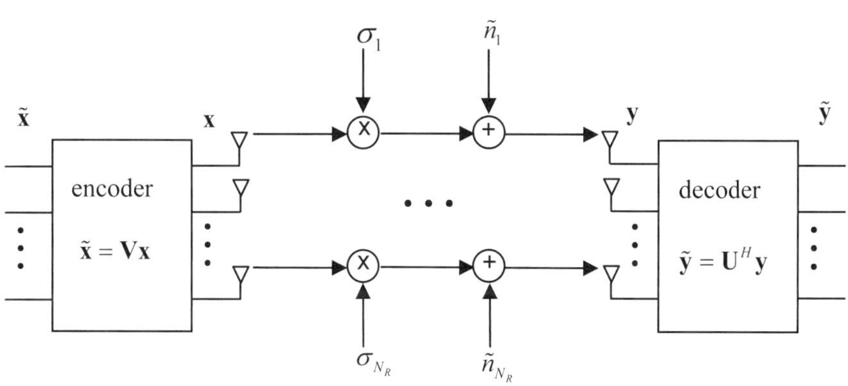

Figure 5.3. Parallel decomposition of a MIMO channel.

distance, it is possible that a M-SISO system maybe superior in performance than a M×M MIMO system. In addition for the M×M MIMO system different feed arrangements are required to properly combine the voltages at each of the M-receive antennas vectorially for the M independent orthogonal spatial modes in hardware.

Based on this decomposition, multiple signals can be transmitted simultaneously, which increases the data rate of the system without changing the bandwidth of the systems. However, the performance of the transmission strictly depends on the singular values of the channel matrix since they represent gains of the decomposed channels. It is also possible that the matrix $\mathbf{H}$ is not full rank. In that case, the decomposed channel will be useful for transmission of data only by the mode with singular values larger than a threshold. This results in a multiplexing gain of K, where $K \leq \min(N_R, N_T)$. When $\mathbf{H}$ is full rank and well conditioned, $K = \min(N_R, N_T)$, the channel is called a *rich scattering* environment. This is where we gain benefits from multiple antenna systems.

To implement such a concept, different assumptions are made about the knowledge of the channel matrix. These assumptions lead to different schemes for MIMO systems. In the next section, we will first discuss channel capacity followed by assumptions about knowledge of the channel and their impact to the MIMO systems.

A moment of reflection at this point will reveal that the concept of the best region of operation of a MIMO in a rich scattering environment is developed from a purely scalar point of view. So from a philosophical view point, it is possible to have some pathological scenarios where $\mathbf{H}$ may be of full rank containing large singular values so that K indeed can be $\min(N_R, N_T)$. However, it is extremely doubtful that for the vector electromagnetic problem, such a scenario can indeed occur as will be illustrated later. If one has a collection of vector electric fields representing the various signal transmission mechanisms then it is not clear at all that by combining all the various vector signals without paying any attention to their relative phases will provide a better solution! Alternately, one can view MIMO system as an over-moded waveguide where simultaneous operation is carried out utilizing the various orthogonal modes. The difficulty in such a system is how to simultaneously excite all the orthogonal modes and then how to extract the signal from each mode at the receiver.

5.5 CHANNEL CAPACITY OF THE MIMO ANTENNA SYSTEMS

In this section, fundamental limit on the spectral efficiency in MIMO antenna systems is discussed. The quantity that represents the maximum error-free transmission rate is called the *channel capacity*. The channel capacity for additive white Gaussian noise channels was first introduced by Claude Shannon [10,11]. This quantity represents the transmission rate per Hertz of bandwidth (bits/s/Hz) of information. To understand how multiple antenna systems can increase the capacity, we first consider the capacity of a SISO system. We will assume the channel to be a narrowband flat fading model unless otherwise stated.

From (5.1) we repeat here that the received signal for a single transmission is given by

$$r(t) = \sqrt{E_S}\alpha e^{-j\phi} s(t) + n(t). \tag{5.10}$$

If E_S is the transmitted signal power and N_0 is the noise power, the channel capacity of this SISO system can be written as

$$C_1 = \log_2\left(1 + \frac{E_S}{N_0}\cdot\alpha^2\right), \qquad \text{bits/s/Hz} \tag{5.11}$$

For MIMO systems with N_R receive and N_T transmit antennas in (5.7), the capacity is defined as follows [12]:

$$C = \max_{Tr(R_{xx})=N_T} \log_2\left|\mathbf{I}_{N_R} + \frac{E_S}{N_T N_0}\mathbf{H}\mathbf{R}_{xx}\mathbf{H}^H\right| \tag{5.12}$$

wherein $\mathbf{I}_{N_R}$ is an identity matrix of dimension $N_R \times N_R$, $\mathbf{R}_{xx} = E\{\mathbf{x}\mathbf{x}^H\}$ is the covariance matrix of the transmitted signal $\mathbf{x}$. $E\{\cdot\}$ is the expected value operator. $|\cdot|$ and $Tr(\cdot)$ denote the determinant and the trace of a matrix, respectively. $Tr(\mathbf{R}_{xx}) = N_T$ is the total average radiated power constraint at the transmitter. This capacity represents the data rate per unit bandwidth that can be transmitted with negligible error over the MIMO channel. Clearly, to achieve the maximum MIMO channel capacity, we need to optimize $\mathbf{R}_{xx}$ that is the correlation of the transmitted signals.

A word of caution at this point is that the channel capacity has been derived under the assumption of the average radiated power. No mention is ever made of the input power to the transmitter. Therefore it is difficult to scientifically compare the performance of various systems solely based on the channel capacity if the value for the input power is not specified and the radiating efficiency of the antennas is not taken into account.

Next we consider the situation when the channel capacity will be maximized. We first assume that the number of transmit and receive antennas are equal, $N_T = N_R = M$. When the transmitted signal is uncorrelated, $\mathbf{R}_{xx} = \mathbf{I}_M$, the channel capacity reduces to

$$C = \log_2\left|\mathbf{I}_M + \frac{E_S}{M N_0}\mathbf{H}\mathbf{H}^H\right|. \tag{5.13}$$

Assuming the channel matrix is orthogonal, we have $\mathbf{H}\mathbf{H}^H = \mathbf{H}^H\mathbf{H} = \mathbf{I}_M$. If we assume that $\alpha_{11} = \alpha_{22} = \ldots = \alpha_{MM}$, this results in the following expression for (5.13) as

$$C = M\log_2\left(1+\frac{E_S\,\alpha_{11}^2}{N_0 M}\right) \tag{5.14}$$

where α_{ii} denotes the channel gain from the ith transmit antenna to the ith receive antenna. The multiplication by the factor of M outside the logarithm in (5.14) has more profound effect than the division by M inside the logarithm, provided, the signal is large compared to that of noise. This is equivalent to increasing the channel capacity by a factor of M when compared with a single channel transmission in (5.11) with the same radiated power. In other words, when the channel matrix is orthogonal and the transmitted signals are uncorrelated, transmitting information through $M \times M$ MIMO channels is equivalent to transmitting signals via completely uncoupled M parallel SISO, where each channel is weighted by the singular value of the channel matrix representing the radiation efficiency of the transmission system. Later it will be shown that by using the parallel decomposition of MIMO channels, an increase in channel capacity can be obtained. In summary, the intriguing feature of (5.14) is that if one has M-barely functional spatial modes and thus the total power supplied to the receiver is significantly less than a single SISO, yet the presence of the factor M outside the *log* really skews up the result for the capacity and thus provides a number which may be large. It is therefore important to interpret the mathematical value for the capacity with great caution from a system perspective. It must be remembered that the numerical value for the capacity is just a mathematical entity just like entropy described in chapter 4 and is a quantitative entity, and not a qualitative one. That means that channel capacity is not something that is fundamentally intuitive, but something that is fundamentally defined via an equation, via mathematics applied to physics. There is no such thing as capacity, without an equation that defines it. This will be illustrated later.

5.6 CHANNEL KNOWN AT THE TRANSMITTER

In practice, the channel information at the receiver can easily be obtained as opposed to at the transmitter, since to have channel information at the transmitter one requires some kind of feedback to send the information from the receiver back to the transmitter. Thus, we will assume the knowledge of the channel at the transmitter is known throughout this chapter. At the receiver side, the knowledge of the channel is also known. When the transmitter knows the channel information, $\mathbf{H}$, (referred to as a closed-loop system) either by a feedback from the receiver or through estimation using reciprocity in the TDD systems [12,13], the MIMO channel decomposition can be utilized to increase the capacity. Assuming that the transmitted signals are uncorrelated, i.e., $\mathbf{R}_{xx} = \mathbf{I}_{N_T}$ and given the eigendecomposition $\mathbf{HH}^H = \mathbf{Q\Lambda Q}^H$, the MIMO channel capacity in (5.12) can be written as

$$C = \log_2 \left| \mathbf{I}_{N_R} + \frac{E_S}{N_0 N_T} \mathbf{Q \Lambda Q}^H \right| \tag{5.15}$$

$$C = \sum_{i=1}^{K} \log_2 \left(1 + \frac{E_S}{N_0 N_T} \psi_i \right) \tag{5.16}$$

where ψ_i is the eigenvalue of $\mathbf{HH}^H$ and K is the rank of the channel matrix $\mathbf{H}$. This channel capacity can be interpreted as the sum of capacities of K channels, each having gain of ψ_i. Since the value of ψ_i is not the same for all the channels, we need to allocate the power for each channel properly to maximize the capacity.

5.6.1 Water-filling Algorithm

Since the power gain for each channel is not equal, the best way to transmit data efficiently is to spread the power input to the transmitter such that each channel contains similar levels of energy. This leads to the well known solution for power allocation known as the *water-filling* algorithm. The philosophy of the algorithm of power allocation follows the following recipe [12,14]:

1. Select the number of parallel channels: $m = K$.
2. Determine the constant: $\mu = \frac{P_T}{m} + \frac{1}{m} \sum_{i=1}^{m} \frac{1}{\psi_i}$, where ψ_i is the eigenvalue and they are placed in the descending order ($\psi_i \geq \psi_j$, for $i < j$). P_T denotes the total transmit power.
3. Determine the power allocated to the ith sub-channel: $p_i = \mu - \frac{1}{\psi_i}$, for $i = 1, \ldots, m$.
4. If $p_m > 0$, then we set $E_{S_i} = p_i$ for $i = 1, \ldots, m$; otherwise, we set $E_{Sm} = 0$ and replace m by $m - 1$. Then we return to step 2.

The water-filling algorithm will allocate the transmitted power determined by the sub-channel gain according to the eigenvalue ψ_i. Whenever the sub-channel gain is low (high attenuation), the algorithm will discard the use of that sub-channel, which in turn reduces the number of parallel channels available in the MIMO system.

It is important to note that many of the conclusions are based on a point source theory for the antennas. The deficiency of such an assumption is that all antennas are considered to be operating in the far field scenario and there is no mutual coupling between the antennas. The result is that one does not make any distinction between the excitation voltages for the point sources or the power radiated by them. In a far field scenario, the power is simply the square of the absolute value of the excitation voltages and is real. However, when using realistic model for antennas it is necessary to compute separately the voltages and the currents and then compute the input power, which is generally a complex number. So information for both the voltages and the currents are necessary in an array environment to evaluate the complex power. When using realistic antennas the real singular values of the channel matrix are no longer then related to the complex value of the transmitted power, and the terminology for this water filling algorithm needs to be modified!

5.7 CHANNEL UNKNOWN AT THE TRANSMITTER

So far, we have discussed the use of MIMO systems and their benefits, of an increase in the channel capacity over the SISO systems when the transmitters have knowledge about the channels. In the absence of this information, it is not possible to allocate transmitted power efficiently over the transmit antennas. The best strategy then is to transmit equal power for each transmit antenna. If the transmitted signals are chosen such that they are uncorrelated, the channel capacity is given in (5.16). As the capacity is a function of the eigenvalues, which are not known to the transmitter in this case, the multiplexing gain may be reduced due to the fact that some eigenvalues may be too small to convey the transmitted information. However, one could utilize the diversity gain for the MIMO systems as well through space-time coding [14, Chap.5]. Next we provide an example of a coding scheme that provides a diversity gain even when there is no information available about the channels at the transmitter.

5.7.1 Alamouti Scheme

With an appropriate coding scheme, the transmit diversity can be obtained even in the absence of channel information. Alamouti [15] has proposed a simple coding scheme that can provide a diversity order of $2N_R$ for the system of 2 transmit antennas and N_R receive antennas. Let us consider a MIMO system with 2×2 antennas and assume that the transmitter has no channel information. The channels are assumed to be frequency flat fading and assumed to remain constant over the transmissions of two symbols. The channel matrix, $\mathbf{H}$, is given by

$$\mathbf{H} = \begin{bmatrix} h_{11} & h_{12} \\ h_{21} & h_{22} \end{bmatrix}. \tag{5.17}$$

In the first symbol period, two different symbols, s_1 and s_2, are transmitted simultaneously from antenna 1 and antenna 2, respectively, with the energy per bit per transmission is $E_S/2$. At the second symbol period, however, the transmitter sends $-s_2^*$ from antenna 1 and s_1^* from antenna 2 instead, where s^* denotes complex conjugate of s. The received signals at the receive antennas over two consecutive symbol periods, $\mathbf{y}_1$ and $\mathbf{y}_2$, can be expressed as

$$\mathbf{y}_1 = \sqrt{\frac{E_S}{2}} \begin{bmatrix} h_{11} & h_{12} \\ h_{21} & h_{22} \end{bmatrix} \begin{bmatrix} s_1 \\ s_2 \end{bmatrix} + \begin{bmatrix} n_1 \\ n_2 \end{bmatrix} \tag{5.18}$$

$$\mathbf{y}_2 = \sqrt{\frac{E_S}{2}} \begin{bmatrix} h_{11} & h_{12} \\ h_{21} & h_{22} \end{bmatrix} \begin{bmatrix} -s_2^* \\ s_1^* \end{bmatrix} + \begin{bmatrix} n_3 \\ n_4 \end{bmatrix} \tag{5.19}$$

where n_1, n_2, n_3, and n_4 are the additive white Gaussian noise (AWGN) samples with zero-mean and power N_0. At the receiver, the received signals are put together to form a signal vector $\mathbf{y} = [\mathbf{y}_1 \quad \mathbf{y}_2^*]^T$. The signal vector $\mathbf{y}$ can be expressed as

$$\mathbf{y} = \sqrt{\frac{E_S}{2}} \begin{bmatrix} h_{11} & h_{12} \\ h_{21} & h_{22} \\ h_{12}^* & -h_{11}^* \\ h_{22}^* & -h_{21}^* \end{bmatrix} \begin{bmatrix} s_1 \\ s_2 \end{bmatrix} + \begin{bmatrix} n_1 \\ n_2 \\ n_3^* \\ n_4^* \end{bmatrix}, \quad \mathbf{y} = \sqrt{\frac{E_S}{2}} \mathbf{H}_A \mathbf{s} + \mathbf{n}, \tag{5.20}$$

where $\mathbf{s} = \begin{bmatrix} s_1 & s_2 \end{bmatrix}^T$ and $\mathbf{n} = \begin{bmatrix} n_1 & n_2 & n_3^* & n_4^* \end{bmatrix}^T$. According to the structure of $\mathbf{H}_A$, it implies that $\mathbf{H}_A$ is orthogonal or $\mathbf{H}_A^H \mathbf{H}_A = \left(|h_{11}|^2 + |h_{12}|^2 + |h_{21}|^2 + |h_{22}|^2\right)\mathbf{I}_{2\times 2}$ is a diagonal matrix. Thus, if the receiver performs $\mathbf{z} = \mathbf{H}_A^H \mathbf{y}$, we will get

$$\mathbf{z} = \sqrt{\frac{E_S}{2}} \left(|h_{11}|^2 + |h_{12}|^2 + |h_{21}|^2 + |h_{22}|^2\right)\mathbf{I}_{2\times 2}\,\mathbf{s} + \tilde{\mathbf{n}}, \tag{5.21}$$

where $\tilde{\mathbf{n}} = \mathbf{H}_A^H \mathbf{n}$ is a complex Gaussian noise vector with zero mean and covariance $E\{\tilde{\mathbf{n}}\tilde{\mathbf{n}}^H\} = \left(|h_{11}|^2 + |h_{12}|^2 + |h_{21}|^2 + |h_{22}|^2\right) N_0 \mathbf{I}_{2\times 2}$. We see that the receiver decouples the two transmitted symbols. Each of which is contained in the components of $\mathbf{z}$,

$$z_i = \sqrt{\frac{E_S}{2}} \left(|h_{11}|^2 + |h_{12}|^2 + |h_{21}|^2 + |h_{22}|^2\right) s_i + \tilde{n}_i, \quad i = 1, 2. \tag{5.22}$$

This gives the SNR for each received symbol as follows

$$\eta = \frac{\left(|h_{11}|^2 + |h_{12}|^2 + |h_{21}|^2 + |h_{22}|^2\right) E_S}{2N_0} \tag{5.23}$$

It yields an increase in the received SNR according to the summation of the channel gains. When the channel is considered as a spatially white channel, i.e., $E\left\{|h_{ij}|^2\right\} = 1$, the Alamouti scheme yields the maximum diversity gain, which is equal to four. Note that the transmission rate does not get improved by using this scheme as the transmission of two symbols requires two symbol periods. In other word, there is no multiplexing gain, but there is diversity gain for this MIMO system. A question that still lingers on, whether this value of four can be reached when using real antennas.

5.8 DIVERSITY-MULTIPLEXING TRADEOFF

In the previous section, we see that even though the multiplexing gain cannot be obtained due to the absence of any knowledge of the channel, but with a proper transmission scheme, the diversity gain can be achieved. Zheng and Tse have shown in [16] that it is not possible to achieve the maximum of both diversity and multiplexing gains at the same time. We briefly review this fundamental tradeoff in this section. According to [16] a scheme is said to achieve the multiplexing gain r if

$$\lim_{SNR\to\infty} \frac{R(SNR)}{\log(SNR)} = r \tag{5.24}$$

and achieve the diversity gain d if

$$\lim_{SNR\to\infty} \frac{\log P_e(SNR)}{\log(SNR)} = -d \tag{5.25}$$

where $P_e(SNR)$ is the probability of error at a given SNR and $R(SNR)$ is the rate of a code scheme as a function of SNR. For a given scheme with block length of $l \geq N_T + N_R - 1$, the optimal tradeoff between the diversity and the multiplexing gains that any scheme can reach in the case of Rayleigh-fading MIMO channel can be given by [16]:

$$d(r) = (N_T - r)(N_R - r), \quad r = 0, 1, \ldots, \min(N_T, N_R) \tag{5.26}$$

where $d(r)$ represents the tradeoff curve which is a piecewise-linear function connecting the points $\{r, d(r)\}$. The tradeoff curves for the case of

$N_T = N_R = 2$ and $N_T = N_R = 3$ is shown in Figure 5.4. It is seen that the Alamouti scheme for the 2×2 system considered in the previous section has the maximum diversity while there is no increase in the transmission rate. However, since this fundamental limit is derived from the channel capacity, constrained for an arbitrarily low bit error rate, when this constraint is relaxed, it is possible to have a full-diversity and full-rate transmission [14,17].

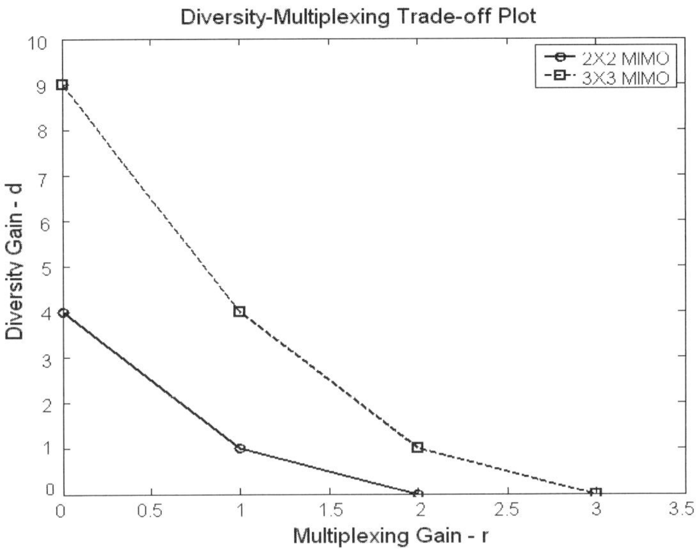

Figure 5.4. Diversity-Multiplexing tradeoff for 2×2 and 3×3 MIMO systems.

5.9 MIMO UNDER A VECTOR ELECTROMAGNETIC METHODOLOGY

So far all the discussions about MIMO has been based on the scalar statistical methodology, which illustrates that use of multiple antennas for transmit and receive have superior performance over a single transmit receive system. In this section, we investigate how the introduction of the vector Maxwellian principles impacts the previous discussions. So, the first departure from the scalar statistical theory lies in using real antennas instead of using point sources, which really does not exist in practice. There are additional hidden subtleties which are missed in the point source model. For example, the entire field radiated by a point source is akin to a far-field electromagnetic radiation and which is real and the power can be computed from either the electric or the magnetic field. But for a finite length dipole there are both near and far fields. In the near field the approximations done in the signal processing literature does not hold. Moreover, there is mutual coupling between the antennas, which can completely alter the nature of the conclusions if their effects are taken into account. Finally, we limit the total input power to the systems so that we can make meaningful performance comparisons between various systems.

Research topics in MIMO systems under an electromagnetic point of view have been recently studied [2–4]. As we know, the signal processing algorithms, to be more specific – array processing, was originally developed for sonar applications. Later on it was adapted for wireless communications where antennas are parts of the systems. Unfortunately, these two arenas are disjoint. While sonar signal is a scalar, the electromagnetic signal is a vector [18,19]. Thus, great care needs to be taken into account in system and/or algorithm development to obtain the best performance. Also for an antenna the space and time variables are not independent but they are related as an antenna is both a temporal filter and also a spatial filter and that is why one uses Maxwell's equations to study antennas as these equations capture the fundamental real life physics which is missing in a scalar statistical analysis.

In this section, we illustrate the vector nature of the MIMO electromagnetic system through a few simulated numerical examples. The actual parameters of the example are not important in the sense that the examples presented will demonstrate that MIMO does not always perform better than a SISO system, from a real system standpoint but may provide numerical values using a statistical methodology which seem to indicate that the performance is better. Hence, it is important to make a distinction between conclusions based on statistical aberrations as opposed to basic physics. Also, due to the existence of complicated multipath environments, which are non-existent in a near-field scenario, the discussion in this section will be based only on numerical simulations, using an accurate numerical electromagnetic analysis computer code [20]. Our goal is to provide some numerical examples to illustrate the basic principles when using a system standpoint.

5.9.1 MIMO versus SISO

As an example, consider two dipole antennas of 15 cm in length and of radius 1 mm separated in free space by 1.5 m. The center-fed dipoles are conjugately matched at 1 GHz by connecting the load of $90.7 - j\,42.7\ \Omega$ at their feed points. For a 1 V excitation of the transmitting dipole we have an excitation current of $5.5 + j\,0.0015$ mA in the transmitter antenna which induces a current of $0.045 + j\,0.14$ mA at the feed of the receiving dipole. Therefore for an input power of 1 W in the transmitting antenna, there will be a received power of 0.36 mW in the load of the receiving antenna. This describes a typical SISO system.

Next we consider two conjugately matched center-fed dipoles as transmitters and two additional conjugately matched center-fed dipoles as receivers replacing the single antenna system to represent a MIMO system. We consider the two half wave dipole antennas at both the transmitter and the receiver to be separated by half a wavelength. The four center-fed dipoles have the same length, radius and loading as before. They are also separated by the same distance of 5 wavelengths or 1.5 m. The basic philosophy is that since there are two antennas each for the transmit receive systems, one can communicate with two spatial orthogonal modes of the system. The two orthogonal modes of the excitation of the antennas will be of 1 V each fed to the transmitting antennas so that they operate in phase. The other orthogonal mode will have a +1 V and –1 V excitations to each of the transmitting antennas so that the excitations are orthogonal. The basic principle of MIMO is to simultaneously use both of these spatial modes for transmission. For a co-phase 1 V excitation of the dipoles we have an excitation current of $6.0 + j\ 1.2$ mA in each of the transmitting antennas which induces a current of $0.014 - j\ 0.33$ mA in each of the receiving dipoles. Therefore for a total input of 1 W of power to the two transmitting antennas, they will produce a total received power of 1.6 mW in the loads of the receiving antennas. For an anti-phase excitation of +1 V and –1 V in each of the transmitting antennas representing the second orthogonal mode we have an excitation current of $4.7 - j\ 0.74$ mA in each of the transmitting antennas which induces a current of $12.9 - j\ 13.3$ μA in each of the receiving dipoles. Therefore for a total of 1 W of input power to the two transmitting antennas will result in a received power of only 6.5 μW in the loads of the receiving antennas. Even though there are two spatial modes, the first mode has a higher radiation efficiency than the second mode, by a factor of 246 approximately. Electromagnetically this second mode will never be used in practice because of its poor radiation efficiency. Communication using the first mode in the antenna literature is called a phased array. Using the first spatial mode in this two antenna system it is possible to get a gain of 4.44 over the SISO system. It is important to note that this value is greater than the number 4 which is the limit obtained from a scalar statistical analysis! In addition, it is not clear how to develop two different corporate feeds which will simultaneously separate out the two voltages due to the two spatial modes from the same antennas as the voltages are vector in nature and even though the two modes are orthogonal their electrical separation at the same frequency is not a trivial problem!

The discussion should generally stop here. However, since a different metric called the channel capacity other than the received power is used to compare the performance between systems, we need to explore what is the system performance under this new metric. The channel capacity is a formula that has been derived from the concept of entropy which is purely philosophical nature and not connected with the basic physics as we have seen in chapter 4. The Shannon channel capacities for the SISO and the MIMO systems for bandwidth B will then be given by

$$C_{SISO} = B \log_2\left(1 + \frac{0.00036}{P_N}\right) \tag{5.27}$$

$$C_{MIMO} = B \log_2\left(1 + \frac{0.0016}{2 \times P_N}\right) + B \log_2\left(1 + \frac{0.0000065}{2 \times P_N}\right) \tag{5.28}$$

where P_N is the thermal noise power. The factor of 2 appearing in the denominator of (5.28) is due to the fact that for the same input power of 1 W for the two spatial modes, we will be feeding 0.5 W to each of the spatial modes so that the total input power remains constant. Now if the denominator of both (5.27) and (5.28) is the thermal noise power, then even the minuscule power received for the second MIMO mode will contribute to the formula for the channel capacity, even though it may be useless from a practical standpoint. One may even argue that by some appropriate coding this mode can be put to use. The physics disappear at this point and the logarithm of the radiated to the thermal noise power appears really very attractive even though its contribution is dismal in a real system! Therefore, a MIMO antenna system is thought of as simultaneously transmitting multiple orthogonal modes in a multi-moded wave guiding system. This type of multi-moded transmission is seldom used in real life because of the dispersion in the system and the logistics involved in simultaneously exciting all the orthogonal modes and combining them in the same waveguide for transmission and then separating the various modes at the receiver, is quite complex. In summary, in the capacity formula the linearity of the addition of the separate channels overwhelms any gain achievable through the hardware of the antenna systems. The thinking appears to be that the logarithmic increase in power is not as relevant as the multiple channels even though they may be unacceptable vehicles for transmission from a hardware point of view! However, if one uses the expression for the Hartley capacity given by (4.9) instead of the Shannon capacity (4.1) then due to the discretization of the induced voltages, one would get a more realistic value for the capacity of the system as the second mode may be weeded out as it may yield a voltage comparable to the first quantization level.

So far so good, and one can relate the physics with mathematics for any system. However, the next step really becomes bizarre if we now pose the problem as follows: In the MMO system that is just described, the direct line of sight creates the more efficient channel and if we take the direct line of sight out then the linearity of the two terms will predominate in (5.28). In actual practice there may seldom be a direct line of sight communication. There could be many mulipaths and the second orthogonal mode whose performance is really dismal in the line of sight operation perhaps may be a viable mode of propagation. Unfortunately, this way of thinking clearly misses the vector nature of the wireless communication problem and is mostly guided by the scalar channel capacity theorem. Let us illustrate the statement by another example.

Let us now encapsulate the transmit-receive antennas described in the previous example in a concentric region so that there will be no direct line of sight of communication. The spacing between the antennas, dimensions and the load remains the same as before. Let us place the SISO system described earlier in a concentric region characterized by two conducting structures as shown in Figure 5.5. The closed inner conducting box has a dimension of 1 m × 1 m × 0.5 m. The outer conducting shell has an inner dimension of 2 m × 2 m × 0.5 m. The thickness of the conducting walls is 6 cm. In this case there is no line-of-sight of

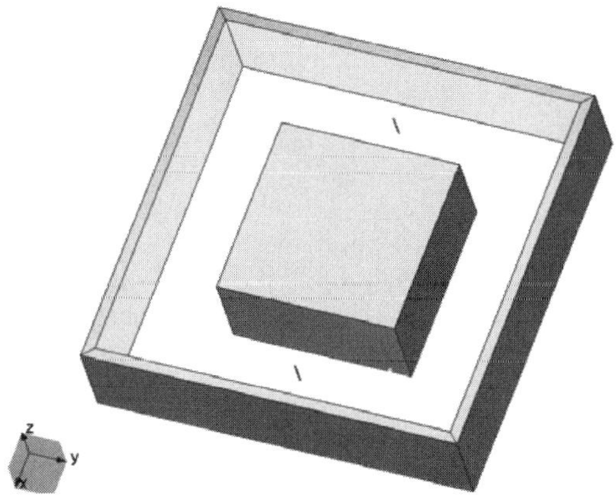

Figure 5.5. A SISO system enclosed in conducting concentric cylinders.

communication as the inner conducting box prevents such a scenario. Also, the conducting structure will guide the signals more to the receiver and therefore if we place multiple antennas, one of them may pickup more signal. Unfortunately, such knave simplistic reasoning does not hold for the vector electromagnetic problem as we will observe next.

For a 1 V excitation of the transmitting dipole in the SISO system will produce an excitation current of $5.2 + j\ 3.4$ mA in the transmit antenna. It will also induce a current of $0.99 + j\ 0.4$ mA in the receiving dipole. Therefore for an input power of 1 W in the transmitting antenna, it will produce a received power of 16.64 mW in the load of the receiving antenna. An increase in the received power over the earlier open-air SISO example is expected as the signals are directed in this case to the receiving antenna by the concentric guiding structure.

Next we consider a 2×2 MIMO antenna systems where the two transmitting and the receiving antennas as discussed before are encapsulated by the concentric cylinders. The situation is depicted in Figure 5.6. The two dipoles as transmitters and two dipoles as receivers are now replacing the single antenna systems to represent a MIMO system. We consider the same configuration of two half wave dipole antennas instead of the single one but separated by half a wavelength. They will have the same length, radius, and loading as before. They are also separated by the same distance of 5 wavelengths. The two antenna transmit receive systems can communicate using two spatial orthogonal modes. One of the orthogonal modes of the antenna systems will be an excitation of 1 V to each to the transmitting antennas so that they will be operating in phase. There are no direct paths linking the transmitting and the receiving antennas. For a co-phase 1 V excitation of the dipoles we have an excitation current of $4.2 + j\ 1.8$ mA induced in each of the transmitting antennas will induce a current of $0.075 + j\ 0.86$ mA in each of the receiving dipoles. Therefore, for a total input power of 1 W to the two transmitting antennas, there will be a total received power of 14.79

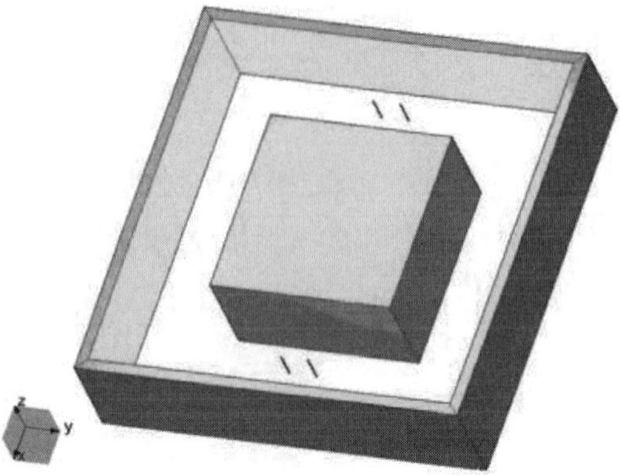

Figure 5.6. A 2 × 2 MIMO system enclosed in a conducting concentric box.

mW in the loads of the receiving antennas. For the second spatial orthogonal mode, an anti-phase excitation of +1 V and –1 V in each of the transmitting antennas will produce an excitation current of 4.7 + j 0.22 mA in the first transmitting antenna and a current of 0.53 + j 0.3 mA in the first receiving dipole.

Therefore for the second orthogonal spatial mode, for a total input power of 1 W to the two transmitting antennas there will be a received power of only 7.15 mW in the loads of the receiving antennas. Even though there are two spatial modes, the first mode is generally more efficient so far as the radiated power is concerned than the second mode, by a factor of only 2.07 approximately, as the line of sight has been eliminated. Electromagnetically it appears that both the two spatial orthogonal modes in this case have poorer radiation efficiency than the SISO case. Yet, if one writes the capacity in this case for the SISO and the MIMO system, one obtains the following two expressions:

$$C_{SISO} = B \log_2 \left(1 + \frac{0.01664}{P_N} \right) \tag{5.29}$$

$$C_{MIMO} = B \log_2 \left(1 + \frac{0.01479}{2 \times P_N} \right) + B \log_2 \left(1 + \frac{0.00715}{2 \times P_N} \right) \tag{5.30}$$

Assuming a background thermal noise floor of about 2 pW, one evaluates the two capacities as

$$C_{SISO} = B \times 32.95 \tag{5.31}$$

$$C_{MIMO} = B \times 31.78 + B \times 30.74 = B \times 62.52 \tag{5.32}$$

Here is the dichotomy. The total power received by the MIMO antenna systems using the two spatial modes is less than the total power received by the SISO system. So, from an electromagnetic system point of view, we have two inferior modes of propagation than over a single antenna system, yet if one were to claim that (5.32) is a better system than (5.31), then it must be based on statistical aberrations and not from a sound physical system point of view. The other point is quite clear, that two independent SISO systems will always be better than a 2 × 2 MIMO system under all conditions and keeping the total amount of input power to the system fixed. In addition, it is not clear how to develop two corporate feeds for the MIMO system which will separate out the two voltages from the same antennas as the voltages are vector in nature and even though the two modes are orthogonal their electrical separation at the same frequency is not a trivial problem! Finally, equations (5.29) and (5.30) indicate that depending on the value of P_N, C_{SISO}, or C_{MISO} will be larger. Hence, there is no guarantee for a general situation that a SISO will be inferior in performance than a MIMO unless the actual system parameters are exactly specified.

In conclusion, the statistical analysis which is responsible for the derivation of the channel capacity does not support basic physics. Examples of such non-intuitive results based on application of probability theory are available in the literature [21].

5.10 MORE APPEALING RESULTS FOR A MIMO SYSTEM

5.10.1 Case Study: 1

To get an intuitive idea for the statistical results, we first look at a simple scenario when a SIMO system of $N_T = 1$ and $N_R = 3$ is considered. The receiver uses the MRC combiner mentioned earlier in section 5.2. Note that it is shown in (5.6) that the diversity gain of the MRC is equal to three in this case (three receive antennas), which results in three times higher output SNR compared to a single-antenna system. Let us consider all the antennas as $\lambda/2$ (half-wavelength long) dipoles operating at $f = 1\,GHz$ with a radius of $\lambda/300$ (1 mm.). All the antennas are centrally loaded with the complex impedance $Z_L = 90.57 - j42.51\ \Omega$, so that they are matched, which implies that they are going to radiate maximum power in free space. For the SISO system, the two antennas are 100 m apart in free space. A power of 1 W is fed to the transmitting antenna and this will produce a received current of $I_{SISO} = 0.017 + j0.023$ mA. The received power is then $P_{SISO} = 78.0$ nW. When three identical antennas are used at the receiver, the receive antennas are 5 m apart from each other. Note that all the antennas are vertically polarized along the z-direction, which is out of the plane of the paper. The simulation setup for this SISO and SIMO case is shown in Figure 5.7.

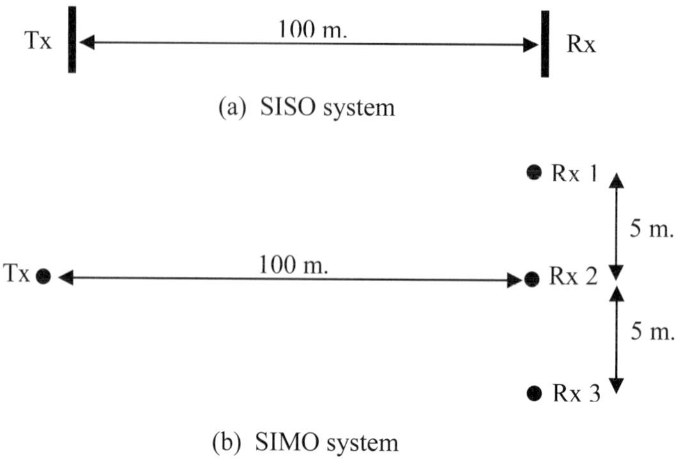

Figure 5.7. SISO and SIMO setups for case 1, all antennas are vertically polarized; (a) SISO system and (b) SIMO system.

The currents induced at the receive antennas in mA are $I_1 = -0.007 - j0.028$, $I_2 = 0.020 + j0.021$, and $I_3 = -0.007 - j0.028$. Note that $I_1 = I_3$ due to the symmetry of the structure. This yields the received power of $P_1 = 75.6$, $P_2 = 76.2$, and $P_3 = 75.6$ nW. It means that under this ideal case, (no noise and losses), $P_{MRC} = \sum_{i=1}^{3} P_i = 227.4$ nW or $2.92 P_{SISO}$. This loss in the diversity gain is due to the fact that there is mutual coupling between antennas and the radiated power is a function of $1/R^2$, where R is the distance between a transmit and a receive antenna. Note that the distances from the transmitter to the antenna 1 and 3 in our case are longer than the one from the transmitter to the antenna 2.

5.10.2 Case Study: 2

For the next example, let us make the situation more complicated by introducing two objects in the scene. This will cause multipath interference at the receiver. The antennas used in this simulation are exactly the same as in the previous case. Note that all the antennas are vertically polarized along the z-direction. The simulation setup is as follows:

1. The transmit antenna is at the origin.
2. Three receivers are placed at the locations given by the following (x, y, z) coordinates as (5,100,0), (0,100,0), and (−5,100,0), respectively, where the unit is in meter.

3. A metallic sphere with diameter of λ or 0.3 m. is located at (−5,60,0).
4. A metallic cube with the dimension of $0.3\,\text{m} \times 0.3\,\text{m} \times 0.3\,\text{m}$ is placed at (10,75,0).

Figure 5.8 shows the setup for this simulation. The received current for the SISO case, when using only Rx2 as a receiver, is $I_{SISO} = 0.020 + j0.021$ mA or the received power is $P_{SISO} = 79.7$ nW. The currents induced at the receive antennas and their received power for the SIMO case are related to $I_1 = -0.006 - j0.028$ mA, $I_2 = 0.021 + j0.021$ mA, $I_3 = -0.006 - j0.027$ mA, and therefore, $P_1 = 73.6$ nW, $P_2 = 78.3$ nW, and $P_3 = 73.7$ nW, respectively. This is similar to the results presented earlier, even though there are multipath fading, the received power from the MRC is $P_{MRC} = 225.6$ nW, which is $2.83\,P_{SISO}$ in this case.

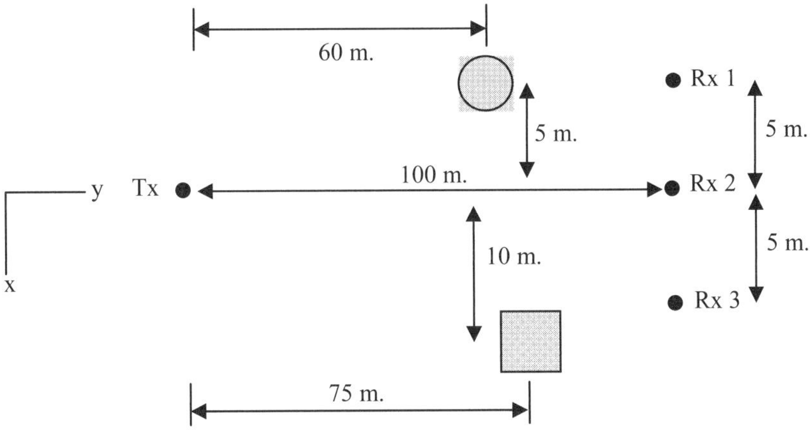

Figure 5.8. SISO and SIMO setups for case 2.

5.10.3 Case Study: 3

In this simulation, we investigate the multiplexing gain of a MIMO system along with its tolerance to a dynamic environment. We, in fact, simulate the multipath fading with some degrees of randomness. We use a MIMO system of $N_T = N_R = 2$ with the same antenna configuration as before and we use the method of parallel decomposition as our transmit scheme. The two transmitters are separated by a wavelength (λ) and the two receivers are also placed λ apart. They are located inside a room of dimension $2\,\text{m} \times 2\,\text{m}$. The room has a metallic wall of height 0.5 m. There is no ceiling and floor for the room considered in this simulation. Inside the room, there is one metallic sphere with a diameter of 0.4 m and one metallic cube with a dimension of $0.4\,\text{m} \times 0.4\,\text{m} \times 0.4\,\text{m}$. If we let the room to be centered at the origin, where the wall starts from $z = 0$ m as shown in Figure 5.9, then the sphere is centered at (−0.2, 0.3, 0.25). The cube is centered at (0.4, −0.1, 0.25). Two transmit antennas are centered at (0.15, −0.75, 0.175) and (−0.15, −0.75, 0.175). Similarly, the receive antennas are centered at (0.15, 0.75, 0.175) and (−0.15, 0.75, 0.175), respectively. In order to create random multipath fading, we vary the locations for the sphere and the cube. However, to keep the fading statistic stationary, the variations of the locations of the two objects are considered as a zero-mean Gaussian random variable with a standard deviation $\underline{\sigma}_x = \underline{\sigma}_y = \underline{\sigma}_z = \underline{\sigma}$, where $\underline{\sigma}_x$ denotes the standard deviation of the change in the location along x direction. Similarly $\underline{\sigma}_y$ and $\underline{\sigma}_z$ denote the standard deviation of the change in the location along y and z directions.

Before introducing the randomness, we evaluate the channel matrix $\mathbf{H}$ (when the sphere and cube are centered at their initial positions) and calculate its SVD. The transmit symbol is $\mathbf{x} = [x_1\ x_2]^T$ where $x_1 = 1$ and $x_2 = j$. These transmitted symbols are assumed to be on the QPSK constellation mapping. By using the SVD, our encoded symbols are $\tilde{\mathbf{x}} = \mathbf{Vx}$ as explained in section 5.4. On receive, the receiver decodes the received signal through $\tilde{\mathbf{y}} = \mathbf{U}^H\mathbf{y}$. We expect that by parallel decomposition the transmitted symbols x_1 and x_2 will be attenuated in proportion to the singular values of $\mathbf{H}$. We note here that the true values for $\mathbf{V}$ and $\mathbf{U}$ will be used through out the simulation, even though there is a variation in the channel matrix $\mathbf{H}$, due to the change in object positions. This is practically true since after the channel matrix is estimated both at the receiver side and also at the transmitter side, the estimated value will be used as

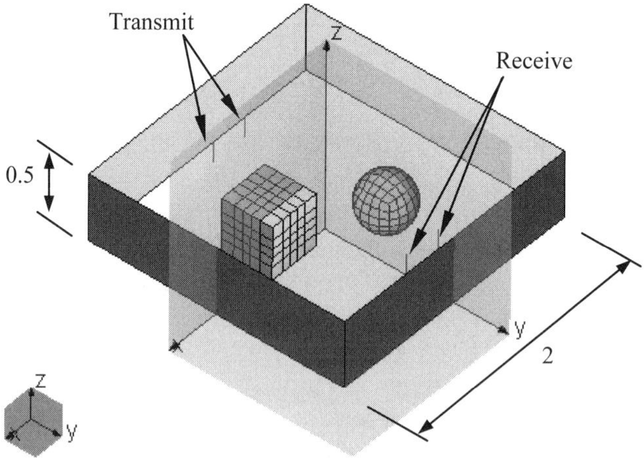

Figure 5.9. (2×2) MIMO simulation for case 3.

long as the channel is considered stationary, which is not necessarily a constant. From our simulation, the exact channel matrix is given by

$$\mathbf{H}=\begin{bmatrix} 0.022+j0.132 & -0.057+j0.042 \\ -0.195-j0.061 & 0.245+j0.587 \end{bmatrix}\cdot 10^{-3}$$

The SVD of $\mathbf{H}$ is then computed as follows

$$\mathbf{U}=\begin{bmatrix} 0.017-j0.042 & 0.131+j0.990 \\ -0.948-j0.315 & -0.045+j0.009 \end{bmatrix},\quad \Sigma=\begin{bmatrix} 0.668 & 0 \\ 0 & 0.148 \end{bmatrix}\cdot 10^{-3},$$

$$\mathbf{V}=\begin{bmatrix} 0.298 & 0.955 \\ -0.627+j0.720 & 0.196-j0.224 \end{bmatrix}.$$

When there is no variation of the scene, the decoded symbol $\tilde{\mathbf{y}}=\begin{bmatrix} 0.668 \\ j0.148 \end{bmatrix}\cdot 10^{-3}$ is exactly what it is expected to be. Note that since the phases of the two received symbols are correct, the messages can be conveyed successfully. It means that the multiplexing gain of 2 can really be obtained for this MIMO system.

Now, let us introduce some variation in the location of the two objects with $\underline{\sigma}=0.01$ m which means the standard deviation in the object position is 1 cm along all the directions. We used 20 simulations and the normalized decoded signal constellation is shown in Figure 5.10. From the simulations, it is clear that with the deviation in object locations there is no effect on the first symbol, and it can be decoded very accurately. For the second symbol, which is attenuated by a factor of 4.5 compared to the first symbol, there is a larger variation in its phase. However, this still can be decoded with high accuracy.

Next, let us increase the standard deviation further to $\underline{\sigma}=0.05$ m. The simulation was run as before with 20 different scenarios. The normalized decoded signal constellation is shown in Figure 5.11. Even when the standard deviation of the object position is only 5 cm. (or 0.17λ), this (2×2) MIMO system cannot resolve the received signals. There is no multiplexing gain in this case as we cannot transmit two symbols simultaneously. Thus, only one symbol can be transmitted at a time or the multiplexing gain is 0. Note that thermal noise is not included in all the simulations. Thus, once the vector nature of the problem is introduced, it is possible that no multiplexing gain can be achieved. It is possible that use of the inverse channel technique based on the reciprocity theorem can provide the maximum multiplexing gain as one can direct the energy to each of the receive antenna separately. This will be discussed in chapter 13.

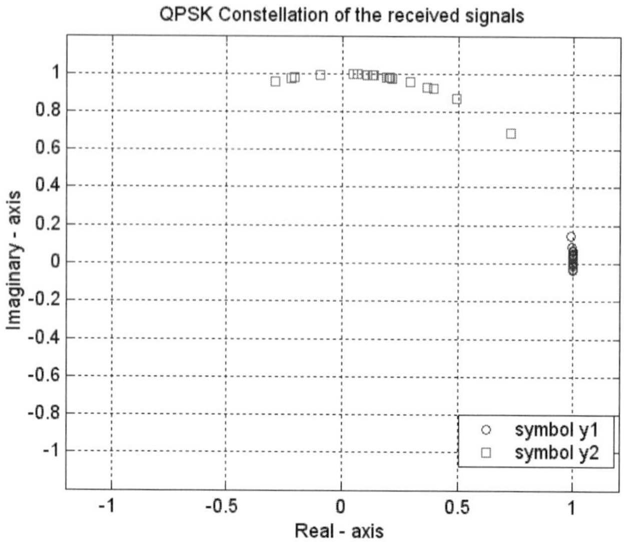

Figure 5.10. QPSK constellation of the received symbols for $\underline{\sigma} = 1$ cm.

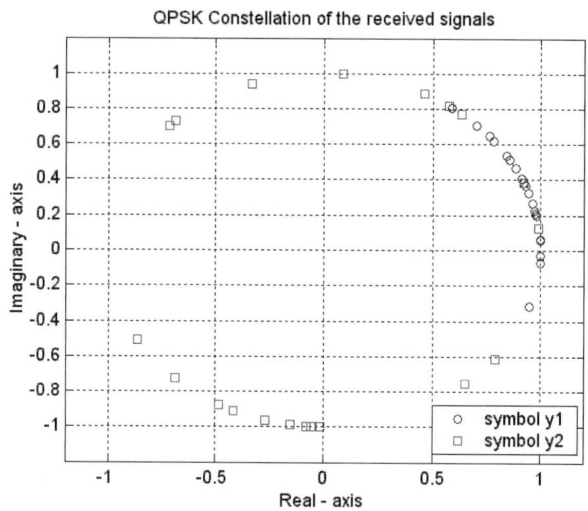

Figure 5.11. QPSK constellation of the received symbols for $\underline{\sigma} = 5$ cm.

5.10.4 Case Study: 4

In this example, we illustrate the diversity-multiplexing tradeoff in a MIMO system. We use the same experimental setup as in case 3; however, the sphere and cube are replaced by a metallic box of dimension $1\ \text{m} \times 1\ \text{m} \times 0.5\,\text{m}$ placed at the origin as shown in Figures 5.12 and 5.13. In this case, there is no direct line-of-sight (LOS) transmission from the transmitter to the receiver. For the (2×2) MIMO case, the setup is shown in Figure 5.12. The results will be compared with the SISO system shown in Figure 5.13, where there is only one antenna for the transmitter and one for the receiver.

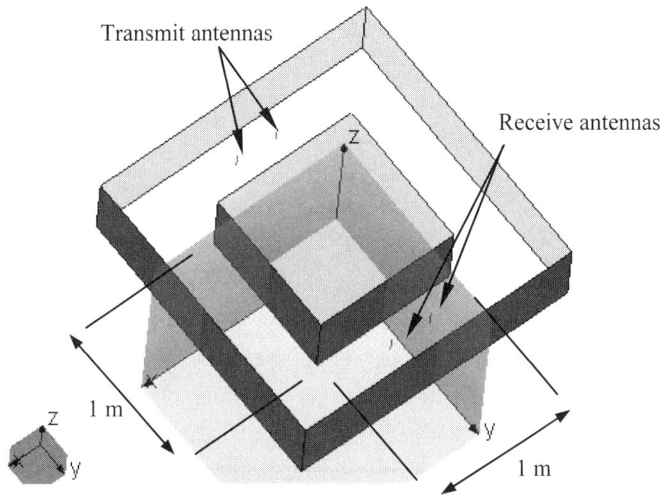

Figure 5.12. (2×2) MIMO simulation for case 4.

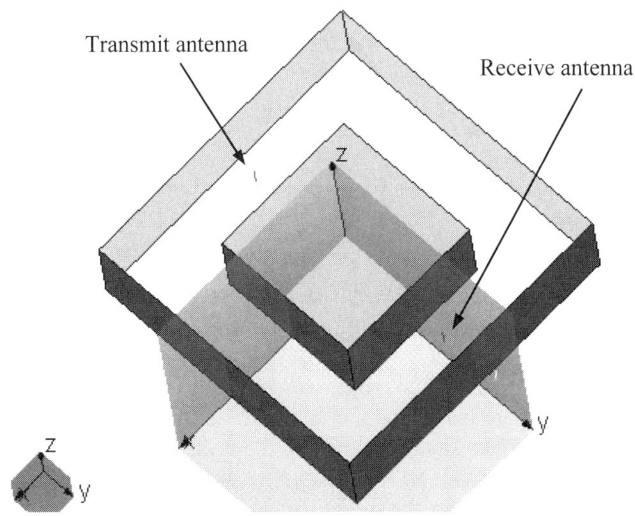

Figure 5.13. SISO simulation for case 4.

Let us consider the SISO case. When the total transmitted power is 1 W, the received current is $I_R = 13.989 + j9.958$ mA. The power received at the receiver for the SISO case is $P_{SISO} = 0.026$ W. For the MIMO case, the channel matrix can be written as

$$\mathbf{H} = \begin{bmatrix} -0.685 + j0.122 & 0.186 + j0.977 \\ 0.186 + j0.977 & -0.685 + j0.122 \end{bmatrix} \cdot 10^{-3}.$$

The SVD of this channel matrix is given by

$$\mathbf{U} = \begin{bmatrix} -0.504 - j0.495 & -0.292 + j0.644 \\ 0.504 + j0.495 & -0.292 + j0.644 \end{bmatrix}, \quad \Sigma = \begin{bmatrix} 0.0012 & 0 \\ 0 & 0.0012 \end{bmatrix},$$

$$\mathbf{V} = \begin{bmatrix} 0.707 & 0.707 \\ -0.707 & 0.707 \end{bmatrix}.$$

The input signal is selected from the QPSK constellation mapping as $\mathbf{x} = [x_1\ x_2]^T$ where $x_1 = 1$ and $x_2 = j$. Thus, the input to the MIMO system is $\tilde{\mathbf{x}} = \mathbf{V}\mathbf{x} = \begin{bmatrix} 0.707 + j0.707 \\ -0.707 + j0.707 \end{bmatrix}$. To compare the results with the SISO case, we scale the input $\tilde{\mathbf{x}}$ such that the total power of 1 W is radiated from the MIMO transmitters. Since this MIMO system transmits two symbols at a time, it is equivalent as the transmitted power is 0.5 W per symbol per transmission. The received currents at the receive antennas are as follows $I_{R1} = -13.791 - j9.472$ mA and $I_{R2} = -1.604 + j2.491$ mA. At the receiver, these currents are decoded by using $\tilde{\mathbf{y}} = \mathbf{U}^H \mathbf{y}$ as explained in section 5.3, where I_{R1} and I_{R2} are the first and the second components of vector $\mathbf{y}$, respectively. The received currents at the output of the decoder are $\tilde{\mathbf{y}} = \begin{bmatrix} 0.012 \\ j0.012 \end{bmatrix}$ A. The received powers for each symbol are $P_{R1} = 0.013$ W and $P_{R2} = 0.013$ W, respectively. Therefore, the total received power for the MIMO case is $P_{MIMO} = 0.026$ W, which is equivalent to 0.013 W per symbol. However, as mentioned earlier that the transmitted power is 0.5 W/symbol/transmission. For 1 W/symbol/transmission the received power for this MIMO example is 0.026 W/symbol, which is the same as that for the SISO case. This shows that there is no diversity gain in this MIMO example since there is no improvement in the output SNR of the system. However, there is a multiplexing gain of 2 instead, as two different symbols can be transmitted at a time. This yields an increase in the overall transmission rate of the system. This example shows the diversity-multiplexing tradeoff needs to be considered in designing a MIMO system. It is important to note that in these examples we have not provided the power balance and the amount of power fed to the transmitting antenna.

5.10.5 Case Study: 5

An example of multiantenna systems is considered in this example to illustrate that great care is needed when deploying a multiantenna system in practice. It has been shown in the previous section that a SIMO and a MISO system would give an improvement in the SNR at the receiver over a SISO system. This is mathematically true when the antennas are ideal point sources and they are operated in free space. This example shows that the improvement in the SNR is not always true especially when the antennas are deployed above a ground plane, which is the case in practice. To make this example more realistic, a transmitter composed of an array of two dipoles is placed 20 m above a perfect electric conductor (PEC) ground plane and a receiving dipole antenna is placed 2 m above the ground. This scenario represents a base station tower (as a transmitter) and a mobile unit (as a receiver). All of the antennas are conjugately matched so that they radiate maximum power. The operating frequency is $f = 1\,\text{GHz}$ and the radius of the dipoles is $\lambda/300$ (1 mm.). The received power for the 2×1 MISO system will be compared with the one that would be obtained from a SISO system when its transmitting antenna is placed at the center position of the MISO transmitter. In this comparison, the spacing between the two transmitting antennas for the MISO system and the distance between the transmitter and the receiver are considered to be the variables. The spacing between the two transmitting antennas is varied from 0.4λ to 50λ and the distance between the transmitter and the receiver is varied from 5λ to 100λ. The scenario of the simulation is shown in Figure 5.14. The received power ratio between the MISO system and the SISO system when they use the same input power to the transmitter is shown in Figure 5.15. From an array processing theory, the gain of 3 dB would be expected from the 2×1 MISO system. However, as shown in the Figure 5.15, there are many areas (the dark areas) that the received power from the MISO system is less than the one from the SISO system. At some points the MISO system receives power as low as -10 dB when compared to the SISO system. Figure 5.16 shows a power ratio when the transmitting antennas are 30λ apart. It is obvious that there are many positions from the transmitter where the mobile unit may be placed so that a SISO system will perform better than a MISO system. Thus, what is predicted from the scalar statistical theory is not always correct as the presence of the ground plane representing earth may skew the results. This is because, an electromagnetic wave is vector in nature and therefore to predict the interference pattern correctly, it is necessary to know the phase between the various vectors.

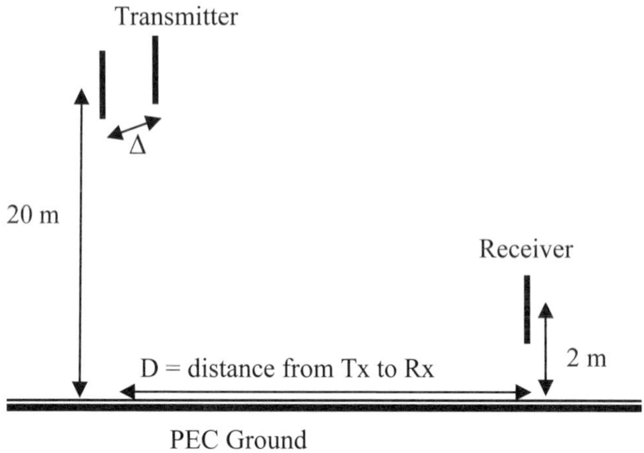

Figure 5.14. MISO system setup; Transmitter is 20 m above ground; Rx is 2 m above ground; Δ = transmitting antenna spacing; D = distance between the transmitter and the receiver. (For the SISO case, the transmitting antenna is placed at the center of the MISO transmitter.)

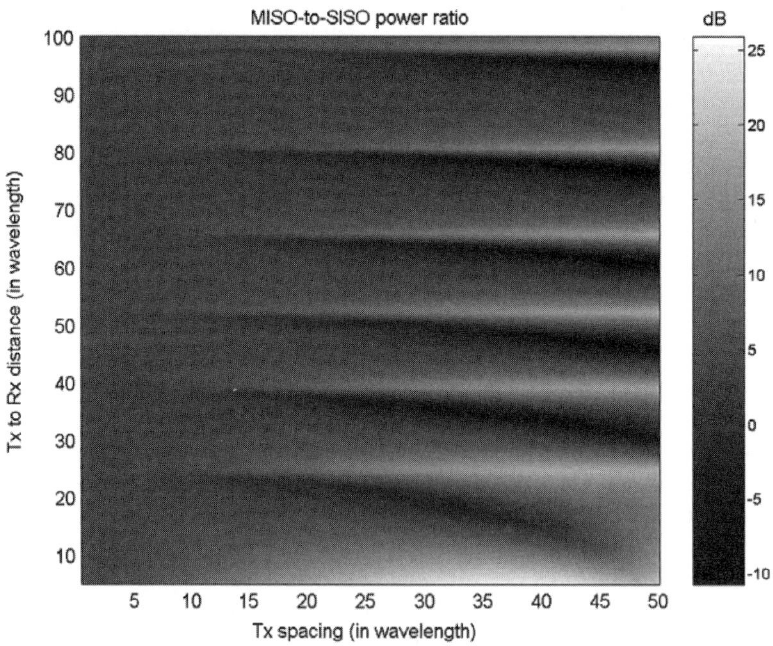

Figure 5.15. Received power ratio between MISO and SISO systems when the transmitter is 20 m and the receiver is 2 m above ground, respectively.

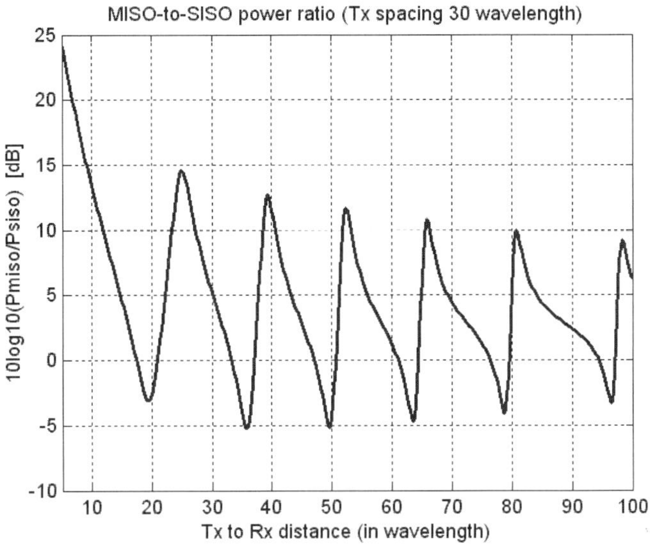

Figure 5.16. Received power ratio when the transmitting antennas are 30λ apart.

These simple examples demonstrate the shortcomings of an exclusive statistical analysis without paying attention to the electromagnetic aspects. That is without considering the electromagnetic scenario; the MIMO system performance can be misinterpreted. Further study on the electromagnetic effects to MIMO systems need to be carried out to establish the credibility of this methodology.

5.11 PHYSICS OF MIMO IN A NUTSHELL

The objective of MIMO is to provide spatial diversity through the use of multiple transmit and receive antennas. So, if there is N transmit and N receives antennas, then one can generate N spatially orthogonal modes to communicate between these transmit-receive systems. The goal in MIMO then is to simultaneously communicate with these N spatial modes using N transmit and N receive antennas. We now look into the system engineering aspects of this deployment and observe what really is possible from a physics perspective by employing the Maxwell's equations for the analysis. We consider several different cases of antenna arrays oriented along the broadside directions and antenna arrays oriented along the end fire directions. In addition we consider the radiation efficiency of the various spatial MIMO modes for a constant input power. Two different scenarios are chosen for the evaluation of the various spatial modes both in the presence and in the absence of a direct line-of-sight path between the transmitter and the receiver.

5.11.1 Line-of-Sight (LOS) MIMO Systems with Parallel Antenna Elements Oriented Along the Broadside Direction

In this study we consider several different cases, where we have 0.5 λ long dipole antenna elements with a radius of 1 mm all oriented vertically. We consider a 1 × 1 MIMO system, which is a SISO, to a 5 × 5 MIMO system, where there are 5 transmit and 5 receive antennas. In an array, both the transmit and the receive antennas are located half a wavelength apart as shown in Figure 5.17a. The transmit and the receive antenna arrays are horizontally separated by 100 m and it is operating at 1 GHz. We consider several scenarios of the transmit and the receive antenna arrays, where the entire array may be located in free space or situated at different heights above a ground plane whereas the received array is situated 2 m above a perfect ground plane as seen in Fig. 5.17b. We use an electromagnetic analysis code, excite one antenna at a time and compute the channel matrix using a voltage excitation to each of the antennas of this MIMO array. Furthermore, each antenna element both in the transmit and receive array are conjugately matched with a complex value of the load impedance so that they can efficiently radiate and receive the various electromagnetic signals. For a N × N MIMO system we compute the voltage channel matrix $[H_v]_{N\times N}$ which will be a N × N square matrix. In this case, N can take any value between 1 and 5. We perform a singular value decomposition of the matrix $[H_v]$ to observe how effectively each spatial mode will radiate with respect to the SISO case. This is accomplished by squaring the ratio of the singular values for the voltage channel matrix scaled by the singular value of the 1 × 1 MIMO system. For these conjugately matched antennas, the square of the voltage singular values will represent how efficiently each spatial mode of a MIMO system is radiating with respect to the SISO case for the same input power.

Table 5.1 provides the square of the ratio of the various singular values with respect to the SISO case. So if we consider the radiation efficiency for the

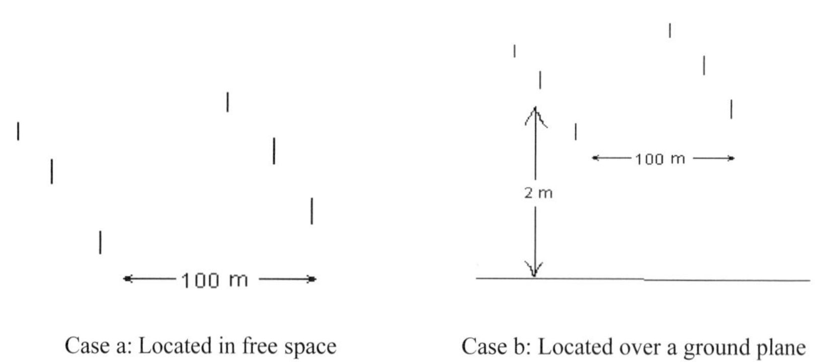

Figure 5.17. A typical 3 × 3 MIMO system consisting of half wave dipoles, half wavelength spaced and separated by 100 m.

SISO case to be unity, a value greater than one will indicate that the radiation efficiently of that particular spatial MIMO mode is better than the SISO case. In that case, the use of this spatial MIMO mode has a definite advantage over the use of a SISO. For this example, there is a direct LOS connection between the transmitter and the receiver.

Table 5.1. Ratio of the Square of the Singular Values for the Various Spatial MIMO Modes with Respect to the SISO Case (Broadside Orientation).

SISO 1×1	MIMO 2×2	MIMO 3×3	MIMO 4×4	MIMO 5×5
1.0	5.21	11.95	22.18	34.85
	3.73×10^{-6}	9.67×10^{-5}	6.26×10^{-4}	2.75×10^{-3}
		2.85×10^{-11}	1.88×10^{-9}	2.40×10^{-8}
			4.28×10^{-16}	5.42×10^{-14}
				9.15×10^{-21}

Table 5.1 presents the results when the transmit and the receive antenna array is operating in free space. We also consider cases, where these antenna arrays are placed over a ground plane with the receive antenna array located at 2 m above a perfect ground plane whereas the height of the transmit antenna is varied from 2 m and 20 m above the ground plane. It is important to note that the ratios of the singular values do not change up to the second place of decimal for different widely varying scenarios in the deployment of the arrays, even when the ratio of the square of the singular values is less than $\leq 10^{-15}$, even though the individual values may change greatly for the different MIMO systems. For the last two rows of Table 5.1 some change is observed in the last two values of the 5 × 5 MIMO case. However, since these spatial modes are so inefficient radiators that it would not be wise to use them at all. There are three observations that can be made from this Table.

1. There is in fact ***only*** one spatial mode that is really useful and provides a real gain over the SISO case. This is the classical broadside, phased array mode when all the antennas are excited in phase.
2. Beside this dominate mode, the other spatial modes are essentially useless for real applications as they are at least 10^{-6} lower than the dominant mode. This implies that if one uses 1 W of power to excite the dominant mode, one needs to put 1 MW to excite the transmitting antenna to produce similar values for the received power using the second spatial mode in the 2 × 2 MIMO system.
3. The power gain of the phased array mode which is the dominant MIMO mode is greater than N^2 over the SISO case.

In summary, with a direct line-of-sight link between the transmitter and the receiver, there is only one spatial mode that is really useful from an engineering perspective as evidenced from the numerical results obtained from the solution of the Maxwell's equations. Next, we consider a different orientation of the transmit-receive system.

5.11.2 Line-of-Sight MIMO Systems with Parallel Antenna Elements Oriented Along the Broadside Direction

In this study we consider the antenna elements for each of the transmit and receive antenna array system to be located as a collinear array as shown in Figure 5.18a, when they are located in free space and in Figure 5.18b when they are located over a perfectly conducting ground plane. In this form of the deployment all the antenna elements are vertically oriented and located one on top of the other. This end fire configuration may be perhaps useful as it has smaller mutual coupling between the elements in the array.

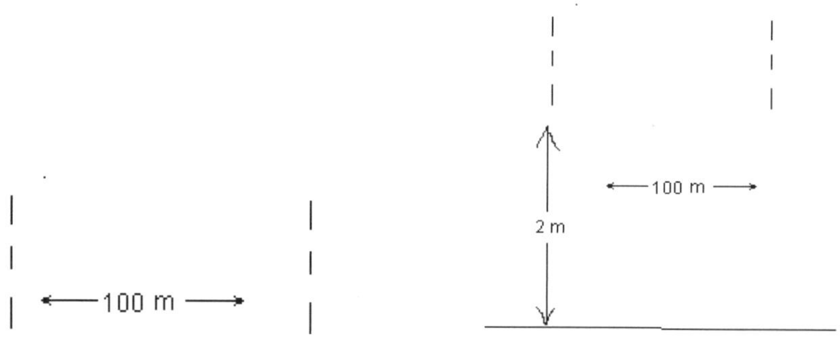

Case a: Located in free space   Case b: Located over a ground plane

Figure 5.18. A typical 3×3 collinear array MIMO system consisting of half wave dipoles, half wavelength spaced and separated by 100 m.

The antenna elements are of the same length and radius as in the previous case. However, in this situation their center to center separation along the vertical direction is 1 λ. They are again conjugately matched with their respective loads so that they radiate in the most efficient manner. Table 5.2 provides the square of the singular values of the transfer voltage matrix normalized with respect to the SISO case. The transmit and the receive array are separated by 100 m.

Table 5.2 provides the results when the transmit and the receive array are operating in free space with a line of sight link existing between them. As before there is one dominant mode, and the second spatial mode is at least down by 10^{-2} over that of a SISO. This mode may perhaps work for N = 4 and 5. This implies that instead of a 1 MW of transmit power for the previous case, one will require approximately 100 W to induce similar received power. Hence, it may not be very useful from a practical stand point. One could again draw similar conclusions as before:

1. The useful spatial mode in an endfire MIMO system has a power gain of slightly greater than N^2 over a SISO system.
2. The second dominant mode is down at least by 10^{-2} from the dominant mode.

Table 5.2. Ratio of the Square of the Singular Values for Various Spatial MIMO Modes with Respect to the SISO Case (Collinear Array Over a Ground Plane).

SISO 1×1	MIMO 2×2	MIMO 3×3	MIMO 4×4	MIMO 5×5
1.0	4.46	10.58	19.45	31.05
	7.96×10^{-5}	1.38×10^{-3}	9.13×10^{-3}	3.79×10^{-2}
		1.22×10^{-8}	4.68×10^{-7}	6.00×10^{-6}
			3.47×10^{-12}	2.33×10^{-10}
				1.60×10^{-15}

However, this endfire array system is more sensitive to the deployment of these antennas over a perfectly conducting ground plane representing moist earth, than when they are deployed in free space. For example, when both the transmitter and the receiver are located 2 m above a perfectly conducting ground plane as shown in Fig 5.18b, the respective ratios of the singular values are now given by Table 5.3.

Table 5.3. Ratio of the Square of the Singular Values for the Various Spatial MIMO Modes for Endfire Arrays with Respect to the SISO Case (Collinear Array Over a Ground Plane).

SISO 1×1	MIMO 2×2	MIMO 3×3	MIMO 4×4	MIMO 5×5
1.0	3.22	5.02	5.66	5.68
	1.12×15^{-3}	3.18×10^{-2}	3.44×10^{-1}	2.16
		7.53×10^{-8}	4.74×10^{-6}	9.53×10^{-5}
			8.66×10^{-12}	1.21×10^{-9}
				1.23×10^{-15}

In the case for the 5 × 5 array, presence of the ground plane may make 2 spatial modes of transmission viable, but the gain of each of the spatial modes over that of the SISO case is quite small. Hence, one need to see whether it is cost effective to deploy a 5 × 5 MIMO system over two SISO systems in this case. In addition, the other spatial modes are not that useful as they are at least down by several orders of magnitude over that of the SISO case. However, a collinear array has a better promise than a broadside oriented array. It appears, the promise of multiple spatial modes really does not pan out for arrays when there is a direct LOS link between the transmitting and the receiving arrays.

Next we consider systems where the LOS is not present between the transmitter and the receiver arrays.

5.11.3 Non-line-of-Sight MIMO Systems with Parallel Antenna Elements Oriented Along the Broadside Direction

In this example, we consider MIMO systems located between two concentric conducting cylinders with no ground plane. Here the antennas are considered to be half wavelength long of radius 1 mm and are conjugately matched to receive and transmit maximum power. Each of the elements in the array are spaced half a wavelength apart. The separation between the transmit and the receive antenna arrays is 100 m as before. The inner conducting cylinder is of dimensions 96 m × 8 m × 0.5 m and the outer conducting cylinder are of dimension 104 m × 10 m × 0.5 m as illustrated in Fig. 5.5. Hence these two conducing cylinders direct the transmitted power to the receiver. In this situation we would like to observe how many of the spatial modes are effective radiators over that of a SISO. It is generally perceived by many practioners that apparently in this situation the MIMO will provide a great benefit over SISO. Our objective is to examine such concepts under a more rigorous physics based analysis to check, if these statements are really true. In Table 5.4, the various ratio of the square of the singular values for the different spatial modes of a MIMO system is normalized with respect to that of a SISO system.

Table 5.4. Ratio of the Square of the Singular Values for the Various Spatial MIMO Modes with Respect to the SISO Case (Broadside Array with No Line-of-sight Communication).

SISO 1×1	MIMO 2×2	MIMO 3×3	MIMO 4×4	MIMO 5×5
1.0	5.87	24.86	33.34	60.07
	7.83×10^{-4}	5.73×10^{-2}	2.70×10^{-1}	9.84×10^{-1}
		9.12×10^{-5}	2.86×10^{-3}	3.52×10^{-2}
			5.35×10^{-6}	6.3×10^{-6}
				2.42×10^{-6}

Table 5.4 presents the ratio of the square of the singular values for the channel voltage transfer function matrix. In this situation, there is no direct LOS communication between the transmitter and the receiver. When there is no LOS, the second mode is about 10^{-2} below the dominant spatial mode in power. In comparison to the SISO case, it appears that for the 5 × 5 MIMO case there is a significant increase in the gain for the dominant mode. There is a potentially useful second mode, but the other spatial modes are at least an order of magnitude lower than the SISO case. So, it appears that even though there is some potential for improvement in the singular values when the energy is channelized towards the receiver, even for the 5 × 5 MIMO case, only 2 spatial modes may be of sufficient value. To look at this situation from an engineering perspective, only one spatial modes is really useful over a SISO and the other spatial mode may or may not be useful depending on the environment. We will look at a different scenario next. If we take away the focusing effect introduced by the outer conducting cylinder by simply removing it and keeping only the inner conducting cylinder, so that it still eliminates the direct line of sight, the results are not very encouraging. Table 5.5 provides the results for the square of the various singular values for the various spatial MIMO modes over the SISO case.

Table 5.5. Ratio of the Square of the Singular Values for the Various Spatial MIMO Modes with Respect to the SISO Case (Broadside Array with No Line-of-sight and no Outer Conducting Cylinder).

SISO 1×1	MIMO 2×2	MIMO 3×3	MIMO 4×4	MIMO 5×5
1.0	7.02	35.74	57.00	119.0
	1.68×10^{-4}	4.39×10^{-3}	1.50×10^{-2}	6.02×10^{-2}
		1.50×10^{-5}	4.92×10^{-5}	4.27×10^{-4}
			5.38×10^{-6}	9.76×10^{-6}
				3.84×10^{-6}

Table 5.5 illustrates that in the absence of line of sight, the dominant spatial mode can be an extremely efficient radiator. However, the second spatial mode leaves very little to be desired as it is at least an order of magnitude less than the SISO case, even when there is no line-of-sight link between the transmitter and the receiver.

By observing the data from Tables 5.1 to 5.5, it is natural to conclude that in a MIMO system only one spatial mode is really useful. The other spatial modes are not better radiators than a SISO system. This has significant practical implications. Even if one is deploying arrays of antennas for the transmit or for the receive system, the only advantage is that it produces a high gain over a SISO system as one would expect from the phased array theory. Even though the other spatial modes for a MIMO system may be proven to exist from a purely theoretical point of view, using a physics based analysis indicate that they are not good carriers of signals from a practical point of view.

Perhaps, it would make more sense from a practical standpoint to deploy a MISO system as it will be as effective as a MIMO system, without the additional cost of deploying a phased array at the receiver and significantly reducing the hardware complexity. This will be relevant for mobile systems where the footprint at the receiver is quite small. In other words, this is the typical phased array scenario that we are back to!

In summary, if there is going to be only a single spatial useful mode for practical application, then it will be worthwhile to deploy a MISO system with adaptive processing. Adaptive processing in this case will have a better potential utility. Hence, in the next chapter 6, we look into the various forms of the optimum filters and select one which will be suitable for our applications. We also address the issue of dealing with multiple uncorrelated receivers in a MISO system and illustrate how the signal can be directed to an intended receiver using the principles of reciprocity. This topic will be addressed in chapter 13.

5.12 CONCLUSION

In this chapter, the MIMO technology is discussed starting from the basic statistical idea to the implementations. By using multiple antennas at the transmitter and/or at the receiver, the performance of a wireless communication system can sometimes be improved. However, from a system point of view, whether the system will actually work in practice or not cannot solely be determined from a numerical value obtained from a statistical analysis. For proper operation of the system, one needs to know the actual power exchanged between the transmitter and the receiver.

Multiple antennas allow us to transmit signals spatially through a number of independent paths caused by multipath fading. It has been shown that the diversity gain or the improvement in the received SNR can be achieved at both ends of the wireless systems via coding schemes. The transmission rate, or multiplexing gain, can also be increased by using multiple antennas depending on whether the knowledge of channels is available or not. With the knowledge of the channels, the MIMO systems provide a number of independent paths for the transmission. Diversity and multiplexing gain tradeoff is discussed as a criterion for consideration in a MIMO system design. The electromagnetic effects to the MIMO channels are also illustrated through numerical simulations. It becomes obvious that without taking into account the electromagnetic effects the expected MIMO system performance may not be realized to its full potential. Great care needs to be taken when designing a wireless system based on an array theory as the vector nature of the electromagnetic fields can provide a completely different picture over the conclusions arrived at by performing a scalar analysis.

With the introduction of a new dimension, i.e., space, to the wireless communications, the MIMO technology is a promising tool to bring communication systems toward the 4G and beyond. However, its success can only be guaranteed if the design is carried out using fundamental physical principles based on a Maxwellian framework.

REFERENCES

[1] G. J. Foschini, "Layered space-time architecture for wireless communication in a fading environment when using multi-element antennas," *Bell Labs Tech. J.*, Vol. 1, No. 2, pp.41-59, 1996.

[2] M. D. Migliore, "An intuitive electromagnetic approach to MIMO communication systems," *IEEE Antenna and Propagation Magazine*, Vol. 48, No. 3, Jun. 2006.

[3] M. D. Migliore, "On the role of the number of degrees of freedom of the field in MIMO channel," *IEEE Trans. on Antenna and Propagation*, Vol. 54, No. 2, pp. 620-628, Feb. 2006.

[4] M. A. Jensen and J. W. Wallace, "A review of antennas and propagation for MIMO wireless communications," *IEEE Trans. on Antenna and Propagation*, Vol. 52, No. 11, pp. 2810-2824, Nov. 2004.

[5] D. G. Brennan, "Linear Diversity Combining Techniques," *Proc. IRE.*, Vol. 47, pp. 1075-1102, June 1959.

[6] J. G. Proakis, *Digital Communications*, McGraw-Hill, New York, 2001.

[7] H. Bölcskei, D. Gesbert, C. B. Papadias, and A.-J. Van Der Veen, *Space-Time Wireless Systems from Array Processing to MIMO Communications*, Cambridge University Press, UK, 2006.

[8] G. M. Calhoun, *Third Generation Wireless Systems, Volume 1: Post-Shannon Signal Architectures*, Artech House, Norwood, MA, 2003.

[9] D. S. Watkins, *Fundamentals of Matrix Computations*, John Wiley, 2nd. Ed., New York, 2002.

[10] C. E. Shannon, "A mathematical theory of communication," *Bell System Tech. J.*, Vol. 27, Jun. 1948.

[11] C. E. Shannon, "Communication in the presence of noise," *Proceeding of the IEEE*, Vol. 86, No. 2, pp. 447-458, Feb. 1998.

[12] A. Paulraj, R. Nabar, and D. Gore, *Introduction to Space-Time Wireless Communications*, Cambridge University Press, United Kingdom, 2003.

[13] S. Hwang, A. Medouri, and T. K. Sarkar, "Signal enhancement in a near-field MIMO environment through adaptivity on transmit," *IEEE Trans. on Antenna and Propagation*, Vol. 53, No. 2, Feb. 2005.

[14] S. Barbarossa, *Multiantenna Wireless Communication Systems*, Artech House, Norwood, MA, 2005.

[15] S. M. Alamouti, "A simple transmit diversity technique for wireless communications," *IEEE J. Sel. Areas. Comm.*, Vol. 16, No. 8, pp. 1451-1458, Oct.1998.

[16] L. Zheng and D. N. C. Tse, "Diversity and multiplexing: A fundamental tradeoff in multiple-antenna channels," *IEEE Trans. on Information Theory*, Vol. 49, No. 5, pp. 1073-1096, May 2003.

[17] X. Ma and G. B. Giannakis, "Full-diversity full-rate complex-field space-time coding," *IEEE Trans. on Signal Processing*, Vol. 49, pp. 2917-2930, Nov. 2003.

[18] T. K. Sarkar, M. Wicks, M. Salazar-Palma, and R. Bonneau, *Smart Antennas*,

John Wiley, New York, 2003.

[19] T. K. Sarkar, S. Burintramart, N. Yilmazer, S. Hwang, Y. Zhang, A. De, and M. Salazar-Palma, "A discussion about some of the principles/practices of wireless communication under a Maxwellian framework," *IEEE Trans. on Antenna and Propagation*, Vol. 54, No. 12, Dec. 2006.

[20] B. M. Kolundzija, J. S. Ognjanovic, and T. K. Sarkar, *WIPL-D: Electromagnetic Modeling of Composite Metallic and Dielectric Structures, Software and User's Manual*, Artech House, Norwood, MA, 2000.

[21] S. Kay, "Can Detectability Be Improved by Adding Noise?," *IEEE Signal Processing Letters*, Vol. 7, No. 1, Jan 2000, pp. 8-10.

6

USE OF THE OUTPUT ENERGY FILTER IN MULTIANTENNA SYSTEMS FOR ADAPTIVE ESTIMATION

6.0 SUMMARY

The goal of deploying multiantenna systems is to adaptively enhance the signal of interest in presence of jammers both desired and undesired, clutter and noise. This is achieved by adaptively weighting the vector values of all the voltages received at all the antennas. The same principle can also be applied on transmit to direct the energy towards a receiver of interest by adaptive weighting of the excitation voltages at each of the antennas. The application of digital beam forming over analog beam forming allows one to cancel closely spaced interferers without increasing the physical size of the antenna arrays. That is why, in this chapter, we study various versions of digital filters for enhancing the signal-to-noise ratio at the receiver or focus the radiated energy along a particular direction in the presence of near field scatterers. The three optimum filters often described in the literature for enhancement of signals are the matched filter, Wiener filter and the output energy filter. They are all termed optimum filters in their respective applications and therefore it is necessary to know under what conditions each of the filters is optimum. This topic is described in section 6.1. The output energy filter is then chosen as it fits our requirements of adaptively extracting a signal in the presence of strong interferers both coherent and non-coherent, and noise. The novelty of this approach is that this procedure can be applied to a single snapshot of the data and therefore is quite suitable to operate in a highly dynamic environment or for cases where one has to deal with blinking jammers. This single snapshot based adaptive procedure is described in section 6.2 and it has exactly the same number of degrees of freedom for the coherent signal case as it would be in a statistical multiple snapshot based methodology. A salient feature of this direct data domain technique is that it does not need any information at all on the statistics of the clutter and solves the estimation problem under the least squares metric. In addition, simultaneous applications of independent multiple implementations of this technique can be used in a real situation where the solution is unknown. Multiple independent estimates of the same solution can therefore increase the level of confidence in the results. The operation counts for these adaptive methods are an order of magnitude lower than the statistical approaches and can be efficiently implemented in a digital processor for real time processing. In section 6.3, this methodology is extended to deal with Space-Time Adaptive Processing (STAP) where the goal is to extract the signal of interest by filtering in Doppler and spatial angle and for a given range cell. Finally in section 6.4, a comparison is made between the direct data domain approaches to STAP with the classical stochastic methodology using real data from the airborne MCARM platform to illustrate the superior performance of the former.

6.1 VARIOUS FORMS OF THE OPTIMUM FILTERS

There are typically three different types of filters [1] which have been or are extensively used in the signal extraction area depending on their desired goals and their performance requirements. All of them interestingly have been termed as *optimum* filters. The term *optimum filter* therefore by itself is quite ambiguous. In this context the term optimum is nebulous, as each filter optimizes different mathematical criteria. And hence it is optimum under those conditions.

Next we describe the three different filters from a mathematical perspective and illustrate which one is suitable for our application following the methodology described in [1]. Consider a signal consisting of two discrete data sequence $\{b_0; b_1\}$, as an example. This signal is fed to a filter. Let the filter coefficients be also given by the two sample discrete sequence $\{a_0; a_1\}$. Then the output of the filter will be given by the sequence $\{c_0; c_1; c_2\}$ which will be the result of the convolution between the input sequences $\{b_0; b_1\}$ and the impulse response of the filter $\{a_0; a_1\}$. If ✪ denotes the convolution between the two sequences then

$$\{c_0;\ c_1;\ c_2\} = \{b_0;\ b_1\} \text{ ✪ } \{a_0;\ a_1\} = \{a_0 b_0;\ a_0 b_1 + a_1 b_0;\ a_1 b_1\}. \tag{6.1}$$

The output sequence of the filter will then be equal to

$$c_0 = a_0 b_0;\ c_1 = a_0 b_1 + a_1 b_0;\ c_2 = a_1 b_1. \tag{6.2}$$

Now we introduce the concept of the three filters and illustrate how they individually tend to modify the output. The three types of filters that are going to be described are [1]:

- A *Matched filter* also called a *cross-correlation filter*
  This filter maximizes the output signal-to-noise ratio at a particular time instance (say at t = 0) at the output of the filter.
- A *Wiener filter*
  The goal of the Wiener filter is to describe a linear filter that will match the output of the filter to a given desired waveform.

- An *output energy filter* also called a *minimum variance filter*
This filter maximizes/minimizes the total output energy at the output of the filter.

Next we illustrate what mathematical criteria are used to design each of these optimum filters and in what sense they are optimum [1].

6.1.1 Matched Filter (Cross-correlation filter) [1]

The objective here is to maximize the specific value at the output c_n of the filter at a particular instance of time $t = n$, which is hopefully large enough so that the presence or absence of the signal at this point can be established. The matched filters were primarily developed for radar applications. In a radar problem, we transmit a pulse and then we try to receive the same shaped pulse after some delay τ. For the radar case, the received signal will have the same shape as the transmitted pulse but reduced in amplitude due to the propagation loss in the media as per the divergence of the waves dictated by the Huygens's principles. The goal here is to detect whether the transmitted radar signal has been reflected from a target and whether it is present at the input of the filter. The presence of the signal will be characterized by a large value of the output signal-to-noise ratio at the time instance $t = \tau$. Whatever happens at other time instances at the output of the filter is of no interest to us! Hence, the goal is: *Maximize the square of the output value* c_1^2, *subject to the unit energy constraints on the filter coefficients* $a_0^2 + a_1^2 = 1$. The unit energy constraints of the filter coefficients are necessary; otherwise, one can make the filter coefficients arbitrarily large to maximize the output signal-to-noise ratio (SNR) at the output of the filter at the particular time instance $t = 1$.

Therefore, by the terms of the problem,

$$c_1^2 = \left(a_0 b_1 + a_1 b_0\right)^2 . \tag{6.3}$$

By using the Cauchy-Schwarz inequality, one can transform the above equality to

$$c_1^2 \leq \left(a_0^2 + a_1^2\right)\left(b_1^2 + b_0^2\right). \tag{6.4}$$

This is true in general and the equality for the above case holds only if $\{a_0; a_1\} = \aleph\{b_1; b_0\}$ where $\aleph$ is a scalar constant. By imposing the unit energy constraints on the filter coefficients results in the value of the scalar constant as

$$\aleph = 1\Big/\sqrt{\left(b_0^2 + b_1^2\right)} . \tag{6.5}$$

Therefore the solution to the matched filter is given by

$$\{a_0; a_1\} = \left\{b_1\Big/\sqrt{\left(b_0^2 + b_1^2\right)};\ b_0\Big/\sqrt{\left(b_0^2 + b_1^2\right)}\right\} . \tag{6.6}$$

The optimum filter given by these coefficients is matched to the signal (the reverse of the signal in time) and hence this filter is called the matched filter [1]. Therefore the matched filter maximizes the output signal-to-noise ratio at a particular time instance ($t = 1$, in this case). This can only occur if the shape of the input signal in time is the reverse of the shape of the temporal response of the filter and thus is useful for radar target detection.

6.1.2 A Wiener Filter [1]

The Wiener filter is optimum in the sense that it is the only linear filter which minimizes the mean squared error between the desired output d_t and the actual output c_t obtained from the filter [1]. However, neither the matched filter nor the output energy filter allows us to control the *shape* of the actual output. Here, the objective is: *Minimize the sum of the square of the differences between the values of the desired waveform d_t and the values obtained from the actual output c_t of the filter due to a given input signal.* In this case, the unit energy constraints on the filter coefficients are not necessary as the desired sample values that the output of the filter needs to match are given explicitly. Therefore by the terms of the problem, the goal is to minimize the least squares error Υ between the given signal waveform values $\{d_0; d_1; d_2\}$ and the actual output $\{c_0; c_1; c_2\}$ obtained from the filter. Hence, the goal is to minimize the functional

$$\begin{aligned}\Upsilon &= (d_0 - c_0)^2 + (d_1 - c_1)^2 + (d_2 - c_2)^2 \\ &= (d_0 - a_0 b_0)^2 + (d_1 - a_0 b_1 - a_1 b_0)^2 + (d_2 - a_1 b_1)^2\end{aligned} \tag{6.7}$$

The filter coefficients which will minimize the error Υ will be given by taking the following partial derivatives and setting them to zero, resulting in

$$\partial\Upsilon/\partial a_0 = 0 \quad \text{and} \quad \partial\Upsilon/\partial a_1 = 0. \tag{6.8}$$

This produces

$$\begin{bmatrix} r_0 & r_1 \\ r_1 & r_0 \end{bmatrix} \begin{pmatrix} a_0 \\ a_1 \end{pmatrix} = \begin{pmatrix} b_0 d_0 + b_1 d_1 \\ b_0 d_1 + b_1 d_2 \end{pmatrix} \tag{6.9}$$

where the autocorrelation sequence of the data is given by

$$r_0 = b_0^2 + b_1^2 \tag{6.10}$$

$$r_1 = b_0 b_1 \tag{6.11}$$

and

$$\begin{pmatrix} a_0 \\ a_1 \end{pmatrix} = \begin{bmatrix} r_0 & r_1 \\ r_1 & r_0 \end{bmatrix}^{-1} \begin{pmatrix} b_0 d_0 + b_1 d_1 \\ b_0 d_1 + b_1 d_2 \end{pmatrix} \tag{6.12}$$

Hence, the filter coefficients are computed by the product of the inverse of a matrix whose entries are the autocorrelation of the input signal with a vector which is given by the cross-correlation between the desired waveform and the input waveform to the filter.

It is seen that, in order to implement a Weiner filter, one essentially has to know the desired waveform that one is going to match the actual signal to. Almost all adaptive algorithms implemented in hardware in the signal processing and communication theory literature are based on the Weiner filter theory. Hence, in all current applications, one needs to carry out a calibration procedure, which is tantamount to solving for the Wiener filter of the system before each and every transmission. This is a serious problem in a real-time implementation. Secondly, when the scene is changing very quickly, it may not be possible to perform the calibration procedure adequately before each transmission. The need to carry out frequent calibration can be minimized for the next type of filters.

6.1.3 An Output Energy Filter (Minimum Variance Filter) [1]

For this filter, the goal is to deal with the total energy in the entire output and not the output value at a particular time instance, as is the case for a matched filter. Hence, in this case one is dealing with the total SNR at the output. The difference between an inverse filter and an output energy filter is that for the inverse filter there will be a delta function at the time origin of the output of the filter. In some situations, for the inverse filter the impulse may not be located at the origin but could be time shifted. Hence, this filter is a generalization of the inverse filter. The goal of this filter is to: *Maximize/Minimize the sum of the squares of all the output values* $c_0^2 + c_1^2 + c_2^2$, *subject to the unit energy constraints on the filter coefficients* $a_0^2 + a_1^2 = 1$.

By the terms of the problem, the objective is to maximize/minimize the output energy $c_0^2 + c_1^2 + c_2^2$ from the filter subject to $a_0^2 + a_1^2 = 1$. The constraints on the filter coefficients are necessary so as to make sure one does not get the trivial solution as the result of the optimization of the output energy of the filter. The above specifications result in maximizing/minimizing the following expression for Ξ involving the Lagrange multiplier λ, so that the following expression is optimized.

$$\Xi = c_0^2 + c_1^2 + c_2^2 - \lambda\left(a_0^2 + a_1^2 - 1\right). \tag{6.13}$$

Therefore, to obtain a stationary point for the output energy is equivalent to setting the partial derivatives of Ξ with respect to a_0 and a_1 equal to zero, and this results in

$$\begin{aligned} r_0 a_0 + r_1 a_1 - \lambda a_0 &= 0 \ ; \\ r_1 a_0 + r_0 a_1 - \lambda a_1 &= 0 \end{aligned} \tag{6.14}$$

with r_0 given by (6.10) and r_1 by (6.11). Hence,

$$\left(\begin{bmatrix} r_0 & r_1 \\ r_1 & r_0 \end{bmatrix} - \lambda \begin{bmatrix} 1 & 0 \\ 0 & 1 \end{bmatrix}\right) \times \begin{pmatrix} a_0 \\ a_1 \end{pmatrix} = \begin{pmatrix} 0 \\ 0 \end{pmatrix}. \tag{6.15}$$

In this equation, the maximum/minimum eigenvalue λ in magnitude provides the signal-to-noise ratio at the output of the filter that needs to be optimized and the eigenvector corresponding to that eigenvalue provides the solution for the coefficients of the output energy filter.

Out of all the three filters, only for this filter can the output waveform be controlled. The advantage of using this filter over the Weiner filter is equivalent to solving an estimation problem rather than a detection problem as is conventionally done in signal processing and communication theory. In an output energy filter the goal is to control the output shape of the filter when a desired signal is present at the input of the filter. In a multipath environment, the goal is not to detect the desired signal as there is plenty of it, but to obtain its proper amplitude so that the system under consideration can properly operate. The output energy filter in a multipath environment can estimate the parameters of the signal of interest, as it is not solving a detection problem. Also, the output energy filter is well suited to be implemented as an adaptive filter in a highly dynamic environment where the signal and interference scenario may change from snapshot to snapshot. In addition, it has the advantage over a Wiener filter in the sense that frequent calibration of the system is not necessary.

6.1.4 Example of the Filters [1]

As an example, consider the following signal which consists of two samples of signal values $\{b_0; b_1\}$ to be numerically equal to $\{3,1\}$. Then the coefficients of the matched filter will be given by the sequence $\{a_0; a_1\}$ which will be equal to $\{b_1; b_0\}$. This will yield a numerical value of $\{1; 3\}$, which, when normalized to unit energy $a_0^2 + a_1^2 = 1$, will result in $\{a_0; a_1\} = \{0.316; 0.948\}$. The output from the matched filter will be given by the sequence [1]

$$\begin{aligned} \{c_0;\ c_1;\ c_2\} &= \{b_0;\ b_1\} ✪ \{a_0;\ a_1\} = \{a_0 b_0;\ a_0 b_1 + a_1 b_0;\ a_1 b_1\} \\ &= \{0.948, 3.160, 0.948\} \end{aligned}. \tag{6.16}$$

The SNR at the output of the matched filter will then be given by $c_1^2 = (a_0 b_1 + a_1 b_0)^2 = 10$. The matched filter by definition will maximize c_1. The response of this matched filter is plotted in Figure 6.1 by the dashed curve. This is also the spectrum of the signal [1].

Next we look at the response of the Wiener filter. For the Wiener filter we need to have the output of the filter $\{c_0; c_1; c_2\}$ match a desired sequence given by $\{d_0; d_1; d_2\}$. We assume the desired sequence to be $\{0; 1; 0\}$. The equations for the filter coefficients for the Wiener filter is given by (6.12) and is obtained as $\{a_0; a_1\} = \{0.067; 0.998\}$. The output of the Wiener filter is given by

$\{c_0;\ c_1;\ c_2\} = \{0.201;\ 3.061;\ 0.998\}$. Clearly the matched filter has better performance than the Wiener filter if we are interested only in the value of the term c_1^2. The response of the Wiener filter for this problem is shown by the dotted line of Figure 6.1.

Finally we look at the output energy filter where the goal is to maximize (let us assume, instead of minimizing) the output energy. Or equivalently the goal is to obtain the largest value for $c_0^2 + c_1^2 + c_2^2$ from the filter subject to $a_0^2 + a_1^2 = 1$. The solution to equation (6.15) is obtained as $\{a_0;\ a_1\} = \{0.707;\ 0.707\}$ and the maximum eigenvalue $\lambda = 13$. Hence, it is clear that out of all the three filter methodologies the output energy filter provides the maximum output energy. The response of the filter is shown by the solid line in Figure 6.1. It is seen that the output energy filter has the sharpest response [1].

Next we apply the output energy filter for the adaptive estimation problem where the goal will be to extract the signal of interest (SOI) in the presence of unknown strong jammers and clutter. We apply the estimation procedure to a single snapshot of the voltages induced in the antenna array at a particular instance of time.

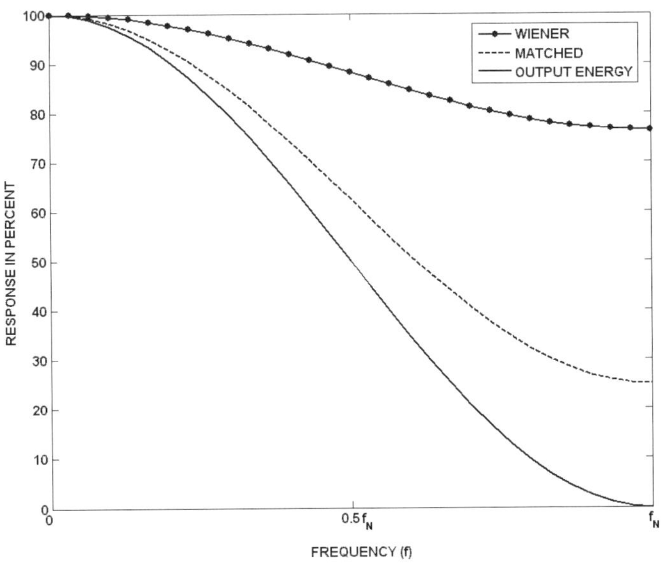

Figure 6.1. Normalized energy density spectra for the various filters.

6.2 DIRECT DATA DOMAIN LEAST SQUARES APPROACHES TO ADAPTIVE PROCESSING BASED ON A SINGLE SNAPSHOT OF DATA

In a conventional adaptive technique, the covariance matrix of the data must first be computed, and then inverted, but in the direct data domain least squares

(D3LS) approach, the SOI is directly estimated from the solution of a matrix equation, which has a Hankel structure. Use of the conjugate gradient and the Fast Fourier Transform techniques in solving such matrix equations make this procedure highly suitable for real-time implementation of these algorithms, as the computational time is one order of magnitude less than the conventional adaptive techniques based on statistical methodologies [2–5]. A least squares technique is directly applied to the data on a snapshot-by-snapshot basis without forming a covariance matrix, hence this method is computationally quite efficient. A snapshot is defined as the voltages measured at the feed points of all the antenna elements in the array at a particular instance of time.

In radar problems, where the direction of arrival (DOA) of the SOI is known, the energy is transmitted along a certain specific direction, and then we attempt to detect targets along that direction from the signal returns. Since the angular direction along which the electromagnetic energy was initially transmitted is known, one has a reasonably good knowledge of the DOA of the reflected energy from the target, if it exists along that specific direction. The development presented here is based on the assumption that each antenna element is an omni-directional point radiator, and uniformly spaced along a line. However, for applications to realistic antenna elements the procedure can be modified to take into account mutual coupling and other near field effects [2]. In addition to the SOI contributing to the received voltages at each antenna element, there are also contributions due to jammers, clutter, and thermal noise. The incoming interferers and the clutter may be coherent with the SOI. Details of how to compensate for the mutual coupling between the antennas and take into account near-field coupling between the antenna and the environment in which it is deployed are available in [2].

Consider a uniformly spaced linear array consisting of N+1 isotropic omni-directional point radiators as shown in Figure 6.2. The voltage X_n induced at the nth antenna element at a particular instance of time will then be given by

$$X_n = s\, e^{\frac{j2\pi n d \cos\varphi_s}{\lambda}} + \sum_{p=1}^{P} A_p\, e^{\frac{j2\pi n d \cos\varphi_p}{\lambda}} + C_n + \xi_n \tag{6.17}$$

where s = complex amplitude of the SOI (to be determined)

φ_s = directional of arrival of the SOI (assumed to be known)

d = spacing between each of the antenna elements (known)

λ = wavelength of transmission (assumed to be known)

P = total number of interferers (unknown)

A_p = complex amplitude of the pth undesired interferer (unknown)

φ_p = directional of arrival of the pth interferer (unknown)

C_n = clutter induced at the nth antenna element (unknown)

ξ_n = thermal noise induced at the nth antenna element (unknown)

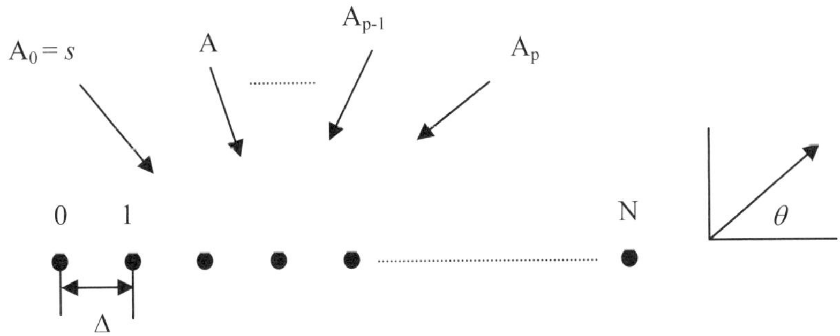

Figure 6.2 A uniform linear uniform array of isotropic radiators.

In a real environment, the DOA of the SOI will be unknown. In that case, one simply assumes a value for the DOA and solves for the amplitude of the unknown signal s. If there is no signal this value will be small. In practice, multiple processors can be employed to estimate the amplitude for the SOI for a given DOA and then try to find the SOI without actually knowing the DOA. Other secondary processing techniques can also be applied which will be presented in chapter 7. In addition to the DOA, other signal parameters can be used to perform adaptive processing like the spectral composition as in a cyclostationarity assumption [2].

The clutter is modeled as a bunch of reflected/diffracted rays bouncing back from the ground or platforms on which the array is mounted and from nearby buildings or trees. The amplitudes and phases of these rays have been determined by two random number generators. Hence, the clutter is modeled by the true physics of an electromagnetic model and is not based on some probability distributions which do not satisfy any known electromagnetic phenomenon. The detailed discussion can be found in [2–8]. The measured voltages X_n for $n = 0, 1, \ldots N$ at the antenna elements are assumed to be known along with θ_s the DOA of the SOI. The goal is to estimate the complex amplitude s for the SOI. Here, we define a single snapshot by the voltages measured at all of the antenna elements at a certain instant of time t_m. It is understood that all the SOI, jammers, clutter, and thermal noise vary as a function of time.

In conventional adaptive processing, it is assumed that a set of weights W_n for $n = 0, 1, \ldots, N$ is connected to each one of the antenna element. Then, a block of data is generated corresponding to M+1 snapshots, i.e., X_n^m for $m = 0, 1, \ldots M$ and $n = 0, 1, \ldots N$. Here the superscript m on X_n denotes that the voltage X_n^m is induced at antenna element n at a specific time instance m. Then, a covariance matrix of this block of data of $(N+1) \times (M+1)$ samples is evaluated and the adaptive weights are given by the Wiener solution, which is related to the inverse of the covariance matrix. The computational load of forming a covariance matrix and its inversion is an $\Theta(N^3)$ operation, where $\Theta(\bullet)$ represents of the order of. Hence, it is difficult to implement this procedure in real time. In addition, the procedure assumes that the data are stationary over these $(M+1)$

samples, i.e., the environment of the SOI, clutter, and jammer scenarios have not changed over the entire data collection process and this also precludes blinking jammers. Because of these disadvantages of the conventional adaptive processing, the D3LS approach using a single snapshot of the data for $M = 0$ has been proposed [2].

Consider the same linear array as in Figure 6.2. Here, we have a single snapshot of the voltages measured at the feed point of the antenna elements. The goal is to estimate the complex amplitude s for a given φ_s. To obtain the complex amplitude of the SOI using a single snapshot of the data, the number of coherent jammers must be less than or equal to $N/2$ in the absence of clutter and noise. It is important to point out that in this procedure no distinction is made between coherent or noncoherent interferers. The classical stochastic based techniques will be able to handle more than $N/2$ noncoherent interferers but no more than $N/2$ coherent interferers. However, the price to be paid for this is that a snapshot of at least $N+1$ voltages is required. Therefore, for these direct data domain methods there may be a significant loss in the number of degrees of freedom for the noncoherent case, when using a single snapshot of the data, however, multiple snapshots can be processed in a similar manner and in that case the degrees of freedom will not be reduced. It will be shown how to go beyond the limitation of the degrees of freedom of $N/2$ for the single snapshot case, in the subsequent sections. When the interferers are coherent with the signal, the number of degrees of freedom is the same both for the direct data domain methods and the stochastic based methods even though the latter is using multiple snapshots as opposed to a single snapshot used by the former.

6.2.1 Eigenvalue Method [2,7,8]

The eigenvalue method is one of the forms of the D3LS method. Under the assumption that L is less than or equal to $N/2$, one can form the matrix pencil $[X]-\alpha[S]$ of dimension $L+1$ (here, α is the estimate of the complex amplitude for the unknown SOI, s, to be solved for), where

$$[X] = \begin{bmatrix} X_0 & X_1 & \cdots & X_L \\ X_1 & X_2 & \cdots & X_{L+1} \\ \vdots & \vdots & \cdots & \\ X_L & X_{L+1} & \cdots & X_N \end{bmatrix}_{(L+1)\times(L+1)} \tag{6.18}$$

$$[S] = \begin{bmatrix} S_0 & S_1 & \cdots & S_L \\ S_1 & S_2 & \cdots & S_{L+1} \\ \vdots & \vdots & \cdots & \\ S_L & S_{L+1} & \cdots & S_N \end{bmatrix}_{(L+1)\times(L+1)} . \tag{6.19}$$

Since we are using a single snapshot for the voltages at a particular instance of time, the superscript on X_n in (6.18) has been suppressed. In addition the elements of the matrix [S] are related to the DOA of the SOI and is given by

$$S_n = e^{\frac{j2\pi\, nd \cos\varphi_s}{\lambda}} \; . \tag{6.20}$$

Thus the elements of matrices [X] and [S] can be determined using (6.17), and (6.20). The elements of the matrices in (6.18) and (6.19) are defined in such a way that the difference $X_n - \alpha S_n$ at each antenna element represents the contribution due to signal multipaths, jammers, clutter, and thermal noise (all the undesired components of the signals except due to the SOI which is s) since it is assumed that S_n is the voltage induced at the nth element due to a signal arriving from the same direction as the SOI, but whose amplitude is unity.

It is assumed that $(N+1) \geq (2L+1)$ and the total number of antenna elements N+1 is always odd. Let us say there are P jammers, and then there are total of $2P$+1 unknowns to deal with. In an adaptive processing methodology, the column vector of weights [W] are chosen in such a way that the contribution from the jammers, clutter, and thermal noise are minimized to enhance the output signal to interference plus noise ratio. If the matrix $[U] = \{[X] - \alpha[S]\}$ is defined in this way, then one gets the following generalized eigenvalue problem,

$$[U]_{(L+1,L+1)}\,[W]_{(L+1)} = \{[X] - \alpha[S]\}_{(L+1)\times(L+1)}\,[W]_{(L+1)\times 1} = 0 \tag{6.21}$$

where α, is the estimate of the complex amplitude for SOI in (6.17) and is obtained from the solution of the generalized eigenvalue problem. The weights [W] are given by the generalized eigenvector. Since there is only one signal arriving from θ_s, the matrix [S] is of rank unity, and hence the generalized eigenvalue equation given by (6.21) has only one eigenvalue and that eigenvalue α provides an estimate for the complex amplitude for the SOI, s. Alternately, one can view the left-hand side of (6.21) as the total noise signal at the output of the adaptive processor due to jammer, clutter, and thermal noise. Hence, the weighted sum of the total interference plus noise voltage is given by

$$\mathrm{Y}_{\text{out}} = [U][W] = \{[X] - \alpha[S]\}[W] \; . \tag{6.22}$$

Therefore, the total noise power would be given by

$$\mathrm{Y}_{\text{power}} = [W]^H \{[X] - \alpha[S]\}^H \{[X] - \alpha[S]\}\, [W] \; . \tag{6.23}$$

The objective is to minimize this noise power by selecting [W] for a fixed signal strength α. This is achieved by differentiating $\mathrm{Y}_{\text{power}}$ with respect to each of the individual weights W_i and setting each one of the partial derivatives equal to zero. When each of the individual equations is assembled together, this results in (6.21). From a computational point of view, an alternate way to solve for α, is to make the determinant of the following matrix equal to zero, i.e., det $\{[X] - \alpha\, [S]\} = 0$, for a suitable value of α. Then equation (6.21) can be rewritten in terms of the following two generalized eigenvalue equations, $[X][W] = \alpha\, [S][W]$, or alternatively $[S][W] = \frac{1}{\alpha}[X][W]$. Even though (6.21) is a matrix of size $(L+1) \times (L+1)$, the matrix $[S]$ is of rank unity and so there is only one eigenvalue and that is the least squares solution to the estimation of the complex amplitude.

The solution of an eigenvalue problem requires an operation count which is of the order of $\theta(L+1)^3$ and hence difficult to implement in real time on a digital signal processing chip. Secondly, the computation of the generalized eigenvalue problem in (6.21) can become unstable as the rank of the matrix $[S]$ is unity. That is why we transform the solution procedure from the solution of a generalized eigenvalue problem to the solution of a matrix equation.

6.2.2 Forward Method [2,6]

The (1,1) and (1,2) elements of the interference plus noise matrix $[U]$ from (6.21) or (6.22) can be written as,

$$U_{1,1} = X_0 - \alpha\, S_0 \tag{6.24}$$

$$U_{1,2} = X_1 - \alpha\, S_1 \tag{6.25}$$

where X_0 and X_1 are the voltages received at antenna elements 0 and 1 due to the signal, jammer, clutter, and noise, whereas S_0 and S_1 are the values of the SOI only at those elements due to a signal of unit strength. Define

$$Z = \exp\left[\, j2\pi\frac{d}{\lambda}\cos\varphi_s \right] \tag{6.26}$$

Then, $U_{1,1} - Z^{-1}U_{1,2}$ contains no components of the SOI, as

$$S_0 = \exp\left[\, j2\pi\frac{(n=0)d}{\lambda}\cos\varphi_s \right] \text{ with } n = 0\ , \tag{6.27}$$

and

$$S_1 = \exp\left[\, j2\pi\frac{(n=1)d}{\lambda}\cos\varphi_s \right] \text{ with } n = 1\,. \tag{6.28}$$

The same is true for $U_{1,2} - Z^{-1}U_{1,3}$, and in general, for $U_{i,j} - Z^{-1}U_{i,\,j+1}$, for $i = 1, \ldots, L+1$, $j = 1, \ldots, L$. Here, we have $L = N/2$. Therefore, one can form a reduced rank matrix $[T]_{L\times(L+1)}$, generated from $[U]$ such that

$$[T]=\begin{bmatrix} X_0-Z^{-1}X_1 & X_1-Z^{-1}X_2 & \cdots & X_L \quad Z^{-1}X_{L+1} \\ \vdots & \vdots & \vdots & \vdots \\ X_{L-1}-Z^{-1}X_L & X_L-Z^{-1}X_{L+1} & \cdots & X_{N-1}-Z^{-1}X_N \end{bmatrix}_{L\times(L+1)} \times \begin{bmatrix} W_0 \\ \vdots \\ W_L \end{bmatrix}_{(L+1)\times 1} = 0 \tag{6.29}$$

In order to restore the signal component in the adaptive processing, we fix the gain of the subarray formed by the $L+1$ elements along the direction θ_s and then evaluate a weighted sum of the voltages $\sum_{i=0}^{L} W_i X_i$. Let us say the gain of the subarray is D along the direction of φ_s. This provides an additional equation resulting in a square matrix

$$\begin{bmatrix} 1 & Z & \cdots & Z^L \\ X_0-Z^{-1}X_1 & & & X_L-Z^{-1}X_{L+1} \\ \vdots & \vdots & & \vdots \\ X_{L-1}-Z^{-1}X_L & X_L-Z^{-1}X_{L+1} & \cdots & X_{N-1}-Z^{-1}X_N \end{bmatrix}_{(L+1)\times(L+1)}$$

$$\times \begin{bmatrix} W_0 \\ W_1 \\ \vdots \\ W_L \end{bmatrix}_{(L+1)\times 1} = \begin{bmatrix} D \\ 0 \\ \vdots \\ 0 \end{bmatrix}_{(L+1)\times 1} \tag{6.30}$$

or, equivalently

$$[F][W]=[Y] \; . \tag{6.31}$$

Once the weights are solved for by using (6.30), the signal component α, estimate for the amplitude of the SOI s, in (6.17) can be evaluated by using

$$\alpha = \frac{1}{D}\sum_{i=0}^{L} W_i X_i \; . \tag{6.32}$$

The proof of (6.32) is available in [5].

6.2.3 Backward Method [2,6]

A second independent estimate for the complex amplitude for the SOI can be obtained by rearranging the same data. This can be accomplished by reversing the data sequence and then complex conjugating each term of that sequence. It is well known in the parametric spectral estimation literature that a sampled sequence which can be represented by a sum of exponentials with purely imaginary argument can be used either in the forward or in the reverse direction resulting in the same value for the exponent [2,6]. From physical considerations, it is known that if a polynomial equation can be solved with the weights W_i evaluated from (6.30) as the coefficients then the roots of the polynomial equation provide the DOA for all the unwanted signals including the interferers. Therefore, whether the snapshot is treated as a forward sequence as presented in the previous section or by a reverse conjugate of the same sequence, the final computed values for W_i must be the same. Hence, for these classes of problems, the data when analyzed either in the forward direction or in the reverse direction provide two independent data sets and hence two independent estimates for the same solution. This is equivalent to creating a virtual array of the same size but located along a mirror symmetry line. Therefore, if the data is conjugated and one forms the reverse sequence, then one gets an independent set of equations similar to (6.30) for the solution of the weights $[W]$. This is represented by

$$\begin{bmatrix} 1 & Z & \cdots & Z^L \\ X_N^* - Z^{-1}X_{N-1}^* & X_{N-1}^* - Z^{-1}X_{N-2}^* & \cdots & X_L^* - Z^{-1}X_{L-1}^* \\ \vdots & \vdots & \cdots & \vdots \\ X_{L+1}^* - Z^{-1}X_L^* & X_L^* - Z^{-1}X_{L-1}^* & \cdots & X_1^* - Z^{-1}X_0^* \end{bmatrix}_{(L+1)\times(L+1)} \tag{6.33}$$

$$\times \begin{bmatrix} W_0 \\ W_1 \\ \vdots \\ W_L \end{bmatrix}_{(L+1)\times 1} = \begin{bmatrix} D' \\ 0 \\ \vdots \\ 0 \end{bmatrix}_{(L+1)\times 1}$$

or equivalently in a matrix form as

$$[\mathrm{B}][\mathrm{W}]=[\mathrm{Y}]. \tag{6.34}$$

The signal strength α can now be determined using (6.33) as

$$\alpha = \left[\frac{Z^L}{D'} \sum_{i=0}^{L} W_i X_{L-i}^* \right]^* \tag{6.35}$$

Note that for both the forward and the backward methods described so far L is equal to $N/2$. Hence, the degrees of freedom are the same for both forward and backward methods.

6.2.4 Forward-Backward Method [2,6]

Both the forward and the backward methods can be combined to double the given data set and thereby increase the number of weights or the degrees of freedom significantly over that of either the forward or the backward method. In the forward-backward model the amount of data is doubled by not only considering the samples of the given data in the forward direction but also conjugating them and reversing the direction of increment of the independent variable. This type of processing can be done as long as the series to be approximated can be fit by exponential functions of purely imaginary argument. This is always true for the adaptive array case. For the forward-backward method, the number of degrees of freedom can be significantly increased by approximately 50%, without increasing the number of antenna elements. The equation that needs to be solved for the weights is given by combining (6.30) and (6.33), with D′ = D, into

$$\begin{bmatrix} 1 & Z & \cdots & Z^V \\ X_0 - Z^{-1}X_1 & X_1 - Z^{-1}X_2 & \cdots & X_V - Z^{-1}X_{V+1} \\ \vdots & \vdots & \vdots & \vdots \\ X_{V-1} - Z^{-1}X_V & X_V - Z^{-1}X_{V+1} & \cdots & X_{N-1} - Z^{-1}X_N \\ X_N^* - Z^{-1}X_{N-1}^* & X_{N-1}^* - Z^{-1}X_{N-2}^* & \cdots & X_V^* - Z^{-1}X_{V-1}^* \\ \vdots & \vdots & \vdots & \vdots \\ X_{V+1}^* - Z^{-1}X_V^* & X_V^* - Z^{-1}X_{V-1}^* & \cdots & X_1^* - Z^{-1}X_0^* \end{bmatrix}_{(V+1)\times(V+1)} \times \begin{bmatrix} W_0 \\ W_1 \\ \vdots \\ W_V \end{bmatrix}_{(V+1)\times 1} = \begin{bmatrix} D \\ 0 \\ \vdots \\ 0 \end{bmatrix}_{(V+1)\times 1} \tag{6.36}$$

or equivalently

$$[FB][W] = [Y] \tag{6.37}$$

The value of V in (6.36) is now much greater than the value of L in equations (6.30) and (6.33). Since in (6.36), the total amount of data is now doubled, the number of degrees of freedom V in this case will be much greater than L. This increase in the degrees of freedom has been achieved by considering both the forward and reverse forms of the data sequence. In summary, in a conventional adaptive technique where there is a weight attached to each element and the processing is done in time, the number of degrees of freedom is $N + 1$, provided that the environment is stationary in time and the interferers are noncoherent. For coherent interferers, whether one is using the conventional methods or this technique, the maximum number of interferers V that can be handled is much greater than L or $N/2$. So for the forward-backward method this proposed spatial processing based on a snapshot-by-snapshot analysis will provide the number of degrees of freedom $V = N/1.5 + 1$. It is important to note that this is the maximum number of degrees of freedom for handling coherent interferers by any method! In addition, this is a least squares-based approach. The advantage of doing snapshot-by-snapshot processing is that the stationarity assumption about the data can be relaxed.

6.2.5 Real-time Implementation of the Adaptive Procedure [9–11]

As noted in [2,4,11], equations (6.30), (6.33), or (6.36) can be solved very efficiently by applying the fast Fourier transform (FFT) and the conjugate gradient method. They have been implemented to operate in real time utilizing a digital signal processing (DSP) chip (the algorithm was actually implemented on a DSP32C chip produced by AT&T) to solve these type of equations [9–11].

For the solution of equations of the form [F][W] = [Y] in (6.30), (6.33), or (6.36), the conjugate gradient method starts with an initial guess $[W]_0$ for the solution and continues with the calculation of

$$[P]_0 = -b_{-1}[F]^H[R]_0 = -b_{-1}[F]^H\{[F][W]_0 - [Y]\} \tag{6.38}$$

where H denotes the conjugate transpose of a matrix. At the kth iteration the conjugate gradient method develops the following:

$$c_k = \frac{1}{\|[F][P]_k\|^2} \tag{6.39}$$

$$[W]_{k+1} = [W]_k + c_k[P]_k \tag{6.40}$$

$$[R]_{k+1} = [R]_k + c_k[F][P]_k \tag{6.41}$$

$$b_k = \frac{1}{\left\|[F]^H[R]_{k+1}\right\|^2} \tag{6.42}$$

$$[P]_{k+1} = [P]_k - b_k[F]^H[R]_{k+1} \tag{6.43}$$

The norm is defined by

$$\|[F][P]_k\|^2 = [P]_k^H[F]^H[F][P]_k \ . \tag{6.44}$$

The equations above are applied in an iterative fashion until the desired error criterion for the residuals $\|[R]_k\|$ is satisfied, where $[R]_k = [F][W]_k - [Y]$. In our case, the error criterion is defined by

$$\frac{\|[R]_k\|}{\|[Y]\|} = \frac{\|[F][W]_k - [Y]\|}{\|[Y]\|} \le 10^{-6} \tag{6.45}$$

The iterative procedure is stopped when the criterion defined above is satisfied. A detailed description of this method along with a few sample computer programs is presented in [2]. The strength of the conjugate gradient method is that the final solution is still going to converge to an acceptable one even if the matrix [F] is exactly singular. Thus, the conjugate gradient method has the advantage of a direct method as it is guaranteed to converge to the exact solution after a finite number of steps barring any numerical errors and for any initial guess [16]. It has also the advantage of an iterative method as round-off and truncation error is limited to the last stage of iteration. Finally, it will converge even when the matrix is exactly singular when the direct methods fail.

The computational bottleneck in the conjugate gradient method is in the evaluation of the matrix-vector products $[F][P]_k$ and $[F]^H[R]_{k+1}$. Typically, matrix vector products in real-time computations can slow down the computational process when they are transported to a digital signal-processing chip. However, in our examples, these computational bottlenecks can be streamlined through exploitation of the block Hankel structure in the matrix [F] as seen from (6.30), (6.33), or (6.36). A block Hankel structure implies that the elements along any diagonal are equal. Under this special circumstance, that the matrix [F] has a block Hankel structure, the matrix-vector products defined by $[F][P]_k$ or $[F]^H[R]_{k+1}$ can be carried out efficiently through use of the fast Fourier transform (FFT) [2,4]. This is accomplished as shown in the next paragraph.

Consider the following matrix-vector product, when the matrix has a block Hankel structure so that we have the expression

$$\begin{bmatrix} f_1 & f_2 & f_3 \\ f_2 & f_3 & f_4 \\ f_3 & f_4 & f_5 \end{bmatrix}_{(r\times r)} \times \begin{bmatrix} w_1 \\ w_2 \\ w_3 \end{bmatrix}_{(r\times 1)} \tag{6.46}$$

In (6.46), the value of $r = 3$. A matrix-vector product is usually accomplished in r^2 operations, where r is the dimension of the matrix. However, since the matrix has a Hankel structure, we can rewrite the matrix-vector product as a result of the convolution of the two sequences $\{f\} ✪ \{w\} = \{f_1\ f_2\ f_3\ f_4\ f_5\} ✪ \{w_3\ w_2\ w_1\ 0\ 0\}$, where ✪ denotes a convolution operation. We observe that the fourth, fifth, and sixth elements of this convolution provide the correct expression for the matrix-vector product. The convolution actually results in more terms than we require for the matrix-vector product. However, that is not relevant. In fact, convolutions can be carried out very efficiently using the FFT. Here, since we have finite sequences, the FFT will provide the correct solution even though it is periodizing both sequences in carrying out the convolution. We take the FFT of the two sequences $\{f\}$ and $\{w\}$. Next, we multiply the two transformed sequences term by term. Then we take an inverse FFT to obtain the results for the matrix-vector product. In this procedure, the total operation count for the operations $\text{FFT}^{-1}[\text{FFT}\{f\} \times \text{FFT}\{w\}]$ will be $3[2r - 1]\log[2r - 1]$. For a value of r greater than 30, this procedure becomes quite advantageous, as the operation count is on the order of $(r \log r)$ as opposed to r^2 for a conventional matrix-vector product. Also, in this new procedure, there is no need to store an array. Thus, the time spent in accessing the elements of the array in the hard disk of the computer is virtually nonexistent, as everything is now one-dimensional and can be stored in the main memory. This procedure is quite rapid and easy to implement in hardware [9–11]. Hence this D3LS method is not only efficient [as the computational count for K iterations is $K \times (L+1)\log(L+1)$ as opposed to $2(L+1)^3$] but can also be implemented efficiently on a DSP chip for accurate solution of the adaptive problem.

Next we apply this methodology to the space-time adaptive processing.

6.3 DIRECT DATA DOMAIN LEAST SQUARES APPROACH TO SPACE-TIME ADAPTIVE PROCESSING

One way to detect small signals of interest in a noisy environment is to have a large array providing sufficient power and a large enough aperture to achieve narrow beams. In addition, the array must have extremely low sidelobes simultaneously on transmit and receive. This is very difficult and expensive to achieve in practice. An electrically large aperture will provide a narrow beam on transmit and receive with which to search, and the low sidelobes would help keep interferers from entering the system through the sidelobes. The problem with this solution is that the manufacture of such an array would be very difficult, as extremely tight mechanical tolerances would be necessary and thus will be expensive to build. In addition, the real estate on airborne platforms is limited. This is the situation with the early warning system airborne platforms called AWACS (Airborne Early Warning and Control System). Another solution, the one we discuss here, is to use space-time adaptive processing [12,13] to suppress the interferers and enable the system to detect potentially weak target signals.

Therefore, instead of using a high-gain antenna with very low sidelobes, we plan to achieve the same goal through space-time adaptive processing (STAP). STAP is carried out by performing two-dimensional filtering on signals which are collected by simultaneously combining signals from the elements of an antenna array (the spatial domain) as well as from the multiple pulses from a coherent radar (the temporal domain). The data collection mechanism is shown in Figure 6.3. The temporal domain thus consists of multiple pulse repetition periods of a coherent processing interval (CPI). By performing simultaneous multidimensional filtering in space and time, the goal is not only to eliminate clutter that arrives at the same spatial angle as the target but also to remove clutter that comes from other spatial angles which has the same Doppler frequency as the target. Hence, STAP provides the necessary mechanism to detect low observables from an airborne radar. The goal of adaptive processing is to weight the received space-time data vectors as seen in Figure 6.3 to maximize the output signal-to-interference plus noise ratio (SINR).The adaptive algorithms presented in the previous section used data from a space snapshot, which consists of samples from across the array at an instant in time (a given pulse at a given range bin). In this section we present four algorithms that operate across pulses and elements, increasing the degrees of freedom over that of the element domain alone. The first processor to be described implements a generalized eigenvalue equation [2,9,10], while the last three processors implement a least squares solution to a linear matrix equation [12,13].

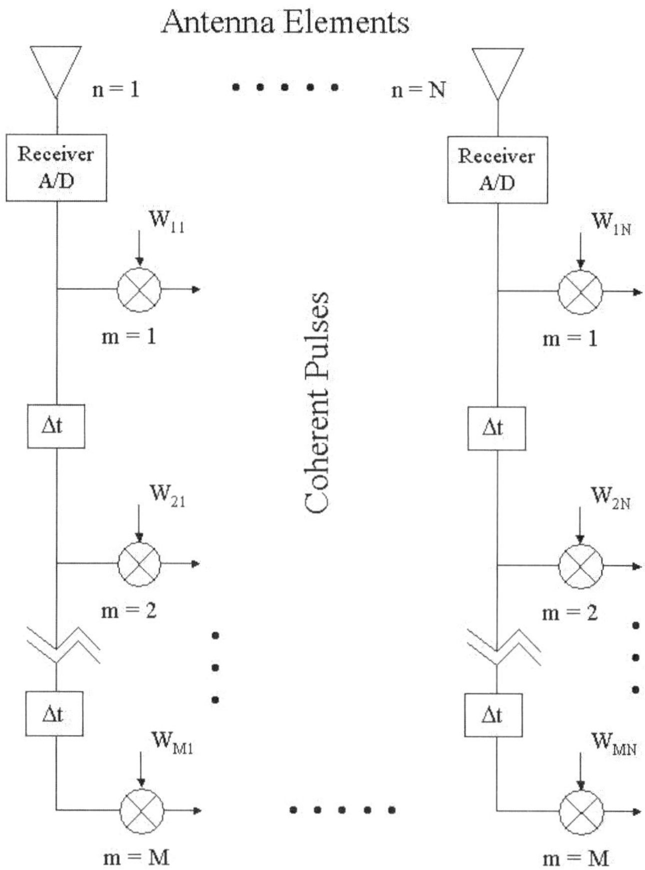

Figure 6.3. Data collection system.

We consider a pulsed Doppler radar situated on an airborne platform which is moving at a constant velocity. The radar consists of an antenna array where each element has its own independent receiver channel. The linear antenna array has $N+1$ elements uniformly spaced by a distance Δ, as shown in Figure 6.2. In this configuration the received voltages at each of the antenna elements are the sampled values of the data as there is a receive channel behind every element in the planar array.

We also assume that the system processes M coherent pulses within a coherent processing interval (CPI) (i.e., the radar transmits a coherent burst of M pulses at a constant pulse repetition frequency) where each pulse repetition interval consists of the transmission of a pulsed waveform of finite bandwidth and the reception of reflected energy captured by the aperture and passed through a receiver with a bandwidth equal to that of the pulse. In the receive chain the signal is down-converted, matched filtered, sampled, and digitized, and the baseband samples are stored. In this manner complex samples are generated at R range bins for M pulses at N elements. To facilitate working with the array output, the baseband samples can be arranged into a three-dimensional matrix commonly referred to as a data cube, as shown in Figure 6.4. The three axes of the data cube correspond to the pulse (M), antenna element (N), and range (R) dimensions. At a particular range r_i, the sheet or slice of the data cube is referred to as a space-time snapshot, indicated by the shaded plane in Figure 6.4.

Therefore, with M pulses and N antenna elements, each having its own independent receiver channels, the received data for a coherent processing interval consists of RMN complex baseband samples. These samples, often referred to as the *data cube*, consist of $R \times M \times N$ complex baseband data samples of the received pulses. The data cube then represents the voltages defined by $V(m; n; r)$ for $m = 1, ..., M$; $n = 1, ..., N$; and $r = 1, ..., R$. These complex baseband measured voltages contain the SOI, jammers, and clutter, including thermal noise. A space-time snapshot then is referred to as MN samples for a fixed range gate value of r.

We assume that the signals entering the array are narrowband and consist of the SOI and interference plus noise. The noise (thermal noise) originates in the receiver and is assumed to be independent across elements and pulses. The interference is external to the receiver and consists of clutter (reflection of the transmitted electromagnetic energy from the earth), jammers, mutual coupling, and multipath (due to the SOI, clutter, and/or jamming). We assume that for each jammer, the energy impinging on the array is confined to a particular DOA and is spread in frequency. The jammers may be blinking or stationary. From the data cube shown in Figure 6.4, we focus our attention to the range cell r and consider the space-time snapshot for this range cell.

We assume that the SOI for this range cell r is incident on the uniform linear array from an angle φ_s and is at Doppler frequency f_s. Our goal is to estimate its amplitude, given φ_s and f_s only. In a surveillance radar, φ_s and f_s set

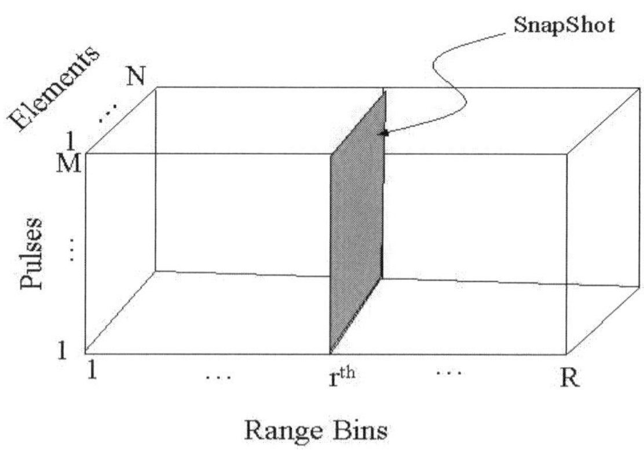

Figure 6.4. Representative data cube.

the look directions and a SOI (target) may or may not be present along this look direction and Doppler. Let us define *S(m; n)* to be the complex voltage received at the qth antenna element corresponding to the pth time for the same range cell r. We further stipulate that the voltage *S(m; n)* is due to a signal of unity magnitude incident on the array from the azimuth angle φ_s corresponding to Doppler frequency f_s. Hence, the signal-induced voltage under the assumed array geometry and a narrowband signal is a complex sinusoidal given by

$$S(m, n) = \exp\left[j \left(\frac{2\pi\Delta n}{\lambda} \cos\varphi_s + \frac{2\pi f_s m}{f_r} \right) \right] \quad (6.47)$$

$$\text{for } m = 1, \ldots, M; \; n = 1, \ldots, N$$

where λ is the wavelength of the radio-frequency radar signal and f_r is the pulse repetition frequency.

Let X $(m; n)$ be the actual measured complex voltages that are in the data cube of Figure 6.4 for the range cell r. The actual voltages X will contain the SOI of amplitude α (α is a complex quantity), jammers which may be due to coherent/incoherent multipaths of the radiated signal, and clutter which is the reflected electromagnetic energy from the ground which has been transmitted through both the main lobe and sidelobes. The interference competes with the SOI at the Doppler frequency of interest. There is also a contribution to the measured voltage from the receiver thermal noise. Hence the actual measured voltages $X(p; q)$ are

$$X(m, n) = \alpha \exp\left[j \left(\frac{2\pi\Delta m}{\lambda} \cos\varphi_s + \frac{2\pi f_s n}{f_r} \right) \right] \quad (6.48)$$

$$+ \text{ clutter } + \text{ jammer } + \text{ thermal noise}$$

The goal is to extract the SOI, α, given these voltages X, the DOA for the SOI, φ_s, and the Doppler frequency, f_s.

The signals entering the array consist of the SOI plus interference (clutter, jammers, multipath, etc.) and noise. For the nth element and mth pulse, at the rth range bin, the complex envelope of the received signal is

$$X(m,n) = \alpha S(m,n) + \text{ interference } + \text{ noise} \quad (6.49)$$

where α is the amplitude of the SOI entering the array. In the D3LS procedures to be described, the adaptive weights are applied to the single space-time snapshot for the range cell r. Here a two-dimensional array of weights numbering N_a N_p is used to extract the SOI for the range cell r. Hence the weights are defined by $w(m; n; r)$ for $m = 1, \ldots, N_p < M$ and $n = 1, \ldots, N_a < N$ and are used to extract the SOI at the range cell r. Therefore, for the D3LS method we essentially perform a high-resolution filtering in two dimensions (space and time) for each range cell [2,12–14].

6.3.1 Two-Dimensional Generalized Eigenvalue Processor

For each pulse-element cell (given a range bin r) the difference equation

$$X(m,n) - \alpha S(m,n) \tag{6.50}$$

removes the SOI from the nth element and mth pulse sample, leaving noise plus interference. It is important to note that α, the amplitude of the SOI, is still an unknown quantity. Based on (6.50) a two-dimensional matrix pencil can be created whose solution will result in a weight vector which will null out the interferers and extract the SOI. The elements of this matrix pencil can be constructed by sliding a window (box) over the space-time snapshot data, as shown by the shaded plane in Figure 6.4. By creating a vector using the elements in the window, each window position generates a row in the S and X matrices as shown next:

$$\begin{array}{r} \text{Box 1} \rightarrow \\ \text{Box 2} \rightarrow \\ \\ S = \\ \text{Box 3} \rightarrow \\ \\ \\ \end{array} \begin{pmatrix} S_{11} & \cdots & S_{1Na} & \cdots & S_{Np1} & \cdots & S_{NpNa} \\ S_{12} & \cdots & S_{1(Na+1)} & \cdots & S_{Np2} & \cdots & S_{Np(Na+1)} \\ \vdots & & \vdots & & \vdots & & \vdots \\ S_{21} & \cdots & S_{2Na} & \cdots & S_{(Np+1)} & \cdots & S_{(Np+1)Na} \\ \vdots & & \vdots & & \vdots & & \vdots \\ S_{(M-Np)(N-Na)} & \cdots & S_{(M-Np)Na} & \cdots & S_{M(N-Na)} & \cdots & S_{MNa} \end{pmatrix} \tag{6.51}$$

$$\begin{array}{r} \text{Box 1} \rightarrow \\ \text{Box 2} \rightarrow \\ \\ X = \\ \text{Box 3} \rightarrow \\ \\ \\ \end{array} \begin{pmatrix} X_{11} & \cdots & X_{1Na} & \cdots & X_{Np1} & \cdots & X_{NpNa} \\ X_{12} & \cdots & X_{1(Na+1)} & \cdots & X_{Np2} & \cdots & X_{Np(Na+1)} \\ \vdots & & \vdots & & \vdots & & \vdots \\ X_{21} & \cdots & X_{2Na} & \cdots & X_{(Np+1)} & \cdots & X_{(Np+1)Na} \\ \vdots & & \vdots & & \vdots & & \vdots \\ X_{(M-Np)(N-Na)} & \cdots & X_{(M-Np)Na} & \cdots & X_{M(N-Na)} & \cdots & X_{MNa} \end{pmatrix} \tag{6.52}$$

The window size along the element dimension is N_a and is N_p along the pulse dimension. Selection of N_a determines the number of spatial degrees of freedom, while N_p determines the temporal degrees of freedom. Typically, for a single domain processing, N_a and N_p must satisfy the equations

$$N_a \leq \frac{N+1}{2} \tag{6.53}$$

$$N_p \leq \frac{M+1}{2} \tag{6.54}$$

The advantage of a joint domain processing is that either of these bounds can be relaxed (i.e., one can exchange spatial degrees of freedom with the temporal degrees of freedom). So, indeed, it is possible to cancel a number of interferers which is greater than the number of antenna elements in a joint domain processing. The total number of degrees of freedom, Q, for any method is

$$Q = N_a \times N_p \,. \tag{6.55}$$

Given the system constraints, most airborne radar systems contain more temporal degrees of freedom than spatial (i.e., $N << M$). Therefore, since the terms $X(m, n) - \alpha S(m, n)$ eliminate the SOI, these elements represent the contribution due to the unwanted signal multipaths, jammers, unwanted signals at the same Doppler, and receiver thermal noise. In D3LS adaptive processing, the goal is to take a weighted sum of these matrix elements defined in (6.50) and extract the SOI, which is going to be α for the range cell r.

The total number of degrees of freedom (DOFs = Q) then represents the total number of weights, and this is the product $N_a N_p$, where N_a is the number of spatial DOFs and N_p is the number of temporal DOFs. Next it is illustrated how a D3LS approach is taken for the extraction of SOI.

The least squares procedure for the 1D case (i.e., with only the subscript n and no m) is available in [2,12–14]. Here the same least squares procedure is extended to two dimensions. Consider the two matrices C_1 and C_2. The elements of C_1 and C_2 are formed by

$$C_1(x\,;y) = S(g+h-1;\ d+e-1) \tag{6.56}$$

$$C_2(x\,;y) = X(g+h-1;\ d+e-1) \tag{6.57}$$

$$1 \le x = g+(d-1)\,N_p \tag{6.58}$$

$$1 \le y = h+(e-1)\,N_p \le N_a N_p \tag{6.59}$$

$$1 \le d \le N - N_a \tag{6.60}$$

$$1 \le e \le N_a \tag{6.61}$$

$$1 \le g \le M - N_p \tag{6.62}$$

$$1 \le h \le N_p \tag{6.63}$$

so that $1 \le x, y \le N_a N_p$. Now if we consider a matrix pencil of size $N_a N_p$,

$$[C_2]_{N_a N_p \times N_a N_p} - \alpha [C_1]_{N_a N_p \times N_a N_p} \tag{6.64}$$

Then (6.64) represents the contribution of the unwanted signals, as the desired SOI have been canceled out. The elements of the matrices $[C_2]$ and $[C_1]$ are created out of the data matrices $[X(m, n)]$ and $[S(m, n)]$, respectively, as defined by (6.47) and (6.48).

Now in the STAP processing, the elements of the weight vector $[W]$ are chosen in such a way that the contribution from the jammers, clutter, and thermal noise is zero. Hence, if we define the generalized eigenvalue problem,

$$[R]\,[W] = \{[C_2] - \alpha\,[C_1]\}\,[W] = 0 \tag{6.65}$$

then α, the strength of the signal, is a generalized eigenvalue and the weights are given by the corresponding generalized eigenvector. Here W is a column vector of length $N_a\, N_p$ for the range cell r. Since we have assumed that only the SOI is arriving from φ_s corresponding to the Doppler f_s, the matrix $[C_1]$ is of rank one, and hence the generalized eigenvalue equation has only one nonzero eigenvalue, which provides an estimate of the complex amplitude of the signal.

Alternatively, one can view the left-hand side of (6.65) as the total noise signal at the output of the adaptive processor due to jammer, clutter, and thermal noise. One is therefore trying to reduce the noise voltage at the output of the adaptive processor, which is given by

$$N_{\text{out}} = [R][W] = \{[C_2] - \alpha[C_1]\}[W]\,. \tag{6.66}$$

The total output noise power then can be obtained as

$$N_{\text{power}} = [W]^H\{[C] - \alpha[C_1]\}^H\{[C_2] - \alpha[C_1]\}[W] \tag{6.67}$$

where H represents the conjugate transpose of a matrix. Our objective is to set the output noise power as small as possible by selecting $[W]$ for a fixed signal strength α. This is done by differentiating the real quantity N_{power} with respect to the elements of $[W]$ and setting each component equal to zero. This yields equation (6.65). The total number of DOFs, N_aN_p, is determined by both M and N. Clearly, we need $N_p < M$ and $N_a < N$ so that enough equations can be generated to form equation (6.65). Generally, $M >> N$, and therefore there are a larger number of temporal DOFs than spatial DOFs. The goal therefore is to extract the SOI at a given Doppler and angle of arrival in a given range cell r by using a two-dimensional filter of size $N_a\, N_p$. The filter is going to operate on the data snapshot depicted in Figure 6.4 of size NM to extract the SOI.

In real-time applications, it is difficult to solve numerically for the generalized eigenvalue problem in real time, particularly if the value N_aN_p representing the total number of weights is large and the matrix $[C_2]$ is highly rank deficient. For this reason, we convert the solution of a generalized eigenvalue problem given by (6.65) to the solution of a linear matrix equation.

6.3.2 Least Squares Forward Processor

The formulation of the direct data domain least squares space-time algorithm [2,12–14] can be obtained through extension of the one-dimensional case. We start by developing the forward case and then present the backward and forward-backward algorithms. As before, the matrix equation to be solved can be defined as

$$[T][W]=0 \tag{6.68}$$

where $[T]$ is the system matrix and $[W]$ is the vector of space-time weights, which have the potential to null the interferers. The system matrix, $[T]$, contains the angle-Doppler look direction of the SOI as well as the cancellation rows, which contain the angle-Doppler information on the interferers. This interferer information is obtained through difference equations similar to equation (6.29), where the contribution of the SOI is removed, leaving information of interferers only. In the two-dimensional case these difference equations are performed with elements offset in space only, time only, and space and time. Define the element-to-element offset of the SOI in space and time, respectively, as

$$Z_1 = e^{j2\pi\frac{d}{\lambda}\cos(\varphi_s)} \tag{6.69}$$

$$Z_2 = e^{j2\pi\frac{f_s}{f_r}} \tag{6.70}$$

Again, SOI has an angle of arrival of φ_s and a Doppler frequency of f_s. The three types of difference equations are then given by

$$X(m,n)-X(m,n+1)Z_1^{-1} \tag{6.71}$$

$$X(m,n)-X(m+1,n)Z_2^{-1} \tag{6.72}$$

$$X(m,n)-X(m+1,n+1)Z_1^{-1}Z_2^{-1} \tag{6.73}$$

Note that in (6.71), the signal component (SOI) is canceled from samples taken from different antenna elements at the same time. Similarly, (6.72) represents signal cancellation from samples taken at the same antenna elements at different time. Finally, (6.73) represents signal cancellation from neighboring samples in both space and time. Therefore, we are performing a filtering operation simultaneously using N_aN_p samples of the space-time data. The cancellation rows of the matrix $[T]$ can now be formed using (6.71)–(6.73) through the application of various windows as shown in Figure 6.5. In this case the dots in Figure 6.5 represent the induced voltages, $X(m, n)$ as defined in (6.48) or (6.49), for a given element-pulse location. Just as was done for the generalized eigenvalue algorithm for the 1D case in the previous section, a space-time window is passed over the data. For each given location of the window function in Figure 6.5, three rows in matrix $[T]$ are formed by implementing (6.71)–(6.73), which remove the SOI. The different rows are formed by performing an element-by-element subtraction between the sampled data inside the windows and then arranging the resulting computations into a row vector, as shown in Figure 6.6. The window is then slid one spacc to the right and three more rows are generated, and so on.

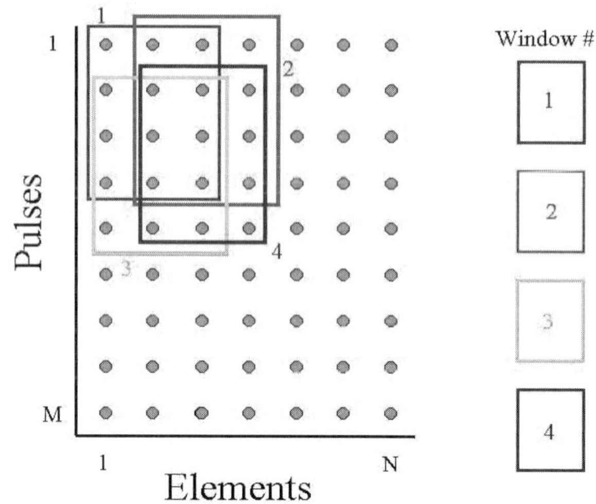

Figure 6.5. Space-time data.

Example: Computing a cancellation row using $N_a = 3$ and $N_p = 4$

$$X_{\text{window1}} - X_{\text{window2}} \bullet Z_1^{-1},$$

$$\Rightarrow \begin{pmatrix} x_{11} & x_{12} & x_{13} \\ x_{21} & x_{22} & x_{23} \\ x_{31} & x_{32} & x_{33} \\ x_{41} & x_{42} & x_{43} \end{pmatrix} - \begin{pmatrix} x_{12} & x_{13} & x_{14} \\ x_{22} & x_{23} & x_{24} \\ x_{32} & x_{33} & x_{34} \\ x_{42} & x_{43} & x_{44} \end{pmatrix} \bullet Z_1^{-1}$$

$$= \begin{pmatrix} x_{11} - x_{12}z_1^{-1} & x_{12} - x_{12}z_1^{-1} & x_{13} - x_{12}z_1^{-1} \\ x_{21} - x_{12}z_1^{-1} & x_{22} - x_{12}z_1^{-1} & x_{23} - x_{12}z_1^{-1} \\ x_{31} - x_{32}z_1^{-1} & x_{32} - x_{12}z_1^{-1} & x_{33} - x_{12}z_1^{-1} \\ x_{41} - x_{42}z_1^{-1} & x_{42} - x_{12}z_1^{-1} & x_{43} - x_{12}z_1^{-1} \end{pmatrix}$$

$$= \begin{pmatrix} \underline{x}_{11} & \underline{x}_{12} & \underline{x}_{13} \\ \underline{x}_{21} & \underline{x}_{22} & \underline{x}_{23} \\ \underline{x}_{31} & \underline{x}_{32} & \underline{x}_{33} \\ \underline{x}_{41} & \underline{x}_{42} & \underline{x}_{43} \end{pmatrix}$$

Converting to a row vector

$$\begin{pmatrix} \underline{x}_{11} & \underline{x}_{12} & \underline{x}_{13} & \underline{x}_{21} & \underline{x}_{22} & \underline{x}_{23} & \underline{x}_{31} & \underline{x}_{32} & \underline{x}_{33} & \underline{x}_{41} & \underline{x}_{42} & \underline{x}_{43} \end{pmatrix}$$

Figure 6.6. Creating a cancellation row.

After this window has reached the second column to the far right and three rows are generated, the window is lowered a row and shifted back to the left side of the data array, and the generation of rows continues. This is repeated until $Q-1$ cancellation rows have been formed. The elements of this row can be obtained by placing an $N_a \times N_p$ window, such as window 1 in Figure 6.5, over data. In order to restore the signal component in the adaptive processing, we fix the gain of the subarray (in both space and time) formed by fixing the first row of the matrix [T]. The elements of the first row are given by

$$T(1;y) = Z_1^{(e-1)} Z_2^{(h-1)} \tag{6.74}$$

where y, e, and h are given by equations (6.59), (6.61), and (6.63), respectively. By fixing the gain of the system for the given Doppler and the DOA, the following row vector can be generated:

$$[1 \;\; Z_1 \;\; Z_1^2 \cdots Z_1^{Na-1} \;\; Z_2 \;\; Z_1 Z_2 \;\; Z_1^2 Z_2 \;\cdots\; Z_1^{Na-1} \;\; Z_2 \;\; Z_2^2 \;\; Z_1 Z_2^2 \;\cdots\; Z_1^{Na-1} \, Z_2^{Np-1}]$$

By setting the product of [T] and [W] equal to a column vector [Y] the matrix equation is completed and it becomes a square system. The first element of [Y] consists of the constraint gain G and the remaining $Q-1$ elements are set to zero in order to complete the cancellation equations. The resulting matrix equation is then given by

$$[T][W] = [Y] = \begin{bmatrix} G \\ 0 \\ \vdots \\ 0 \end{bmatrix} \tag{6.75}$$

where G is a complex constant. In solving this equation one obtains the weight vector [W], which places space-time nulls in the direction of the interferers while maintaining gain along the direction of the SOI. The amplitude of the SOI can be estimated using

$$\hat{\alpha} = \frac{1}{G} \sum_{m=1}^{N_p} \sum_{n=1}^{N_a} W(n+m-1)\, X(m,n) \tag{6.76}$$

The analysis above was conducted for a single constraint. As in the 1D case, the SOI for the 2D space-time case could arrive at the array slightly off the look direction, either in angle or Doppler or both. In order to keep the processor from nulling the SOI, multiple constraints can be implemented [2]. The added constraints would reduce the number of degrees of freedom, but given the antenna beam width and Doppler filter width of a real system, the constraints could help maintain the system gain over this finite look direction extent. In a manner similar to the single constraint, L constraints can be implemented using L row constraints where the look direction of the ℓth row is determined by φ_ℓ and f_ℓ. For the ℓth constraint,

$$Z_{\ell 1} = e^{j 2\pi \frac{d}{\lambda} \cos(\varphi_\ell)} \tag{6.77}$$

$$Z_{\ell 2} = e^{j 2\pi \frac{f_\ell}{f_r}} \tag{6.78}$$

and the ℓth row of $[T]$, denoted as $T(\ell, :)$, becomes

$$T(\ell,:) =$$
$$\begin{bmatrix} 1 & Z_{\ell 1} & Z_{\ell 1}^2 & \cdots & Z_{\ell 1}^{Na-1} & Z_{\ell 2} & Z_{\ell 1}Z_{\ell 2} & Z_{\ell 1}^2 Z_{\ell 2} & \cdots & Z_{\ell 1}^{Na-1}Z_{\ell 2} & Z_{\ell 2}^2 & Z_{\ell 1}Z_{\ell 2}^2 & \cdots & Z_{\ell 1}^{Na-1} & Z_{\ell 2}^{Np-1} \end{bmatrix} \tag{6.79}$$

The L constraints provide a more accurate solution when there is some uncertainty associated with either the Doppler or the DOA. This technique will be utilized in chapter 18 for broadband processing of signals.

6.3.3 Least Squares Backward Processor

A second direct least squares space-time processor can be implemented by conjugating the element-pulse data and processing this data in reverse [2,12–14]. It is well known in the parametric spectral estimation literature that a sampled sequence consisting of a sum of complex exponentials can be estimated by observing it in either the forward or reverse direction. If we now conjugate the data and form the reverse sequence, we obtain an equation similar to (6.75) for the weights. In this case the first rows of $[T]$ and $[Y]$ are the same as before, as in (6.75). The remaining equations of (6.75) now have to be modified. Under the present circumstances, one would obtain the three consecutive rows of the $[T]$ matrix by taking a weighted difference between the neighboring elements to form

$$\begin{aligned} T(x, y) = {} & X^*(M-h-g+2;\ N-d-e+2) \\ & - Z_1^{-1} X^*(M-h-g+2;\ N-d-e+1) \end{aligned} \tag{6.80}$$

$$\begin{aligned} T(x+1;\ y) = {} & X^*(M-h-g+2;\ N-d-e+2) \\ & - Z_2^{-1} X^*(M-h-g+1;\ N-d-e+2) \end{aligned} \tag{6.81}$$

$$\begin{aligned} T(x+2;\ y) = {} & X^*(M-h-g+2;\ N-d-e+2) \\ & - Z_1^{-1} Z_2^{-1} X^*(M-h-g+1;\ N-d-e+1) \end{aligned} \tag{6.82}$$

for any row number x, and column y; the variables h, g, d, and e have been defined in (6.60)–(6.63). The row number increases by multiples of three, and

$$1 \le y = N_p(e-1) + h \le N_a N_p \tag{6.83}$$

The form of this linear matrix equation is similar to that of the forward algorithm, resulting in

$$[B][W]=[Y] \tag{6.84}$$

where the matrices $[B]$, $[W]$, and $[Y]$ are of size $Q \times Q$, $Q \times 1$, and $Q \times 1$, respectively. The constraint rows in $[B]$ are implemented in the same manner as the constraint rows in $[T]$. The difference between $[B]$ and $[T]$ is in the cancellation equations. For the backward method, these equations are formed by first conjugating the space-time snapshot given in Figure 6.5. Then using a windowing procedure similar to the forward case, three cancellation rows are generated for each position of the window, except now the window starts in the lower right corner of the space-time snapshot, as shown in Figures 6.5 and 6.6. This window is then moved to the right and up the snapshot. The three difference equations that are used to cancel the SOI are given by

$$X^*(m,n)-X^*(m,n-1)Z_1^{-1} \tag{6.85}$$

$$X^*(m,n)-X^*(m-1,n)Z_2^{-1} \tag{6.86}$$

$$X^*(m,n)-X^*(m-1,n-1)Z_1^{-1}Z_2^{-1} \tag{6.87}$$

Using equations (6.85)–(6.87), the SOI is removed from the windowed data. Once the weights are solved for by solving a system of equations similar to (6.84), the strength of the desired signal at range cell r is estimated from [2]

$$\alpha \approx \frac{1}{G}\sum_{e=1}^{N_a}\sum_{h=1}^{N_p} W\{N_p(e-1)+h\}\, X^*(M-h+1;\, N-e+1) \tag{6.88}$$

Thus the backward procedure provides a second independent realization of the same solution. In a practical environment, where the real solution is unknown, generation of two independent sets of solutions may provide some degree of confidence in the final results.

For systems where the DOA and Doppler frequency of the SOI are not known exactly, but are known approximately (e.g., within the mainbeam of the antenna), multiple constraints can be implemented to preserve the SOI. This is discussed in chapter 7. The procedure for doing this is identical to that of the forward method. For each additional constraint an additional row replaces a cancellation equation in $[B]$ and the corresponding amplitude is placed in $[Y]$. The constraint equations are determined using (6.79).

6.3.4 Least Squares Forward-Backward Processor

A system of equations may be formed by combining the forward and backward solution procedure as described for the 1D case [2,12–14]. Since, in this process, one is doubling the amount of data available by considering it in both the forward and reverse directions, one can essentially do one of the following two things: either increase the number of equations and solve an equation similar to (6.75) or (6.84) in a least squares fashion, or equivalently, increase the number of weights by combining the data of the forward and backward methods resulting in an increase in the number of degrees of freedom (DOFs) by as much as 50%. The second alternative is extremely attractive if one is processing the data on a snapshot-by-snapshot basis in a highly nonstationary environment. In this way one can effectively deal with a situation where the number of data samples may be too few to perform any other processing.

Since the total number of data points is $2MN$, the number of degrees of freedom can be increased from the two cases presented earlier. Hence the number of degrees of freedom in this case will be $N_a^{'}N_p^{'}$ where $N_a^{'} > N_a$ and $N_p^{'} > N_p$. The increase in the number of degrees of freedom depends on the number of antennas N and the time samples M. But clearly, $N_a^{'}N_p^{'}$ is significantly greater than N_aN_p. This increase is by a factor of approximately 2 when dealing with a data cube where $N = 22$ (the number of antennas in the array) and $M = 128$ (the number of time samples). By using the samples from N antenna elements and M pulses, we formulate the following matrix equation:

$$[FB][W] = [Y] \tag{6.89}$$

Here again the system matrix $[FB]$ consists of constraint rows and cancellation rows. The constraint rows preserve the SOI during the adaptive process. There is at least one constraint row, while multiple constraints may be used just as in (6.79) to maintain the gain of the array toward the SOI, which may possess a slightly different look direction φ_s and Doppler frequency f_s along the look direction. The remaining rows in [FB] consist of cancellation equations that are formed in both the forward and backward directions.

We now apply the various direct data domain STAP algorithms described so far to real experimental data to study the performance of each method. In addition, the result for the statistical based STAP method is also presented to illustrate the superiority in performance of this new direct data domain least squares techniques, over conventional statistical methods.

6.4 APPLICATION OF THE DIRECT DATA DOMAIN LEAST SQUARES TECHNIQUES TO AIRBORNE RADAR FOR SPACE-TIME ADAPTIVE PROCESSING

The Multichannel Airborne Radar Measurement (MCARM) program had as its objective the collection of multiple spatial channel airborne radar data for the development and evaluation of STAP algorithms for future Airborne Early Warning (AEW) systems. The airborne MCARM testbed, a BAC1-11 aircraft, used for these measurements is shown in Figure 6.7. The phased array is hosted in an aerodynamic cheek-mounted, and placed just forward of the left wing of the aircraft. The L-band (1.24 GHz) active array consists of 16 columns, with each column having two four-element subarrays, shown in Figure 6.8. The elements are vertically polarized, dual-notch reduced-depth radiators. These elements are located on a rectangular grid with azimuth spacing of 4.3 inches and elevation spacing of 5.54 inches. There is a 20-dB Taylor weighting across the eight elevation elements, resulting in a 0.25-dB elevation taper loss for both transmit and receive. The total average radiated power for the array was approximately 1.5 kW. A 6-dB modified trapezoid weighting for the transmit azimuthal illumination function is used to produce a 7.5° beamwidth pattern along the boresight with –25 dB RMS sidelobes. This pattern can be steered up to ±60°. Of the 32 possible channels, only 24 receivers were available for the data collection program. Two of the receivers were used for analog sum and azimuthal difference beams. There are therefore 22 (N = 22) digitized channels, which in this work are arranged as a rectangular 2 × 11 array. Each CPI comprises 128 (M = 128) pulses at a pulse repetition interval of 1984 Hz [15].

Figure 6.7. The MCARM test bed.

Figure 6.8. The MCARM antenna array.

In the following examples the beam was pointed downward in elevation by about 5°. The flight path of the phased array was over the Delmarva peninsula (as defined by the landmass between the Chesapeake Bay and the Atlantic Ocean). In this experiment, the phased array on the BAC1-11 is trying to locate a Saberliner approaching the BAC1-11 in the presence of sea, urban, and land clutter. The data cubes generated from these measurements, which are available from the Air Force Research Laboratory Web site (*http://sunrise.deepthought.rl.af.mil*), are used to analyze the validity of the algorithms presented in this chapter. The details are available in [15]. The geographical region for the flight is shown in Figure 6.9 and the regions of ground clutter return in Figure 6.10. This indicates that it is possible to have urban, land, and sea clutters simultaneously.

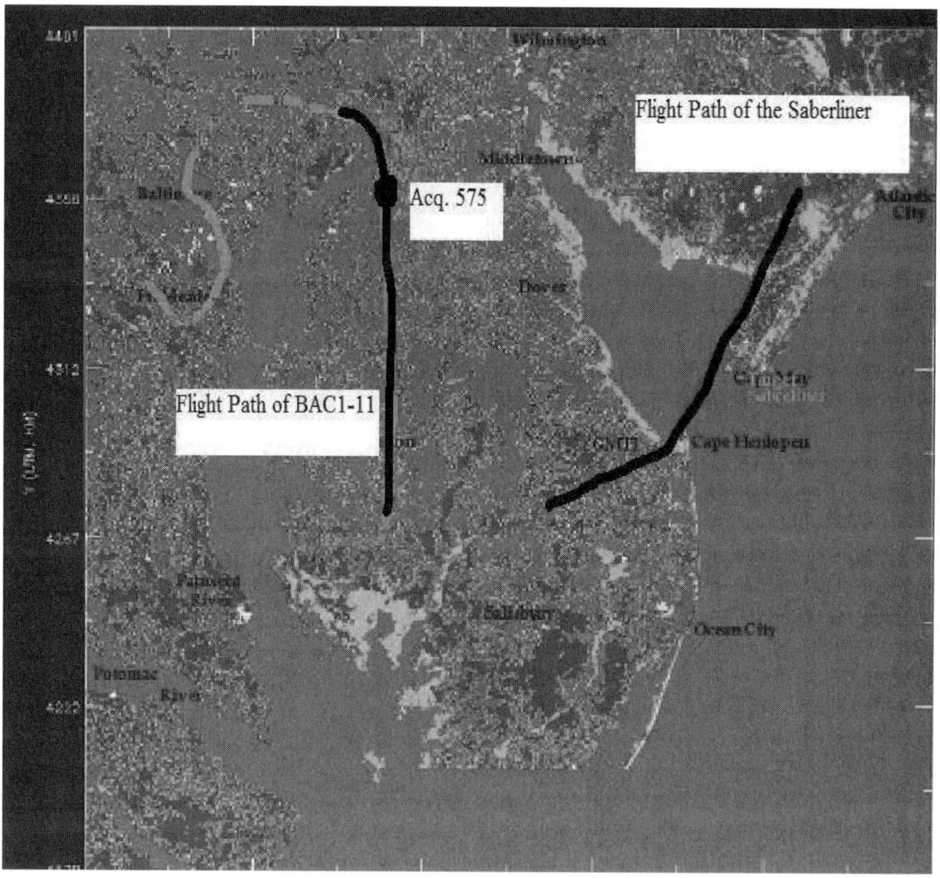

Figure 6.9. Flight paths of the BAC1-11 and the Saberliner over the Delmarva peninsula.

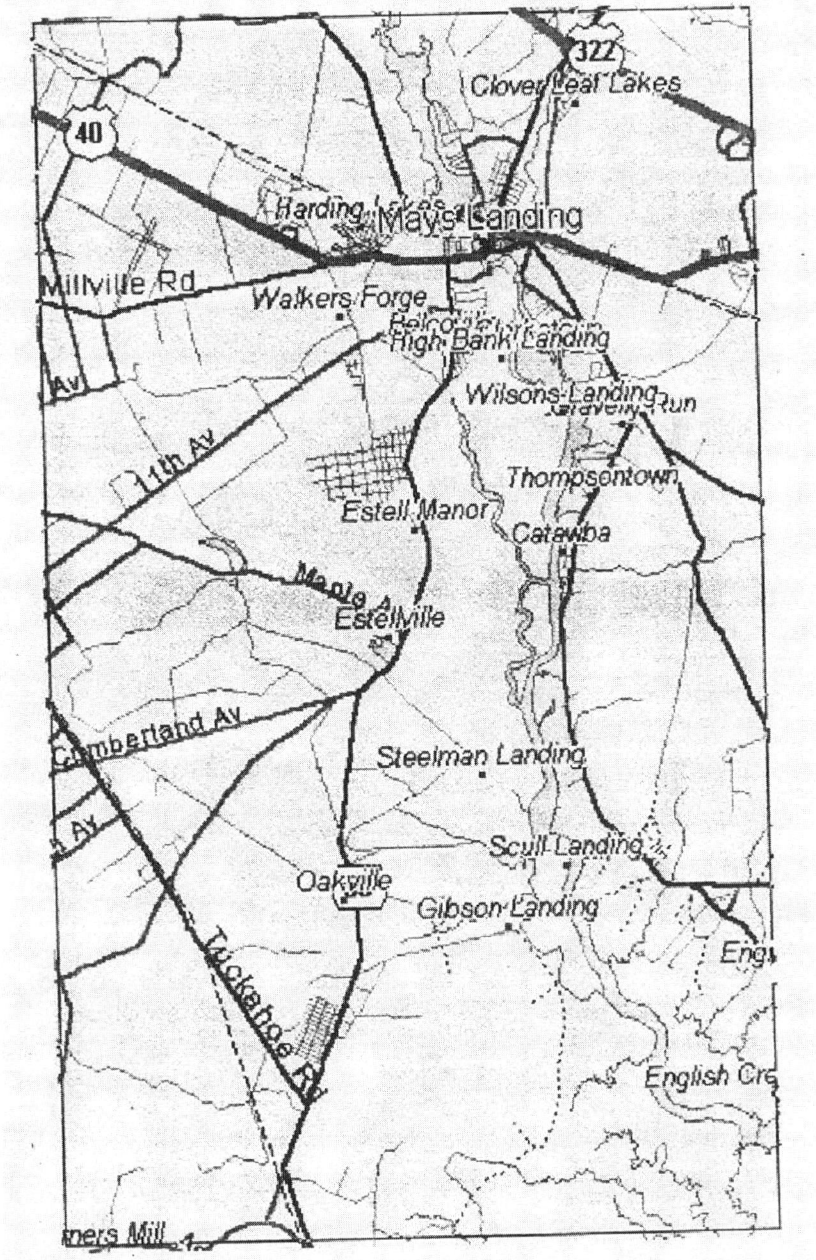

Figure 6.10. Scene of the region of ground clutter.

We next apply the D3LS techniques to the analysis of MCARM data set RL050575.dat. This deals with an actual target buried in clutter. The data was collected by an airborne antenna array. The antenna array had 22 channels in addition to the sum and the difference channel. For each channel, the data in the time domain was sampled at 1984 Hz and there are 128 time samples ($M = 128$). The third dimension of the data set corresponds to the range profile and there are 630 range bins. The 3-dB beamwidth of the antenna is approximately 7.8°. The data was gathered over a flight path over the Delmarva peninsula. The flight path of the down-looking phased array of the BAC1-11 is shown by the left curve in Figure 6.9, on which the particular data set was collected. Its position when it took the data is marked by the circle on the left-hand side. In addition, there is a Sabreliner flying toward the BAC1-11 in a slanted fashion, as shown by the second curve in Figure 6.9. The position of the Sabreliner is marked on the right-hand side. From data collected from geostationary satellites, it appears that the target is at 91° in azimuth and corresponds to the range cell at 318 (this is the second ambiguous range cell, namely 630 + 318) and corresponding to a Doppler frequency of approximately f_s = 520 Hz. Also, this data set contains received land, sea, and urban clutter from the regions shown in Figure 6.10. It also had signal return from highways which may have had some cars traveling by at that time. In the current analysis it is assumed that the signal is coming from $\varphi_s = 90°$ (i.e., broadside).

Before the super resolution D3LS signal processing algorithms can be applied, one needs to compensate for the various electromagnetic effects as shown by the uneven steering vector in Fig. 6.11 pointing to the broadside direction. The compensation is accomplished by coupling an electromagnetic analysis with the signal processing methodology. The various compensation techniques for the various electromagnetic effects have been described in [2] and will not be repeated here. .

The superresolution D3LS analysis is now applied to the real data which is equivalent to a two-dimensional filtering technique is applied to each range cell as described by the forward method. The order of the filter required to identify the signals in the presence of clutter consisted of 17 weights or filter taps in space ($N_a = 17$) and a 39-order filter in time ($N_p = 39$), so that the total number of degrees of freedom is $N_a N_p$ and is 663. This filter was applied to each range bin, corresponding to the signal of arrival from the broadside direction (i.e., φ_s = 90°) and the Doppler frequency was swept from 380 to 600 Hz in steps of 10 Hz. The range cells are swept from 300 to 350. Figure 6.12 represents the contour plot of the estimated signal return utilizing the forward method with weights of $N_a = 17$ and $N_p = 39$. It is seen that there are some activities around Doppler 500 Hz near the range cells of 308, 330, and 347.

Next the backward method is used to analyze the same data set using the same number of degrees of freedom ($17 \times 39 = 663$). The results are shown in Figure 6.13. Again large returns around 500 Hz Doppler frequency are observed in the range cells of 305, 320, 330, and 338. The application of the forward-

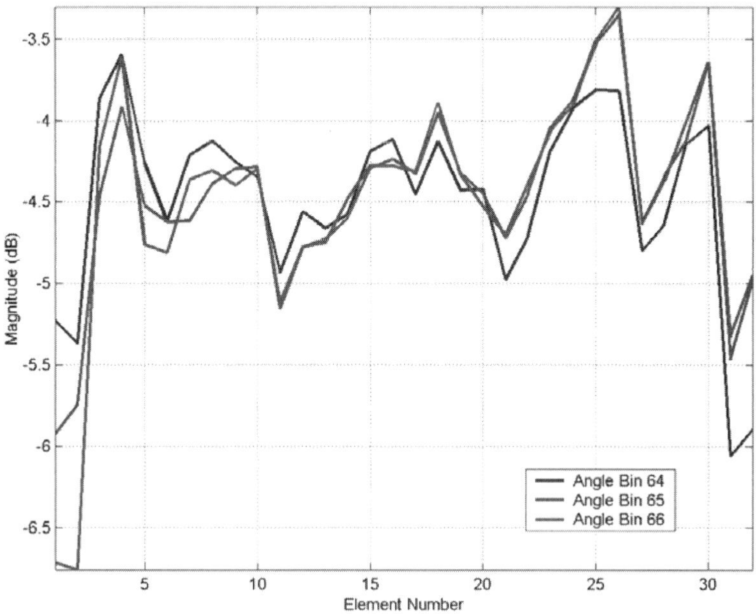

Figure 6.11. Magnitude of MCARM steering vectors.

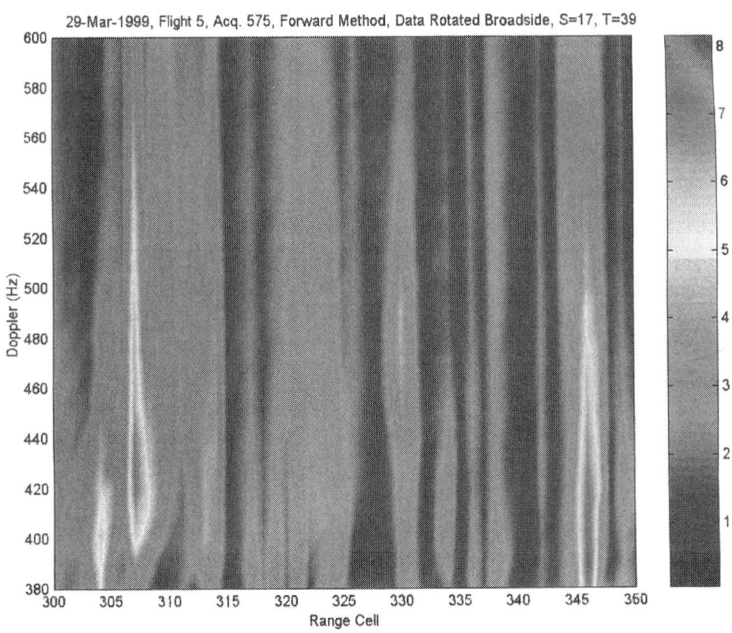

Figure 6.12. Application of the forward method to RL050575.dat.

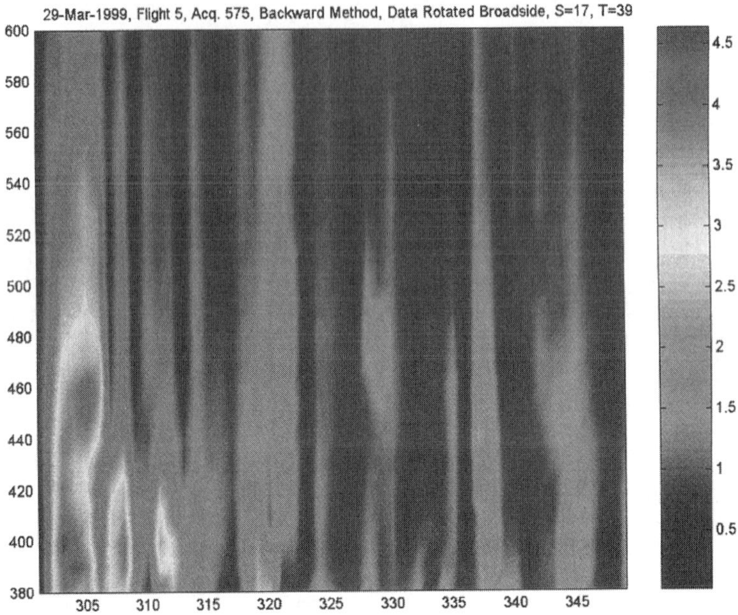

Figure 6.13. Application of the backward method to RL050575.dat.

backward method with the following weights of order ($N_a = 19$, $N_p = 61$) results in signal returns as shown in Figure 6.14. The number of degrees of freedom of the forward-backward method is $19 \times 61 = 1159$. This is nearly a two fold increase in the number of degrees of freedom used in either the forward or backward method. It is seen that the return is dominant at the range cells of 308 and 330.

By comparing the results of the three graphs, one could say with confidence which strong signal is a true return corresponding to a particular Doppler and range cell. This is because the three methods are analyzing the same data set in three independent ways, and hence it makes sense to compare the three results. Therefore, simultaneous use of all three methods would provide a reliable estimate of the signal and will minimize the probability of false alarm. However, the forward-backward method uses significantly more weights than either the forward or the backward method and therefore is expected to yield better results over either the forward or the backward method.

Typical running time for each data point using the forward or backward method is less than a minute on a Pentium PC with a CPU clock of 450 MHz. The forward-backward method takes slightly more time than either the forward or backward method. It is important to note that each range cell/Doppler/look angle can be processed in parallel. Hence the computational requirements are very modest for real-time applications. In all the computations, the value of the gain factor G in equations (6.75), (6.84), and (6.89) has been chosen as unity.

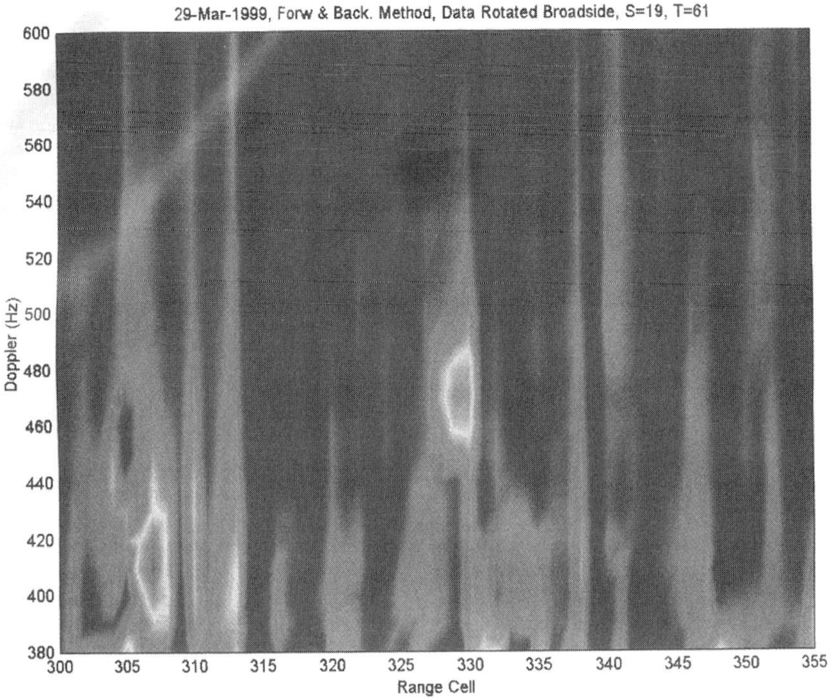

Figure 6.14. Application of the forward-backward method to RL050575.dat.

Next a conventional stochastic method is used to estimate the signal strengths. The application of a 9 × 9 covariance matrix utilizing a JDL stochastic approach [15] is also used to estimate the signals. This is after the data has been Fourier transformed into the Doppler domain utilizing a Kaiser–Bessel window and a 128-point FFT. The result is shown in Figure 6.15, where some weak activity can be seen in range cells 308 and 330 around the Doppler of 500 Hz. It is quite clear that the direct data domain methods provide a better presentation of the results in the Doppler-range space than the statistical method. However, without any "ground truth," it is difficult to predict which signal return is actually the Saberliner in all of these! This is because there are channel mismatches in the measurements and various uncertainties, such as the crab angle of the two aircrafts. The actual results show some deviations from the theoretical estimates. There may be several factors of uncertainty in the measured data such as the velocity of each aircraft, its elevation, and its direction of travel. However, all the methods predicted returns around the Doppler of 500 Hz in the range bin of 330. This slight discrepancy in Doppler and range can happen due to various factors, as outlined. Some shift may occur due to the matched filter processing if the target is not exactly at $\varphi_s = 90°$, or due to errors in the array calibration

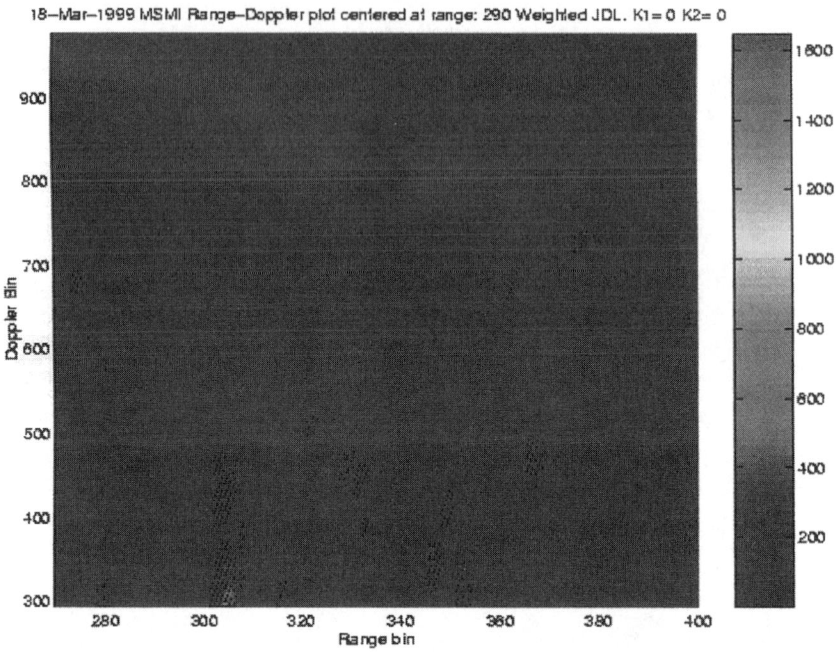

Figure 6.15. Application of the stochastic method to RL050575.dat.

introduced by mutual coupling and near-field coupling with the aircraft frame which were not fully accounted for in the analysis.

Similar conclusions regarding the statistical methods were also reached in [15]. Some of the reasons for the visible differences between the direct data domain least squares methods and the statistical based techniques will be discussed in chapter 11.

6.5 CONCLUSION

A D3LS method based on the spatial samples of a single snapshot of data is presented. In this approach the adaptive analysis is done on a snapshot-by-snapshot basis, and therefore nonstationary environments can be handled quite easily, including coherent multipaths. Associated with adaptive processing is the same *a priori* knowledge about the nature of the signal, which in this case is the DOA. The assumption that the target signal is coming from an exactly known direction will probably never be met in any real array. In communication systems the location of the transmitter may be known only approximately, or the propagation of the signal through the atmosphere may distort the wavefront such that it appears to be coming from a slightly different direction. For example, diffraction could cause enough error for the determination of an elevation angle to be important for some systems. Or the adaptive receive array may be surveyed into a location with small errors, and thus the angle to the transmitter from the broadside of the array will be in error. Other applications of adaptive arrays will also have at least small errors in the DOA of the SOI. In this approach additional constraints can be placed to correct these imperfections as discussed next in chapter 7.

The advantage of the D3LS approach based on spatial processing of the array data may prove beneficial over conventional adaptive techniques utilizing time averaging of the data. This will be quite relevant in a non-stationary environment. A D3LS method has also been presented to carry out space-time adaptive processing. Limited examples have been presented to illustrate the applicability of this technique to deal with real airborne platform data.

REFERENCES

[1] S. Treitel and E. A. Robinson, "Optimum Digital Filters for Signal to Noise Ratio Enhancement," *Geophysical Prospecting*, 1969, pp. 380-425.

[2] T. K. Sarkar, M. C. Wicks, M. Salazar-Palma and R. J. Bonneau, *Smart Antennas*, Wiley-IEEE Press, 2003.

[3] T. K. Sarkar, J. Koh, R. S. Adve, R. A. Schneible, M. C. Wicks, M. Salazar-Palma, and S. Choi, "A Pragmatic Approach to Adaptive Antennas," *IEEE Antennas and Propagation Magazine*, Vol. 42, No. 2, pp. 39–55, Apr. 2000.

[4] T. K. Sarkar, E. Arvas, and S. M. Rao, "Application of FFT and the Conjugate Gradient Method for the Solution of Electromagnetic Radiation from Electrically Large and Small Conduction Bodies," *IEEE Transactions on Antennas and Propagation*, Vol. 34, No. 5, pp. 635–640, 1986.

[5] T. K. Sarkar and N. Sangruji, "An Adaptive Nulling System for a Narrowband Signal with a Look Direction Constraint Utilizing the Conjugate Gradient Method," *IEEE Transactions on Antennas and Propagation*, Vol. 37, No. 7, pp. 940–944, 1989.

[6] T. K. Sarkar, S. Park, J. Koh, and R. A. Schneible, "A Deterministic Least Square Approach to Adaptive Antennas," *Digital Signal Processing: A Review Journal*, Vol. 6, pp. 185–194, 1996.

[7] S. Park and T. K. Sarkar, "Prevention of Signal Cancellation in Adaptive Nulling Problem," *Digital Signal Processing: A Review Journal*, Vol. 8, No. 2, pp. 95–102, Apr. 1995.

[8] S. Park, T. K. Sarkar, and Y. Hua, "A Singular Value Decomposition Based Method for Solving a Deterministic Adaptive Problem," *Digital Signal Processing: A Review Journal*, Vol. 9, pp. 57–63, 1999.

[9] R. Brown and T. K. Sarkar, "Real Time Deconvolution Utilizing the Fast Fourier Transform and the Conjugate Gradient Method," in *5th Acoustic Speech and Signal Processing Workshop on Spectral Estimation and Modeling*, Rochester, NY, 1990.

[10] M. G. Bellanger, *Adaptive Digital Filters and Signal Analysis*, Marcel Dekker, New York, 1987.

[11] T. K. Sarkar, *Application of the Conjugate Gradient Method to Electromagnetics and Signal Analysis*, Vol. 5, Progress in Electromagnetics Research, Elsevier, New York 1990.

[12] J. Carlo, T. K. Sarkar and M. C. Wicks, "Application of Deterministic Techniques

to STAP," In *Applications of Space-Time Adaptive Processing*, edited by R. Klemm, IEE Press, 2004, pp. 375-411.

[13] T. K. Sarkar, H. Wang, S. Park, J. Koh, R. Adve, K. Kim, Y. Zhang, M. C. Wicks and R. D. Brown, "A Deterministic Least squares Approach to Space Time Adaptive Processing (STAP)," *IEEE Trans. Antennas and Propagation*, Vol. 49, January 2001, pp. 91-103.

[14] T. K. Sarkar, S. Nagaraja and M. C. Wicks, "A Deterministic Direct Data Domain Approach to Signal Estimation Utilizing Nonuniform and Uniform 2-D Arrays," *Digital Signal Processing – A Review Journal,* Vol. 8, 114-125 (1998).

[15] P. Sanyal, "STAP Processing Monostatic and Bistatic MCARM Data," AFRL-SN-RS-TR-1999-197, Final Technical Report, Air Force Research Laboratory, Sensors Directorate, Rome Research Site, Sept 1999.

[16] H. Chen, T. K. Sarkar, S. A. Dianat, and J. D. Brule, "Adaptive Spectral Estimation by the Conjugate Gradient Method", *IEEE Transactions on Acoustics, Speech & Signal Processing*, Vol. ASSP-34, No. 2, pp. 272-284, Apr. 1986.

7

MINIMUM NORM PROPERTY FOR THE SUM OF THE ADAPTIVE WEIGHTS IN ADAPTIVE OR IN SPACE-TIME PROCESSING

7.0 SUMMARY

In most adaptive algorithms, it is generally assumed that one knows the direction of arrival (DOA) of the signal of interest (SOI) through the steering vector of the array, and the goal is to estimate its complex amplitude in the presence of jammer, clutter and noise. In space-time adaptive processing (STAP) the goal is to seek for a target located along a certain look direction and at a particular Doppler frequency through a given steering vector. Therefore, the accuracy of the computed results in either case is based on the reliability of this *a priori* assumption of the steering vector. It is possible that, due to mechanical vibrations, calibration errors, or atmospheric refractions of the incident electromagnetic waves, the assumed DOA may not be very accurate or that the assumed value of the Doppler frequency is not appropriate. In either of these cases, the adaptive algorithm treats the SOI as an interferer and nulls it out. This perennial problem of signal cancellation is an open problem for adaptive algorithms. In this chapter we propose a secondary processing scheme for the direct data domain least squares (D3LS) method to illustrate on how to refine the estimate for the steering vector. It is shown that the proper steering vector occurs at the minimum of the sum of the norm of the adaptive weights and can be used as an indicator to refine the estimate of the DOA of the SOI in adaptive algorithms or both the DOA or/and the Doppler frequency in STAP. Examples are presented to illustrate that the secondary processing outlined in this chapter may provide a refined estimate for the true DOA or/and Doppler frequency for the SOI in the presence of interference, clutter, and noise.

7.1 INTRODUCTION

Recently, a D3LS algorithm for solving both adaptive [1] and space-time adaptive problems [2–5] has been proposed to overcome the drawbacks of a statistical technique. This methodology has been presented in Chapter 6. In that approach one adaptively minimizes the interference power while maintaining the gain of the antenna array along the direction of the SOI through the proper choice of the steering vector. For the 1-D case, the steering vector corresponds to the voltages induced at the antenna elements for a given DOA for the SOI. Not having to estimate a covariance matrix leads to an enormous savings in memory and computer processor time and makes it possible to carry out an adaptive process in real time [3]. One of the open problems is how to refine the *a priori* information available for the assumed DOA of the SOI through the steering vector. Here, we address the RADAR problem, where we know *a priori* along what direction the transmitted energy was sent and therefore we know approximately through what angles we should expect the target return. In real life, there are always some uncertainties associated with this assumed DOA for RADAR problems. This may be due to mechanical vibrations of the antenna array, incorrect calibration, or atmospheric refraction of the incident electromagnetic wave. The goal of this chapter is to illustrate that some secondary processing may be done to improve the initial assumption about the DOA. It is shown that the norm of the adapted weights may provide a refined estimate for the actual DOA of the SOI and therefore improve the accuracy for the assumed steering vector when there are uncertainties associated with their initial estimates. The norm of the adapted weights is simply the sum of the absolute value of the weights. The existence of a minimum in the norm of the weights within the assumed beam width of the transmitter can be used to obtain a refined target return angle. It could also be used to perform the detection process as well. So this chapter represents the minimum norm properties of the optimum weights, when the true DOA of the signal coincides with the assumed DOA of the SOI in the presence of jammers, interferers, clutter, and thermal noise. This could lead to a more accurate estimation of the DOA of the SOI or on a detection process, when a good estimate for the DOA information is not available *a priori*. For STAP processing [4,5] we search for a signal with a particular Doppler shift and DOA. Again, if the assumed values of the DOA and /or the Doppler frequency are not close enough to the real ones then the adaptive algorithm treats the SOI as an interferer and cancels it. The minimum value of the norm of the adaptive weights can also be used to provide a refined estimate for the Doppler and the DOA of the SOI. For the STAP problem, it becomes an additional search in Doppler as we have some *a priori* information about the DOA for this RADAR problem. Here, as we solve an estimation problem, the estimated signal strength for the SOI can also be correlated with the norm of the weights to get a refined estimate.

The presentation is organized as follows. First we provide a brief review of the direct data domain least squares approach for the 1-D adaptive problems in section 7.2 and for the 2-D space-time adaptive case in section 7.3. In section 7.4 we present the minimum norm property of the optimum weights when the assumed DOA coincides with the actual one. The proof is given based on induction. In section 7.5, numerical simulations illustrate this property. Finally, in section 7.6 we present some conclusions.

7.2 REVIEW OF THE DIRECT DATA DOMAIN LEAST SQUARES APPROACH

Let us assume that the SOI is coming from the angular direction θ_s and our objective is to estimate its complex amplitude while simultaneously rejecting all other interferences and noise. The signal arrives at each antenna element at different times dependent on the DOA of the SOI and the geometry of the array. We make the narrowband assumption for all the signals including the interferers. At each of the N antenna elements, the received signal is a sum of the SOI, interference, and thermal noise. It is important to note that here we treat the antenna elements as idealized point sources, for illustration. Use of realistic antennas in adaptive processing has been addressed in [3]. The interference may consist of coherent multipaths of SOI along with clutter and thermal noise. Using the complex envelope representation for a uniform linear array, the $N \times 1$ complex vectors of phasor voltages $\left[X(t)\right]$ received by the antenna elements at a single time instance t can be expressed by

$$[X(t)] = \begin{bmatrix} X_1(t) \\ X_2(t) \\ \vdots \\ X_N(t) \end{bmatrix} = \sum_{m=1}^{M} [a(\theta_m)]\, s_m(t) + [n(t)] \tag{7.1}$$

where s_m and θ_m are the amplitude and DOA, respectively, of the mth source incident on the array at the time instance t. $[a(\theta_m)]$ denotes the steering vector of the array toward direction θ_m and $[n(t)]$ denotes the noise vector at each of the N antenna elements. We now analyze the data using a single snapshot of the voltages measured at the antenna terminals [1,3]. As we are using the phasor notation the functional dependence on t can be dropped from the remaining equations as we are looking at the complex voltages induced at the feed point of the antenna elements at a particular instance of time.

Let us assume that the SOI is coming from the angular direction θ_s and our objective is to estimate its complex amplitude while simultaneously rejecting all other interferences and noise. The arrival of the signal at each antenna element occurs at different times. It depends on the DOA of the SOI and the geometry of the array. We make the narrowband assumption for all the signals including the interferers. At each of the N antenna elements, the received signal given by (7.1) is a sum of the SOI, interference, and thermal noise. The interference may consist of coherent multipaths of the SOI along with clutter and thermal noise.

Therefore, by suppressing the time dependence in the phasor notation, we can reformulate (7.1) as

$$[X] = \begin{bmatrix} X_1 \\ X_2 \\ \vdots \\ X_N \end{bmatrix} = \alpha_s \begin{bmatrix} a_1(\theta_s) \\ a_2(\theta_s) \\ \vdots \\ a_N(\theta_s) \end{bmatrix} + \sum_{m=1}^{M-1} \mathrm{s_m} \begin{bmatrix} a_1(\theta_m) \\ a_2(\theta_m) \\ \vdots \\ a_N(\theta_m) \end{bmatrix} + \begin{bmatrix} n_1 \\ n_2 \\ \vdots \\ n_N \end{bmatrix} \tag{7.2}$$

where α_s is the complex amplitude of the SOI, to be determined. The column vectors in (7.2) explicitly show the various components of the signal induced in each of the N antenna elements. a_n represents the voltage induced at the nth antenna element due to signal of unity amplitude arriving from a particular direction θ. For a conventional adaptive array system, using each of the K weights W_k, we can now estimate the SOI through y by using the following weighted sum

$$y = \sum_{k=1}^{K} W_k \, X_k \tag{7.3}$$

or in a compact matrix form as

$$y = [W]^T [X'] = [X']^T \, [W] \tag{7.4}$$

where $[X']$ contains K of the N elements of $[X]$, and the superscript T denotes the transpose of a matrix. Since we are using a single snapshot of the received voltages in which there may be possible coherent interferers, then K is less than N as illustrated in [3]. Depending on the type of the direct data domain processing chosen, the relationship between K and N can be chosen as either $K \cong N/2$ or can be $0.66N$ as explained in section 6.2 and described in detail in [3].

Let us define

$$Z = \exp\left[j2\pi \frac{d}{\lambda} \cos\theta_s \right] \tag{7.5}$$

where θ_s is the angle of arrival corresponding to the desired signals. Then $X_1 - Z^{-1} X_2$ contains no components of the SOI. Since

$$a_1(\theta_s) = \exp\left[j2\pi \frac{(n-1)\,d}{\lambda} \cos\theta_s \right] \quad \text{with } n = 1 \tag{7.6}$$

$$\mathrm{a}_2(\theta_s) = \exp\left[j2\pi \frac{(n-1)\,d}{\lambda} \cos\theta_s \right] \quad \text{with } n = 2 \tag{7.7}$$

Therefore, one can form a reduced rank matrix $[T]$ where the weighted sum of all its elements would be zero [3]. The matrix $[T]$ is given by

$$[T] = \begin{bmatrix} X_1 - Z^{-1}X_2 & X_2 - Z^{-1}X_3 & \cdots & X_K - Z^{-1}X_{K+1} \\ \vdots & \vdots & & \vdots \\ X_{K-1} - Z^{-1}X_K & X_K - Z^{-1}X_{K+1} & \cdots & X_{N-1} - Z^{-1}X_N \end{bmatrix}_{(K-1)\times K} \quad (7.8)$$

In order to make the matrix full rank, we fix the gain of the subarray by forming the weighted sum $\sum_{k=1}^{K} W_k Z^{k-1}$ along the DOA of the SOI to a prespecified value. Let us say the gain of the subarray is C along the direction of θ_s. This provides an additional equation resulting in

$$\begin{bmatrix} 1 & \cdots & Z^{K-1} \\ X_1 - Z^{-1}X_2 & \cdots & X_K - Z^{-1}X_{K+1} \\ \vdots & \vdots & \vdots \\ X_{K-1} - Z^{-1}X_K & \cdots & X_{N-1} - Z^{-1}X_N \end{bmatrix}_{K\times K} \begin{bmatrix} W_1 \\ W_2 \\ \vdots \\ W_K \end{bmatrix}_{K\times 1} = \begin{bmatrix} C \\ 0 \\ \vdots \\ 0 \end{bmatrix}_{K\times 1} \quad (7.9)$$

$$\text{Or, equivalently, } [F][W] = [C] \quad (7.10)$$

There exist many DOA estimation methods like MUSIC, ESPRIT, and maximum likelihood method, which are based on using multiple snapshots in time to evaluate the DOA's of the signals [6–8]. These methods form a covariance matrix and they estimate the DOA's from the properties of the covariance matrix. However, when there are coherent signals, additional processing needs to be done with these techniques to estimate the DOA's. Also these techniques do not make any distinction between the SOI and the interferers. In this section, our goal is not to determine the DOA of all the signals including interferers, but to refine the estimate of the DOA of the SOI that we have. In the current approach, even though we are dealing with a single snapshot of the data, there is no problem associated with refining the estimate of the DOA of the SOI when it is coherent with its undesired multipaths. For the conventional methods, it will be difficult to estimate the DOA's of the signals, when there is clutter present in the measured response.

In summary, this chapter does not propose an algorithm to find the DOA of all the signals impinging on the array but rather a methodology for refining the given estimates through secondary processing for the DOA for the 1-D case of adaptive processing and the DOA and the Doppler frequency for the 2-D STAP case, to be outlined next.

7.3 REVIEW OF SPACE-TIME ADAPTIVE PROCESSING BASED ON THE D3LS METHOD

In a uniformly spaced linear array consisting of isotropic elements, the complex envelope of the received SOI with unity amplitude, for the pth pulse and q^{th} antenna element, can be described as [2–5]

$$S_{p,q} = \exp\left[j2\pi \left\{ \frac{(q-1)d}{\lambda}\cos(\theta_s) + \frac{(p-1)f_s}{f_r} \right\} \right] \tag{7.11}$$

$$\text{for } p = 1,2,\ldots,P\,;\ q = 1,2,\ldots,Q$$

where d is the spacing between the antenna elements, λ is the wavelength, θ_s is the angle of arrival, f_s is the Doppler frequency of SOI, and f_r is the pulse repetition frequency. For the pth pulse at the qth antenna, the complex envelope of the received signal is given by

$$X_{p,q} = \alpha_s\, S_{p,q} + \text{Interferers (Clutter plus coherent/noncoherent jammers)} + \text{Noise} \tag{7.12}$$

where α_s is the complex amplitude of the SOI.

The received data for a given space-time snap shot can be arranged as

$$[X] = \begin{bmatrix} X_{11} & X_{12} & \cdots & X_{1Q} \\ X_{21} & X_{22} & \cdots & X_{2Q} \\ \vdots & \vdots & \cdots & \vdots \\ X_{P1} & X_{P2} & \cdots & X_{PQ} \end{bmatrix} \tag{7.13}$$

For a particular row of (7.13) the column-to-column phase difference, due to the SOI, is

$$Z_1 = \exp(j2\pi \frac{d}{\lambda}\cos\theta_s) \tag{7.14}$$

The row-to-row phase difference in a given column, due to the Doppler of the SOI, is given by

$$Z_2 = \exp(j2\pi \frac{f_s}{f_r}) \tag{7.15}$$

Differencing is performed with elements offset in space, time, and jointly in space and time for the two-dimensional case. The three types of difference equations are then given by

$$X_{p,q} - Z_1^{-1} \cdot X_{p,q+1} \tag{7.16}$$

$$X_{p,q} - Z_2^{-1} \cdot X_{p+1,q} \tag{7.17}$$

$$X_{p,q} - Z_1^{-1} \cdot Z_2^{-1} \cdot X_{p+1,q+1} \tag{7.18}$$

Now we can form the cancellation rows in matrix [F] using equations (7.16)–(7.18). And the elements of the first row of matrix [F], similar to (7.10), is given by [2–5]

$$\begin{bmatrix} 1 & Z_1 & Z_1^2 & \ldots & Z_1^{N_a-1} & Z_2 & Z_1Z_2 & Z_1^2Z_2 & \ldots & Z_1^{N_a-1}Z_2 & Z_2^2 & Z_1Z_2^2 & \ldots & Z_1^{N_a-1}Z_2^{N_p-1} \end{bmatrix} \tag{7.19}$$

where the number of the antenna weights $N_a \leq (Q+1)/2$ and the number of temporal weights $N_p \leq (P+1)/2$. The total number of weights in space-time then will be $R = N_a \times N_p$. It is important to note that it is not a factored space time methodology.

The resulting matrix equation for the STAP case is then given by

$$[F]_{R\times R}\,[W]_{R\times 1} = [Y]_{R\times 1} = \begin{bmatrix} C & 0 & \ldots & 0 \end{bmatrix}_{1\times R}^{T} \tag{7.20}$$

7.4 MINIMUM NORM PROPERTY OF THE ADAPTIVE WEIGHTS AT THE DOA OF THE SOI FOR THE 1-D CASE AND AT DOPPLER FREQUENCY AND DOA FOR STAP

What we show in this section is that if we sum up all the squared values of the weights obtained for the solution of the 1-D problem for an assumed DOA of the SOI, then that sum is a minimum when the actual DOA of the SOI coincides with the assumed DOA. The norm of the weights for the vector W can be defined by $\|W\| = \sqrt{[|W_1|^2 + |W_2|^2 \;\ldots\ldots\; + |W_K|^2]}$. So, for a scalar quantity, the norm represents its square of the absolute value. For the 2-D case, the minimum of the sum of the squared absolute value of the weights will occur when the assumed Doppler and the DOA for the SOI coincides with the actual ones. The norm of the weights then succinctly defines the sum of the squared absolute value of the weights.

The minimum of the sum of the squared values of the adaptive weights, typically referred to as the minimum norm occurs at the true value for the angle of arrival of the SOI irrespective of what the interference scenario is.

Similarly one can observe the same property for the 2-D space-time adaptive processing (STAP), when there is an additional Doppler frequency term. So if we find the minimum of the norm of the weights, we can refine the estimate for the assumed steering vector and therefore of the actual Doppler frequency and DOA of the SOI in STAP.

The proof of this section is illustrated through induction for the 1-D case. First we develop the proof for the 1-D case with three antenna elements. It can then be extended to the more complex cases and for the 2-D STAP cases.

If we consider the voltages X_n induced at a set of N equispaced antenna elements separated by a distance d due to a set of M incident waves arriving from an angle θ_m and amplitude A_m, then we know we can write

$$X_n = \sum_{m=1}^{M} A_m \exp\left[j\frac{2\pi(n-1)d}{\lambda}\sin\theta_m \right] \quad \text{for } n = 1, \ldots, N \tag{7.21}$$

Prony's method [9, 10] gives us a recipe to compute not only the value for M, but also the angles θ_m and A_m given the set of voltages X_n. We rewrite the above equation in the following form as

$$X_n = \sum_{m=1}^{M} A_m \exp[j\, n\, z_m] \tag{7.22}$$

where

$$z_m = 2\pi d \sin\theta_m / \lambda \tag{7.23}$$

The solution procedure for z_m and A_m in Prony's method starts in the following way by assuming $M = 2$ and we have no noise in the data. The proof can then be extended to any other value of M. Next, we solve for the set of coefficients W_i which satisfies

$$\begin{bmatrix} X_1 & X_2 & X_3 \\ X_2 & X_3 & X_4 \\ X_3 & X_4 & X_5 \end{bmatrix} \begin{bmatrix} W_1 \\ W_2 \\ W_3 \end{bmatrix} = [0] \tag{7.24}$$

Since (7.24) is a homogeneous equation, the solution to it is not unique. To generate a unique solution we can set any of the four constraints

$$W_1 = 1, \text{ or } W_2 = 1 \text{ or } W_3 = 1 \text{ or } \sqrt{W_1^2 + W_2^2 + W_3^2} = 1 \tag{7.25}$$

We then solve for the matrix equation of interest (7.24) for W_i, with any one of the four constraints of (7.25). It is clear that by observing (7.24), it is seen that the matrix $[W]$ is the eigenvector corresponding to the zero eigenvalue of the Hankel matrix formed by the data. In this way there is similarity between the MUSIC method of which Pisarenko's method is a special case and the Prony method. Prony's method deals directly with the data whereas the other two techniques deal with the covariance matrix of the data. To continue with the development of the Prony's method, the unknown poles or the exponents z_m in (7.22) are obtained from the solution of the polynomial equation

$$W_1 + z\, W_2 + z^2\, W_3 = 0 \tag{7.26}$$

where the values of W_i are obtained from the solution of (7.24). The two roots of (7.26) provide the directions of arrival of the two signals through z_m defined in (7.23). Let us assume that the two solutions are z_1 and z_2.

One could rewrite (7.24) by multiplying the second row by $\mathbb{Z}^{-1}$ (the value of $\mathbb{Z}$ in this case is known) and subtracting it from the first row. It is assumed that $\mathbb{Z} \neq z_1$ or $\mathbb{Z} \neq z_2$, i.e., $\mathbb{Z}$ does not assume any of the specific values of the exponents z_1 or z_2. In that case (7.24) becomes

$$\begin{bmatrix} X_1 - \mathbb{Z}^{-1}X_2 & X_2 - \mathbb{Z}^{-1}X_3 & X_3 - \mathbb{Z}^{-1}X_4 \\ X_2 - \mathbb{Z}^{-1}X_3 & X_3 - \mathbb{Z}^{-1}X_4 & X_4 - \mathbb{Z}^{-1}X_5 \end{bmatrix} \begin{bmatrix} W_1 \\ W_2 \\ W_3 \end{bmatrix} = [0] \tag{7.27}$$

If instead of using any of the constraints of (7.25) we use the following constraint

$$W_1 + \mathbb{Z}\, W_2 + \mathbb{Z}^2\, W_3 = 1 \tag{7.28}$$

then the solution procedure is similar to what we are solving in (7.9). If we use the solutions W_1, W_2 and W_3 of (7.27) to satisfy a polynomial equation of the form

$$W_1 + \mathbb{Z}\, W_2 + \mathbb{Z}^2\, W_3 = 1 \tag{7.29}$$

then the roots of this equation will be the DOA of the two SOI, namely z_1 and z_2, as was first discovered by Prony [9].

Now let us assume that $\mathbb{Z} = z_1$. In that case the difference

$$X_m - \mathbb{Z}^{-1}\, X_{m+1} \tag{7.30}$$

will cancel all the components of the signal which has a DOA of z_1 from the elements of the matrix of (7.27). For this case, we will have only the signal components corresponding to the DOA of the signal z_2 and the matrix of (7.27) will be rank-deficient. However, one can still find the minimum norm solution for the weights of (7.27) which will provide the DOA of the signal z_2. In fact, this is the procedure advocated by Kumaresan and Tufts [9,10] for the modified Prony method to find the system poles which relate to the DOA of the SOI when the system is rank deficient and the proper weights, W_i, need to be solved for to find the DOA of the signals arriving at the array. Kumaresan and Tufts addressed the spectral estimation or the DOA estimation problem. However, in the proof of Kumeresan and Tufts, the rank deficiency in (7.24) is related to the noise subspace. So, in conclusion, when $\mathbb{Z} = z_1$, we have a rank deficient system and when we solve for the minimum norm solution for W_i , then the roots of the polynomial equation similar to (7.29) will provide the DOA of the signal components remaining in the data matrix of (7.27). Now, in our case, we are interested in the complementary problem. As long as $\mathbb{Z} \neq z_1$ or z_2 then (7.27) along with the constraint (7.28) will be of full rank. When $\mathbb{Z} = z_1$, then the system will be rank deficient. This is the desired property that we are interested in. At the true DOA, $\mathbb{Z} = z_1$ the weights which are the solution of (7.27) has to be the result of the minimum norm solution as the system is rank deficient. Numerically, we can determine the minimum norm solution by using the conjugate gradient method to solve this equation as it is one of the few methods that can handle singular systems. If one solves (7.27) with the constraint (7.28) with the conjugate gradient method starting with a zero initial guess, then it is well known that the solution method will converge to a minimum norm solution, which is the solution procedure advocated in Chapter 6 and expanded in [3] for the 1-D and the 2-D adaptive problems.

Here, we are interested in the complementary problem than that is usually addressed in the signal processing literature. We therefore use the complementary property of [9,10] so that when $\mathbb{Z} = z_1$, the coefficients W_i can be computed from a minimum norm solution. The roots of the polynomial equation formed by the W_i will provide the DOA of the other signals remaining in the nonzero terms of the elements of matrix W_i in (7.27). Or conversely, if we change the value of $\mathbb{Z}$ in (7.27) then when $\mathbb{Z}$ takes the value of z_1 we will obtain a minimum norm solution for the weights W_i as the system will be rank deficient. The roots of that polynomial equation in (7.29) will yield in our case the DOA of the remainder of the signals other than z_1. If we perform a number of simulations varying the parameter $\mathbb{Z}$, then when the particular value of $\mathbb{Z}$ exactly coincides with the DOA of the actual signal, (z_1 in this case) the norm of the weights will display a minimum as then the data matrix will be rank deficient. Hence through this property of the minimum norm of the weights we can get an accurate estimate of the DOA of the SOI to refine our initial estimate. When $\mathbb{Z}$ does not coincide with any of the system poles or roots of the polynomial equation, the data matrix is full rank and in that case the weights have a unique solution.

In summary, when the value of $\mathbb{Z}$ exactly coincides with the true DOA of the signal z_1, then the sum of the norm of the weights will be a minimum as the system is then rank deficient. This property is often used for the spectral estimation problem to estimate the proper dimension of the model in order to correctly estimate the dimension of the noise subspace [9,10]. But in this chapter, we utilize a similar property to get an accurate estimate of the DOA for the SOI. We now illustrate this principle through numerical simulations.

The proof has been presented under the assumption that the dimension of the noise subspace is unity. However, the proof becomes very complex when the dimension of the subspace is greater than one, which is what happens in the real situation, but the procedure still works.

7.5 NUMERICAL EXAMPLES

The first example deals with a conventional adaptive algorithm using a twenty-one-element antenna array. In addition to the SOI and thermal noise there are also three jammers and their parameters are listed in Table 7.1.

Table 7.1. Parameters for the SOI and Interference

	Magnitude	Phase	DOA
Signal	1.0 V/m	0.0	91.5°
Jammer #1	1000.0 V/m	0.0	60.0°
Jammer #2	100.0 V/m	0.0	110.0°
Jammer #3	100.0 V/m	0.0	130.0°

The number of adaptive weights chosen for our simulation is thirteen [1]. One of the jammers is 60 dB stronger and the other two are 40 dB stronger than the SOI. The actual location of the SOI is at 91.5°. If in this adaptive algorithm we assume that the SOI is arriving from 90° instead of the actual DOA of 91.5° and carry out the adaptive processing, the SOI will be cancelled, as it will be treated as an interferer. So the question is what to do? First we fix the beam width of the adaptive receive antenna pattern which can be related to the beam width of the transmitter antenna for the RADAR problem. This is achieved using the five-constraint algorithm [1] with a pre-specified beam width of the adapted receive beam pattern to be formed. The five constraints are used for the adapted receive beam pattern so that the adaptive algorithm does not form any pattern null in the main beam of the receive adaptive antenna pattern.

In each case, the simulation is repeated for five different samples of noise at each antenna element. For each noise sample, we now solve the adaptive problem a number of times assuming different values of DOA for the SOI. In this case, we solve the problem a hundred different times by assuming each time that the signal is arriving from any one of the hundred angles covering from 85° to 95° at every 0.1° intervals. Here, we have assumed that the actual DOA of the SOI is between 85° and 95°. We scan this sector and demonstrate that the norm of the weights has a minimum at the actual DOA of the SOI. This 10° scan is assumed based on the *a priori* information of the transmitter beam width. We compute the sum of the norm of the adapted weights for each simulation, corresponding to each one of the assumed DOA for the SOI. It is quite possible that the output from the adaptive algorithm may yield a zero value for the estimate of the amplitude of the SOI when we assume that the SOI is arriving from say 93°, when the actual DOA of the SOI is 91.5°. But still we carry out the computations for all of these 100 different assumed angles of arrival. Then we plot the sum of the norm of the weights as a function of the assumed DOA, which in this case varies from 85° to 95°. The actual location of the target however is at 91.5° for all the cases. Figures 7.1a and 7.1b show the norm of the weights for different values of the background thermal noise. We observe that the norm of the sum of the weights is a minimum, and this minimum occurs when the actual DOA of the SOI coincides with the assumed DOA for the SOI. The minimum norm is then referred to the minimum of the sum of the weights. The only difference is that the shape of the null is different for different values of the thermal noise level. Figure 7.1a shows the results for a very strong target return (20dB signal-to-noise ratio (SNR) at each element). The minimum of the sum of the weights occurs very close to the true target direction for all the five simulations with different receiver noise. The five different curves in the figures represent five different simulations of the problem. The jammers are effectively nulled in all cases and the only effect of receiver noise is that the target return angle is estimated to be in between 91.4° and 91.6° instead of the true 91.5°. The location of the minimum norm of the weights is easy to identify across the entire mainbeam orientation. Hence, this approach has successfully refined the estimate of the DOA of the SOI. Figure 7.1b shows the results for a weak target return (10 dB SNR at each element). The effects of receiver noise are more significant in this case. The estimates of the target location have greater spread (the estimate varies between 91° and 92°). The location of the minimum is still easy to identify but the ratio between the maximum and the minimum is now smaller than for the 20 dB case. All the three interferers have also been nulled out.

Similarly, Figures 7.2a and 7.2b deal with the backward method for 20 db and 10 dB signal-to-noise ratio, respectively. Finally, Figures 7.3a and 7.3b deal with the forward-backward method for 20 db and 10 dB signal-to-noise ratio, respectively. These sets of figures illustrate that the minimum norm property of the adaptive weights are maintained at the true DOA irrespective of the presence of interferers and noise.

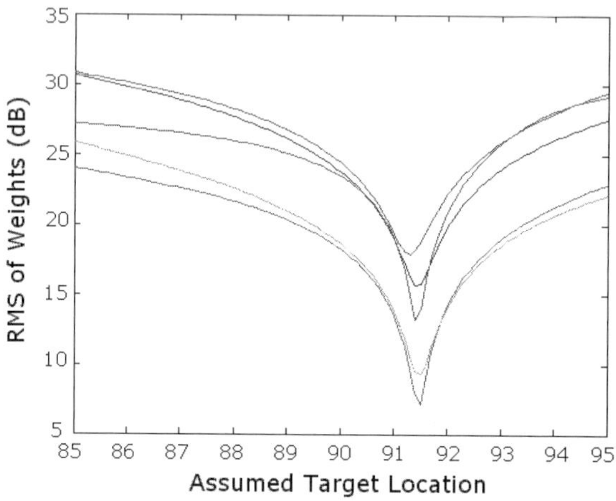

Figure 7.1a. Norm of the weights for 1-D case using the forward method with 20 dB noise.

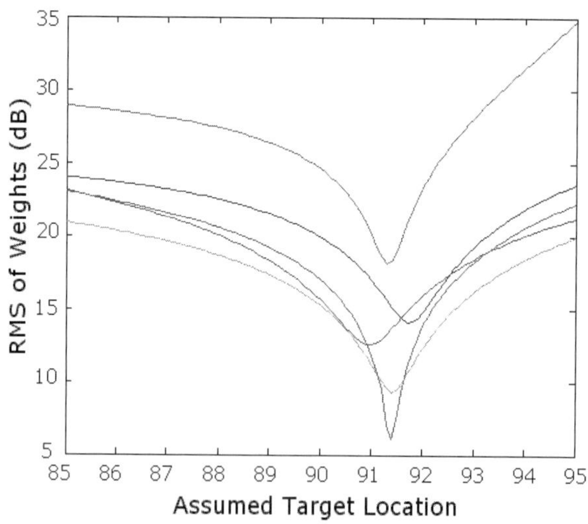

Figure 7.1b. Norm of the weights for 1-D case using the forward method with 10 dB noise.

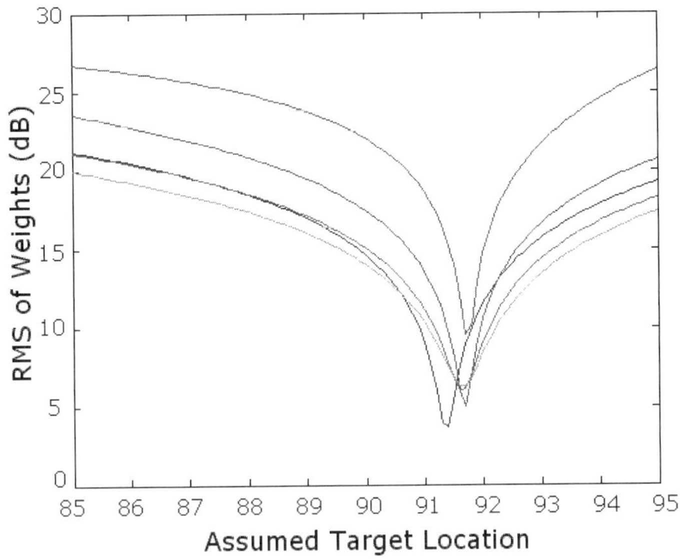

Figure 7.2a. Norm of the weights for 1-D case using the backward method with 20 dB noise.

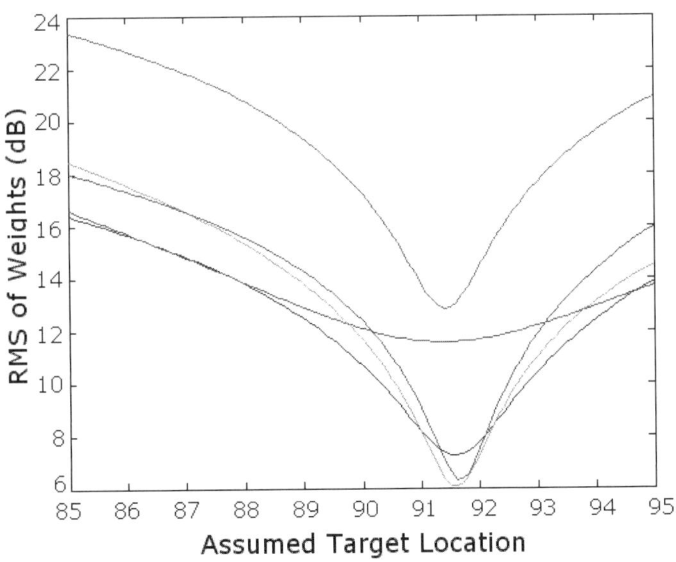

Figure 7.2b. Norm of the weights for 1-D case using the backward method with 10 dB noise.

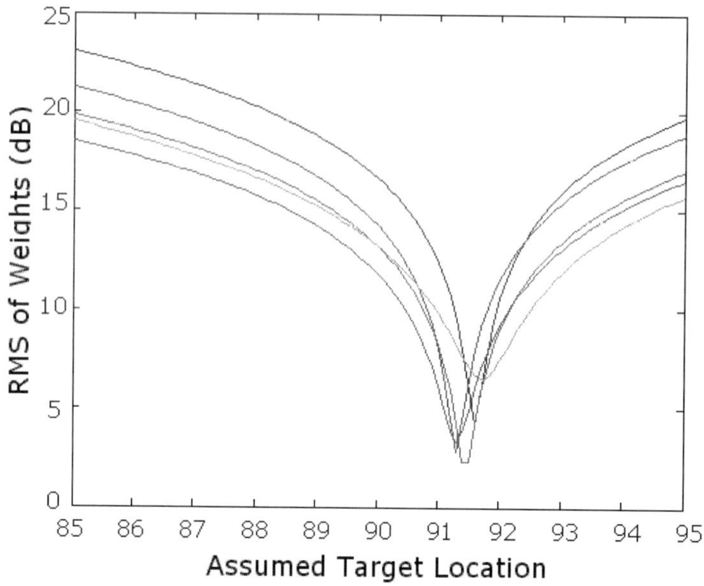

Figure 7.3a. Norm of the weights for the 1-D case using the forward-backward method with 20 dB noise.

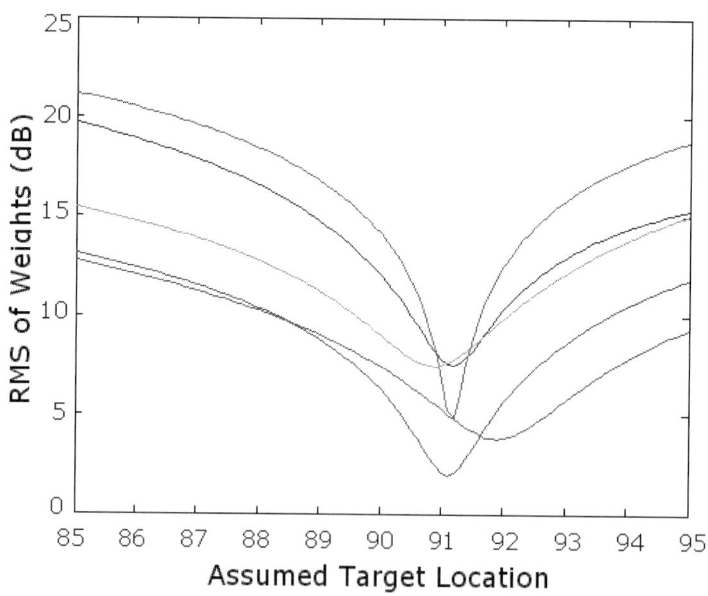

Figure 7.3b. Norm of the weights for the 1-D case using the forward-backward method with 10 dB noise.

For the second example, we simulate a 2-D space-time adaptive processing scenario. The parameters of the input signal are shown in Figure 7.4. In these figures, the small '∘' mark represents the locations of the discrete interferers and the large '∘' denotes clutter and '+' denotes the target location. As seen in this figure there is a strong clutter ridge near the SOI. Numerical values for the SOI and the various interferences including clutter are summarized in Table 7.2.

Here the clutter is generated by assuming multiple plane waves that are arriving from 60° to 66° at intervals of every 0.1°. The Doppler spread of the clutter is from 788 Hz to 970 Hz, every 2 Hz apart. The complex amplitudes of the plane waves constituting the clutter patch are determined by two random number generators, one for the amplitude and the other for the phase. Figure 7.5 shows the target location in 2-D, Doppler frequency and the angle of arrival using five constraints, which are described in Table 7.3. Again, the constraints are used so that no null is produced by the adaptive algorithm in the main beam so as to cancel the SOI. The five constraints in Table 7.3 correspond to the shaping of the received adapted beam in two dimensions. The peak of the adapted beam will occur at 65° and at the Doppler of 1300 Hz. The other four points marked in the Figure 7.5 are the –3 dB points of the adapted receive pattern in the angle-Doppler space. In this figure, the '•' marks the *a priori* beam constraint points of the adapted receive pattern and the '+' mark represents the actual location of the target. In this example we have used different values for the SNR, i.e., 30 dB, 20 dB and 10 dB. The results are shown in Figures 7.6–7.10.

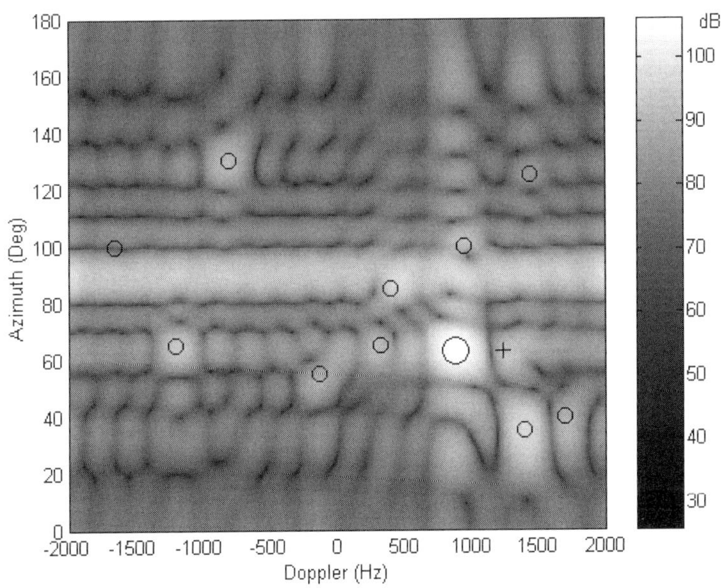

Figure 7.4. Parameters of the Input Signal for Example 2.

Table 7.2. Parameters for the SOI and Interference

	AOA (degree)	Doppler (Hz)	
Signal	63°	1240	
Discrete Interferers	85°, 130°, 40° 35°, 65°, 100° 100°, 65°, 55°, 125°	400, −800, 1700 1400, 325, −1650 950, −1200, −125, 1450	SDIR = −9.5dB
Jammer	90°	For all Doppler frequencies of interest	SJR = −15.5dB
Clutter	60° : 0.1° : 66°	788 : 2 : 970	SCR = −14.7dB

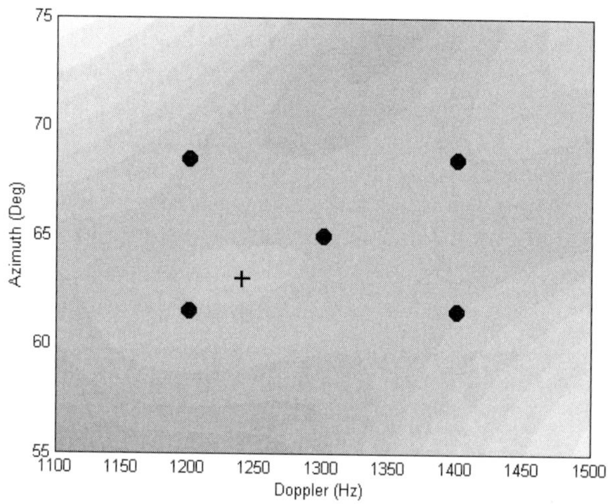

Figure 7.5. Angle and Doppler frequency values of interest.

Table 7.3. Doppler Frequency and Angle Arrival of the Signal and the Five *A Priori* Constraints for the Adapted Receive Pattern

	Angle (Degree)	Doppler (Hz)
Signal	63.0°	1240 Hz
Constraints	65.0°	1300 Hz
	68.5°	1400 Hz
	68.5°	1200 Hz
	61.5°	1400 Hz
	61.5°	1200 Hz

Figure 7.6a shows the minimum norm property with strong target return using the forward method. In these figures, the ' ∘ ' marks represent the constraint points and the '+' mark represent the location of the target. As seen in this figure, the norm of the weights is a minimum at the actual target angle and Doppler frequency. Figures 7.6b and 7.6c repeat the simulation results using different values of SNR, namely 20 dB and 10 dB. The sum of the norm of the weights has a minimum at the estimated angle of arrival and the Doppler frequency. The shape of the minimum is sharper for the higher SNR case and the difference between the maximum and minimum values of the norm of the weights is also larger for the higher SNR case.

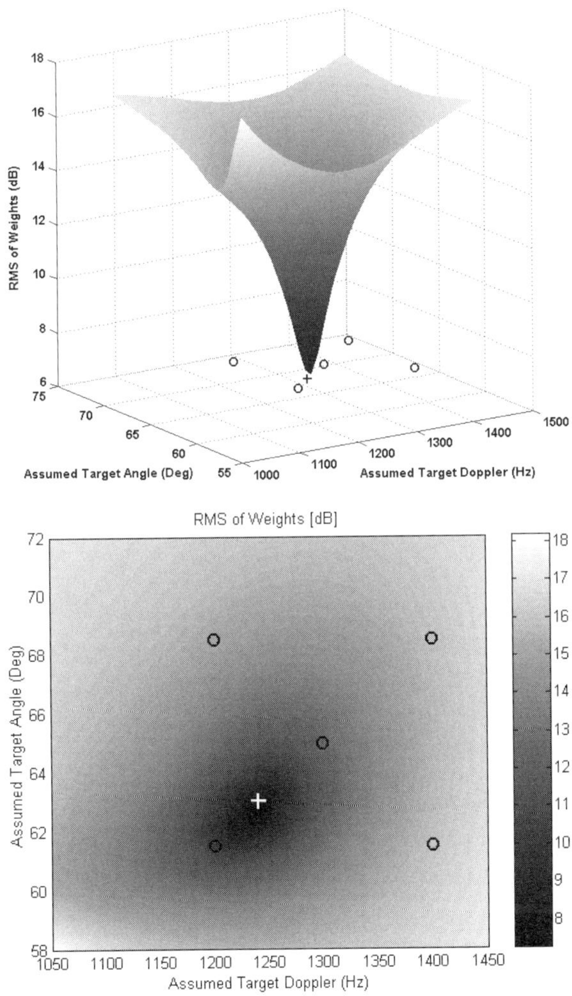

Figure 7.6a. Norm of the weights with 30 dB noise using the forward method.

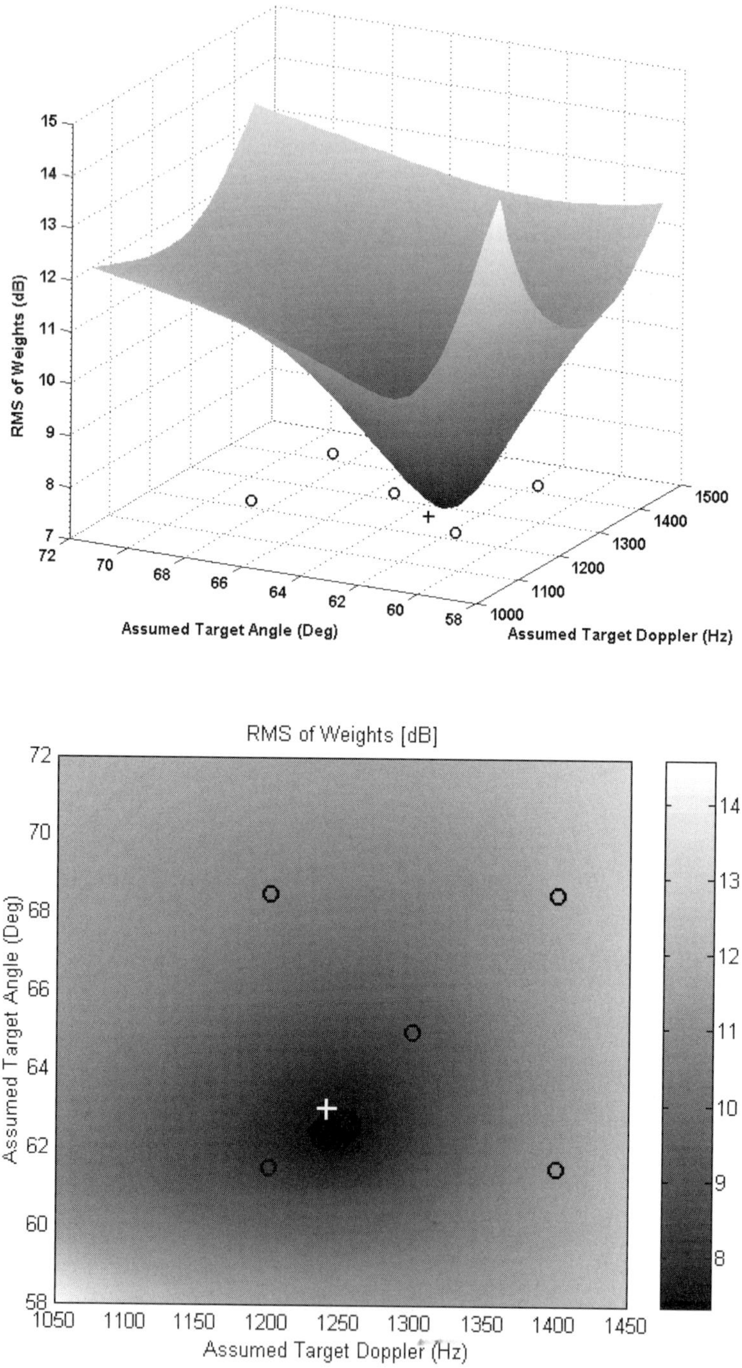

Figure 7.6b. Norm of the weights with 20 dB noise using the forward method.

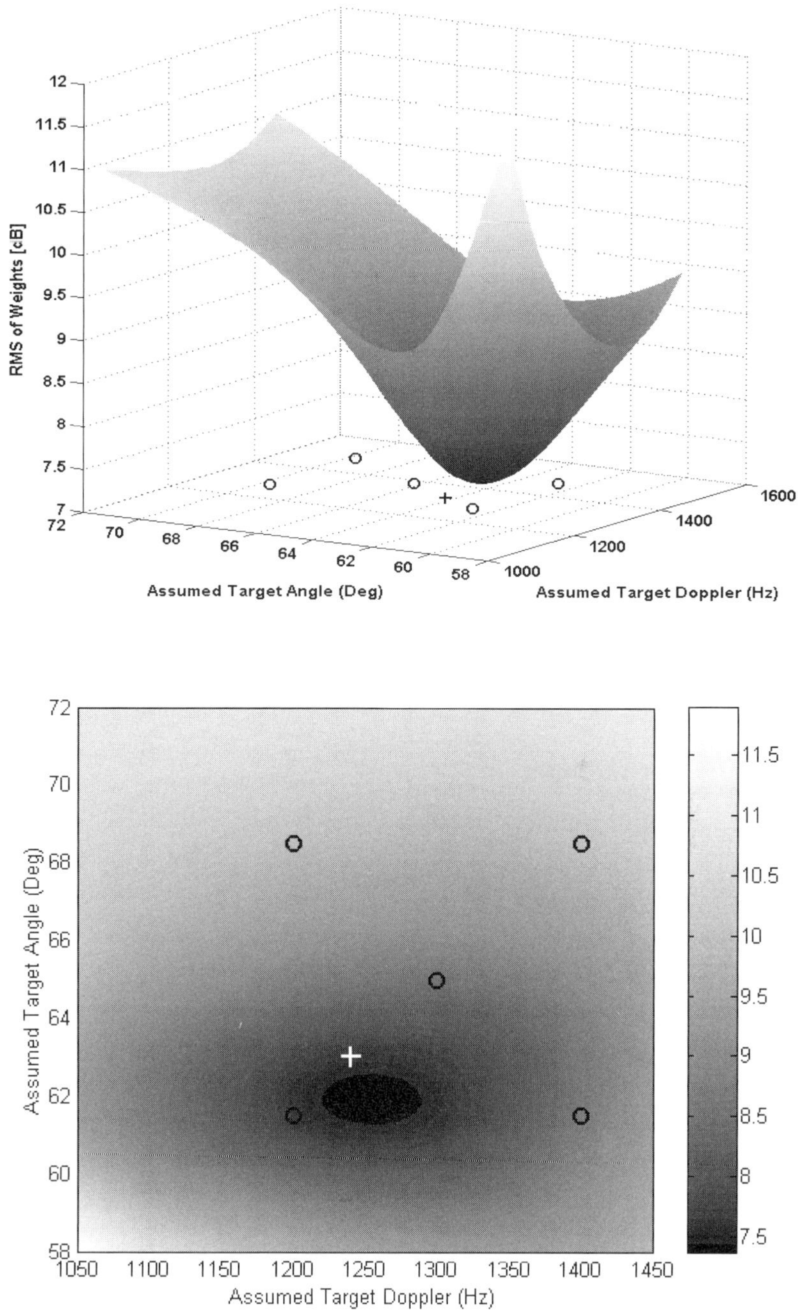

Figure 7.6c. Norm of the weights with 10 dB noise using the forward method.

Figures 7.7a, 7.7b, and 7.7c repeat the simulation results using the backward method for different values of SNR starting with 30 dB and then reducing it to 20 dB and 10 dB, respectively. The sum of the norm of the weights has a minimum at the estimated angle of arrival and the Doppler frequency. The shape of the minimum is sharper for the higher SNR case and the difference between the maximum and minimum values of the norm of the weights is also larger for the higher SNR case.

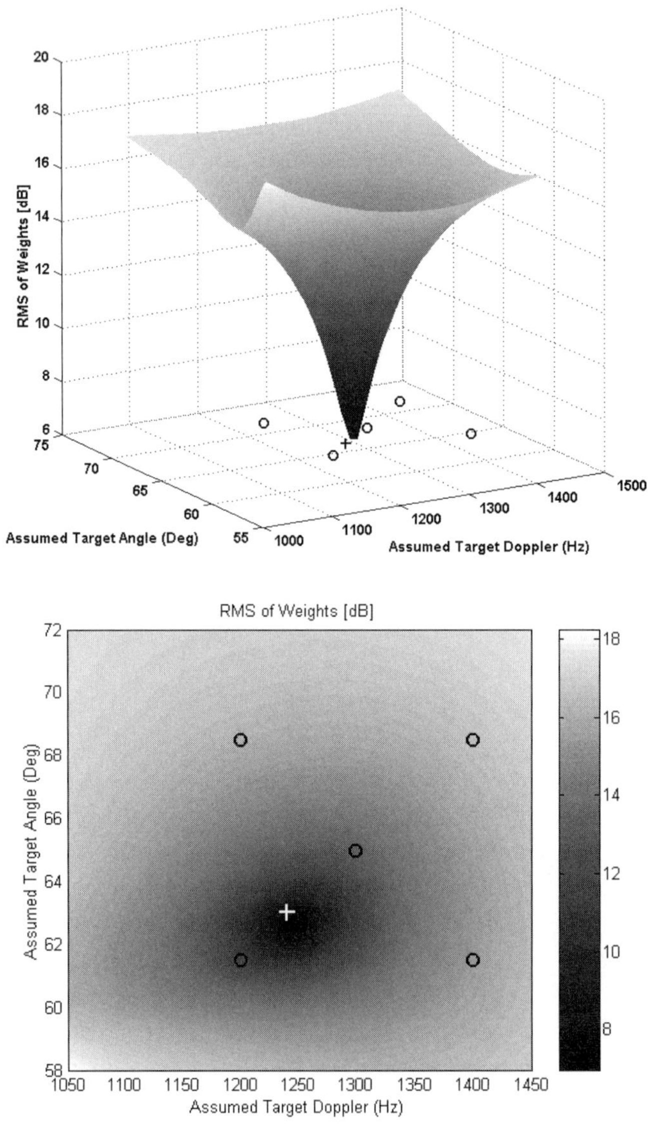

Figure 7.7a. Norm of the weights with 30 dB noise using the backward method.

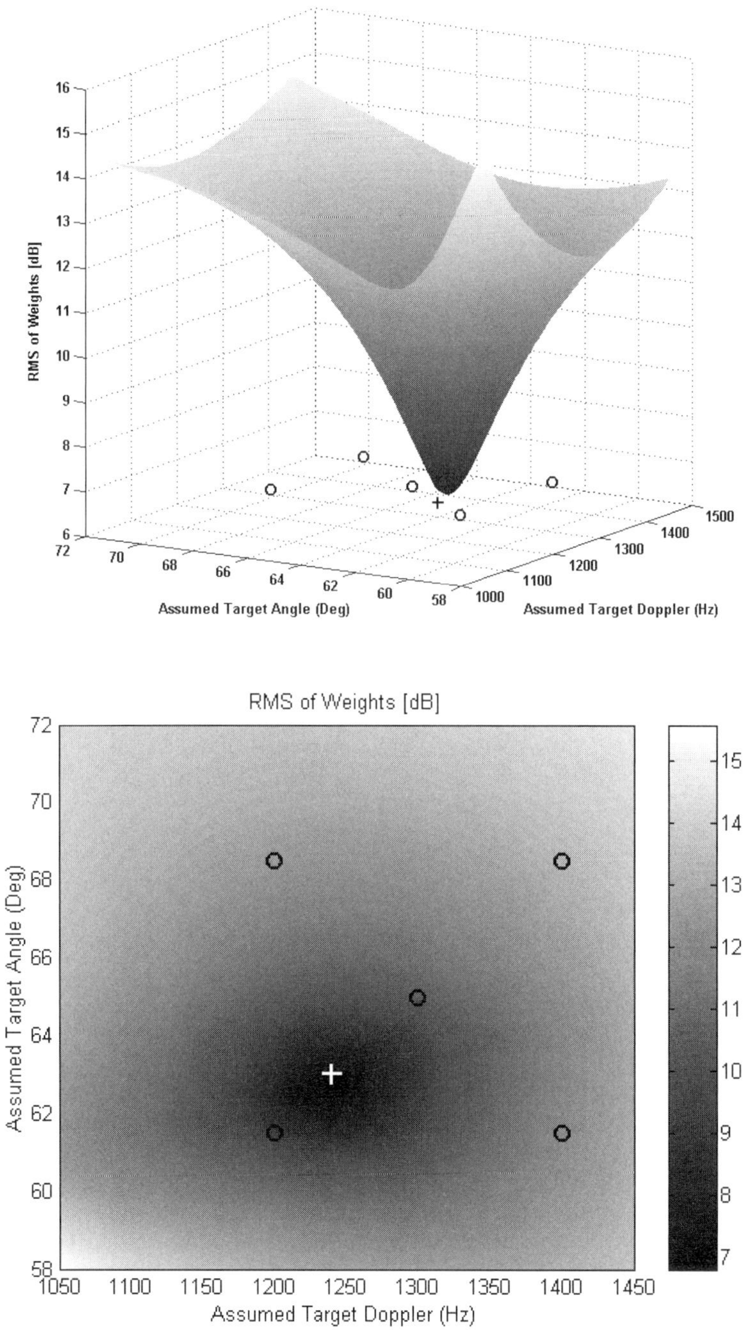

Figure 7.7b. Norm of the weights with 20 dB noise using the backward method.

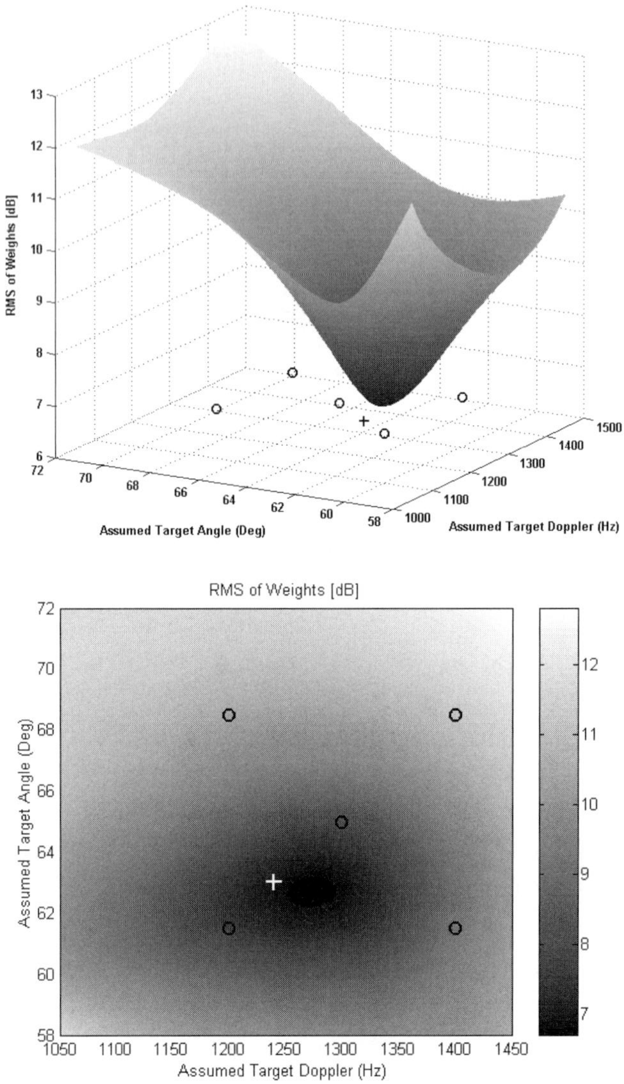

Figure 7.7c. Norm of the weights with 10 dB noise using the backward method.

Finally, Figures 7.8a, 7.8b, and 7.8c repeat the simulation results using the forward-backward method for different values of SNR starting with 30 dB and then reducing it to 20 dB and 10 dB, respectively. The sum of the norm of the weights has a minimum at the estimated angle of arrival and the Doppler frequency. The shape of the minimum is sharper for the higher SNR case and the difference between the maximum and minimum values of the norm of the weights is also larger for the higher SNR case.

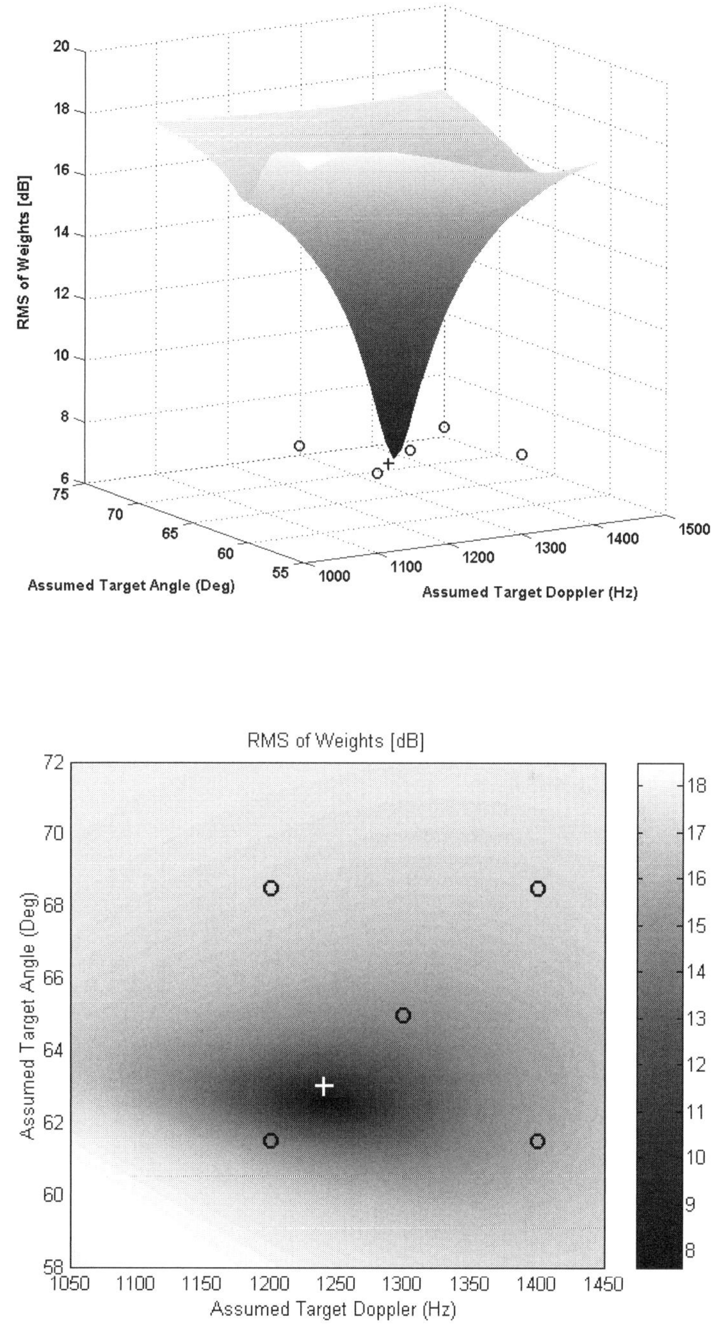

Figure 7.8a. Norm of the weights with 30 dB noise using the forward-backward method.

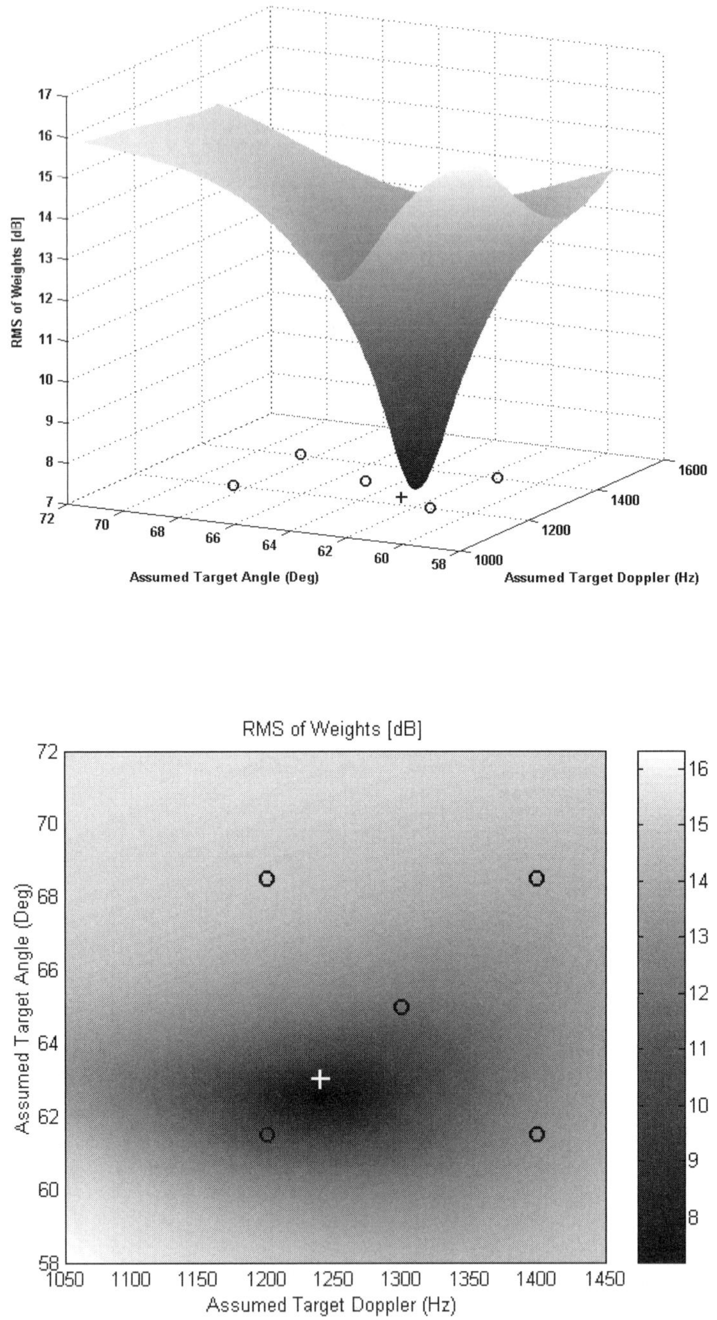

Figure 7.8b. Norm of the weights with 20 dB noise using the forward-backward method.

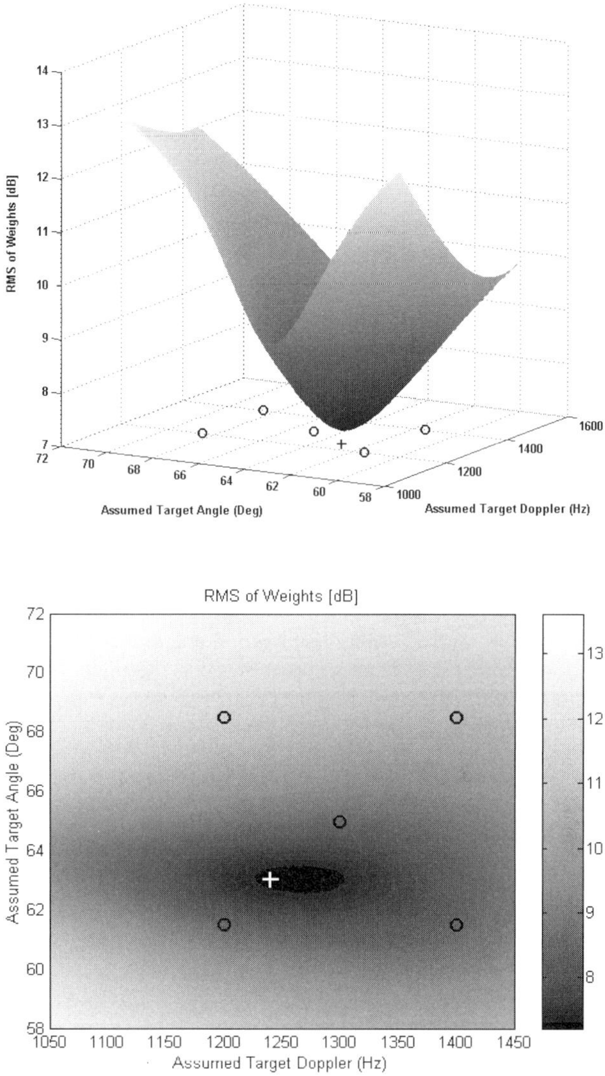

Figure 7.8c. Norm of the weights with 10 dB noise using the forward-backward method.

7.6 CONCLUSION

The existence of a minimum in the sum of the norm of the weights can be used to further refine the estimate of the angle of arrival for 1-D adaptive problem and both the angle of arrival and the Doppler of the target return in 2-D space-time adaptive processing. In this way, the method can also prevent the problem of signal cancellation for an adaptive algorithm due to inaccurate *a priori* information about the DOA and the Doppler for RADAR problems. This technique can also be used to perform the detection process as well. If there is a large ratio between the minimum and the maximum values of the sum of the norm of the weights across the scan of interest then that is an indication that a target is present. The strongest linear progression of the random noise samples sets a lower limit on that detection process.

The sum of the adaptive weights varies as the value of the assumed target direction is varied across the mainbeam forming a minimum when it coincides with the actual direction. This secondary processing can enhance the estimate of the DOA of the SOI further in adaptive processing. When a strong target is present the ratio between the maximum and minimum values of the norm of the weights is large. If there is no target present, the ratio between the maximum and the minimum values of the norm of the weights are small. This could lead to a more accurate estimation of the DOA of the signal and the target Doppler, when an accurate estimation of them is not available *a priori*. In addition, the ratio between the maximum and the minimum values in the norm of the weights can also be used in detecting the presence of the target.

REFERENCES

[1] W. Choi and T. K. Sarkar, "Minimum Norm Property for the Sum of the Adaptive Weights for a Direct Data Domain Least Squares Algorithm", *IEEE Transactions on Antennas and Propagation*, Volume 54, Issue 3, Mar. 2006 , pp. 1045-1050.

[2] T. K. Sarkar, H. Wang, S. Park, J. Koh, R. Adve, K. Kim, Y. Zhang, M. C. Wicks and R. D. Brown, "A Deterministic Least Squares Approach to Space Time Adaptive Processing (STAP)," *IEEE Trans. Antennas and Propagation*, Vol. 49, January 2001, pp. 91-103.

[3] T. K. Sarkar, M. C. Wicks, M. Salazar-Palma and R. J. Bonneau, *Smart Antennas*, Wiley-IEEE Press, Hoboken, NJ, 2003.

[4] J. Carlo, T. K. Sarkar and M. C. Wicks, "Application of Deterministic Techniques to STAP", In *Applications of Space-Time Adaptive Processing*, edited by R. Klemm, IEEE Press, 2004, pp. 375-411.

[5] J. Carlo, T. K. Sarkar and M. C. Wicks, "A Least Square Multiple Constraint Direct Data Domain Approach for STAP", *Applications of Space-Time Adaptive Processing*, In Vol. 2 edited by R. Klemm, IEEE Press, 2004 and also in *Proceedings of 2003 IEEE Conference on Radar,* pp. 431-438, 2003.

[6] D. H. Johnson and D. E. Dudgeon, *Array Signal Processing*, Prentice Hall, Englewood Cliffs, NJ, 1993.

[7] S. Haykin, *Adaptive Filter Theory*, 4th ed., Prentice Hall, Upper Saddle River, NJ, 2002.

[8] H. L. Van Trees, *Optimum Array Processing*, Wiley, New York, 2002.

[9] P. Stoica and R. Moses, *Introduction to Spectral Analysis*, Prentice Hall, Upper Saddle River , NJ, 1997.

[10] D. W. Tufts and R. Kumaresan, "Estimation of Frequencies of Multiple Sinusoids: Making Linear Prediction Perform Like Maximum Likelihood", *Proc. of IEEE*, Vol. 70, No. 9, 1982, pp. 975-989.

8

USING REAL WEIGHTS IN ADAPTIVE AND SPACE-TIME PROCESSING

8.0 SUMMARY

In traditional adaptive signal processing algorithms one changes both the amplitude and phase of the weight vectors associated with an array at each of the antenna elements. The use of complex weights offers greater control over the array response at the expense of system complexity. However, it is easier if one requires only amplitude variation with a fixed phase for all the weight vectors associated with the antenna elements. Because one uses only real arithmetic operations to find the amplitude of the weights connected to the antenna, the computational complexity is reduced considerably. In this chapter, we first address the use of real weights in an adaptive system. Next we extend this methodology to space-time adaptive processing (STAP) using a single snapshot-based direct data domain least squares (D3LS) approach. The D3LS STAP method is applied to data collected by an antenna array utilizing space and time (Doppler) diversity. Here the weights involved in amplitude-only STAP systems are designed to also have a fixed phase. Because one uses only real arithmetic operations to find the amplitude of the adaptive weights in the array, the computational complexity is reduced considerably. This technique may be useful for real time implementation of the D3LS method on a chip.

8.1 INTRODUCTION

Adaptive array signal processing has been used in many applications in such fields as radar, sonar, wireless mobile communication, and so on. One principle advantage of an adaptive array is the ability to recover the desired signal while also automatically placing deep pattern nulls along the direction of the interference.

Generally, adaptive antenna array systems perform by changing the adaptive weights having complex weights, i.e., magnitudes and phases. For large array systems, the computational complexity becomes quite large. For this reason, several authors have proposed phase-only weight control [1–4] and amplitude-only weight adaptive algorithms [5–10]. The emphasis of these algorithms is on faster response and simpler software and hardware design. For amplitude-only weights in an adaptive algorithm, several methods have been proposed, such as symmetric amplitude-only control (SAOC) [8–10] and real amplitude-only nulling algorithm (RAMONA) [5–7]. But algorithms based on statistical approaches require independent, identically distributed secondary data to estimate the covariance matrix of the interference. The formation of the covariance matrix is quite time consuming and so is the evaluation of its inverse. Also, because one requires several snapshots of the data to generate a covariance matrix, it is assumed that the environment remains stationary during that process. However, for a dynamic environment the present D3LS methods may be more suitable as one processes the information based on a single snapshot. A single snapshot is defined as the complex voltages measured across the elements of the array simultaneously at a particular instance of time.

Another related problem is space-time adaptive processing (STAP). STAP, as available in the published literature deals with the statistical treatment of clutter and this involves estimating a covariance matrix of the interference using data over the range cells [11–18]. The statistical procedures require secondary data for processing and this is in short supply for a nonstationary environment. In addition, the formation of the covariance matrix and the computation of its inverse are not only computationally intensive but also break down under highly nonstationary environments, particularly when the clutter scenario changes from land to urban to sea clutter and when there are blinking jammers and hot clutter. For airborne radars it is necessary to detect targets in the presence of clutter, jammers and thermal noise. The airborne radar scenario has been described in [11–14]. It is necessary to suppress the levels of the undesired interferers well below the weak desired signal. For airborne radar surveillance systems the detection of airborne and ground targets is complicated by many factors in the radar signal environment. In general, STAP algorithms perform by changing the adaptive weights having different complex weights. The use of complex weights offers greater control over the array response at the expense of system complexity. However, it is easier if one requires only amplitude variation with a fixed phase for all the weight vectors associated with all the antenna elements. Because one uses only real arithmetic operations to find the amplitude of the weights at the antenna, computational complexity is reduced considerably. Hence, this chapter addresses the amplitude-only adaptive systems whose weights have fixed phases. In this chapter we describe a new amplitude-only adaptive method based on a D3LS approach, which utilizes only a single snapshot of the data for adaptive processing.

Recently a direct data domain least squares (D3LS) approach has been proposed to the computational issues to deal with single snapshot of data for both the adaptive processing and STAP [19–24]. A direct data domain approach has certain advantages related associated with the adaptive array signal processing problem, which adaptively analyzes the data by snapshots as opposed to forming a covariance matrix of the data from multiple snapshots, then solving for the weights utilizing that information. Another advantage of the D3LS approach is that when the direction of arrival (DOA) of the signal is not known precisely, additional constraints can be applied to fix the main beamwidth of the receiving array *a priori* (i.e., specifying the 3 dB beamwidth of the adapted pattern before it is formed) and thereby reduce the signal cancellation problem as discussed in chapter 7. Here, we present the amplitude-only adaptive processing based on a D3LS technique. In section 8.2 we formulate the problem. We present three different independent formulations of the same procedure. Generation of three independent estimates for the same solution provides a higher level of confidence for the unknown solution. Section 8.3 presents some simulation results illustrating the performance of the proposed method. In section 8.4, we present the amplitude-only adaptive systems for STAP whose weights have fixed phases. Section 8.5 describes simulation results illustrating the performance of the proposed method. Finally, in section 8.6 we present some conclusion followed by a list of selected references, which is by no means complete on the statistical processing schemes.

8.2 FORMULATION OF A DIRECT DATA DOMAIN LEAST SQUARES APPROACH USING REAL WEIGHTS

8.2.1 Forward Method

Consider an array composed of $N + 1$ antennas separated by a distance Δ as shown in Figure 6.2. We assume that narrowband signals consisting of the desired signal plus possibly coherent multipaths and jammers with the same center frequency f_0 are impinging on the array from various angles θ, with the constraint $0 \le \theta \le 180^\circ$. In addition, there can be strong interferers in the main beam and thermal noise. For sake of simplicity we assume that the incident fields are coplanar and that they are located in the far field of the array. However, this methodology can easily be extended to the non-coplanar case without any problem including the added polarization diversity.

The problem formulation has been described in detail in [19] and has been summarized in section 6.2 for completeness. In this discussion we assume modeling the antennas by point sources. However, how to deal with real antennas operating in the presence of near-field scatterers and mutual coupling between the elements are described in details in [19].

Here we consider that we have a single snapshot of the voltages measured at the feed point of the antenna elements (i.e., at a time $t = t_m$, we have the voltages X_n, for $n = 0, 1, \ldots, N$ measured at the feed points of all $N + 1$ antenna elements). The complex voltage X_n induced at the nth antenna element at a particular instance of time will then be given by

$$X_n = s \exp\left[\frac{j 2\pi n \Delta \cos\theta_s}{\lambda}\right] + \sum_{p=1}^{P} A_p \exp\left[\frac{j 2\pi n \Delta \cos\theta_s}{\lambda}\right] + C_n + \xi_n \tag{8.1}$$

where s = A_0 = complex amplitude of the SOI (to be determined)
θ_s = direction of arrival of the SOI (assumed to be known)
Δ = spacing between each of the antenna elements (known)
λ = wavelength of transmission (here it is assumed that we are dealing with narrowband signals) – (known)
P = total number of interferers (unknown)
A_p = complex amplitude of the pth undesired interferer (unknown)
θ_p = direction of arrival of the pth interferer (unknown)
C_n = clutter induced at the nth element (in some cases, this may be diffused electromagnetic radiation from distant mountains, lands, buildings, water, and so on) (unknown)
ξ_n = thermal noise induced at the nth antenna element (unknown)

Here we model the clutter as a bunch of reflected/diffracted rays bouncing back from the ground, platforms on which the array is mounted, and from nearby buildings, trees, or uneven terrains. The amplitude and phase of these rays have been determined by two random number generators. Hence, the clutter is modeled by a true physics of an electromagnetic model and not based on some probability distributions which do not satisfy any known electromagnetic phenomenon. For the array shown in Figure 6.2, the measured voltages X_n for $n = 0, 1, \ldots, N$, at the antenna elements are assumed to be known along with θ_s, the DOA of the SOI. The goal is to estimate the complex amplitude s for the SOI. Here we define a single snapshot by the voltages X_n measured at the nth element at a certain instant of time t_m. It is understood that all the SOI, jammers, clutter, and thermal noise vary as a function of time.

Our goal is to estimate s given θ_s and X_n. The methodology on how to extract the SOI in the presence of coherent jammers and clutter has been presented in section 6.2. We also know that in order to obtain the SOI, the number of coherent jammers must be $\leq N/2$ in the absence of clutter and noise. It is important to point out that in this procedure we do not make any distinction between coherent or noncoherent interferers. The classical techniques based on the statistical methodology will be able to handle more than $N/2$ noncoherent interferers but no more than $N/2$ coherent interferers. However, the price to be paid for this is that we require at least $N+1$ snapshots of voltages. The forward direct data domain method has been described in section 6.2.2 and the equation implementing that methodology is described by (6.30).

The extraction of the SOI in the presence of the undesired signals can be carried out by solving for the weights, $[W]$, as the first step, by obtaining the following equation similar to (6.30). Therefore, one obtains

$$Z = \exp\left[j\, 2\, \pi\, \Delta \cos(\theta_s) / \lambda \right] \tag{8.2}$$

$$\begin{bmatrix} 1 & Z & \cdots & Z^L \\ X_0 - Z^{-1}X_1 & \cdots & \cdots & X_L - Z^{-1}X_{L+1} \\ \vdots & \vdots & \vdots & \vdots \\ X_{N-L-1} - Z^{-1}X_{N-L} & X_{N-L} - Z^{-1}X_{N-L+1} & \cdots & X_{N-1} - Z^{-1}X_N \end{bmatrix}_{(L+1)\times(L+1)} \times \begin{bmatrix} W_0 \\ W_1 \\ \vdots \\ W_L \end{bmatrix}_{(L+1)\times 1} = \begin{bmatrix} C \\ 0 \\ \vdots \\ 0 \end{bmatrix}_{(L+1)\times 1} \tag{8.3}$$

In the amplitude-only adaptive processing, the weight vector $[W]$ should be real numbers. But use of the real weights will make a symmetric antenna beam pattern. This creates the problem of signal cancellation if the SOI is incident from $(90 + \theta)°$ and the interferer arrives from $(90 - \theta)°$ and vice versa. Then, due to the symmetry property of the antenna pattern caused by a set of real weights, both the signal and the interferer would be canceled. So before carrying out any processing, we need to shift the incoming data, $[X]$, in angle. We propose to shift the input data so that the direction of arrival of the SOI is now 90° instead of θ_s by using the following transformation matrix

$$[\text{Tr}] = \text{diag}\left(\frac{\text{steering vector of } 90°}{\text{steering vector of DOA of SOI}} \right) \tag{8.4a}$$

Then transfer the input vector, $[X]$, to the new vector, $[\overline{X}]$, by multiplying it with the matrix in (8.4a) to form

$$[\overline{X}] = [\text{Tr}] \times [X] \tag{8.4b}$$

With the new data matrix $[\overline{X}]$ substituted in (8.3) one obtains

$$\begin{bmatrix} 1 & \cdots & Z^L \\ \overline{X}_0 - Z^{-1}\overline{X}_1 & \cdots & \overline{X}_L - Z^{-1}\overline{X}_{L+1} \\ \vdots & \vdots & \vdots \\ \overline{X}_{N-L-1} - Z^{-1}\overline{X}_{N-L} \cdots & & \overline{X}_{N-1} - Z^{-1}\overline{X}_N \end{bmatrix}_{(L+1)\times(L+1)} \begin{bmatrix} W_0 \\ W_1 \\ \vdots \\ W_L \end{bmatrix}_{(L+1)\times 1} = \begin{bmatrix} C \\ 0 \\ \vdots \\ 0 \end{bmatrix}_{(L+1)\times 1} \tag{8.5}$$

$$\text{or equivalently, } [F][W] = [Y] \tag{8.6}$$

where C is assumed to be a real number.

Once the weights are solved by using (8.5), the signal component α, an estimate for the true value s, may be estimated from

$$\alpha = \frac{1}{C} \sum_{i=0}^{L} W_i \overline{X}_i \tag{8.7}$$

The proof of (8.7) is available in [19]. It is also possible to estimate α from any of the following $L+1$ equations

$$\alpha_v = \frac{1}{C Z^v} \sum_{i=0}^{L} W_i \overline{X}_{i+v} \ , \quad for\ v = 0, 1, \cdots, L \tag{8.8}$$

or by averaging any one of the equations given by the set of $L+1$ equations in (8.8). However, it is interesting to note that because of (8.5), averaging $L+1$ estimates of α obtained from (8.8) is no better than using (8.7).

Now to compute a set of real weights $[\Re]$, we need to separate the matrix [F] into real and imaginary parts with $K \cong 2L$ as

$$\begin{bmatrix} \mathrm{Re}(F_{11}) & \mathrm{Re}(F_{12}) & \cdots & \mathrm{Re}(F_{1K}) \\ \mathrm{Re}(F_{21}) & \mathrm{Re}(F_{22}) & \cdots & \mathrm{Re}(F_{2K}) \\ \vdots & \vdots & \vdots & \vdots \\ \mathrm{Re}(F_{K1}) & \mathrm{Re}(F_{K2}) & \cdots & \mathrm{Re}(F_{KK}) \\ \mathrm{Im}(F_{11}) & \mathrm{Im}(F_{12}) & \cdots & \mathrm{Im}(F_{1K}) \\ \mathrm{Im}(F_{21}) & \mathrm{Im}(F_{22}) & \cdots & \mathrm{Im}(F_{2K}) \\ \vdots & \vdots & \vdots & \vdots \\ \mathrm{Im}(F_{K1}) & \mathrm{Im}(F_{K2}) & \cdots & \mathrm{Im}(F_{KK}) \end{bmatrix}_{2K \times K} \times \begin{bmatrix} \Re_1 \\ \Re_2 \\ \vdots \\ \Re_K \end{bmatrix}_{K \times 1} = \begin{bmatrix} C \\ 0 \\ \vdots \\ 0 \end{bmatrix}_{2K \times 1} \tag{8.9}$$

Then we can use the Conjugate gradient method to solve (8.9). The conjugate gradient method starts with an initial guess $[\Re]_0$ for the solution and continues with the calculation of the following [19,25]

$$[P]_0 = -b_{-1}[F]^T [\mathbb{R}]_0 = -b_{-1}[F]^T \{[F][\Re]_0 - [Y]\} \ , \tag{8.10}$$

Where the superscript T denotes the transpose of a matrix. At the kth iteration the conjugate gradient method develops the following:

$$c_k = \frac{1}{\| [F][P]_k \|^2} \tag{8.11}$$

$$[\Re]_{k+1} = [\Re]_k + c_k [P]_k \tag{8.12}$$

$$[\mathbb{R}]_{k+1} = [\mathbb{R}]_k + c_k [F][P]_k \tag{8.13}$$

$$b_k = 1./\| [F]^T [\mathbb{R}]_{k+1} \|^2 \tag{8.14}$$

$$[P]_{k+1} = [P]_k - b_k [F]^T [\mathbb{R}]_{k+1} \tag{8.15}$$

The norm is defined by

$$\left\| [F]\,[P]_k \right\|^2 = [P]_k^T\,[F]^T\,[F]\,[P]_k \tag{8.16}$$

The above equations are applied in an iterative fashion until the desired error criterion for the residuals $\left\| [\mathbb{R}]_k \right\|$ is satisfied, where $[\mathbb{R}]_k = [F]\,[\Re]_k - [Y]$. In our case, the error criterion is defined by

$$\frac{\left\| [\mathbb{R}]_k \right\|}{\left\| [Y] \right\|} = \frac{\left\| [F][\Re]_k - [Y] \right\|}{\left\| [Y] \right\|} \le 10^{-6} \tag{8.17}$$

Once the real set of weights $[\Re]$ is solved for, they are substituted in (8.8) for W_i to obtain an estimate for the complex value of s.

8.2.2 Backward Method

Next we reformulate the problem using the same data to obtain a second independent estimate for the solution. This is achieved by reversing the data sequence and then complex conjugating each term of that sequence [19, 22] as has been discussed in section 6.2.3.

It is well known in the parametric spectral estimation literature that a sampled sequence, which can be represented by a sum of exponentials with purely imaginary argument, can be used in either the forward or in the reverse direction, resulting in the same values for the exponent fitting that sequence. From physical considerations we know that if we solve a polynomial equation with the weights W_i as the coefficients, its roots provide the DOA for all the unwanted signals, including the interferers. Therefore, whether we look at the snapshot as a forward sequence as presented in the last section or by a reverse conjugate of the same sequence, the final results for W_i must be the same. Hence for these classes of problems, we can observe the data in either the forward or reverse direction. This is equivalent to creating a virtual array of the same size but located along a mirror symmetry line. So, if we now conjugate the data and form the reverse sequence, we get an independent set of equations similar to (8.5) for the solution of the weights [W]. This is represented by [19, 22] and is similar to (6.34)

$$\begin{bmatrix} 1 & \cdots & Z^L \\ \overline{X}_N^* - Z^{-1}\overline{X}_{N-1}^* & \cdots & \overline{X}_{N-L}^* - Z^{-1}\overline{X}_{N-L-1}^* \\ \vdots & \vdots & \vdots \\ \overline{X}_{L+1}^* - Z^{-1}\overline{X}_L^* & \cdots & \overline{X}_1^* - Z^{-1}\overline{X}_0^* \end{bmatrix}_{(L+1)\times(L+1)} \times \begin{bmatrix} W_0 \\ W_1 \\ \vdots \\ W_L \end{bmatrix}_{(L+1)\times 1} = \begin{bmatrix} C' \\ 0 \\ \vdots \\ 0 \end{bmatrix}_{(L+1)\times 1} \tag{8.18}$$

where the superscript * denotes the complex conjugate. Equivalently in a matrix form as

$$[B][W] = [Y]$$

The signal strength α can again be determined from

$$\alpha_v = \left[\frac{Z^{L+v}}{C'} \sum_{i=0}^{L} W_i X^*_{L-i+v} \right]^* \quad \textit{for} \;\; v = 0, 1, ..., L \tag{8.19}$$

To guarantee a real set of weight vectors, we split (8.18) into real and imaginary parts resulting in an equation similar to (8.9). Once the real set of weights [$\Re$] are solved for, they are substituted in (8.19) to obtain an estimate for the complex value of *s*.

Note that for both the forward and the backward methods described in Sections 8.2.1 and 8.2.2 we have $L = N/2$. Hence the degrees of freedom are the same for both the forward and backward methods. However, we have two independent solutions for the same adaptive problem. In a real situation when the solution is unknown, two different estimates for the same unknown may provide a level of confidence on the quality of the solution.

8.2.3. Forward-Backward Method

Finally, in this section we combine the forward and backward methods to double the given data and thereby increase the number of weights or the degrees of freedom significantly over that of either the forward or backward method alone. In the forward-backward model we double the amount of data not only by considering the data in the forward direction but also conjugating it and reversing the direction of increment of the independent variable. This type of processing can be done as long as the series to be approximated can be fit by exponential functions of purely imaginary argument. This is always true for the adaptive array case. So by considering the data set simultaneously in both the forward and backward directions, denoted by the sequences $\{X_n\}$ and $\{X^*_{-n}\}$, respectively, we have essentially doubled the amount of data without any penalty, as these two data sets for our problem are linearly independent [19,22], has been presented in section 6.2.4.

An additional benefit accrues in this case. For both the forward and backward methods, the maximum number of weights we can consider is given by $N/2$, where $N + 1$ are the number of antenna elements. Hence, even though all the antenna elements are being utilized in the processing, the number of degrees of freedom available for this approach is essentially $N/2$. For the forward-backward method, the number of degrees of freedom can be increased significantly without increasing the number of antenna elements. This is accomplished by considering the forward and backward versions of the array data. For this case, the number of degrees of freedom can reach $Q = N/1.5 + 1$. Hence Q is larger than L used in the forward and the backward method, as we have now doubled the data set. The equation that needs to be solved for in this case for the complex weights is given by combining (8.5) and (8.18), into

$$\begin{bmatrix} 1 & \cdots & Z^{Q} \\ \overline{X}_0 - Z^{-1}\overline{X}_1 & \cdots & \overline{X}_Q - Z^{-1}\overline{X}_{Q+1} \\ \vdots & \vdots & \vdots \\ \overline{X}_{N-Q-1} - Z^{-1}\overline{X}_{N-Q} & \cdots & \overline{X}_{N-1} - Z^{-1}\overline{X}_N \\ \overline{X}_N^{*} - Z^{-1}\overline{X}_{N-1}^{*} & & \overline{X}_{N-Q}^{*} - Z^{-1}\overline{X}_{N-Q-1}^{*} \\ \vdots & \vdots & \vdots \\ \overline{X}_{Q+1}^{*} - Z^{-1}\overline{X}_Q^{*} & \cdots & \overline{X}_1^{*} - Z^{-1}\overline{X}_0^{*} \end{bmatrix}_{(Q+1)\times(Q+1)} \begin{bmatrix} W_0 \\ W_1 \\ \vdots \\ W_Q \end{bmatrix}_{(Q+1)\times 1} = \begin{bmatrix} \mathbb{C} \\ 0 \\ \vdots \\ 0 \end{bmatrix}_{(Q+1)\times 1}$$

or equivalently as, $[FB]\,[W] = [Y]$ (8.20)

To obtain a real set of weight vectors we split (8.20) into the real and the imaginary parts and then solve for the real set of weights.

8.3 SIMULATION RESULTS FOR ADAPTIVE PROCESSSING

For the first example, consider a signal of unit amplitude arriving from $\theta_s = 90°$. We consider a 7-element array with an element spacing of $\lambda/2$. The magnitude of the incident signal is varied from 1 V/m to 10.0 V/m in steps of 0.1 V/m for each of the 100 data snapshots, while maintaining the jammer intensities constant, which are arriving from 70° and 130°. All the signal intensities and their directions of arrival are summarized in Table 8.1. The signal-to-noise ratio at each of the antenna elements is set at 30 dB. The number of weights is 5 for either the forward or the backward methods and it is 6 for the forward-backward method. All the weights are real. Here, we assume that we know the DOA of the signal but need to estimate its complex amplitude. If the jammers have been nulled correctly and the signal recovered properly, it is expected that the recovered signal at each snapshot will have a linear relationship with respect to the intensity of the estimated signal. Figure 8.1a plots the estimated amplitude in volts for the SOI recovered at each snapshot. Figure 8.1b plots the estimated phase in degrees of the SOI at each snapshot. The phase varies within a very small range. The adapted beam pattern associated with this example for each of the three methods is shown in Figure 8.2. For the antenna pattern we set the magnitude of the desired signal to be 1 V/m and the other parameters are as given in Table 8.1. Figure 8.2a plots the beam pattern for the forward method, Figure 8.2b for the backward method and Figure 8.2c for the forward-backward method. As expected, the nulls are deep and occur along the correct directions. Use of the forward-backward method produces an antenna pattern with lower sidelobes as it has more degrees of freedom. The use of amplitude-only weight produces a symmetric pattern for all the three cases.

Table 8.1. Parameters for the SOI and Interference.

	Magnitude	Phase	DOA
Signal	1.0–10.0 V/m	0.0	90°
Jammer #1	1000.0 V/m	0.0	70°
Jammer #2	1000.0 V/m	0.0	130°

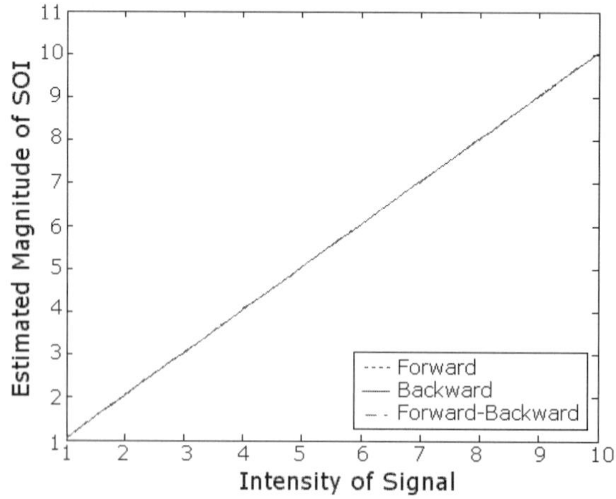

Figure 8.1a. Estimated amplitude (in volts) of the SOI in the presence of jammers and noise.

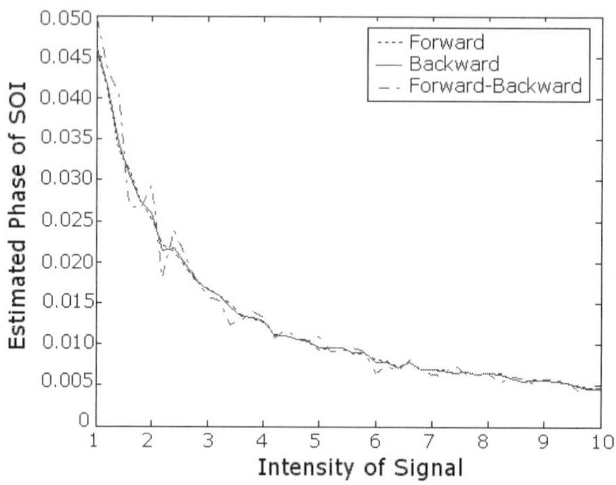

Figure 8.1b. Estimated phase (in degrees) of the SOI in the presence of jammers and noise.

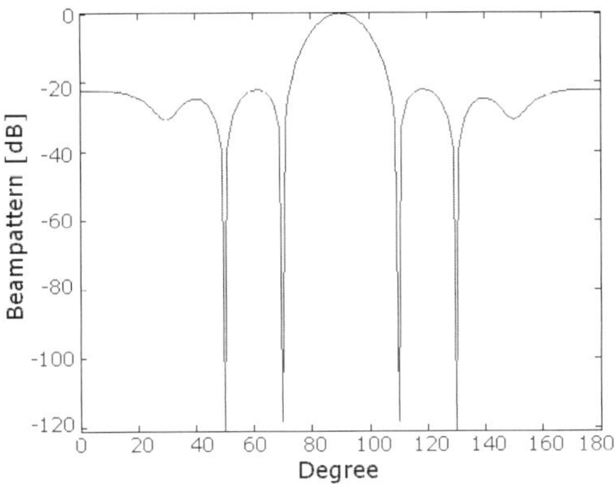

Figure 8.2a. Adaptive beam pattern in the presence of jammers and noise using the forward method.

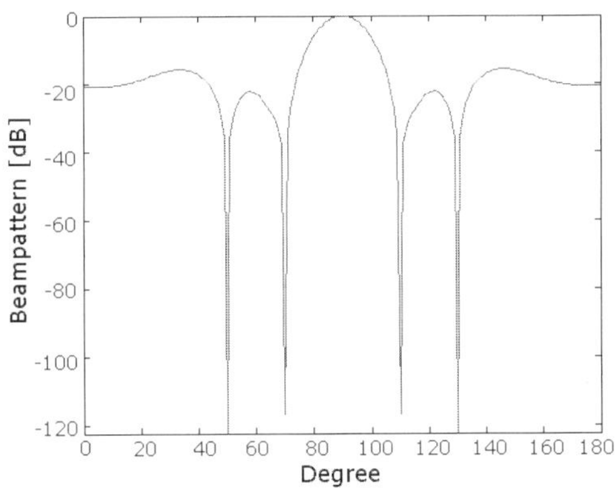

Figure 8.2b. Adaptive beam pattern in the presence of jammers and noise using the backward method.

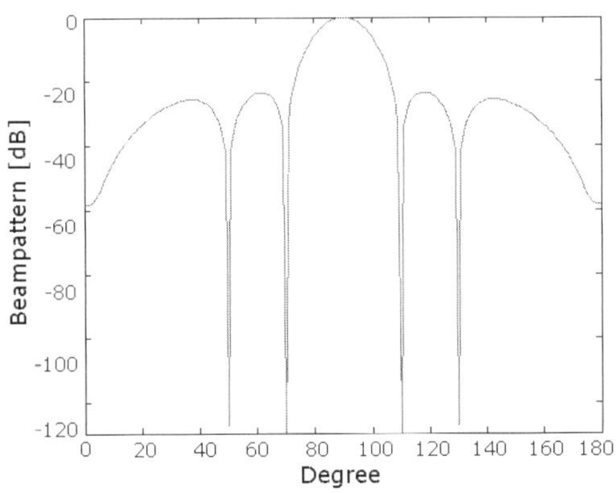

Figure 8.2c. Adaptive beam pattern in the presence of jammers and noise using forward-backward method.

For the second example, we consider three jammers. One of the jammers is incident from an angle symmetric with respect to 90° along with the SOI. All the signals intensities and the directions of arrival are summarized in Table 8.2. In this simulation we have added 30 dB noise at each of the antenna elements. In this simulation we use an 11-element antenna array. The number of weights used for the forward and the backward methods is 8 and the number of weights is 9 for the forward-backward method. All the weights are real. We increase the amplitude of the SOI from snapshot to snapshot. Figure 8.3a plots the recovered amplitude and Figure 8.3b plots the estimated phase of the SOI from snapshot-to-snapshot for all the three methods.

The beam pattern associated with this adaptive system for recovering the SOI of 1 V/m is shown in Figure 8.4 for all the three different methods (forward method in Figure 8.4a, backward method - Figure 8.4b and the forward-backward method in Figure 8.4c). Even though the jammer comes from an angle which is symmetrical with respect to broadside with that of the SOI, it has been nulled using an amplitude-only adaptive algorithm.

Table 8.2. Parameters for the SOI and Interference.

	Magnitude	Phase	DOA
Signal	1.0–10.0 V/m	0.0	100°
Jammer #1	100.0 V/m	0.0	70°
Jammer #2	500.0 V/m	0.0	80°
Jammer #3	1000.0 V/m	0.0	120°

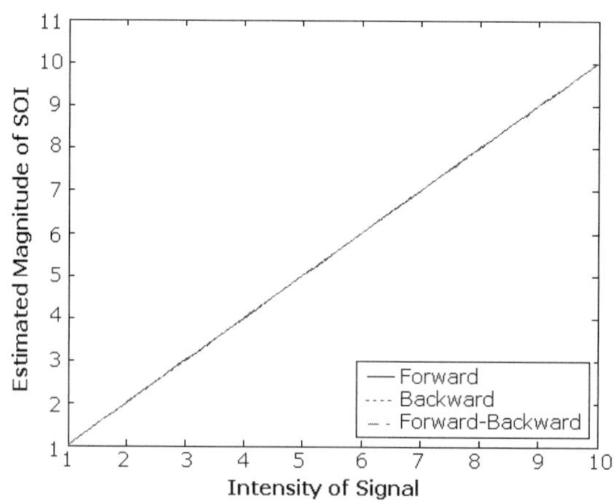

Figure 8.3a. Estimated amplitude (in volts) of the SOI in the presence of jammers and noise.

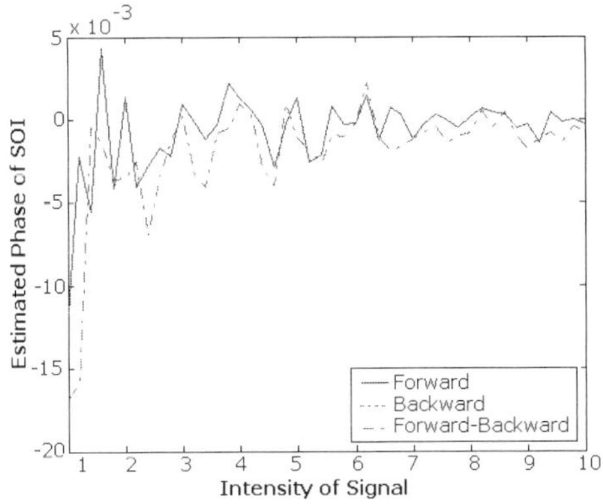

Figure 8.3b. Estimated phase (in degrees) of the SOI in the presence of jammers and noise.

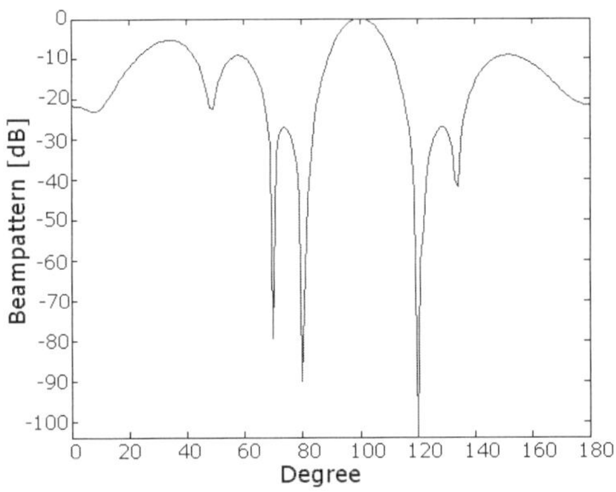

Figure 8.4a. Adaptive beam pattern in the presence of jammers and noise using the forward method.

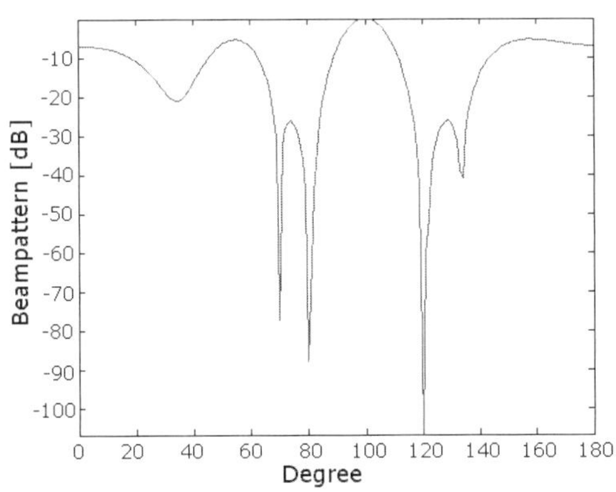

Figure 8.4b. Adaptive beam pattern in the presence of jammers and noise using the backward method.

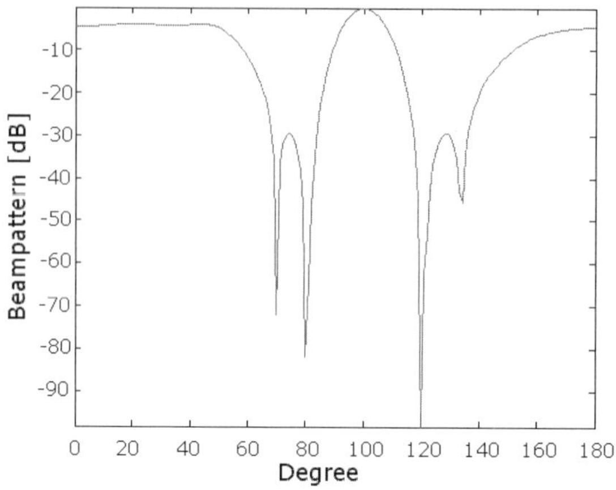

Figure 8.4c. Adaptive beam pattern in the presence of jammers and noise using the forward-backward method.

Using the forward-backward method for the third example, we observe the performance of the proposed algorithm when there are interferers spatially close to the SOI. The SOI arrives from 95° and the DOA of one of the interferers is varied from 121° to 97° in steps of 2° from snapshot to snapshot. The detailed parameters are summarized in Table 8.3. Here we consider a 7-element antenna array equally spaced half-a-wavelength apart and therefore the beamwidth will be approximately $60°/L_\lambda = 20°$, where L_λ is the length of the aperture in wavelengths. Hence, one of the interferers can be in the main lobe. The signal-to-noise ratio is set at each antenna element to be 30 dB. As shown In Figure 8.5, the estimated result of the proposed approach degrades as the interferer comes very close to the SOI as we would expect. This method breaks down for this problem when the separation between the SOI and the interferer is approximately one quarter of the main beamwidth.

Table 8.3. Parameters for the SOI and Interference.

	Magnitude	Phase	DOA
Signal	3.0 V/m	0.0	95°
Jammer #1	10 V/m	0.0	60°
Jammer #2	1 V/m	0.0	121°–97°

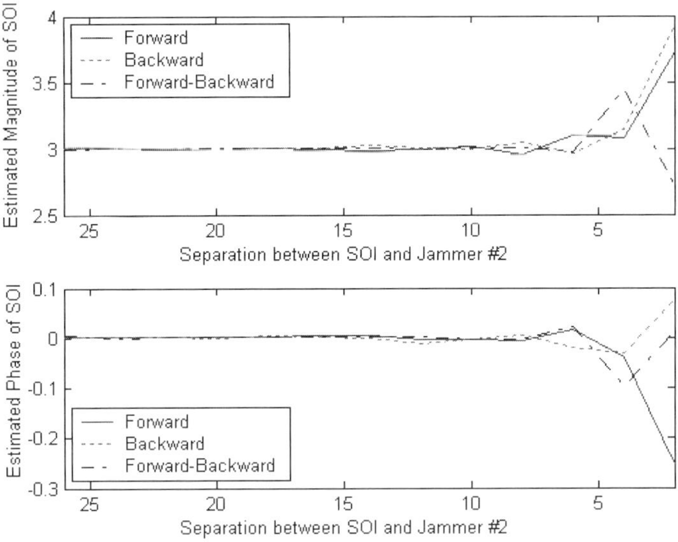

Figure 8.5. Estimation of the complex amplitude (in volts for the amplitude and in degrees for the phase) of the SOI for a close separation in degrees between an interferer and the SOI.

8.4 FORMULATION OF AN AMPLITUDE-ONLY DIRECT DATA DOMAIN LEAST SQUARES SPACE-TIME ADAPTIVE PROCESSING

In this section we describe the application of the real weights to space-time adaptive processing.

8.4.1 Forward Method

We assume that the signals entering the array are narrowband and consist of the SOI and interferences plus noise. We assume that for each jammer, the energy impinging on the array is confined to a particular DOA and is spread in frequency.

In this case, the jammers may be blinking or stationary. The scenario has been described in detail in section 6.3. We focus our attention to the range cell r and consider the space-time snapshot for this range cell. Let $X(p; q)$ be the actual measured complex voltages at the qth antenna element at the pth instance of time, that are in the data cube of Figure 6.4 for the range cell r as explained in section 6.3. Hence the actual measured voltages $X(p; q)$ are

$$X(p,q) = \alpha \exp\left\{ j \left(\frac{2\pi q \Delta}{\lambda} \cos\varphi_s + \frac{2\pi f_s p}{f_r} \right) \right\} + \text{Clutter} + \text{Jammer} + \text{Thermal noise} \tag{8.21}$$

The goal is to extract the SOI, α, given these voltages, the DOA for the SOI, φ_s, f_r is the pulse repetition frequency and the Doppler frequency is f_s.

In the D3LS procedures to be described, the adaptive weights are applied to the single space-time snapshot for the range cell r. Here a two-dimensional array of weights numbering $N_a N_t$ is used to extract the SOI for the range cell r. Hence the weights are defined by $w(p; q; r)$ for $p = 1, ..., N_t < P$ and $q = 1, ..., N_a < Q$ and are used to extract the SOI at the range cell r. Therefore, for the D3LS method we essentially perform a high-resolution filtering in two dimensions (space and time) for each range cell as illustrated in section 6.3. P is the total number of time samples and Q is the number of antenna elements.

In the amplitude-only adaptive processing, this weighting vector should be a real number. But these real weights make symmetric beam pattern, so it can have an ambiguity if signal of interest comes $(90 + \varphi)°$ when interferer comes $(90 - \varphi)°$ because of symmetry property. So before separating the matrix [T], we need to shift the incoming data $X(t)$ because, if signal of interest comes from 90°, then there is no symmetry in angle between 0° to 180°. We can shift the input data by using the following transformation matrix

$$[\mathrm{Tr}] = \mathrm{diag}\left(\frac{\text{steering vector of } 90°}{\text{steering vector of DOA of SOI}}\right) \tag{8.22}$$

Then we modify the input data matrix $[X]$ to a new matrix $[\overline{X}]$ by multiplying it with the transformation matrix.

$$[\overline{X}] = [\mathrm{Tr}] \times [X] \tag{8.23}$$

The window size along the element dimension is N_a, and N_t along the pulse dimension. Selection of N_a determines the number of spatial degrees of freedom, while N_t determines the temporal degrees of freedom. Typically for a single domain processing, N_a and N_t must satisfy the following equations:

$$N_a \le (Q + 0.5)/1.5 \tag{8.24}$$

$$N_t \le (P + 0.5)/1.5 \tag{8.25}$$

In conventional D3LS STAP algorithm maximum number of N_a and N_t is $(Q + 1)/2$ and $(P + 1)/2$. So, more degrees of freedom can be used in an amplitude-only D3LS STAP algorithm. And the advantage of a joint domain processing is that either of these bounds can be relaxed, i.e., one can exchange spatial degrees of freedom with the temporal degrees of freedom. So, indeed it is possible to cancel a number of interferers, which is greater than the number of antenna elements in a joint domain processing. The total number of degrees of freedom, R, for any method is

$$R = N_a \times N_t \tag{8.26}$$

In conventional D3LS STAP algorithm we let the element to element off set of the SOI in space and time, respectively, as

$$Z_1 = \exp\left[j\, 2\, \pi \frac{d}{\lambda} \cos(\varphi_s) \right] \tag{8.27}$$

$$Z_2 = \exp\left[j\, 2\, \pi \frac{f_s}{f_r} \right] \tag{8.28}$$

Again, SOI has an angle of arrival of φ_s and a Doppler frequency of f_s. But in amplitude-only D3LS STAP algorithm we need to change φ_s to 90° and f_s to zero frequency.

And we form a reduced rank matrix $[T]$ of dimension $(N_a \times N_t - 1) \times (N_a \times N_t)$ from the elements of the matrix $\overline{X}$. This will result in an equation similar to (6.75) and will be given by

$$[T][W] = [Y] = \begin{bmatrix} C & 0 & \ldots & 0 \end{bmatrix}^T \tag{8.29}$$

where C is a complex constant. In solving this equation one obtains the weight vector $[W]$, which places space-time nulls in the direction of the interferers while maintaining gain in the direction of the SOI. However, an equation similar to (8.9) needs to be set up to solve for the real valued weights. The complex amplitude of the SOI can be estimated using

$$\alpha \approx \frac{1}{C} \sum_{e=1}^{N_a} \sum_{h=1}^{N_t} W\left\{N_t(e-1)+h\right\} \overline{X}(h\,;e) \tag{8.30}$$

8.4.2 Backward Method

Next we reformulate the problem using the same data to obtain a second independent estimate for the solution. This is achieved by reversing the data sequence and then complex conjugating each term of that sequence as illustrated in section 6.3.3.

It is well known in the parametric spectral estimation literature that a sampled sequence which can be represented by a sum of exponentials with purely imaginary argument can be used either in the forward or in the reverse direction resulting in the same value for the exponent. From physical considerations we know that if we solve a polynomial equation with the weights W_i as the coefficients then its roots provide the DOA for all the unwanted signals including the interferers. Therefore whether we look at the snapshot as a forward sequence as presented in the last section or by a reverse conjugate of the same sequence the final results for W_i must be the same. Hence for these classes of problems we can observe the data either in the forward direction or in the reverse direction. This is equivalent to creating a virtual array of the same size but located along a mirror symmetry line. Therefore, if we now conjugate the data and form the reverse sequence, then one gets an independent set of equations similar to (8.29) and equivalent to (6.84). Once we get the weights by solving a system of equations similar to (8.29), the strength of the desired signal at range cell r is estimated from

$$\alpha \approx \frac{1}{C} \sum_{e=1}^{N_a} \sum_{h=1}^{N_t} W\{N_t(e-1)+h\} \overline{X}^* (M-h+1; N-e+1) \tag{8.31}$$

Note that for both the forward and the backward methods described in Sections 8.4.1 and 8.4.2, we have maximum number of N_a and N_t given by $(Q+0.5)/1.5$ and $(P+0.5)/1.5$. An additional benefit accrues in this case of dealing with the direct data sequence. For the conventional method, the maximum number of N_a and N_t we can consider is given by $(Q+1)/2$ and $(P+1)/2$. In a real situation when the solution is unknown two different estimates for the same solution may provide a level of confidence on the quality of the solution.

8.4.3 Forward-Backward Method

Finally, in this section we combine both the forward and the backward method to double the given data and thereby increase the number of weights or the degrees of freedom over that of either the forward or the backward method. In the forward-backward model we double the amount of data by not only considering the data in the forward direction but also conjugating it and reversing the direction of increment of the independent variable. This is always true for the adaptive array case. So by considering the data set $X(n)$ and $X^*(-n)$ we have essentially doubled the amount of data without any penalty, as these two data sets for our problem are linearly independent as discussed in section 6.3.4 and the governing equations are given by (6.89). This results in

$$[FB][W] = [Y]. \tag{8.32}$$

Once the real weights are solved, the complex amplitude of the signal can be obtained as outlined for the previous cases presented in sections 8.4.2 and 8.4.3.

8.5 SIMULATION RESULTS

For the first example we simulated an amplitude only D3LS STAP algorithm. We consider a signal arriving from Azimuth angle (φ_s) is 65° and Doppler frequency (f_s) is 1300 Hz in this simulation. We used 11 elements array (Q = 11) spacing of λ/2 and 19 pulses (P = 19). And we consider 4 discrete interferers and jammer and clutter. In this simulation we used 30 dB noise (thermal noise). For this example the PRF (Pulse Repetition Frequency) is 4 KHz. All values of the signal and the interferers are summarized in Table 8.4.

Table 8.4. Parameters for the SOI and Interference.

	AOA (degree)	Doppler (Hz)	
Signal	65°	1300	SNR − 30 dB
Discrete Interferers	115°, 40° 145°, 35°	−1300, −200 400, 1400	SDIR = −8.5 dB
Jammers	90°		SJR = −14 dB
Clutter	58°: 0.2° :61°	923: 5 :1073	SCR = −13 dB

Here, we assume that we know the DOA of the signal but need to estimate its complex amplitude using a single snapshot of the data. If the jammers have been nulled correctly then the signal can be recovered with high accuracy. In this simulation we choose one of the discrete interferers to be located symmetrically to the SOI. We can see the input signal spectrum (signal of interest and interferers) in Figure 8.8. And Figure 8.9 plots the results of using amplitude-only STAP method based on the direct data domain approach presented in section 8.4. Results are shown for the three different methods (forward method, backward method and forward-backward method). In this figure the '+' mark denotes the signal of interest, small 'o' marks denote the discrete interferers, and the large 'o' marks shows the location of clutter. Input signal interferer plus noise ratio (input SINR) of this example is –17.117 dB and output signal to interferer plus noise ratio (output SINR) is 43.6 dB using the forward method and 43.5 dB using the backward method and 45.6 dB using the forward-backward method. For the forward and the backward methods, the values for the weights in space and time are $N_a = 6$, $N_t = 10$. For the forward-backward method, the values for the weights in space and time are $N_a = 7$, $N_t = 13$.

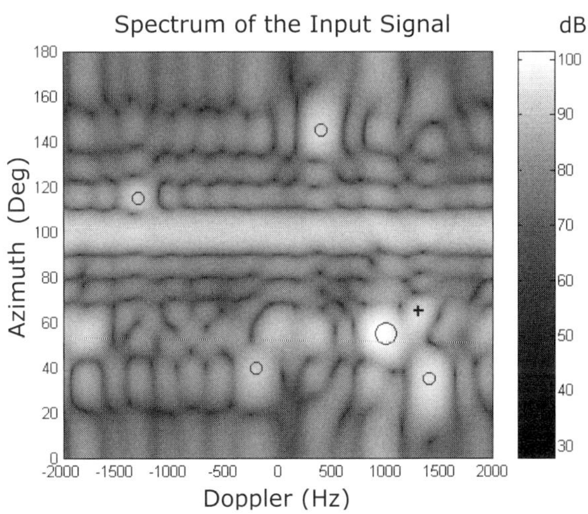

Figure 8.8. Input signal spectrum for the simulation.

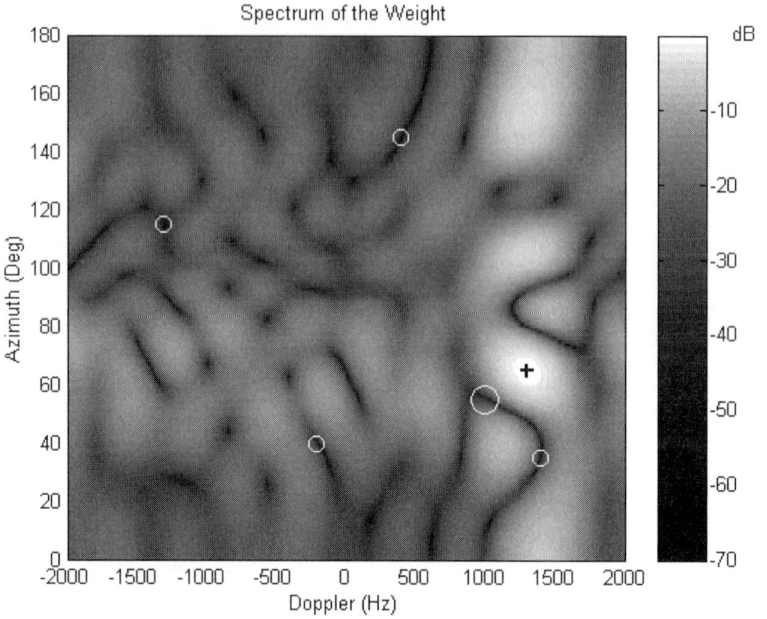

Figure 8.9a. Spectrum of the weights using the forward method.

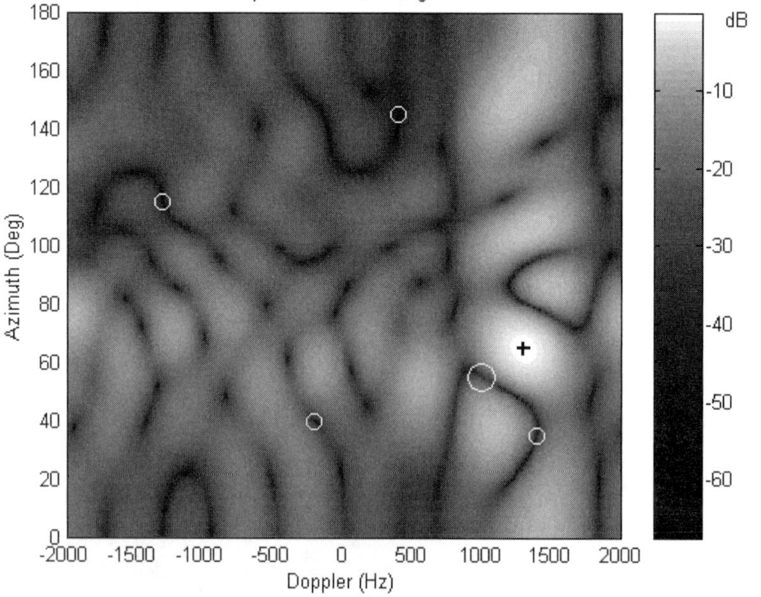

Figure 8.9b. Spectrum of the weights using the backward method.

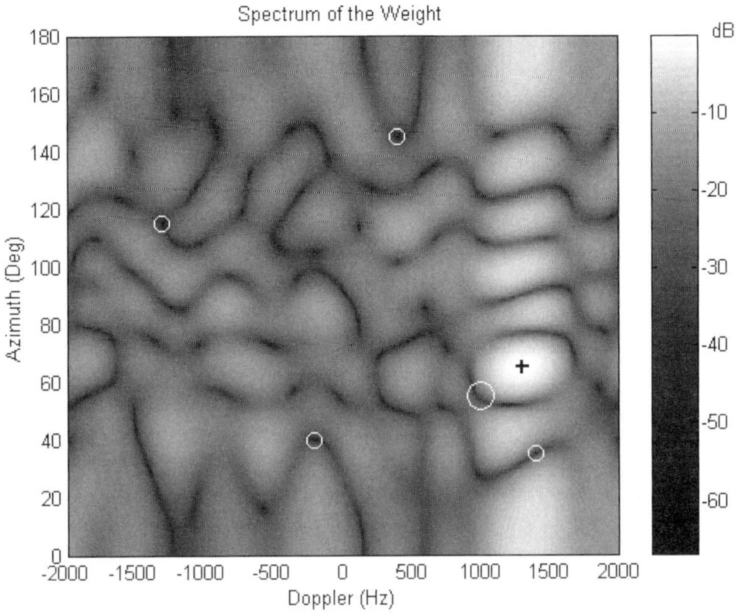

Figure 8.9c. Spectrum of the weights using the forward-backward method.

In Figure 8.10 we show the beam pattern at the Doppler frequency of each of the discrete interferers. Here, we can see that the nulls are deep and occur along the correct directions.

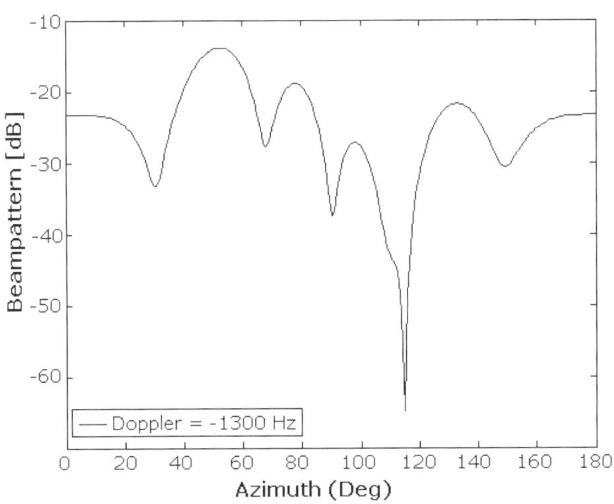

Figure 8.10a. Adaptive beam pattern to illustrate the nulling of the interference at the Doppler frequency.

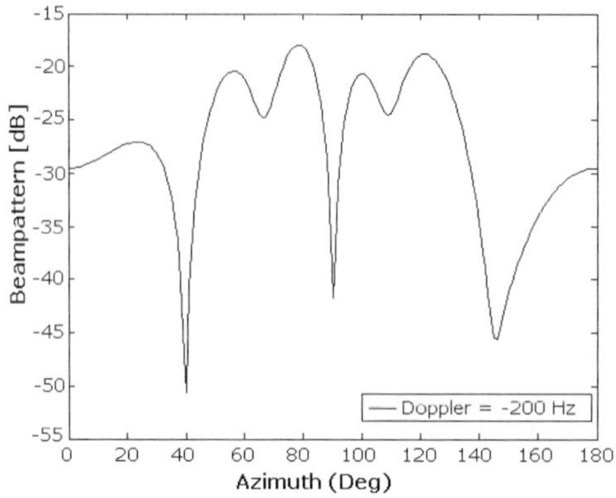

Figure 8.10b. Adaptive beam pattern to illustrate the nulling of the interference at the Doppler frequency.

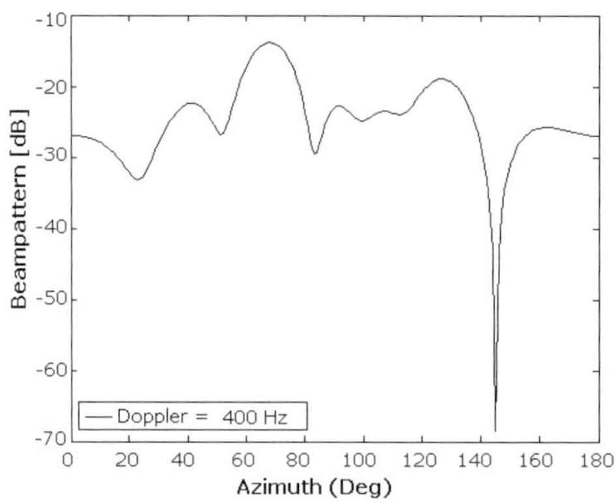

Figure 8.10c. Adaptive beam pattern to illustrate the nulling of the interference at the Doppler frequency.

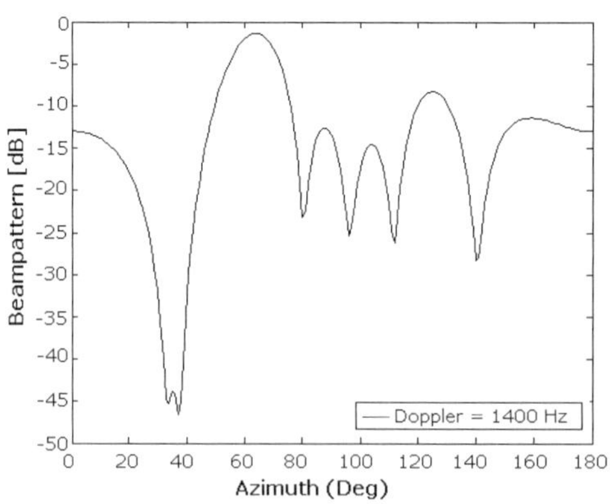

Figure 8.10d. Adaptive beam pattern to illustrate the nulling of the interference at the Doppler frequency.

For the second example, we simulate a more complicated situation (more interferers). We consider a signal arriving from the same azimuth angle and the same Doppler frequency as in the first example. And we used the same number of elements [11-element array ($Q = 11$) spacing of $\lambda / 2$ and 19 pulses ($P = 19$)]. But in this example there are 10 discrete interferers. We use the same PRF (4 KHz). All the signals intensities and the directions of arrival are summarized in Table 8.5. The clutter is distributed from 58° to 61° by a number of signals separated from each other in angle by 0.2°. The amplitude and the phase of these signals are generated by random numbers so that the signal to clutter ratio is approximately –13 dB. The input signal spectrum is shown in Figure 8.11.

Table 8.5. Parameters for the SOI and Interference.

	AOA (degree)	Doppler (Hz)	
Signal	65°	1300	SNR = 30 dB
Discrete Interferers	35°, 40°, 55° 65°, 65°, 85° 100°, 105°,120°, 125°	1400, 1700, –125 –1200, 325, 400 950, –1300, –800, 1450	SDIR = –10 dB
Jammers	90°		SJR = –14 dB
Clutter	58° : 0.2° : 61°	923: 5 :1073	SCR = –13 dB

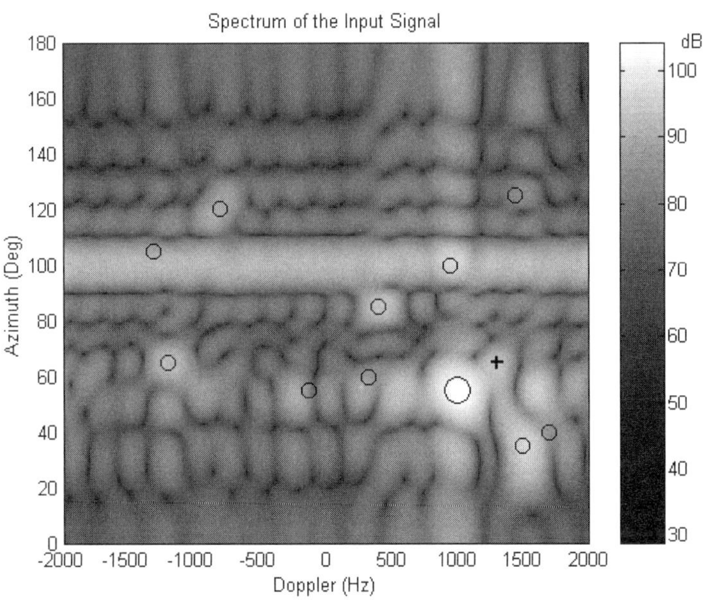

Figure 8.11. Input signal spectrum of the simulation.

Figure 8.12 shows the results of using the amplitude-only STAP method. Results are also shown for the three different methods (forward method, backward method and forward-backward method). In this example we used a more complex situation, but we can see that the nulls are still deep and occur along the correct directions and Doppler frequency. Input SINR of this example is –17.3 dB and output signal to noise and interference ratio is 36.561 dB from the forward method, 36.231dB from the backward method, and 38.057 dB from the forward-backward method.

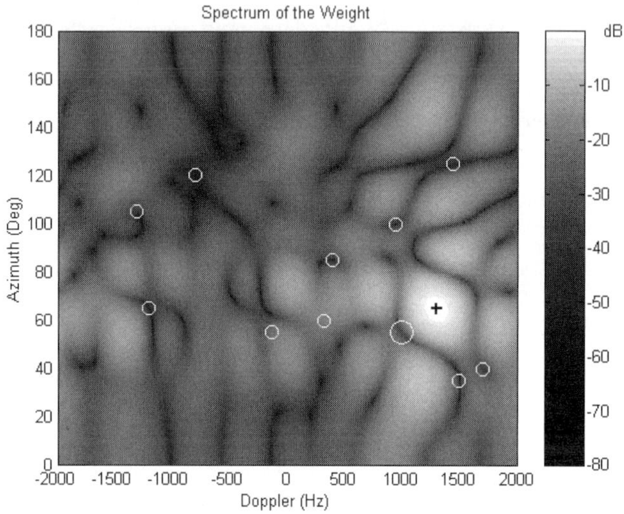

Figure 8.12a. Spectrum of the weights using the forward method.

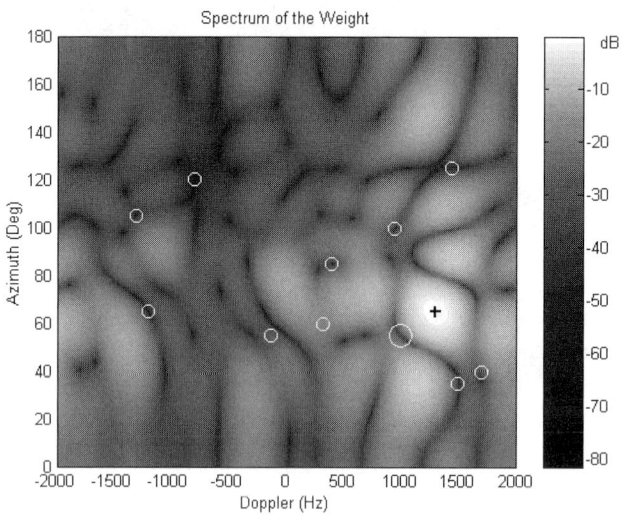

Figure 8.12b. Spectrum of the weights using the backward method.

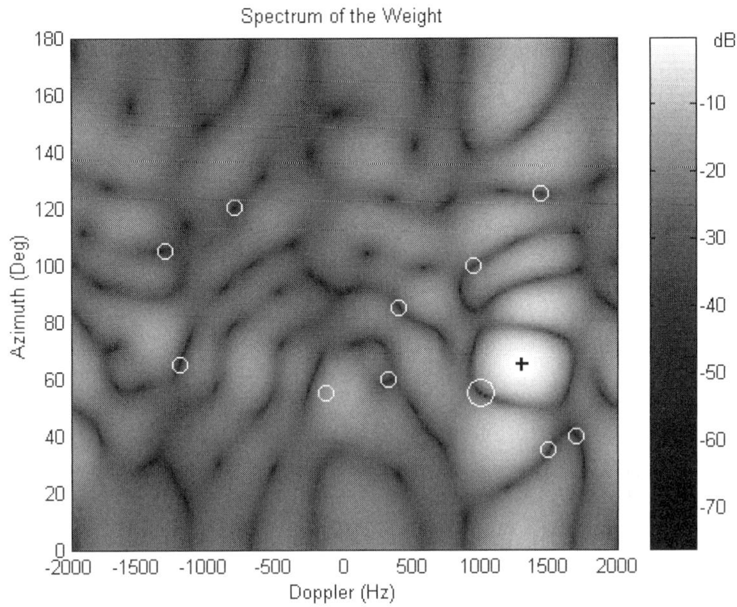

Figure 8.12c. Spectrum of the weights using the forward-backward method.

8.6 CONCLUSION

An adaptive processing technique utilizing a real set of weights based on a direct data domain least squares approach is presented. In this amplitude-only weights-based adaptive algorithm, since no statistical methodology is employed, there is no need to compute a covariance matrix. Therefore, this procedure can be implemented on a general-purpose digital signal processor for real-time implementations with ease. As shown through numerical examples, this proposed technique can cancel strong coherent interferers even though it is processing the data on a snapshot-by-snapshot basis and hence is quite useful in a dynamic environment. In this approach, one uses only real arithmetic operations to find the amplitude of the weights of the array. The use of real weights produces a symmetric antenna pattern and a shifting technique is used to prevent signal cancellation when the interferer is at a symmetric location with respect to the SOI. This method may be useful for a real-time implementation of the algorithm. Finally, the amplitude-only direct data domain least squares approach is applied to the space-time adaptive processing and uses only real arithmetic operations to find the amplitude of the weights in the array. As shown through numerical examples, even in the presence of strong interferers and clutter, the proposed amplitude-only STAP algorithm eliminates the undesired interferers and obtains the correct complex amplitude of the desired signal.

REFERENCES

[1] S. T. Smith, "Optimum Phase-only Adaptive Nulling", *IEEE Trans. Signal Processing*, Vol. 47, pp. 1835-1843, July 1999.

[2] R. L. Haupt, "Phase-only Adaptive Nulling with a Genetic Algorithm", *IEEE Trans. Antennas and Propagation*, Vol. 45, pp. 1009-1015, June 1998.

[3] M. K. Leavitt, "A Phase Adaptation Algorithm", *IEEE Trans. Antennas and Propagation*, Vol. 24, pp. 754– 756, July 1976.

[4] C. A. Baird and G. G. Rassweiler, "Adaptive Sidelobe Nulling Using Digitally Controlled Phase-shifters", *IEEE Trans. Antennas and Propagation*, Vol. 24, Sept. 1976, pp. 638-649.

[5] G. Yanchang and L. Jianxin, "Speed Up and Optimization of RAMONA for Adaptive Digital Beamforming", *Antennas and Propagation*, 1993, Eighth International Conference on, APS. Digest, Vol. 1, pp. 508-511.

[6] G. Yanchang and L. Jianxin, "Real Amplitude-only Nulling Algorithm (RAMONA) for Adaptive Digital Beamforming", *Antennas and Propagation Society International Symposium*, 1999, APS. Digest, Vol. 1, May 1990, pp. 206-209

[7] Y. Guo and J. Li, "Real Amplitude-only Nulling Algorithm (RAMONA) for Adaptive Sum and Difference Patterns", *Antennas and Propagation Society International Symposium*, 1999, AP-S. Digest , Vol. 1, pp. 94-97.

[8] T. Gao, Y. Guo and J. Li, "A Fast Beamforming Algorithm for Adaptive Sum and Difference Patterns in Conformal Array Antennas", *Antennas and Propagation Society International Symposium*, 1999, AP-S Digest, Held in conjunction with: URSI Radio Science Meeting and Nuclear EMP Meeting, Vol. 1, pp. 450-453.

[9] K. W. Lo, "Adaptivity of a Real-symmetric Array by DOA Estimation and Null Steering", *Radar, Sonar and Navigation, IEE Proceedings*, Vol. 144, Issue 5, pp. 245 –251.

[10] T. Vu, "Simultaneous Nulling in Sum and Difference Patterns by Amplitude Control", *IEEE Transactions on Antennas and Propagation*, Vol. 34, Issue 2, Feb 1986, pp. 214-218.

[11] L. E. Brennan and I. S. Reed, "Theory of Adaptive Radar", *IEEE Trans. Aerosp. Electron. Syst.*, Vol. AES-9, pp. 237-252, Mar. 1973.

[12] L. E. Brennan, J. D. Mallet, and I. S. Reed, "Adaptive Arrays in Airborne MTI Radar", *IEEE Trans. Antennas Propagat.*, Vol. AP-24, pp. 605-615, Sept. 1976.

[13] J. Ward, "Space Time Adaptive Processing for Airborne Radar", Lincoln Lab., Lexington, MA, *Tech. Rep.* 1015, Dec. 1994.

[14] R. Klemm, "Adaptive Clutter Suppression for Airborne Phased Array Radars", *Proc. IEEE*, pt. F and H, Vol. 130, No. 2, pp. 125-131, Feb.1983.

[15] E. C. Banle, R. C. Fante, and J. A. Torres, "Some Limitations on the Effectiveness of Airborne Adaptive Radar", *IEEE Trans. Aerosp. Electron. Syst.*, Vol. AES-28, pp.1015-1032, Oct. 1992.

[16] H. Wang and L. Cai, "On Adaptive Spatial-temporal Processing for Air-borne Surveillance Radar Systems", *IEEE Trans. Aerosp. Electron. Syst.*, Vol. 30, pp. 660-669, July 1994.

[17] J. Ender and R. Klemm, "Airborne MTI via Digital Filtering", *Proc. IEEE*, pt. F, Vol. 136, No. 1, pp. 22-29, Feb. 1989.

[18] F. R. Dickey Jr., M. Labitt, and F. M. Standaher, "Development of Air-borne Moving Target Radar for Long Range Surveillance", *IEEE Trans. Aerosp. Electron. Syst.*, Vol. 27, pp. 959–971, Nov. 1991.

[19] T. K. Sarkar, M. C. Wicks, M. Salazar-Palma, and R. J. Bonneau, *Smart Antennas*, John Wiley-IEEE Press, Hoboken, NJ, 2003.

[20] J. Carlo, T. K. Sarkar, and M. C. Wicks, "Application of Deterministic Techniques to STAP", In *Applications of Space-Time Adaptive Processing*, edited by R. Klemm, London, UK, IEE Press, pp. 375-411, 2004.

[21] T. K. Sarkar, S. Park, J. Koh, and R. A. Schneible, "A Deterministic Least Squares Approach to Adaptive Antennas", *Digital Signal Processing - A Review Journa*l, Vol. 6, pp. 185-194, 1996.

[22] W. Choi, T. K. Sarkar, H. Wang, and E. L. Mokole, " Adaptive Processing Using Real Weights Based on a Direct Data Domain Least Squares Approach", *IEEE Transactions on Antennas and Propagation*, Vol. 54, No. 1, pp. 182-191, Jan 2006.

[23] T. K. Sarkar, S. Nagaraja and M. C. Wicks, "A Deterministic Direct Data Domain Approach to Signal Estimation Utilizing Nonuniform and Uniform 2-D Arrays", *Digital Signal Processing – A Review Journal*, Vol. 8, 114-125. 1998.

[24] T. K. Sarkar, H. Wang, S. Park, J. Koh, R. Adve, K. Kim, Y. Zhang, M. C. Wicks, and R. D. Brown, "A Deterministic Least Squares Approach to Space Time Adaptive Processing (STAP)", *IEEE Trans. Antennas and Propagation*, Vol. 49, pp. 91-103, Jan 2001.

[25] J. H. Mathews and K. D. Fink, *Numerical Methods Using Matlab*, Third edition, Upper Saddle River, NJ, Prentice Hall.

9

PHASE-ONLY ADAPTIVE AND SPACE-TIME PROCESSING

9.0 SUMMARY

Generally, adaptive antenna arrays operate by changing the complex adaptive weights consisting of both magnitudes and phases applied at each of the antenna elements. However, it is easier to require the adaptive weights to have only phase variation with a fixed amplitude at each of the antenna elements. Hence, this chapter addresses the phase only adaptive systems whose weights have fixed amplitude and its phase is adjusted through a new phase only adaptive method based on a direct data domain least squares approach (D3LS), which utilizes only a single snapshot of the data for adaptive processing. This technique can also be applied to space-time adaptive processing (STAP).

9.1 INTRODUCTION

Traditionally, adaptive signal processing algorithms apply both amplitude and phase weighting at each of the antenna elements in an array as weight vectors. However, some existing antenna systems possess capability only of changing the phase at the antenna elements to mitigate the undesired interference while preserving simultaneously the desired signal. Hence, several authors have proposed phase perturbation algorithms whose resulting beam pattern place nulls to cancel out the interferences along some directions [1–3]. These approaches are fast but require knowledge of the directions of arrival (DOA) of the interference, which are not known in a real situation. Another class of algorithms based on statistical approaches adjusts the phase of the elements in an array to reduce the total output power from the array [4–6]. However, they require independent identically distributed secondary data to estimate the covariance matrix of the interference. The formation of the covariance matrix is quite time consuming and so is the evaluation of its inverse, particularly when the system matrix is nearly singular.

Recently, a D3LS algorithm has been proposed [7–10] and described in chapter 6. A D3LS approach has certain advantages related to the computational issues associated with the adaptive array processing problem as it adaptively analyzes the data at each snapshot as opposed to forming a covariance matrix of the data from multiple snapshots, and then solving for the weights utilizing that information. A single snapshot in this context is defined as the array of complex voltages measured at the feed point of the antenna elements. Another advantage of the D3LS approach is that when the DOA of the signal is not known precisely, additional constraints can be applied to fix the main beam width of the receiving array *a priori* and thereby reduce the signal cancellation problem. Therefore, in this chapter we represent the phase-only adaptive processing based on a D3LS approach to overcome the drawbacks of statistical approaches.

The chapter is organized as follows. In section 9.2 we formulate the problem. In section 9.3 we present simulation results illustrating the performance of the proposed method. Section 9.4 presents the phase-only STAP algorithms followed by some numerical results in section 9.5. Finally, in section 9.6 we present some conclusions.

9.2 FORMULATION OF THE DIRECT DATA DOMAIN LEAST SQUARES SOLUTION FOR A PHASE-ONLY ADAPTIVE SYSTEM

9.2.1 Forward Method

Consider an array composed of N sensors separated by a distance d as shown in Figure 9.1. We assume that narrowband signals consisting of the desired signal plus possibly coherent multipaths and jammers with center frequency f_0 are impinging on the array from various angles θ, with the constraint $0 \leq \theta \leq 180^\circ$. For sake of simplicity we assume that the incident fields are coplanar and that they are located in the far field of the array. However, this methodology can easily be extended to the non-coplanar case without any problem including the added polarization diversity.

Using the complex envelope representation, the $N \times 1$ complex vectors of phasor voltages $[\underline{x}(t)]$ received by the antenna elements at a single time instance t can be expressed by

$$[\underline{x}(t)] = \begin{bmatrix} x_1(t) \\ x_2(t) \\ \vdots \\ x_N(t) \end{bmatrix} = \sum_{m=1}^{M} [\underline{a}(\theta_m)]\, s_m(t) + [\,n(t)\,] \tag{9.1}$$

where $s_m(t)$ denotes that the incident signal from the mth source directed towards the array at the instance t. $[\underline{a}(\theta)]$ denotes the steering vector of the array toward direction θ and $[n(t)]$ denotes the noise vector at each of the antenna

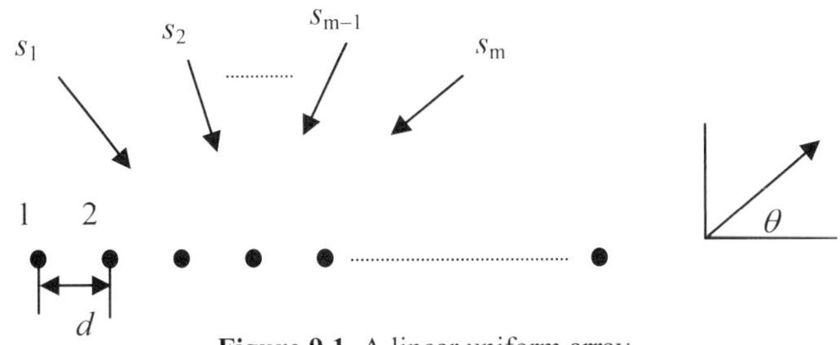

Figure 9.1. A linear uniform array.

elements. We now analyze the data using a single snapshot of the voltages measured at the antenna terminals.

Using a matrix notation, (9.1) becomes

$$[\underline{x}(t)] = [A(\theta)]\,[s(t)] + [n(t)] \tag{9.2}$$

where $[A(\theta)]$ is the $N \times M$ matrix of the steering vectors, referred to as the array manifold

$$[A(\theta)] = [a(\theta_1) \quad a(\theta_2) \;.\;.\;.\; a(\theta_M)] \tag{9.3}$$

Here $[s(t)]$ is a $M \times 1$ vector representing the various signals incident on the array at time instance t. In practice, there are mutual couplings between the antenna elements in the array, which undermine the performance of any conventional adaptive signal processing algorithm. But in this chapter we assume that the elements of the antenna are omni-directional point radiators in the plane of the radiation of interest. The mutual coupling effects and the presence of near-field scatterers can be taken into account as outlined in [10].

Hence, our problem can be stated as follows: Given the sampled data vector snapshot $[\underline{x}(t)]$ at a specific instance of time (single snapshot), how do we recover the desired signal arriving from a given look direction while simultaneously rejecting all other interferences and clutter which may be coherent.

Let us assume that the signal is coming from θ_d and our objective is to estimate its amplitude while simultaneously rejecting all other interferences. The signal arrives at each sensor at different times dependent on the direction of arrival of the target and the geometry of the array. At each of the N antennas, the received signal (9.1) is the sum of the signal of interest (SOI), interference, clutter, and thermal noise. The interference may consist of multiple delayed copies of the SOI which may be coherent multipaths. Therefore, we can reformulate (9.1) as

$$[\underline{x}(t)] = s_d\left[a(\theta_d)\right] + \sum_{m=1}^{M-1} s_m\left[a(\theta_m)\right] + [n(t)] \tag{9.4}$$

where s_m and θ_m are the amplitude and direction of arrival (DOA) of the mth interference signals. s_d and θ_d are the amplitude and direction of arrival of the SOI. Here we assume that we know θ_d, the DOA of the SOI and we need to estimate its complex amplitude.

We can represent the received voltage at the time instance t solely due to the desired signal at the kth sensor element by

$$s_d = \alpha \exp\left[j\psi(\theta_d)\right] \tag{9.5}$$

Here, since we are using a single snapshot to extract the SOI from jammers and interferences, the array of N elements is partitioned into a number of segments of length k as determined in [10].

The strength of the SOI, $s_d(t)$, is the desired unknown parameter which will be estimated for the given snapshot at the time instance t. $\psi(\theta_d)$ does not provide a linear phase regression along the elements of the real array, when the elements deviate from isotropic omni directional point sensors. This deviation from phase linearity undermines the capabilities of the various signal-processing algorithms. For a standard adaptive weighting system we can now estimate the SOI by a weighted sum given by

$$y(t) = \sum_{k=1}^{K} w_k \, x_k(t) \tag{9.6}$$

or in a compact matrix form as

$$[y(t)] = [W]^T[X] = [X]^T[W] \tag{9.7}$$

where the superscript T denotes the transpose of a matrix and K is equal to number of weights, so $K = (N+1)/2$ in this case. Also K has to be greater than the number of interferers $M-1$, i.e., $K \geq M$. The two vectors $[W]$ and $[X]$ are given by

$$[W]^T = \left[e^{j\varphi_1} \;\; e^{j\varphi_2} \; \ldots\ldots \; e^{j\varphi_K} \right] \tag{9.8}$$

$$[X]^T = \left[x_1 \;\; x_2 \; \ldots\ldots \; x_K \right] \tag{9.9}$$

Let $[V]$ be a matrix whose elements comprise the complex voltages measured at a single time instance t at all the N elements of the array simultaneously. The received signals may also be contaminated by thermal noise. Let us define another matrix $[S]$ whose elements comprise the complex voltages received at the antenna elements due to a signal of unity amplitude coming from the desired direction θ_d. Then the elements of this matrix contain the elements s_{dm}, where m represents the induced voltage at antenna element m due to the SOI with an assumed amplitude of 1 V. However, the actual complex amplitude of the SOI is not 1 V but α which is to be determined. Then if we form the matrix pencil using these two matrices, then

$$[V] - \alpha [S] \tag{9.10}$$

where

$$[V] = \begin{bmatrix} x_1 & x_2 & \cdots & x_K \\ x_2 & x_3 & \cdots & x_{K+1} \\ \vdots & & & \\ x_K & x_{K+1} & \cdots & x_M \end{bmatrix}_{K \times K} \tag{9.11}$$

$$[S] = \begin{bmatrix} s_{d1} & s_{d2} & \cdots & s_{dK} \\ s_{d2} & s_{d3} & \cdots & s_{dK+1} \\ \vdots & & & \\ s_{dK} & s_{dK+1} & \cdots & s_{dM} \end{bmatrix}_{K \times K} \tag{9.12}$$

represents a matrix of only the undesired signals. This difference at each of the antenna elements consisting of $\{[V] - \alpha [S]\}$ represents the contribution of all the undesired signals due to interferences which may be coherent or noncoherent multipath components, clutter and thermal noise (i.e., all the undesired components except the SOI). One could form the undesired noise power from (9.10) and estimate a value of α by using a set of weights $[W]$, which minimizes the noise power. Next, in an adaptive processing methodology, the column vectors of the weights $[W]$ are chosen in such a way that the contribution from the jammers, clutter, and thermal noise are minimized to enhance the output signal-to-interference plus noise ratio. Hence, if we define the matrix $[U] = \{[V] - \alpha[S]\}$, then one gets the following generalized eigenvalue problem, which is a least squares solution to the estimation of the SOI for that snapshot [6–8]

$$[U][W] = \{[V] - \alpha[S]\}[W] = 0 \tag{9.13}$$

Note that the (1,1) and (1,2) elements of the interference plus noise matrix, $[U]$ as defined in (9.13), are given by

$$U_{1,1} = X_1 - \alpha s_{d1} \tag{9.14}$$

$$U_{1,2} = X_2 - \alpha s_{d2} \tag{9.15}$$

where X_1 and X_2 are the voltages received at antenna elements 1 and 2 due to the signal, jammer, clutter, and noise whereas s_{d1} and s_{d2} are the values of the SOI only at those elements due to a signal of unit strength

$$Z = \exp\left[j 2\pi \frac{d}{\lambda} \cos\theta_d \right] \tag{9.16}$$

where θ_d is the angle of arrival corresponding to the desired signals. Then, $U_{1,1} - Z^{-1} U_{1,2}$ contains no components of SOI, as

$$s_{d1} = \exp\left[j 2\pi \frac{id}{\lambda} \cos\theta_d \right], \quad \text{with } i = 1 \tag{9.17}$$

$$s_{d2} = \exp\left[j 2\pi \frac{id}{\lambda} \cos\theta_d \right], \quad \text{with } i = 2. \tag{9.18}$$

Therefore one can form a reduced rank matrix $[T]_{(K-1)\times K}$, generated from $[U]$ such that

$$[T] = \begin{bmatrix} X_1 - Z^{-1}X_2 & \cdots & X_K - Z^{-1}X_{K+1} \\ \vdots & \vdots & \vdots \\ X_{K-1} - Z^{-1}X_K & \cdots & X_{N-1} - Z^{-1}X_N \end{bmatrix}_{(K-1)\times K} \begin{bmatrix} e^{j\varphi_1} \\ \vdots \\ e^{j\varphi_k} \end{bmatrix}_{K\times 1} = \begin{bmatrix} 0 \\ \vdots \\ 0 \end{bmatrix}_{(K-1)\times 1} \tag{9.19}$$

In order to make the matrix full rank, we fix the gain of the array by forming a weighted sum of the voltages $\sum_{k=1}^{K} W_k X_k$ along the direction of arrival of the SOI. Let us say the gain of the array is C in the direction of θ_d. This provides an additional equation resulting in

$$\begin{bmatrix} 1 & \cdots & Z^{K-1} \\ X_1 - Z^{-1}X_2 & \cdots & X_K - Z^{-1}X_{K+1} \\ \vdots & \vdots & \vdots \\ X_{K-1} - Z^{-1}X_K & \cdots & X_{N-1} - Z^{-1}X_N \end{bmatrix}_{K\times K} \begin{bmatrix} e^{j\varphi_1} \\ e^{j\varphi_2} \\ \vdots \\ e^{j\varphi_k} \end{bmatrix}_{K\times 1} = \begin{bmatrix} C \\ 0 \\ \vdots \\ 0 \end{bmatrix}_{K\times 1} \tag{9.20}$$

or, equivalently,

$$[F]\,[W] = [Y] \tag{9.21}$$

Once the weights are solved by using (9.13), the signal component α may be estimated from

$$\alpha = \frac{1}{C} \sum_{k=1}^{K} W_k X_k \tag{9.22}$$

The proof of (9.22) is available in [10].

Consider the effect of small phase perturbations on the phase-only weight vector. If each phase φ_k is perturbed by the small amount Δ_k, i.e., $\varphi_k = \varphi_k + \Delta\varphi_k$, then the phase-only weight vector is perturbed as

$$\left[e^{j\varphi_1}\ e^{j\varphi_2} \cdots e^{j\varphi_K} \right] \rightarrow \left[e^{j(\varphi_1+\Delta\varphi_1)}\ e^{j(\varphi_2+\Delta\varphi_2)} \cdots e^{j(\varphi_K+\Delta\varphi_K)} \right] \tag{9.23}$$

It will be convenient to represent this perturbation via the matrix exponential of real n-by-n matrix with diagonal entries $\Delta\varphi_k$, $k=1, \cdots, K$.

$$w \;\to\; \exp\left[j\Delta\varphi_k\right] w \tag{9.24}$$

To update phase-only variant weight vector (9.23) in the nonlinear equation (9.20), we use the version of the Conjugate Gradient (CG) method that is described in [11,12].

The CG method can be used not only to find the minimum point of a quadratic form, but can also minimize any continuous function f for which the gradient f' can be computed. The iterative formula for the phase only weighting algorithm is given by

$$w_{k+1} = w_k + \lambda_k d_k \tag{9.25}$$

where λ_k is a step-length and d_k is the line search direction defined by

$$d_k = \begin{cases} -g_k & \text{for } k=1 \\ -g_k + \beta_k d_{k-1} & \text{for } k \geq 2 \end{cases} \tag{9.26}$$

where β_k is a scalar and g_k is a gradient of the cost function f.

As with the linear CG, a value of λ_k that minimize the cost function $f(w_k + \lambda_k d_k)$ is found by ensuring that the gradient is orthogonal to the search direction. We can use any algorithms that find the zeros of the expression $[f'(w_k + \lambda_k d_k)]^T d_k$. And the best-known formulas for β_k are the following Fletcher-Reeves and Polak-Ribière formulas, which are given by

$$\beta_k^{FR} = \|g_k\|^2 / \|g_{k-1}\|^2 \tag{9.27}$$

$$\beta_k^{PR} = g_k^T (g_k - g_{k-1}) / \|g_{k-1}\|^2 \tag{9.28}$$

Convergence of the Polak-Ribière method can be guaranteed by choosing $\beta = \max\{\beta^{PR}, 0\}$. Using this value is equivalent to restarting CG if $\beta^{PR} < 0$. Here is the outline of the nonlinear CG method. Assume,

$$d_1 = -g_1 \tag{9.29}$$

Find λ_k that minimizes $f(w_k + \lambda_k d_k)$

$$w_{k+1} = w_k + \lambda_k d_k \tag{9.30}$$

$$d_k = -g_k \tag{9.31}$$

$$\beta_k = \|g_k\|^2 / \|g_{k-1}\|^2 \text{ or } \beta_k = \max\{g_k^T (g_k - g_{k-1}) / \|g_{k-1}\|^2, 0\} \tag{9.32}$$

The performance of the phase-only adaptive system based on a D3LS approach in this way will be considered in section 9.3.

9.2.2 Backward Method

Next we reformulate the problem using the same data to obtain a second independent estimate for the solution. This is achieved by reversing the data sequence and then complex conjugating each term of that sequence. It is well known in the parametric spectral estimation literature that a sampled sequence which can be represented by a sum of complex exponentials with purely imaginary argument can be used either in the forward or in the reverse direction resulting in the same value for the exponent. From physical considerations we know that if we solve a polynomial equation with the weights W_i as the coefficients then its roots provide the direction of arrival for all the unwanted signals including the interferers. Therefore whether we look at the snapshot as a forward sequence as presented in the last section or by a reverse conjugate of the same sequence the final results for W_i must be the same. Hence for these classes of problems we can observe the data either in the forward direction or in the reverse direction. This is equivalent to creating a virtual array of the same size but located along a mirror symmetry line. Therefore, if we now conjugate the data and form the reverse sequence, then one gets an independent set of equations similar to (9.20) for the solution of the weights $[W]$. This is represented by

$$\begin{bmatrix} 1 & \cdots & Z^{K-1} \\ X_N^* - Z^{-1}X_{N-1}^* & \cdots & X_K^* - Z^{-1}X_{K-1}^* \\ \vdots & \vdots & \vdots \\ X_{K+1}^* - Z^{-1}X_K^* & \cdots & X_2^* - Z^{-1}X_1^* \end{bmatrix}_{K\times K} \times \begin{bmatrix} e^{j\varphi_1} \\ e^{j\varphi_2} \\ \vdots \\ e^{j\varphi_K} \end{bmatrix}_{K\times 1} = \begin{bmatrix} C' \\ 0 \\ \vdots \\ 0 \end{bmatrix}_{K\times 1} \tag{9.33}$$

or equivalently in a matrix form as

$$[B][W] = [Y] \tag{9.34}$$

The signal strength α can again be determined by (9.22), once (9.33) is solved for weights. C' is the gain of the antenna array along the direction of the arrival of the signal.

Note that for both the forward and the backward methods described in sections 9.2.1 and 9.2.2, we have $K = (N + 1)/2$. Hence the degrees of freedom are the same for both the forward and the backward method. However, we have two independent solutions for the same adaptive problem. In a real situation when the solution is unknown, two different estimates for the same solution may provide a level of confidence on the quality of the solution.

9.2.3 Forward-Backward Method

Finally, in this section we combine both the forward and the backward method to double the given data and thereby increase the number of weights or the degrees of freedom significantly over that of either the forward or the backward method. This thus provides a third independent solution. In the forward-backward model we double the amount of data by not only considering the data in the forward direction but also conjugating it and reversing the direction of increment of the independent variable. This type of processing can be done as long as the series to be approximated can be fit by exponential functions of purely imaginary argument. This is always true for the adaptive array case. So by considering the data set $X(n)$ and $X^*(-n)$ we have essentially doubled the amount of data without any penalty, as these two data sets for our problem, are linearly independent.

An additional benefit accrues in this case. For both the forward and the backward method, the maximum number of weights we can consider is given by $(N+1)/2$, where N is the number of the antenna elements. Hence, even though all the antenna elements are being utilized in the processing, the number of degrees of freedom available for this approach is essentially $(N+1)/2$. For the forward-backward method, the number of degrees of freedom can be significantly increased without increasing the number of antenna elements. This is accomplished by considering the forward and backward versions of the array data. For this case, the number of degrees of freedom can reach $(N + 0.5)/1.5$. This is approximately equal to 50% more weights or number of degrees of freedom than the two previous cases. The equation that needs to be solved for the weights is given by combining (9.20) and (9.33), with $C' = C$, into

$$\begin{bmatrix} 1 & \cdots & Z^{V-1} \\ X_1 - Z^{-1}X_2 & \cdots & X_V - Z^{-1}X_{V+1} \\ \vdots & \vdots & \vdots \\ X_{V-1} - Z^{-1}X_V & \cdots & X_{N-1} - Z^{-1}X_N \\ X_N^* - Z^{-1}X_{N-1}^* & \cdots & X_V^* - Z^{-1}X_{V-1}^* \\ \vdots & \vdots & \vdots \\ X_{V+1}^* - Z^{-1}X_V^* & \cdots & X_2^* - Z^{-1}X_1^* \end{bmatrix} \times \begin{bmatrix} e^{j\varphi_1} \\ e^{j\varphi_2} \\ \vdots \\ e^{j\varphi_V} \end{bmatrix} = \begin{bmatrix} C \\ 0 \\ \vdots \\ 0 \end{bmatrix} \tag{9.35}$$

or equivalently

$$[FB][W] = [Y]. \tag{9.36}$$

9.3 SIMULATION RESULTS

As a first example consider a signal of unit amplitude arriving from $\theta_s = 100°$. We consider an 11-element array with an element spacing of $\lambda/2$ as shown in Figure 9.1. The magnitude of the incident signal is varied from 1 V/m to 10.0 V/m in the steps of 0.1 V/m while maintaining the jammer intensities constant, which are arriving from 70° and 110°. The amplitude of the signal of interest is increased by steps of 0.1 V/m from snapshot to snapshot. So the signal intensity is changing from snapshot to snapshot. This amounts to 100 snapshots. Signal-to-thermal noise ratio at each antenna element is set at 30 dB. All signal intensities and directions of arrival are summarized in Table 9.1.

Table 9.1. Parameters of the Incident Signals.

	Magnitude	Phase	DOA
Signal	1.0 - 10.0 V/m	0.0°	100°
Jammer #1	1.0 V/m	0.0°	70°
Jammer #2	1.0 V/m	0.0°	110°

The amplitude of the SOI is changed from snapshot to snapshot, however, the intensity of the jammer remains fixed. Here, we assume that we know the DOA of the signal but need to estimate its complex amplitude. In addition, we do not know the complex amplitudes or the DOA of the interferers nor any probabilistic description of the thermal noise. If the jammers have been nulled correctly and the signal is recovered properly, it is expected that the recovered signal will have a linear relationship with respect to the intensity of the incident signal for each snapshot. Figures of 9.2 plot the results of using phase-only adaptive weights using the D3LS presented in section 9.2. The amplitude and the phase of the recovered signal are shown in Figs. 9.2a and 9.2b, respectively. The output signal-to-interference-plus-noise ratio (SINR) for the forward, backward and the forward-backward methods are plotted in Fig. 9.2c. As can be seen, this amplitude displays the expected linear relationship and the estimated phase varies within a very small value. The beam pattern associated with this example is shown in Figure 9.3. The adapted beam pattern for the forward method is shown in Fig. 9.3a, for the backward method in Fig. 9.3b, and for the forward-backward method in Fig. 9.3c. The nulls are deep and occur along the correct directions. In Figure 9.3d, we show the comparison of beam patterns between the phase-only adaptive and the conventional adaptive algorithms described in section 6.2. As seen in this figure, the conventional adaptive algorithms using both the amplitude and phase as variables, produce deeper nulls in the beam pattern at the location of jammers. So the output signal-to-interference plus noise ratio is higher in the conventional case. But for the present phase-only algorithm, the nulls are also deep and occur along the correct directions of the interferers. We can also estimate correctly the magnitude and phase information of the SOI using this phase-only adaptive algorithm.

As a second example consider a signal of unit amplitude arriving from $\theta_s = 100°$. We consider a 13-element array with a spacing of $\lambda/2$ as shown in Figure 9.1. And we use three jammers, which is close to the DOA of the SOI. These jammers are arriving from 80°, 105°, and 120°. An interferer is close to the SOI. All the signal intensities and their DOA are summarized in Table 9.2. Figure 9.4 shows the results of thc adapted beam patterns using the three phase-only adaptive methods based on the D3LS approach. All the three methods, the forward method, the backward method, and the forward-backward method, produce similar results. The adapted beam produce deep pattern nulls at the locations of the interferers and the SOI has been recovered correctly at 0 dB.

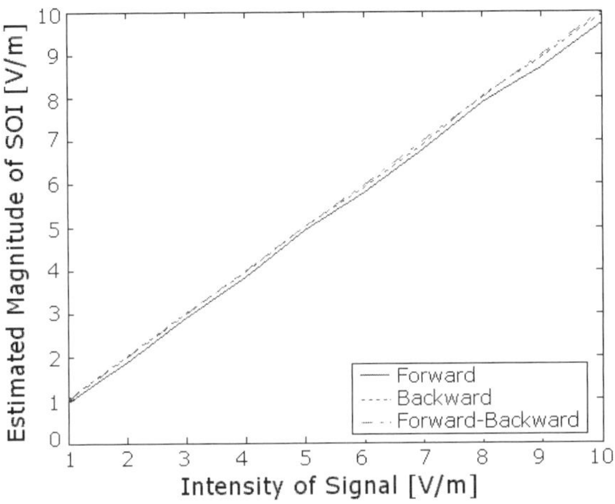

Figure 9.2a. Estimated amplitude in volts of the SOI in the presence of jammers and thermal noise.

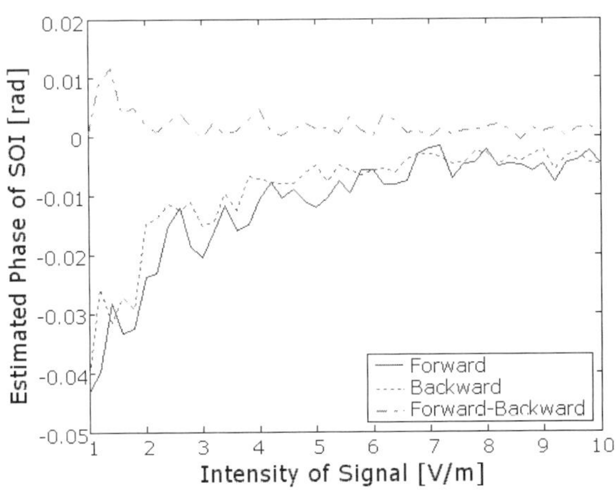

Figure 9.2b. Estimated phase in degrees of the SOI in the presence of jammers and noise.

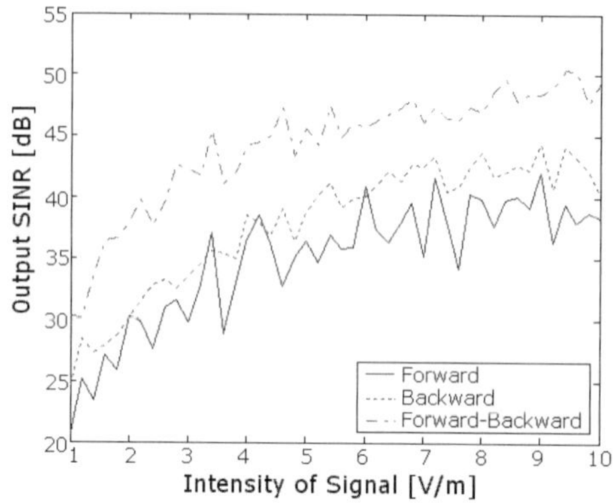

Figure 9.2c. Output signal-to-interference plus noise ratio in dB in the presence of jammers and noise.

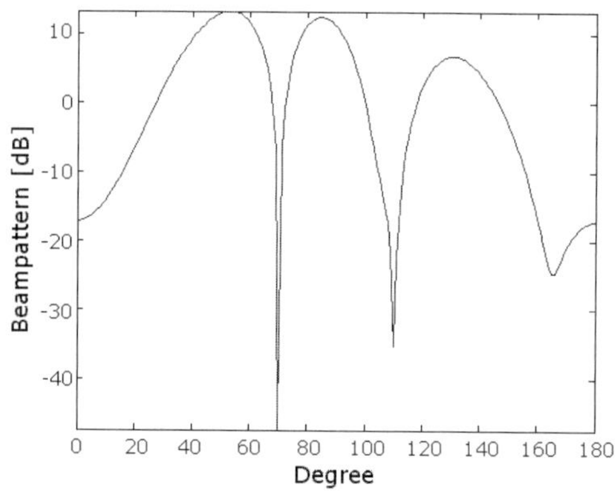

Figure 9.3a. Adaptive Beam Pattern in the presence of jammers and noise using the forward method.

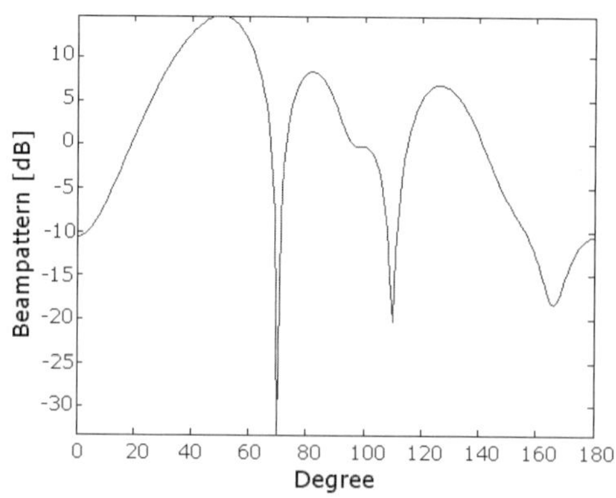

Figure 9.3b. Adaptive Beam Pattern in the presence of jammers and noise using the backward method.

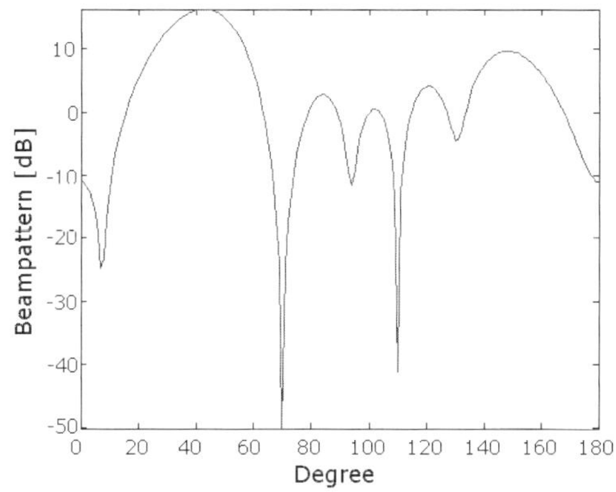

Figure 9.3c. Adaptive Beam Pattern in the presence of jammers and noise using the forward-backward method.

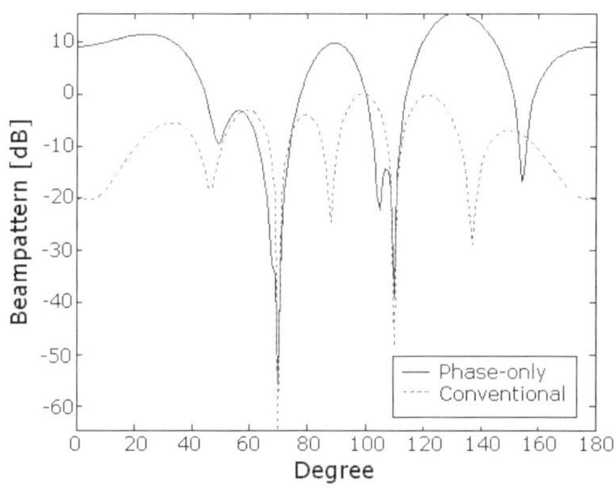

Figure 9.3d. Comparison of the beam pattern produced by the phase-only and the conventional adaptive algorithm.

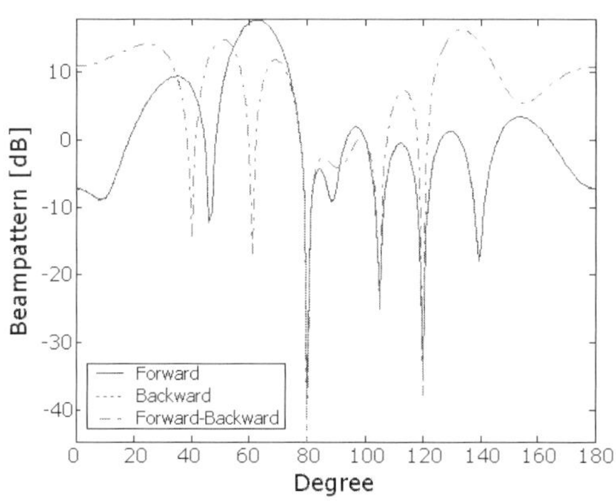

Figure 9.4. Adaptive Beam Pattern in the presence of jammers located close to the SOI.

Table 9.2. Parameters of the Incident Signals.

	Magnitude	Phase	DOA
Signal	1.0 V/m	0.0	100°
Jammer #1	1.0 V/m	0.0	80°
Jammer #2	1.0 V/m	0.0	105°
Jammer #3	1.0 V/m	0.0	120°

For the third example, we use three jammers. One of the jammer is very close to the DOA of the signal of interest. In this example, the jammer intensities are constant which are arriving from 85°, 102°, and 120°. All the signals intensities and DOA are summarized in Table 9.3. We can see the adapted beam pattern in Figure 9.5 for all the three methods. As seen in the figure, the pattern nulls are deep and occur along the correct directions to null out the interferers.

Table 9.3. Parameters of the Incident Signals.

	Magnitude	Phase	DOA
Signal	1.0 V/m	0.0	100°
Jammer #1	1.0 V/m	0.0	85°
Jammer #2	1.0 V/m	0.0	102°
Jammer #3	1.0 V/m	0.0	120°

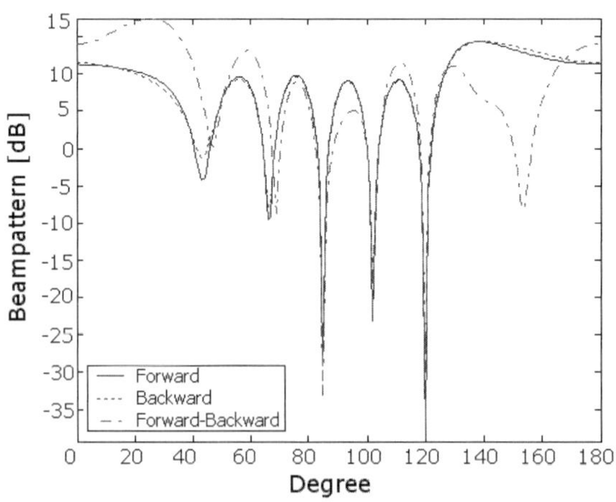

Figure 9.5. Adaptive Beam Pattern in the presence of jammers very close to the SOI.

For the final example we consider the convergence properties of the phase-only adaptive method. The scenario used in this example deals with three different arrays with different number of antenna elements which are 21, 31, and 41 with the SOI arriving along 95° and the two interferers are incident from 60°and 110°. All signals intensities and DOA are summarized in Table 9.4. As shown in Figure 9.6, we observe that an increase in the number of antenna elements in the array increases, the rate of convergence of the phase-only adaptive system decreases and it becomes little slower as expected, but this decrease is not too much. In this figure we plot the logarithm to the base 10 of the $\mathcal{L}_2$ norm of the errors. We also compare the CPU-time of this simulation using the three different methods, namely the forward, backward, and forward-backward method. These results are summarized in Table 9.5.

Table 9.4. Parameters for the SOI and Interference.

	Magnitude	Phase	DOA
Signal	1.0 V/m	0.0	95°
Jammer #1	1.0 V/m	0.0	60°
Jammer #2	1.0 V/m	0.0	110°

Table 9.5. CPU-Time Comparison.

# of Elements	Forward	Backward	Forward-Backward
11	0.2700 (sec)	0.2600 (sec)	0.2900 (sec)
21	0.3640 (sec)	0.3310 (sec)	0.6500 (sec)
31	0.9100 (sec)	0.8510 (sec)	1.2420 (sec)
41	1.4920 (sec)	1.4620 (sec)	2.0030 (sec)

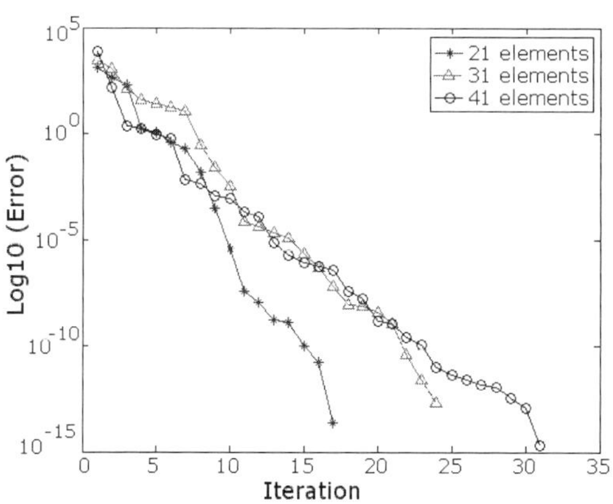

Figure 9.6. Convergence rate of the adaptive algorithm in terms of number of array.

9.4 FORMULATION OF A PHASE-ONLY DIRECT DATA DOMAIN LEAST SQUARES SPACE-TIME ADAPTIVE PROCESSING

9.4.1 Forward Method

We assume that the signals entering the array are narrowband and consist of the SOI and interferences plus noise. We assume that for each jammer, the energy impinging on the array is confined to a particular DOA and is spread in frequency. The jammers may be blinking or stationary. From the data cube shown in Figure 6.4, we focus our attention to the range cell r and consider the space-time snapshot for this range cell.

In the D3LS procedures as described in section 6.3 for the STAP problem, the adaptive weights are applied to the single space-time snapshot for the range cell r. Here a two-dimensional array of weights numbering $N_a\ N_t$ is used to extract the SOI for the range cell r. Hence the weights are defined by $w(p; q; r)$ for $p = 1, ..., N_t < P$ and $q = 1, ..., N_a < Q$ and are used to extract the SOI at the range cell r. Therefore, for the D3LS method we essentially perform a high resolution filtering in two dimensions (space and time) for each range cell.

The window size along the element dimension is N_a, and N_t along the pulse dimension. Selection of N_a determines the number of spatial degrees of freedom, while N_t determines the temporal degrees of freedom. Typically for a single domain processing, N_a and N_t must satisfy (6.53) and (6.54). In conventional D3LS STAP algorithm, the maximum number of N_a and N_t is $(Q + 1)/2$ and $(P + 1)/2$. So, more degrees of freedom can be used in a Phase-only D3LS STAP algorithm. The resulting matrix equation is then given by a equation similar to (6.75). In solving this equation one obtains the weight vector $[W]$, which places space-time nulls in the direction of the interferers while maintaining gain in the direction of the SOI. The amplitude of the SOI can be estimated using (6.76). However, the weights are of constant amplitude.

9.4.2 Backward Method

Next we reformulate the problem using the same data to obtain a second independent estimate for the solution as described in section 6.3.3. This is achieved by reversing the data sequence and then complex conjugating each term of that sequence. Therefore, if we now conjugate the data and form the reverse sequence, then one gets an independent set of equations similar to (6.84). Using equations (6.85)–(6.87) the SOI is removed from the windowed data. Once we get the weights by solving a system of equations similar to (6.84), the strength of the desired signal at range cell r is estimated from (6.88).

9.4.3 Forward-Backward Method

Finally, in this section we combine both the forward and the backward method to double the given data and thereby increase the number of weights or the degrees of freedom over that of either the forward or the backward method as illustrated in 6.3.4. In the forward-backward model we double the amount of data by not only considering the data in the forward direction but also conjugating it and reversing the direction of increment of the independent variable. This is always true for the adaptive array case. So by considering the data set $X(n)$ and $X^*(-n)$ we have essentially doubled the amount of data without any penalty, as these two data sets for our problem are linearly independent. The equation to be solved in this case is similar to (6.89).

9.5 SIMULATION RESULTS

As an example we simulate a phase-only D3LS STAP algorithm. We consider a signal arriving from Azimuth angle (φ_s) is 65° and the Doppler frequency (f_d) is 1300 Hz in this simulation. We used 11 elements array ($Q = 11$) spacing of $\lambda/2$ and 19 pulses ($P = 19$). And we consider four discrete interferers and jammer and clutter. In this simulation we the signal is 30 dB above noise (thermal noise), and so the SNR is 30 dB. In this example the PRF (Pulse Repetition Frequency) is 4 KHz. Parameters for all the signals and interferers are summarized in Table 9.6.

Table 9.6. Parameters for the SOI and Interference.

	AOA (degree)	Doppler (Hz)	
Signal	–25°	1300	SNR = 30dB
Discrete Interferers	–5°, 50°, 5°, 30°, 10°, –25°	–400, 200, 325, –1650, –950, –1200	SDIR = –9dB
Clutter	18°: 0.2° : 31°	–923: –5 : –1073	SCR = –13dB

Here, we assume that we know the DOA of the signal but need to estimate its complex amplitude. If the jammers have been nulled correctly then the signal will be recovered properly. We can see the input signal spectrum (signal of interest and interferers) in Figure 9.7. Here 0° is assumed to be the broadside direction of the array. Figures 9.8a–9.8c plot the results of using phase-only STAP method based on the direct data domain approach. Results are shown for the three different methods (forward method, backward method, and forward-backward method). In this figure, the '+' mark denotes the SOI, the small 'o' marks denote discrete interferers, and the large 'o' mark shows the location of clutter. For the forward and the backward method, the values for the weights in space and time are $N_a = 6$, $N_t = 10$. For the forward-backward method, the values for the weights in space and time are $N_a = 7$, $N_t = 13$.

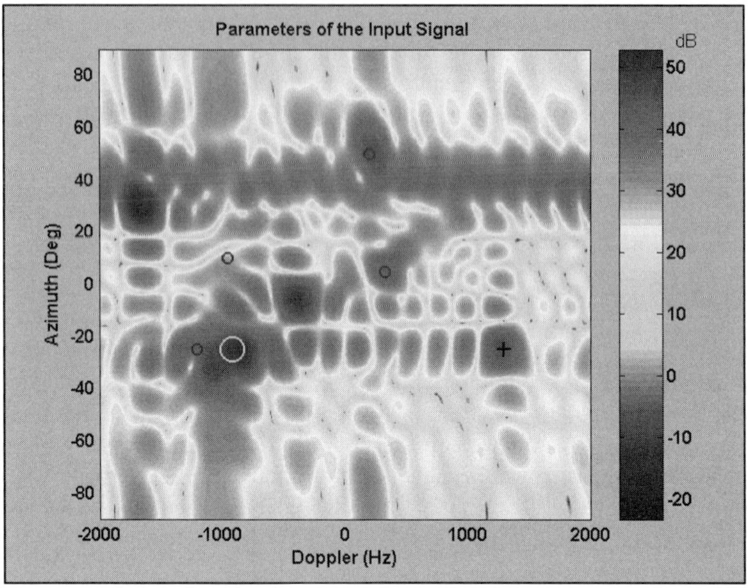

Figure 9.7. Input signal spectrum of simulation. Here 0° is assumed to be the broadside direction of the array.

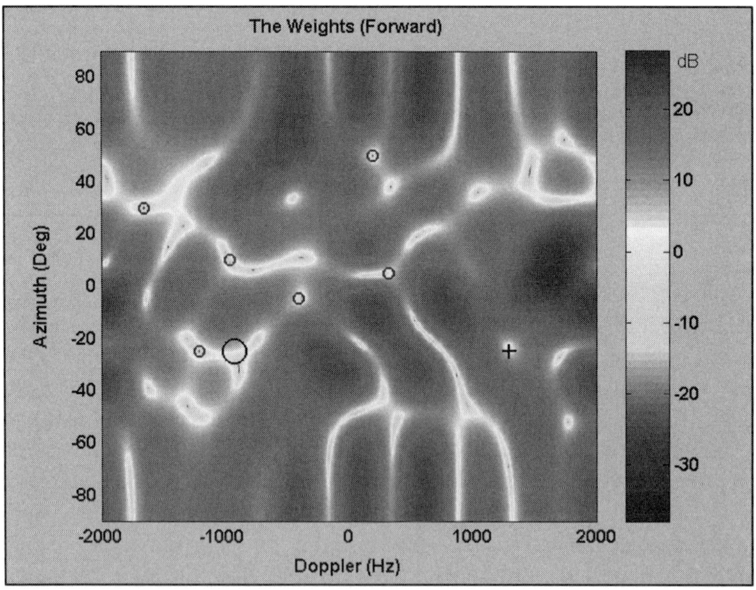

Figure 9.8a. Spectrum of the weights using the forward method. Here 0° is assumed to be the broadside direction of the array.

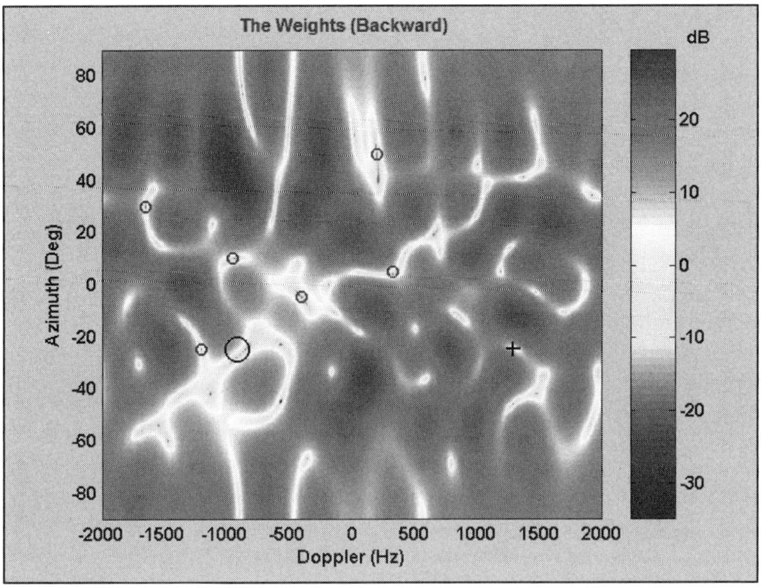

Figure 9.8b. Spectrum of the adaptive weights using the backward method. Here 0° is assumed to be the broadside direction of the array.

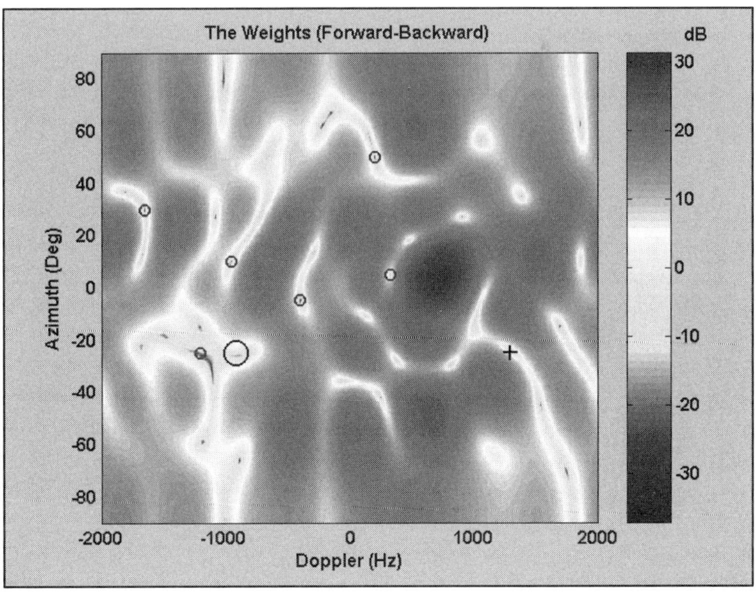

Figure 9.8c. Spectrum of the weights using the forward-backward method. Here 0° is assumed to be the broadside direction of the array.

9.6 CONCLUSION

The phase-only adaptive processing based on a direct data domain least squares approach using the Nonlinear Conjugate Gradient method has been presented in this chapter. In the proposed phase-only adaptive algorithm since no statistical methodology is employed, there is no need to compute a covariance matrix. Therefore, this procedure can be implemented on a general-purpose digital signal processor for real time implementations. As shown by the numerical examples that even in the presence of very closely spaced interferer, the phase-only adaptive processing based on a direct data domain least squares approach obtain an accurate estimate of the complex amplitude of the desired signal. Furthermore, we observe that when we increase the number of elements of the array the rate of convergence of the Nonlinear Conjugate Gradient method in solving a nonlinear equation slows down, but not too much.

REFERENCES

[1] M. K. Leavitt, "A Phase Adaptation Algorithm," *IEEE Trans. Antennas and Propagation*, Vol. 24, July 1976, pp. 754-756.

[2] C. A. Baird and G. G. Rassweiler, "Adaptive Sidelobe Nulling Using Digitally Controlled Phase-Shifters", *IEEE Trans. Antennas and Propagation*, Vol. 24, Sept. 1976, pp. 638- 649.

[3] H. Steyskal, "Simple Method for Pattern Nulling by Phase Perturbation", *IEEE Trans. Antennas and Propagation*, Vol. 31, January 1983, pp. 163- 166.

[4] S. T. Smith, "Optimum Phase-only Adaptive Nulling", *IEEE Trans. Signal processing*, Vol. 47, July 1999, pp. 1835 – 1843.

[5] R. L. Haupt, "Phase-only Adaptive Nulling with a Genetic Algorithm", *IEEE Trans. Antennas and Propagation*, Vol. 45, June 1998, pp. 1009 – 1015.

[6] D. H. Johnson and D. E. Dudgeon, *Array Signal Processing*, Prentice Hall, Englewood Cliffs, NJ, 1993.

[7] T. K. Sarkar, S. Park, J. Koh and R. A. Schneible, "A Deterministic Least Squares Approach to Adaptive Antennas", *Digital Signal Processing - A Review Journal*, Vol. 6, 185-194 (1996).

[8] W. Choi and T. K. Sarkar, "Phase-only Adaptive Processing Based on a Direct Data Domain Least Squares Approach Using the Conjugate Gradient Method", *IEEE Trans. Antennas and Propagation*, Vol. 52, No. 12, Dec. 2004, pp. 3265-3272

[9] T. K. Sarkar, H. Wang, S. Park, J. Koh, R. Adve, K. Kim, Y. Zhang, M. C. Wicks and R. D. Brown, "A Deterministic Least Squares Approach to Space Time Adaptive Processing (STAP)", *IEEE Trans. Antennas and Propagation*, Vol. 49, January 2001, pp. 91-103.

[10] T. K. Sarkar, M. Wicks, M. Salazar-Palma and R. Bonneau, *Smart Antennas*, John Wiley & Sons, Hoboken, NJ, 2003.

[11] Y. H. Dai and J. Han, "Convergence Properties of Nonlinear Conjugate Gradient Methods", *SIAM J. Optim.*, Vol. 10, No. 2, pp. 345-356.

[12] Y. H. Dai and Y. Yuan, "A Nonlinear Conjugate Gradient Method with a Strong Global Convergence Property", *SIAM J. Optim.*, Vol. 10, No. 1, pp. 177-182.

10

SIMULTANEOUS MULTIPLE ADAPTIVE BEAMFORMING

10.0 SUMMARY

This chapter presents a multiple adaptive beamforming technique based on a Direct Data Domain Least Squares (D3LS) approach. The D3LS algorithm has been proposed to deal with a single snapshot of data. Direct data domain approaches have certain advantages associated with the adaptive array signal processing problem, which adaptively analyzes the data by snapshots as opposed to forming a covariance matrix of the data from multiple snapshots, then solving for the weights utilizing that information. But conventional adaptive algorithms generally deal with a single main beam. Here we extend the D3LS approach for simultaneously generating multiple received beams. Using this new algorithm one can receive multiple signals of interest (SOI) at the same time. In addition, we present a new multiple beamforming Space-Time Adaptive Processing (STAP) based on a D3LS approach. The D3LS STAP algorithm has been proposed to deal with a single snapshot of data as conventional STAP algorithms can handle only one SOI. This new technique can handle multiple SOI by forming simultaneous multiple beams using the D3LS STAP approach.

10.1 INTRODUCTION

The principal advantage of an adaptive array is the ability to electronically steer the mainlobe of the antenna to any desired direction while also automatically placing deep pattern nulls along the specific directions of interferences. Recently, a direct data domain least squares (D3LS) algorithm has been proposed [1–2] and presented in chapter 6. The D3LS approach has certain advantages related to the computational issues associated with the adaptive array processing problem as it analyzes the data for each snapshot as opposed to forming a covariance matrix of the data using multiple snapshots, and then solving for the weights utilizing that information. Conventional adaptive algorithms based on statistical approaches require independent identically distributed secondary data to estimate the covariance matrix of the interference. The formation of the covariance matrix is quite time consuming and so is the evaluation of its inverse. Also because one requires several snapshots of the data to generate a covariance matrix, it is assumed that the environment remains stationary during that process. However, for a dynamic environment the present method may be more suitable since the direct data domain approach is based on a single snapshot of the data. Here, the term single snap shot implies that we process the space-time voltages by range cells, so that all the range cells can be simultaneously processed in parallel.

But the conventional D3LS algorithm can handle only one signal of interest (SOI) at a time. If there are two SOI, only one SOI can be handled and the other SOI can be treated as an interferer. The scenario is that even though the two SOI have the same carrier frequency, they may have different codes, so that they can be separated at the receiver. So, a new technique is presented for multiple beamforming using the D3LS approach. Using this new algorithm one can handle multiple SOI and estimate the complex amplitudes of multiple SOI using a single processing scheme.

The D3LS methodology has also been extended to deal with multiple SOI for different directions of arrival and Doppler frequencies for Space-Time Adaptive Processing (STAP) [2–4] using a single snap shot of the data over a single range cell without requiring secondary data sets as explained in section 6.3.

In section 10.2 the problem of simultaneous formation of multiple beams in an adaptive problem is described. Three different independent formations of the same procedure are presented. Generation of three independent estimates provides a level of confidence for the data. In section 10.3 some simulation results illustrating the performance of the proposed method to the adaptive problem are described. Section 10.4 discusses the issues related to simultaneous estimation of multiple targets in STAP. Here also three different independent solutions of the same mathematical problem. Generation of three independent estimates provides a level of confidence for the data. In section 10.5 some simulation results illustrate the performance of the proposed method for STAP followed by conclusions in section 10.6.

10.2 FORMULATION OF A DIRECT DATA DOMAIN APPROACH FOR MULTIPLE BEAMFORMING

10.2.1 Forward Method

Consider an array composed of $N+1$ antenna elements separated by a distance Δ as shown in Figure 10.1. We assume that narrowband signals consisting of the desired two signals plus possibly coherent multipaths and jammers. In addition, there can be strong interferers in the main beam and thermal noise. The phasor voltage X_n (for $n = 0, 1, \ldots, N$) induced at the nth antenna element at a particular instance of time will then be given by

$$X_n = \alpha_1 \, e^{\frac{j 2\pi n \Delta \cos\theta_{s1}}{\lambda}} + \alpha_2 \, e^{\frac{j 2\pi n \Delta \cos\theta_{s2}}{\lambda}} + \text{Undesired interferers} + \xi_n \tag{10.1}$$

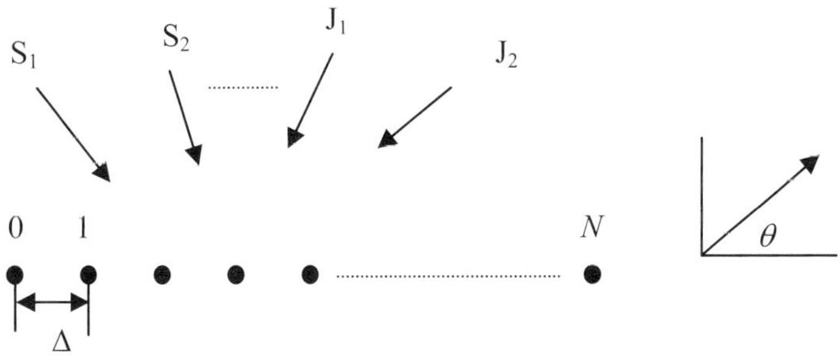

Figure 10.1 A linear uniform array.

where:

α_1	=	complex amplitude of the SOI #1 (to be determined)
α_2	=	complex amplitude of the SOI #2 (to be determined)
θ_{s1}	=	direction of arrival of the SOI #1 (known)
θ_{s2}	=	direction of arrival of the SOI #2 (known)
Δ	=	Spacing between each of the antenna elements
λ	=	wavelength of transmission (here it is assumed that we are dealing with narrowband signals) (unknown)
ξ_n	=	Thermal noise induced at the nth antenna element (unknown)

So, the goal is to estimate α_1 and α_2 simultaneously. If we define

$$Z_1 = \exp\left[\frac{j2\pi\Delta\cos\theta_{s1}}{\lambda}\right] \tag{10.2}$$

$$Z_2 = \exp\left[\frac{j2\pi\Delta\cos\theta_{s2}}{\lambda}\right] \tag{10.3}$$

then $\left(X_0 - Z_2^{-1}X_1\right) - Z_1^{-1}\left(X_1 - Z_2^{-1}X_2\right)$ contains no components of the SOI. Therefore one can form a reduced rank matrix where the weighted sum of all its elements would be zero [2]. Then, the reduced rank matrix $[T]_{(L-1)\times(L+1)}$, which contains no SOI, is formed as

$$[T]=\begin{bmatrix} \left(X_0-Z_2^{-1}X_1\right)-Z_1^{-1}\left(X_1-Z_2^{-1}X_2\right) & \cdots & \left(X_L-Z_2^{-1}X_{L+1}\right)-Z_1^{-1}\left(X_{L+1}-Z_2^{-1}X_{L+2}\right) \\ \vdots & & \vdots \\ \left(X_{N-L-2}-Z_2^{-1}X_{N-L-1}\right)-Z_1^{-1}\left(X_{N-L-1}-Z_2^{-1}X_{N-L}\right) & \cdots & \left(X_{N-2}-Z_2^{-1}X_{N-1}\right)-Z_1^{-1}\left(X_{N-1}-Z_2^{-1}X_N\right) \end{bmatrix} \tag{10.4}$$

In order to restore the signal component in this adaptive processing, we fix the gain of the subarray formed by the $L + 1$ elements along the direction θ_{s1} and θ_{s2}. Let us say that the gain of the subarray is C along the direction of θ_s. This provides an additional equation, resulting in a square matrix:

$$\begin{bmatrix} 1 & \cdots & Z_1^L \\ 1 & \cdots & Z_2^L \\ \left(X_0 - Z_2^{-1}X_1\right) - Z_1^{-1}\left(X_1 - Z_2^{-1}X_2\right) & \cdots & \left(X_L - Z_2^{-1}X_{L+1}\right) - Z_1^{-1}\left(X_{L+1} - Z_2^{-1}X_{L+2}\right) \\ \vdots & \cdots & \vdots \\ \left(X_{N-L-2} - Z_2^{-1}X_{N-L-1}\right) - Z_1^{-1}\left(X_{N-L-1} - Z_2^{-1}X_{N-L}\right) & \cdots & \left(X_{N-2} - Z_2^{-1}X_{N-1}\right) - Z_1^{-1}\left(X_{N-1} - Z_2^{-1}X_N\right) \end{bmatrix}$$

$$\times \begin{bmatrix} w_0 \\ w_1 \\ \vdots \\ w_L \end{bmatrix} = \begin{bmatrix} C \\ C \\ \vdots \\ 0 \end{bmatrix} \tag{10.5}$$

or in a matrix form

$$[F][W] = [Y] \tag{10.6}$$

Once the weights are solved by using (10.5), the complex amplitude of SOI α may be estimated from

$$\alpha = (\alpha_1 + \alpha_2) = \frac{1}{C} \sum_{i=0}^{L} W_i X_i \tag{10.7}$$

And complex amplitude α_1 for SOI #1 can be individually estimated using weights which can be found by solving the following equation,

$$[F][W_1] = [C \quad 0 \quad 0 \quad 0 \quad 0]^T \tag{10.8}$$

where superscript T denotes the transpose.

Similarly, complex amplitude α_2 for SOI #2 can be estimated using weights which can be found by solving the following equation separately,

$$[F][W_2] = [0 \quad C \quad 0 \quad \cdots \quad 0]^T \tag{10.9}$$

or from

$$[W_2] = [W] - [W_1]. \tag{10.10}$$

Alternately, they can be separated through the use of separate codes. So that

$$\alpha_1 = \frac{1}{C} \sum_{i=0}^{L} [W_1]_i X_i \tag{10.11}$$

and

$$\alpha_2 = \frac{1}{C} \sum_{i=0}^{L} [W_2]_i X_i \tag{10.12}$$

For the solution of $[F][W]=[Y]$ in (10.6), the conjugate gradient method is used as illustrated in section 6.2.5 [1–3].

10.2.2 Backward Method

Next we reformulate the problem using the same data to obtain a second independent estimate for the solution. This is achieved by reversing the data sequence and then complex conjugating each term of that sequence. It is well known in the parametric spectral estimation literature that a sampled sequence, which can be represented by a sum of exponentials with purely imaginary argument, can be used in either the forward or in the reverse direction, resulting in the same values for the exponent fitting that sequence. Backward matrix equation can be written as

$$\begin{bmatrix} 1 & \cdots & Z_1^L \\ 1 & \cdots & Z_2^L \\ \left(X_N^*-Z_2^{-1}X_{N-1}^*\right)-Z_1^{-1}\left(X_{N-1}^*-Z_2^{-1}X_{N-2}^*\right) & \cdots & \left(X_{N-L}^*-Z_2^{-1}X_{N-L-1}^*\right)-Z_1^{-1}\left(X_{N-L-1}^*-Z_2^{-1}X_{N-L-2}^*\right) \\ \vdots & & \vdots \\ \left(X_{L+2}^*-Z_2^{-1}X_{L+1}^*\right)-Z_1^{-1}\left(X_{L+1}^*-Z_2^{-1}X_L^*\right) & \cdots & \left(X_2^*-Z_2^{-1}X_1^*\right)-Z_1^{-1}\left(X_1^*-Z_2^{-1}X_0^*\right) \end{bmatrix} \times \begin{bmatrix} w_0 \\ w_1 \\ \vdots \\ w_L \end{bmatrix} = \begin{bmatrix} C \\ C \\ \vdots \\ 0 \end{bmatrix} \tag{10.13}$$

where the superscript * denotes the complex conjugate. Or in a matrix form

$$[B][W]=[Y] \tag{10.14}$$

And complex amplitude α_1 for SOI #1 can be individually estimated using weights which can be found by the following equation,

$$[B][W_1]=[C \quad 0 \quad 0 \quad \cdots \quad 0]^T. \tag{10.15}$$

Similarly, complex amplitude α_2 for SOI #2 can be estimated using weights which can be found by the following equation,

$$[B][W_2]=[0 \quad C \quad 0 \quad \cdots \quad 0]^T. \tag{10.16}$$

Or through appropriate coding or through

$$[W_2]=[W]-[W_1]. \tag{10.17}$$

10.2.3 Forward-Backward Method

Finally, in this section both the forward and the backward methods are combined to double the given data and thereby increase the number of weights or the degrees of freedom significantly over that of either the forward or the backward method. This provides the third independent solution. In the forward-backward model we double the amount of data by not only considering the data in the forward direction but also conjugating it and reversing the direction of increment of the independent variable. This type of processing can be done as long as the series to be approximated can be fitted by exponential functions of purely imaginary arguments. This is always true for the adaptive array case.

An additional benefit accrues in this case. For both the forward and the backward method, the maximum number of the value L we can consider is given by $N/2$, where $N + 1$ is the number of antenna elements. For the forward-backward method, the number of degrees of freedom can be significantly increased without increasing the number of antenna elements. For this case, the number of degrees of freedom can reach $(N/1.5) + 1$.

$$[FB][W] = [Y] \tag{10.18}$$

10.3 SIMULATION RESULTS

For the first simulation, consider an 11-element array with an element spacing of $\lambda/2$ as shown in Figure 10.1. Two signals of interest that are arriving from 60° and 110° are chosen. And four strong interferers were considered in this simulation. Signal-to-noise ratio is set at each antenna element to be 20 dB. All of the signal intensities and their DOA are summarized in Table 10.1.

Table 10.1. Parameters for the SOI and Interference.

	Magnitude	Phase	DOA
SOI #1	1.0 V/m	0.0	60°
SOI #2	2.0 V/m	0.0	110°
Jammer #1	1000.0 V/m	0.0	40°
Jammer #2	1000.0 V/m	0.0	80°
Jammer #3	1000.0 V/m	0.0	95°
Jammer #4	1000.0 V/m	0.0	150°

Results are shown for all three different methods (forward method, backward method and forward-backward method) in Fig. 10.2. The number of weights is 6 for either the forward or the backward methods and it is 7 for the forward-backward method. Figure 10.2a plots the beam pattern for the forward method, Figure 10.2b for the backward method and Figure 10.2c for the forward-backward method. As expected, the nulls are deep and occur along the correct directions. And also the two SOI are recovered properly along the correct angles. The estimated complex amplitude of the SOI #1 (α_1) and the SOI #2 (α_2) is summarized in Table 10.2.

Table 10.2. Estimated Complex Amplitude of SOI.

	Forward	Backward	Forward-Backward
Estimated α_1	$1.01 + j\,0.09$	$1.01 + j\,0.09$	$1.01 + j\,0.06$
Estimated α_2	$2.07 - j\,0.10$	$2.07 - j\,0.10$	$2.03 - j\,0.09$

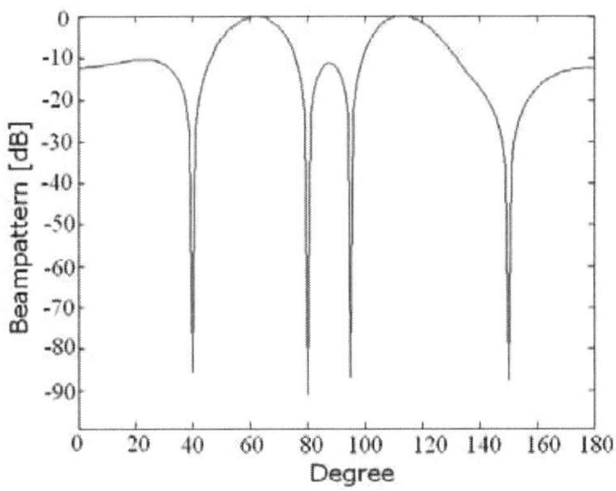

Figure 10.2a. Adaptive beam pattern in the presence of jammers and thermal noise using the forward method.

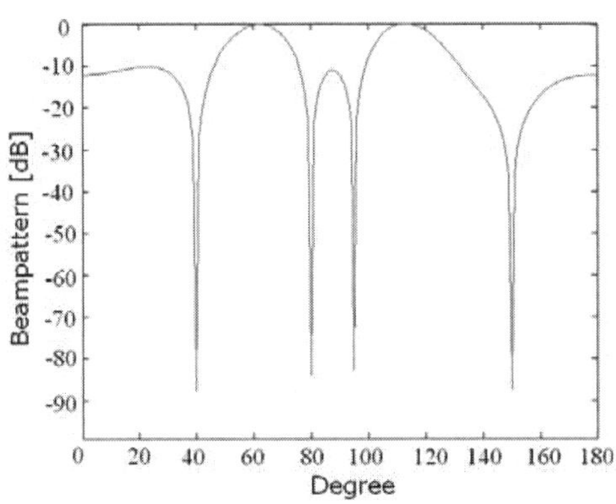

Figure 10.2b. Adaptive beam pattern in the presence of jammers and thermal noise using the backward method.

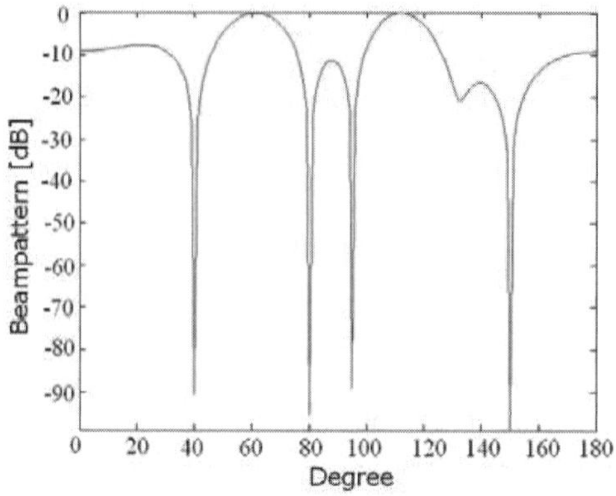

Figure 10.2c. Adaptive beam pattern in the presence of jammers and thermal noise using the forward-backward method.

For the second example, the angles of arrival of two SOI are chosen to be close to each other. An 11-element array with an element spacing of $\lambda/2$ is considered. The two SOI are arriving from 75° and 95°. And four strong interferers are chosen in this simulation. One of the interferer is located between the two SOI. Signal-to-noise ratio is set at each antenna element to be 20 dB. Since the beamwidth for this array will be approximately $60°/L_\lambda \cong 60°/5 = 12°$, one of the interferers and the two SOI's are in the periphery of the main beam. All the signal intensities and their directions of arrival are summarized in Table 10.3.

Table 10.3. Parameters for the SOI and Interference.

	Magnitude	Phase	DOA
SOI #1	1.0 V/m	0.0	75°
SOI #2	2.0 V/m	0.0	95°
Jammer #1	1000.0 V/m	0.0	50°
Jammer #2	1000.0 V/m	0.0	85°
Jammer #3	1000.0 V/m	0.0	120°
Jammer #4	1000.0 V/m	0.0	150°

Results are shown for all the three different methods (forward method, backward method, and forward-backward method). The number of weights is also 6 for either the forward or the backward methods and it is 7 for the forward-backward method. Figure 10.3a plots the beam pattern for the forward method, Figure 10.3b for the backward method and Figure 10.3c for the forward-backward method. As expected, the nulls are deep and occur along the correct directions and the SOI are recovered at 0 dB.

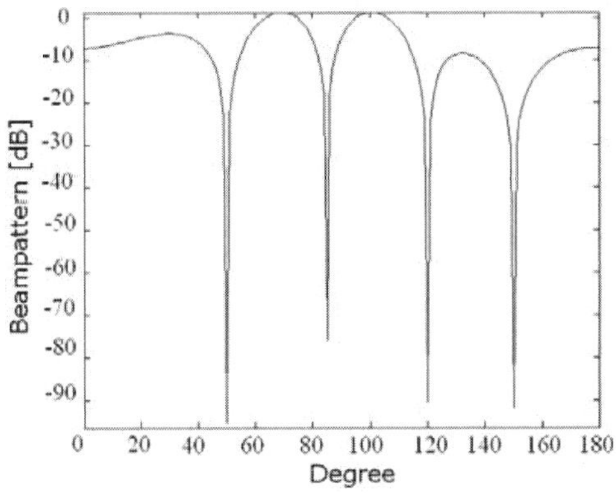

Figure 10.3a. Adaptive beam pattern in the presence of jammers and thermal noise using the forward method.

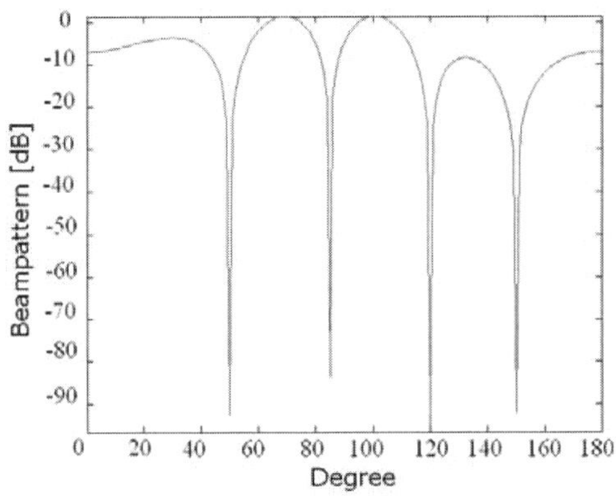

Figure 10.3b. Adaptive beam pattern in the presence of jammers and thermal noise using the backward method.

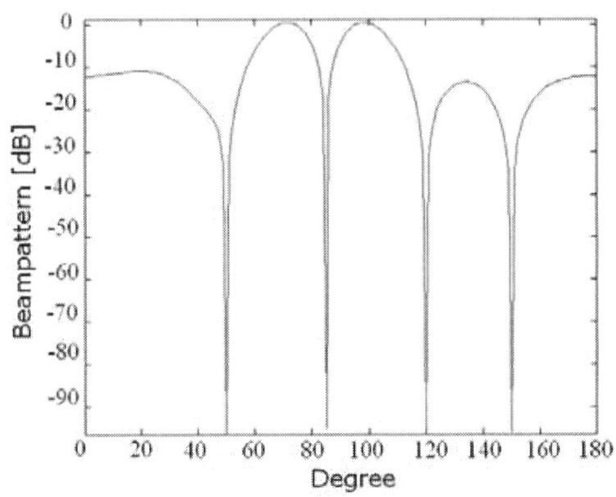

Figure 10.3c. Adaptive beam pattern in the presence of jammers and thermal noise using the forward-backward method

As seen from the computations, the two SOI are recovered properly along with their correct angles of arrival. The estimated complex amplitude of the SOI #1 (α_1) and the SOI #2 (α_2) is summarized in Table 10.4.

Table 10.4. Estimated Complex Amplitude of SOI.

	Forward	Backward	Forward-Backward
Estimated α_1	$0.90 - j\,0.02$	$0.90 - j\,0.02$	$0.95 - j\,0.15$
Estimated α_2	$2.11 - j\,0.06$	$2.11 - j\,0.06$	$2.07 + j\,0.10$

10.4 FORMULATION OF A DIRECT DATA DOMAIN LEAST SQUARES APPROACH FOR MULTIPLE BEAMFORMING IN SPACE-TIME ADAPTIVE PROCESSING

10.4.1 Forward Method

Next, this procedure is applied to the STAP problem. The SOI for this STAP problem is considered to be located at range cell r and is incident on the uniform linear array from an angle φ_s and is at Doppler frequency f_s. Our goal is to estimate its complex amplitude, given φ_s and f_s only. In a surveillance radar, φ_s and f_s set the look directions and a SOI (target) may or may not be present along this look direction and Doppler. Let us define $S(p,q)$ to be the complex voltage received at the qth antenna element corresponding to the pth time instance for the same range cell r. It is further stipulated that the voltage $S(p,q)$ is due to two signals of unity magnitude incident on the array from the azimuth angle φ_{s1} corresponding to Doppler frequency f_{s1} for SOI #1 and the azimuth angle φ_{s2} corresponding to Doppler frequency f_{s2} for SOI #2. Hence, the signal-induced voltage under the assumed array geometry and a narrowband signal is a complex sinusoidal given by

$$
\begin{aligned}
S(p,q) &= \exp\left\{ j\left(\frac{2\pi\,\Delta(q-1)}{\lambda}\cos\varphi_{s1} + \frac{2\pi f_{s1}(p-1)}{f_r} \right) \right\} \\
&+ \exp\left\{ j\left(\frac{2\pi\,\Delta(q-1)}{\lambda}\cos\varphi_{s2} + \frac{2\pi f_{s2}(p-1)}{f_r} \right) \right\}
\end{aligned}
\tag{10.19}
$$

for $p=1, \ldots, P$ and $q=1, \ldots, Q$. And λ is a wavelength of the radio frequency radar signal, Δ is the spacing between each of the antenna elements and f_r is the pulse repetition frequency.

Let $X(p,q)$ be the actual measured complex voltages that are in the data cube of Figure 6.4 for the range cell r. The actual voltages X will contain the signal of interest of amplitude α (α is a complex quantity), jammers which may be due to coherent multipaths both in the mainlobe and in the sidelobes, and clutter which is the reflected electromagnetic energy from the ground. The interference competes with the SOI at the Doppler frequency of interest. There is also a contribution to the measured voltage from the receiver thermal noise. Hence the actual measured voltages $X(p,q)$ are

$$
\begin{aligned}
X(p,q) = {} & \alpha_1 \exp\left\{ j \left(\frac{2\pi\,\Delta\,(q-1)}{\lambda} \cos\varphi_{s1} + \frac{2\pi f_{s1}(p-1)}{\mathrm{f_r}} \right) \right\} + \\
& \alpha_2 \exp\left\{ j \left(\frac{2\pi\,\Delta(q-1)}{\lambda} \cos\varphi_{s2} + \frac{2\pi f_{s2}(p-1)}{f_r} \right) \right\} + \\
& + \text{Clutter} + \text{Jammer} + \text{Thermal noise}
\end{aligned}
\tag{10.20}
$$

The goal is to extract the SOI, α_1, the direction of arrival for the SOI #1, φ_{s1}, and the Doppler frequency, f_{s1}, and α_2, the direction of arrival for the SOI #2, φ_{s2}, and the Doppler frequency, f_{s2}, given the various voltages for the space-time snapshot.

In the D3LS procedures to be described, the adaptive weights are applied to the single space-time snapshot for the range cell r. Here a two-dimensional array of weights numbering N_a, N_t is used to extract the SOI from the range cell r. Hence the weights are defined by $W(m,n,r)$ for $p=1, \cdots, N_t < P$ and $q=1, \cdots, N_a < Q$ and are used to extract the SOI at the range cell r. Therefore, for the D3LS method we essentially perform a high resolution filtering in two dimensions (space and time) for each range cell. At a particular range, r, the sheet or slice of the data cube is referred to as a space-time snapshot as marked by the shaded plane in Figure 6.4.

The window size along the element dimension is N_a, and N_t along the pulse dimension. Selection of N_a determines the number of spatial degrees of freedom, while N_t determines the temporal degrees of freedom. Typically for a single domain processing, N_a and N_t must satisfy the following equations

$$N_a \le (Q+1)/2 \tag{10.21}$$

$$N_t \le (P+1)/2 \tag{10.22}$$

And the advantage of a joint domain processing is that either of these bounds can be relaxed, i.e., one can exchange spatial degrees of freedom with the temporal degrees of freedom. So, indeed it is possible to cancel a number of interferers, which is greater than the number of antenna elements in a joint domain processing. The total number of degrees of freedom, R, for any method is

$$R = N_a \times N_t \tag{10.23}$$

In the D3LS STAP algorithm we let the element to element offset of the SOI in space and time, respectively, as

$$Z_{a1} = \exp\left[j2\pi \frac{\Delta}{\lambda} \cos\varphi_{s1} \right] \tag{10.24}$$

$$Z_{t1} = \exp\left[j2\pi \frac{f_{s1}}{f_r} \right] \tag{10.25}$$

and

$$Z_{a2} = \exp\left[j2\pi \frac{\Delta}{\lambda} \cos\varphi_{s2} \right] \tag{10.26}$$

$$Z_{t2} = \exp\left[j2\pi \frac{f_{s2}}{f_r} \right] \tag{10.27}$$

And we form a reduced rank matrix $[F]$ of dimension $(N_a \times N_t - 2) \times (N_a \times N_t)$ from the elements of the matrix X. So as to obtain

$$\begin{aligned} F(x; y) = &\left\{ X(g+h-1; d+e-1) - Z_{a1}^{-1} X(g+h-1; d+e) \right\} \\ &- Z_{a2}^{-1} \left\{ X(g+h-1; d+e) - Z_{a1}^{-1} X(g+h-1; d+e+1) \right\} \end{aligned} \tag{10.28}$$

$$\begin{aligned} F(x+1; y) = &\left\{ X(g+h-1; d+e-1) - Z_{t1}^{-1} X(g+h; d+e-1) \right\} \\ &- Z_{t2}^{-1} \left\{ X(g+h; d+e-1) - Z_{t1}^{-1} X(g+h+1; d+e-1) \right\} \end{aligned} \tag{10.29}$$

$$\begin{aligned} F(x+2; y) = &\left\{ X(g+h-1; d+e-1) - Z_{a1}^{-1} Z_{t1}^{-1} X(g+h; d+e) \right\} \\ &- Z_{a2}^{-1} Z_{t2}^{-1} \left\{ X(g+h; d+e) - Z_{a1}^{-1} Z_{t1}^{-1} X(g+h+1; d+e+1) \right\} \end{aligned} \tag{10.30}$$

where

$$1 \le x = g + (d-1) N_t \tag{10.31}$$

$$1 \leq y = h + (e-1)\ N_t \leq N_a N_t \tag{10.32}$$

$$1 \leq d \leq Q - N_a \tag{10.33}$$

$$1 \leq e < N_a \tag{10.34}$$

$$1 \leq g \leq P - N_t \tag{10.35}$$

$$1 \leq h \leq N_t \tag{10.36}$$

so that $1 \leq x, y \leq N_a N_t$.

Note that in (10.28), the signal component (SOI) is canceled from samples taken from different antenna elements at the same time. Similarly (10.29) represents signal cancellation from samples taken at the same antenna elements at different time instances. Finally, (10.30) represent signal cancellation from neighboring samples in both space and time. Therefore, we are performing a filtering operation simultaneously using $N_a N_t$ samples of the space-time data. The cancellation rows of the matrix $[F]$ can now be formed using (10.28)–(10.30). This is similar to windowing of the data in Fig. 6.5. In this case the dots in Figure 6.5 represent the induced voltages, $X(p,q)$ as defined in (10.20), for a given element-pulse location. For each given location of the window function as illustrated in Fig. 6.5, 3 rows in matrix $[F]$ are formed by implementing (10.28) – (10.30), which remove the SOI. The rows are formed by performing an element by element subtraction between the elements of the windows and then arranging the resulting data into a row vector. The window is then slided one space to the right and 3 more rows are generated, and so on. After this window has reached the second column to the far right and 3 additional rows are generated as seen in Fig. 6.5, the window is lowered a row and shifted back to the left side of the data array, and the generation of rows continues. This is repeated until $(R-2)$ cancellation rows have been formed. The elements of this row can be obtained by placing a $N_a \times N_t$ window, such as window #1 as shown in Figure 6.5, over data. In order to restore the signal component in the adaptive processing, we fix the gain of the subarray (in both space and time) formed by fixing the first row of the matrix $[F]$. The elements of the first row are given by

$$F(1;y) = Z_{a1}^{(e-1)}\ Z_{t1}^{(h-1)} \tag{10.37}$$

$$F(2;y) = Z_{a2}^{(e-1)}\ Z_{t2}^{(h-1)} \tag{10.38}$$

where y, e and h are given by equations (10.31)–(10.36). So, the elements of the first and second rows are given by

$$\Big[1\ \ Z_{a1}\ \ Z_{a1}^2\ \ \ldots\ \ Z_{a1}^{N_a-1}\ \ Z_{t1}\ \ Z_{a1}Z_{t1}\ \ Z_{a1}^2 Z_{t1}\ \ \ldots\ \ Z_{a1}^{N_a-1}Z_{t1}\ \ \ldots\ \ Z_{t1}^2\ \ Z_{a1}Z_{t1}^2\ \ \ldots\ \ Z_{a1}^{N_a-1}Z_{t1}^{N_t-} \tag{10.39}$$

and

$$\begin{bmatrix} 1 & Z_{a2} & Z_{a2}^2 & \ldots & Z_{a2}^{N_a-1} & Z_{t2} & Z_{a2}Z_{t2} & Z_{a2}^2 Z_{t2} & \ldots & Z_{a2}^{N_a-1}Z_{t2} & \ldots & Z_{t2}^2 & Z_{a2}Z_{t2}^2 & \ldots & Z_{a2}^{N_a-1}Z_{t2}^{N_t-1} \end{bmatrix} \tag{10.40}$$

respectively.

The resulting matrix equation is then assembled as

$$[F][W] = [Y] = \begin{bmatrix} C & C & 0 & \cdots & 0 \end{bmatrix}^T \tag{10.41}$$

where C is a complex constant and the superscript T denotes the transpose of a matrix. In solving this equation one obtains the weight vector $[W]$, which places space-time nulls in the direction of the interferers while maintaining gain along the directions of the SOI. The amplitude of the SOI can be estimated using

$$\alpha = (\alpha_1 + \alpha_2) \approx \frac{1}{C} \sum_{e=1}^{N_a} \sum_{h=1}^{N_t} W\{N_t(e-1)+h\}\ X(h;e) \tag{10.42}$$

and the complex amplitude α_1 for SOI #1 can be individually estimated using weights which can be found by solving the following equation,

$$[F][W_1] = \begin{bmatrix} C & 0 & 0 & \cdots & 0 \end{bmatrix}^T \tag{10.43}$$

where superscript T denotes the transpose.

Similarly, complex amplitude α_2 for SOI #2 can be estimated using weights which can be found by solving the following equation separately,

$$[F][W_2] = \begin{bmatrix} 0 & C & 0 & \cdots & 0 \end{bmatrix}^T \tag{10.44}$$

or from

$$[W_2] = [W] - [W_1] \tag{10.45}$$

Alternately, they can be separated through the use of codes. So that

$$\alpha_1 = \frac{1}{C} \sum_{e=1}^{N_a} \sum_{h=1}^{N_t} W_1\{N_t(e-1)+h\} X(h;e) \tag{10.46}$$

and

$$\alpha_2 = \frac{1}{C} \sum_{e=1}^{N_a} \sum_{h=1}^{N_t} W_2\{N_t(e-1)+h\} X(h;e) \tag{10.47}$$

10.4.2 Backward Method

Next, the problem using the same data to obtain a second independent estimate for the solution is described. This is achieved by reversing the data sequence and then complex conjugating each term of that sequence. It is well known in the parametric spectral estimation literature that a sampled sequence, which can be represented by a sum of exponentials with purely imaginary argument, can be used in either the forward or in the reverse direction, resulting in the same values for the exponent fitting that sequence.

Therefore, if the data is conjugated and a reverse sequence is formed, then one gets an independent set of equations similar to (10.28)–(10.30). This is represented by

$$\begin{aligned} B(x;y) = &\left\{ X^*(M-g-h+2;\, N-d-e+2) - Z_{a1}^{-1} X^*(M-g-h+2;\, N-d-e+1) \right\} \\ &- Z_{a2}^{-1} \left\{ X^*(M-g-h+2;\, N-d-e+1) - Z_{a1}^{-1} X^*(M-g-h+2;\, N-d-e) \right\} \end{aligned} \tag{10.48}$$

$$\begin{aligned} B(x+1;y) = &\left\{ X^*(M-g-h+2;\, N-d-e+2) - Z_{t1}^{-1} X^*(M-g-h+1;\, N-d-e+2) \right\} \\ &- Z_{t2}^{-1} \left\{ X^*(M-g-h+1;\, N-d-e+2) - Z_{t1}^{-1} X^*(M-g-h;\, N-d-e+2) \right\} \end{aligned} \tag{10.49}$$

$$\begin{aligned} B(x+2;y) = &\left\{ X^*(M-g-h+2;\, N-d-e+2) - Z_{a1}^{-1} Z_{t1}^{-1} X^*(M-g-h+1;\, N-d-e+1) \right\} \\ &- Z_{a2}^{-1} Z_{t2}^{-1} \left\{ X^*(M-g-h+1;\, N-d-e+1) - Z_{a1}^{-1} Z_{t1}^{-1} X^*(M-g-h;\, N-d-e) \right\} \end{aligned} \tag{10.50}$$

where the superscript * denotes the complex conjugate.

Using equations (10.48)–(10.50) the SOI are removed from the windowed data. And the resulting matrix equation is then given by

$$[B][W] = [Y] = \begin{bmatrix} C & C & 0 & \cdots & 0 \end{bmatrix}^T \tag{10.51}$$

and the complex amplitude α_1 for SOI #1 can be individually estimated using weights which can be found by following equation,

$$[B][W_1] = \begin{bmatrix} C & 0 & 0 & \cdots & 0 \end{bmatrix}^T \tag{10.52}$$

Similarly, complex amplitude α_2 for SOI #2 can be estimated using weights which can be found by following equation,

$$[B][W_2] = \begin{bmatrix} 0 & C & 0 & \cdots & 0 \end{bmatrix}^T \tag{10.53}$$

Or through appropriate coding or through

$$[W_2] = [W] - [W_1]. \tag{10.54}$$

10.4.3 Forward-Backward Method

Finally, in this section, both the forward and the backward method are combined to double the given data and thereby increase the number of weights or the degrees of freedom over that of either the forward or the backward method. In the forward-backward model we double the amount of data by not only considering the data in the forward direction but also conjugating it and reversing the direction of increment of the independent variable. This is always true for the adaptive array case. So by considering the data set $X(p,q)$ and $X^*(p,q)$ we have essentially doubled the amount of data without any penalty, as these two data sets for our problem are linearly independent.

$$[FB][W]=[Y]=\begin{bmatrix}C & C & 0 & \cdots & 0\end{bmatrix}^T \tag{10.55}$$

It is important to note that, for this single snapshot Forward-Backward method, the number of degrees of freedom are really the maximum that can be achieved by any method when the interferers are coherent.

10.5 SIMULATION RESULTS

For the first example, an array of 11 elements (Q =11) is used with an element spacing of $\lambda/2$. In addition, 19 pulses (P = 19) are considered to from the data set. Two SOI are considered along with some undesired interferers in this simulation. The azimuth angle (φ_{s1}) is 65° and Doppler frequency (f_{s1}) is 1300Hz for SOI #1 and φ_{s2} is 130° and f_{s2} is 1000 Hz for SOI #2. And we also considered four discrete interferers and jammer and clutter in this simulation. And signal-to-noise ratio is set at each antenna element to be 30 dB. In this example, the PRF (Pulse Repetition Frequency) is 4 KHz. Here clutter is modeled by discrete scatterers located every 0.1° apart in azimuth and approximately every 3.2 Hz apart in Doppler. The magnitude for the point-source clutter returns is determined by a uniformly distributed random number generator, with values distributed between 0 and 1. The phase is also determined by another uniformly distributed random number generator, with values between 0 and 2π. The signal- to-clutter ratio (SCR) is –10 dB. And the jammer is modeled as a broadband noise signal that arrives from 100° in azimuth, and covers all Doppler frequencies of interest. The signal-to-jammer ratio (SJR) is –13 dB. All of the signal intensities and their directions of arrival are summarized in Tables 10.5 and 10.6.

In this simulation, 6 weights or filter taps in space (N_a = 6) and a 10 order filter in time (N_t = 10) for the forward method and the backward method has been used. And N_a = 7 and a N_t = 13 is assumed for the forward-backward method.

Here, the DOA of the signal of interests are assumed to be known and the goal is to estimate their complex amplitudes. If the undesired interferers have been nulled correctly and the signal recovered properly, it is expected that the recovered signal will be clearly visible at each snapshot. The input signal spectrum of this simulation is shown in Figure 10.4. And Figure 10.5 plots the results of using multiple beam STAP algorithm based on the D3LS approach presented in section 10.4. Results are shown for the three different methods (forward method, backward method, and forward-backward method). In this figure '+' marks denote the SOI and small 'o' marks denote the discrete interferers and large 'o' mark shows the location of clutter. As expected, the nulls are deep and occur along the correct directions.

Table 10.5. Parameters of the SOI and Interference.

	AOA (degree)	Doppler (Hz)	
SOI #1	65°	1300	
SOI #2	130°	1000	
Discrete Interferers	145°, 115°, 40°, 35°	400, –1300, –200, 1400	SDIR = –14dB
Jammer	100°	Covers all Doppler frequencies of interest	SJR = –13dB
Clutter	62° ~ 68°	783 ~ 980	SCR = –10dB

Table 10.6. Magnitude and Phase of the SOI.

	Magnitude	Phase
SOI #1 (α_1)	80.0 V/m	0.0°
SOI #2 (α_2)	120.0 V/m	0.0°

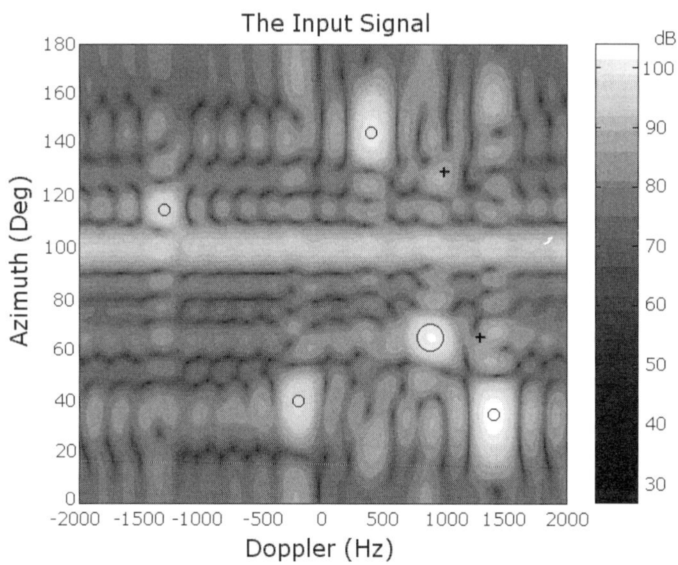

Figure 10.4. Input signal of simulation 1.

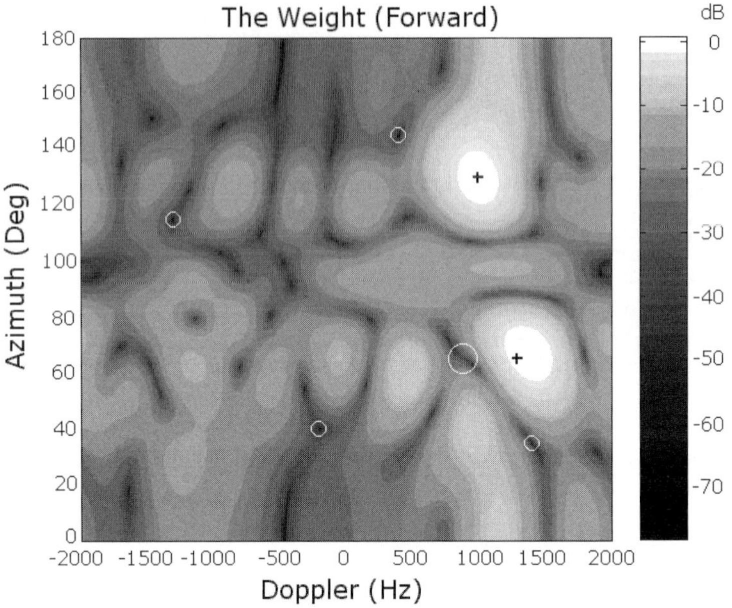

Figure 10.5a. The weight spectrum using the forward method.

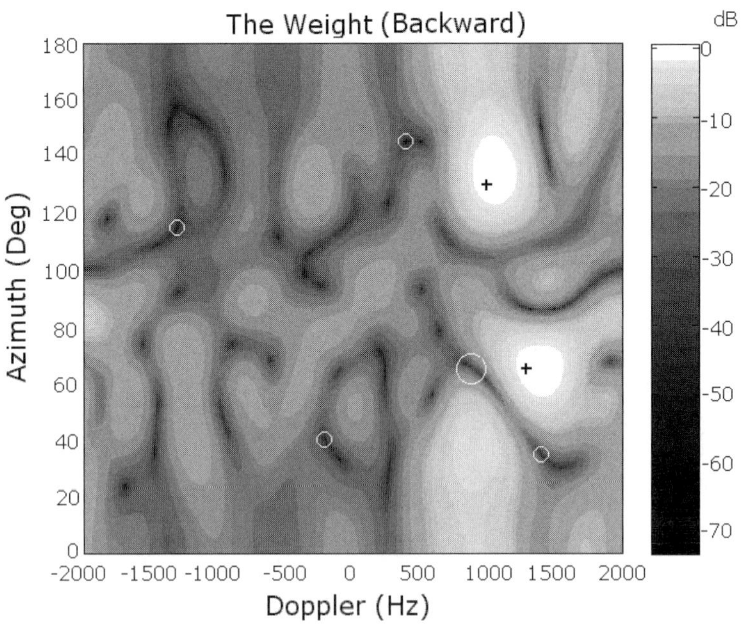

Figure 10.5b. The weight spectrum using the backward method.

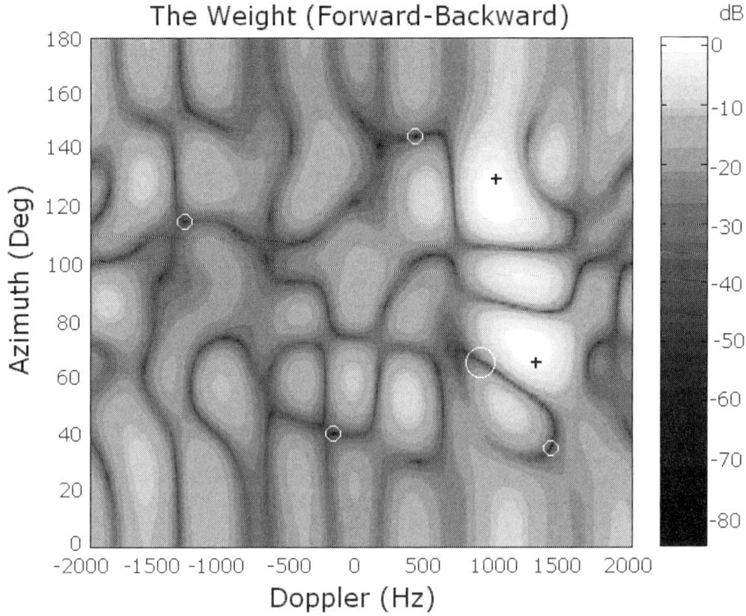

Figure 10.5c The weight spectrum using the forward-backward method

The spectrum of the polynomial containing the weights for the forward method is shown in Fig. 10.5a, for the backward method in Fig. 10.5b, and for the forward-backward method in Fig. 10.5c. Also the two SOI have been recovered properly along the correct angles and Doppler frequency as shown in all the three figures by the mark '+'. The estimated complex amplitude of the SOI #1 (α_1) and the SOI #2 (α_2) is summarized in Table 10.7 as compared to the true values given in Table 10.6.

Table 10.7. Estimated Complex Amplitude of the SOI.

	Forward	Backward	Forward-Backward
Estimated α_1	$81.17 - j\,0.05$	$81.49 + j\,0.58$	$80.58 + j\,0.27$
Estimated α_2	$119.93 - j\,0.23$	$119.78 - 0.j\,14$	$119.90 - j\,0.16$

For the second example a more complex situation is considered. We also used an 11-element array ($Q = 11$) with a spacing of $\lambda/2$ and 19 pulses ($P = 19$) in this simulation. We also considered two SOI and undesired interferers in this simulation. But in this simulation the angles of arrival of two SOI are closer than the first simulation. The azimuth angle (φ_{s1}) is 65° and Doppler frequency (f_{s1}) is 1300 Hz for SOI #1 and φ_{s2} is 60° and f_{s2} is 350 Hz for SOI #2. And we considered 10 discrete interferers and jammer and clutter. SNR is set at each antenna element to be 30 dB. In this example PRF (Pulse Repetition Frequency) is 4 KHz. The clutter is modeled as point scatterers placed every 0.1° apart in azimuth and approximately every 3.2 Hz apart in Doppler. The complex amplitudes for the point-source clutter returns are generated by two uniformly distributed random number generators as mentioned before. The signal to clutter ratio, SCR, is –10 dB. And the jammer is modeled as a broadband noise signal that arrives from 100° in azimuth, and covers all Doppler frequencies of interest. All the signal intensities and their DOA are summarized in Tables 10.8 and 10.9.

Table 10.8. Parameters of the SOI and Interference.

	AOA (degree)	Doppler (Hz)	
SOI #1	65°	1300	SNR = 30 dB
SOI #2	60°	350	
Discrete Interferers	85°,120°,40°,35° 60°,105°,100° 50°,55°,125°	400, –800, 1700, 1500 –1000, –1300, 950 –1200, –125, 1450	SDIR = –15 dB
Jammer	100°	Covers all Doppler frequencies of interest	SJR = –13 dB
Clutter	62° ~ 68°	783 ~ 980	SCR = –10 dB

Table 10.9. Magnitude and Phase of the SOI.

	Magnitude	Phase
SOI #1 (α_1)	80.0 V/m	0.0°
SOI #2 (α_2)	120.0 V/m	0.0°

In this simulation we used 6 weights or filter taps in space ($N_a = 6$) and a 10 order filter in time ($N_t = 10$) for the forward and the backward methods. And $N_a = 7$ and a $N_t = 13$ is used for the forward-backward method.

Figure 10.6 shows the locations of the input signals in this simulation. And Figure 10.7a plots the weight spectrum for the forward method, Figure 10.7b for the backward method and Figure 10.7c for the forward-backward method. Even though the angles of arrival of two SOI are closer and there are more interferers, the interferers are eliminated at the correct angles and also the two SOI's are recovered properly at the correct angles and Doppler frequencies. The estimated complex amplitude of the SOI #1 (α_1) and the SOI #2 (α_2) is summarized in Table 10.10.

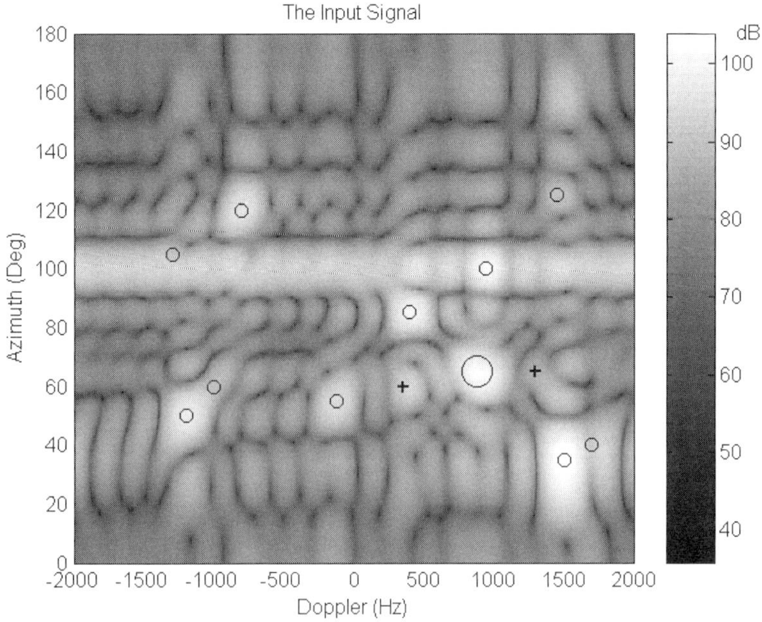

Figure 10.6. Input signal parameters of simulation 2.

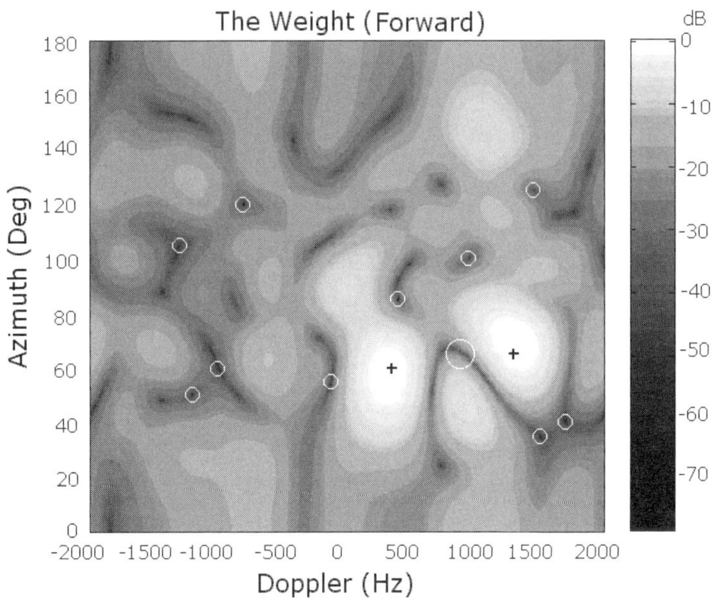

Figure 10.7a. The weight spectrum using the forward method.

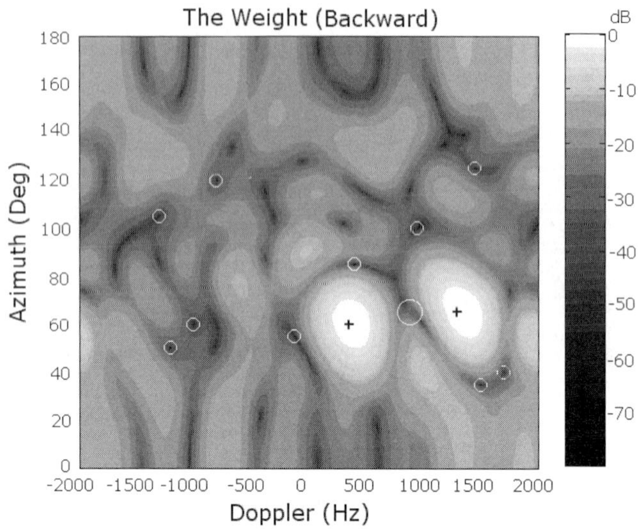

Figure 10.7b. The weight spectrum using the backward method.

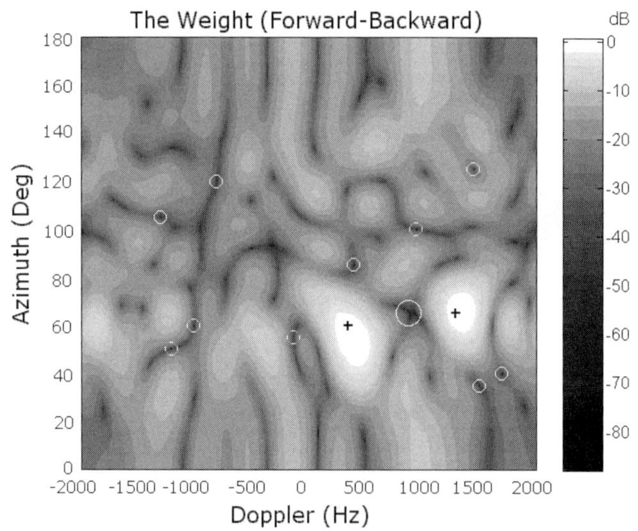

Figure 10.7c. The weight spectrum using the forward-backward method.

Table 10.10. Estimated Complex Amplitude of the SOI.

	Forward	Backward	Forward-Backward
Estimated α_1	$80.05 - j\,0.01$	$80.21 - j\,0.94$	$80.06 - j\,0.69$
Estimated α_2	$119.79 + j\,1.02$	$119.83 + j\,0.49$	$119.83 + j\,0.49$

10.6 CONCLUSION

A new multiple beamforming technique using the D3LS approach is presented. In a conventional adaptive processing, only one SOI can be handled at a time. So, when multiple SOI are presented, the other SOI can be treated as an interferer. Then it can be eliminated during the adaptive processing like other interferers. But using this new algorithm one can handle multiple SOI, simultaneously while the interferers are nulled along the correct angles at the same time. Onc can also estimate the complex amplitudes of the multiple SOI using this algorithm.

A new multiple beamforming STAP algorithm based on a D3LS approach also has been described. In a conventional STAP algorithm, only one SOI can be handled at a time. So, when multiple SOI are presented, the other SOI can be treated as an interferer. Then it can be eliminated during the adaptive processing like other interferers. But using this new STAP algorithm one can handle multiple SOI in angles and Doppler frequencies, simultaneously while the interferers are nulled along the correct angles at the same time. Even though the angles of arrival of the two SOI may be close, the interferers are eliminated along the correct angle and also the two SOI are recovered properly along the correct angles and Doppler frequencies. Again, one can also estimate the complex amplitudes of the multiple SOI using this algorithm.

REFERENCES

[1] T. K. Sarkar, S. Park, J. Koh, and R. A. Schenieble, "A Deterministic Least Square Approach to Adaptive Antennas, " *Digital Signal Processing* 6, pp. 185-194, 1996

[2] T. K. Sarkar, M. C. Wicks, M. Salazar-Palma, and R. J. Bonneau, *Smart Antennas*, John Wiley-IEEE Press, Hoboken, NJ, 2003.

[3] T. K. Sarkar, H. Wang, S. Park, J. Koh, R. Adve, K. Kim, Y. Zhang, M. C. Wicks and R. D. Brown, "A Deterministic Least-squares Approach to Space Time Adaptive Processing (STAP)", *IEEE Trans. Antennas and Propagation*, Vol. 49, January 2001, pp. 91-103.

[4] J. Carlo, T. K. Sarkar, M. C. Wicks, "A Least Square Multiple Constraint Direct Data Domain Approach for STAP", *Proceedings of 2003 IEEE Conference on Radar,* pp. 431-438, 2003.

11

PERFORMANCE COMPARISON BETWEEN STATISTICAL-BASED AND DIRECT DATA DOMAIN LEAST SQUARES SPACE-TIME ADAPTIVE PROCESSING ALGORITHMS

11.0 SUMMARY

In the situation where an airborne radar platform is moving very fast, the number of training data used in space-time adaptive processing (STAP) is a major concern. Less number of training data sets is preferred in this situation. In this chapter, four versions of the statistical-based and direct data domain least squares (D3LS) STAPs are discussed and compared by performance when the number of training data is varied. The four statistical-based methods are the full-rank statistical method, the relative importance of the eigenbeam (RIE) method, the principle component generalized sidelobe canceller (GSC) method, and the cross-spectral GSC method. The performance of these four statistical methods is compared with the D3LS approach, which utilizes only one snapshot of data (space and time corresponding to one range cell only). The channel mismatch is also introduced to all methods to evaluate their performance. It is found that to make the statistical-based methods work one needs to know the rank of the interference covariance matrix. The D3LS performs better than the statistical-based methods when the number of training data available is less than the rank of the interference covariance matrix.

11.1 INTRODUCTION

For airborne (or spaceborne) radars, the detection of moving targets is a primary objective. During detection, the radar encounters the effect of strong clutter return around a weak return from the target. If the radar platform is stationary, the effect of the clutter can easily be removed by Doppler filtering. However, when the platform is nonstationary, the space-time adaptive processing (STAP) has been developed to address these applications. Many STAP algorithms have been discussed in the literature. Guerci summarized the taxonomy of STAP in [1]. Most of the STAP methods are based on statistical properties of the interferences (clutter, jammer, and thermal noise) while the Direct Data Domain Least Squares (D3LS) method [2,3] uses the actual interference information at that particular space-time snapshot in the cancellation of the interference itself. For statistical-based methods, the interference covariance matrix is used to calculate a weight vector that maximizes the signal-to-interference plus noise ratio (SINR) at the receiver output. However, this covariance matrix is not known beforehand, and needs to be estimated from the secondary data or training data using adjacent snapshots. The statistical properties of these training data must be the same as that of the cell under test from which the data is gathered. The number of training data that gives a good estimation of the covariance matrix is related to the degrees of freedom (DOF) which in this case is the number of weights used in the receiver. Because the appropriate DOF determines the ability of the system to null the interference, the number of training data used to calculate the covariance matrix is intimately related to the interference scenario of that particular radar platform. However, if the platform is moving relatively fast compared to the interference scenario, the assumption of a stationary interference in the training data may be destroyed and the result is an inaccurate estimation of the interference covariance matrix and an improper nulling of the interference. On the other hand, the D3LS method calculates the weight vector on a snapshot by snapshot basis and the training data are not needed for canceling interferences in this technique.

The goal of this chapter is to evaluate the performance of these statistical and direct data domain techniques. This is achieved by a comparison between the numbers of training data required and the accuracy of the estimation for the signal of interest. In section 11.2, the mathematical description of the various signals used for both the direct data domain and the statistical methods is developed. Section 11.3 introduces the statistical-based STAP approaches and section 11.4 describes the D3LS methodology. The channel mismatch for evaluating the robustness of these two methods is applied in section 11.5. Simulation results are presented in section 11.6. Section 11.7 offers some conclusions based on the various numerical simulations.

11.2. DESCRIPTION OF THE VARIOUS SIGNALS OF INTEREST

There are many signals involved in the radar return and it is not possible to model each signal individually. However, the radar return can be accurately modeled by identifying four components. These are the target signal, called the "signal of interest" (SOI), the clutter return, intentionally transmitted jammers, and discrete interferers. In this section, the mathematical modeling of these four important signals is described.

11.2.1 Modeling of the Signal-of-Interest

Assuming that the SOI impinges on an antenna array of N elements from one direction θ_s relative to the boresight direction, the signal spatial steering vector is given by:

$$\boldsymbol{a} = \left[1, \ \exp(j2\pi\frac{d}{\lambda}\sin\theta_s), \ \cdots, \ \exp(j2\pi\frac{d}{\lambda}(N-1)\sin\theta_s \right]^T \tag{11.1}$$

where d is the interelement spacing between the antenna elements and the superscript T denotes the transpose of a matrix. λ is the wavelength of the SOI. Since the antenna platform is moving, there is also a Doppler shift, f_d, in the received signal. With M pulses received by a single antenna element, the Doppler steering vector of the received signal can be written as:

$$\boldsymbol{b} = [\ 1, \ \exp(j2\pi f_d), \ \cdots, \ \exp(j2\pi(M-1)f_d)]^T \tag{11.2}$$

Therefore, the SOI space-time steering $(NM \times 1)$ vector can be written in terms of the Kronecker product as follows:

$$\boldsymbol{s} = \boldsymbol{b} \otimes \boldsymbol{a} \tag{11.3}$$

or,

$$\boldsymbol{s} = [\ \boldsymbol{a}^T, \ \exp(j2\pi f_d)\ \boldsymbol{a}^T, \ \cdots, \ \exp(j2\pi(M-1)f_d)\ \boldsymbol{a}^T\ ]^T \tag{11.4}$$

where $\otimes$ denotes the Kronecker product. The received SOI with an unknown complex amplitude, α , is then equal to $\alpha\boldsymbol{s}$. Note that α will be estimated and it will determine the level of performance of the five STAP techniques compared in this chapter.

11.2.2 Modeling of the Clutter

In general the clutter can be considered as a collection of small patches of scatterers; therefore, it can be modeled in a similar way as the SOI. We first define $\tilde{\gamma}_i$ as a random complex amplitude of the received signal from the clutter patch i. The received signal, X_c, corresponding to a group of clutter patches can be modeled as follows:

$$X_c = \sum_{i=1}^{N_c} \tilde{\gamma}_i \boldsymbol{v}_i \tag{11.5}$$

where, $\boldsymbol{v}_i$ is a space-time steering vector of *i-th* clutter patch, and N_c is the number of clutter patches.

11.2.3 Modeling of the Jammer

In general, the jammer signal is an intentionally transmitted signal to saturate the receiver. The jammer signal usually contains a wide range of frequencies which covers all Doppler frequencies of interest; thus, it depends only on the angle-of-arrival (AOA). The jammer signal is also uncorrelated from pulse-to-pulse; therefore, the model for a single received jammer signal is:

$$J = \boldsymbol{p} \otimes \boldsymbol{q} \tag{11.6}$$

where $\boldsymbol{p} = [\,\tilde{\beta}_1,\ \tilde{\beta}_2,\ \ldots,\ \tilde{\beta}_M\,]^T$ is a vector of random complex intensity of the jammer signal. $\tilde{\beta}_k$ is a random complex amplitude of the jammer for the *k-th* pulse. And $\boldsymbol{q}$ is the spatial steering vector of the jammer signal for angle θ_j.

The condition of multiple jammer signals is modeled as $\sum_{j=1}^{N_j} \boldsymbol{p}_j \otimes \boldsymbol{q}_j$ where N_j is the total number of jammers in the target environment.

11.2.4 Modeling of the Discrete Interferers

The interference from any stationary transmitter such as television and radio stations or even coherent multipaths of the transmitted signal can be considered as sources of discrete interference. The amplitude and Doppler frequency of the discrete interferers are assumed to be stationary over the coherent processing interval (CPI) which consists of M pulses received from N antennas with fixed pulse repetition interval. Therefore, the discrete interference can be modeled in the same way as the clutter in (11.5) as follows:

$$D = \sum_{i=1}^{N_d} \tilde{\kappa}_i \, \boldsymbol{d}_i \tag{11.7}$$

where $\tilde{\kappa}_i$ is a random complex amplitude of the *i-th* interference and $\boldsymbol{d}_i$ is a space-time steering vector of the *i-th* interference with its corresponding AOA and Doppler frequency.

From the above signal modeling, the $(NM \times 1)$ received signal, or the space-time signal snapshot, from the array antenna can be formulated as:

$$\boldsymbol{x} = \alpha \boldsymbol{s} + \sum_{i_c=1}^{N_c} \tilde{\gamma}_{i_c} \boldsymbol{v}_{i_c} + \sum_{j=1}^{N_j} \boldsymbol{p}_j \otimes \boldsymbol{q}_j + \sum_{l=1}^{N_d} \tilde{\kappa}_l \, \boldsymbol{d}_l + \boldsymbol{n} \tag{11.8}$$

where N_c, N_j, and N_d are the number of clutter patches, jammers, and discrete interferences, respectively. The additive component $\boldsymbol{n}$ is the snapshot of the thermal noise which is uncorrelated with the other signals.

11.3 STATISTICAL-BASED STAP ALGORITHMS

Several statistical-based STAP methods have been developed to address the moving radar platform effect. In this section, four versions of them will be discussed. The optimum STAP which is full-rank will be mentioned first followed by the reduced-rank STAP based on the relative importance of the eigenbeam method. Then, two versions of the generalized sidelobe canceller based reduced-rank STAP will be discussed at the end of this section.

11.3.1 Full-Rank Optimum STAP

In order to improve the detection capability of the system in the presence of interferences, i.e. clutter, jammer, discrete interferers, and thermal noise, an adaptive beamformer is applied to the signal space-time snapshot. The key is to introduce the weight factor, w_n , that maximizes the output SINR to each received signal in the snapshot of the data vector $\boldsymbol{x}$. The optimum weight vector $\boldsymbol{w} = [w_1 \; w_2 \; \dots \; w_{NM}]^T$ can be obtained from [1, 4]:

$$\boldsymbol{w} = \eta \boldsymbol{R}^{-1} \boldsymbol{s} \tag{11.9}$$

where η is a scalar, and $\boldsymbol{R}$ is the covariance matrix of the interference. However, in general, $\boldsymbol{R}$ is not known *a priori* and needs to be estimated. One common way to estimate the covariance matrix is as follows [4]:

$$\hat{\boldsymbol{R}} = \frac{1}{L} \sum_{k=1}^{L} \boldsymbol{x}_k \boldsymbol{x}_k^H \tag{11.10}$$

where $\boldsymbol{x}_k$ is called the secondary or training data for the *k-th* range cell where the SOI is absent. And it is assumed that all other interferences are present and stationary in these L training data. Figure 11.1 shows a space-time data cube with the training data used in estimating the covariance matrix. In order to get a good estimation for $\boldsymbol{R}$, the number of training data, L, is recommended to be $L \approx 2NM$ [1, 4]. However, in some situations, this large number of training data is not available since the interference scenario changes rapidly particularly for an airborne platform. The number of training data needed to calculate $\hat{\boldsymbol{R}}$ has to be less. One of the methods that require less number of training data is the Reduced-Rank STAP and it will be discussed in the next section.

In order to estimate the complex amplitude of the SOI, the unit gain constraint is applied to the weight vector along the direction of the SOI or $\boldsymbol{w}^H \boldsymbol{s} = 1$. We define the output SINR based on the estimated amplitude so that we can compare the output with the D3LS method in section 11.4 as [2]:

$$SINR_{output} \triangleq 20 \log \left| \frac{\alpha}{\alpha - \hat{\alpha}} \right| \tag{11.11}$$

where $\hat{\alpha}$ is the estimated complex amplitude of the SOI.

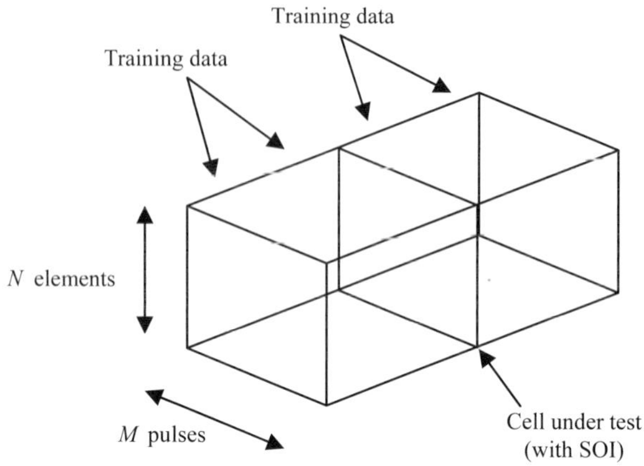

Figure 11.1. Space-time data cube with cell under test and training data cells.

11.3.2 Reduced-Rank STAP (Relative Importance of the Eigenbeam Method)

Usually the number of dominant interferences is less than the degrees of freedom (DOF) of the system. The DOF is determined by the number of weights used in beamforming, $DOF = NM$, in our case. Thus, it is possible to reduce the number of training data according to the number of dominant interferences. This number can be determined from the rank of the interference covariance matrix $\boldsymbol{R}$. However, as mentioned above, $\boldsymbol{R}$ is unknown; therefore, the rank of the interference covariance matrix has to be estimated from $\hat{\boldsymbol{R}}$ instead. There are many Reduced-Rank methods proposed in the literature [1,4]. We first start with the Relative Importance of the Eigenbeam (RIE) method [1]. For the RIE, the effective covariance inverse (ECI), $\hat{\boldsymbol{R}}_{eff}^{-1}$, is used instead of $\hat{\boldsymbol{R}}^{-1}$ for the weight calculation. First, for exact covariance inverse matrix, $\hat{\boldsymbol{R}}_{eff}^{-1}$, can be found from the following equation:

$$\boldsymbol{R}_{eff}^{-1} = \left(\boldsymbol{I} - \sum_{i=1}^{k} \frac{\lambda_i - \lambda_{\min}}{\lambda_i} \boldsymbol{u}_i \boldsymbol{u}_i^T \right) \tag{11.12}$$

where λ_i and $\boldsymbol{u}_i$ denote the *i-th* eigenvalue and eigenvector of the $\boldsymbol{R}$. However, $\boldsymbol{R}$ is unknown, thus, $\hat{\lambda}_i$ and $\hat{\boldsymbol{u}}_i$ have to be estimated from $\hat{\boldsymbol{R}}$. Then $\hat{\boldsymbol{R}}_{eff}^{-1}$ follows by substituting the estimated $\hat{\lambda}_i$ and $\hat{\boldsymbol{u}}_i$ in (11.12) instead of using the exact λ_i and $\boldsymbol{u}_i$. Only k interference subspaces are used in estimating the covariance matrix. It is expected that a reduced number of training data will be needed instead of using the full rank given by (11.10) yet preserving the same nulling ability.

11.3.3 Reduced-Rank STAP (Based on the Generalized Sidelobe Canceller)

In this subsection the Generalized Sidelobe Canceller (GSC) based STAP will be discussed starting with the full-rank method and followed by the reduced-rank method. The idea of the GSC is to find a signal blocking matrix that blocks the SOI and passes the interference through the system. Next, the weight vector that maximizes the output SINR due to the interference only is calculated. Then, at the final step, the result for the SOI and the interference are subtracted to obtain the output from the beamformer. The GSC beamformer is shown in Fig. 11.2 where $\boldsymbol{x}$ is the received signal vector defined in (11.8) and $\boldsymbol{s}$ is the space-time steering vector as in (11.4). The signal blocking matrix $\boldsymbol{B}$ is of full-row rank and of size $(NM-1)\times NM$. The signal blocking matrix satisfies the constraint $\boldsymbol{B}\boldsymbol{s} = \mathbf{0}$. There are several methods to compute the matrix $\boldsymbol{B}$, but two simple methods based on the Singular Value Decomposition (SVD) and the QR decomposition is mentioned in the Appendix of [5].

We can see that the weight vector in this beamformer is of dimension $(NM-1)\times 1$. In other words, the matrix $\boldsymbol{B}$ reduces the dimension of the weight vector by 1. This concept will lead to the reduced-rank GSC in the following discussion. Since the received signal vector $\boldsymbol{x}$ is transformed by the matrix $\boldsymbol{B}$, the covariance matrix $\boldsymbol{R}_b$ can be obtained from:

$$\mathbf{R}_b = \mathcal{E}\left[\boldsymbol{b}\boldsymbol{b}^H\right] = \boldsymbol{B}\boldsymbol{R}\boldsymbol{B}^H \tag{11.13}$$

where $\mathcal{E}[\bullet]$ is an expected value operator and the transformed signal or noise vector $\boldsymbol{b} = \boldsymbol{B}\boldsymbol{x}$. Note that since $\boldsymbol{B}$ blocks all the SOI component $\boldsymbol{b}$ can be implied as an interference only vector. Now the weight vector $\boldsymbol{w}_{gsc}$, which maximizes the SINR (due to interference only), can be found from the standard Wiener solution [4–6] as:

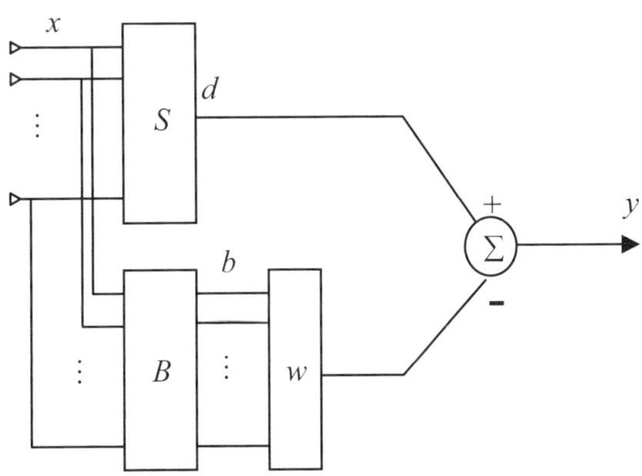

Figure 11.2. Full-rank GSC beamformer.

$$\boldsymbol{w}_{gsc} = \boldsymbol{R}_b^{-1}\boldsymbol{r}_{bd} \tag{11.14}$$

where $\boldsymbol{r}_{bd} = \boldsymbol{B}\,\boldsymbol{R}\,\boldsymbol{s}$ (for a detailed derivation see [4]).

Next, we apply the Eigen decomposition to the full-rank GSC beamformer in order to map the interference or noise vector onto its corresponding eigenvectors as shown in Fig. 11.3. First the covariance matrix $\boldsymbol{R}_b$ is decomposed as:

$$\boldsymbol{R}_b = \boldsymbol{U}\Lambda\boldsymbol{U}^H \tag{11.15}$$

where $\boldsymbol{U}$ is a unitary matrix composed of $NM-1$ eigenvectors, and Λ is the diagonal matrix of their associated eigenvalues. Using $\boldsymbol{U}$, we transform the noise vector $\boldsymbol{b}$ to a principle coordinate vector $\boldsymbol{p}$ as follows:

$$\boldsymbol{p} = \boldsymbol{U}^H\boldsymbol{b} \tag{11.16}$$

Then the estimated covariance matrix $\boldsymbol{R}_p$ is defined as:

$$\boldsymbol{R}_p = \mathcal{E}\left[\boldsymbol{p}\,\boldsymbol{p}^H\right] = \boldsymbol{U}^H\boldsymbol{R}_b\boldsymbol{U} = \Lambda \tag{11.17}$$

Now the full-rank transformed GSC weight vector , $\boldsymbol{w}_{pc}$, can be calculated from the following equation:

$$\boldsymbol{w}_{pc} = \boldsymbol{R}_p^{-1}\,\boldsymbol{r}_{pd} \tag{11.18}$$

where $\boldsymbol{r}_{pd} = \boldsymbol{U}^H\boldsymbol{r}_{bd}$. Since the rank of the interference covariance matrix, in general, is less than the DOF of the receiver, a number of eigenvectors corresponding to the dominant eigenvalues are selected. Instead of using full rank of the matrix $\boldsymbol{U}$ and Λ, we use only the K largest eigenvalues for the matrix $\tilde{\Lambda}$ and their corresponding eigenvectors for the matrix $\tilde{\boldsymbol{U}}$ where

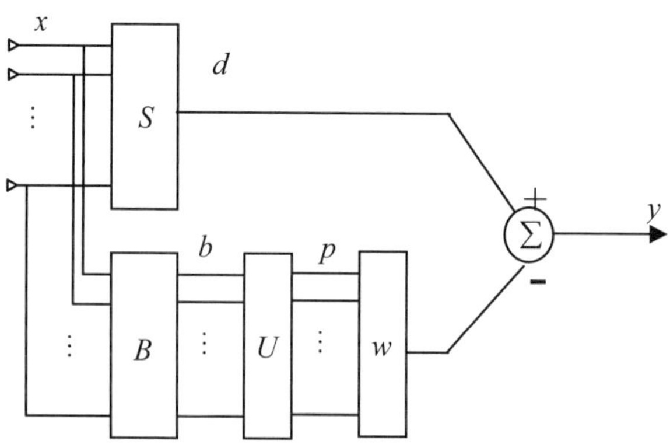

Figure 11.3. Full-rank transformed GSC beamformer.

$K < (NM - 1)$ in the transformation. By performing this operation, the dimension of the weight vector is reduced to be $K \times 1$. And it is supposed to reduce the number of training data required for the covariance matrix estimation. This reduced-rank method by choosing the K largest eigenvalues and their corresponding eigenvectors is called the principle component generalized sidelobe canceller (PC-GSC) technique. Another similar method, called cross-spectral GSC (CS-GSC), where the K eigenvalues and eigenvectors are selected by some other criteria will be discussed next. Figure 11.4 shows the reduced-rank GSC beamformer in general.

For CS-GSC, instead of choosing K largest eigenvectors, we choose them according to the largest cross-spectral energy projected along the eigenvector. The metric of cross-spectral energy is defined as [5]:

$$\left| \frac{\boldsymbol{u}_i^H \boldsymbol{r}_{bd}}{\sqrt{\lambda_i}} \right|^2 \tag{11.19}$$

where $\boldsymbol{u}_i$ is the *i-th* eigenvector corresponding the *i-th* eigenvalue λ_i. For the GSC-based beamformer to estimate the signal complex amplitude, it is necessary to impose the constraint of unit gain for the beamformer output, $\boldsymbol{y}$, when the input signal is the SOI with unit amplitude or $\boldsymbol{s}^H \boldsymbol{s} - \boldsymbol{W}^H \tilde{\boldsymbol{U}}^H \boldsymbol{B} \boldsymbol{s} = 1$. Since from the definition of the blocking matrix, $\boldsymbol{B}$, the second term of the left-hand side is zero, the only constraint left is $\boldsymbol{s}^H \boldsymbol{s} = 1$, which will effect the calculation of $\boldsymbol{r}_{bd}$ in (11.14). Once this constraint is satisfied, the amplitude can be estimated simply by solving for $\boldsymbol{y}$.

The different statistical based STAP algorithms discussed so far will be used in our performance comparison. In the next section, the direct data domain least squares method will be presented in this context.

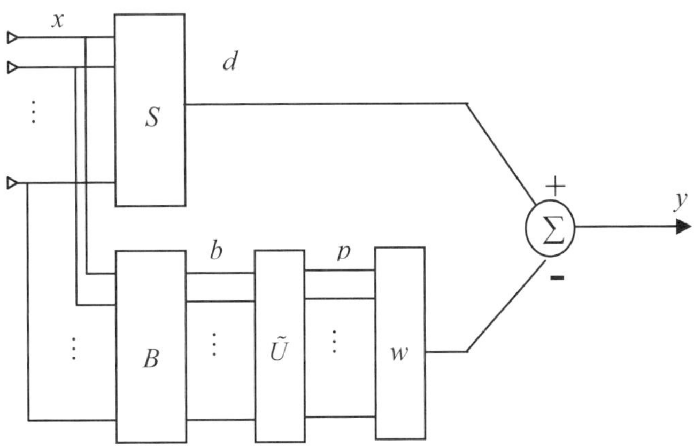

Figure. 11.4. Reduced-rank GSC beamformer.

11.4 DIRECT DATA DOMAIN LEAST SQUARES STAP ALGORITHMS

In order to overcome the problem of the training data needed in the statistical based STAP, a deterministic approach that utilizes only one space-time snapshot of data has been introduced to STAP problem [2,3,7–10]. By using the D3LS method, the weight vector needs to be evaluated on a snapshot by snapshot basis with a known direction of the SOI. The D3LS method has three different approaches namely forward, backward, and forward-backward approaches [2]. The forward method has been described in detail in section 6.3.2. The backward method has been illustrated in section 6.3.3. Finally, the forward-backward method has been presented in section 6.3.4.

However, if the actual signal comes from a slightly different direction than the assumed SOI, the three methods will not preserve the SOI, as it will be treated as an interferer and will get cancelled. Hence, a multiple constraint direct data domain technique has been developed and has been described in details in [2,10]. These same principles have been used in Chapter 10 to simultaneously form multiple beams. Therefore, the cost of using multiple constraints reduces the DOF of the system.

In the next section, the channel mismatch will be applied to the system in order to evaluate the performance of the STAPs.

11.5 CHANNEL MISMATCH

In an actual array antenna, the channel mismatch between the channels always occurs. One category of the channel mismatch is an angle-dependence. The angle-dependent channel mismatches, which are very complex, are usually due to the electromagnetic effects such as near-field scatterings or multipath effects [1]. Angle-independent channel mismatch is generally due to the transfer function between the antenna and the receiver of each channel. Even though the array antenna can be calibrated, this transfer function can be slightly changed. Since we are evaluating the performance of the STAP techniques, we will strictly limit ourselves to only the angle-independent channel mismatch for the assumption of narrowband signal modeling.

For a uniformly spaced linear array, the channel mismatch can be considered as introducing a random transfer function to each channel as shown in Fig. 11.5. We assume the mismatch is stable over a CPI, and it is randomly changed for each space-time snapshot. For a spatial steering signal vector at the *i-th* CPI, $\boldsymbol{a}_i$, we can define the received spatial steering signal vector with the effect of channel mismatch as:

$$\hat{\boldsymbol{a}}_i = \boldsymbol{a}_i \odot \boldsymbol{t}_i \tag{11.20}$$

where $\boldsymbol{t}_i = \left[\varepsilon_1 e^{j\varphi_1},\ \varepsilon_2 e^{j\varphi_2}, \cdots, \varepsilon_N e^{j\varphi_N} \right]^T$ with ε_k and φ_k are the amplitude and the phase errors at the *k-th* channel, respectively, and $\odot$ denotes the

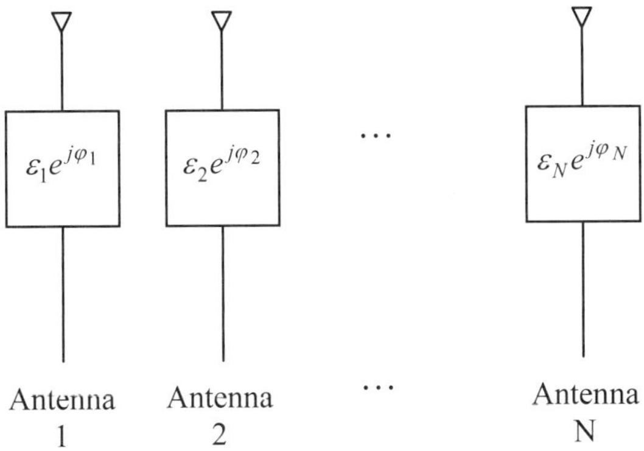

Figure 11.5. Angle-independent channel mismatch modeling as a transfer function of each channel for the narrowband signal.

elementwise multiplication. The amplitude error represents the variation in gain of each channel which lies in the range of $0 \le \varepsilon_k \le 1$ but is usually less than 0.1 dB or effectively is bounded by $0.99 \le \varepsilon_k \le 1$. The phase error is generally less than 5° [1].

In the next section, the channel mismatch defined above will be applied to our numerical simulations for both statistical-based and the D3LS STAPs in order to compare the tolerances of each algorithm. In addition, the performance of all the three direct data domain methods is compared with the statistical based approaches using the same data sets.

11.6 SIMULATION RESULTS

In this section, the details of our numerical simulations and their results will be discussed. The receiver has 10 omnidirectional point radiators with 16 coherent pulses $(N = 10, M = 16)$. The antenna spacing d is half wavelength at the operating frequency. The radar pulse repetition frequency (PRF) is 4000 *Hz*. The SOI is coming from 80° with a Doppler frequency of 900 *Hz*. The complex amplitude of the SOI is $\alpha = 5 + j2$. Thermal noise is 20 dB SNR. A jammer is located at 40° with the signal-to-jammer ratio of –20 dB. There are two different sets of discrete interferers and clutters in our simulations. The details of the interferers and the clutters are shown in Table 11.1. For both the cases, the Signal-to-Discrete Interferer ratio and the Signal-to-Clutter ratio is –10 dB. For the D3LS method, we used $N_a = 7, N_p = 9$ for the forward and backward methods, and $N_a' = 8, N_p' = 9$ for the forward-backward method. Figure 11.6 shows the input beampattern of the simulation in case I. The SOI is marked by a triangle in the figure while the discrete interferers are marked as circles. The center of the clutter region is marked with an asterisk $(*)$ symbol.

Table 11.1. Discrete Interference and Clutter Information (Case I).

Discrete Interferences										
AOA (degree)	85°	120°	40°	35°	65°	120°	100°	65°	55°	125°
Doppler Freq. (*Hz*)	400	–800	1700	1400	325	–1650	950	–1200	–125	1450

Clutter Characteristics		
Number of Clutter Patches	AOA (degree)	Doppler Freq. (*Hz*)
100	75°–85°	950–1450

For the first simulation, the output beam patterns of the full-rank statistical method and the reduced-rank RIE are plotted in Figs. 11.7 and 11.8. Figures 11.9 and 11.10 show the output beam patterns for the PC-GSC and CS-GSC methods. Then the output beam patterns for the forward, backward, and forward-backward D3LS methods are plotted in Figures 11.11–11.13, for the forward, backward, and the forward-backward methods, respectively. In this simulation, the number of training data used is $L = N\,M = 160$ with the number of weights 160 for all the statistical-based methods and when using only one snapshot (number of data in one snapshot = 160) the number of weights for the D3LS method 63 for both the forward and the backward methods, and 72 for the forward-backward method. Noting that the D3LS method cannot use all the DOF of the data available in one snapshot since it needs to form the cancellation matrix with an appropriate dimension in order to carry out a matrix inversion for the weight vector. The number of eigenvalues and eigenvectors used in rank reduction is 30 in each case. As seen, all the methods can null the interference and the jammer properly. Since the number of principal eigenvalues of the estimated covariance matrix $\hat{\boldsymbol{R}}$ is about 30 as shown in Fig. 11.14, it implies that the rank of the interference covariance matrix of this simulation scenario is about 30. Therefore, when the number of training data is greater than the interference rank, it is possible to obtain a proper weight vector that nulls the interference for the statistical-based methods.

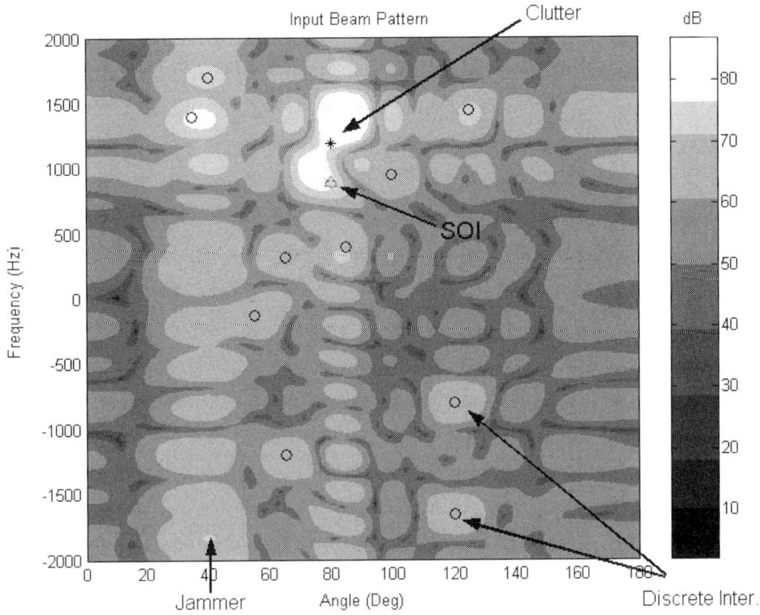

Figure 11.6. Input beam pattern for Case I.

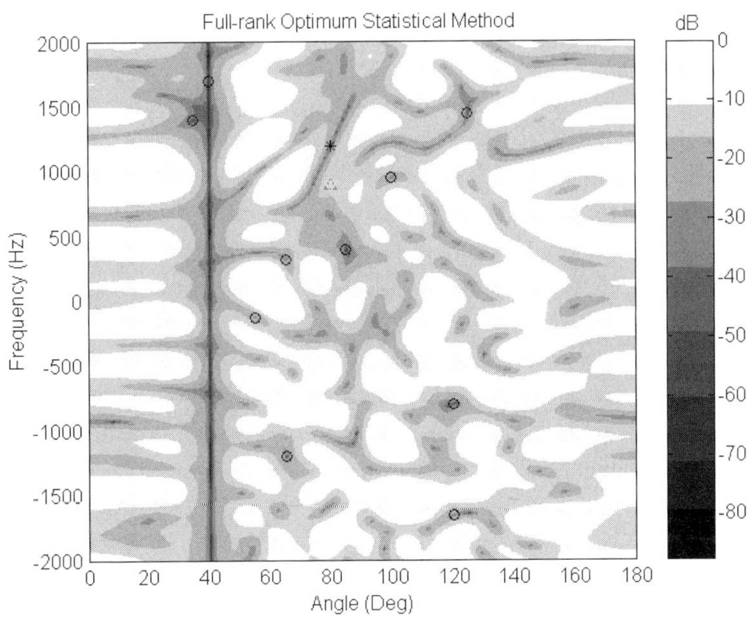

Figure 11.7. Output beam pattern for full-rank optimum statistical method (Case I).

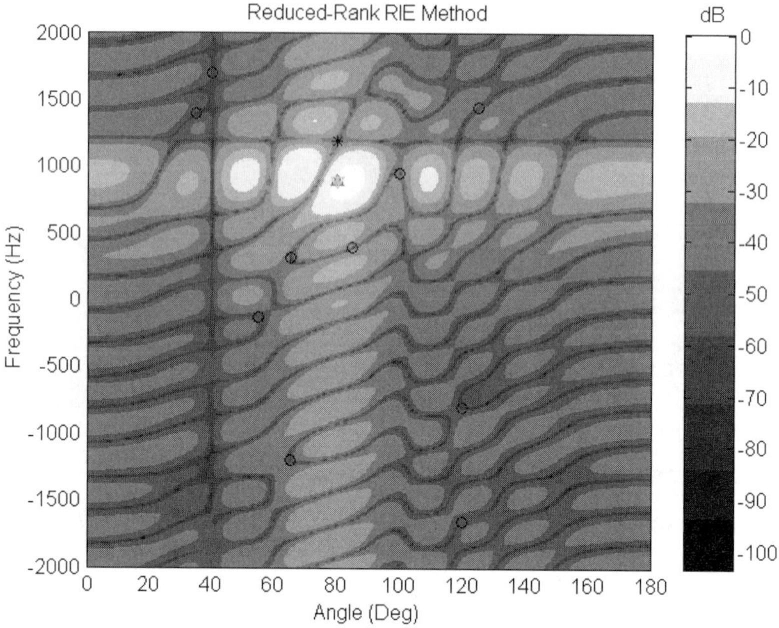

Figure 11.8. Output beam pattern for reduced-rank RIE method (Case I).

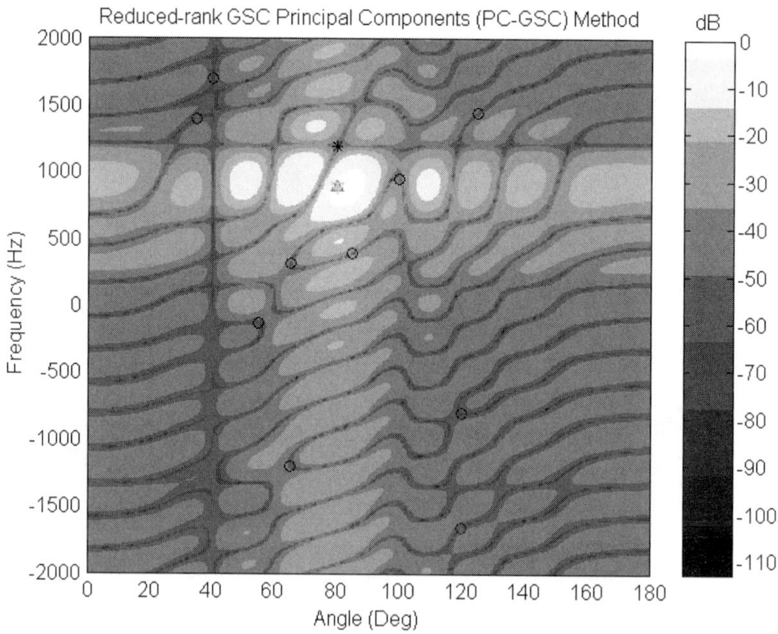

Figure 11.9. Output beam pattern for PC-GSC method (Case I).

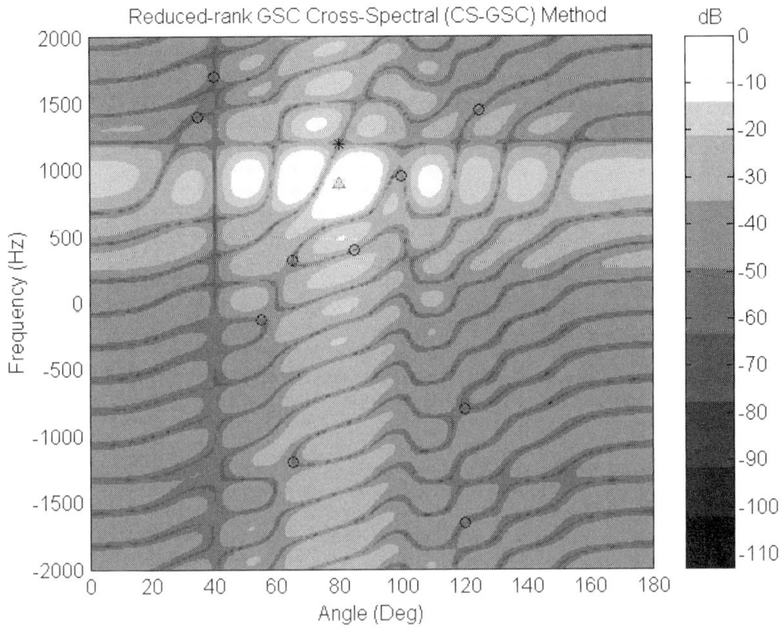

Figure 11.10. Output beam pattern for CS-GSC method (Case I).

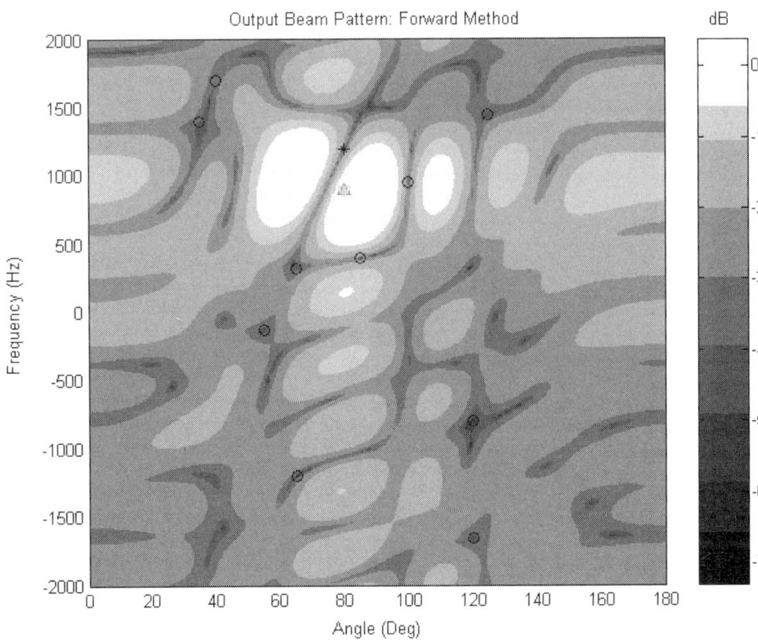

Figure 11.11. Output beam pattern for D3LS forward method (Case I).

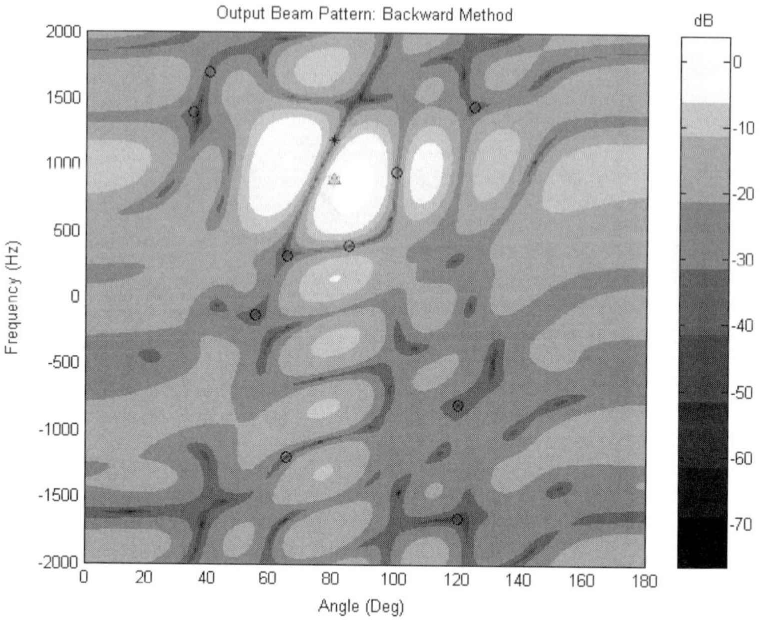

Figure 11.12. Output beam pattern for D3LS backward method (Case I).

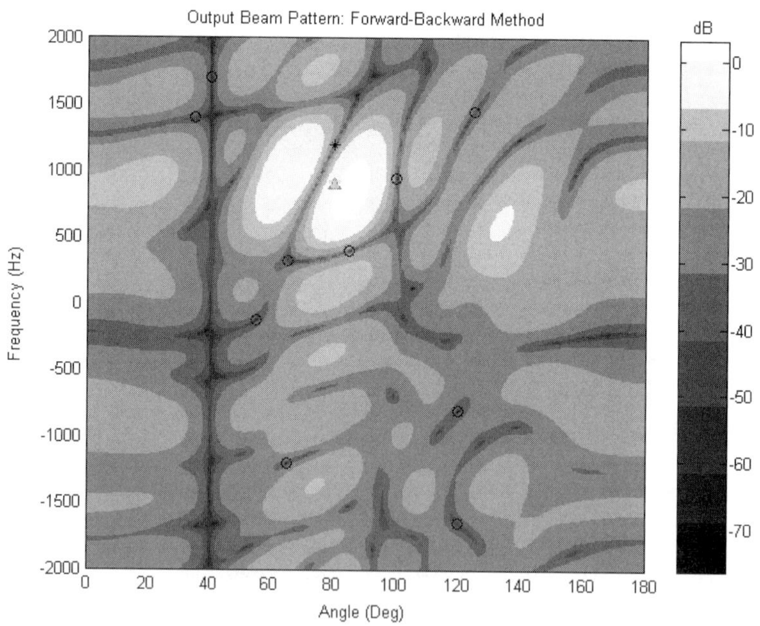

Figure 11.13. Output beam pattern for D3LS forward-backward method (Case I).

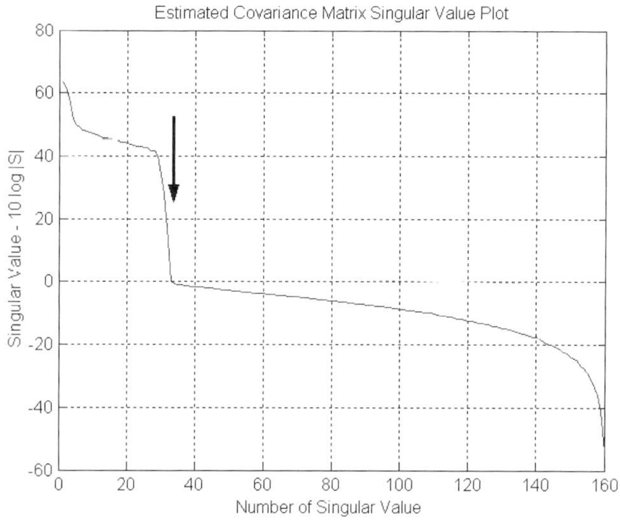

Figure 11.14. Singular value plot of the estimated covariance matrix shows that the number of dominant singular values at about 30.

In the next simulation, it will be shown that once the number of training data becomes small, the performance of the system based on the $SINR_{output}$ defined in (11.11) is also reduced. However, because the D3LS method utilizes only one snapshot of the data in evaluating the weight vector, there is no effect of reduced number of data for the D3LS method. For clarity, in Fig. 11.15 only the result of the forward-backward method, using the averaged amplitude with $N_a^{'} = 8, N_p^{'} = 9$ or $DOF = 72$, is plotted and compared with other methods.

It is obvious that the full-rank statistical method requires L to be at least equal to the DOF (= 160 in this case) in order to keep its performance above the acceptable limit. 100 independent runs were simulated to get the averaged $SINR_{output}$ in this comparison. The RIE, PS-GSC, and the CS-GSC give similar results since they use the same principle eigenvectors of $\hat{\boldsymbol{R}}$ in evaluating the weight vector. For the D3LS method, it is seen that the $SINR_{output}$ does not depend on L since it uses only one snapshot in the estimation. According to Figures 11.14 and 11.15, it is interesting to see that even though the performance of the reduced-rank methods (PC-GSC and CS-GSC) is quite good, it reduces dramatically once the number of training data becomes less than the rank of the interference covariance matrix.

In the next simulation, the interference and clutter in Table 11.2 will be used in order to increase the rank of the interference covariance matrix. Figure 11.16 shows the output SINR for the different methods. Figure 11.17 plots the magnitude of the singular values of the covariance matrix for the signal parameters given by Table 11.2.

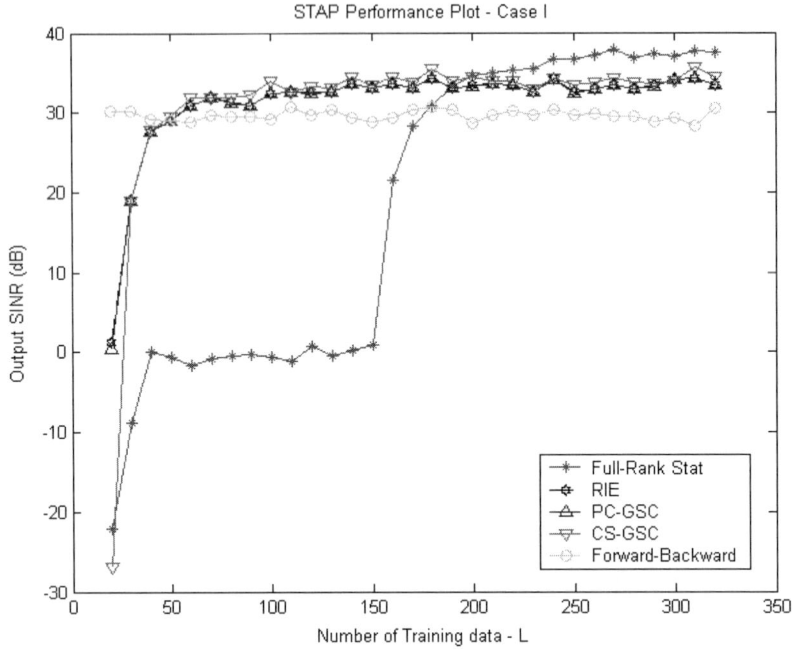

Figure 11.15. Output SINR of various methods are plotted with the number of training data used in the estimation.

Table 11.2. Discrete Interference and Clutter Information (Case II).

Discrete Interference										
AOA (degree)	85°	120°	40°	35°	65°	120°	100°	65°	55°	125°
Doppler Freq. (*Hz*)	400	−800	1700	1400	325	−1650	950	−1200	−125	1450
AOA (degree)	150	110°	70°	75°	90°	120°	140°	35°	60°	170°
Doppler Freq. (*Hz*)	1300	250	1450	800	−1100	−1400	600	−300	−1800	−1500

Clutter		
Number of Clutter Patches	AOA (degree)	Doppler Freq. (*Hz*)
300	65°–95°	450–1950

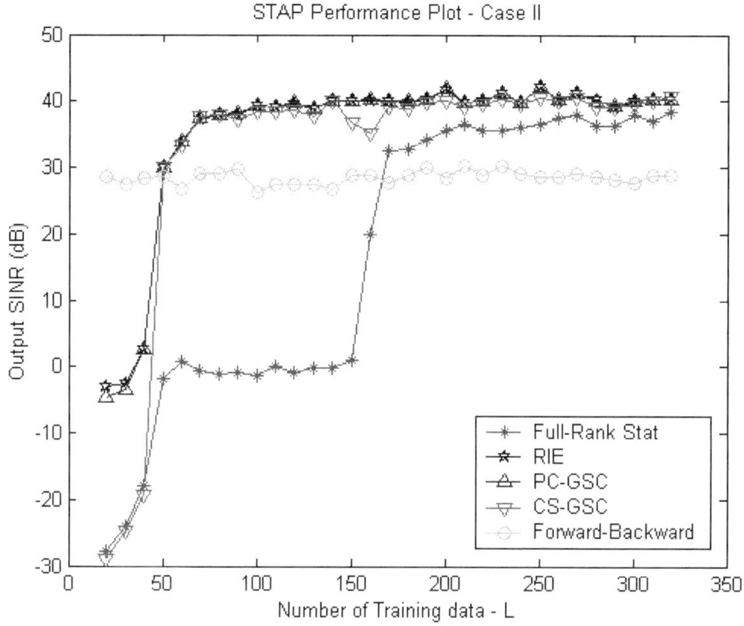

Figure 11.16. Output SINR of Case II where the interference rank is about 50.

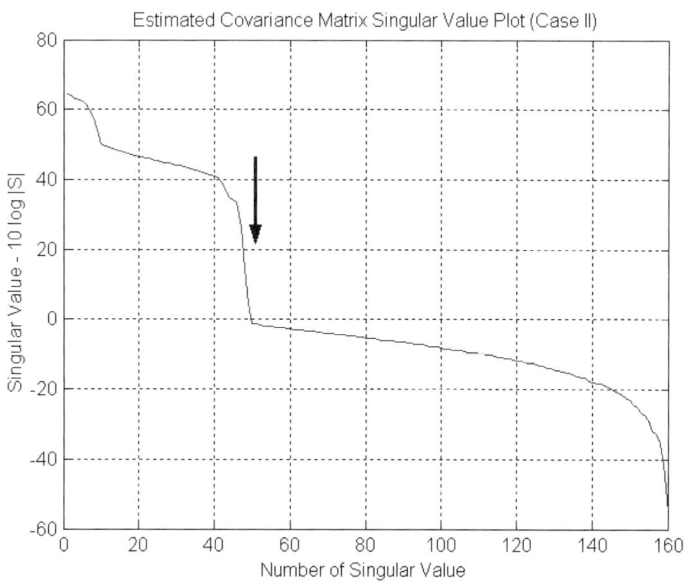

Figure 11.17. Singular value plot of the estimated covariance matrix for Case II shows the number of dominant singular values at about 50.

As seen, the output SINRs of the reduced-rank methods decrease when L is getting lower than the dominant singular values of the covariance matrix. However, the output SINR of the D3LS method is low compared to the statistical based method because the DOF of the forward-backward method is only 72 while the interference rank is about 50. The nulling capability of the method is reduced and yields a lower performance in the estimation of the complex amplitude of the SOI.

In the next simulation, we show that one needs to have a rough idea about the rank of the interference covariance matrix for the reduced-rank statistical methods; otherwise, the performance will be degraded dramatically. The interference scenario as presented in Case I is used in this simulation. The dominant rank of the interference for this case is about 30; however, the selected rank or eigenvectors (and/or eigenvalues) in rank reduction process is 15. The effect of selecting a lower rank in the process than the actual interference rank is shown in Fig. 11.18.

Compared with Fig. 11.15, it is obvious that when the rank in the method is less than the actual one, the performance of the reduced-rank methods decreases dramatically. Therefore, when using the reduced-rank STAP, the rank of the interference covariance matrix must be estimated in an appropriate

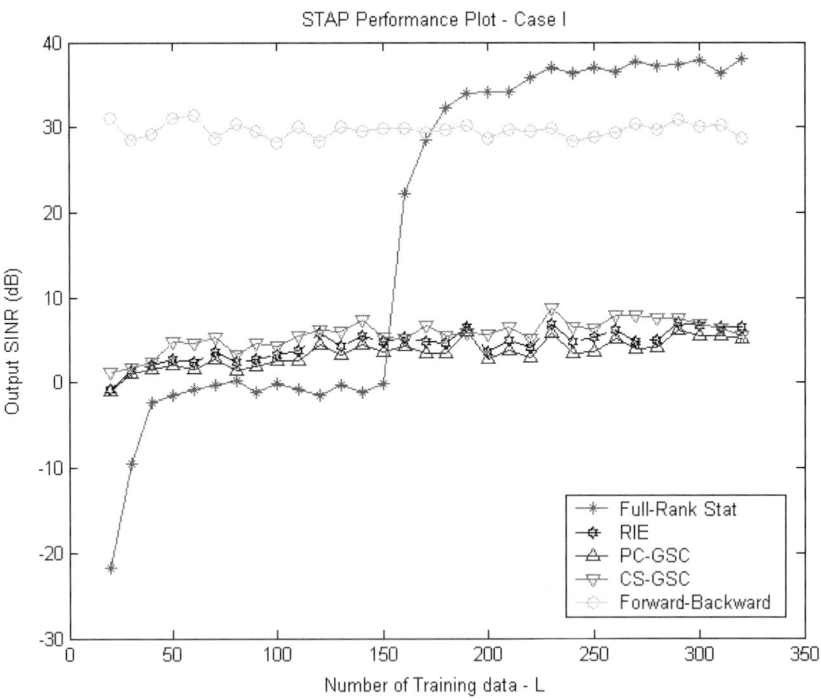

Figure 11.18. Output SINR for Case I when the selected rank in the methods is less than the actual interference rank.

fashion. However, a good estimation of this rank can be determined from $\hat{\boldsymbol{R}}$, which implies that there should be enough number of training data to get a good estimation of $\hat{\boldsymbol{R}}$. However, the D3LS method does not require this additional information and its performance remains unchanged irrespective of whether one has an estimate of the rank of the interference covariance matrix or the length of the secondary data.

Next, we are going to evaluate the tolerance of each method in the presence of channel mismatch as explained in Section 11.5. In this simulation, the amplitude error is uniformly distributed and is set to be less than 0.1 dB, or $0.99 \leq \varepsilon_k \leq 1$ and the phase error for each channel is also uniformly distributed with the maximum phase error of 3° or $-3^\circ \leq \varphi_k \leq +3^\circ$. The interference and the clutter as described in the first example are used in these simulations.

From Fig. 11.19, with the present channel mismatch, all the STAP methods gave similar results when compared with the simulation shown before without the channel mismatch. The only difference is that the $SINR_{output}$ is about -15 dB lower than the perfectly matched case for all the methods.

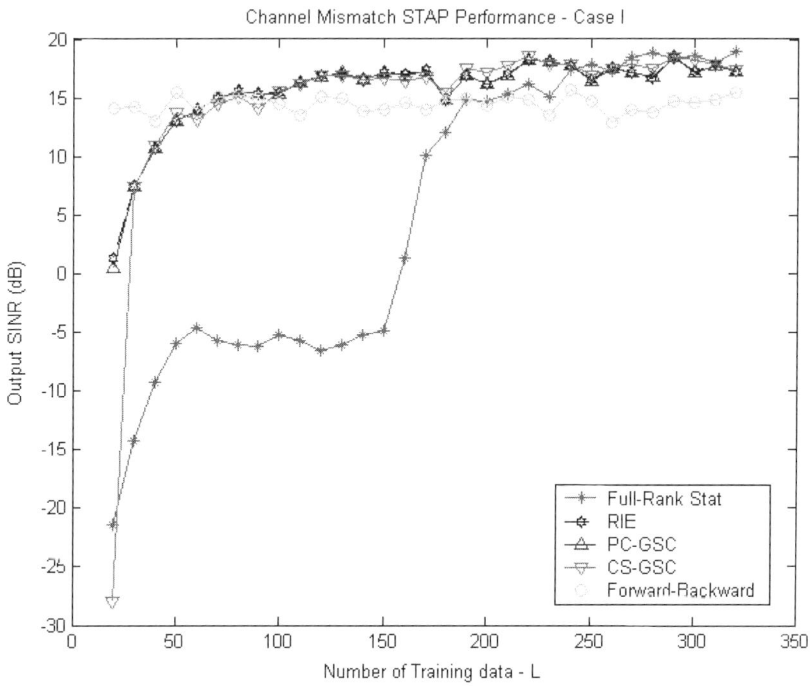

Figure 11.19. Output SINR in the presence of channel mismatch for various STAP methods.

11.7 CONCLUSION

Four statistical based STAP methods and the direct data domain least squares approach are described in this chapter. A performance comparison is made by the amplitude estimation of the SOI. The number of training data needed in the statistical methods is the major consideration in this analysis since in many applications it is not possible to obtain a large number of homogeneous training data. While the statistical-based methods require the training data, the D3LS method does not. The benefit of not requiring the training data comes at the cost of a reduced *DOF* of the system. On the other hand, if enough number of training data is available, the reduced-rank statistical-based methods provide good performances. For the reduced-rank methods, the estimation of rank of the interference covariance matrix is needed and it also requires a number of training data in the covariance matrix estimation. The effect of channel mismatch is introduced in the various methods to evaluate their tolerances. It does not alter the number of training data required for the methods, but it decreases the output SINR of the system evaluated by the methods.

From our simulation results, we recognize that there are two key factors: the number of training data available and the rank of the interference covariance matrix. As observed in the plots of the simulation results, it is clear that the D3LS method has superior performance when the number of training data is less than the rank of the interference covariance matrix. However, we cannot know *a priori* the rank of this covariance matrix unless a large number of training data is available. Therefore, the D3LS method will offer more stable and consistent performance when a large number of training data is not available. Each application will determine whether a large amount of training data is available for the STAP processing.

REFERENCES

[1] J. R. Guerci, *Space-Time Adaptive Processing for Radar*, Artech House, Norwood, MA, 2003.

[2] T. K. Sarkar, M. C. Wicks, M. Salazar-Palma, and R. J. Bonneau, *Smart Antennas*, John Wiley & Sons, Hoboken, NJ, 2003.

[3] T. K. Sarkar, H. Wang, S. Park, J. Koh, R. Adve, K. Kim, Y. Zhang, M. C. Wicks and R. D. Brown, "A Deterministic Least squares Approach to Space Time Adaptive Processing (STAP)", *IEEE Trans. Antennas and Propagation*, Vol. 49, January 2001, pp. 91-103.

[4] W. L. Melvin, "A STAP Overview", *IEEE A&E Systems Magazine*, Vol. 19, No. 1, Jan. 2004.

[5] J. S. Goldstein and I. S. Reed, "Theory of Partially Adaptive Radar", *IEEE Trans. on Aerospace and Electronic Systems*, Vol. 33, No. 4, Oct. 1997.

[6] S. Haykin, *Adaptive Filter Theory*, Prentice Hall, 4th ed., Upper Saddle River, NJ, 2002.

[7] A. Luthra, "A Solution to the Adaptive Nulling Problem with a Look-Direction Constraint in the Presence of Coherent Jammers", *IEEE Trans. on Antenna and Propagation*, Vol. 34, pp. 702-710, May 1986.

[8] T. K. Sarkar and N. Sangruji, "An Adaptive Nulling System for a Narrow-Band Signal with a Look-Direction Constraint Utilizing the Conjugate Gradient Method", *IEEE. Trans. on Antenna and Propagation*, Vol. 37, pp. 940-944, July 1989.

[9] R. Schneible, *A Least Squares Approach for Radar Array Adaptive Nulling*, Ph.D. dissertation, Syracuse University, Syracuse, NY, May 1996.

[10] J. T. Carlo, *Multiple Constraint Space-Time Direct Data Domain Approach Using Nonlinear Arrays*, Ph.D. dissertation, Syracuse University, Syracuse, NY, August 2003.

12

APPROXIMATE COMPENSATION FOR MUTUAL COUPLING USING THE IN SITU ANTENNA ELEMENT PATTERNS

12.0 SUMMARY

This chapter presents a new technique for an approximate compensation of the effects of mutual coupling among the antenna elements of an array using the computed or the measured in-situ element patterns in a direct data domain least squares adaptive algorithm. In this chapter, we consider the antenna elements in the phased array to have finite dimensions, i.e., they are not omnidirectional radiators. Hence, the antenna elements sample and reradiate the incident fields resulting in mutual coupling between the antenna elements. Mutual coupling not only destroys the linear wavefront assumption for the signal of interest but also for all the interferers impinging on the array. Thus, we propose a new direct data domain approach that partly compensates for the effect of mutual coupling, specifically when the jammer strengths are comparable to that of the signal. It is based on using the in-situ either measured or computed individual antenna element patterns. For strong interferers, a more accurate compensation for the mutual coupling is necessary using the transformation matrix through the formation of a uniform linear virtual array. Numerical results are presented to illustrate the principles of this new technique.

12.1 INTRODUCTION

The principal advantage of an adaptive array is the ability to electronically steer the mainlobe of the antenna to any desired direction while also automatically placing deep pattern nulls along the specific directions of interferences. Recently, a direct data domain least squares (D3LS) algorithm has been proposed [1–3]. A D3LS approach [3] has certain advantages related to the computational issues associated with the adaptive array processing problem as it analyzes the data for each snapshot as opposed to forming a covariance matrix of the data using multiple snapshots, and then solving for the weights utilizing that information. A single snapshot in this context is defined as the array of complex voltages measured at the feed point of the antenna elements. Another advantage of the D3LS approach is that when the direction of arrival (DOA) of the signal is not known precisely, additional constraints can be applied to fix the mainlobe beam width of the receiving array *a priori* and thereby reduce the signal cancellation problem.

Most adaptive algorithms assume that the elements of the receiving array are independent isotropic point sensors that sample, but do not reradiate, the incident fields. They further assume that the array is isolated from its surroundings. In a real system, however, each array element has physical dimensions. Therefore, the antenna elements spatially sample and reradiate the incident fields. The reradiated fields interact with the other antenna elements causing the sensors to be mutually coupled. The effect of mutual coupling may provide erroneous results for the estimated strength of the signal of interest (SOI) and DOA of the signal. Several authors have proposed different algorithms to eliminate the effects of mutual coupling in adaptive processing.

Gupta and Ksienski [4] analyzed and compensated for the effects of mutual coupling using a statistical adaptive algorithm. In references [5–7], examples have been presented to compensate for the degradation in the performance of the D3LS adaptive algorithms if the mutual couplings are not accounted for. In [7] a transformation matrix is used to correct the voltages induced at the antenna elements to compensate for mutual coupling.

In this chapter, we propose a new technique for an approximate compensation for the effect of mutual coupling between the elements of an array using the D3LS algorithms. In these algorithms we use the measured voltages across the loads connected to the antenna elements in the array without using a transformation matrix to compensate for the mutual coupling and other near field effects, as presented in [7]. This is equivalent to using the reciprocity theorem to link the measured in-situ far field element pattern to the voltage we are now measuring at the element terminals. We test the new algorithm using an array of dipoles. The mutual coupling among the dipole elements and the in-situ antenna element patterns are computed using an electromagnetic analysis code [8]. However, the proposed method of using the in-situ antenna element patterns to compensate for the mutual coupling in the array becomes inaccurate when the intensity of the jammers increases. The reason for such graceful degradation of this method has also been presented. The performance of the proposed method is also investigated by adding additional dummy elements at the end of the antenna array.

This chapter is organized as follows. In section 12.2 we formulate the problem. In section 12.3 we present simulation results illustrating the performance of the proposed method. The reason for limited region of validity of this technique is described in section 12.4. Finally, in section 12.5 we present our conclusions.

12.2 FORMULATION OF THE NEW DIRECT DATA DOMAIN LEAST SQUARES APPROACH APPROXIMATELY COMPENSATING FOR THE EFFECTS OF MUTUAL COUPLING USING THE IN SITU ELEMENT PATTERNS

12.2.1 Forward Method

Using the complex envelope representation for a uniform linear array where all the antenna elements are equally spaced, the $N \times 1$ complex vectors of phasor voltages $[V(t)]$ received by the antenna elements at a single time instance t can be expressed by

$$[V(t)] = \begin{bmatrix} V_1(t) \\ V_2(t) \\ \vdots \\ V_N(t) \end{bmatrix} = \sum_{m=1}^{M} [a(\theta_m)] s_m(t) + [n(t)] \tag{12.1}$$

where s_m and θ_m are the amplitude and DOA, respectively, of the mth source incident on the array at the instance t, while $[a(\theta_m)]$ is the steering vector of the array toward direction θ_m and $[n(t)]$ is the noise vector at each of the antenna elements. We now analyze the data using a single snapshot of the voltages measured at the antenna terminals.

Let us assume that the signal is coming from the angular direction θ_s and our objective is to estimate its amplitude while simultaneously rejecting all other interferences. The signal arrives at each sensor at different times dependent on the DOA of the target and the geometry of the array. At each of the N antennas, the received signal is the sum of the SOI, interference, clutter, and thermal noise. The interference may consist of coherent multipaths of the SOI along with clutter and thermal noise. Therefore, by suppressing the time dependence in phasor notation, we can reformulate (12.1) as

$$[V] = \begin{bmatrix} V_1 \\ V_2 \\ \vdots \\ V_N \end{bmatrix} = \alpha_s \begin{bmatrix} a_1(\theta_s) \\ a_2(\theta_s) \\ \vdots \\ a_N(\theta_s) \end{bmatrix} + \sum_{m=1}^{M-1} s_m \begin{bmatrix} a_1(\theta_m) \\ a_2(\theta_m) \\ \vdots \\ a_N(\theta_m) \end{bmatrix} + [n] \tag{12.2}$$

where α_s is the complex amplitude of the SOI, to be determined. The column vectors in this equation explicitly show the various components of the signal induced in each of the N antenna elements. In (12.2), a_n represents the voltage induced at the nth antenna element due to a signal of unit volt amplitude arriving from the particular direction θ. There are $M - 1$ undesired signal components in addition to the SOI. For the conventional adaptive array system, using each of K weights W_k, we can estimate the SOI through the following weighted sum

$$y = \sum_{k=1}^{K} W_k V_k \tag{12.3}$$

or in compact matrix form as

$$y = [W]^T [V] \tag{12.4}$$

where the superscript T denotes the transpose of a matrix and K is equal to number of weights. In the present method, $K = (N + 1)/2$ [3]. Also K has to be greater than the number of interferers $M - 1$, i.e., $K \geq M$.

Let us define another matrix $[S]$

$$[S] = \begin{bmatrix} S_1 \\ S_2 \\ \vdots \\ S_N \end{bmatrix} = \begin{bmatrix} a_1(\theta_s) \\ a_2(\theta_s) \\ \vdots \\ a_N(\theta_s) \end{bmatrix} \tag{12.5}$$

where a_n represents the voltage induced at the nth antenna element due to the SOI only, with an assumed amplitude of 1V. However, the actual complex amplitude of the SOI is not 1V but α_s which is to be determined. This SOI is arriving from the particular direction θ_s. So the value of S_n in the absence of mutual coupling is

$$S_n = \exp\left[j2\pi \frac{nd}{\lambda} \cos\theta_s \right], \quad n = 1, 2, \ldots, N \tag{12.6}$$

Then, $V_2/S_2 - V_1/S_1$ will have no components of the SOI, moreover, there are only undesired signal components left [9]. In a real environment, however, there is mutual coupling among the antenna elements. In this case the elements of the vector $[S]$ should be the measured voltages due to the SOI in the antenna array with an assumed amplitude of 1V. So, if we use the actual voltages from the real antenna array for the vector $[S]$ and $[V]$, then $V_2/S_2 - V_1/S_1$ contains undesired signal components and mutual coupling due to both the SOI and the undesired signals.

Therefore one can form a reduced rank matrix $[T]_{(K-1)\times K}$, generated from the vector $[V]$ and $[S]$, such that

$$\begin{bmatrix} \frac{V_2}{S_2} - \frac{V_1}{S_1} & \frac{V_3}{S_3} - \frac{V_2}{S_2} & \cdots & \frac{V_{K+1}}{S_{K+1}} - \frac{V_K}{S_K} \\ \frac{V_3}{S_3} - \frac{V_2}{S_2} & \frac{V_4}{S_4} - \frac{V_3}{S_3} & \cdots & \frac{V_{K+2}}{S_{K+2}} - \frac{V_{K+1}}{S_{K+1}} \\ \vdots & \vdots & \vdots & \vdots \\ \frac{V_K}{S_K} - \frac{V_{K-1}}{S_{K-1}} & \frac{V_{K+1}}{S_{K+1}} - \frac{V_K}{S_K} & \cdots & \frac{V_N}{S_N} - \frac{V_{N-1}}{S_{N-1}} \end{bmatrix}_{(K-1)\times K} \begin{bmatrix} W_1 \\ W_2 \\ \vdots \\ W_K \end{bmatrix}_{K\times 1} = \begin{bmatrix} 0 \\ 0 \\ \vdots \\ 0 \end{bmatrix}_{(K-1)\times 1} \tag{12.7}$$

So if we find the weighting vectors which satisfy the above matrix equation, we can then eliminate all the undesired signals. Mutual coupling due to the undesired signals is also partially compensated for when we use the actual measured voltage from the real antenna array for the elements of vector $[V]$ and $[S]$. To achieve a perfect compensation we have to use a transformation matrix [3] that transforms the measured voltages $[V]$ to an equivalent set of voltages that is induced in a uniform linear virtual array consisting of isotropic point radiators radiating in free space. This transformation takes care of the effect of the dissimilarity in the values in the self terms of the port admittance matrix of the array which is reduced to an identity matrix when dealing with isotropic point radiators operating in free space.

In order to make the matrix full rank, we fix the gain of the array to be C along the direction of θ_s. This provides an additional equation resulting in

$$\begin{bmatrix} 1 & 1 & \cdots & 1 \\ \frac{V_2}{S_2} - \frac{V_1}{S_1} & \frac{V_3}{S_3} - \frac{V_2}{S_2} & \cdots & \frac{V_{K+1}}{S_{K+1}} - \frac{V_K}{S_K} \\ \frac{V_3}{S_3} - \frac{V_2}{S_2} & \frac{V_4}{S_4} - \frac{V_3}{S_3} & \cdots & \frac{V_{K+2}}{S_{K+2}} - \frac{V_{K+1}}{S_{K+1}} \\ \vdots & \vdots & \vdots & \vdots \\ \frac{V_K}{S_K} - \frac{V_{K-1}}{S_{K-1}} & \frac{V_{K+1}}{S_{K+1}} - \frac{V_K}{S_K} & \cdots & \frac{V_N}{S_N} - \frac{V_{N-1}}{S_{N-1}} \end{bmatrix}_{K\times K} \begin{bmatrix} W_1 \\ W_2 \\ \vdots \\ W_K \end{bmatrix}_{K\times 1} = \begin{bmatrix} C \\ 0 \\ \vdots \\ 0 \end{bmatrix}_{K\times 1} \tag{12.8}$$

or, in short,

$$[F][W] = [Y] \tag{12.9}$$

Once the weights are solved by using (12.9), the complex amplitude of SOI α_s may be estimated from

$$\alpha_s = \frac{1}{C} \sum_{k=1}^{K} \frac{V_k}{S_k} W_k \tag{12.10}$$

For the solution of $[F][W] = [Y]$ in (12.9), the conjugate gradient method starts with an initial guess $[W]_0$ for the solution and continues with the calculation of the following [1–3] and has been outlined in details in section 6.2.5.

The above equations are applied in an iterative fashion till the desired error criterion for the residuals $\|[R]_k\|$, is satisfied, where $[R]_k = [F][W]_k - [Y]$. In our case, the error criterion is defined by

$$\frac{\|[R]_k\|}{\|[Y]\|} = \frac{\|[F][W]_k - [Y]\|}{\|[Y]\|} \leq 10^{-6} \; . \tag{12.11}$$

12.2.2 Backward Method

Next we reformulate the problem using the same data to obtain a second independent estimate for the solution. This is achieved by reversing the data sequence and then complex conjugating each term of that sequence. It is well known in the parametric spectral estimation literature that a sampled sequence which can be represented by a sum of complex exponentials with purely imaginary argument can be used either in the forward or in the reverse direction resulting in the same value for the exponent. From physical considerations we know that if we solve a polynomial equation with the weights W_i as the coefficients then its roots provide the DOA for all the unwanted signals including the interferers. Therefore, whether we look at the snapshot as a forward sequence as presented in the last section or by a reverse conjugate of the same sequence, the final results for W_i must be the same. Hence for these classes of problems we can observe the data either in the forward direction or in the reverse direction. This is equivalent to creating a virtual array of the same size but located along a mirror symmetry line. Therefore, if we now conjugate the data and form the reverse sequence, then one gets an independent set of equations similar to (12.8) for the solution of the weights $[W]$. This is represented by

$$\begin{bmatrix} 1 & 1 & \cdots & 1 \\ \frac{V_N^*}{S_N^*} - \frac{V_{N-1}^*}{S_{N-1}^*} & \frac{V_{N-1}^*}{S_{N-1}^*} - \frac{V_{N-2}^*}{S_{N-2}^*} & \cdots & \frac{V_K^*}{S_K^*} - \frac{V_{K-1}^*}{S_{K-1}^*} \\ \frac{V_{N-1}^*}{S_{N-1}^*} - \frac{V_{N-2}^*}{S_{N-3}^*} & \frac{V_{N-2}^*}{S_{N-2}^*} - \frac{V_{N-3}^*}{S_{N-3}^*} & \cdots & \frac{V_{K-1}^*}{S_{K-1}^*} - \frac{V_{K-2}^*}{S_{K-2}^*} \\ \vdots & \vdots & \vdots & \vdots \\ \frac{V_{K+1}^*}{S_{K+1}^*} - \frac{V_K^*}{S_K^*} & \frac{V_K^*}{S_K^*} - \frac{V_{K-1}^*}{S_{K-1}^*} & \cdots & \frac{V_2^*}{S_2^*} - \frac{V_1^*}{S_1^*} \end{bmatrix}_{K \times K} \begin{bmatrix} W_1 \\ W_2 \\ \vdots \\ W_K \end{bmatrix}_{K \times 1} = \begin{bmatrix} C' \\ 0 \\ \vdots \\ 0 \end{bmatrix}_{K \times 1} \tag{12.12}$$

where the superscript * denotes the complex conjugate. It can be now written in a matrix form as

$$[B][W] = [Y] \tag{12.13}$$

and the complex amplitude of signal α_s can again be determined by

$$\alpha_s = \left[\frac{1}{C'} \sum_{k=1}^{K} W_k \frac{V^*_{K-k+1}}{S^*_{K-k+1}} \right]^* \tag{12.14}$$

Note that for both the forward and the backward methods described in sections 12.2.1 and 12.2.2, we have $K = (N + 1)/2$. Hence the degrees of freedom are the same for both the forward and the backward method. However, we have two independent solutions for the same adaptive problem. In a real situation when the solution is unknown two different estimates for the same solution may provide a level of confidence on the quality of the solution.

12.2.3 Forward-Backward Method

Finally, in this section we combine both the forward and the backward method to double the sizes of the given data set and thereby increase the number of weights or the degrees of freedom significantly over that of either the forward or the backward method. This provides a third independent solution. In the forward-backward model we double the amount of data by not only considering the data in the forward direction but also conjugating it and reversing the direction of increment of the independent variable. This type of processing can be done as long as the series to be approximated can be fit by exponential functions of purely imaginary arguments. This is always true for the adaptive array case. So by considering the data set V_n and V^*_{-n} we have essentially doubled the amount of data without any penalty, as these two data sets for our problem are linearly independent.

An additional benefit accrues in this case. For both the forward and the backward method, the maximum number of weights we can consider is given by $(N + 1)/2$, where N is the number of the antenna elements. Hence, even though all the antenna elements are being utilized in the processing, the number of degrees of freedom available for this approach is essentially $(N + 1)/2$. For the forward-backward method, the number of degrees of freedom can be significantly increased without increasing the number of antenna elements. This is accomplished by considering the forward and backward versions of the data from the array. For this case, the number of degrees of freedom Q, can reach $(N + 0.5)/1.5$. This, Q, is approximately equal to 50% more weights or number of degrees of freedom than the two previous cases of K. The equation that needs to be solved for the weights is given by combining (12.8) and (12.12), with $C' = C$, into

$$\begin{bmatrix} 1 & 1 & \cdots & 1 \\ \frac{V_2}{S_2}-\frac{V_1}{S_1} & \frac{V_3}{S_3}-\frac{V_2}{S_2} & \cdots & \frac{V_{Q+1}}{S_{Q+1}}-\frac{V_Q}{S_Q} \\ \frac{V_3}{S_3}-\frac{V_2}{S_2} & \frac{V_4}{S_4}-\frac{V_3}{S_3} & \cdots & \frac{V_{Q+2}}{S_{Q+2}}-\frac{V_{Q+1}}{S_{Q+1}} \\ \vdots & \vdots & \vdots & \vdots \\ \frac{V_Q}{S_Q}-\frac{V_{Q-1}}{S_{Q-1}} & \frac{V_{Q+1}}{S_{Q+1}}-\frac{V_Q}{S_Q} & \cdots & \frac{V_N}{S_N}-\frac{V_{N-1}}{S_{N-1}} \\ \frac{V_N^*}{S_N^*}-\frac{V_{N-1}^*}{S_{N-1}^*} & \frac{V_{N-1}^*}{S_{N-1}^*}-\frac{V_{N-2}^*}{S_{N-2}^*} & \cdots & \frac{V_Q^*}{S_Q^*}-\frac{V_{Q-1}^*}{S_{Q-1}^*} \\ \frac{V_{N-1}^*}{S_{N-1}^*}-\frac{V_{N-2}^*}{S_{N-3}^*} & \frac{V_{N-2}^*}{S_{N-2}^*}-\frac{V_{N-3}^*}{S_{N-3}^*} & \cdots & \frac{V_{Q-1}^*}{S_{Q-1}^*}-\frac{V_{Q-2}^*}{S_{K-2}^*} \\ \vdots & \vdots & \vdots & \vdots \\ \frac{V_{Q+1}^*}{S_{Q+1}^*}-\frac{V_Q^*}{S_Q^*} & \frac{V_Q^*}{S_Q^*}-\frac{V_{Q-1}^*}{S_{Q-1}^*} & \cdots & \frac{V_2^*}{S_2^*}-\frac{V_1^*}{S_1^*} \end{bmatrix}_{(2Q-1)\times Q} \begin{bmatrix} W_1 \\ W_2 \\ \vdots \\ W_Q \end{bmatrix}_{Q\times 1} = \begin{bmatrix} C \\ 0 \\ \vdots \\ 0 \end{bmatrix}_{(2Q-1)\times 1} \tag{12.15}$$

or in matrix form as

$$[FB][W]=[Y] \tag{12.16}$$

12.3 SIMULATION RESULTS

As a first example, consider a signal of unit amplitude arriving from $\theta_s = 95°$. We form a 7-element uniform linear array (ULA) consisting of dipoles which are half wavelengths long and centrally loaded with 50 Ω as shown in Figure 12.1. The parameters for the dipole antenna array are given in Table 12.1. We consider the mutual coupling in this array of dipoles and analyze the antenna array using an electromagnetic analysis code [8]. The goal is to recover the complex amplitude of the SOI using the proposed method, in the presence of mutual coupling using the in-situ measured element patterns.

However, for numerical simulation we use the voltages measured at the loads of the antenna elements as the in-situ element patterns are related to these voltages through reciprocity. The proposed algorithm tries to maintain the gain of the array along the direction of SOI while automatically placing nulls along the directions of the interferences. In this simulation we take the SOI of unit amplitude to be arriving from $\theta_s = 95°$. Two jammers are present at 80° and 110°. The intensities of these jammers are varied from 1 to 30 V/m. We assume

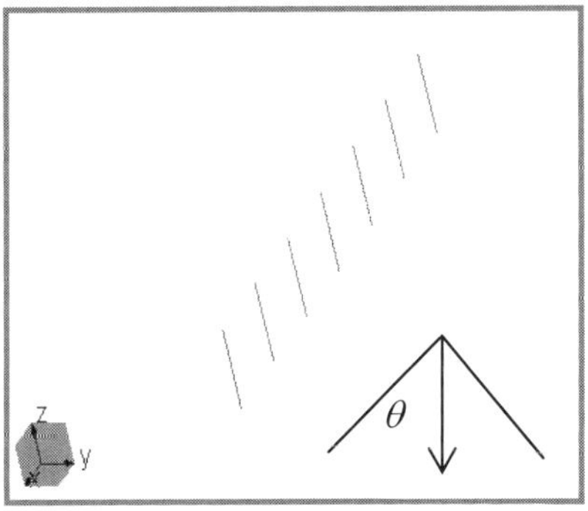

Figure 12.1. A uniform linear array of dipoles.

Table 12.1. Parameters Defining the Elements of the Dipole Array.

Number of elements in array	7
Length of z-directed wires	$\lambda/2$
Radius of wires	$\lambda/200$
Spacing between wires	$\lambda/2$
Loading at the center	50 Ω

that we know the DOA of the signal but we need to estimate its complex amplitude. In addition, we do not know the complex amplitudes or the DOA of the interferers nor do we have any probabilistic description of the thermal noise. We assume that we have 20 dB of signal-to-noise ratio at the antenna elements. The output signal-to-interference-plus-noise ratio (output SINR) is shown in Figure 12.2. The output SINR is an indicator of the accuracy of our estimate. It is defined as

$$\mathrm{SINR}_{\mathrm{out}} = 20\log\left|\frac{\alpha_s}{\alpha_s - \alpha_{est}}\right| \tag{12.18}$$

where the numerator α_s provides the true value for the complex amplitude of the desired signal and α_{est} is the output providing the estimated complex amplitude. The denominator term $\alpha_s - \alpha_{est}$ then provides the residual interference plus noise error, which resulted from the processing. Results are shown for the three different methods (forward method, backward method, and the forward-backward method) in Figure 12.2. It is seen from Figure 12.2 that as the intensity of the interferer increases the results become inaccurate if we do not correct for the mutual coupling.

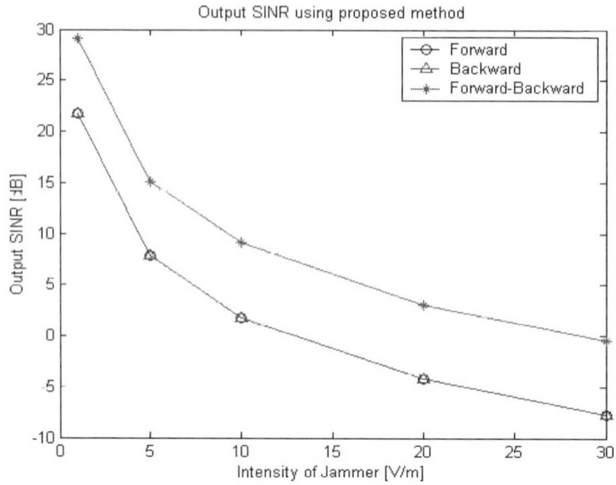

Figure 12.2. Output SINR as a function of the intensity of the jammer for a 7-element ULA of half-wave dipoles without correcting for the mutual coupling.

Figure 12.3a presents the results for the D3LS method which does account for the mutual coupling between the antenna elements using the transformation matrix and forming a uniform linear virtual array as illustrated in [3]. Figure 12.3a shows the improvement in the performance of the output SINR for the forward method and Figure 12.3b for the forward-backward method as the results for the backward method is similar to that of the forward method. As seen in Figure 12.3, the proposed method deteriorates gracefully as the intensity of the jammers increase and a more accurate analysis, through the use of a uniform linear virtual array to compensate for the mutual coupling [7] between the antenna elements, is then necessary.

For the second example, we study the performance of the proposed method as we increase the number of antenna elements both with and without an accurate compensation for the mutual coupling between the elements of the array. In this simulation we use the same geometry of the antenna array as in the first simulation. The DOA of the SOI and the jammers are also the same. The intensities of the jammers are 30 V/m. The number of antenna elements in the array is increased from 7 to 31. In Figure 12.4 we present the output SINR as a function of the number of antenna elements using the proposed method. It is seen that if we significantly increase the number of antenna elements then it is not necessary to perform an accurate compensation for the mutual coupling.

Figures 12.5a and 12.5b present the output SINR when we do and do not perform an accurate compensation for the mutual coupling between the elements in the array. It is seen that accurately compensating for the mutual coupling provides a more accurate estimate for the SOI.

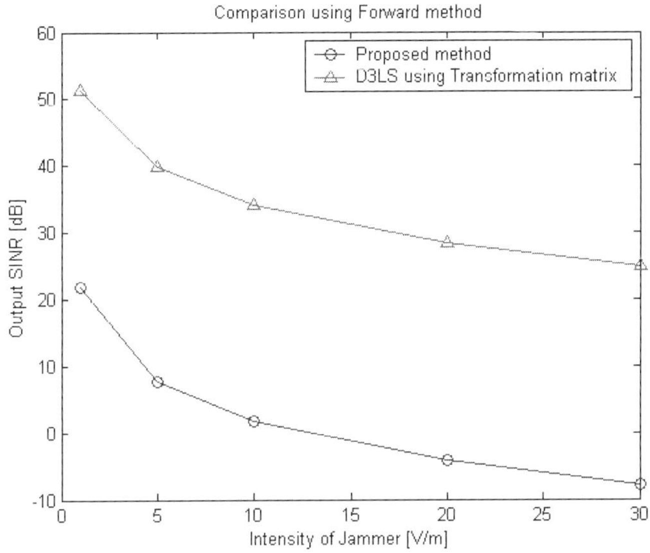

Figure 12.3a. Comparison of the output SINR using the forward method between the proposed method and the use of a more accurate treatment of mutual coupling through the transformation matrix.

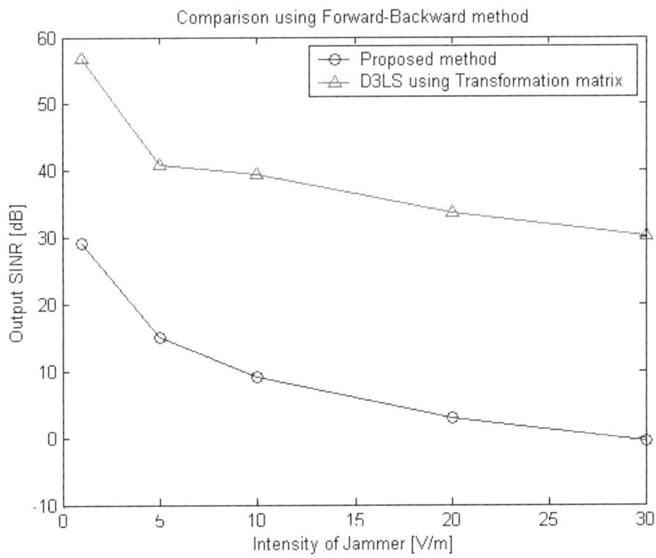

Figure 12.3b. Comparison of the output SINR using the forward-backward method between the proposed method and the use of a more accurate treatment of mutual coupling through the transformation matrix.

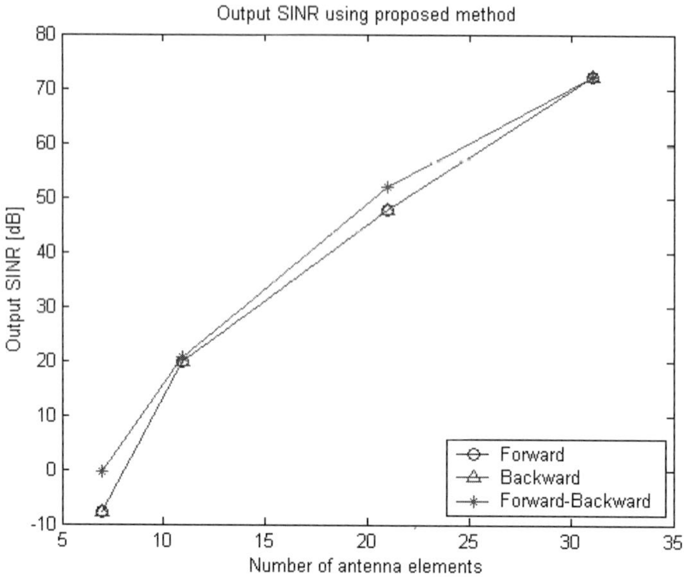

Figure 12.4. Output SINR as a function of the number of antenna elements in the array using the proposed methods.

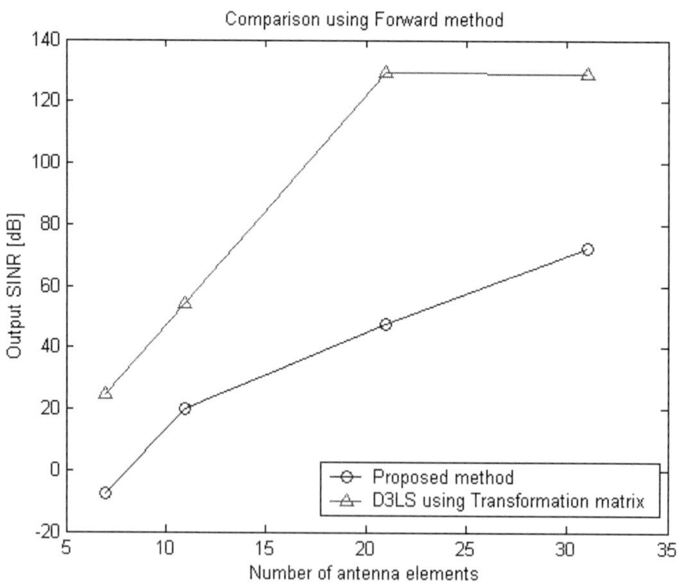

Figure 12.5a. Comparison of the output SINR between the proposed method and a more accurate compensation for mutual coupling using the forward method.

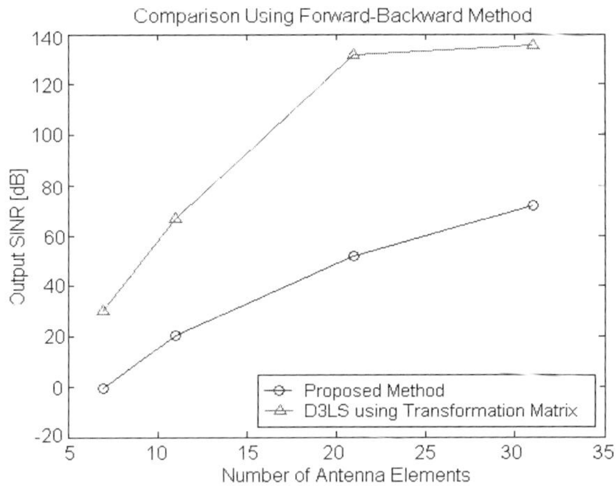

Figure 12.5b. Comparison of the output SINR between the proposed method and a more accurate compensation for mutual coupling using the forward-backward method.

For the third example, we simulate the same array but put some additional dummy antenna elements at the ends of the antenna array. In this case, the measured element pattern will be approximately the same for all the elements of the array including the ones at the end. We receive the same signal and the interferers using this modified array, which has dummy elements at the end, and apply the proposed adaptive algorithm to the voltages received by this modified array. In this simulation we take the SOI of unit amplitude to be arriving from $\theta_s = 90°$. Two jammers are present at 70° and 120°. And the intensity of the SOI is 1V/m and the intensity of the interferers is varied from 30 V/m to 1000 V/m. Simulation results are shown in Tables 12.2 through 12.5. As shown in the tables, the estimation for the SOI can be improved significantly when we use additional dummy elements at the end of the array. However, as the intensity of the interferer increases, a more accurate compensation method described in [7] may be necessary.

Table 12.2. Output SINR as a Function of the Number of Dummy Elements at the Two Ends of the Array (Intensity of SOI: 1 V/m, Intensity of Jammers: 30 V/m).

# of Antenna Elements		Number of Dummy Elements (Each Side)			
		0	1	2	3
7	$(\text{SINR})^{\alpha}_{\text{OUT}}$	$1.81 - j\,1.12$ (−2.80 dB)	$1.26 - j\,0.27$ (8.4145 dB)	$0.87 + j\,0.23$ (11.52 dB)	$1.00 + j\,0.03$ (29.26 dB)
11	$(\text{SINR})^{\alpha}_{\text{OUT}}$	$1.06 + j\,0.38$ (8.25 dB)	$1.04 + j\,0.05$ (23.44 dB)	$0.99 - j\,0.01$ (37.02 dB)	$1.01 + j\,0.01$ (36.85 dB)
21	$(\text{SINR})^{\alpha}_{\text{OUT}}$	$1.00 + j\,0.00$ (51.92 dB)	$1.00 + j\,0.00$ (55.22 dB)	$1.00 + j\,0.00$ (56.57 dB)	$1.00 - j\,0.00$ (64.97 dB)

Table 12.3. Output SINR as a Function of the Number of Dummy Elements at the Two Ends of the Array (Intensity of SOI: 1 V/m, Intensity of Jammers: 100 V/m).

# of Antenna Elements		Number of Dummy Elements (Each Side) 0	1	2	3
7	$(\mathrm{SINR})_{\mathrm{OUT}}^{\alpha}$	$3.70 - j\,3.73$ (−13.26 dB)	$1.88 - j\,0.90$ (−2.01 dB)	$0.57 + j\,0.77$ (1.06 dB)	$1.02 + j\,0.11$ (18.80 dB)
11	$(\mathrm{SINR})_{\mathrm{OUT}}^{\alpha}$	$1.19 + j\,1.28$ (−2.21 dB)	$1.14 + j\,0.18$ (12.99 dB)	$0.98 - j\,0.04$ (26.56 dB)	$1.04 + j\,0.03$ (26.39 dB)
21	$(\mathrm{SINR})_{\mathrm{OUT}}^{\alpha}$	$0.99 + j\,0.00$ (41.46 dB)	$0.99 + j\,0.00$ (44.76 dB)	$0.99 + j\,0.00$ 46.12 dB)	$1.00 - j\,0.00$ (54.52 dB)

Table 12.4. Output SINR as a Function of the Number of Dummy Elements at the Two Ends of the Array (Intensity of SOI: 1 V/m, Intensity of Jammers: 500V/m).

# of Antenna Elements		Number of Dummy Elements (Each Side) 0	1	2	3
7	$(\mathrm{SINR})_{\mathrm{OUT}}^{\alpha}$	$14.51 - j\,8.63$ (−27.24 dB)	$5.41 - j\,4.51$ (−15.99 dB)	$-1.17 + j\,3.85$ (−12.92 dB)	$1.09 + j\,0.57$ (4.82dB)
11	$(\mathrm{SINR})_{\mathrm{OUT}}^{\alpha}$	$1.94 + j\,6.39$ (−16.20 dB)	$1.69 + j\,0.88$ (−0.99 dB)	$0.92 - j\,0.22$ (12.58 dB)	$0.92 - j\,0.22$ (12.41 dB)
21	$(\mathrm{SINR})_{\mathrm{OUT}}^{\alpha}$	$0.96 + j\,0.01$ (27.48 dB)	$0.97 + j\,0.01$ (30.78 dB)	$0.98 + j\,0.01$ (32.14 dB)	$1.00 - j\,0.01$ (40.54 dB)

Table 12.5. Output SINR as a Function of the Number of Dummy Elements at the Two Ends of the Array (Intensity of SOI: 1 V/m, Intensity of Jammers: 1000 /m).

# of Antenna Elements		Number of Dummy Elements (Each Side) 0	1	2	3
7	$(\mathrm{SINR})_{\mathrm{OUT}}^{\alpha}$	$28.02 - j\,7.27$ (−33.26 dB)	$9.82 - j\,9.01$ (−22.01 dB)	$-3.35 + j\,7.71$ (−18.94 dB)	$1.18 + j\,1.13$ (−1.20 dB)
11	$(\mathrm{SINR})_{\mathrm{OUT}}^{\alpha}$	$2.88 + j\,12.76$ (−22.21 dB)	$2.39 + j\,1.76$ (−7.01 dB)	$0.83 - j\,0.44$ (6.56 dB)	$1.40 + j\,0.26$ (6.39 dB)
21	$(\mathrm{SINR})_{\mathrm{OUT}}^{\alpha}$	$0.92 + j\,0.03$ (21.46 dB)	$0.94 + j\,0.02$ (24.76 dB)	$0.95 + j\,0.02$ (26.12 dB)	$1.00 - j\,0.02$ (34.52 dB)

For the final example, we consider a semi-circular array (SCA). The SCA is analyzed using an numerical electromagnetic code. The signal intensity is then recovered using the proposed method, by using the voltages induced in the elements of the array. The antenna elements have the same dimension as presented in Table 12.1. In this simulation we consider the SOI of unit amplitude is arriving from $\theta_s = 100°$. There are two jammers which are coming from 60° and 120°. The signal intensity is set to 1 V/m and the intensities of these coherent jammers are varied from 1 to 50 V/m. In this case we observe the performance of the forward method as a function of the number of antenna elements in the circular array. As the number of elements increases the array dimension also increases as shown in Table 12.6. Table 12.7 provides the output SINR for the circular array as a function of the number of antenna elements for the forward method. It is clear that a more accurate compensation for the mutual coupling is necessary when the intensity of the jammer increases, as explained in [3,7,10].

Again the performance of the adaptive algorithm improves as the number of elements in the array is increased. In addition, when the jammers are much stronger than the signal a more accurate method for treating mutual coupling between the antenna elements presented in [3] should be employed.

Table 12.6. Radius of the SCA.

# of Elements	Radius
7	1.115 m
11	1.752 m
21	3.344 m
31	4.936 m

Table 12.7. Output SINR as a Function of the Number of Antenna Elements in the SCA Using the Forward Method (Intensity of Signal: 1 V/m).

# of Antenna Elements		Intensities of the Two Jammers 1 V/m	30 V/m	50 V/m
7	$(\mathrm{SINR})_{\mathrm{OUT}}^{\alpha}$	$1.67 + j\,0.11$ (3.28 dB)	$21.26 + j\,3.44$ (–26.26 dB)	$34.77 + j\,5.74$ (–30.69 dB)
11	$(\mathrm{SINR})_{\mathrm{OUT}}^{\alpha}$	$1.30 - j\,0.17$ (9.19 dB)	$10.13 - j\,5.00$ (–20.35 dB)	$16.22 - j\,8.33$ (–24.79 dB)
21	$(\mathrm{SINR})_{\mathrm{OUT}}^{\alpha}$	$0.92 - j\,0.05$ (20.69 dB)	$-1.27 - j\,1.58$ (–8.85 dB)	$-2.79 - j\,2.64$ (–13.29 dB)
31	$(\mathrm{SINR})_{\mathrm{OUT}}^{\alpha}$	$0.99 - j\,0.03$ (30.38 dB)	$0.68 - j\,0.85$ (0.83 dB)	$0.47 - j\,1.4$ (–3.60 dB)

12.4 REASON FOR A DECLINE IN THE PERFORMANCE OF THE ALGORITHM WHEN THE INTENSITY OF THE JAMMER IS INCREASED

The performance of the proposed method for compensating for mutual coupling using the in-situ antenna element patterns deteriorates when one increases the intensity of the jammers with respect to the SOI. This is because when one is using the embedded element pattern, it is equivalent to use of the measured voltages at the loads of the antenna elements due to the SOI only. This value is affected by the port admittance matrix of the array. When mutual coupling is present, the port admittance matrix is not diagonal. Because of these additional non-diagonal terms the equations used in (12.8), (12.12), and (12.15) do not exactly cancel the signal when we take the difference between the two ratios in the elements of the matrix. That is why when the jammer intensity increases it is necessary to introduce a more accurate compensation methodology using the transformation matrix. The use of a SCA provides worse results than a linear array as the influence of the port admittance matrix is more dominant in the circular array as the influence from neighboring elements are increased in a SCA over that of a linear array.

12.5 CONCLUSION

We have presented a new technique that partially compensates for the effect of mutual coupling among the elements of an array based on a direct data domain least squares algorithm using the in-situ embedded element patterns only. Since no statistical methodology is employed in the proposed adaptive algorithm, there is no need to compute a covariance matrix. Therefore, this procedure can be implemented on a general-purpose digital signal processor for real-time implementations. As shown in the numerical examples, the proposed method provides a good estimate for the complex amplitude of the signal of interest when the jammer intensities are not high. We also investigated the relationship between the number of antenna elements in the array and the estimate of the signal, both in the presence and absence of dummy elements at end of the array. When we increase the number of elements of the array we can get a higher output signal-to-interference plus noise ratio (output SINR), even for a non-linear array like a semi-circular array.

REFERENCES

[1] T. K. Sarkar, S. Park, J. Koh and R. A. Schneible, "A Deterministic Least Squares Approach to Adaptive Antennas", *Digital Signal Processing - A Review Journal*, Vol. 6, pp. 185-194, 1996.

[2] T. K. Sarkar and R. S. Adve, "Space Time Adaptive Processing Using Circular Arrays", IEEE Antennas and Propagation Magazine, Vol. 43, No. 1, pp. 138-143, Feb. 2001.

[3] T. K. Sarkar, M. Wicks, M. Salazar-Palma and R. Bonneau, *Smart Antennas*, John Wiley & Sons, Hoboken, NJ, 2003.

[4] I. J. Gupta and A. A. Ksienski, "Effect of Mutual Coupling on the Performance of Adaptive Arrays", *IEEE Trans. Antennas and Propagation*, vol. AP-31, pp. 785–791, Sept. 1983.

[5] R. S. Adve and T. K. Sarkar, "Compensation for the Effects of Mutual Coupling in Adaptive Algorithms", *IEEE Trans. Antennas and Propag.*, Vol. 48, No. 1, pp. 86–94, Jan 2000.

[6] W. S. Choi and T. K. Sarkar, "Phase-Only Adaptive Processing Based on a Direct Data Domain Least Squares Approach Using the Conjugate Gradient Method", *IEEE Trans. Antennas and Propag.*, Vol. 52, No. 12, pp. 3265–3272, Dec 2004.

[7] K. Kim, T. K. Sarkar, and M. Salazar Palma, "Adaptive Processing Using a Single Snapshot for a Nonuniformly Spaced Array in the Presence of Mutual Coupling and Near-Field Scatterers", *IEEE Transactions on Antennas and Propagation*, Vol. 50, No. 5, pp. 582-590, May 2002.

[8] B. M. Kolundzija, J. S. Ognjanovic and T. K. Sarkar, *WIPL-D, Electromagnetic Modeling of Composite Metallic and Dielectric Structures, Software and User's Manual*, Artech House, Norwood, MA, 2000.

[9] T. K. Sarkar, S. Nagaraja and M. C. Wicks, "A Deterministic Direct Data Domain Approach to Signal Estimation Utilizing Nonuniform and Uniform 2-D Arrays", *Digital Signal Processing – A Review Journal,* Vol. 8, 114-125 (1998).

[10] K. Kim, T. K. Sarkar, H. Wang and M. Salazar Palma, "Direction of Arrival Estimation Based on Temporal and Spatial Processing Using a Direct Data Domain Approach", *IEEE Transactions on Antennas and Propagation*, Vol. 52, Issue: 2, Feb. 2004, pp. 533-541.

13

SIGNAL ENHANCEMENT THROUGH POLARIZATION ADAPTIVITY ON TRANSMIT IN A NEAR-FIELD MIMO ENVIRONMENT

13.0 SUMMARY

In this chapter polarization adaptivity on transmit has been used to enhance the received signals directed to a pre-selected receiver in a near-field multi-input multi-output (MIMO) environment. The objective here is to select a set of weights on the transmitting antennas adapted to individual receivers based on the principles of reciprocity. Using the polarization properties, when the number of receiving antennas is greater than the number of transmitting antennas, the transmitted signal may be directed more to a particular receiver location while simultaneously minimizing the reception signal strength at other receivers. Wideband performance of this system is also studied. Numerical simulations have been made to illustrate the novelty of the proposed approach.

13.1 INTRODUCTION

Many methods have been developed in recent years, to enhance reception of signals in a multi-input multi-output (MIMO) environment [1–10]. For mobile communication, the development of a methodology that mitigates the deleterious effects of multipath fading, near-field scatterers (buildings, trees, and platforms), etc., is necessary for improved reception. How to counteract multipath fading in an adaptive antenna, which can significantly improve the performance of a system, has also been discussed [11–15]. In this chapter, an alternative way of directing the signals from the transmitter to a specific receiving antenna is discussed. Spatial diversity on transmit is obtainable at a base station, if that station has multiple antennas and the signals being fed to each transmitting antenna corresponding to a particular receiver are weighted. At a mobile receiver there is generally no spatial diversity. In this way, the transmitted signals would be received at the designated receiver, when weighted appropriately while it would be canceled at the other locations. By providing spatial diversity on transmit it is possible to mitigate the effects of multipath fading, as the directed energy from the transmitted antennas would combine vectorially at the selected receiving antenna element either to produce a maximum or a minimum. This methodology is based on the reciprocity theorem [16,17] and is applied to a collection of receiving and transmitting antennas. Unlike conventional MIMO, which requires an array of receiving antennas, this method is more like MISO, where the signal is directed to a particular receiver. This proposed methodology is different from the transmit diversity techniques based on reciprocity presented in [3–10], as the effects of antenna mutual coupling and other near-field effects are taken into account in the present methodology. The present methodology works over a finite bandwidth which may cover the frequency corresponding to the up and the down links. The goal here is to select a set of weights that will induce large currents at a specific receiver, while simultaneously the currents induced on the other receivers would be minimized. This is very convenient for mobile systems where it may be difficult to place multiple antennas for receptions.

The objective of the present method is to enhance the signal strength rather than the power because we may be dealing with near-field environments. In a near-field environment, it is necessary to deal with the Poynting vector, which requires knowledge of both the electric and the magnetic fields or, in other words, the voltage and the current at the feed points of the elements of the transmit antenna array. In a near-field environment, since the port currents at the antenna are determined by the electromagnetic environment, optimization of the received power is not possible for a realistic environment, as it changes with time.

This methodology is always applicable as long as the number of transmitting antennas N is greater than or equal to the number of receivers M. On the other hand, if the number of receivers exceeds the number of transmitting antennas ($M > N$), it may not be possible to achieve a perfect match (namely simultaneously maximizing the induced current at pre-selected receivers, with practically zero current induced on the rest of the receivers). These statements will be illustrated through numerical examples. Also, using the polarization properties, one can enhance the signal strength at a particular receiver while simultaneously minimizing it at the other receivers.

In section 13.2, the principles of the signal enhancement methodology through adaptivity on transmit is reviewed. Section 13.3 discusses the polarization properties related to the transmit-receive systems. In section 13.4 some numerical examples which have been analyzed using an electromagnetic simulation code are presented. This chapter does not address how the actual communication takes place and what type of modulation and pulse widths are used, but rather treat the feasibility of the concept. Conclusions follow in section 13.5.

13.2 SIGNAL ENHANCEMENT METHODOLOGY THROUGH ADAPTIVITY ON TRANSMIT

In this procedure, one simultaneously employs the principle of reciprocity and the concept of adaptivity on transmit. The spatial diversity of fixed proximate transmitting antennas permits signals transmitted from a base station to be directed to a pre-selected mobile station without worrying about the presence of other near-field scatterers or the existence of a multipath environment. To illustrate the methodology for N transmitting and M independent disjoint mobile receiving antennas operating at the same frequency f_0, consider the communication between a base station with two transmitting antennas (T_1 and T_2) and two independent mobile receiving antennas (R_1 and R_2) that is depicted in Figure 13.1. Let $V_{T_1}^{R_1}$ and $V_{T_2}^{R_1}$ be the induced voltages on the load resistances located at the feed points of the antennas placed on transmitters T_1 and T_2, when the antenna on receiver R_1 transmits with an excitation voltage of 1V. Here, the superscript R_1 corresponds to the situation when receiver R_1 is transmitting and the antennas on the two transmitters T_1 and T_2 are operating in the receive mode. Thus the subscript denotes the receiving element whereas the superscript specifies the transmitting element. The known load impedances at the two

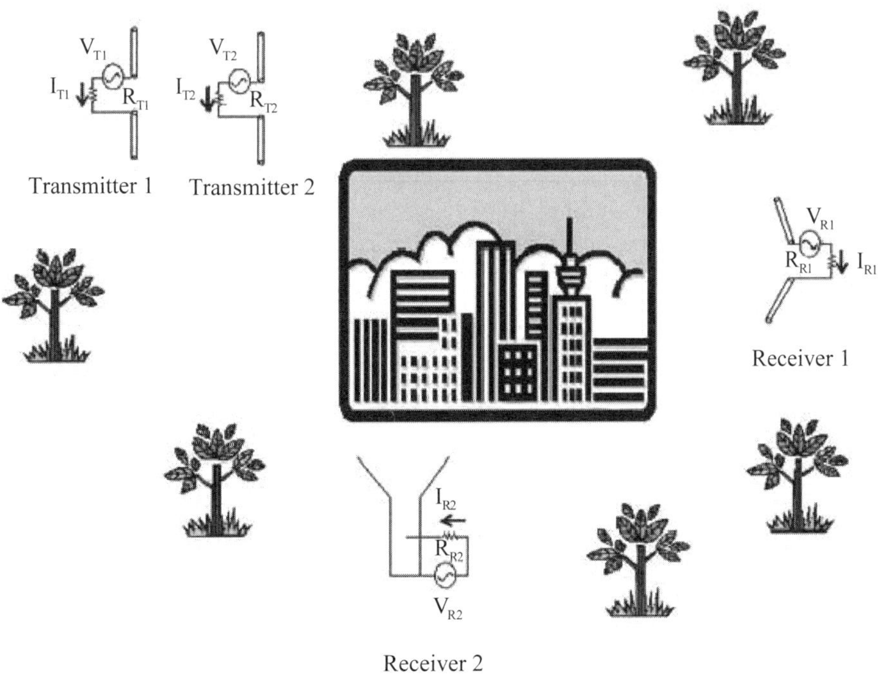

Figure 13.1. A multiple-user transmit/receive scenario.

transmit antennas are Z_{T_1} and Z_{T_2}. These loads conjugately match the transmit antennas for maximum power transfer. Then the currents induced at the feed points of the two transmit antennas are $I_{T_1}^{R_1} = V_{T_1}^{R_1} / Z_{T_1}$ and $I_{T_2}^{R_1} = V_{T_2}^{R_1} / Z_{T_2}$.

Similarly, when the receiving antenna R_2 transmits with an applied input voltage of 1V, then the respective voltages $V_{T_1}^{R_2}$ and $V_{T_2}^{R_2}$ induced in antennas located on transmitters T_1 and T_2 are known. The superscript R_2 corresponds to the case when the receiving antenna R_2 is transmitting. So now the currents induced at the two transmit antennas are $I_{T_1}^{R_2} = V_{T_1}^{R_2} / Z_{T_1}$ and $I_{T_2}^{R_2} = V_{T_2}^{R_2} / Z_{T_2}$. Using these four currents ($I_{T_1}^{R_1}, I_{T_2}^{R_1}, I_{T_1}^{R_2}$, and $I_{T_2}^{R_2}$), it is possible to find a set of corresponding weighted excitations ($W_{R_1}^{T_1}, W_{R_1}^{T_2}, W_{R_2}^{T_1}$, and $W_{R_2}^{T_1}$) to be applied at the transmitting antennas so that the induced current can be simultaneously maximized at the antenna located on receiver R_1 and the induced current minimized in the antenna located on receiver R_2 and vice versa based on the principles of reciprocity. When these weights are used in pairs as excitations on each of the two transmitting antennas, then one will observe that excitations with the following weightings, $W_{R_1}^{T_1}$ and $W_{R_1}^{T_2}$, will enhance the induced current at the antenna located on receiver R_1 while minimizing the induced current at the antenna located on receiver R_2. The subscript on each W specifies the receiving antenna element at which the induced current is enhanced. Similarly, if $W_{R_2}^{T_1}$ and $W_{R_2}^{T_2}$ are used as excitations on the two transmitting antennas, then this will enhance the induced current at the antenna located on receiver R_2 while minimizing the induced current at the antenna located on receiver R_1.

Through the application of the reciprocity principle [16,17], it is now illustrated how this can be accomplished. Consider the system represented in Figure 13.1 and assume that the antenna located on receiver R_1 is transmitting with a 1 V excitation. Then the induced currents at the loads located at the feed points of the two transmitting antennas will be $I_{T_1}^{R_1}$ and $I_{T_2}^{R_1}$. The excitation at the antenna located on receiver R_1 will also induce a voltage at the antenna located on receiver R_2, which is ignored because it is superfluous to this discussion. This does not mean that this induced voltage is small, but it does not enter into the theory! Now exciting antenna on transmitter T_1 with 1 V will induce currents $I_{R_1}^{T_1}$ and $I_{R_2}^{T_1}$, respectively, at the loads located at the feed points of the two receiving antennas. The superscript T_1 indicates that in this case only transmitter T_1 is active. Moreover, the 1 V excitation of transmitter T_1 also induces a current at transmitter T_2 that is not germane to this development. Now if one applies the principle of reciprocity between the excitation voltages and currents flowing at the two feed ports corresponding to antennas located on transmitter T_1 and receiver R_1, one observes that the respective currents are related by

$$I_{T_1}^{R_1} = I_{R_1}^{T_1}, \tag{13.1}$$

for 1 V excitations. Similarly, if the reciprocity principle is applied to the feed ports of antennas on transmitter T_1 and receiver R_2, then

$$I_{T_1}^{R_2} = I_{R_2}^{T_1}. \tag{13.2}$$

Therefore, exciting the antenna on transmitter T_1 with voltage W^{T_1} in (13.1) and (13.2) induces currents equal to $W^{T_1} I_{T_1}^{R_1}$ at antenna on receiver R_1 and $W^{T_1} I_{T_1}^{R_2}$ at antenna on receiver R_2, by reciprocity applied to the respective ports of the transmitting and receiving antennas.

If we now excite the antenna on receiver R_2 with 1V, then currents $I_{T_1}^{R_2}$ and $I_{T_2}^{R_2}$ are induced at the antennas on transmitters T_1 and T_2, respectively, and the induced current in the antenna on receiver R_1 is ignored, as it is not germane to the present discussions. Recall that the superscript R_2 on the currents implies that antenna on receiver R_2 is transmitting. Now exciting antenna on transmitter T_2 with 1 V induces currents $I_{R_1}^{T_2}$ and $I_{R_2}^{T_2}$, respectively, at the loads located at the feed points of the two receiving antennas, as well as an inconsequential current at the inactive transmitting antenna T_1. If one now applies the principle of reciprocity between the excitation voltages and currents at the two feed ports corresponding to antennas on transmitter T_2 and receiver R_1, then the respective currents are related by

$$I_{T_2}^{R_1} = I_{R_1}^{T_2}, \tag{13.3}$$

when antennas on transmitter T_2 and receiver R_1 are excited with 1 V. Similarly, if we apply the same principle of reciprocity between the feed ports of antennas located on transmitter T_2 and receiver R_2, then we will obtain

$$I_{T_2}^{R_2} = I_{R_2}^{T_2}. \tag{13.4}$$

Therefore, exciting antenna on transmitter T_2 with voltage W^{T_2} and applying the principle of reciprocity to the respective ports of the transmitting and the receiving antennas will induce currents equal to $W^{T_2} I_{T_2}^{R_1}$ at the antenna on receiver R_1 and $W^{T_2} I_{T_2}^{R_2}$ at the antenna on receiver R_2.

Next, the principle of superposition is applied to modify some of these induced currents. Suppose antennas on transmitters T_1 and T_2 are excited with voltages $W_{R_1}^{T_1}$ and $W_{R_1}^{T_2}$, respectively. The subscript R_1 symbolically specifies the goal of maximizing the induced current in antenna on receiver R_1 while ensuring the current in antenna on receiver R_2 is practically zero. Assume that this maximum current is 1A. Under these conditions, the desired total currents induced in antennas on receivers R_1 and R_2, respectively, are

$$W_{R_1}^{T_1} I_{T_1}^{R_1} + W_{R_1}^{T_2} I_{T_2}^{R_1} = 1, \tag{13.5}$$

$$W_{R_1}^{T_1} I_{T_1}^{R_2} + W_{R_1}^{T_2} I_{T_2}^{R_2} = 0. \tag{13.6}$$

Similarly, it is possible to choose a set of excitations, $W_{R_2}^{T_1}$ and $W_{R_2}^{T_2}$, to be applied to the antennas on transmitters T_1 and T_2, such that no current is induced in antenna on receiver R_1 and the current induced at antenna on receiver R_2 is 1A. Under these conditions, the total current induced in antenna on receivers R_1 and R_2 will be

$$W_{R_2}^{T_1} I_{T_1}^{R_1} + W_{R_2}^{T_2} I_{T_2}^{R_1} = 0, \tag{13.7}$$

$$W_{R_2}^{T_1} I_{T_1}^{R_2} + W_{R_2}^{T_2} I_{T_2}^{R_2} = 1. \tag{13.8}$$

Equations (13.5)–(13.8) can be written more compactly in matrix form as

$$\begin{pmatrix} I_{T_1}^{R_1} & I_{T_2}^{R_1} \\ I_{T_1}^{R_2} & I_{T_2}^{R_2} \end{pmatrix} \times \begin{pmatrix} W_{R_1}^{T_1} & W_{R_2}^{T_1} \\ W_{R_1}^{T_2} & W_{R_2}^{T_2} \end{pmatrix} = \begin{pmatrix} 1 & 0 \\ 0 & 1 \end{pmatrix} \tag{13.9}$$

By using (13.9), one can solve for an a priori set of excitations that will direct the signal to a pre-selected receiver by vectorially combining the signal from the two transmitting receivers. The excitations are obtained by inverting the current matrix in (13.9) to yield

$$\begin{pmatrix} W_{R_1}^{T_1} & W_{R_2}^{T_1} \\ W_{R_1}^{T_2} & W_{R_2}^{T_2} \end{pmatrix} = \begin{pmatrix} I_{T_1}^{R_1} & I_{T_2}^{R_1} \\ I_{T_1}^{R_2} & I_{T_2}^{R_2} \end{pmatrix}^{-1} \begin{pmatrix} 1 & 0 \\ 0 & 1 \end{pmatrix} \tag{13.10}$$

The caveat here is if antennas on transmitters T_1 and T_2 are excited by $W_{R_1}^{T_1}$ and $W_{R_1}^{T_2}$, respectively, then the total induced current due to all the electromagnetic signals will be vectorially additive at the load, which is located at the feed point of antenna on receiver R_1, and would be destructive at the feed point of antenna on receiver R_2. In contrast, if we apply $W_{R_2}^{T_1}$ and $W_{R_2}^{T_2}$ to antennas on transmitters T_1 and T_2, then the received electromagnetic signal will be vectorially destructive at the load, which is located at the feed point of antenna on receiver R_1, and will be vectorially additive at the feed point of antenna on receiver R_2, generating a large value for the induced load current.

In short, by knowing the voltages that are induced in each of the transmitting antennas by every receiver, it is possible to select a set of weights based on reciprocity that will induce large currents at a specific receiving antenna. This relationship, based on the principles of reciprocity and superposition, can be applied only at the terminals of the transmitting and receiving antennas. This also assumes that there exists a two-way link between the transmitter and the receiver. Furthermore, this principle of directing the signal energy to a pre-selected receiver is independent of the sizes and shapes of the receiving antennas and the near field environments.

13.3 EXPLOITATION OF THE POLARIZATION PROPERTIES IN THE PROPOSED METHODOLOGY

Polarization diversity at the transmit antennas can also be utilized as shown in Figure 13.2. Either an antenna transmitting different linear polarizations can be switched on or one can be dealing with circular polarization. However, the principles are the same in both cases. Using the polarization properties, one can also enhance the signal strength at a particular receiver while simultaneously minimizing it at the other receivers. For the example in Figure 13.2, each of the receivers can have arbitrary polarization whereas the transmitter can switch to either vertical or horizontal polarization. Again the mathematical principles to direct the signal to a pre-selected receiver and producing zero signals at the others follow exactly as presented in the previous section. These principles are now explained though numerical simulations.

13.4 NUMERICAL SIMULATIONS

For the first example, three transmitting and three receiving antennas operating in free space is considered. Two kinds of antennas are chosen as possible candidates for either transmitting or receiving; helical and biconical antennas. Helices produce circular polarization but they are not as broadband as the bicones. However, the bicones produce linear polarization. For the second example, a triad of three transmitting helical antennas and a triad of three receiving helical antennas which are placed inside a finite conducting cylinder are used. In the final example, four receiving antennas and two transmitting antennas are chosen. Two cases are considered for this example. In the first case these antennas are placed in free space, and in the second case they are put inside a finite conducting cylinder with a metal partition so that there is no line of sight propagation of the direct path.

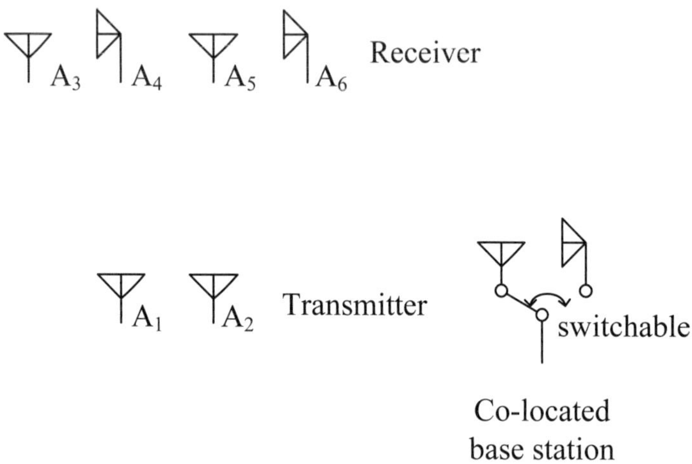

Figure 13.2. Configurations for multiple-users using linear polarizations.

13.4.1 Example 1

Consider three transmitting helical antennas A_1, A_2, and A_3, located at a base station. Each helical antenna has a circumference C = 0.3 m. The parameter of each of the antenna is: diameter of the helix D = C/ π = 0.0955 m, pitch angle α = 13°, the spacing between turns S = C $\times$ tan (α), length of one turn L = $\sqrt{C^2 + S^2}$ = 0.3079 m, number of turns n = 10, and the axial length A = n*S = 0.6926 m. The operating frequency is 1 GHz. Next the three receiving antennas marked as A_4, A_5, and A_6 in Figure 13.3 are considered. Dimension of the receiving antennas are the same and they are separated from the transmitters by a distance of 6 m. All of the six antennas are loaded with 140 Ω at the feed point.

First consider maximizing the currents induced in the antenna on receiver A_4. The antenna on receiver A_4 is excited with 1V, which induces currents in antennas of receivers A_5 and A_6 and antennas on transmitters A_1, A_2, and A_3. These currents are computed by an electromagnetic analysis code. In turn, these induced currents generate voltages across the loads of the other five loaded helices. The induced currents at the antennas of transmitters A_1, A_2, and

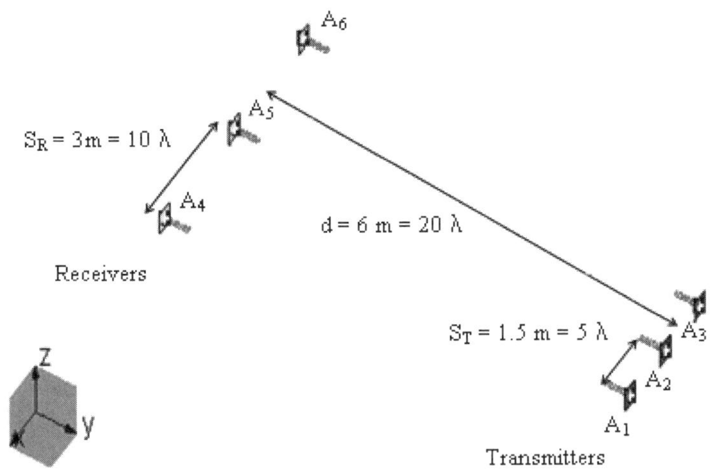

Figure 13.3. A six helical antenna transmit/receive system.

A_3 are $I_{A_1}^{A_4}$, $I_{A_2}^{A_4}$, and $I_{A_3}^{A_4}$, respectively. As noted earlier, the currents induced in antennas of A_5 and A_6 are not considered, as they are not relevant in the present discussions. The induced currents have been obtained using the electromagnetic analysis code [18].

Next, when antenna on receiver A_5 is excited with 1V, it induces the currents $I_{A_1}^{A_5}$, $I_{A_2}^{A_5}$, and $I_{A_3}^{A_5}$ on antennas located on transmitters A_1, A_2, and A_3, respectively. Similarly, exciting antenna of receiver A_6 with 1V induces currents $I_{A_1}^{A_6}$, $I_{A_2}^{A_6}$, and $I_{A_3}^{A_6}$ in antennas of transmitters A_1, A_2, and A_3. Based on the available information, the claim is that one can choose a set of complex voltages $\{W_{A_i}^{A_1}$, $W_{A_i}^{A_2}$ and $W_{A_i}^{A_3}\}$, for i = 4, 5, or 6, which when exciting the three antennas located on each of the transmitters will result in an additive vectorial combination of the electromagnetic fields at antenna on receiver A_i while inducing zero currents at the antennas on the other two receivers. The currents in the antennas located on the receivers then would be

$$I^{A_4} = W_{A_4}^{A_1} I_{A_1}^{A_4} + W_{A_4}^{A_2} I_{A_2}^{A_4} + W_{A_4}^{A_3} I_{A_3}^{A_4} \quad \text{(at receiver } A_4\text{)}$$

$$I^{A_5} = W_{A_4}^{A_1} I_{A_1}^{A_5} + W_{A_4}^{A_2} I_{A_2}^{A_5} + W_{A_4}^{A_3} I_{A_3}^{A_5} \quad \text{(at receiver } A_5\text{)}$$

$$I^{A_6} = W_{A_4}^{A_1} I_{A_1}^{A_6} + W_{A_4}^{A_2} I_{A_2}^{A_6} + W_{A_4}^{A_3} I_{A_3}^{A_6} \quad \text{(at receiver } A_6\text{)}$$

The objective now is to select the excitations $W_{A_j}^{A_i}$ for each j of {1, 2 or 3} in such a fashion that the received currents are maximal at antenna of receiver A_j and zero at the other antennas of the other receivers.

To determine the weight vectors that should induce the maximum current to antenna of only one receiver A_i ($i = 4$, 5 and 6), one can solve

$$\begin{pmatrix} I_{A_1}^{A_4} & I_{A_2}^{A_4} & I_{A_3}^{A_4} \\ I_{A_1}^{A_5} & I_{A_2}^{A_5} & I_{A_3}^{A_5} \\ I_{A_1}^{A_6} & I_{A_2}^{A_6} & I_{A_3}^{A_6} \end{pmatrix} \begin{pmatrix} W_{A_4}^{A_1} & W_{A_5}^{A_1} & W_{A_6}^{A_1} \\ W_{A_4}^{A_2} & W_{A_5}^{A_2} & W_{A_6}^{A_2} \\ W_{A_4}^{A_3} & W_{A_5}^{A_3} & W_{A_6}^{A_3} \end{pmatrix} = \begin{pmatrix} 1 & 0 & 0 \\ 0 & 1 & 0 \\ 0 & 0 & 1 \end{pmatrix} \tag{13.11}$$

to obtain

$$\begin{aligned} [\mathbf{W}_4, \mathbf{W}_5, \mathbf{W}_6] &= \begin{pmatrix} W_{A_4}^{A_1} & W_{A_5}^{A_1} & W_{A_6}^{A_1} \\ W_{A_4}^{A_2} & W_{A_5}^{A_2} & W_{A_6}^{A_2} \\ W_{A_4}^{A_3} & W_{A_5}^{A_3} & W_{A_6}^{A_3} \end{pmatrix} \\ &= \begin{pmatrix} 0.323 + 4.636j & -0.368 + 0.509j & 0.318 - 0.631j \\ 2.217 + 1.405j & -2.302 - 1.432j & 2.188 + 1.390j \\ 0.322 - 0.634j & 0.360 + 0.504 & 0.309 + 4.613j \end{pmatrix} \end{aligned} \tag{13.12}$$

To demonstrate the feasibility of this methodology, one can use these voltages of $\mathbf{W}_4$ as excitation inputs in the electromagnetic analysis code [18] to compute the induced currents on the antennas located on receivers A_4, A_5, and A_6 as

$$I^{A4} = 1 - 0\cdot001j, \quad I^{A5} = -0\cdot001 + 0\cdot001j, \quad I^{A6} = 0\cdot001 - 0\cdot001j. \tag{13.13}$$

where $R_{A_1} = R_{A_2} = R_{A_3} = 140\Omega$ and all currents are multiplied by 10^{-3}A. Clearly, all the electromagnetic signals are vectorially additive at the antenna on receiver A_4, and the currents are practically zero at the feeds of the antennas of receivers A_5 and A_6. Similarly, to direct the signals to the antenna of receiver A_5, one can use the computed voltages $\mathbf{W}_5$ in the electromagnetic analysis code to find the currents induced at the feed point of the antennas located on the various receivers as

$$I^{A4} = 0, \quad I^{A5} = 1.0, \quad I^{A6} = 0, \tag{13.14}$$

which clearly shows that the induced energy can be directed to antenna of receiver A_5 while producing no appreciable induced currents at the other two antennas located on receivers A_4 and A_6. Finally, to direct the signal from the transmitting antennas to the antenna located on receiver A_6, one can use the computed voltages $\mathbf{W}_6$ in the electromagnetic analysis code to find the feed currents at the antennas located on the receivers as

$$I^{A4} = -0.001, \quad I^{A5} = -j0.001, \quad I^{A6} = 1.0 \tag{13.15}$$

This clearly demonstrates that by appropriately choosing the complex values of the excitations at the different transmitting antennas it is possible to direct the signal so that it vectorially adds up at the antenna of a pre-selected receiver. No electromagnetic characterization of the environment is necessary.

Next, the behavior of the magnitude of the currents on antennas located on receivers A_4, A_5, and A_6 as a function of frequency is analyzed to observe what the useful bandwidth of the proposed methodology is. So one fixes the weights at the transmitting antennas which have been evaluated at 1 GHz and then one can use the same weights at other frequencies to observe how well this methodology works. The induced currents at each of the receivers are simulated in Figs. 13.4–13.6 over the 12.0% bandwidth from 0.94 GHz to 1.06 GHz when using the three set of frequency independent voltages $\{\mathbf{W}_4, \mathbf{W}_5, \mathbf{W}_6\}$ obtained for 1 GHz. As indicated in Figure 13.4 for $\mathbf{W}_4$, the induced currents at the antennas of the other receivers are down by a factor of 4.3 at the lower frequency and by a factor of 4.8 at the upper frequency. For the middle receiver, the induced currents at the antennas of the other receivers are down by a factor of 10 at the higher frequency end and by a factor of 12 at the lower frequency end points as shown in Figure 13.5. Finally the result in Figure 13.6 is just the reverse of Figure 13.4 as the six helical antennas have an axis of symmetry.

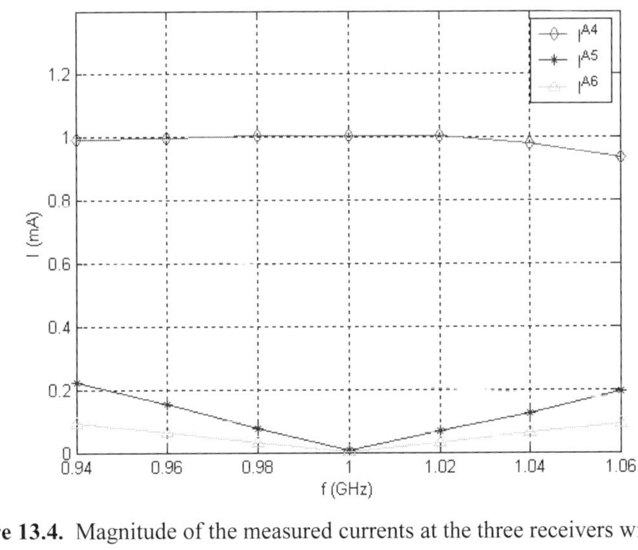

Figure 13.4. Magnitude of the measured currents at the three receivers with $\mathbf{W}_4$.

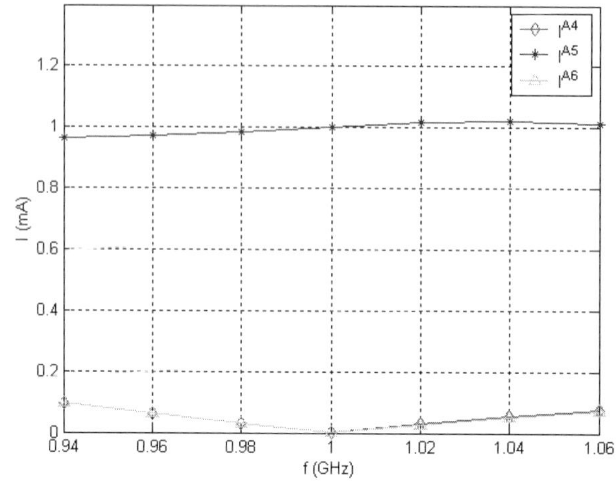

Figure 13.5. Magnitude of the measured currents at the three receivers with $\mathbf{W}_5$.

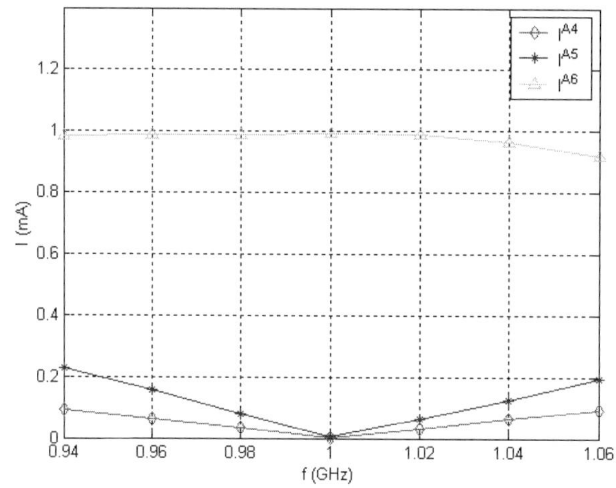

Figure 13.6. Magnitude of the measured currents at the three receivers with $\mathbf{W}_6$.

The useful bandwidth of this methodology can be increased by using biconical antennas instead of helices, as demonstrated next. The six helical antennas of Figure 13.3 are replaced by six biconical antennas as shown in Figure 13.7. The bicone antenna has an angle of 90° so that its input impedance is approximately 106 Ω. In this case, one can now sweep the frequency from 1.5 GHz to 1.7 GHz generating a 12.5% bandwidth, and the excitations are solved for in a similar fashion as outlined above.

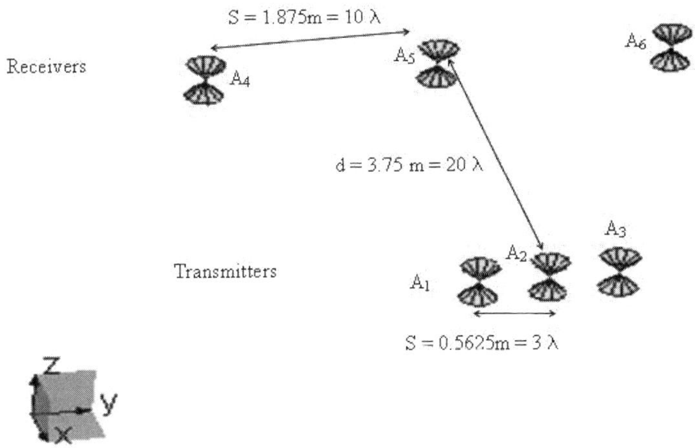

Figure 13.7. A six bicone antenna transmit/receive system.

As indicated in Figure 13.8, for a 12.5% bandwidth centered at 1.6 GHz, the induced currents at the antennas of the other receivers are down by a factor of 3 at the lower frequency end and by a factor of 5.7 at the upper frequency end. For the center receiver, the induced currents at the other receivers are down by a factor of 3 at the higher and by a factor of 6 at the lower frequency end points as shown in Figure 13.9. Finally the result for the third receiver in Figure 13.10 is identical to that of Figure 13.8 as the six antennas have an axis of symmetry.

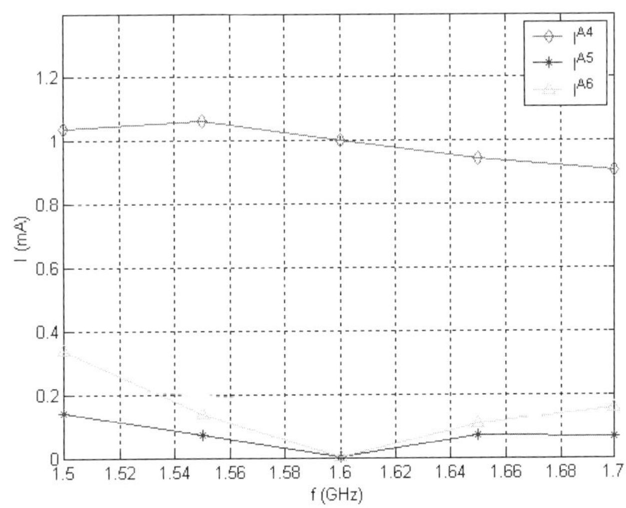

Figure 13.8. Magnitude of the measured currents at the three receivers with $\mathbf{W}_4$.

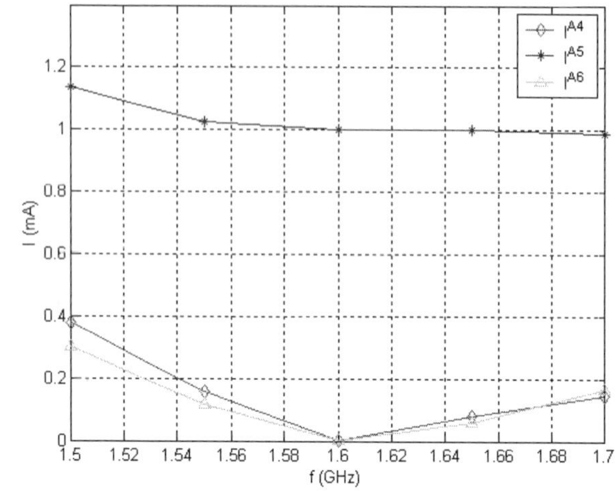

Figure 13.9. Magnitude of the measured currents at the three receivers with $\mathbf{W}_5$.

13.4.2 Example 2

Now, the six helical antennas are placed inside a concentric perfectly conducting cylinder as shown in Figure 13.11. The dimension of the conducting cylinder is 7.8 m × 4.8 m × 1.2 m. The transmitting and the receiving antenna sets are separated by 6 m and the inter-element spacing within each set is 1.5 m. All the six antennas are loaded by 140 Ω.

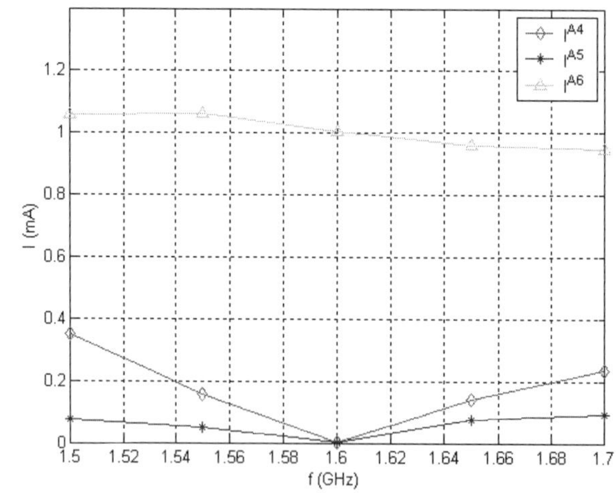

Figure 13.10. Magnitude of the measured currents at the three receivers with $\vec{\mathbf{W}}_6$.

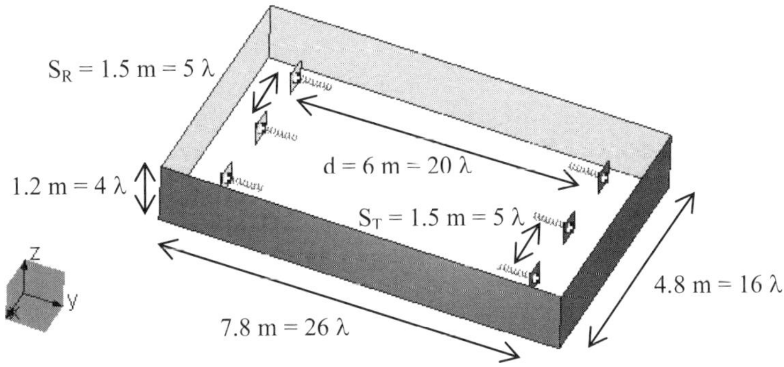

Figure 13.11. A six helical antenna transmit/receive system inside a conducting rectangular cylinder.

Along the same line, one can also choose a set of excitation voltages as follows

$$\left[\mathbf{W}_4, \mathbf{W}_5, \mathbf{W}_6\right] = \begin{pmatrix} W_{A4}^{A1} & W_{A5}^{A1} & W_{A6}^{A1} \\ W_{A4}^{A2} & W_{A5}^{A2} & W_{A6}^{A2} \\ W_{A4}^{A3} & W_{A5}^{A3} & W_{A6}^{A3} \end{pmatrix} \tag{13.16}$$
$$= \begin{pmatrix} -1.359 - 0.995j & -0.803 + 0.326j & 0.333 - 0.394j \\ -0.805 + 0.328j & -0.989 - 0.256j & -0.831 + 0.335j \\ 0.336 - 0.397j & -0.830 + 0.341j & -1.473 - 0.994j \end{pmatrix}.$$

so that the signals can be directed to a preselected receiver. For the three sets of excitation voltages Table 13.1 shows that the signal can be directed to each receiver.

Table 13.1. Complex Values of the Currents Measured at the Three Antennas Located on the Three Receivers Marked as 4, 5, and 6 for Three Different Choices of Excitations Given by (13.16).

By the excitation voltages	I^{A4}	I^{A5}	I^{A6}
$\mathbf{W}_4$	$1.0013 - j0.0008$	$-0.0023 - j0.0010$	$0.0004 + j0.0009$
$\mathbf{W}_5$	$-0.0004 + j0.0017$	$1.0016 + j0.0001$	$-0.0009 - j0.0029$
$\mathbf{W}_6$	$-0.0022 - j0.0010$	$0.0031 + j0.0042$	$0.9970 + j0.0007$

Next, the total power fed to each transmit antenna is calculated which will result in the received power to a single receiving antenna. The various input and output power requirements are presented in Table 13.2. (The numbers are rounded to 4 decimal places). The first column displays the magnitude of the total input power required at the three transmitting antennas and the second column provides the received power at the individual receiving antennas caused by the three sets of excitation voltages. The received power for each excitation is practically the same, 50 μW, because the induced current at the receiving antenna is 1 mA.

Table 13.2. The Magnitude of the Input Power at the Transmitting Antennas and the Received Power at the Receiving Antennas for the Three Choices of the Excitations.

Excitation Voltages	Input Power [mW]	Received Power [μW]
$\mathbf{W}_4$	19.0218	50.1352
$\mathbf{W}_5$	12.2887	50.1637
$\mathbf{W}_6$	20.8739	49.7052
TOTAL	52.1844	150.0041

Next the eigenvectors of the transfer function in this transmit/receive system for a conventional MIMO system is determined. The various transfer functions can be defined as

$$\begin{pmatrix} I_{A_4} \\ I_{A_5} \\ I_{A_6} \end{pmatrix} = \begin{pmatrix} H_{11} & H_{12} & H_{13} \\ H_{21} & H_{22} & H_{23} \\ H_{31} & H_{32} & H_{33} \end{pmatrix} \begin{pmatrix} V_{A_1} \\ V_{A_2} \\ V_{A_3} \end{pmatrix} \tag{13.17}$$

$$[\mathrm{H}] = \begin{pmatrix} H_{11} & H_{12} & H_{13} \\ H_{21} & H_{22} & H_{23} \\ H_{31} & H_{32} & H_{33} \end{pmatrix}$$
$$= \begin{pmatrix} \frac{I_{A_4}^{A_1}}{V_{A_1}}\Big|_{V_{A_2}=V_{A_3}=0} & \frac{I_{A_4}^{A_2}}{V_{A_2}}\Big|_{V_{A_1}=V_{A_3}=0} & \frac{I_{A_4}^{A_3}}{V_{A_3}}\Big|_{V_{A_1}=V_{A_2}=0} \\ \frac{I_{A_5}^{A_1}}{V_{A_1}}\Big|_{V_{A_2}=V_{A_3}=0} & \frac{I_{A_5}^{A_2}}{V_{A_2}}\Big|_{V_{A_1}=V_{A_3}=0} & \frac{I_{A_5}^{A_2}}{V_{A_3}}\Big|_{V_{A_1}=V_{A_2}=0} \\ \frac{I_{A_6}^{A_1}}{V_{A_1}}\Big|_{V_{A_2}=V_{A_3}=0} & \frac{I_{A_6}^{A_2}}{V_{A_2}}\Big|_{V_{A_1}=V_{A_3}=0} & \frac{I_{A_6}^{A_2}}{V_{A_3}}\Big|_{V_{A_1}=V_{A_2}=0} \end{pmatrix} \tag{13.18}$$

$I_{A_4}^{A_1}$ is the induced current at the feed points of the antenna on receiver A_4, when the antenna on transmitter A_1 operates with an excitation voltage of V_{A_1} V. Thus the subscript denotes the receiving element whereas the superscript specifies the transmitting element.

The various transfer function can then be obtained as

$$[H] = \begin{pmatrix} \frac{I_{A_4}^{A_1}}{1} & \frac{I_{A_4}^{A_2}}{1} & \frac{I_{A_4}^{A_3}}{1} \\ \frac{I_{A_5}^{A_1}}{1} & \frac{I_{A_5}^{A_2}}{1} & \frac{I_{A_5}^{A_2}}{1} \\ \frac{I_{A_6}^{A_1}}{1} & \frac{I_{A_6}^{A_2}}{1} & \frac{I_{A_6}^{A_2}}{1} \end{pmatrix}$$

$$= \begin{pmatrix} -0.2777 + 0.2807j & -0.1698 - 0.2768j & 0.2368 + 0.0957j \\ -0.1688 - 0.2753j & -0.4407 + 0.4592j & -0.1552 - 0.2733j \\ 0.2351 + 0.0962j & -0.1527 - 0.2734j & -0.2734 + 0.2619j \end{pmatrix}$$

In a conventional MIMO, three independent orthogonal propagation mechanisms are generated by performing a singular value decomposition (SVD) of [H] to form $[\mathrm{U}]^H[\mathrm{H}][\mathrm{V}] = [\Sigma]$. Therefore,

$$[\mathrm{H}] = [\mathrm{U}][\Sigma][\mathrm{V}]^H \tag{13.19}$$

where [U] is a 3 × 3 unitary matrix whose columns are the eigenvectors of $[\mathrm{H}][\mathrm{H}]^H$, $[\Sigma]$ is the 3 × 3 diagonal matrix with the singular values, σ_i^2, of [H] written through

$$[\Sigma] = \begin{pmatrix} 0.8834 & 0 & 0 \\ 0 & 0.5417 & 0 \\ 0 & 0 & 0.4082 \end{pmatrix}$$

and $[\mathrm{V}]$ is the 3 × 3 unitary matrix whose columns are the eigenvectors of $[\mathrm{H}]^H[\mathrm{H}]$ and are given by

$$[\mathrm{V}] = \begin{pmatrix} 0.3739 & 0.7547 & 0.5391 \\ -0.6471 + 0.5484j & 0.0392 - 0.0208j & 0.3939 - 0.3512j \\ 0.3750 - 0.0040j & -0.6540 + 0.0284j & 0.6554 - 0.0370j \end{pmatrix}.$$

The superscript H denotes the conjugate transpose of a matrix. Then the eigenvectors of $[\mathrm{H}]^H[\mathrm{H}]$ will be used as excitation voltage sets. Table 13.3 shows the total input power and received power for the eigenmodes when the eigenvector voltage sets are used as excitations. Note that the singular values and the received power for each of the eigenmode in a conventional MIMO caused by the eigenvector sets have a relationship given by

$$\frac{\sigma_2^2}{\sigma_1^2}, \frac{\sigma_3^2}{\sigma_1^2} = \frac{P_{reveived\ by\ 2^{nd}\ \text{eigenvector}}}{P_{reveived\ by\ 1^{st}\ \text{eigenvector}}}, \frac{P_{reveived\ by\ 3^{rd}\ \text{eigenvector}}}{P_{reveived\ by\ 1^{st}\ \text{eigenvector}}}$$

Therefore, to deliver the same amount of the received power 50 μW through each propagation mode in a conventional MIMO, the amount of input power needed for each mode is presented in Table 13.3. For the identical power of 50 μW received at each of the individual receivers delivered by the current procedure, the requirement for the total input power at the transmitter is also presented in Table 13.3. It is seen that using reciprocity, the dynamic range of the input power associated with each mode of transmission is much smaller than the ones in a conventional MIMO.

Table 13.3. The Magnitude of the Input Power at the Transmitting Antennas and the Received Power at the Receiving Antennas for Three Choices of Excitations Using the Six Helical Antenna Transmit/Receive System Inside of the Cylinder.

Using the Eigenvector Excitation Sets in MIMO			By Using Reciprocity, the Excitation Sets		
Excitation sets	Input Power [mW]	Received Power [μW]	Excitation Sets	Input Power [mW]	Received Power [μW]
1st eigenvector	6.62	50	$\mathbf{W}_4$	18.97	50
2nd eigenvector	17.22	50	$\mathbf{W}_5$	12.25	50
3rd eigenvector	28.37	50	$\mathbf{W}_6$	21.00	50
TOTAL	52.22	150	TOTAL	52.22	150

13.4.3 Example 3

Consider a situation where the number of receiving antennas is greater than the number of transmitting antennas, which is different from the previous examples where they have been equal. How the polarization properties are useful for the signal enhancement in this situation is examined.

Consider two transmitters and four receivers. All of them have the same polarization that is a right-hand circular polarization. The dimensions of the antennas are same as in the previous examples operating at 1 GHz. Antennas are located as described in Figure 13.12. The transmitting and the receiving antenna sets are separated by 12 m, the inter-element spacing of the receiving antennas is 1.5 m and the inter-element spacing of the transmitting antennas is 1.8 m. Since the number of transmitters is less than the number of receivers, it may not be possible to direct the signals to a particular receiver and cancel it at the other receivers. Table 13.4 presents the magnitudes of the currents at the four receiving antennas when the two transmitting antennas are simultaneously radiating energy. From Table 13.4, it is seen that there is approximately a 2.5 dB difference in the signal levels between the desired receivers and the neighboring one.

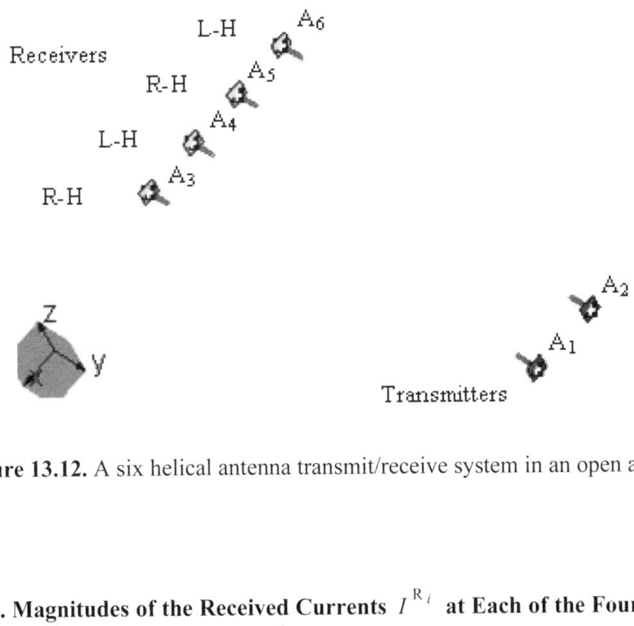

Figure 13.12. A six helical antenna transmit/receive system in an open area.

Table 13.4. Magnitudes of the Received Currents I^{R_i} at Each of the Four Receiving Antennas without Exploiting Polarization.

Excitation Voltages	$\lvert I^{A_3}\rvert$	$\lvert I^{A_4}\rvert$	$\lvert I^{A_5}\rvert$	$\lvert I^{A_6}\rvert$
$\mathbf{W}_3$	0.4654	0.3448	0.0651	0.3492
$\mathbf{W}_4$	0.3473	0.5382	0.3532	0.0709
$\mathbf{W}_5$	0.0684	0.3559	0.5338	0.3402
$\mathbf{W}_6$	0.3598	0.0741	0.3427	0.4625

Now the concept of polarization is introduced. Some of the receivers have left-handed helical antennas and some of them have right-handed antennas. For an axial mode helical antenna, the direction of the windings determines the sense of polarization. So, in this example, the transmission of two orthogonally polarized waves is examined. There are two types of antennas, which are co-located at the base station, and these antennas can be switched as illustrated in Fig. 13.2. In addition, users A_3 and A_5 use a right-hand polarization antenna and A_3 and A_6 has a left hand polarization antenna.

When the transmitters are using the R-H polarization mode, it will excite each mobile receiving antenna which has an R-H polarization, A_3 and A_5. If we measure the currents that are induced at the transmitting antennas located in the base station, when the two receivers are transmitting independently, then, one can choose a set of complex voltages $\{W_{A_i}^{A_1}$ and $W_{A_i}^{A_2}\}$, for $i = 3$, or 5, to direct the transmitted signal to A_3 or A_5. By using the polarization property, the induced currents at A_4 and A_6, which have L-H polarization, should be zero. The excitations to direct the signal to the receivers 3 and 5 (R-H receivers) are as follows,

$$[\mathbf{W}_3, \mathbf{W}_5] = \begin{bmatrix} W_{A_3}^{A_1} & W_{A_5}^{A_1} \\ W_{A_3}^{A_2} & W_{A_5}^{A_2} \end{bmatrix} = \begin{bmatrix} 0.4382 - j1.6414 & 0.9224 - j\,0.5240 \\ 1.2530 - j\,0.8170 & -1.5067 - j\,0.3129 \end{bmatrix}.$$

When the transmitters are using the L-H polarization mode of operation, one can choose a set of complex voltages $\{W_{A_i}^{A_1}$ and $W_{A_i}^{A_2}\}$, for $i = 4$, or 6, to direct the transmitted signals to A_4 or A_6. These weights are for the L-H receivers A_4 and A_6 as follows,

$$[\mathbf{W}_4, \mathbf{W}_6] = \begin{bmatrix} W_{A_4}^{A_1} & W_{A_6}^{A_1} \\ W_{A_4}^{A_2} & W_{A_6}^{A_2} \end{bmatrix} = \begin{bmatrix} -1.5106 - j0.3137 & 1.2336 - j\,0.8324 \\ 0.9273 - j\,0.5202 & 0.4306 - j\,1.6395 \end{bmatrix}.$$

Using these weights, when the polarization of the desired receiving antenna is the same as that of the transmitting antennas, the received currents at each of the four receiving antennas are shown in Table 13.5.

As seen from Table 13.5, using the specified excitations, when the transmitter is simultaneously transmitting simultaneously the two polarization modes, one obtains 1 mA at the antenna of receiver 3, 0.005 mA at antenna of receiver 5, and 0.08 mA at the antennas of receivers 4 and 6 which have a different polarization. The reason that the current at the antenna of receivers 4 and 6 are not zero is that the axial ratio of these helical antennas is 1.074. Theoretically, the formula for axial ratio is

$$|AR| = \frac{2\Upsilon + 1}{2\Upsilon}, \tag{13.20}$$

where Υ is the number of turns. So for this helical antenna AR is 1.05 theoretically. It thus indicates that the quality of the circular polarization improves with the number of turns.

Table 13.5. Magnitudes of the Received Currents I^{A_i} at Each of the Four Receiving Antennas.

Excitation Voltages	$\lvert I^{A_3}\rvert$	$\lvert I^{A_4}\rvert$	$\lvert I^{A_5}\rvert$	$\lvert I^{A_6}\rvert$
$\mathbf{W}_3$	1.0025	0.0794	0.0049	0.0766
$\mathbf{W}_4$	0.0915	0.9999	0.0938	0.0030
$\mathbf{W}_5$	0.0034	0.0901	0.9996	0.0900
$\mathbf{W}_6$	0.0745	0.0056	0.0741	0.9972

Next, all the antennas on the mobile receivers, A_3, A_4, A_5, and A_6, are simultaneously excited. The currents that are induced at the loads of the transmitting antennas in the base station are measured which allows one to determine the appropriate weight vectors to be applied at the transmitters so the signal can be directed to a particular receiver. For directing the signals to the jth receiver, for j = 3 or 5, which has the same polarization will require a set of complex weights which will be different when the signal is to be directed to receivers 4 and 6. However, if the weight vectors are appropriately chosen the signal can be directed to the appropriate receiver simultaneously also using the polarization properties as seen in Table 13.6.

Table 13.6. Magnitudes I_{R_i} of the Induced Currents at Each of the Four Receiving Antennas.

$\lvert I^{A_3}\rvert$	$\lvert I^{A_4}\rvert$	$\lvert I^{A_5}\rvert$	$\lvert I^{A_6}\rvert$
0.9910	0.0784	0.0054	0.0757
0.0900	0.9832	0.0922	0.0035
0.0044	0.0886	0.9834	0.0886
0.0737	0.0055	0.0732	0.9857

Next, one considers four receiving and two transmitting antennas that are located inside a conducting cylinder of finite height, with a partition in the middle as shown in Figure 13.13. In this example there is no direct path of communication and the interaction takes place either through the guided waves or the diffracted waves. Transmitters and receivers are all situated in the near field of the obstacles where beam forming is not possible. Table 13.7 shows the currents induced at the receivers. Using the L-H polarization one can get similar results. When one uses no polarization, the results are not as good as when the polarization properties are exploited.

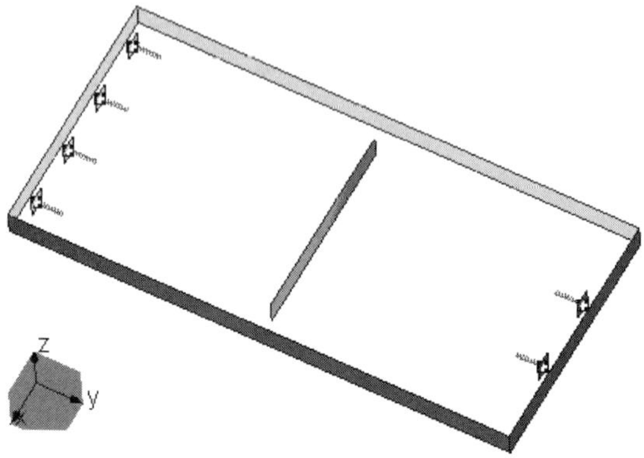

Figure 13.13. A six helical antenna transmit/receive system inside of the cylinder with a wall.

Table 13.7. Magnitudes I^{R_i} at Each of the Four Receiving Antennas Due to W_3 When Using No Polarization Diversity, as Opposed to Choice of Different Polarizations.

Polarization	$\left\|I^{A_3}\right\|$	$\left\|I^{A_4}\right\|$	$\left\|I^{A_5}\right\|$	$\left\|I^{A_6}\right\|$
No polarization	0.5814	0.2920	0.1533	0.3671
R-H polarization	1.0069	0.2463	0.0031	0.5987
L-H polarization	0.7241	0.1292	0.1216	0.4110

13.5. CONCLUSION

A new method is presented for directing the signal to a particular receiving antenna by choosing the appropriate excitations for the transmitting antennas, thereby resulting in adaptivity on transmit. Polarization properties are also exploited. In this way, transmitted signals can be directed to a pre-selected receiver using a finite bandwidth in the presence of coherent multipaths and near-field scatterers exploiting the principle of reciprocity. Several numerical results using an electromagnetic simulation tool have been presented to illustrate the applicability of this novel approach based on the principles of reciprocity and superposition while simultaneously utilizing the polarization properties.

REFERENCES

[1] G. B. Giannakis, Y. Hua, P. Stoica and L. Tong, *Signal Processing Advances in Wireless & Mobile Communications*, Vol. 1, Prentice Hall PTR, Upper Saddle River, NJ, 2000.

[2] L. Setian, *Antennas with Wireless Applications*, Prentice Hall PTR, Upper Saddle River, NJ, 1998.

[3] W. C. Jakes, *Microwave Mobile Communications*, IEEE Press, Lucent Technologies, 1974.

[4] S. P. Morgan, "Interaction of Adaptive Antenna Arrays in an Arbitrary Environment", *Bell System Tech. J.* 44, Jan. 1965.

[5] Y. S. Yeh, "An Analysis of Adaptive Retransmission Arrays in a Fading Environment", *Bell System Tech. J.* 44, Oct. 1970.

[6] S. M. Alamouti, "A Simple Transmit Diversity Technique for Wireless Communications", *IEEE Journal on Select Areas in Communicaions*, Vol. 16, No. 8, Oct. 1998.

[7] H. Lee, M. Shin and C. Lee, "An Eigen-based MIMO Multiuser Scheduler with Partial Feedback Information", *IEEE Communications Letters*, Vol. 9, No. 4 Apr. 2005.

[8] H. Sampath, P. Stoica, and A. Paulraj, "Generalized Linear Precoder and Decoder Design for MIMO Channels Using the Weighted MMSE Criterion", *IEEE Trans. on Communications*, Vol. 49, No. 12, Dec. 2001.

[9] G. Lebrun, J. Gao, and M. Faulkner, "MIMO Transmission Over a Time-Varying Channel Using SVD", *IEEE Trans. on Wireless Communications*, Vol. 4, No. 2, Mar. 2005.

[10] T. Dahl, N. Christophersen, and D. Gesbert, "Blind MIMO Eigenmode Transmission Based on the Algebraic Power Method", *IEEE Trans. on Signal Processing*, Vol. 52, No. 9, Sept. 2004.

[11] S. Choi and D. Yun, "Design of Adaptive Antenna Array for Tracking the Source of Maximum Power and Its Applications to CDMA Mobile Communications", *IEEE Trans. on Antennas and Propagation*, Vol. 45, No. 9, Sept. 1997.

[12] S. Choi, D. Shim and T. K. Sarkar, "A Comparison of Tracking-Beam Arrays and Switching-Beam Arrays Operating in a CDMA Mobile Communication Channel", *IEEE Antennas and Propagation Magazine*, Vol. 41, No. 6, pp. 10-22, Dec. 1999.

[13] R. C. Qui and I. T. Lu, "Multipath Resolving with Frequency Dependence for Wide-Band Wireless Channel Modeling," *IEEE Trans. on Vehicular Technology*, Vol. 48, No. 1, Jan. 1999.

[14] G. D. Durgin and T. S. Rappaport, "Theory of Multipath Shape Factors for Small-scale Fading Wireless Channels", *IEEE Trans. on Antennas and Propagation*, Vol. 48, No. 5, May 2000.

[15] T. K. Sarkar, M. Wicks, M. Salazar-Palma and R. Bonneau, *Smart Antennas*, John Wiley and Sons, Hoboken, NJ, 2003.

[16] R. F. Harrington, *Time-Harmonic Electromagnetic Fields*, McGraw-Hill, New York, NY, 1961.

[17] S. Hwang, A. Medouri, and T. K. Sarkar, "Signal Enhancement in a Near-Field MIMO Environment Through Adaptivity on Transmit", *IEEE Trans. on Antennas and Propagation*, Vol. 53, Issue 2, pp. 685-693, Feb. 2005.

[18] B. M. Kolundzija, J. S. Ognjanovic and T. K. Sarkar, *WIPL-D, Electromagnetic Modeling of Composite Metallic and Dielectric Structures, Software and User's Manual*, Artech House, Norwood, MA, 2004.

14

DIRECTION OF ARRIVAL ESTIMATION BY EXPLOITING UNITARY TRANSFORM IN THE MATRIX PENCIL METHOD AND ITS COMPARISON WITH ESPRIT

14.0 SUMMARY

In this chapter, a unitary transform is applied to the computations involved in the Matrix Pencil Method (MPM) for the direction of arrival (DOA) estimation using real arithmetic. For this modified two-dimensional (2-D) MPM the goal is to compute the 2-D poles related to the azimuth and elevation angles of the various fields incident on a 2-D antenna array. In this formulation, the MPM uses a single snapshot of the data and processes it directly without forming a covariance matrix. Using real computations through the unitary transformation in the 2-D MPM leads to a very efficient computational methodology for real-time implementation on a DSP chip. The numerical simulation results are provided to observe the performance of the method. The simulation results show that for low signal-to-noise ratio (SNR) cases the new 2-D Unitary MPM outperforms the current implementation of the 2-D MPM.

In addition, the performance of ESPRIT (Estimation of Signal Parameters using Rotational In-variance Technique) and MPM under varying number of snapshots are compared. MPM works well under the correlated signal case, as opposed to ESPRIT, a statistical subspace-based estimation technique which requires additional spatial smoothing techniques. Simulation results are provided to show better performance of the MPM over ESPRIT when the number of snapshots available is small.

14.1 INTRODUCTION

The problem of estimating the direction of arrival (DOA) of the various sources impinging on an antenna array has received considerable attention in many fields, including radar, sonar, radio astronomy, and mobile communications.

In this chapter, a unitary transform [1–4] for the one-dimensional (1-D) matrix pencil method (MPM) [5] has been extended to the two-dimensional (2-D) case for estimating the DOA of the signals incident on a 2-D array [6–10]. The azimuth and elevation angles are evaluated using an efficient computational procedure in which the complexities of the computations are reduced significantly by using a unitary matrix transformation. Hence, only real valued computations are carried out in this methodology. In many applications, such as radar imaging [11] and nuclear magnetic resonance imaging [12] or wave number estimation [13] require the estimation of 2-D poles using 2-D data. Unitary transform can convert the complex matrix generally used in the computations to a real matrix along with their eigenvectors and thereby reducing the computational cost at least by a factor of four without sacrificing accuracy. This reduction in the number of computations is achieved by using a transformation, which maps Centro-Hermitian matrices to real matrices [1,3,4]. It is very important to increase the resolution of the DOA estimation as well as to reduce their computational complexity. In the MPM, based on the spatial samples of the data, the analysis is done on a snapshot-by-snapshot basis, and therefore non-stationary environments can be handled easily [14–16]. Unlike the conventional covariance matrix based techniques, the MPM can find DOA easily in the presence of coherent multi-path signals without performing additional processing of spatial smoothing.

Increasing the accuracy of DOA estimation as well as reducing the computational complexity is vital in real-time systems. Capon's minimum variance technique [17] attempts to overcome the poor resolution problems associated with the delay-and-sum method. More advanced approaches are so-called super-resolution techniques that are based on the eigen-structure of the input covariance matrix including MUSIC (Multiple Signal Classification), Root-MUSIC [18], and ESPRIT (Estimation of Signal Parameters via Rotational Invariance Techniques) [19] provides the high-resolution DOA estimation. Music algorithm proposed by Schmidt [18] returns the pseudo-spectrum at all frequency samples. Root-MUSIC [10] returns the estimated discrete frequency spectrum, along with the corresponding signal power estimates. Root-MUSIC is one of the most useful approaches for frequency estimation of signals made up of a sum of exponentials embedded in white Gaussian noise.

The conventional signal processing algorithms using the covariance matrix work on the premise that the signals impinging on the array are not fully correlated or coherent. Under uncorrelated conditions, the source covariance matrix satisfies the full rank condition, which is the basis of the eigen-decomposition. Many techniques involve modification of the covariance matrix through a preprocessing scheme called spatial smoothing [13]. Hua and Sarkar [6,14,15] utilized the matrix pencil to get the DOA of the signals in a coherent multi-path environment. In MPM a 2-D estimation problem has been reduced to two 1-D problems. Each one of the poles is estimated separately and these poles are paired [6] to get the correct pair to estimate the frequencies or elevation and azimuth angles for DOA problems.

On the other hand, some efforts have been made to reduce the computational complexity of the calculations. Huang and Yeh [2] have developed a unitary transform, which can convert a complex matrix to a real matrix along with their eigenvectors. Their simple transformation reduces the processing time by dealing with only real valued computations. The processing time could be reduced almost four times, since the complex multiplication cost four times more than that of real multiplications. More work has been done by Haardt, and Nossek [8], and they applied the method to ESPRIT [19] to successfully reduce the computational burden. In this work, the unitary transform is applied to 2-D MPM to reduce the computational complexity for DOA estimation problems greatly.

In section 14.2, the unitary transform and the related theorems are given. In section 14.3 the signal model for the 2-D case is presented, and the 2D-MPM estimation technique and the pole pairing is introduced. Section 14.4 summarizes the methodology for the 1-D case. The unitary 2-D MPM is introduced in section 14.5. The computer simulation is provided in section 14.6.

Next, we compare the performances between the ESPRIT method [19] and the MPM for dealing with multiple snapshots of the signals. ESPRIT is a high-resolution DOA estimation method and is very similar to MPM so far as the basic philosophy of the methods are concerned. However, ESPRIT is a statistical based technique and thus requires the formation of a covariance matrix of the data. This poses a problem in dealing with coherent signals unless additional sub-aperture based processing is used. Barring this inconvenience of forming a reliable covariance matrix which requires additional computation time, the goal here is to study when multiple snapshots of the data is available which of this method provides a more accurate result. The MPM method can work on a single snap shot of the data, where as ESPRIT requires multiple snapshot of the data to form a covariance matrix. A short overview about ESPRIT is presented in section 14.7. The multiple snapshot-based MPM is described in section 14.8 followed by some numerical simulations in 14.9. Conclusions are presented in section 14.10.

14.2 THE UNITARY TRANSFORM

A square matrix, $B_{N\times N}$, is called a unitary matrix, if it satisfies $BB^H = I$. The superscript H denotes the complex conjugate transpose of a matrix, where I is the identity matrix of dimensions $N \times N$.

Any matrix A, where $A \in C^{P\times S}$, is called Centro-Hermitian [1,2], if it satisfies

$$A = \prod_P A^* \prod_S \tag{14.1}$$

$\prod_P$ is called the exchange matrix and defined as

$$\Pi_P = \begin{bmatrix} 0 & \cdots & 0 & 0 & 1 \\ 0 & \cdots & 0 & 1 & 0 \\ 0 & \cdots & 1 & 0 & 0 \\ \vdots & ⋰ & \vdots & \vdots & \vdots \\ 1 & \cdots & 0 & 0 & 0 \end{bmatrix}_{P \times P} \tag{14.2}$$

Here, Π_P is a $P \times P$ square matrix, and A^* is conjugate of A.

Theorem 1: If the matrix A is Centro-Hermitian, then $Q_P{}^H AQ_S$ is a real matrix. Here, the matrix Q is unitary, whose columns are conjugate symmetric and has a sparse structure [2]. When P is even, we have

$$Q_P = \frac{1}{\sqrt{2}} \begin{bmatrix} I & jI \\ \Pi & -j\Pi \end{bmatrix} \tag{14.3}$$

Here, I and Π are matrices that have the dimension of $P/2$ and $j = \sqrt{-1}$.

When P is odd, we have

$$Q_P = \frac{1}{\sqrt{2}} \begin{bmatrix} I & O & jI \\ O & \sqrt{2} & O \\ \Pi & O & -j\Pi \end{bmatrix} \tag{14.4}$$

Here I and Π are matrices that have the dimension of $(P-1)/2$, and $\boldsymbol{O}$ is a $(P-1)/2 \times 1$ vector whose elements are 0.

Proof: Using $\Pi_P \Pi_P = I$, the conjugate of $Q_P{}^H AQ_S$ is

$$\left(Q_P{}^H AQ_S\right)^* = Q_P{}^T A^* Q_S{}^* = Q_P{}^T \Pi_P \Pi_P A^* \Pi_S \Pi_S Q_S{}^* \tag{14.5}$$

Since $\Pi Q^* = Q$, and $\Pi_P A^* \Pi_S = A$, then we obtain

$$\left(Q_P{}^H AQ_S\right)^* = Q_P{}^H AQ_S. \tag{14.6}$$

Therefore, $Q_P{}^H AQ_S$ is a real matrix. Other related theorems can be found in [1] and [5].

14.3 1-D UNITARY MATRIX PENCIL METHOD REVISITED

Let us consider a Uniform Linear Array (ULA) consisting of N isotropic omni directional antenna elements, and M impinging signals incident on the array. The received voltages at the antenna elements are $\{x(0),x(1),\ldots,x(N-1)\}$. This column data vector can be written in the form of a Hankel Matrix to obtain the matrix Y [5].

$$Y=\begin{bmatrix} x(0) & x(1) & \cdots & x(L) \\ x(1) & x(2) & \cdots & x(L+1) \\ \vdots & \vdots & \ddots & \vdots \\ x(N-L-1) & x(N-L) & \cdots & x(N-1) \end{bmatrix}_{(N-L)\times(L+1)} \tag{14.7}$$

L is the pencil parameter. The Matrix Pencil extracted from the Hankel Matrix Y can be written as

$$J_2 Y-\lambda J_1 Y \tag{14.8}$$

The matrices J_1 and J_2 are called selection matrices and are defined as follows,

$$J_1=\begin{bmatrix} 1 & 0 & \cdots & 0 & 0 \\ 0 & 1 & \cdots & 0 & 0 \\ \vdots & \vdots & \ddots & \vdots & \vdots \\ 0 & 0 & \cdots & 1 & 0 \end{bmatrix}_{(N-L-1)\times(N-L)} \quad ; \; J_2=\begin{bmatrix} 0 & 1 & 0 & \cdots & 0 \\ 0 & 0 & 1 & \cdots & 0 \\ \vdots & \vdots & \vdots & \ddots & \vdots \\ 0 & 0 & \cdots & 0 & 1 \end{bmatrix}_{(N-L-1)\times(N-L)} \tag{14.9}$$

The matrices J_1 and J_2 are used to select the first and last $(N-1)$ components of the matrix Y as discussed in [5]. Then (14.8) can be written as

$$Q^H J_2 Q Q^H Y Q=\lambda Q^H J_1 Q Q^H Y Q \tag{14.10}$$

Note that $QQ^H=I$, and $Q^H Y Q=X_r$ is real [1], since the matrix Y is centro-hermitian. Then

$$Q^H J_2 Q X_r=\lambda Q^H J_1 Q X_r \tag{14.11}$$

It can be shown that $\prod_P \prod_P = I$, $\prod_P Q_P = Q^*$, $Q_P{}^H \prod_P = Q^T$, and $\prod_P J_2 \prod_{P+1} = J_1$, and therefore, (14.11) can be rewritten as

$$Q^H \prod\prod J_2 \prod\prod Q X_r = Q^H \prod J_1 \prod Q X_r = Q^T J_1 Q^* X_r = \left(Q^H J_1 Q\right)^* X_r \tag{14.12}$$

Hence, equation (14.12) can be converted to

$$\left(Q^H J_1 Q\right)^* X_r=\lambda Q^H J_1 Q X_r \tag{14.13}$$

Therefore,

$$\tan\left(\frac{\omega_i}{2}\right)\operatorname{Re}\left(Q^H J_1 Q\right)X_r=\operatorname{Im}\left(Q^H J_1 Q\right)X_r \tag{14.14}$$

The singular value decomposition (SVD) of X_r can be written as

$$X_r = U\Sigma V^T \tag{14.15}$$

Here, U , and V are orthogonal matrices whose elements are the eigenvectors, $\boldsymbol{\Sigma}$ is a diagonal matrix with its element σ_i , which are called the singular values of the matrix X_r . So,

$$U = [u_1 \ u_2 \ \dots \ u_M \dots]$$

$$V = [v_1 \ v_2 \ \dots \ v_M \dots]$$

$$\boldsymbol{\Sigma} = diag(\sigma_1 \geq \sigma_2 \geq \cdots \sigma_M \geq \sigma_{M+1}, \cdots)$$

Let us define the matrix $E_S = [u_1 \ u_2 \ \dots \ u_M]$, such that the elements of E_s are the first M singular vectors corresponding to the largest singular values $\sigma_1, \sigma_2, \cdots, \sigma_M$. M is number of the signals and can be estimated from the singular values based on the criteria defined in [6]. In order to reduce the effect of the noise, one can write (14.14) as

$$\tan\left(\frac{\omega_i}{2}\right)\mathrm{Re}\left(Q^H J_1 Q\right)E_s = \mathrm{Im}\left(Q^H J_1 Q\right)E_s, i = 1, \dots, M \tag{14.16}$$

So, $\tan\left(\frac{\omega_i}{2}\right)$ can be identified with the generalized eigenvalue of the matrix pair $\left\{\mathrm{Im}\left(Q^H J_1 Q\right)E_S, \mathrm{Re}\left(Q^H J_1 Q\right)E_S\right\}$. The solution to this problem can be reduced to an ordinary eigenvalue problem, where $\tan\left(\frac{\omega_i}{2}\right)$ is the eigenvalue of $\left[\mathrm{Re}\left(Q^H J_1 Q\right)E_S\right]^{-1} \mathrm{Im}\left(Q^H J_1 Q\right)E_S$. In this algorithm all computations are made using real numbers and therefore, no variable is complex including the eigenvalues and the eigenvectors in this procedure.

One should state that premultiplying and postmultiplying Y by Q^H and Q require only additions and scaling. For example, in the case of N antenna elements, and for a given Pencil parameter L , the computations done for the unitary transformation, $X_r = Q^H YQ$, requires around $(N-L) \times 2(L+1)$ real additions. This is negligible when compared to the computations required in an Eigen decomposition. Eigen-structure-based methods for estimating DOA of the sources impinging on a ULA requires complex calculations in computing the eigenvectors and the eigenvalues. The MPM, in addition, requires the computation of a Singular Value Decomposition (SVD) of the complex-valued data. It should be stated that eigen-decomposition with complex-valued data matrix is quite computation intensive. The eigen-decomposition process consists of a large portion of the whole computational load. To reduce the computational complexity during eigen-decomposition, application of a unitary transformation is proposed for DOA estimation by using a real-valued SVD. Computing the eigen-components of the unitary transformed data matrix requires only real computations.

The Unitary MPM (UMPM) is thus a completely real-valued algorithm, as it requires only real-valued computations. Apart from finding the singular values and vectors, the rest of the calculations are also real computations as opposed to the ones done in the conventional Matrix Pencil Method. A big portion of the computational load is occupied by the multiplication operations, so transforming the data can save a noticeable amount of computations and the processing time is reduced greatly.

14.4 SUMMARY OF THE 1-D UNITARY MATRIX PENCIL METHOD

The algorithm can be summarized as follows:

1. Convert the complex data matrix into the real matrix X_r, by using $X_r = Q^H Y Q$.
2. Compute the SVD of X_r and calculate E_S, which contains the M principal singular vectors of X_r.
3. Evaluate $\mathrm{Re}\left(Q^H J_1 Q\right)$, and $\mathrm{Im}\left(Q^H J_1 Q\right)$.
4. Calculate the generalized eigenvalues, $\lambda_1, \lambda_2, \cdots, \lambda_M$ of $\mathrm{Im}\left(Q^H J_1 Q\right) E_S$, and $\mathrm{Re}\left(Q^H J_1 Q\right) E_S$.
5. Calculate $\omega_i = 2\tan^{-1}\left(\lambda_i\right)$, $i = 1, \ldots, M$.

14.5 THE 2-D UNITARY MATRIX PENCIL METHOD

In many applications, such as radar imaging and nuclear magnetic resonance imaging or wave number estimation require the estimation of two-dimensional (2-D) poles in 2-D data. Let us consider the 2-D uniform rectangular array (URA). The noiseless data $z(m,n)$ measured at the feed points of all the omni directional antennas are now defined as

$$z(m,n) = \sum_{p=1}^{P} a_p \, \mathrm{e}^{j\varphi_p} \, \mathrm{e}^{j\frac{2\pi m}{\lambda}\Delta_x \sin\theta_p \cos\phi_p} \, \mathrm{e}^{j\frac{2\pi n}{\lambda}\Delta_y \sin\theta_p \sin\phi_p},$$

$$0 \le m \le M-1 \quad \text{and} \quad 0 \le n \le N-1 \tag{14.17}$$

This equation can be simplified to

$$z(m,n)=\sum_{p=1}^{P}\alpha_p x_p{}^m y_p{}^n \tag{14.18}$$

where $\alpha_p = a_p\,\mathrm{e}^{j\varphi_p}$, and $x_p = \mathrm{e}^{j\frac{2\pi}{\lambda}\Delta_x \sin\theta_p \cos\phi_p}$, and $y_p = \mathrm{e}^{j\frac{2\pi}{\lambda}\Delta_y \sin\theta_p \sin\phi_p}$, $x_p = \mathrm{e}^{j\omega_x}$, $y_p = \mathrm{e}^{j\omega_y}$, $\omega_x = \dfrac{2\pi}{\lambda}\Delta_x \sin\theta_p \cos\phi_p$, and $\omega_y = \dfrac{2\pi}{\lambda}\Delta_y \sin\theta_p \sin\phi_p$.

Basically, in the 2-D Matrix Pencil Method, the 2-D problem is subdivided into a number of 1-D cases, and solves the poles for each dimension and pairs them together to find the corresponding DOA. The data matrix $z(m,n)$ can be enhanced and written in a Block Hankel Matrix structure as defined in [6]. The distance between the antenna elements along the x and y directions are equal and is given by $\Delta_x = \Delta_y = \lambda/2$. The signal model has P, 2-D exponential signals, where a_p, and φ_p are the magnitudes and the phases, respectively. The objective is to find the (x_p, y_p) pairs, which correspond to the azimuth and elevation angles for each of the signals. The formulation of the 2-D MPM is discussed in detail in [6]. The noiseless data corresponding to (14.18) can be written in matrix form as,

$$D=\begin{bmatrix} z(0;0) & z(0;1) & \cdots & z(0;N-1) \\ z(1;0) & z(1;1) & \cdots & z(1;N-1) \\ \vdots & \vdots & \ddots & \vdots \\ z(M-1;0) & z(M-1;1) & \cdots & z(M-1;N-1) \end{bmatrix}. \tag{14.19}$$

Basically, in the 2-D MPM, the problem is divided into a number of 1-D MPM, and it is solved for each of the dimension separately and these estimated poles are then paired to find the corresponding DOA. The data matrix $z(m,n)$ can be enhanced and written using a Block Hankel Matrix structure as,

$$D_e=\begin{bmatrix} D_0 & D_1 & \cdots & D_{M-\chi} \\ D_1 & D_2 & \cdots & D_{M-\chi+1} \\ \vdots & \vdots & \ddots & \vdots \\ D_{\chi-1} & D_\chi & \cdots & D_{M-1} \end{bmatrix}. \tag{14.20}$$

The original data matrix $z(m,n)$ is a $M \times N$ complex matrix. Extending the original data matrix by stacking results in a Hankel structure D_e in (14.20), where each element of D_e is also a Hankel matrix, which is obtained by windowing the rows of the original data matrix $z(m,n)$. In addition,

$$D_m = \begin{bmatrix} z(m;0) & z(m;1) & \cdots & z(m;N-\zeta) \\ z(m;0) & z(m;1) & \cdots & z(m;N-\zeta+1) \\ \vdots & \vdots & \ddots & \vdots \\ z(m;\zeta-1) & z(m;\zeta) & \cdots & z(m;N-1) \end{bmatrix}. \tag{14.21}$$

The χ and ζ are the window pencil parameters used to obtain the Hankel matrix in (14.20).

The matrix D_m can be written as $D_m = Y_\zeta G X_d{}^m Y_R$ Here,

$$Y_\zeta = \begin{bmatrix} 1 & 1 & \cdots & 1 \\ y_1 & y_2 & \cdots & y_P \\ \vdots & \vdots & \ddots & \vdots \\ y_1{}^{(\zeta-1)} & y_2{}^{(\zeta-1)} & \cdots & y_P{}^{(\zeta-1)} \end{bmatrix}_{(\zeta\times P)} \tag{14.22}$$

$$Y_R = \begin{bmatrix} 1 & y_1 & \cdots & y_1{}^{(N-\zeta)} \\ 1 & y_2 & \cdots & y_2{}^{(N-\zeta)} \\ \vdots & \vdots & \ddots & \vdots \\ 1 & y_P & \cdots & y_P{}^{(N-\zeta)} \end{bmatrix}_{P\times(N-\zeta+1)} \tag{14.23}$$

$X_d = diag(x_1, x_2, \cdots x_P)$, and G is the diagonal matrix, where $G = diag(a_1, a_2, \cdots a_P)$, D_m is an $\zeta \times (N-\zeta+1)$ Hankel matrix. The extended matrix D_e could be written as

$$D_e = E_\zeta G E_R \tag{14.24}$$

which is a $\chi \times (M-\chi+1)$ block Hankel matrix, so that

$$E_\zeta = \begin{bmatrix} Y_\zeta \\ Y_\zeta X_d \\ \cdots \\ Y_\zeta X_d{}^{\chi-1} \end{bmatrix} \tag{14.25}$$

$$E_R = [Y_R, X_d Y_R, \cdots, X_d Y_R] \tag{14.26}$$

By multiplying E_ζ by the Shuffling matrix P gives $E_{\zeta P}$, where

$$E_{\zeta P} = \begin{bmatrix} X_\zeta \\ X_\zeta Y_d \\ \cdots \\ X_\zeta Y_d^{\chi-1} \end{bmatrix} \tag{14.27}$$

$$X_\zeta = \begin{bmatrix} 1 & 1 & \cdots & 1 \\ x_1 & x_2 & \cdots & x_P \\ \vdots & \vdots & \ddots & \vdots \\ x_1^{(\zeta-1)} & x_2^{(\zeta-1)} & \cdots & x_P^{(\zeta-1)} \end{bmatrix}_{(\zeta \times P)} \qquad Y_d = diag\left(y_1, y_2, \cdots y_P\right) \tag{14.28}$$

As can be seen, we need to repeat this procedure for each of the dimensions, such as x, and y directions in rectangular coordinates. For each pole or exponent, we need to compute the Singular Value Decomposition (SVD). So, we need to compute the SVD twice. We can solve the problem by computing the SVD only once and then shuffling the signal subspace. Let us introduce the following shuffling matrix S as

$$S = \begin{bmatrix} s(1) \\ s(1+\zeta) \\ \cdots \\ s(1+(\chi-1)\zeta) \\ s(2) \\ s(2+\zeta) \\ \cdots \\ s(2+(\chi-1)\zeta) \\ \cdots \\ \cdots \\ s(\zeta) \\ s(\zeta+\zeta) \\ \cdots \\ s(\zeta+(\chi-1)\zeta) \end{bmatrix} \tag{14.29}$$

where, $s(i)$ is $1 \times \chi\zeta$ row vector with one at the ith position and zero elsewhere. The eigen-structure of the matrix D_e is found by taking the Singular Value Decomposition

$$D_e = U_s \Sigma_s V_s^H + U_n \Sigma_n V_n^H \tag{14.30}$$

The U_s, Σ_s, and V_s^H are in the signal subspace corresponding to the P principal components whereas U_n, Σ_n, and V_n^H are in the noise subspace. Finding the poles x_i, by knowing that U_s and E_ζ are in the same column space, we can write

$$range(U_s) = range(E_\zeta)$$

with $U_s = E_\zeta T$, where T is a nonsingular matrix. The matrix pencil can be written along the x direction as

$$U_{x2} - \lambda U_{x1} \tag{14.31}$$

$$U_{x1} = U_s \text{ with last } \zeta \text{ rows deleted} \tag{14.32}$$

$$U_{x2} = U_s \text{ with first } \zeta \text{ rows deleted} \tag{14.33}$$

The rank reducing numbers of the matrix pencils, $U_{x2} - \lambda U_{x1}$, are the poles of $X_d = diag(x_1, x_2, \cdots x_P)$. So the parameter u_p can be estimated from the eigenvalues of the matrix $(U_{x2})^+ U_{x1}$. $(U_{x2})^+$ is called the Pseudo-inverse and is defined as

$$(U_{x2})^+ = \left(U_{x2}^H U_{x2}\right)^{-1} U_{x2}^H = pinv(U_{x2}) \tag{14.34}$$

To compute the poles along the y direction, we can define,

$$U_{SY} = SE_\zeta T \tag{14.35}$$

So, the matrix pencil can be written as

$$U_{y2} - \lambda U_{y1} \tag{14.36}$$

$$U_{y1} = U_{SY}, \text{ with last } \zeta \text{ rows deleted} \tag{14.37}$$

$$U_{y2} = U_{SY}, \text{ with first } \zeta \text{ rows deleted} \tag{14.38}$$

The rank reducing numbers of the matrix pencil, $U_{y2} - \lambda U_{y1}$, are the poles of $Y_d = diag(y_1, y_2, \cdots y_P)$. So the parameter v_p can be estimated from the matrix $(U_{y2})^+ U_{y1}$. $(U_{y2})^+$ is called the Pseudo-inverse of a matrix and is defined as $(U_{y2})^+ = \left(U_{y2}^H U_{y2}\right)^{-1} U_{y2}^H = pinv(U_{y2})$. The necessary conditions for choosing the pencil parameters χ, and C are pointed out in [6]. They are

$$\begin{aligned}&(\chi-1)\zeta \geq P\\&\chi(\zeta-1) \geq P\\&(M-\chi+1)(N-\zeta+1) \geq P\end{aligned} \tag{14.39}$$

After finding the poles x_P ,and y_P , which are not in pair, they need to be paired to get the correct association so that they correspond to the true azimuth and elevation angles. By exploiting the property that the signal space is orthogonal to the noise subspace, which implies that $e_L \perp U_n$, the pairs can be matched together by maximizing the criterion below [6]:

$$J_{sp}(i,j) = \sum_{p=1}^{P} \left\| u_p{}^H e_L(x_i, y_j) \right\|^2, \quad p=1,2,\ldots P \tag{14.40}$$

$$e_L = x_L \otimes y_L \tag{14.41}$$

$\otimes$ is the Kronecker product. And the variables are $x_L = \left[1, x, \ldots, x^{\chi-1}\right]^T$, and $y_L = \left[1, y, \ldots, y^{\zeta-1}\right]^T$. Here $\left\{u_p;\, i=1,2,\ldots P\right\}$ are the principle eigenvectors.

We now extend the 1-D UMP method to the 2-D UMP method. So by using the unitary transform, the complex valued data matrix can be converted to a real matrix. By using the Centro-symmetry of the antenna arrays or any complex matrix could be written into a Centro-Hermitian matrix form. So the data matrix D_e can be used to compute

$$D_{ch} = \left[D_e \,\vdots\, \Pi\, D_e^*\, \Pi\right] \tag{14.42}$$

By using theorem 1, we can write

$$Q^H D_{ch} Q = D_r, \tag{14.43}$$

D_r is a real valued matrix. Let us introduce the selection matrices that will be used to write the matrix pencils. The matrices J_3, J_4, J_5, and J_6 are called selection matrices and used to select the rows of the real matrix D_r in order to write the matrix pencil along the x and y directions in rectangular coordinates.

$$\begin{aligned}J_3&=\left[I_{(\chi\zeta-\zeta)\times(\chi\zeta-\zeta)} \,\vdots\, 0_{(\chi\zeta-\zeta)\times(\zeta)}\right]_{(\chi\zeta-\zeta)\times(\chi\zeta)}\\J_4&=\left[0_{(\chi\zeta-\zeta)\times(\chi\zeta-\zeta)} \,\vdots\, I_{(\chi\zeta-\zeta)\times(\zeta)}\right]_{(\chi\zeta-\zeta)\times(\chi\zeta)}\end{aligned} \tag{14.44}$$

$$\begin{aligned}J_5&=\left[I_{(\chi\zeta-\chi)\times(\chi\zeta-\chi)} \,\vdots\, 0_{(\chi\zeta-\zeta)\times(\chi)}\right]_{(\chi\zeta-\zeta)\times(\chi\zeta)}\\J_6&=\left[0_{(\chi\zeta-\chi)\times(\chi)} \,\vdots\, I_{(\chi\zeta-\zeta)\times(\chi\zeta-\chi)}\right]_{(\chi\zeta-\zeta)\times(\chi\zeta)}\end{aligned} \tag{14.45}$$

One can easily write the matrix pencil for the 2-D case along the x direction as $(J_4)(D_e)-\lambda(J_3)(D_e)$. It is reduced to the form below

$$\tan\left(\frac{\omega_{xp}}{2}\right)\ \mathrm{Re}\left(Q^H J_3 Q\right)\ E_{sx}=\mathrm{Im}\left(Q^H J_3 Q\right)\ E_{sx} \tag{14.46}$$

for $i=1,\ldots,P$, where P is number of impinging signals arriving to the URA, E_{sx} is the left singular eigenvector corresponding to P principal eigenvalues. So $\tan\left(\omega_{xp}/2\right)$ can be solved as the generalized eigenvalue of the matrix pair $\left\{\mathrm{Im}\left(Q^H J_3 Q\right) E_{sx},\ \mathrm{Re}\left(Q^H J_3 Q\right)\ E_{sx}\right\}$. In a similar way, one can write the matrix pencil for the 2-D case along the y-direction as $(J_6)(S)(D_e)-\lambda(J_5)(S)(D_e)$. It is reduced to the form below

$$\tan\left(\omega_{yp}/2\right)\ \mathrm{Re}\left(Q^H J_5 Q\right)\ E_{sy}=\mathrm{Im}\left(Q^H J_5 Q\right) E_{sy} \tag{14.47}$$

to find the poles, y_p, along the y direction. E_{sy} is the left singular eigenvector corresponding to P principal eigenvalues which is found by applying the shuffling matrix to the D_r.

So $E_{sy}=P\times E_{sx}$ and $\tan\left(\omega_{yp}/2\right)$ can be solved as the generalized eigenvalue of the matrix pair $\left\{\mathrm{Im}\left(Q^H J_5 Q\right) E_{sy},\ \mathrm{Re}\left(Q^H J_5 Q\right) E_{sy}\right\}$.

SVD of D_r can be written as $D_r=U\Sigma V^T$ where $U=\left[u_1\ u_2\ \ldots\ u_P\ldots\right]$, and $V=\left[v_1\ v_2\ \ldots\ v_P\ldots\right]$ are orthogonal matrices. $\mathbf{\Sigma}$ is a diagonal matrix with its elements σ_i, $\sigma_1\geq\sigma_2\geq\cdots\sigma_p\geq\sigma_{p+1},\cdots$, which are called the singular values. Let us define $E_{SX}=\left[u_1\ u_2\ \ldots\ u_P\right]$, such that the elements of E_{SX} are the first P singular vectors corresponding to the largest singular values $\sigma_1,\sigma_2,\cdots,\sigma_P$. P is the number of signals and can be estimated from the dominant singular values.

14.5.1 Pole Pairing for the 2-D Unitary Matrix Pencil Method

After finding the poles x_p and y_p, it needs to be paired to get the correct association, which corresponds to the true azimuth and elevation angles. This is accomplished by exploiting the property that the signal space is orthogonal to the noise subspace, which implies that $e_L\perp U_n$. The two pairs can be matched together by maximizing the criterion below. But the difference here again is that the pairing algorithm is real valued which reduces the computation time as well.

$$J_{sp}(i,j) = \sum_{p=1}^{P} \left\| u_p{}^H e_L(x_i, y_j) \right\|^2, \quad p = 1,2,\ldots P \tag{14.48}$$

Here, $e_L = x_L \otimes y_L$, $\otimes$ is the Kronecker product and the variables are $x_L = \left[1, x, \ldots, x^{B-1}\right]^T$, and $y_L = \left[1, y, \ldots, y^{C-1}\right]^T$. $\{u_p; i = 1,2,\ldots P\}$ are the principles eigenvectors.

14.5.2 Computational complexity

In the 2-D MPM, most of the computational load is associated with the computation of the SVD, formation of the matrix pencil, and correct pairing of the poles. In the 2D-MPM all the computations are complex valued, but for the 2-D UMP, all these computations are done with real valued computations, which reduce the complexity. The difference between the 2-D UMP and the 2-D MPM comes into play at this point, because all the computations of SVD are real-valued in UMP unlike for the MP method, which reduces the computational complexity greatly. As an example, the multiplication of two complex numbers requires four real multiplications and two additions. Computing the SVD, and evaluating the left singular vectors of a matrix $[D_e]_{BC\times(M-B+1)(N-C+1)}$ requires $17/3\left(B^3C^3\right) + B^2C^2(M-B+1)(N-C+1)$ multiplications. The computation of $Q^H D_{ch} Q = D_r$ can be reduced to $M \times 2N$ real additions. For the pairing method, the number of computations becomes $0.5 \times \left(P^3BC\right)$. The computation required for the matrix pencil for the two of them is $3P^2BC$. For the 2-D UMP method, all these computations are done with real valued functional evaluations as opposed to the 2-D MPM where these computations are complex valued operations. So the computational complexity is reduced greatly.

14.5.3 Summary of the 2-D Unitary Matrix Pencil Method

The algorithm can be summarized as follows:

1. Convert the complex data matrix into the real matrix D_r, by using $D_r = Q^H D_{ch} Q$.
2. Use the Singular Value Decomposition of D_r, to calculate E_{sx}, and E_{sy} which are the P principal singular vectors of D_r.
3. Solve for the generalized eigenvalues

$$\tan\left(\omega_{xp}/2\right) \operatorname{Re}\left(Q^H J_3 Q\right) E_{sx} = \operatorname{Im}\left(Q^H J_3 Q\right) E_{sx}$$

4. Solve for the generalized eigenvalues

$$\tan\left(\omega_{yp}/2\right)\ \mathrm{Re}\left(Q^H J_5 Q\right)\ E_{sy}\ =\ \mathrm{Im}\left(Q^H J_5 Q\right)\ E_{sy}$$

5. Pair the generalized eigenvalues by using the criterion of (14.48).

14.6 SIMULATION RESULTS RELATED TO THE 2-D UNITARY MATRIX PENCIL METHOD

In this section, illustrative computer simulation results are provided to demonstrate the performance of this technique. The noise contaminated signal model is formulated as $\tilde{z}(m,n) = z(m,n) + w(m,n)$, where $0 \le m \le M-1$, and $0 \le n \le N-1$, $w(m,n)$ is a zero mean Gaussian white noise with variance σ^2. A two-dimensional rectangular array of omnidirectional isotropic point sensors are considered in this study. The distance between the antenna elements along the *x*- and *y*-directions are $\Delta_x = \Delta_y = \lambda/2$. It is assumed that there are three signals that are impinging on the URA. The signals have a phase of $\gamma_i = 0$ degrees. The number of the antenna elements along each coordinate direction is $M = N = 20$. The DOA of the three signals are $\{(\phi_i,\theta_i)$; (15°,40°), (25°,30°), (35°,20°)} and are summarized in Table 14.1. The scatter plot of the estimated elevation and azimuth angles are shown in Figures 14.1–14.3 for different signal-to-noise (SNR) ratios of SNR = 5 dB, SNR = 15 dB, and SNR = 20 dB. The results are based on 200 Monte Carlo simulations. As expected, when the SNR increases, the estimated values approach to its true values in the scatter plot.

Table 14.1. Summary of Signal Features Incident on the Antenna Array.

	Signal 1	Signal 2	Signal 3
ϕ	15°	35°	25°
θ	40°	20°	30°

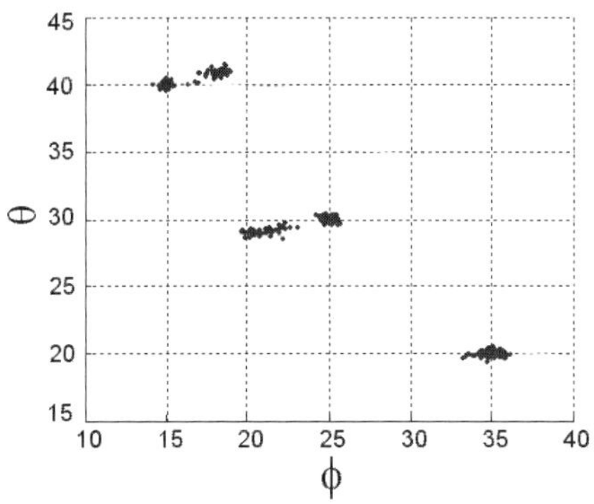

Figure 14.1. The scatter plot of 3 impinging signals, SNR = 5 dB, 200 Monte Carlo simulations.

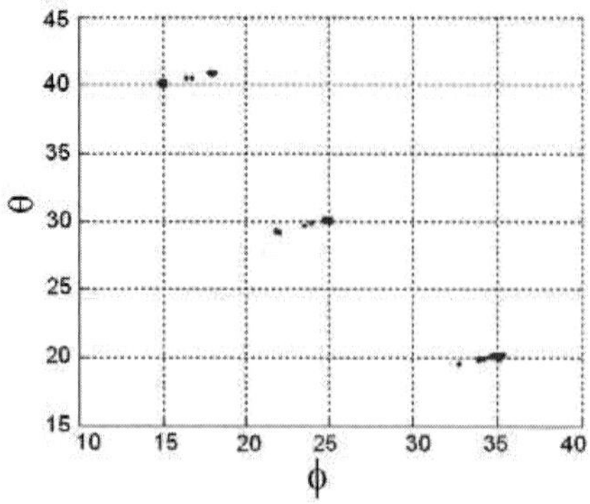

Figure 14.2. The scatter plot of 3 impinging signals, SNR = 15 dB, 200 Monte Carlo simulations.

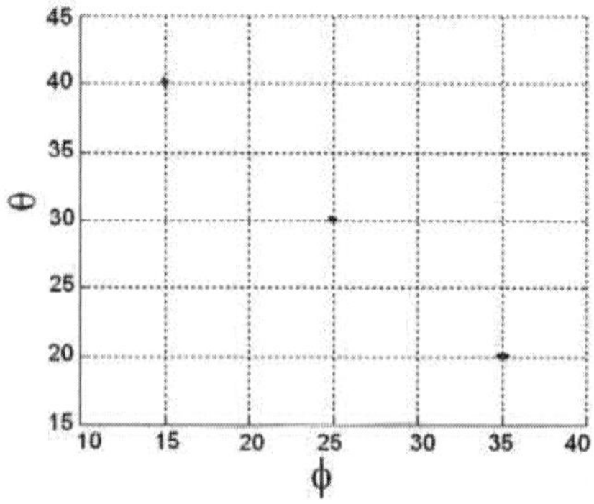

Figure 14.3. The scatter plot of 3 impinging signals, SNR = 20 dB, 200 Monte Carlo simulations.

The inverse of the sample variance of the estimates of ϕ (azimuth angle), θ (elevation angle), for the 2-D UMP is compared against the 2-D MPM versus signal-to-noise ratio (SNR) of the incoming signals and are plotted in Figures 14.4a–14.4f. Different values of SNR are plotted along the x- axis and the variance of the estimated azimuth, elevation angles are in the logarithmic domain, as shown along the y-axis. For low values of SNR, the new 2-D UMP method performs better than the 2-D MPM and has a lower variance. The results are based on 500 Monte Carlo simulations.

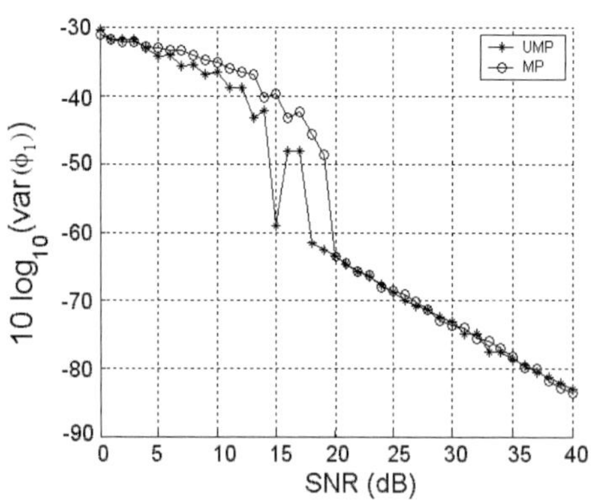

Figure 14.4a. Variance $10\log_{10}\left(\mathrm{var}\left(\phi_1\right)\right)$ 2-D UMP, MP are plotted against SNR.

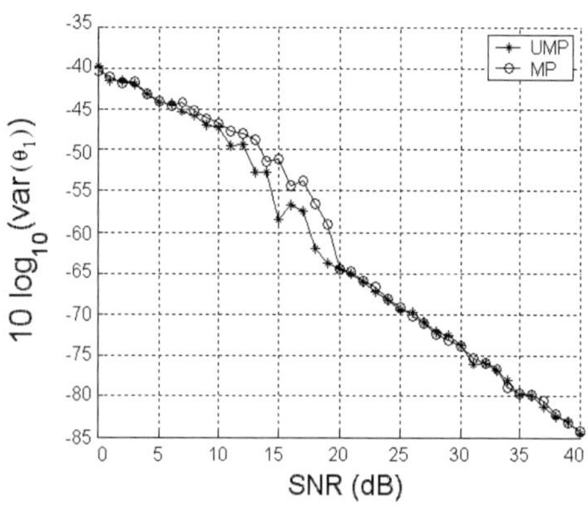

Figure 14.4b Variance $10\log_{10}\left(\mathrm{var}\left(\theta_1\right)\right)$ 2-D UMP, MP are plotted against SNR.

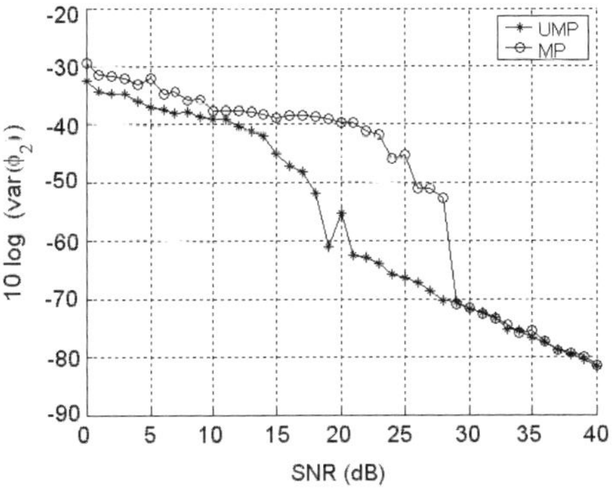

Figure 14.4c. Variance $10\log_{10}\left(\mathrm{var}\left(\phi_2\right)\right)$ 2-D UMP, MP are plotted against SNR.

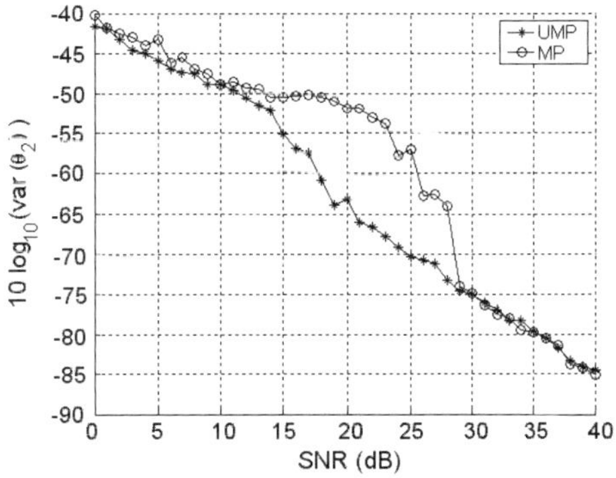

Figure 14.4d. Variance $10\log_{10}\left(\mathrm{var}\left(\theta_2\right)\right)$ 2-D UMP, MP are plotted against SNR.

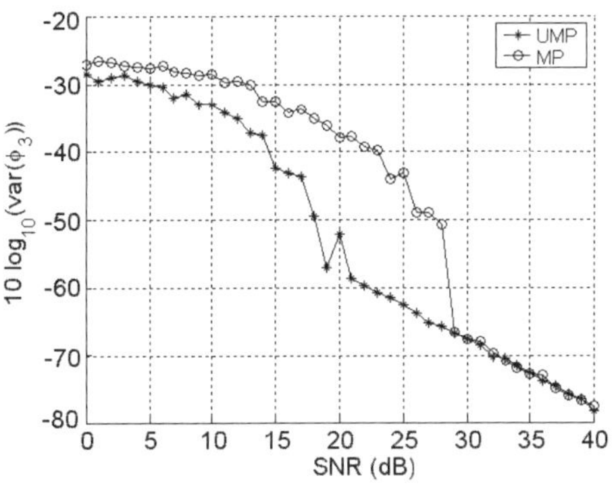

Figure 14.4e. Variance $10\log_{10}\left(\mathrm{var}\left(\phi_3\right)\right)$ 2-D UMP, MP are plotted against SNR.

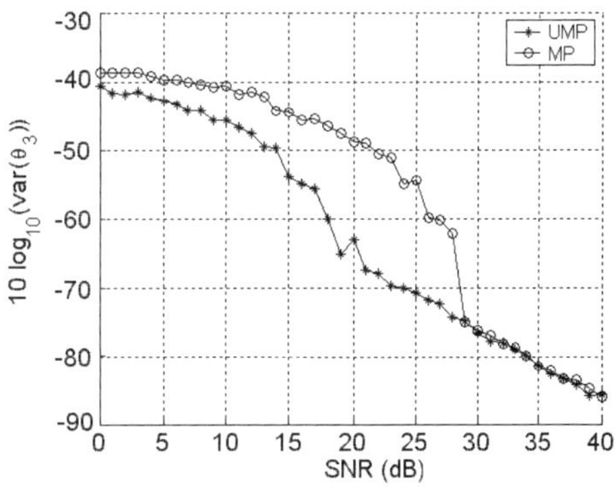

Figure 14.4f. Variance $10\log_{10}\left(\mathrm{var}\left(\theta_3\right)\right)$ 2-D UMP, MP are plotted against SNR.

The bias of the estimator has also been studied to see the efficiency of the new method. For the computed elevation and azimuth angles, the bias of the estimator has also been evaluated. The bias is calculated as

$$bias\left(\phi\right) = \mathcal{E}\left(\hat{\phi}\right) - \phi \tag{14.49}$$

$$bias\left(\theta\right) = \mathcal{E}\left(\hat{\theta}\right) - \theta \tag{14.50}$$

where $\mathcal{E}\left(\bullet\right)$ denotes the expected value, and $\hat{\phi}$ and $\hat{\theta}$ are the estimates of ϕ and θ, respectively. The bias (in dB) of the estimator for azimuth and elevation angles, versus SNR is shown in Figures 14.5 and 14.6, respectively. The bias for the first signal is given since the other two have the similar characteristics. As expected, for higher values of the SNR, the bias of the estimator decreases.

Next we compare the performance of the MPM with ESPRIT for a few snapshots of the data.

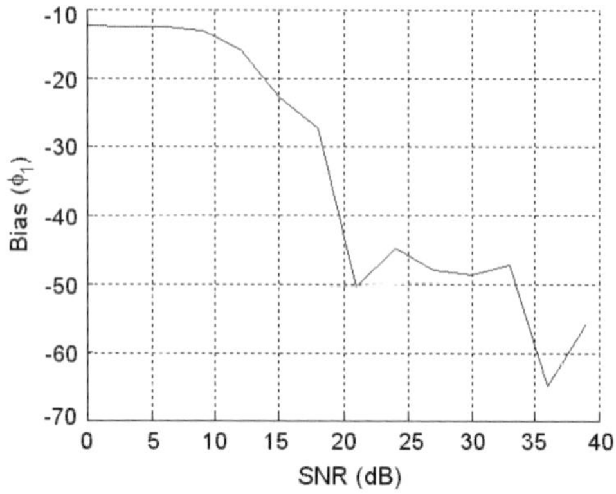

Figure 14.5. Bias of the estimator (in dB) for ϕ_1 (azimuth angle) versus SNR.

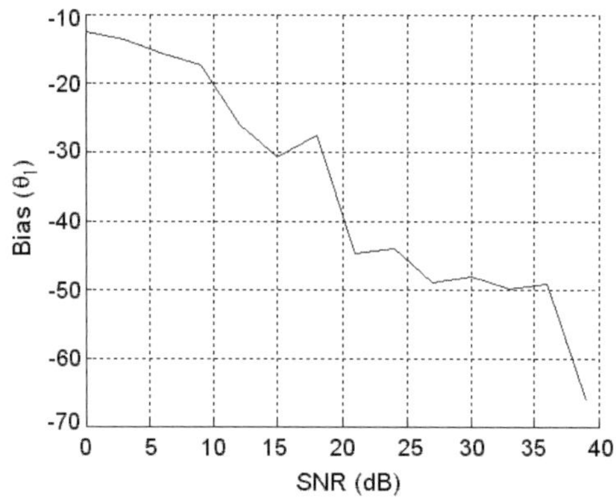

Figure 14.6. Bias of the estimator (in dB) for θ_1 (elevation angle) versus SNR.

14.7 THE ESPRIT METHOD

The ESPRIT method is based on a statistical methodology. Our goal is to compare the computational efficiency and the accuracy between the ESPRIT method and the MPM. ESPRIT is another parameter estimation technique similar to the MPM. Based on the fact that in the steering vector, the signal at one antenna element has a constant phase shift from the previous antenna element, ESPRIT develops a subspace-based estimation technique where the poles are computed using a Covariance matrix [19].

The data model is assumed to consist of M signals and sampled at N points. By using the following data model for the 1-D case, we get

$$y(p) = \sum_{i=1}^{M} A_i z_i^{\,p} + n(p), \text{ for } p = 0, 1, \cdots, N-1 \tag{14.51}$$

As all statistically based techniques form a covariance matrix, the correlation matrix R of y can be written as

$$R = E\left[yy^H\right] = E\left[(x+n)(x+n)^H\right] \tag{14.52}$$

or, in a compact form as

$$R = AR_s A^H + R_n \tag{14.53}$$

where A is defined as

$$A = \left[a(\theta_1) \quad a(\theta_2) \quad \cdots a(\theta_M)\right], \tag{14.54}$$

and

$$a(\theta) = \left[1 \quad e^{j\theta} \quad \cdots e^{j(M-1)\theta}\right]. \tag{14.55}$$

Assume that the signal and noise are uncorrelated. Then R_s is the signal covariance matrix and its rank is M, equal to the number of the signals coming to the array. It is computed using

$$R_s = \frac{1}{N}\sum_{n=1}^{N} x(n)x^H(n). \tag{14.56}$$

In addition, the noise covariance matrix R_n is generally assumed to be

$$R_n = \sigma^2 \mathrm{I}. \tag{14.57}$$

Where I is an identity matrix. In real applications, we have only a limited number of time samples, so the covariance matrix can be estimated from these limited numbers of samples as

$$R \approx \frac{1}{N}\sum_{n=1}^{N} y(n)y^H(n). \tag{14.58}$$

Let E_s contain the M largest eigenvectors corresponding to the signal subspace of the matrix R. E_s and A have the same range, so we can write $E_s = AT$, where the matrix T is non-singular. Let us define two matrices E_1, and E_2, by deleting the first and last row of the covariance matrix respectively, one obtains

$$E_2 = E_1 \Phi \tag{14.59}$$

where $\Phi = T^{-1}\Omega T$. The eigenvalues of Φ will provide the poles for which we are looking. Both MPM and ESPRIT have some similarities, in that respect. Both algorithms estimate a diagonal matrix whose entries are the poles of the system. The major difference is that ESPRIT works with the signal subspace as defined by the correlation matrix, whereas MPM works with the data directly. MPM thus has a significant saving in terms of computation load and very suitable for real time applications as it can also deal with a single snapshot of the data. Next, we present the MPM when it utilizes multiple snapshots. In that way one can compare the performance of both the methods for a particular data set.

14.8 MULTIPLE SNAPSHOT-BASED MATRIX PENCIL METHOD

The formulation for MPM is generally given for the single snapshot case [14]. For the case of multiple snapshots, the Hankel matrix is constructed for each snapshot. Let us say, there are K snapshots, then for each snapshot the Hankel matrix can be constructed as

$$Y_k = \begin{bmatrix} x_k(0) & x_k(1) & \cdots & x_k(L-1) \\ x_k(1) & x_k(2) & \cdots & x_k(L) \\ \vdots & \vdots & \ddots & \vdots \\ x_k(N-L) & x_k(N-L+1) & \cdots & x_k(N-1) \end{bmatrix}_{(N-L+1)\times(L)}, \quad k = 1, 2, \cdots, K \tag{14.60}$$

These Hankel matrices are now appended side by side to form,

$$Y_E = \begin{bmatrix} Y_1 & Y_2 & \cdots & Y_K \end{bmatrix}_{(N-L+1)\times(KL)} \tag{14.61}$$

The size of the matrix Y_E is $(N-L+1)\times(KL)$. This matrix is used to find the poles and the corresponding DOA. This extended data matrix will increase the computational complexity in the computation of the Singular Value Decomposition over the single snapshot-based procedure.

14.9 COMPARISON OF ACCURACY AND EFFICIENCY BETWEEN ESPRIT AND THE MATRIX PENCIL METHOD

In this section, computer simulation results are provided to make a meaningful comparison between the performances of these two popular methods. The noise contaminated signal y given by (14.51) is used. $n(k)$ is treated as a zero mean Gaussian white noise with variance σ^2. An array of uniformly spaced omni-directional isotropic point sensors are considered in this study. The distance between any two elements of the ULA is half a wavelength. $y(k)$ is the voltage induced at each of the antenna elements, for $k = 0, 1, ..., N-1$. The SNR of the data is fixed at 12 dB. It is assumed that there are eight antenna elements and two signals are impinging on the array which has amplitudes $A_1 = A_2 = 1$. The signals are coming from 30° and 45°. Noise for each run is independent of each other. The optimum value for the pencil length, L, is chosen to be 4 for efficient noise filtering.

The histogram plot is given for 1000 trials using the same noisy data containing two signals. Figures 14.7 and 14.8 provide the histogram plot for the MPM for the directions of arrival of the two signals using a single snapshot. Figures 14.9 and 14.10 provide the histogram plot for the ESPRIT method for the directions of arrival of the two signals using a single snapshot.

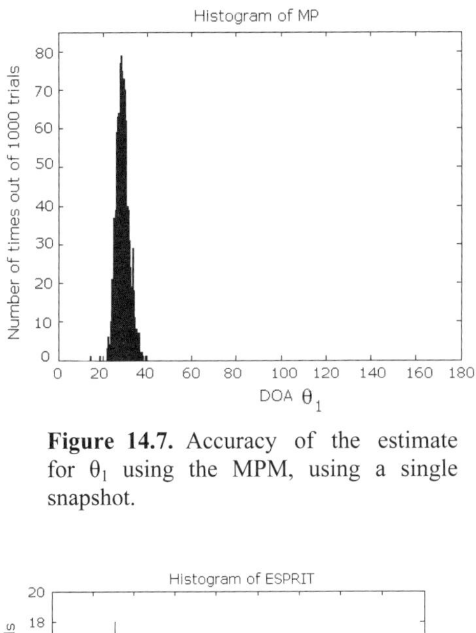

Figure 14.7. Accuracy of the estimate for θ_1 using the MPM, using a single snapshot.

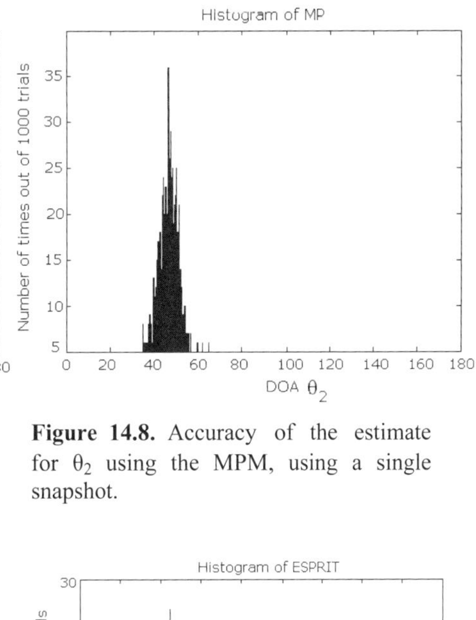

Figure 14.8. Accuracy of the estimate for θ_2 using the MPM, using a single snapshot.

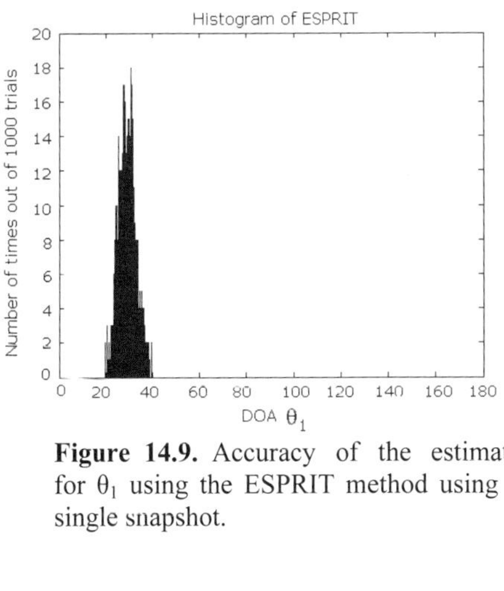

Figure 14.9. Accuracy of the estimate for θ_1 using the ESPRIT method using a single snapshot.

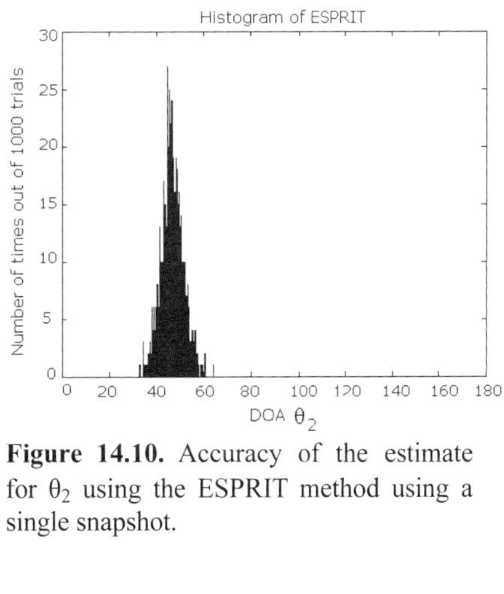

Figure 14.10. Accuracy of the estimate for θ_2 using the ESPRIT method using a single snapshot.

The performance of these two estimation methods is compared under multiple snapshots, eight to be exact. Figures 14.11 and 14.12 provide the histogram plot for the MPM for the directions of arrival of the two signals using eight snapshots. Figures 14.13 and 14.14 provide the histogram plot for the ESPRIT method for the directions of arrival of the two signals using eight snapshots.

The variances of the estimators are plotted in Figure 14.15 using three different scenarios. First, using a single snapshot, next using 8 snapshots and finally using 16 snapshots. Five hundred Monte Carlo simulations are conducted for each case. As can be seen from Figure 14.15, the MPM has a lower variance than the ESPRIT for all the three cases considered.

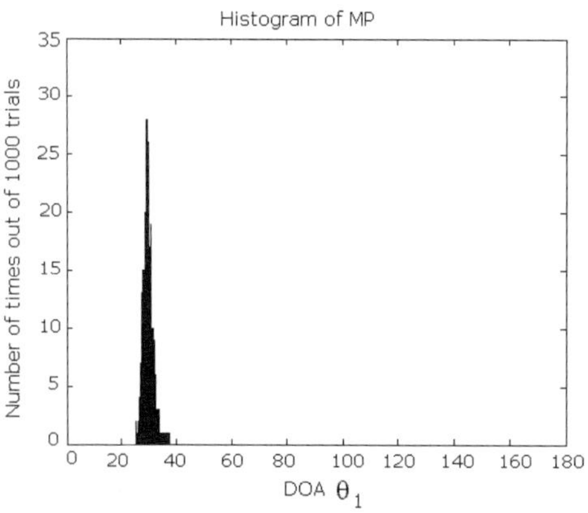

Figure 14.11. Accuracy of the estimate for θ_1 using the MPM using eight snapshots.

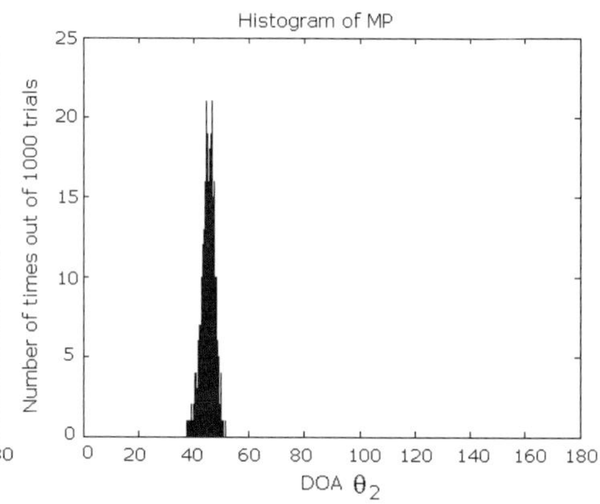

Figure 14.12. Accuracy of the estimate for θ_2 using the MPM using eight snapshots.

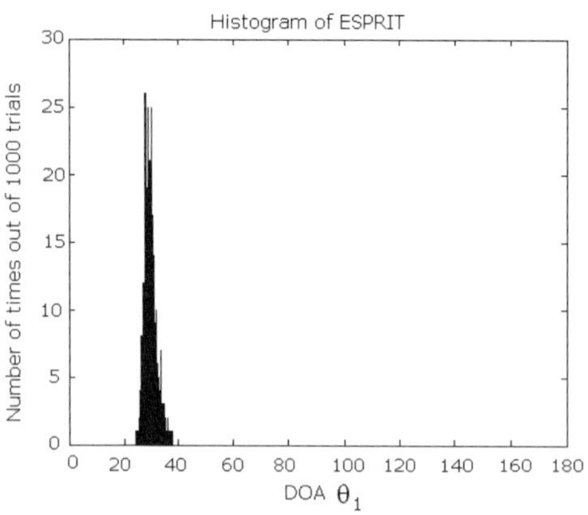

Figure 14.13. Accuracy of the estimate for θ_1 using the ESPRIT method using eight snapshots.

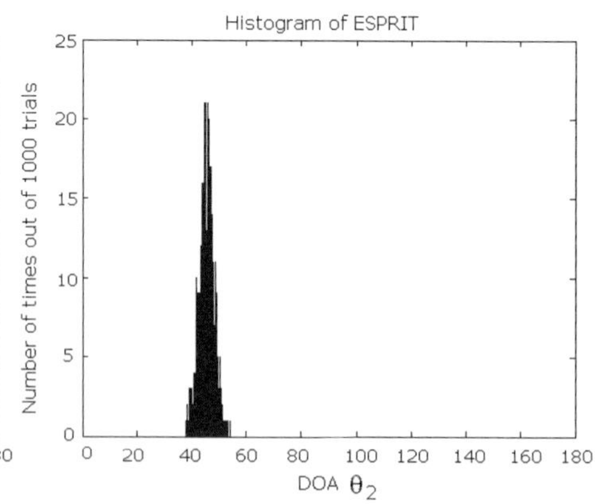

Figure 14.14. Accuracy of the estimate for θ_2 using the ESPRIT method using eight snapshots.

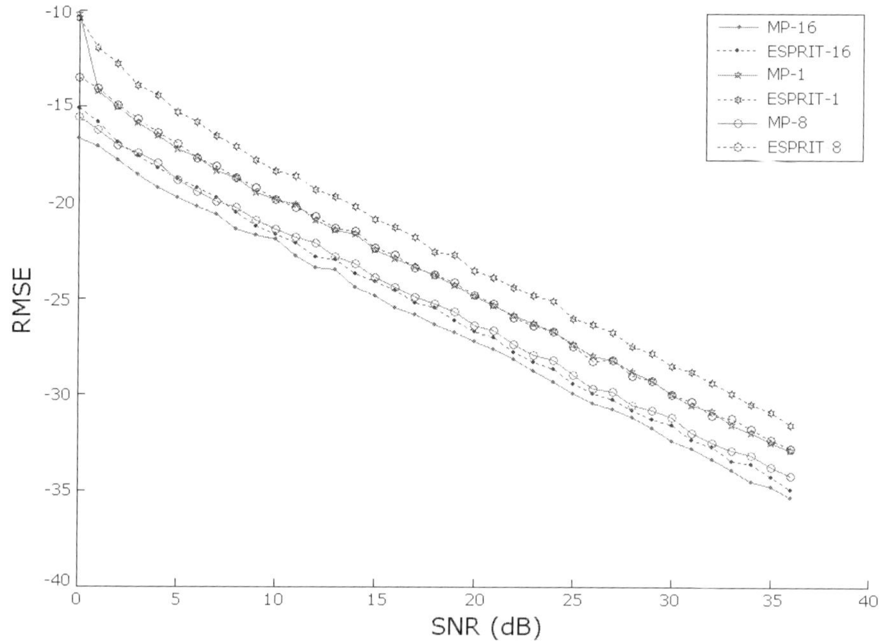

Figure 14.15. Comparing of the performance of the Matrix Pencil and ESPRIT methods for different number of snapshots.

14.10 CONCLUSION

A Unitary Transform for the 2-D MPM has been presented and utilized to convert the complex data matrix to a real matrix, hence reducing the computational complexity significantly for DOA estimation problems. It is seen that for lower SNR of the data, the 2-D Unitary MPM performs better than the 2-D MPM. Both the MP and the new UMP method can be used to model a given data set by a sum of complex exponentials and the UMP can be implemented on a DSP chip using only real arithmetic.

In addition, it has been shown that as we take more samples for the MPM, the variance of the estimator gcts smaller. The MPM results are compared with the statistical ESPRIT method under limited number of samples scenario. It is seen that the MPM outperforms ESPRIT in the accuracy of the solution. Also, as the MPM does not form a covariance matrix, the number of computations is smaller. The computational load however increases for the MPM for the multi snapshot case, as the matrix size is increased by appending the various Hankel matrices corresponding to the different snapshots.

REFERENCES

[1] A. Lee, "Centrohermitian and Skew-centrohermitian Matrices," *Linear Algebra and its Applications*, Vol. 29, pp. 205-210, 1980.

[2] K. C. Huang and C. C. Yeh, "Unitary Transformation Method for Angle of Arrival Estimation", *IEEE Transactions on Signal Processing*, Vol. 39, No. 4, pp. 975-977, 1991.

[3] G. Xu, R. H. Roy, and T. Kailath, "Detection of Number of Sources via Exploitation of Centro-Symmetric Property", *IEEE Transactions on Signal Processing*, Vol. 42, No. 1, pp. 102-111, Jan 1994.

[4] M. Haardt and J. A. Nossek, "Unitary ESPRIT: How to Obtain Increased Estimation Accuracy with a Reduced Computational Burden", *IEEE Transactions on Signal Processing*, Vol. 43, No. 5, pp. 1232-1242, 1995.

[5] N. Yilmazer, J. Koh and T. K. Sarkar, "Utilization of a Unitary Transform for Efficient Computation in the Matrix Pencil Method to Find the Direction of Arrival", *IEEE Transactions on Antennas and Propagation*, Vol. 54, No. 1, pp. 171-181, Jan 2006.

[6] Y. Hua, "Estimating Two Dimensional Frequencies by Matrix Enhancement and Matrix Pencil", *IEEE Transactions on Signal Processing*, Vol. 40, No. 9, pp. 2267-2280, Sep 1992.

[7] A.J. van der Veen, P. Ober, and E. F. Deprettere, "Azimuth and Elevation Computation in High Resolution DOA Estimation," *IEEE Transactions on Signal Processing*, Vol. 40, No. 7, pp. 1828-1832, July 1992.

[8] M. Haardt, K. Hueper, J. B. Moore, and J A Nossek, "Simultaneous Schur Decomposition of Several Matrices to Achieve Automatic Pairing in Multidimensional Harmonic Retrieval Problem", *Proceedings on EUSIPCO*, pp. 531-534, 1996.

[9] Y. Hua and K. Ahed-Meraim, "Techniques of Eigenvalues Estimation and Association", *Digital Signal Processing*, Vol. 7, No. 4, pp. 253-259, Oct 1997.

[10] A. J. Barabell, "Improving the Resolution Performance of Eigen-structure-based Direction Finding Algorithm", *Proceedings of the IEEE Int'l Conf. on Acoustics, Speech, and Signal Processing*, pp. 336-339, 1983.

[11] L. Datta and D. M. Salvatore, "Some Results on Matrix Symmetries and a Pattern Recognition Application", *IEEE Transactions on Acoustic, Speech, and Signal Processing*, Vol. ASSP-34, No. 4, pp. 992-993, Aug 1986.

[12] Y. Li, J. Razavilar, and K. J. Liu, "A High Resolution Technique for Multi-dimensional NMR Spectroscopy", *IEEE Transactions on Biomedical Engineering*, Vol. 45, No: 1, pp. 78-86, Jan 1998

[13] R. Bachl, "The Forward and Backward Averaging Technique Applied to TLS-ESPRIT Processing", *IEEE Transactions on Signal Processing*, Vol. 43, No. 11, pp. 2691-2699, Nov 1995.

[14] T. K. Sarkar and O. Pereira, "Using the Matrix Pencil Method to Estimate the Parameters of a Sum of Complex Exponentials", *IEEE Antennas and Propagation Magazine*, Vol. 37, No. 1, pp. 48-54, 1994.

[15] Y. Hua and T. K. Sarkar, "Generalized Pencil-of-Function Method for Extracting Poles of an EM System from Its Transient Response", *IEEE Transactions on Antennas and Propagation*, Vol. 37, No. 2, pp. 229-234, 1989.

[16] T. K. Sarkar, M. C. Wicks, and M. Salazar-Palma, *Smart Antennas*, John Wiley and Sons, Hoboken, NJ, 2003.

[17] J. Capon, "Maximum Likelihood Spectral Estimation", *Nonlinear Methods of Spectral Analysis*, pp. 155-179, 1979.

[18] R. O. Schmidt, "Multiple Emitter Location and Signal Parameter Estimation", *IEEE Transactions on Antennas and Propagation*, Vol. 34, No. 3, pp. 276-280, 1986.

[19] R. Roy and T. Kailath, "ESPRIT-Estimation of Signal Parameters Via Rotational Invariance Techniques", *IEEE Transactions on Signal Processing*, Vol. 37, No. 7, pp. 984-995, Jul 1989.

15

DOA ESTIMATION USING ELECTRICALLY SMALL MATCHED DIPOLE ANTENNAS AND THE ASSOCIATED CRAMER-RAO BOUND

15.0 SUMMARY

In this chapter, a methodology is presented for the direction of arrival (DOA) estimation using the induced voltages that are measured at the loads connected to electrically small matched dipole antenna arrays illuminated by the signal of interest (SOI). A matched antenna transmits/receives the maximum power. The Matrix Pencil method is applied directly to the induced voltages to estimate the DOA of the various signals. Using electrically small matched antennas can be advantageous as they can be placed in close proximity of each other saving the real estate and thus making it possible to deploy phased arrays on small footprints. When dealing with closely spaced matched electrically small antennas, it may be necessary to use the transformation matrix to compensate for the strong mutual coupling that may exist between the antenna elements. The transformation matrix converts the voltages that are induced at the loads corresponding to the feed point of the array operating in the presence of mutual coupling and other near field scatterers to an equivalent set of voltages that will be induced by the same incident wave in an uniform linear virtual array (ULVA) consisting of omnidirectional isotropic point radiators equally spaced and operating in free space. It is important to note that the coupling (in terms of power transfer) between small and between half-wavelength dipole elements is nearly identical if they are connected to conjugately matched loads. For any given incident field, the open circuit voltage developed across the feed-point of the small dipole will always be less than the open circuit voltage developed across the feed-point of the half-wavelength dipole. The difference is in the voltage developed across the loads connected to the dipole's feed-point. With the small dipole antenna, the voltage developed across the load impedance will be orders or magnitude greater than the voltage developed across the load connected to the half-wavelength dipole, even though the power captured by any conjugately matched antenna is approximately the same irrespective of their lengths. Performance of the various techniques in the presence of noise and mutual coupling is studied and compared to the Cramer-Rao bound (CRB) for the DOA. The variance of the DOA estimation results for the conjugately matched antenna array is close to the Cramer-Rao bound of an ideal array consisting of omnidirectional isotropic point radiators. Three different scenarios are presented to illustrate the methodology. We first consider resonant dipole elements spaced a half wavelength apart, next electrically small matched antenna elements spaced a half wavelength apart are considered, and lastly we look at electrically small matched antenna elements placed in close proximity of each other to reduce the footprint without affecting the performance of a phased array. In addition, we consider the possibility of DOA estimation using a combination of different types of electrically small antennas both uniformly and non-uniformly spaced. Numerical examples are presented to illustrate the principles of this methodology.

15.1 INTRODUCTION

With the advent of small airborne platforms it may be necessary to deploy phased arrays that may contain electrically small antennas [1–4] that are spaced very close to each other. One of the interesting properties to note for wire antennas is that when they are tuned with appropriate loading, the small overall electrical length of the structure does not significantly degrade the quality of their performance as a receive antenna. In fact, it will be shown that for the same incident field the voltage induced at the load connected to a conjugate-matched electrically small dipole antenna could be significantly larger than the voltages that are induced at the loads of the resonant half wave dipole antennas. For example, if we consider a half wave dipole antenna of length 0.5 λ and radius 0.0001λ, then the load required at the feed point to tune the half wave dipole will be $80.44 - j46.11\ \Omega$. The dipoles considered are quite thin. If this loaded, tuned dipole is irradiated by an incident field of 1 V/m incident from the broadside direction, then the current (in mA) at the feed point will be given by $2.08 - j0.099$. The complex voltage induced at the load located at the feed point is then given by $0.172 + j0.087$ V. Next we consider an electrically small dipole of length 0.1λ and radius 0.0001λ. We now need the complex impedance of $1.84 + j1910.15$ to be connected at the feed point to tune this electrically short dipole. If the same incident field of 1 V/m is incident broadside to this electrically small conjugate matched dipole antenna then the induced current at the feed point will be given by $13.13 + j0.148$ mA. When we compute the voltage induced across this complex load impedance, the value is given by $0.052 + j25.08$ V. The most interesting part is that the voltage induced at the load of the electrically small dipole is approximately two orders of magnitude larger than the voltage that is induced at the load of the resonant dipole. The more remarkable part, which is often not mentioned in classical antenna theory, is that the power captured by both the resonant and tuned antennas is approximately the same even though their electrical sizes are quite different. For the half-wave dipole antenna the power absorbed from the incident field is 0.348 mW whereas for the conjugately matched electrically small antenna is 0.317 mW. Hence, there is a difference of only 0.4 dB. Therefore, use of electrically small resonant dipoles has significant potentials as phased array elements, when dealing with arrays on a small footprint, if one is interested in carrying out digital beam forming. This will be particularly useful for airborne or other related platforms, which have a small real estate, and use of electrically small dipoles will be of absolute necessity. The radiation efficiency of small wire-type antennas can be increased by using the larger diameter wires. However, such a choice makes the numerical simulation quite complicated as the thin-wire antenna theory no longer holds due to the circumferential variation of the currents over electrically thick wire structures.

In section 15.2 we review the application of the transformation matrix, which can compensate for the various mutual coupling effects between the antennas. The Matrix Pencil method is then applied to some simulated examples to illustrate the accuracy and efficiency of this procedure. In section 15.3, an expression of the Cramer-Rao bound (CRB) for the DOA estimation is presented. The numerical variances of the DOA estimation results of the array are compared to the CRB for the same array. For the numerical examples in section 15.4, we consider three different types of phased array systems. The first one consists of antennas of length 0.1 λ and spaced 0.1 λ apart. The second array consists of elements that are also 0.1 λ long but spaced 0.5 λ apart. The third one consists of antennas that are 0.5 λ long and spaced 0.5 λ apart. All the elements of the array are appropriately loaded so that they are matched at the operating frequency. In section 15.5 we consider the possibility of DOA estimation using a combination of different types of electrically small antennas both uniformly and non-uniformly spaced taking the effects of mutual coupling into account through a transformation matrix. The effect of applying the transformation matrix on this array is also presented. Finally, the conclusion is given in section 15.6.

15.2 DOA ESTIMATION USING A REALISTIC ANTENNA ARRAY

First, we describe the procedure of using a transformation matrix to compensate for the mutual coupling between the elements of the array.

15.2.1 Transformation Matrix Technique

Consider a phased array composed of $N+1$ antenna elements. Assume that $P+1$ narrow-band sources impinge on the array from $P+1$ distinct directions $\varphi_0, \cdots, \varphi_P$. Using the complex envelope representation, a single snapshot of the voltages represents a $(N+1)\times 1$ vector of phasor voltages $\left[\overline{x}(t)\right]$ received by the elements of the actual array at a particular time instance t and can be expressed by

$$\left[\bar{x}(t)\right] = \begin{bmatrix} x_0(t) \\ x_1(t) \\ \vdots \\ x_N(t) \end{bmatrix} = \sum_{i=0}^{P} s_i(t)\left[\bar{a}(\varphi)_i\right] + \left[\bar{\xi}(t)\right] \tag{15.1}$$

where $s_i(t)$ denotes the signal at the elements of the array from the ith source, for $i = 0, 1, \ldots, P$. $\left[\bar{a}(\varphi)\right]$ denotes the far-field pattern of the array toward the direction φ and $\left[\bar{\xi}(t)\right]$ denotes the noise vector at each of the loaded antenna elements. By using a matrix representation (15.1) becomes

$$\left[\bar{x}(t)\right] = \left[A(\varphi)\right]\left[\bar{s}(t)\right] + \left[\bar{\xi}(t)\right] \tag{15.2}$$

where $\left[A(\varphi)\right]$ is the $(N+1)\times(P+1)$ matrix, referred to as the array manifold corresponding to each one of the incident signals of unity amplitude and is represented by

$$\left[A(\varphi)\right] = \left[\bar{a}(\varphi_0),\ \bar{a}(\varphi_1),\ \ldots,\ \bar{a}(\varphi_P)\right] \tag{15.3}$$

In a typical array calibration methodology, a far field source $s_i(t)$ is placed along the angular direction φ_i and then $\left[\bar{x}(t)\right]$ is the voltage measured at the loads of the antenna elements at the time instance t. Equation (15.2) is applicable to antenna arrays which consist of isotropic omni-directional point radiators, located in free space. However, this equation needs to be modified for a real array where there are mutual coupling between the other near field scatterers. In that situation, there would no longer exist such a linear relationship between the incoming signals and the measured snapshot of the voltages. In practice, the array manifold information in a real array related to these induced voltages is contaminated by the effects of non-uniformity in the spacing and mutual coupling between the elements of the array, and the presence of near field scatterers will undermine the performance of a conventional adaptive signal processing algorithm.

By using a transformation matrix, one can compensate for all the undesired electromagnetic effects [5,6]. The voltages induced in an actual array are then transformed to a set of voltages that would be induced in an ULVA consisting of omni-directional isotropic point radiators radiating in free space. It is based on transforming the real array into an ULVA operating in the absence of mutual coupling and other undesired electromagnetic effects. Hence, we are going to select the best-fit transformation, $[\Im]$, between the real array manifold, $[A(\varphi)]$, and the virtual array manifold corresponding to an ULVA, $[A_v(\varphi)]$ such that

$$[\Im]\,[A(\varphi)] = [A_v(\varphi)] \tag{15.4}$$

for all of the azimuth angles φ within a predefined sector.

First we define a set of uniformly defined angles to cover a sector, the azimuth angles spanning $[\varphi_q, \varphi_{q+1}]$:

$$\left[\Phi_q\right] = [\varphi_q, \varphi_q + \phi, \varphi_q + 2\phi, \ldots, \varphi_{q+1}] \tag{15.5}$$

where the angle ϕ represents the azimuth incremental step size. Then we measure the steering vectors associated with the set $[\Phi_q]$ of the real array. The measured real array manifold is defined by

$$[A(\Phi_q)] = [a(\varphi_q),\ a(\varphi_q + \phi),\ \ldots\ ,\ a(\varphi_{q+1})] \tag{15.6}$$

This can include all the undesired electromagnetic coupling effects. Next, we calculate the virtual array manifold, $[A_v(\Phi_q)]$, corresponding to the same set of angles $[\Phi_q]$:

$$[A_v(\Phi_q)] = [a_v(\varphi_q),\ a_v(\varphi_q + \phi),\ \ldots\ ,\ a_v(\varphi_{q+1})] \tag{15.7}$$

where $a_v(\varphi,\theta)$ is a set of theoretical steering vectors corresponding to the uniformly spaced virtual array. Then the transformation matrix is obtained using a least square solution as follows

$$[\Im_q] = [A_v(\Phi_q)][A(\Phi_q)]^H \{[A(\Phi_q)][A(\Phi_q)]^H\}^{-1} \tag{15.8}$$

where the superscript H represents the conjugate transpose of a complex matrix. This transformation matrix needs to be computed only once *a priori* for the defined sector and this computation can be done off-line. Hence, once $[\Im]$ is known we can compensate for the various undesired electromagnetic effects, in real time by carrying out a single matrix vector multiplication. Finally, using Equation (15.8), we can obtain the processed input voltages in which the effects of non-uniformity in the spacing, the mutual coupling effects and the effects of various near field scatterers have been eliminated. The compensated voltages $\left[\bar{x}_c(t)\right]$ will then be given by

$$\left[\bar{x}_c(t)\right] = [\Im]\left[\bar{x}(t)\right] \tag{15.9}$$

After (15.9) is used to process the measured voltages, we apply the Matrix Pencil method to the preprocessed set of voltages $[\bar{x}_c(t)]$ without any significant loss of accuracy to obtain the DOA of the various signals even if they are all coherent.

It is important to note that after the compensation matrix is applied, any DOA estimation method can be used to find the DOA of the various signals. The Matrix Pencil method is one example, but it is not the only method. However, we use this method as we carry out the processing using a single snapshot which may be useful in a highly dynamic environment as illustrated in chapter 14.

Next, we apply the Matrix Pencil method to the processed set of voltages induced at the terminals of the antenna. The details of the Matrix Pencil method are available in chapter 14.

15.3 CRAMER-RAO BOUND FOR DOA ESTIMATION

We now present the Cramer-Rao bound (CRB) for this DOA estimation method. The performance of this DOA estimation technique in the presence of noise is compared to the CRB. We derive a Fisher information matrix which is in such a form that some relations between the CRB and signal parameters are easily seen. From Eq. (15.2), after normalizing the induced voltages by the antenna characteristic factor, α_n, we can write it in a vector form as

$$\boldsymbol{y} = \boldsymbol{g} + \boldsymbol{w} \tag{15.10}$$

where $\boldsymbol{y} = [y_0, y_1, \cdots, y_N]^T$, $\boldsymbol{g} = [g_0, g_1, \cdots, g_N]^T$, and $\boldsymbol{w}$ is a noise vector. The probability density function (pdf) of $\boldsymbol{w}$ is a normal distribution, i.e., $\Lambda(\mathbf{0}, 2\sigma^2 \mathbf{I}_{N+1})$. Here $\mathbf{I}$ is an identity matrix. Therefore the pdf of $\boldsymbol{y}$ is $\Lambda(\boldsymbol{g}, 2\sigma^2 \mathbf{I}_{N+1})$. It is clear that the mean vector $\boldsymbol{g}$ depends on the parameter vector $\boldsymbol{\theta}$ defined as

$$\boldsymbol{\theta} = [\boldsymbol{\theta}_0^T, \ \cdots, \ \boldsymbol{\theta}_P^T]^T \tag{15.11}$$

$$\boldsymbol{\theta}_t = [A_t, \ \gamma_t, \ \varphi_t]^T \tag{15.12}$$

Let $\boldsymbol{\theta}_t$ denotes the tth element of $\boldsymbol{\theta}$. For unbiased estimates, the CRB states that if $\hat{\boldsymbol{\theta}}$ is an unbiased estimate of $\boldsymbol{\theta}$, the variance of each element of $\hat{\boldsymbol{\theta}}$ can be no smaller than the corresponding diagonal term in the inverse of the Fisher information matrix, $\boldsymbol{J}$. Then the (t, j)th element of the Fisher information matrix , $\boldsymbol{J}$, can be written [5] as

$$(\mathbf{J})_{t,j} = \frac{1}{\sigma^2} \sum_{k=0}^{N} \text{Re}\left\{ \frac{\partial g_k}{\partial \theta_t} \frac{\partial g_k^*}{\partial \theta_j} \right\} \tag{15.13}$$

where $\partial(\ )/\partial\theta_t$ is a partial derivative and $\text{Re}\{\bullet\}$ denotes the real part. But $\boldsymbol{J}$ can be partitioned as

$$\boldsymbol{J}=\{\boldsymbol{J}_{t,j};\, t, j=1, 2,\cdots, P+1\} \tag{15.14}$$

where $\boldsymbol{J}_{t,j}$ is a 3 × 3 (t, j)th block matrix of $\boldsymbol{J}$, and by using (15.13) it can be shown to be of the form

$$\boldsymbol{J}_{t,j}=\frac{1}{\sigma^2}\sum_{k=0}^{N}\operatorname{Re}\begin{Bmatrix} \frac{\partial g_k}{\partial A_t}\frac{\partial g_k^*}{\partial A_j} & \frac{\partial g_k}{\partial A_t}\frac{\partial g_k^*}{\partial \gamma_j} & \frac{\partial g_k}{\partial A_t}\frac{\partial g_k^*}{\partial \varphi_j} \\ \frac{\partial g_k}{\partial \gamma_t}\frac{\partial g_k^*}{\partial A_j} & \frac{\partial g_k}{\partial \gamma_t}\frac{\partial g_k^*}{\partial \gamma_j} & \frac{\partial g_k}{\partial \gamma_t}\frac{\partial g_k^*}{\partial \varphi_j} \\ \frac{\partial g_k}{\partial \varphi_t}\frac{\partial g_k^*}{\partial A_j} & \frac{\partial g_k}{\partial \varphi_t}\frac{\partial g_k^*}{\partial \gamma_j} & \frac{\partial g_k}{\partial \varphi_t}\frac{\partial g_k^*}{\partial \varphi_j} \end{Bmatrix} \tag{15.15}$$

Therefore,

$$\boldsymbol{J}_{t,j}=\frac{1}{\sigma^2}\boldsymbol{A}_t\operatorname{Re}\left\{\exp\left[\,j(\gamma_t-\gamma_j)\right]\boldsymbol{H}_{t,j}\right\}\boldsymbol{A}_j \tag{15.16}$$

in which

$$\boldsymbol{A}_t=\operatorname{diag}\{1,\, A_t,\, A_t\},\ \boldsymbol{A}_j=\operatorname{diag}\{1,\, A_j,\, A_j\} \tag{15.17}$$

$$\boldsymbol{H}_{t,j}=\begin{bmatrix} \beta_{t,j,0} & (-j)\beta_{t,j,0} & j2\pi\Delta\sin\varphi_j\,\beta_{t,j,1} \\ j\beta_{t,j,0} & \beta_{t,j,0} & -2\pi\Delta\sin\varphi_j\,\beta_{t,j,1} \\ -j2\pi\Delta\sin\varphi_t\,\beta_{t,j,1} & -2\pi\Delta\sin\varphi_t\,\beta_{t,j,1} & (2\pi\Delta)^2\sin\varphi_t\sin\varphi_j\beta_{t,j,2} \end{bmatrix} \tag{15.18}$$

$$\beta_{t,j,0}=\sum_{k=0}^{N}\exp[\,j2\pi\Delta k(\cos\varphi_t-\cos\varphi_j)] \tag{15.19}$$

$$\beta_{t,j,1}=\sum_{k=0}^{N}k\exp[\,j2\pi\Delta k(\cos\varphi_t-\cos\varphi_j)] \tag{15.20}$$

$$\beta_{t,j,2}=\sum_{k=0}^{N}k^2\exp[\,j2\pi\Delta k(\cos\varphi_t-\cos\varphi_j)] \tag{15.21}$$

The 3 diagonal elements of $\boldsymbol{J}_{t,t}$, which is the 3 × 3 (t, j) th block matrix of $\boldsymbol{J}^{-1}$, are the CRB for A_t, γ_t, and φ_t, respectively.

15.4 DOA ESTIMATION USING 0.1 λ LONG ANTENNAS

As the first example, consider five antenna elements as shown in Figure 15.1. The radius of the wire is 0.00001λ, where λ = 1 m. It is a rather unusually thin

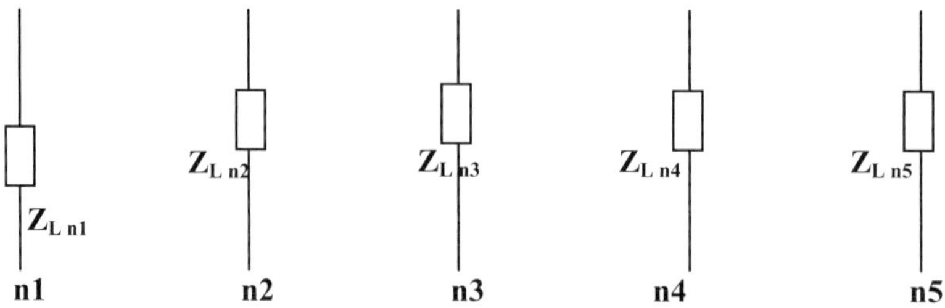

Figure 15.1: The five-antenna array system with loadings.

wire. The length of the dipole antennas are 0.1 λ and the spacing between each of the antenna elements is 0.1 λ. The reference point is chosen at the center of the array. We consider two different cases for this array. For the first case, which we term non-resonant, we use a load at each of the dipole antenna elements which will make them resonant when operating in free space in an isolated fashion. So when taken separately, each of the dipole antenna by itself are matched. However, when these five dipole antennas are placed in an array, due to the mutual coupling between them, the terminating loads are not the same which will make them matched when they are operating in an array environment. The terminating loads need to be tuned depending on how the antennas are placed in the array. The antenna input impedance for the 0.1λ length antenna is $Z_{in\ 0.1\lambda} = 0.19\times10^1 - j\ 0.27\times10^4\ \Omega$. To make the antenna resonant or matched it is then necessary to terminate it with the complex conjugate of $Z_{in\ 0.1\lambda}$ to make it the terminating load when the antenna is operating by itself in free space. However, when we place five of them together in an array, the terminating loads on each of the antenna will differ from this matched value due to the mutual coupling between the antenna elements. Using these two different types of loading we perform a DOA estimation. We consider a signal of amplitude 1 V/m is incident on the array from a single azimuth angle φ at a time covering from $60°$ to $90°$. The estimated DOA is shown in Fig. 15.2. In Fig. 15.2 the resonant case implies that the load at each antenna has been fine tuned so that the mutual coupling effects have been accounted for, in their evaluation, whereas for the nonresonant case, the loads which make them individually tuned in free space have been used.

In Fig 15.3, we plot the Cramer-Rao bound for the DOA estimation given by (15.18). For this case, the incident signals are assumed to arrive from $70°$ with E_θ = 1 V/m. Then, 500 trials were simulated using different noise parameters to calculate the variance of the estimate of the azimuth angle $\hat{\varphi}$ given by both the arrays. The variance of $\hat{\varphi}$ is plotted as $-10\log_{10}\mathrm{var}(\hat{\varphi})$. It is seen that the matched array performs better than the unmatched case. On the same plot we present the result of an ideal array where the elements are not dipoles but isotropic omni directional point radiators. It is seen that the variance of the estimated DOA due to the ideal array and the array with conjugately matched antenna elements, are approximately the same and they are quite close to the Cramer-Rao bound.

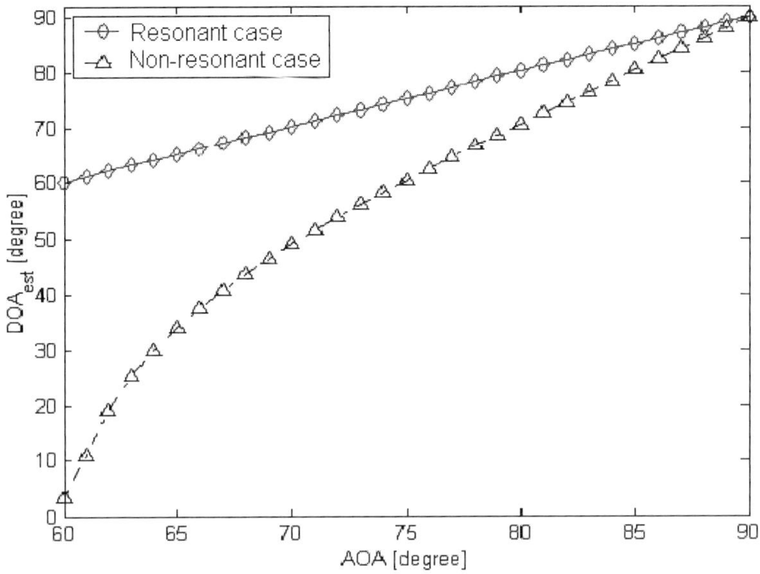

Figure 15.2. The DOA estimation results of the resonant and non-resonant case when the signal arrives at each angle which is 60° to 90°.

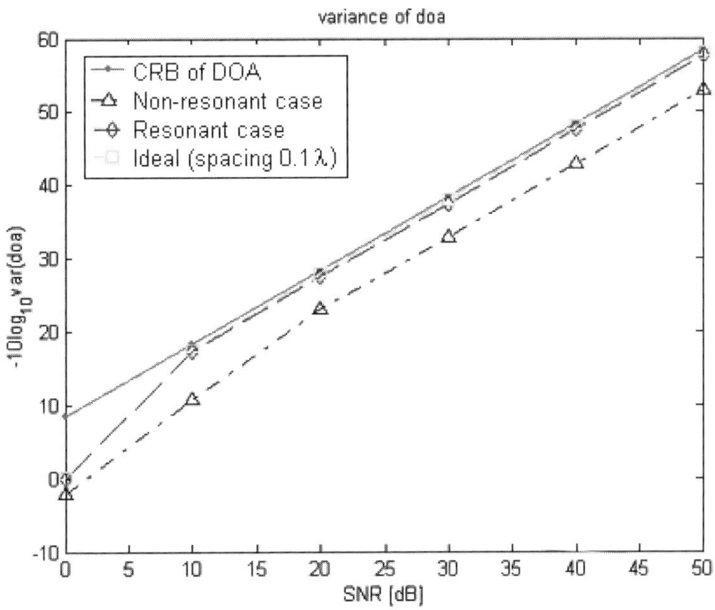

Figure 15.3. Cramer-Rao bound for φ and the variance of $\hat{\varphi}$ for the non-resonant case, the resonant case and the ideal case when the signal arrives from 70°.

15.5 DOA ESTIMATION USING DIFFERENT ANTENNA ARRAY CONFIGURATIONS

In the first part of this section, we consider two different cases as shown in Table 15.1. For each case, we have five antenna elements composed of electrically small tuned dipole wire antennas working as an array as shown in Figure 15.1. The difference between Case I and Case II is that for Case I the same loading which makes a single antenna resonant operating in free space is applied to all the five elements without considering the fact that the presence of mutual coupling will change their input impedances and so the loading required to make each antenna element resonant may depend on its location in the phased array. For Case I the values of the impedance found in section 15.4 was used. For the second case, the terminating impedances were fine tuned for each antenna elements individually to make each one approximately resonant. Hence, the values for the load impedances were different depending on the position (whether the antenna element was in the middle of the array or at the end positions) of the element in the array. For both the cases, three different array configurations (A, B, and C array system) have been used. The three different arrays are all operating at 300 MHz. The radius of each of the wire is 0.001λ. The length of the dipole antenna for the A-array system is 0.5λ and the spacing between the antennas is 0.5λ. For the B-array system, the length of the dipole antenna is 0.1λ and the spacing is 0.1λ. The C-array system has dipole antennas that are 0.1λ long and are spaced 0.5λ apart. This has been summarized in Table 15.1. All the simulations have been carried out using the electromagnetic software simulator [7].

Table 15.1. The Two Cases Consisting of Three Different Array Configurations.

		Loadings in an Array.	Length of Dipole Antenna.	Spacing Between the Antennas.
Case I	A-array	Same loadings on all the antenna elements, irrespective of their locations in the array	$0.5\ \lambda$	$0.5\ \lambda$
	B-array		$0.1\ \lambda$	$0.1\ \lambda$
	C-array		$0.1\ \lambda$	$0.5\ \lambda$
Case II	A-array	Different loadings on the antenna elements based on their locations in the array	$0.5\ \lambda$	$0.5\ \lambda$
	B-array		$0.1\ \lambda$	$0.1\ \lambda$
	C-array		$0.1\ \lambda$	$0.5\ \lambda$

For Case I, we consider that all the dipoles are loaded by the same impedance values. For configuration A that value is $Z_L = 76.74 - j\,41.89\ \Omega$. For configurations B and C that value is $Z_L = 1.97 + j\,3614.57\ \Omega$. These loadings are applied to all the antenna elements. For all the three configurations, the input impedances for each of the antenna elements in the phased array systems are given by Table 15.2. In Table 15.3, the current at the feed point is provided when the array is illuminated from the broadside by an incident wave of 1 V/m. It is seen that the structure is approximately resonant.

In this simulation a signal of amplitude 1 V/m arrives from within the sector of $\varphi = 60°$ to $90°$ and $\theta = 90°$. The signal impinges on the five-element array. When we estimate the DOA and the complex amplitude of the signals using the Matrix Pencil method, it is assumed that the antenna elements are omni-directional isotropic sensors operating in free space. So, for this example, we do not compensate for the mutual coupling between the antenna elements. However, we normalize all the voltage levels with respect to the voltage that will be induced in a single resonant antenna element with a broadside incidence of 1 V/m. Hence the normalization factor is different for each of the three cases. The normalization is required to estimate the correct amplitude of the incoming signal. The normalized voltage in this case is obtained from the ratio

$$\frac{\text{Induced Voltage}}{\text{Normalization Factor (Normalization Voltage)}} \tag{15.22}$$

Table 15.2. The Input Impedance for Each Antenna with the Same Load Connected to Each One (Unit [Ω]).

	Case I-A	Case I-B	Case I-C
$Z_{\text{Element 1}}$	$1.480\times10^{2}-j\,2.350\times10^{1}$	$1.115\times10^{1}+j\,1.911\times10^{0}$	$3.858\times10^{0}-j\,6.531\times10^{-1}$
$Z_{\text{Element 2}}$	$1.224\times10^{2}-j\,4.191\times10^{1}$	$5.804\times10^{0}-j\,1.405\times10^{1}$	$3.204\times10^{0}-j\,1.202\times10^{0}$
$Z_{\text{Element 3}}$	$1.402\times10^{2}-j\,3.608\times10^{1}$	$3.036\times10^{0}-j\,8.368\times10^{0}$	$3.667\times10^{0}-j\,1.050\times10^{0}$
$Z_{\text{Element 4}}$	$1.224\times10^{2}-j\,4.191\times10^{1}$	$5.804\times10^{0}-j\,1.405\times10^{1}$	$3.204\times10^{0}-j\,1.202\times10^{0}$
$Z_{\text{Element 5}}$	$1.480\times10^{2}-j\,2.350\times10^{1}$	$1.115\times10^{1}+j\,1.911\times10^{0}$	$3.858\times10^{0}-j\,6.531\times10^{-1}$

Table 15.3. The Induced Currents on Each Antenna for Case I When the Signal Comes From $\theta = 90°$ with $E_\theta = 1$V/m (Unit [mA]).

	Case I-A	Case I-B	Case I-C
$I_{\text{Element 1}}$	$2.143\times10^{0}+j\,2.886\times10^{-1}$	$4.357\times10^{0}-j\,7.473\times10^{-1}$	$1.259\times10^{1}+j\,2.133\times10^{0}$
$I_{\text{Element 2}}$	$2.368\times10^{0}+j\,7.713\times10^{-1}$	$1.255\times10^{0}+j\,3.039\times10^{0}$	$1.368\times10^{1}+j\,5.131\times10^{0}$
$I_{\text{Element 3}}$	$2.170\times10^{0}+j\,5.098\times10^{-1}$	$-1.955\times10^{0}+j\,5.279\times10^{0}$	$1.260\times10^{1}+j\,3.606\times10^{0}$
$I_{\text{Element 4}}$	$2.368\times10^{0}+j\,7.713\times10^{-1}$	$1.255\times10^{0}+j\,3.039\times10^{0}$	$1.368\times10^{1}+j\,5.131\times10^{0}$
$I_{\text{Element 5}}$	$2.143\times10^{0}+j\,2.886\times10^{-1}$	$4.357\times10^{0}-j\,7.473\times10^{-1}$	$1.259\times10^{1}+j\,2.133\times10^{0}$

Hence, for the case I-A, we normalize the measured voltages at each of the five antenna elements by the voltage measured at the load located at the feed point of a single 0.5λ antenna. For cases I-B and I-C, the normalization factor is given by the voltage measured at the load located at the feed point of the array. These normalization voltages are given in Table 15.4.

Table 15.4. The Normalized Voltages for Each Configuration.

Case	Normalization Voltage
Case I-A	$0.161 - j\,0.091$
Case I-B	$0.063 + j\,43.375$
Case I-C	$0.063 + j\,43.375$

Now, we add noise which is uniformly distributed in amplitude, and the phase is chosen between 0 and 2π so that the signal-to-noise ratio is +26 dB at each of the antenna elements. Figure 15.4 provides the results of the DOA estimation for a signal with noise estimated by the three different array configurations. The DOA estimation has correctly been obtained for the resonant 0.5λ array spaced 0.5λ apart and the 0.1λ long dipole array spaced 0.5λ apart. However, the estimates given by the 0.1λ long dipole array spaced 0.1λ apart are disturbed by the strong mutual coupling that exists between the elements of the closely spaced array, even though the antenna elements are electrically small.

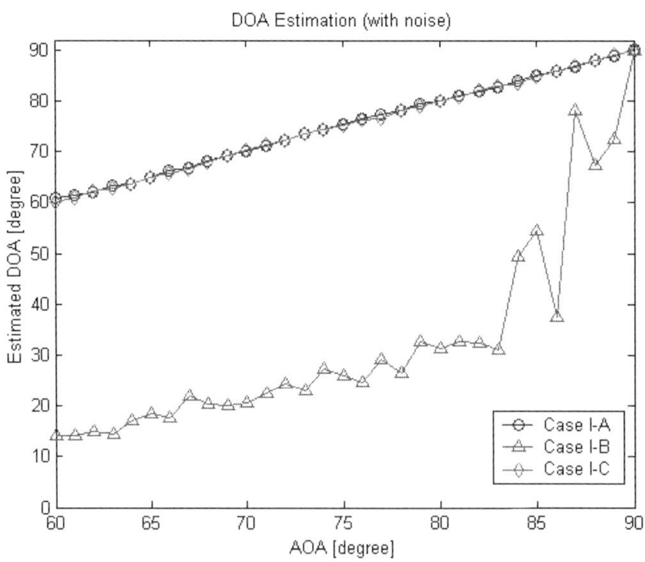

Figure 15.4. DOA estimation for the three configurations of Case I when the signal impinges on the array between ϕ = 60° and 90° with a noise of 26 dB SNR. No compensation for mutual coupling.

Figures 15.5 and 15.6 provide the estimates for the amplitude and the phase of the estimated signal. Again the results for case I-A and I-C are of engineering accuracy. We next compensate for the mutual couplings between the elements of the array using the transformation matrix. The corresponding results are given in Figs. 15.7–15.9. As expected, all the three configurations now yield quite accurate results for the estimate of the SOI paramcters in the presence of noise.

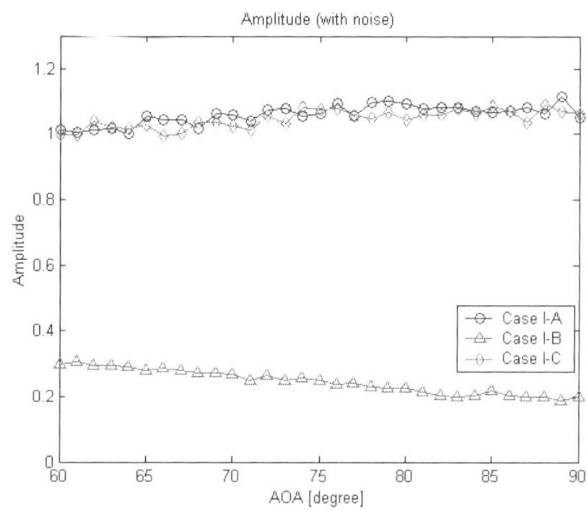

Figure 15.5. Estimate for the amplitude of the signal for all the three configurations of Case I when the signal impinges on the array anywhere between $\phi = 60°$ and 90° with a noise of 26 dB SNR. No compensation for mutual coupling.

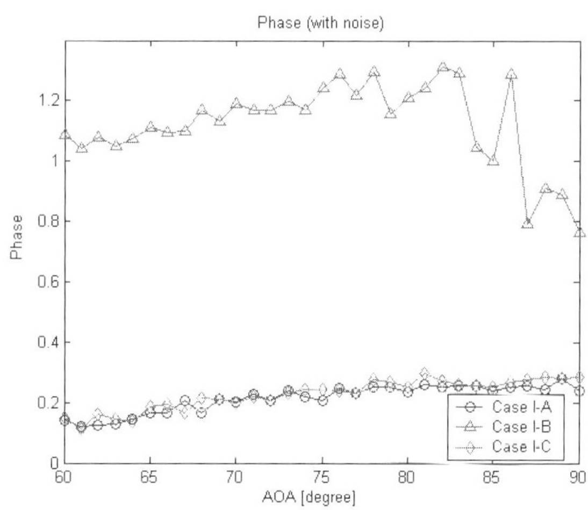

Figure 15.6. Estimate for the phase (in radians) of the signal for all the three configurations of Case I when the signal impinges on the array anywhere between $\phi = 60°$ and 90° with a noise of 26 dB SNR. No compensation for mutual coupling.

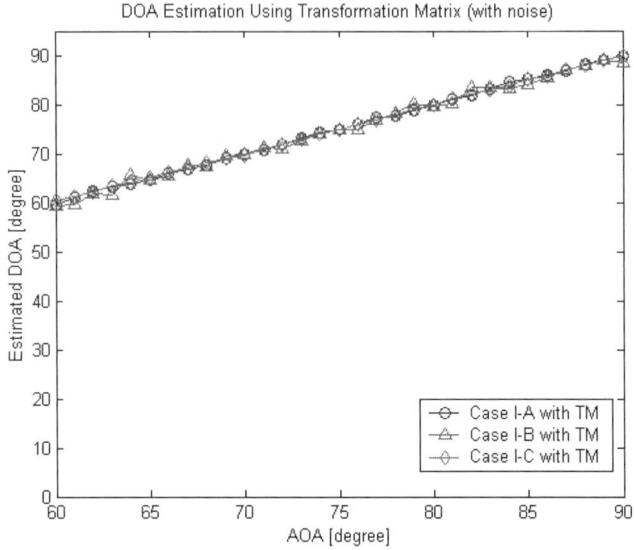

Figure 15.7. DOA estimation for the three configurations of Case I when the signal impinges on the array anywhere between $\phi = 60°$ and $90°$ with a noise of 26 dB SNR. Complete compensation for the mutual coupling for all the three configurations using the transformation matrix has been implemented.

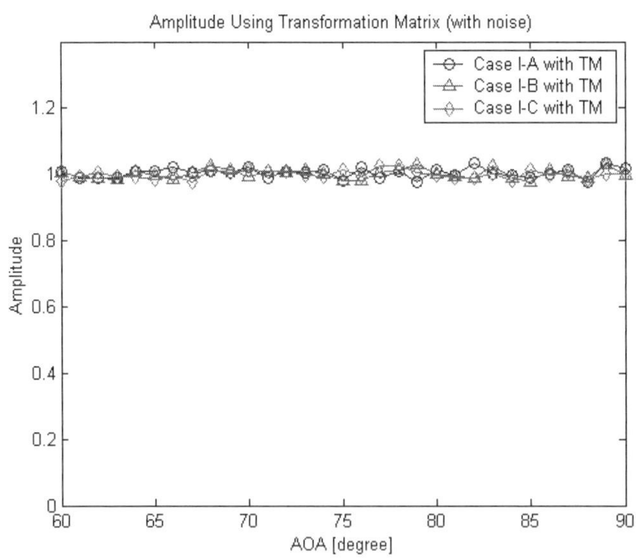

Figure 15.8. Estimate for the amplitude of the signal for all the three configurations of Case I when the signal impinges on the array between $\phi = 60°$ and $90°$ with a noise of 26 dB SNR. Complete compensation for the mutual coupling for all the three configurations using the transformation matrix has been implemented.

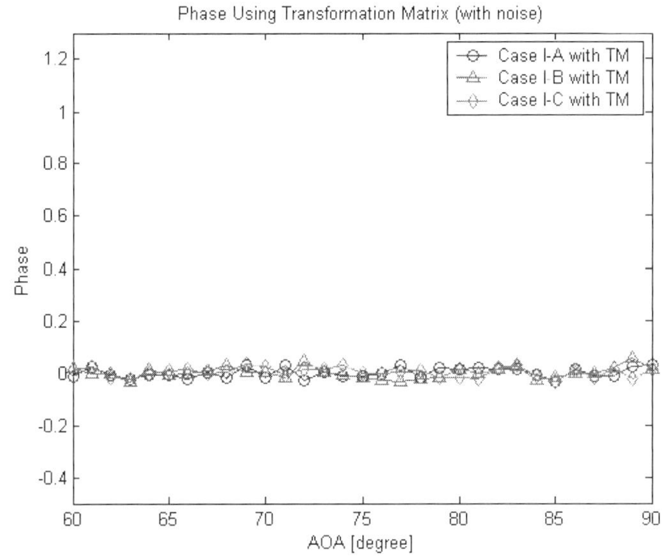

Figure 15.9. Estimate for the phase (in radians) of the signal for all the three configurations of Case I when the signal impinges on the array between $\phi = 60°$ and $90°$ with a noise of 26 dB SNR. Complete compensation for the mutual coupling for all the three configurations using the transformation matrix has been implemented.

For Case II, we consider different loadings for each of the antenna elements as the mutual coupling has different effects on each antenna depending on its position in the array as shown in Fig. 15.1. Table 15.5 provides the possible values for the terminating impedances that will make each of the five antenna elements resonant when operating in an array environment. The induced currents are then given in Table 15.6 for a broadside incident signal.

Table 15.5. Loading for Each Antenna When Different Loadings are Applied (Unit [Ω]).

	Case II-A	Case II -B	Case II –C
$Z_{\text{Element 1}}$	$6.725\times10^1 - j\,1.368\times10^1$	$6.182\times10^0 + j\,3.625\times10^3$	$1.730\times10^0 + j\,3.615\times10^3$
$Z_{\text{Element 2}}$	$5.145\times10^1 - j\,2.870\times10^0$	$7.816\times10^0 + j\,3.631\times10^3$	$1.406\times10^0 + j\,3.616\times10^3$
$Z_{\text{Element 3}}$	$5.757\times10^1 - j\,3.028\times10^0$	$8.403\times10^0 + j\,3.631\times10^3$	$1.521\times10^0 + j\,3.615\times10^3$
$Z_{\text{Element 4}}$	$5.145\times10^1 - j\,2.870\times10^0$	$7.816\times10^0 + j\,3.631\times10^3$	$1.406\times10^0 + j\,3.616\times10^3$
$Z_{\text{Element 5}}$	$6.725\times10^1 - j\,1.368\times10^1$	$6.182\times10^0 + j\,3.625\times10^3$	$1.730\times10^0 + j\,3.615\times10^3$

Table 15.6. The Induced Currents at the Loads for Each of the Antenna Elements When Appropriately Loaded. The Incident Signal is Coming From the Broadside Direction with E_θ= 1V/m (Unit [mA]).

	Case II -A	Case II -B	Case II -C
$I_{\text{Element 1}}$	$2.441\times10^{0}-j\,1.099\times10^{-1}$	$3.656\times10^{0}-j\,4.550\times10^{-1}$	$1.489\times10^{1}+j\,2.820\times10^{-1}$
$I_{\text{Element 2}}$	$3.101\times10^{0}+j\,3.401\times10^{-2}$	$3.374\times10^{0}+j\,1.480\times10^{-1}$	$1.689\times10^{1}-j\,5.621\times10^{-1}$
$I_{\text{Element 3}}$	$2.811\times10^{0}-j\,1.080\times10^{-1}$	$3.219\times10^{0}+j\,3.959\times10^{0}$	$1.702\times10^{1}+j\,4.030\times10^{-1}$
$I_{\text{Element 4}}$	$3.101\times10^{0}+j\,3.401\times10^{-2}$	$3.374\times10^{0}+j\,1.480\times10^{-1}$	$1.689\times10^{1}-j\,5.620\times10^{-1}$
$I_{\text{Element 5}}$	$2.441\times10^{0}-j\,1.099\times10^{-1}$	$3.656\times10^{0}-j\,4.550\times10^{-1}$	$1.489\times10^{1}+j\,2.824\times10^{-1}$

For Case II, the DOA estimation using the measured voltages at the loads of each of the antenna for the three configurations are carried out using the Matrix Pencil method after all the voltages are normalized using the voltage factors as presented in Table 15.7. The reason the normalization factors in Table 15.7 are different for each antenna element is because each antenna is loaded differently. Then the Matrix Pencil method is applied to estimate the DOA of the signal. Figure 15.10 shows the DOA estimation of the signal without applying the transformation matrix with different loadings that make each of the antennas in the array to be resonant in situ.

Table 15.7. The Normalized Voltages for Each Antenna Configuration for Case II.

Elements	Case II-A	Case II-B	Case II-C
Element 1	$0.1626-j\,4.125\times10^{-2}$	$1.672+j\,.325\times10^{1}$	$-9.936+j\,5.384\times10^{1}$
Element 2	$0.1596-j\,7.151\times10^{-3}$	$-0.5110+j1.225\times10^{1}$	$2.056+j\,6.108\times10^{1}$
Element 3	$0.1615-j\,1.473\times10^{-2}$	$-14.35+j\,1.172\times10^{1}$	$-1.431+j\,6.155\times10^{1}$
Element 4	$0.1596-j\,7.151\times10^{-3}$	$-0.5110+j\,1.225\times10^{1}$	$2.056+j\,6.108\times10^{1}$
Element 5	$0.1626-j\,4.125\times10^{-2}$	$1.672+j\,1.325\times10^{1}$	$-0.9936+j\,5.384\times10^{1}$

By comparing Figures 15.4 and 15.10 it is seen that by terminating the electrically small dipoles by their respective conjugately matched loads the compact array may actually minimize the effects of strong mutual coupling between the elements as evidenced by a more accurate estimate for the DOA of the signal. However, we have not actually compensated for the mutual coupling at all in this case. Figures 15.11 and 15.12 show the amplitude and phase of the estimated signal, respectively. By comparing these results with those in Figs. 15.5 and 15.6, it is seen that the estimate of the amplitude of the signal obtained using the electrically small array with conjugately matched elements has a much better performance even though the phase is off.

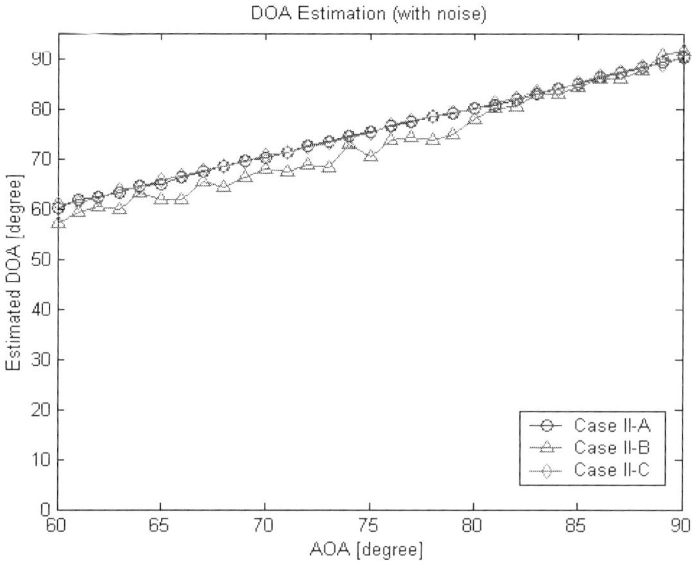

Figure 15.10. DOA estimation for the three configurations of Case II when the signal impinges on the array between $\phi = 60°$ and $90°$ with a noise of 26 dB SNR. No compensation for mutual coupling.

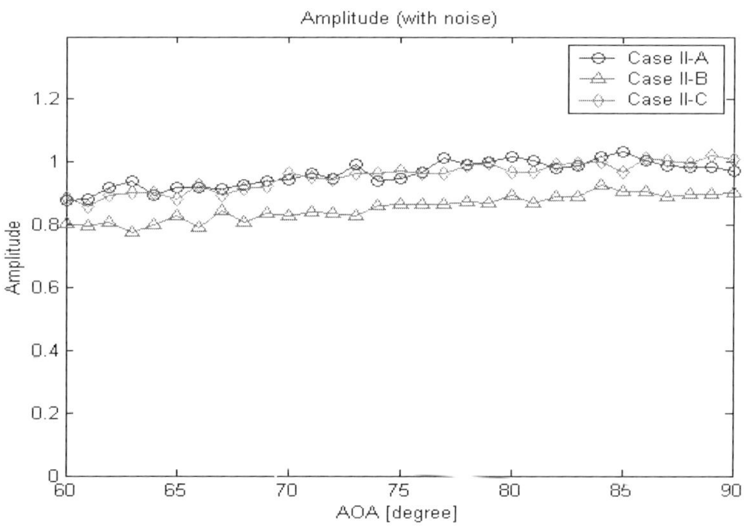

Figure 15.11. Estimate for the amplitude of the signal for all the three configurations of Case II when the signal impinges on the array between $\phi = 60°$ and $90°$ with a noise of 26 dB SNR. No compensation for mutual coupling.

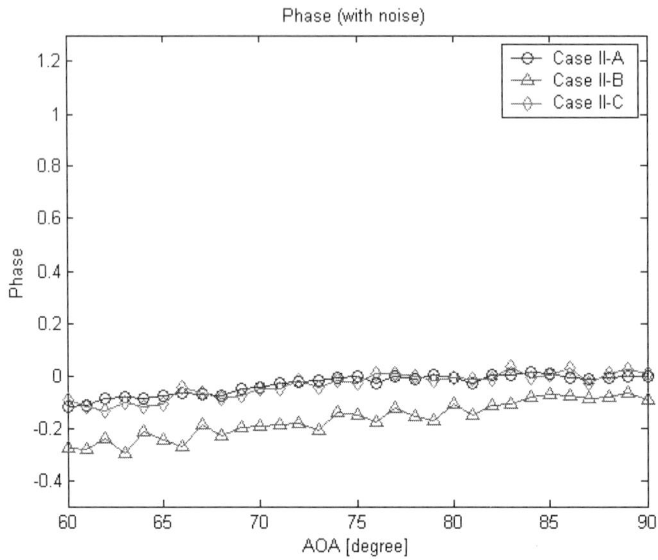

Figure 15.12. Estimate for the phase (in radians) of the signal for all the three configurations of Case II when the signal impinges on the array between $\phi = 60°$ and 90° with a noise of 26 dB SNR. No compensation for mutual coupling.

Next we completely compensate for the mutual coupling in the above two examples by using the transformation matrix as outlined in section 15.2. We now multiply the measured voltages by the transformation matrix to produce an equivalent set of voltages that will be induced in a uniform linear array consisting of isotropic omnidirectional point radiators radiating in free space with the same element spacing as the original array setup. This procedure can be done in real-time, as, given the array geometry one can precalculate the transformation matrix and then one just needs to perform a matrix vector multiplication to get the actual voltages.

Figure 15.13 provides the DOA estimate of the same signal outlined in Fig. 15.10 for Case II, but first we compensate [6] for the mutual coupling between the elements using the transformation matrix. Figures 15.14 and 15.15 provide the estimate for the amplitude and the phase of the signal. It is thus clear by comparing Figs 15.7–15.9 and Figs. 15.13–15.15 that, if we compensate for the mutual coupling between the antenna elements through the use of the transformation matrix, then the effect of loading on the antenna elements is not too critical. It is thus not very important whether the elements are truly tuned or not. However, making them resonant will actually enhance the voltages induced in the array.

Now, to check the performance of this DOA estimation, the CRB for the DOA estimation is compared to the variance of the DOA obtained through numerical simulations for the cases IIA, IIB, and IIC array systems. The mutual coupling between the antenna elements has not been accounted for through the use of the transformation matrix. However, the antenna elements have been matched in their respective locations in the array. Figure 15.16 plots the CRB for φ for the case IIA, IIB, and IIC, and the variance of 500 trials for the computed estimate $\hat{\varphi}$ for the case IIA, IIB, and IIC, when the signal arrives from 70° with $E_\theta = 1$ V/m. As we observe in the plot, there is a difference between the CRB for φ between the cases of IIA and IIC, and the CRB for φ for the case IIB. This difference in the value is approximately −13.9794 dB. It is also exactly equal to the value of $-10 \log_{10} (0.5^2/0.1^2)$. This clearly illustrates that the variance of the DOA estimate increases when the spacing between the antenna elements in the array decreases It is seen that the CRB for cases IIA and IIC are visually indistinguishable.

Also, the variance of $\hat{\varphi}$ for the 500 trials when we use an array consisting of ideal five omnidirectional isotropic point radiators with a spacing of 0.5λ between the elements is plotted in Figure 15.17. Along with the variance of using ideal antenna elements, we plot the variance of the results for cases IIA and IIC, where the spacing is 0.5λ. Figure 15.17 again shows that the CRB for cases IIA and IIC and the variance of the DOA estimation results when using omnidirectional elements are essentially the same. This again enforces what we said in the previous paragraph, namely, the variance of the estimate for DOA is determined primarily by the antenna element spacing and not by their length.

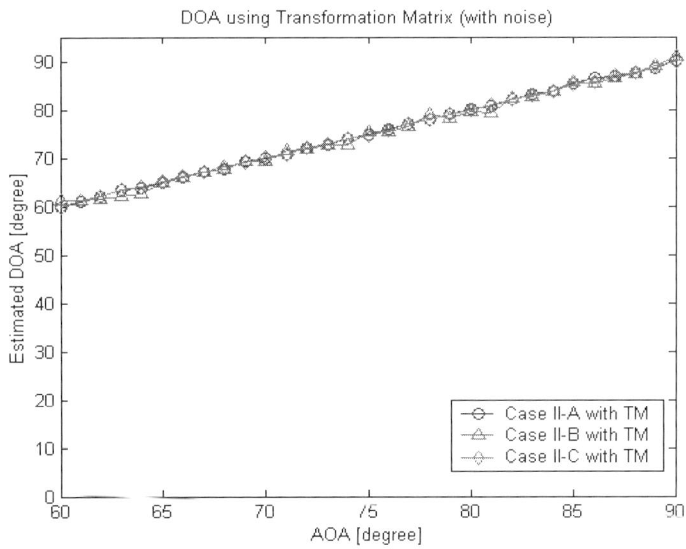

Figure 15.13. DOA estimation for the three configurations of Case II when the signal impinges on the array between $\phi = 60°$ and 90° with a noise of 26 dB SNR. Complete compensation for the mutual coupling for all the three configurations using the transformation matrix has been implemented.

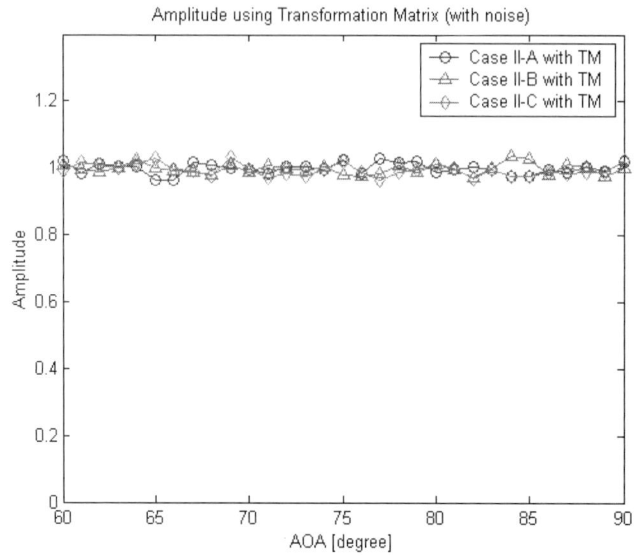

Figure 15.14. Estimate for the amplitude of the signal for all the three configurations of Case II when the signal impinges on the array between $\phi = 60°$ and $90°$ with a noise of 26 dB SNR. Complete compensation for the mutual coupling for all the three configurations using the transformation matrix has been implemented.

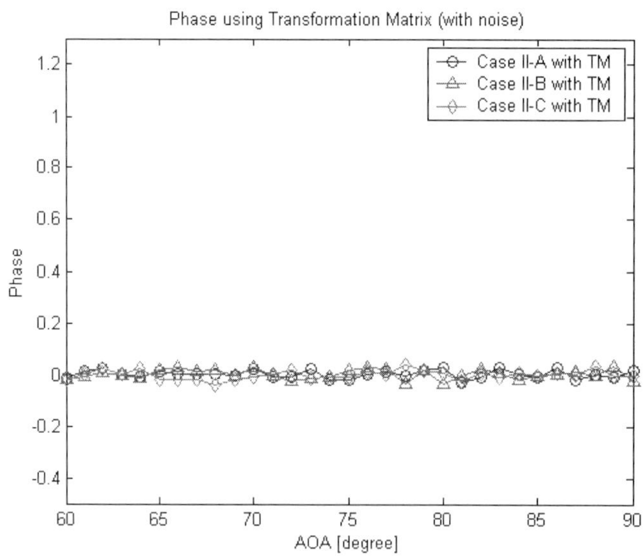

Figure 15.15. Estimate for the phase (in radians) of the signal for all the three configurations of Case II when the signal impinges on the array between $\phi = 60°$ and $90°$ with noise of 26 dB SNR. Complete compensation for the mutual coupling for all the three configurations using the transformation matrix has been implemented.

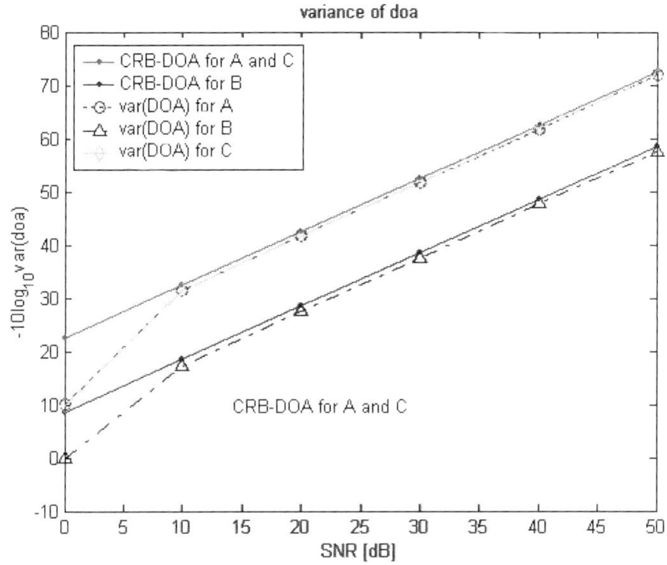

Figure 15.16. CRB for φ for the cases IIA, IIB, and IIC, and the variance of the estimate for $\hat{\varphi}$ for the cases IIA, IIB, and IIC, when the signal arrives from 70°.

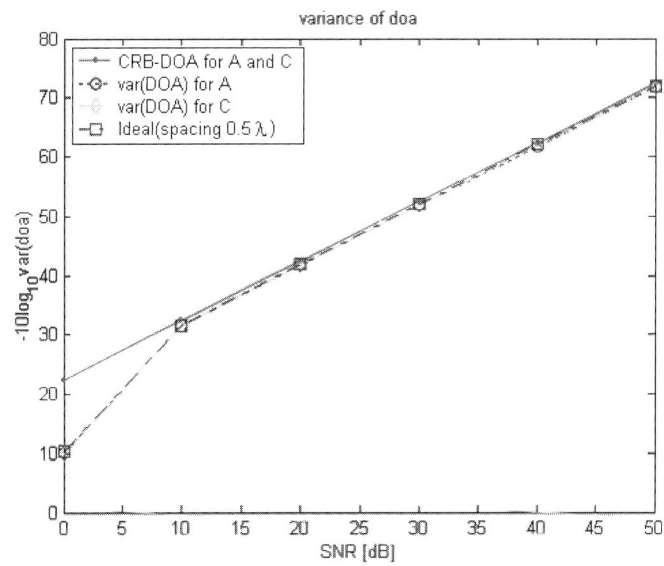

Figure 15.17. CRB for φ for the cases IIA and IIC, and the variance of the estimate for $\hat{\varphi}$ for the cases IIA, IIC, and an array consisting of ideal omnidirectional point radiators. The signal arrives from 70°.

For the final example, we consider an array composed of dissimilar antenna elements which may or may not be non-uniformly spaced. Each of the antenna array is composed of a horn, a bicone, a spiral, and two dipole antennas. They are used to estimate the DOA of two coherent signals of unity amplitude impinging on the array from 45° and 130°. All the antennas are chosen to have their maximum dimension of 0.1λ, i.e., they are electrically small and are tuned by appropriate loading. The value of the terminating impedance will also depend on their position in the array, in addition to the different shapes of the antennas. Each antenna is matched in its location in the array such that it resonates as in Case II.

First, we consider antenna elements with a uniform spacing of 0.5λ as shown in Fig. 15.18. For the second case, the antennas are nonuniformly spaced, but their average spacing satisfies the Nyquist sampling criteria of 0.5λ as shown in Fig. 15.19. For the nonuniform array, the spacing between the first dipole and the spiral is 0.3λ. The spacing between the spiral and the horn is 0.2λ. Spacing between the horn and the bicone and between the bicone and the last dipole are 0.5λ and 0.25λ, respectively. In these simulations, the Matrix Pencil method is used to estimate the directions of two incident signals and their corresponding amplitudes from the voltages induced at the load of these electrically small five antennas. Noise of +26 dB SNR is also added to the simulations. Tables 15.8 and 15.9 show the simulation results with and without using the transformation matrix. When the transformation matrix is applied, the array is calibrated from 0° to 180° to cover one-half of the observation domain. As seen, the uniformly spaced array gives similar results for both with and without using the transformation matrix as the mutual coupling effect is reduced when the antenna has small dimensions and properly matched, indicating that the mutual coupling effects are quite small. However, for the nonuniformly spaced array, the transformation matrix is needed.

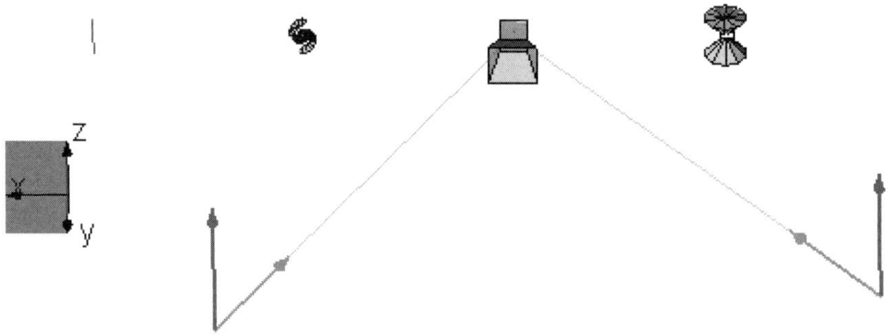

Figure 15.18. An equispaced antenna array composed of electrically small tuned horn, bicone, spiral, and two dipoles. They are spaced by 0.5λ. The arrows show the directions of arrivals of the signal.

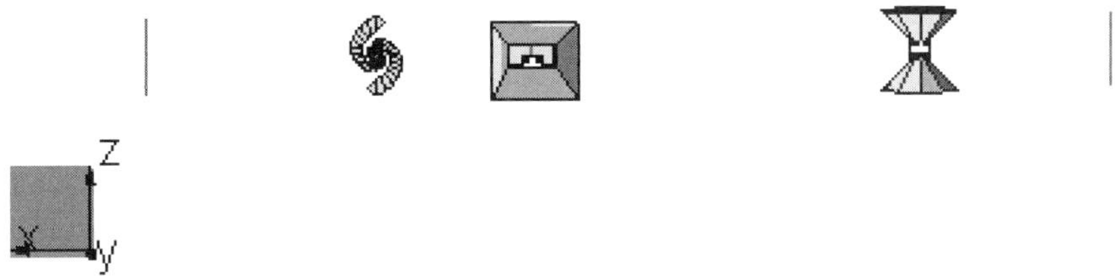

Figure 15.19. A nonuniformly spaced antenna array composed of electrically small tuned horn, bicone, spiral, and two dipoles. The spacing from left to right between the elements are 0.3λ, 0.2λ, 0.5λ, and 0.25λ, respectively.

Table 15.8. DOA Estimation Using the Five Element Array Consisting of Dissimilar Antennas Without Using the Transformation Matrix Calibrated Over 180°.

	DOA (degree)	Estimated DOA (degree)	Amplitude	Estimated Amplitude
Uniformly Spaced Array	45°	44.8°	$1 + j\,0$	$0.95 - j\,0.14$
	130°	129.4°	$1 + j\,0$	$0.96 - j\,0.13$
Nonuniformly Spaced Array	45°	59.9°	$1 + j\,0$	$0.79 - j\,0.76$
	130°	116.1°	$1 + j\,0$	$0.83 + j\,0.37$

15.6 CONCLUSION

In this chapter, we have illustrated that the use of electrically small conjugately matched antenna elements may be used to produce compact phased arrays. These arrays can be used for DOA estimation of signals. This methodology will work in a highly dynamic environment as it is based on a single snapshot of data even for coherent signals as long as the signal-to-noise ratio exceeds 15 dB. We have shown that making the electrically small antenna elements conjugately matched in a phased array may actually yield a higher value for the voltage at the feed points of the array than using resonant half wave elements. Thus, providing a larger value of the signal voltage will actually improve the signal-to-noise ratio and will improve the dynamic range of this methodology. Also, using electrically small resonant elements spaced half wavelength apart actually could provide an accurate estimation for not only the DOA but also the amplitudes and the phases of the signals. Mutual coupling compensation may not be necessary in that case. However, compensation of mutual coupling between the elements of the array may provide a more accurate result irrespective of the loading of the antenna and the environment in which they are operating. It is hoped that this methodology will be useful in deploying phased arrays on small footprints without sacrificing accuracy of the results.

In addition, the Cramer-Rao bound for the estimates has been evaluated to observe how close the computed estimate in the presence of noise will come near the true value. The variance of the DOA estimation for the resonant case is quite close to the CRB for φ as the SNR increases, but the variance of DOA estimate for the non-resonant antenna elements used in an array is higher than those for the resonant case. The CRB is more dependent on the spacing between the antenna elements in the array than on their lengths. Using the transformation matrix, one can compensate for the undesired electromagnetic effects by transforming the induced voltages into the ULVA correctly. However, we need to design the virtual array for the transformation matrix to be well conditioned. Thus one could perform DOA estimation using a single snapshot of the data.

Table 15.9. DOA Estimation Using the Five-element Array Consisting of Dissimilar Antennas with Using the Transformation Matrix Calibrated Over 180°.

	DOA (degree)	Estimated DOA (degree)	Amplitude	Estimated Amplitude
Uniformly Spaced Array	45°	44.9°	$1+j\,0$	$0.98-j\,0.01$
	130°	129.3°	$1+j\,0$	$0.99-j\,0.01$
Nonuniformly Spaced Array	45°	44.7°	$1+j\,0$	$0.97+j\,0.11$
	130°	131.6°	$1+j\,0$	$1.02-j\,0.13$

REFERENCES

[1] S. R. Best, "A Comparison of the Resonant Properties of Small Space-Filling Fractal Antennas", *Antennas and Wireless Propagation Letters*, Vol. 2, pp. 197-200, 2003.

[2] S. R. Best, "A Discussion on the Properties of Electrically Small Self-Resonant Wire Antennas", *IEEE Antennas and Propagation Magazine*, Vol. 46, Issue 6, pp. 9–22, Dec 2004.

[3] H. A. Wheeler, "Fundamental Limitation of Small Antenna", *Proceedings of the IRE*, Vol. 35, pp. 1479-1488, Dec. 1947.

[4] E. E. Altshuler "Electrically Small Self-Resonant Wire Antennas Optimized Using a Genetic Algorithm", *IEEE Transaction Antennas and Propagation*, Vol. 50, pp. 297-300, Mar. 2002.

[5] T. K. Sarkar, M. Wicks, M. Salazar-Palma and R. Bonneau, *Smart Antennas*, John Wiley and Sons, Hoboken, NJ, 2003.

[6] S. Hwang, S. Burintramart, T. K. Sarkar, and S. R. Best, "Direction of Arrival (DOA) Estimation Using Electrically Small Tuned Dipole Antennas", *IEEE Transactions on Antennas and Propagation*, Vol. 54, Issue 11, Part 1, Page(s):3292-3301, Nov 2006.

[7] B. M. Kolundzija, J. S. Ognjanovic and T. K. Sarkar, *WIPL-D, Electromagnetic Modeling of Composite Metallic and Dielectric Structures, Software and User's Manual*, Artech House, Norwood, MA, 2000.

16

NON-CONVENTIONAL LEAST SQUARES OPTIMIZATION FOR DOA ESTIMATION USING ARBITRARY-SHAPED ANTENNA ARRAYS

16.0 SUMMARY

In this chapter, we present an optimization technique based on the non-conventional least squares approximation to the direction-of-arrival (DOA) estimation problem. In contrast to the conventional least squares problem where the number of equations is greater than the number of unknowns, in this non-conventional optimization procedure, the number of unknowns is much greater than the number of equations and hence it is a very under-determined problem. The proposed method utilizes a signal steering vector as a function of azimuth angles similar to the discrete Fourier transform (DFT) concept. Various electromagnetic effects, such as mutual coupling between the elements of the array, antenna element failures, the use of dissimilar antenna elements, the use of non-planar and non-uniformly spaced array elements and coupling from near-field scatterers can be automatically taken into account in the preprocessing. After carrying out the electromagnetic optimization through a preprocessing, the DOA estimation results in a simple matrix multiplication, which reduces the computational complexity in the estimation. This is a single snapshot based estimation method. Hence, this procedure is ideally suited for deployment in a complex environment and the entire computation can be done in real time. Sample numerical results are presented to demonstrate the performance and accuracy of this procedure. This is a good procedure for the DOA estimation but not very accurate in estimation of the amplitude due to the classical picket fence effect produced by a DFT-based methodology.

16.1 INTRODUCTION

Parametric direction-of-arrival (DOA) estimation methods like ESPRIT [1] and MUSIC [2], which require generally multiple snapshots of the data, or the single snapshot-based Matrix Pencil [3–5] are computationally expensive. On the other hand, DFT-based techniques require less computation since they do not need to find the signal and noise subspaces. However, the drawback of a DFT based method compared to the others is its resolution in separating signals coming from close angular directions. The term single snapshot implies that the complex voltages induced in the antenna elements of the array are measured at a single instance of time.

In this chapter, we use a single snapshot-based methodology along with a non-conventional least squares optimization technique to the DFT method for the DOA estimation. The method, in addition, can take into account various electromagnetic effects through proper calibration. Incorporation of the various electromagnetic effects can be computationally intensive but can be done beforehand. So, the DOA estimation problem is reduced to only a matrix multiplication, which can be done in real time. The models for the signals used are presented in section 16.2. A DFT-based DOA estimation technique is described in section 16.3. Section 16.4 presents the DOA estimation based on a non-conventional least squares solution. Examples in section 16.5 illustrate that the proposed technique works well even with complex array structures. The robustness of the method due to instantaneous changes in the environment is demonstrated through numerical simulations followed by conclusions in section 16.6.

16.2 SIGNAL MODELING

Let us consider an antenna array of N antenna elements and let $x(n)$ be the received signal at the port of the antenna 'n'. An $N \times 1$ vector $\boldsymbol{x} = [x(0) \ldots x(N-1)]^T$ of voltages received at the terminals of the antennas can be written as:

$$\boldsymbol{x} = \sum_{m=1}^{M} a(\theta_m) s_m + \boldsymbol{w} = \boldsymbol{A}(\theta)\boldsymbol{s} + \boldsymbol{w} \tag{16.1}$$

where M is the number of plane waves incident on the array. The bold letters denote matrices. For the mth plane wave, let the $N \times 1$ $a(\theta_m)$ be the true steering vector when the wave is incident from direction θ_m with a complex amplitude s_m. $\boldsymbol{w}$ is an $N \times 1$ vector of noise, which contaminates the signal, measured at the feed points of the antenna elements in the array. The $N \times M$ matrix $\boldsymbol{A}(\theta)$ is given by $\boldsymbol{A}(\theta) = [a(\theta_1) \ldots a(\theta_M)]$. $\boldsymbol{s} = [s_1 \ldots s_M]^T$ is a $M \times 1$ vector of complex amplitudes of the signals. Given the received signal voltages, $\boldsymbol{x}$, at the terminals of the antennas, we want to estimate the DOA of the incident waves $\boldsymbol{\theta} = [\theta_1 \ldots \theta_M]^T$. To estimate θ, a matrix of true steering vectors, $\tilde{\boldsymbol{A}}(\boldsymbol{\theta}_c)$, is first measured or computed *a priori* for the given array configuration. The tilde sign '~' indicates that this matrix is the collection of the various received signals when the directions of the incident waves are known. We call these known directions the calibration angles and represent them by a vector

$\boldsymbol{\theta}_c = \left[\tilde{\theta}_1 \ldots \tilde{\theta}_Q\right]^T$ where Q is the total number of calibration angles. The calibration involves putting a far-field source of unit amplitude along a certain angular direction and measuring the voltages at the feed ports of the various antennas in the presence of mutual couplings and all other near-field effects. These directions should be selected to cover the entire region over which we propose to estimate the DOA. For example, a uniform linear array has an observation region from 0° to 180° relative to the end-fire direction of the array. The calibration angles, $\boldsymbol{\theta}_c$, may be chosen to be 0° to 180° with an angular stepping of 2° in the placement of the far field sources, or equivalently $\boldsymbol{\theta}_c = [\,0^\circ\ 2^\circ\ \ldots\ 180^\circ]^T$. The size of $\tilde{\boldsymbol{A}}(\boldsymbol{\theta}_c)$ is $N \times Q$ where in general $Q \gg N$ since the number of calibration angles is usually greater than the number of antennas in the array. During calibration, it is assumed that the array is operating in its natural environment in the presence of all the electromagnetic effects. This implies that $\tilde{\boldsymbol{A}}(\boldsymbol{\theta}_c)$ includes all the electromagnetic effects in the calibration. Once $\tilde{\boldsymbol{A}}(\boldsymbol{\theta}_c)$ is obtained, it will be used to map the received signal to estimate the DOA.

16.3 DFT-BASED DOA ESTIMATION

For a uniform linear array consisting of N isotropic, omni-directional point radiators with element spacing d, a received signal steering vector $\boldsymbol{a}(\theta_m)$ can be written as:

$$\boldsymbol{a}(\theta_m) = \left[\begin{matrix} 1 & \exp(j\frac{2\pi}{\lambda} d\cos\theta_m) & \ldots & \exp\{j\frac{2\pi}{\lambda}(N-1)d\cos\theta_m\} \end{matrix}\right]^T \tag{16.2}$$

where λ is the wavelength of the signal and θ_m is the DOA of the *m-th* signal with respect to the end-fire direction of the array. If $d = \lambda/2$ and θ varies from 0° to 180°, the matrix $\tilde{\boldsymbol{A}}(\boldsymbol{\theta}_c) = \left[a(\theta_1)\ldots a(\theta_M)\right]$ is clearly in the form of a matrix that performs a DFT. Therefore, the complex amplitudes of the signal corresponding to each angle of arrival can be obtained by solving the following equation:

$$\hat{\boldsymbol{s}} = \tilde{\boldsymbol{A}}^H(\boldsymbol{\theta}_c)\boldsymbol{x} \tag{16.3}$$

$\tilde{\boldsymbol{A}}^H$ denotes the conjugate transpose of matrix $\tilde{\boldsymbol{A}}$ and $\hat{\boldsymbol{s}}$ is the estimated complex amplitudes of the signal. The DOA will then correspond to the signal amplitudes of $\hat{s}_m$ which are non-zeros or have significant values. When the columns of $\tilde{\boldsymbol{A}}(\boldsymbol{\theta}_c)$ are chosen such that they are orthogonal to each other, then $\tilde{\boldsymbol{A}}^H(\boldsymbol{\theta}_c)$ is in fact the inverse of the matrix $\tilde{\boldsymbol{A}}(\boldsymbol{\theta}_c)$, as it is the property of the DFT matrix. Thus, given a received signal vector $\boldsymbol{x}$ in (16.1), we can solve for $\hat{\boldsymbol{s}}$ by a simple matrix inversion.

However, due to the finite length of the signal and finite dimension of the array, (16.3) cannot be solved in a straightforward fashion. This is because we do not have a square matrix to perform the DFT operation and hence the number of unknowns in this case is much greater than the number of equations. This is a highly under-determined system and we use the non-conventional least squares optimization to solve this optimization problem. In a classical least squares the number of equations is greater than the number of unknowns.

In addition, a real antenna element cannot be an ideal point radiator. The signal steering vector, $\boldsymbol{a}(\theta_m)$, along a particular direction will not be in the form of (16.2) as it will include various electromagnetic interferences from other nearby antennas and near-field scatterers. When calibrating the array, the matrix $\tilde{\boldsymbol{A}}(\boldsymbol{\theta}_c)$ will no longer be in the form of the DFT. Therefore, we need to seek a simple technique in solving for the DOA by using similar concepts based on the DFT-based estimation.

16.4 NON-CONVENTIONAL LEAST SQUARES OPTIMIZATION

The various effects of the electromagnetic interference make the steering vector different from the exact one. To obtain a correct steering vector when electromagnetic effects are present, we need to replace (16.3) by

$$\hat{\boldsymbol{s}} = \tilde{\boldsymbol{A}}^{\dagger}(\boldsymbol{\theta}_c)\boldsymbol{x} \tag{16.4}$$

where $\tilde{\boldsymbol{A}}^{\dagger}(\boldsymbol{\theta}_c)$ denotes the pseudoinverse of $\tilde{\boldsymbol{A}}(\boldsymbol{\theta}_c)$. However, since the number of calibration angles is usually greater than the number of antennas, $\tilde{\boldsymbol{A}}(\boldsymbol{\theta}_c)$ is not square. This requires an optimization, as it must be done in a least squares sense.

In a classical least squares problem, the null space of the operator is usually empty, however, the null space of the adjoint operator is not empty. Hence, there may be elements $\boldsymbol{x}$ that cannot be obtained by a simple mapping using elements from its domain. In our problem, the null space of the adjoint operator is empty, but the null space of the operator is not. Thus, there is the possibility of having infinite number of solutions and the requirement of the optimization is to seek the solution with the minimum norm. Therefore, the solution for this non-classical least squares problem is quite different. The solution for the complex amplitude of the signal for this non-classical least squares problem can be written as [6]:

$$\hat{\boldsymbol{s}}_{ncls} = \tilde{\boldsymbol{A}}^{\dagger}(\boldsymbol{\theta}_c)\boldsymbol{x} \tag{16.5}$$

where $\tilde{\boldsymbol{A}}^{\dagger}(\boldsymbol{\theta}_c) = \tilde{\boldsymbol{A}}^{H}(\boldsymbol{\theta}_c)\left[\tilde{\boldsymbol{A}}(\boldsymbol{\theta}_c)\tilde{\boldsymbol{A}}^{H}(\boldsymbol{\theta}_c)\right]^{-1}$ is the pseudoinverse of $\tilde{\boldsymbol{A}}(\boldsymbol{\theta}_c)$. We denote the least squares solution of $\hat{\boldsymbol{s}}$ by $\hat{\boldsymbol{s}}_{ncls}$. It will be illustrated through numerical simulation that indeed $\hat{\mathbf{s}}_{ncls}$ provides the proper solution for the DOA estimation problem. So the proper optimization of the element in the null space of the operator has been identified. The computation of $\left[\tilde{A}(\theta_c)\tilde{A}^H(\theta_c)\right]^{-1}$ is performed by the singular value decomposition (SVD) [7]. To apply the SVD, we first decompose the matrix $\left[\tilde{A}(\theta_c)\tilde{A}^H(\theta_c)\right]$ as follows:

$$\left[\tilde{A}(\theta_c)\tilde{A}^H(\theta_c)\right] = \tilde{U}\tilde{\Sigma}\tilde{V}^H \qquad (16.6)$$

where U and V are unitary matrices whose column vectors are the associated left and right singular vectors of $\left[\tilde{A}(\theta_c)\tilde{A}^H(\theta_c)\right]$, respectively. Σ is a diagonal matrix whose elements are the singular values of $\left[\tilde{A}(\theta_c)\tilde{A}^H(\theta_c)\right]$. By selecting k dominant singular values of Σ and discarding the remaining singular values, the inversion of $\left[\tilde{A}(\theta_c)\tilde{A}^H(\theta_c)\right]$ can be obtained as:

$$\left[\tilde{A}(\theta_c)\tilde{A}^H(\theta_c)\right]^{-1} = \tilde{V}\tilde{\Sigma}^{-1}\tilde{U}^H \qquad (16.7)$$

where $\tilde{V}$ and $\tilde{U}$ are obtained form V and U using the first k columns, respectively. Similarly, $\tilde{\Sigma}$ is obtained from Σ by extracting the first k dominant singular values. After obtaining the matrix inversion in (16.7), the pseudoinverse of $\tilde{A}(\theta_c)$ can be found. We note that $\tilde{A}^\dagger(\theta_c)$ can be calculated offline when setting up the antenna in an operating environment. Then the DOA estimation problem in real time is reduced to only a matrix multiplication of $\tilde{A}^\dagger(\theta_c)x$. This is how the non-conventional least squares are numerically implemented so that the proper signal and noise subspaces are appropriately optimized. The important point to note is that this time-consuming key optimization step can be done offline and in a real environment, the final result can be computed in real time through a matrix multiplication, which needs to be done only once to obtain the final result.

In the next section, we illustrate the performance of the proposed technique through computer simulations.

16.5 SIMULATION RESULTS

In this section, the DOA estimation is carried out using the non-conventional least squares solution given by (16.5). The simulations of the incident waves incorporating all the electromagnetic effects have been carried out using the electromagnetic software modeling code [8]. All simulated voltages received at the antenna terminals have been contaminated with a thermal noise of 20 dB signal-to-noise ratio. The estimated DOA is selected by detecting the peak values of the signal complex amplitude $\hat{s}_{ncls}$ corresponding to each direction of arrival.

16.5.1 An Array of Linear Uniformly Spaced Dipoles

As a first example, consider an array of 20 thin $\lambda/2$ long dipoles with a $\lambda/2$ spacing. The radius of the dipoles is $a = 0.01\lambda$. The calibration angle for the array covers from 1° to 180° with an angular stepping of 0.5°. This is represented in the matrix form by $\boldsymbol{\theta}_c = [1°: 0.5° : 180°]^T$. The DOA is measured relative to the end-fire direction of the array or the x-axis as shown in Figure 16.1 where the signals are incident along 60° and 135°. The arrow in the xy-plane shows the DOAs of the signals while the arrows parallel to the z-axis show the signal polarizations. The absolute values of the estimated complex amplitudes of the signals corresponding to the various DOAs are plotted in Figure 16.2. It is shown that the maximum values do correspond to the actual DOA of the signals.

However, the estimated amplitudes of the signal are not correct due to the windowing effect of the finite length of the array and due to the picket fence effect introduced by the DFT. The picket fence effect is due to the fact that if the true DOA does not exactly coincide with the actual DOA used in the calibration matrix then one will observe the results in contiguous DFT bins and it will be smeared by the aperture size of the array. The reason it is called the picket fence

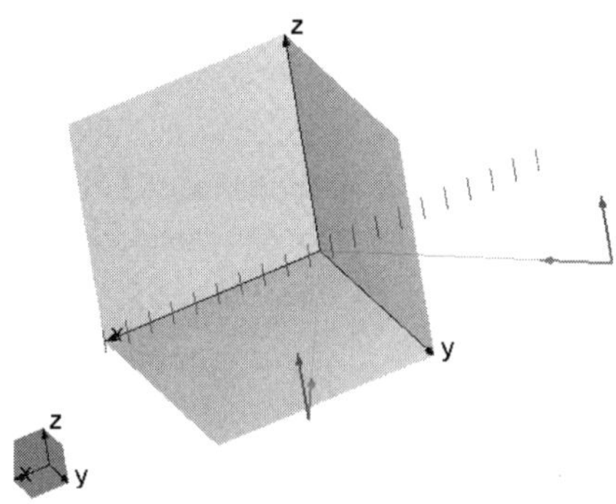

Figure 16.1. An array of 20 half-wave dipoles with 2 signals coming from 60° and 135° from the end-fire direction of the array.

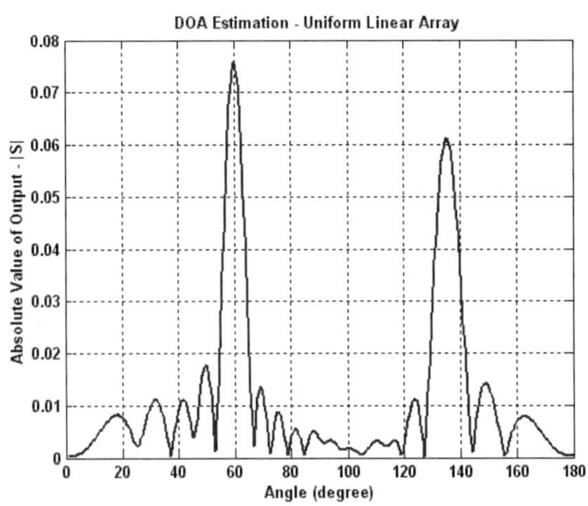

Figure 16.2. Estimated DOAs are at 60° and 135.5°.

effect is that it is as if one is observing the true scene through a picket fence where the opening corresponds to the spatial sampling locations of the array.

Because the cosine function is nonlinear in nature, the steering vectors of the signals coming close to the end-fire directions of the array are not properly weighted. Therefore, when the received signal comes close to the end-fire directions, the estimation is not accurate as shown in Figures 16.3 and 16.4 where the signals are incident along 20° and 135°. However, the problem of estimating the signal close to the end-fire direction can be addressed accurately by selecting the calibration angles, $\boldsymbol{\theta}_c$, such that $\cos(\theta_m)$ are uniformly distributed over the entire span of interest.

For the next example, the calibration angles are uniformly distributed not in angle from 1° and 180° but in the set $\cos(\theta_m)$ space. The simulation result for samples uniformly distributed in the set $\cos(\theta_m)$ space is shown in Fig. 16.5 and compared to the one obtained from uniform $\boldsymbol{\theta}_c$ in Fig. 16.4. It is obvious that by using a uniform cosine calibration, the peaks of the complex amplitudes of the signal can be easily detected when the signal comes close to the end-fire direction of the array.

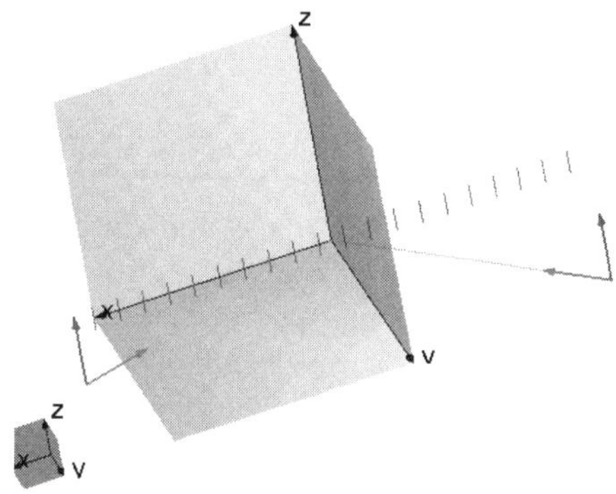

Figure 16.3. An array of 20 half-wave dipoles with 2 signals coming from 20° and 135° with respect to the end-fire direction of the array.

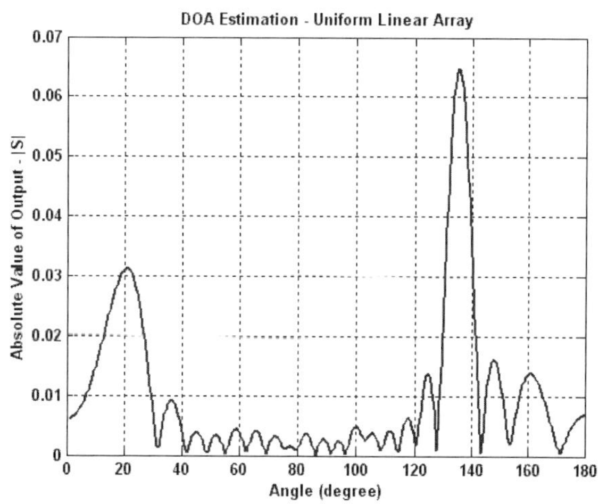

Figure 16.4. Estimated DOAs are at 20° and 135°.

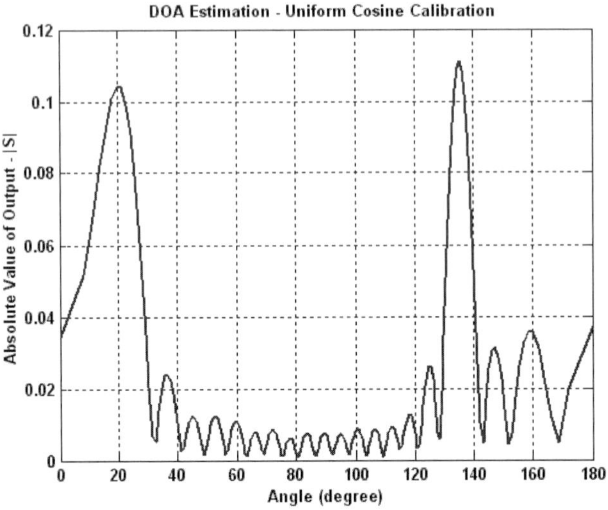

Figure 16.5. Calibration angles are selected such that the cosine function is uniformly discretised. The estimated DOAs are at 20° and 135.25°.

16.5.2 An Array of Linear Non-Uniformly Spaced Dipoles

Next, we deal with a linear non-uniformly spaced array of 20 thin half-wave dipoles. Each of the dipoles have a length of $\lambda/2$ and a radius of $a = 0.01\lambda$. In this structure, the antenna spacing along x-direction is $\lambda/2$, while the spacing along the y-direction is a wavelength. Thus the element spacing is 1.1λ. Even though the antenna spacing is greater than $\lambda/2$ in this case, the average spacing of the elements for this problem is still less than $\lambda/2$. Therefore, estimation is still possible without any ambiguity. The array is calibrated over the range $\boldsymbol{\theta_c} = [1° : 0.5° : 180°]$. Figure 16.6 shows the antenna structure and the incoming signals. The estimated DOA is shown in Fig. 16.7.

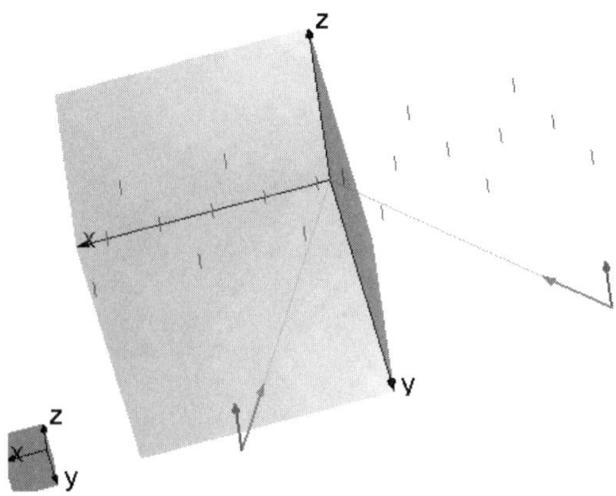

Figure 16.6. An array of 20 half-wave dipoles non-uniformly spaced with 2 signals coming from 60° and 135°.

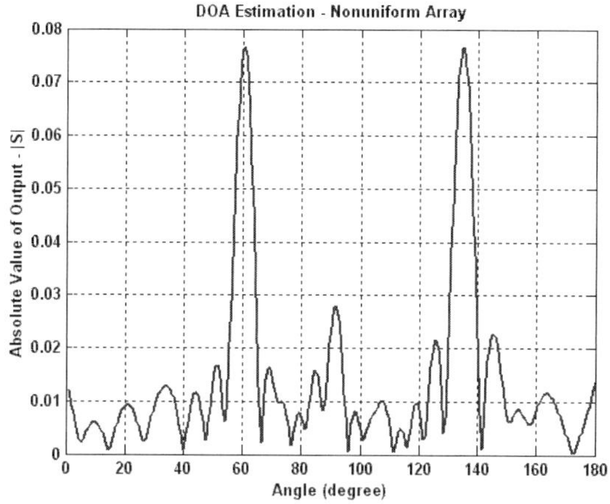

Figure 16.7. Estimated DOA over the entire range of the non-uniform array. The estimated DOAs are at 60°, and 135°.

16.5.3 An Array Consisting of Mixed Antenna Elements

Here, we use different types of antennas in the same array. This will increase the complexity of the electromagnetic effects between the various antenna elements. Four horn antennas with different orientations are used in the array. Each horn antenna has its feed waveguide dimension of $8 \text{ cm} \times 8 \text{ cm} \times 4.8 \text{ cm}$ and its aperture is $0.115 \text{ m} \times 0.128 \text{ m}$. The slant dimension of the horn aperture is 0.089 m. The feed probe for the horn has a length of 0.04 m and is placed at the center of the feed waveguide. Two horn antennas are pointed along 90° (y-axis). Two others are oriented along 45° and 135° as shown in Fig. 16.8. In addition, sixteen dipoles are of 0.1 m in length and radius of 2 mm is randomly put together as a non-planar array. This is shown in Fig. 16.8. The operating frequency is 1.5 GHz. The calibration angle is same as before which is $\theta_c = [1° : 0.5° : 180°]$ and the signals are coming from 60° and 135°. Fig. 16.9 presents the estimated DOAs. It is seen that the DOAs of the two signals have been properly identified.

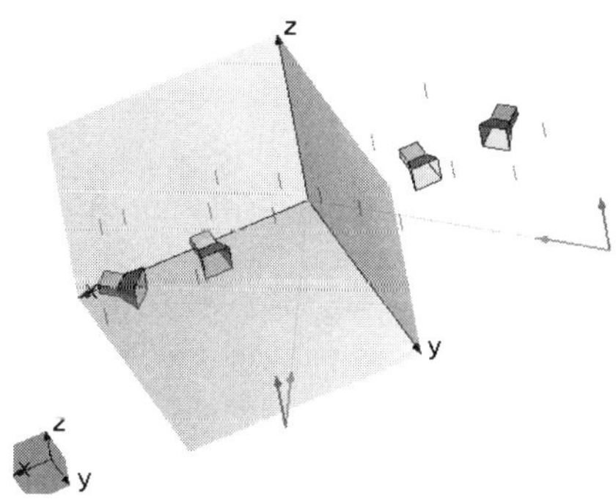

Figure 16.8. An array of dipoles and horns with signals coming at 60° and 135°.

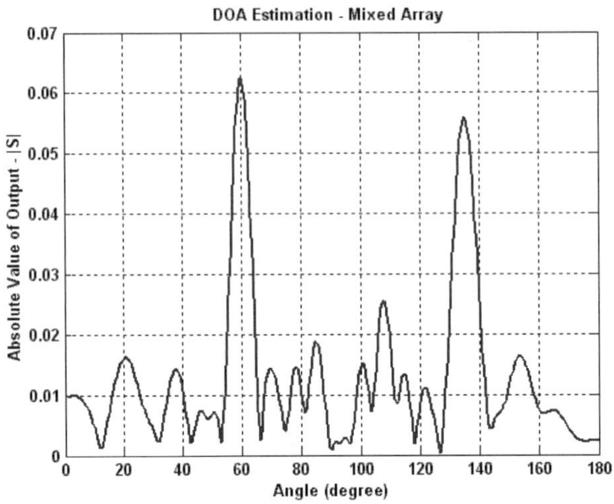

Figure 16.9. Estimated DOA over the entire range for the array of mixed elements. The estimated DOAs are at 60° and 135°.

16.5.4. An Antenna Array Operating in the Presence of Near-Field Scatterers

We assume that the array calibration is performed in an operating environment where there are near-field scatterers. One of the near-field scatterers blocks the direct line-of-sight of one of the incident signals. We use the array of 20 half-wave dipoles; each has a length of 0.1 m and radius 2 mm. The antenna spacing is 0.1 m. Figure 16.10 shows the array structure with two near-field scatterers: a cube and a sphere. The cube has a dimension of $0.2\text{ m} \times 0.2\text{ m} \times 0.2\text{ m}$ and its center is located in the *xy*-plane at the coordinate $(x, y, z) =$ $(-0.5\text{ m}, 0.5\text{ m}, 0\text{ m})$ or 2.5λ away from the array. The closest surface of the cube is 0.4 m or 2λ from the array. Note that the cube is placed along the line-of-sight of a signal, which is incident along 135°. The 0.2 m-diameter sphere is centered at $(0.5\text{ m}, 0.85\text{ m}, 0\text{ m})$ that oriented along the direction of 53.9°. The sphere is located roughly a distance of 4λ or 0.8 m away from the array. There are three signals impinging on the array from 45°, 80°, and 135°. The calibration angle in this simulation has been chosen to be the same as in the previous examples. The estimated DOA for this case is shown in Figure 16.11. It is seen that all the incoming signals have been properly identified even though a near-field scatterer has blocked the line-of-sight of one of the signals. It is therefore clear that proper electromagnetic calibration is necessary.

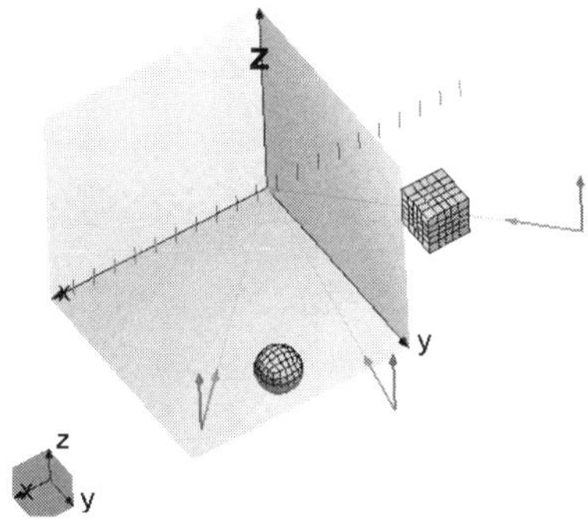

Figure 16.10. An array of 20 half-wave dipoles with two near-field scatterers. Signals are coming from 45°, 80°, and 135°.

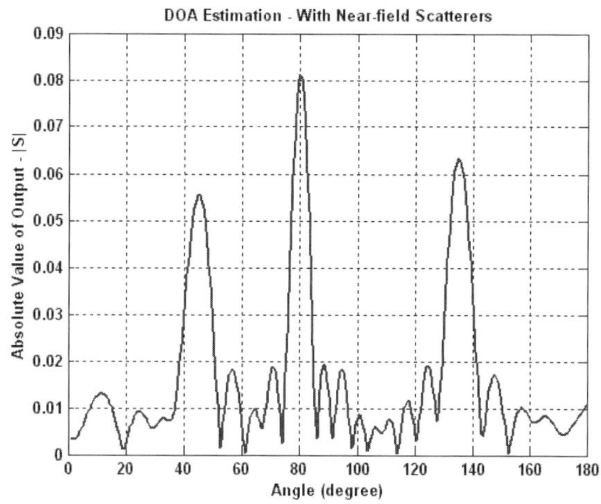

Figure 16.11. Estimated DOA over the entire range in the presence of near-field scatterers. Estimated DOAs are 45°, 80°, and 135°.

16.5.5 Sensitivity of the Procedure Due to a Small Change in the Operating Environment.

Next, we study the sensitivity of the method due to changing of the environment during operation. In this case, we use an array of 20 half-wave dipoles (as shown in Fig. 16.10). However, this time the array is calibrated in free space without any near-field scatterers present, i.e., without the conducting cube and the sphere. In the actual mode of operation, however the three signals coming from 45°, 80°, and 135° encounter two near-field scatterers. We want to see whether this method can still estimate the DOAs of the signals even when the calibration environment has changed. As the sizes of the scatterers are relatively small compared to the size of the entire array, it is seen in Fig. 16.12 that one can still resolve the three signals with an 1° error in the DOA estimation even though the near-field scatterers were not accounted for during the calibration procedure. The amplitude of the estimation is reduced due to the mismatch in the environment.

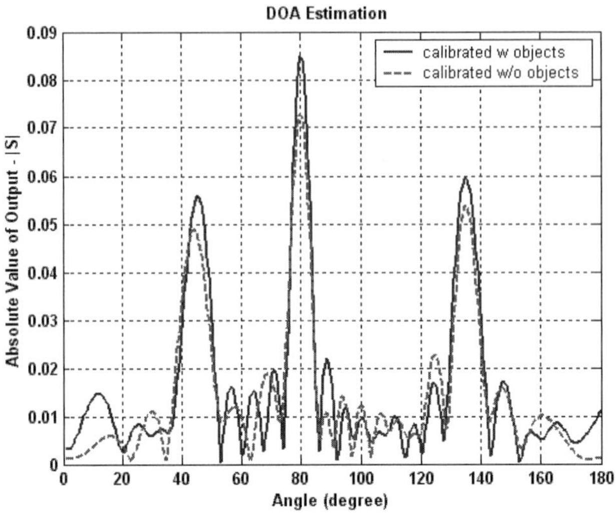

Figure 16.12. Comparison between the results when the array is calibrated both with and without the near-field scatterers; Calibrated *with* scatterers (solid line); calibrated *without* scatterers (dashed line); Estimated DOAs when calibrated *with* scatterers –45°, 80°, and 135°; Estimated DOAs when calibrated *without* scatterers –44°, 80°, and 135°.

16.5.6 Sensitivity of the Procedure Due to a Large Change in the Operating Environment.

We study the consequences that may occur if there is a relatively large change in the operating environment that has not been accounted for during the calibration procedure. In this simulation, the array consists of 20 dipoles, the same as in example 16.5.4, and is calibrated in free space without any of the near-field scatterers. When the array is operated in the presence of three signals coming from 45°, 80°, and 135°, the same as in the previous simulations, we introduce two large near-field scatterers, i.e., a conducting cube and a sphere. The cube has dimension of 2λ along each side and the sphere is 2λ in diameter. The cube and the sphere are placed at $(-0.5\text{ m}, 0.5\text{ m}, 0\text{ m})$ and $(0.5\text{ m}, 0.85\text{ m}, 0\text{ m})$, respectively. And, the cube is still along one of the signal directions $(135°)$ and the sphere is located along 53.9°. The closest face of the cube is only 1.5λ away from the array and the surface of the sphere is 3.25λ from the array. Figure 16.13 shows the array configuration with the near-field scatterers and the incoming signals. The results are somewhat degraded since the electromagnetic effects are much bigger than in the previous example. Figure 16.14 shows that in addition to the initial three signals there is the fourth signal arriving from 93.5°. This may be a reflected signal from the face of the cube. However, if the environment were appropriately calibrated then this reflected signal might have been absent, which would have presented the correct scenario.

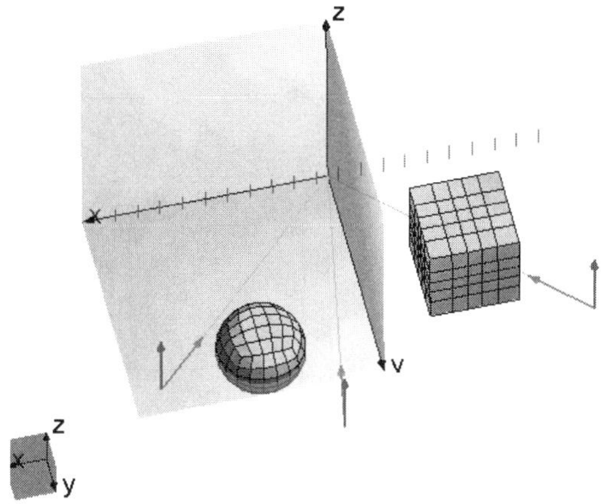

Figure 16.13. An array of 20 half-wave dipoles with two near-field scatterers, which are twice as large as in the previous example. Signals are coming from 45°, 80°, and 135°.

The proposed method thus is quite robust to reasonable changes in the environment, and the system degrades gracefully rather than in a catastrophic fashion. If the array is calibrated in an environment that is as close as possible to the operating environment, then one can obtain reliable estimates for the various DOAs of the different signals even when using a complex antenna array consisting of dissimilar antenna elements.

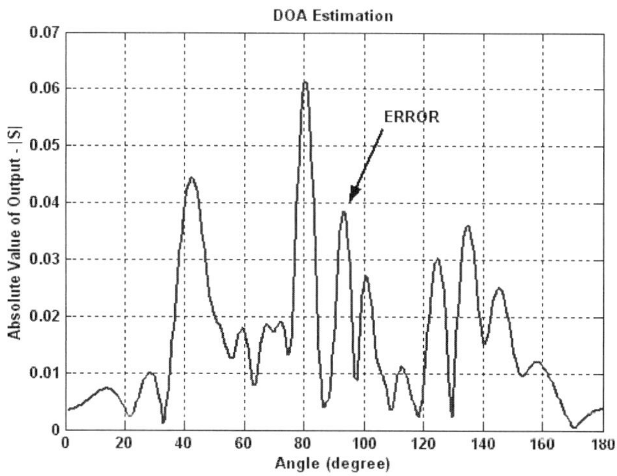

Figure 16.14. DOA estimation when the array is calibrated without the near-field scatterers. Estimated DOAs are 42.5°, 80.5°, and 134.5° with a spurious signal at 93.5°.

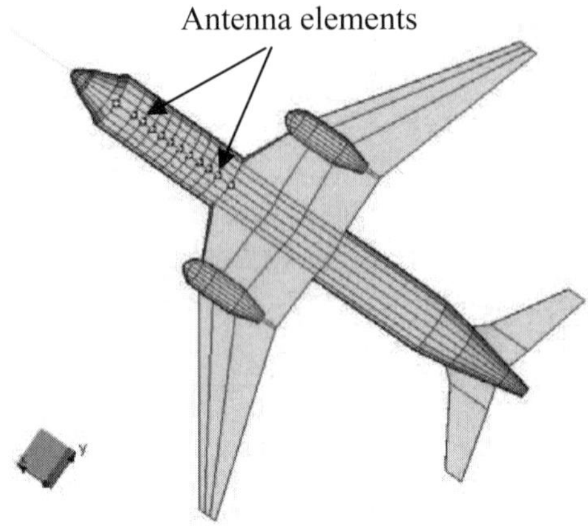

Figure 16.15. An array of 12 monopoles mounted underneath the aircraft body. The dark dots represent the antenna elements.

16.5.7 An Array of Monopoles Mounted Underneath an Aircraft

In this example, the proposed method is used under severe mutual coupling from an array platform. A linear array of 12 monopoles is mounted underneath an aircraft body as shown in Figure 16.15. The operating frequency is 400 MHz. The antenna spacing from antenna 2 to 11 is 1.33λ while the spacing between antenna 1 and 2 is 2.4λ , and the spacing between antenna 11 and 12 is 1.78λ . In this case, as the antenna spacing is larger than half wavelength of the operating frequency, the only visible region of the array (70°– 110°) is considered for the DOA estimation. Thus, the calibration angle is $\boldsymbol{\theta}_c$ = [70°:0.5°:110°]. Figures 16.16 and 16.17 show incident signals from 75° and 95° and the estimates of the DOAs. In this example, even though there are severe mutual couplings between the antennas and the aircraft body including wings and engines, the method shows a very good performance.

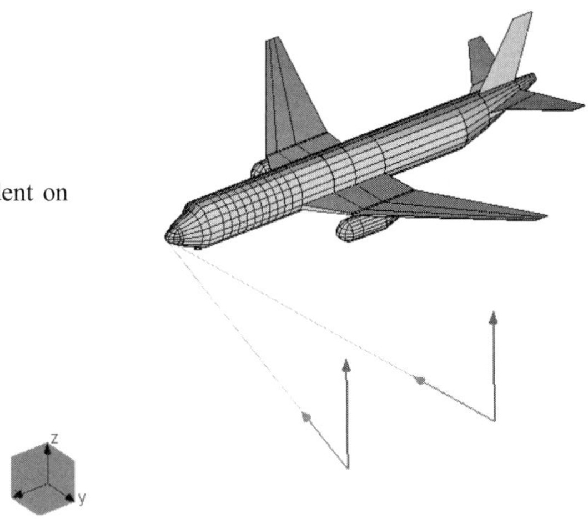

Figure 16.16. Two signals incident on the array from 75° and 95°.

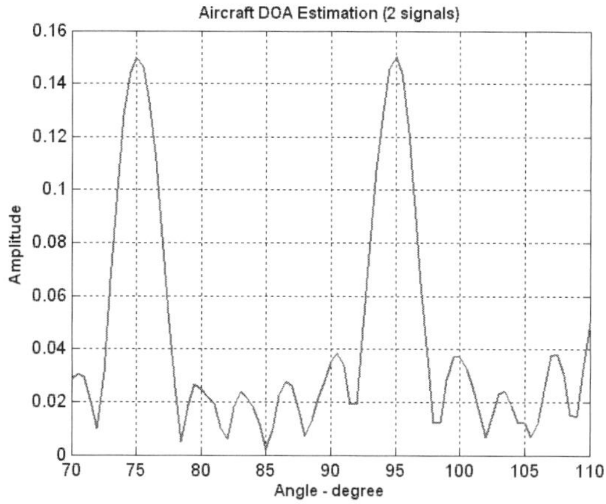

Figure 16.17. DOA estimation for a linear array mounted underneath the aircraft.

16.5.8. A Non-uniformly Spaced Nonplanar Array of Monopoles Mounted Under an Aircraft

Finally, in this last example, a non-uniformly spaced nonplanar array of 12 monopoles is mounted underneath the aircraft in a similar fashion as in the previous example. A non-uniformity in the antenna spacing is introduced in order to reduce the ambiguity of the DOA estimation due to the symmetry in a linear array. As a result, this array is able to estimate the DOA from 0° to 360°. Figure 16.18 shows the antenna positions underneath the aircraft. Since the antenna position is not linear, to preserve the Nyquist sampling criteria, the operating frequency is selected to be 100 MHz as the averaged spacing along the inline antenna elements (along the aircraft axis) is 1.103 m. The calibration angle is $\boldsymbol{\theta_c}$ = [0°:0.5°:360°]. There are two signals coming along 130° and 270°. Thermal noise of 10 dB SNR is added to the received signal at each antenna element. It is seen from Figure 16.19 that two incident signals can be correctly identified.

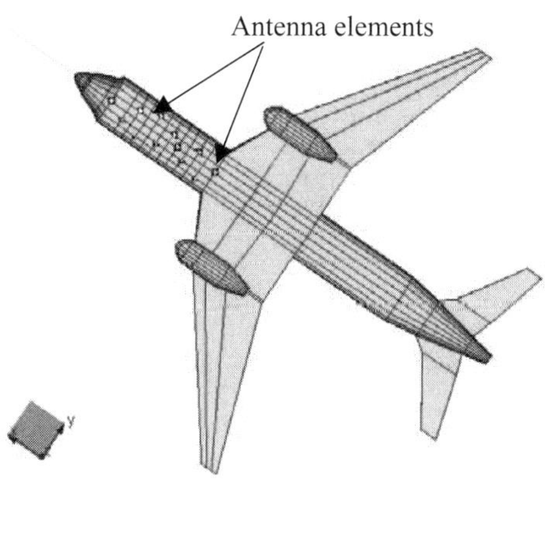

Figure 16.18. An array of non-uniform nonlinearly spaced 12 monopoles mounted underneath the aircraft body. The dark dots represent the antenna elements.

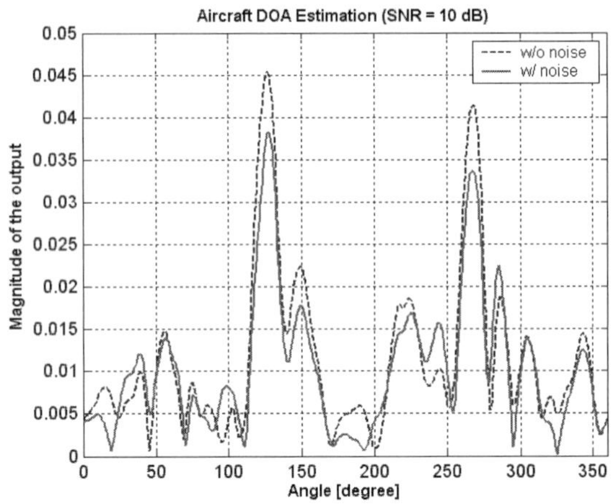

Figure 16.19. DOA estimation for a non-uniform nonlinearly spaced array mounted underneath the aircraft. The signals incident from 130° and 270° with noise of 10 dB SNR.

Since the proposed method is based on the DFT methodology, it has a similar limitation regarding its resolution as the DFT is not a high-resolution technique. Thus, the method can estimate two closely incident signals only when they are well separated. This in turn limits the maximum number of signals can be estimated. Figure 16.20 shows the maximum number of signals that the method can estimate. Figure 16.21 plots the DOA when the method breaks down because too many signals are incident on the array. In Figure 16.21 there are seven signals incident on the array. Note that there are seven peaks, same as the number of signals incident on the array. For the actual signal at 110°, the method has a peak at 119° which is considered too large an error. And the level of peaks is getting closer to the noise floor. Thus, we consider six signals as the maximum number of signals that can be detected for this array configuration. The maximum number of signals can be estimated totally depends on the array configuration and the platform on which the antennas are mounted.

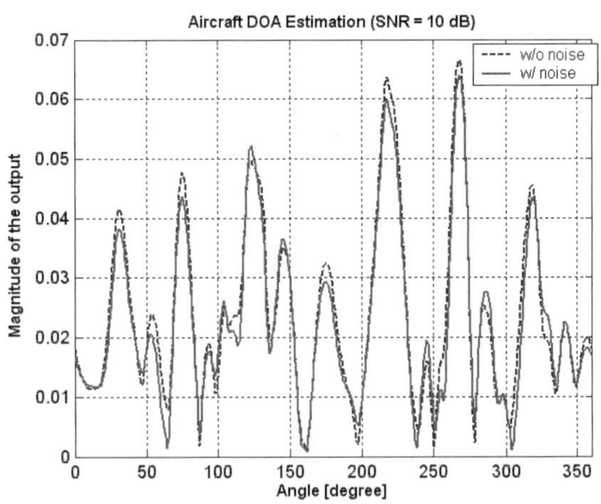

Figure 16.20. The maximum number of signals that can be estimated is 6 in this array structure. The actual signals come from 30°, 80°, 130°, 220°, 270°, and 320°.

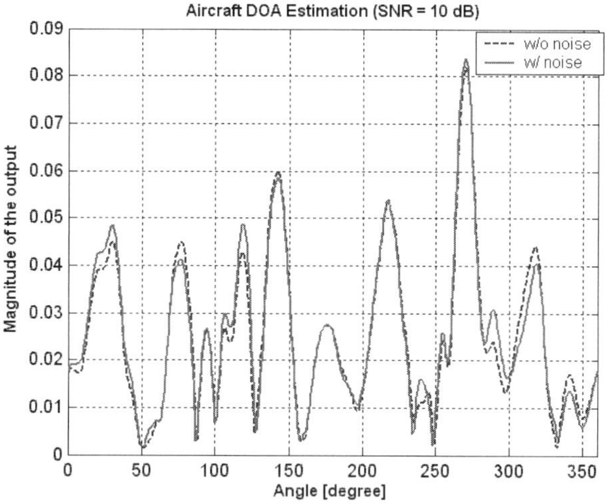

Figure 16.21. The DOA estimate when the method breaks down due to too many signals incident on the array. The 7 actual signals come from 30°, 80°, 110°, 140°, 220°, 270°, and 320°.

16.6 CONCLUSION

In this chapter, a non-conventional least squares procedure is presented for DOA estimation of the signals using a DFT-based method and is implemented using the SVD. The method works even when any arbitrary-shaped antenna array consisting of dissimilar antenna elements is operating in a near-field environment with strong mutual couplings. All of the calibration processes and the optimizations can be done offline, once, before operating the array. Then, the DOA estimation can be simply obtained through a matrix multiplication, which can be done in real time. The validity of this method is illustrated through computer simulations, even when there are strong electromagnetic effects. The calibration technique using a uniformly partitioned cosine function is introduced to reduce the error in the estimation when the signals are arriving close to the end-fire direction of the array. It is also shown that the method is robust to small changes in the environment. This method may be quite suitable for deployment in unmanned aerial vehicles (UAV) where the array may be conformal and it will operate in the presence of various near-field scatterers. In conclusion, this chapter provides a fast way in estimating the DOA of various signals impinging on an array applying the non-classical least squares optimization procedure.

REFERENCES

[1] R. Roy and T. Kailath, "ESPRIT – Estimation of Signal Parameters via Rotational Invariance Techniques", *IEEE Trans. Acoust., Speech, Signal Processing*, Vol. 37, No. 7, pp. 984-995, July 1989.

[2] R. O. Schmidt, "Multiple Emitter Location and Signal Parameter Estimation", *IEEE Trans. on Antennas and Propagation*, Vol. 34, No. 3, pp. 276-280, March 1986.

[3] Y. Hua and T. K. Sarkar, "Matrix Pencil Method for Estimating Parameters for Exponentially Damped/Undamped Sinusoids in Noise", *IEEE Trans. Acoust. Speech, Signal Processing*, Vol. 36, No. 5, pp. 814-824, May 1990.

[4] Y. Hua, A. B. Gershman, and Q. Cheng, *High-Resolution and Robust Signal Processing*, New York, NY: Marcel Dekker, Inc. 2004.

[5] T. K. Sarkar, M. C. Wicks, M. Salazar-Palma, and R. J. Bonneau, *Smart Antennas*, Hoboken, NJ: John Wiley & Sons, 2003.

[6] C. N. Dorny, *A Vector Space Approach to Models and Optimization*, New York, NY: Krieger, 1980.

[7] G. H. Golub and C. F. Van Loan, *Matrix Computations*, 3rd. Ed., Baltimore, MD: Johns Hopkins University Press, 1996.

[8] B. M. Kolundzija, J. S. Ognjanovic, and T. K. Sarkar, *WIPL-D: Electromagnetic Modeling of Composite Metallic and Dielectric Structures*, Norwood, MA: Artech House, 2000.

17

BROADBAND DIRECTION OF ARRIVAL ESTIMATIONS USING THE MATRIX PENCIL METHOD

17.0 SUMMARY

In this chapter we describe a method for simultaneously estimating the direction of arrival (DOA) of the signals along with their unknown frequencies. In a typical DOA estimation problem it is often assumed that all the signals are arriving at the antenna array at the same frequency which is assumed to be known. The antenna elements in the array are then placed 0.5λ apart, where λ is the wavelength at the frequency of operation. However, in practice seldom all the signals arrive at the antenna array at a single pre-specified frequency, but at different frequencies. The question then is what to do when there are signals at multiple frequencies which are unknown. This chapter presents an extension of the Matrix Pencil (MP) method to simultaneously estimate the DOA along with the operating frequency of each of the signal. This novel approach involves pole estimation of the voltages that are induced in a three-dimensional antenna array. Finally, we illustrate how to carry out the broadband DOA estimation procedure using realistic antenna elements. Some numerical examples are presented to illustrate the applicability of this methodology in the presence of noise.

17.1 INTRODUCTION

In contemporary literature, most of the efforts have primarily been directed for estimating the two-dimensional spatial frequencies (namely, the azimuth angle ϕ and the elevation angle θ) of plane waves that are arriving at an antenna array. It is generally assumed that all the signals have the same frequency of operation even though they are arriving from different directions and that the antenna elements in a linear array that are uniformly spaced at 0.5λ, where λ is the wavelength at the frequency of operation. We now describe a methodology for simultaneously estimating the frequency of operation and the DOA of the signals using a three-dimensional antenna array. The voltages induced in the antenna elements of the three-dimensional antenna array are used to estimate the frequency of operation and the DOA of the signal simultaneously using the Matrix Pencil method.

The problem is described in section 17.3, preceded by a short summary of the Matrix Pencil method in section 17.2. Section 17.4 provides the Cramer-Rao bound for the parameters of interest including the DOA and the frequency of interest followed by some numerical examples in section 17.5. In section 17.6 examples are presented using realistic antenna elements and then the conclusion follows in section 17.7.

17.2 BRIEF OVERVIEW OF THE MATRIX PENCIL METHOD

The Matrix Pencil method (MPM) is a direct data domain method [1] for estimation of the direction of arrival (DOA) of various signals impinging on an antenna array. Since, this method does not form a covariance matrix of the data, it is capable of identifying DOA's of coherent signals which is not possible using the conventional techniques like MUSIC and ESPRIT [2–4], unless additional processing is done through the use of subapertures, which is built in the Matrix Pencil approach. This implies that one can deal with coherent signals using a single snapshot (i.e., the data voltages received at the elements of the array at a particular instant of time) of the data [5,6].

The objective is to estimate the DOA of several signals using the MPM in one dimension and then we illustrate how to extend this methodology for the 2D case [7, 8]. If we have a uniformly spaced array of omnidirectional isotropic point sensors located along the z-axis and the distance between any two of them is Δ, one can write the voltage $\vartheta(n)$ induced in each of the n antenna elements, for $n = 0, 1, 2, \ldots, N$, as

$$\vartheta(n) = \sum_{s=1}^{S} A_s \exp(j\gamma_s + \frac{j2\pi}{\lambda}\Delta n \sin\theta_s) \qquad 0 \leq n \leq N \tag{17.1}$$

where A_s, γ_s, and θ_s are the amplitude, phase, and direction of arrival of each of the s plane waves incident on the array, and Δ is the spatial sampling interval, i.e., the spacing between two consecutive antenna elements. Here, we assume all the signals S that are impinging on the array have the same frequency of operation at wavelength λ. $\vartheta(n)$ represents the voltages measured at each of the $N+1$ antenna elements of the linear array at a particular time instance and is assumed to be known. Our goal is to find S and the characteristics of each of the s signals (i.e., to obtain the values for A_s, γ_s and θ_s). Equivalently, one can write (17.1) as

$$\vartheta(n) = \sum_{s=1}^{S} W_s\, a_s^n \text{ ; with } W_s = A_s \exp(j\gamma_s) \text{ and } a_s = \exp\left(\frac{j2\pi}{\lambda}\Delta \sin\theta_s\right) \tag{17.2}$$

or equivalently in a matrix form as

$$[G] = [Y][W] \text{ with; } [G] = \begin{bmatrix} \vartheta(0) \\ \vartheta(1) \\ \vdots \\ \vartheta(N) \end{bmatrix}; [Y] = \begin{bmatrix} 1 & 1 & \cdots & 1 \\ a_1 & a_2 & \cdots & a_S \\ \vdots & \vdots & \ddots & \vdots \\ a_1^N & a_2^N & \cdots & a_S^N \end{bmatrix}; [W] = \begin{bmatrix} W_1 \\ W_2 \\ \vdots \\ W_S \end{bmatrix} \tag{17.3}$$

As the first step in the MPM, $\vartheta(n)$ is partitioned to exploit the structure inherent in a DOA estimation problem using the concept of subapertures to deal with coherent signals. Therefore, we consider the following matrix:

$$[\vartheta] = \begin{bmatrix} \vartheta(0) & \vartheta(1) & \cdots & \vartheta(N-L+1) \\ \vartheta(1) & \vartheta(2) & \cdots & \vartheta(N-L+2) \\ \vdots & \vdots & \ddots & \vdots \\ \vartheta(L-1) & \vartheta(L) & \cdots & \vartheta(N) \end{bmatrix} \tag{17.4}$$

$[\vartheta]$ is an $L \times (N - L + 2)$ Hankel matrix, and each column of $[\vartheta]$ is a windowed segment of the original data $\{\vartheta(0)\ \vartheta(1) \cdots \vartheta(N)\}$ with a window length L. The parameter L, often called the *pencil parameter*, must satisfy the following bounds:

$$N + 1 - S \geq L \geq S + 1 \tag{17.5}$$

Next, a singular value decomposition (SVD) of the matrix $[\vartheta]$ is performed:

$$[\vartheta] = [U][\Sigma][V]^H \tag{17.6}$$

where $[U]$ is the $L \times L$ unitary matrix whose columns are the eigenvectors of $[\vartheta][\vartheta]^H$, $[\Sigma]$ is the $L \times (N - L + 2)$ diagonal matrix with singular values of $[\vartheta]$ located along its main diagonal in descending order $\sigma_1 \geq \sigma_2 \geq \cdots \geq \sigma_{\min}$ and $[V]$ is the $(N-L+2) \times (N-L+2)$ unitary matrix whose columns are the eigenvectors of $[\vartheta]^H[\vartheta]$. The superscript H denotes the conjugate transpose of a matrix.

If we consider that the data $\vartheta(n)$ are not contaminated by any noise, only the first S singular values are nonzero. Hence, $\sigma_i > 0$ for $i = 1, \ldots, S$ and $\sigma_i = 0$ for $i = M + 1, \ldots, L$.

If the data are corrupted by additive noise, the parameter S is estimated through M from observing the ratio of the various singular values to the largest one as defined by

$$\frac{\sigma_M}{\sigma_{\max}} \approx 10^{-w} \tag{17.7}$$

where w is the number of accurate significant decimal digits of the data $\vartheta(n)$ [1]. Hence for noise contaminated data $\sigma_i \geq 10^{-w}\sigma_{\max}$ for $i = 1, \ldots, M$ and $\sigma_i < 10^{-w}\sigma_{\max}$ for $i = M+1, \ldots, L$.

Next, we define the following three submatrices based on the first S dominant singular values:

- $[U]_M$: the first M columns of $[U]$
- $[\Sigma]_M$: the first M columns of $[\Sigma]$
- $[V]_M$: the first M columns of $[V]$

Finally, we form the matrix pencil to estimate the parameters of the problem. We now define the following matrices:

- $[U_1] = [U]_M$ with the last row deleted.
- $[U_2] = [U]_M$ with the first row deleted

and form

$$\left([U_2] - \lambda[U_1]\right)[X] = 0 \tag{17.8}$$

$$[U_1]^H [U_2][X] = \lambda [U_1]^H [U_1][X]$$

$$\left([U_1]^H [U_1]\right)^{-1} [U_1][U_2][X] = \lambda [X]$$

$$[U_1]^{\dagger} [U_2][X] = \lambda [X] \tag{17.9}$$

The eigenvalues of $[U_1]^{\dagger}[U_2]$ provide values for the exponents $\{a_m$: $m = 1, \ldots, M\}$, where $[U_1]^{\dagger}$ is the Moore–Penrose pseudo-inverse of $[U_1]$ and is defined by

$$[U_1]^{\dagger} = \left([U_1]^H [U_1]\right)^{-1}[U_1] \tag{17.10}$$

The directions of arrival θ_m are obtained from the values of the exponents as

$$\theta_m = \text{arc sin}\left[\frac{\lambda \times \ln(a_m)}{j2\pi\Delta}\right] \tag{17.11}$$

The amplitudes and phases of the M signals can be obtained by solving (17.3) using the principle of least squares as

$$[W] = \left([Y]^H [Y]\right)^{-1} [Y]^H [G] \tag{17.12}$$

Estimation of the DOA in two dimensions of several signals that simultaneously impinge on a two-dimensional planar array can also be estimated using the MPM as explained in chapter 14. As an example, consider an array consisting of omnidirectional isotropic point radiators that are uniformly spaced in a two-dimensional grid along the *x,* and *y*-axes. The total number of elements is of dimensions $P \times Q$. Let the spacing between two antenna elements that lie parallel to the *x*-axis be Δ_x and the spacing between the antenna elements that lie parallel to the *y*-axis be Δ_y. The total number of signals impinging on the array is *S*. Each of the *S* signals has associated with them an azimuth and an elevation angle of incidence. They are ϕ_s and θ_s, representing the azimuth and elevation angle, respectively, of the sth signal. Hence, the voltage $\vartheta\,(p;\,q)$ induced at the feed point of the antenna elements will be given by

$$\vartheta(p;\,q) = \sum_{s=1}^{S} A_s \exp\left[\begin{array}{l} j\gamma_s + \dfrac{j2\pi}{\lambda}(p-1)\Delta_y \sin\theta_s \sin\phi_s + \\ \dfrac{j2\pi}{\lambda}(q-1)\Delta_x \sin\theta_s \cos\phi_s \end{array}\right] \tag{17.13}$$

for $1 \leq p \leq P$ and $1 \leq q \leq Q$

We assume that all the *S* signals impinging on the array have the same frequency of operation. Here $\vartheta\,(p;\,q)$, representing the voltages induced in each element of the two-dimensional array, is assumed to be known. The goal is to estimate the parameters that define the azimuth and elevation DOA, ϕ_s and θ_s, respectively, along with their amplitudes and phases, A_s and γ_s, respectively. The number of signals *S* is also to be estimated.

In matrix form (17.13) can be written as

$$\begin{aligned} &\vartheta(p;q) = \sum_{s=1}^{S} W_s\, a_s^{p-1}\, b_s^{q-1}\,; \quad \text{with} \\ &W_s = A_s \exp(j\gamma_s);\; a_s = \exp\!\left(\frac{j2\pi}{\lambda}\Delta_y \sin\theta_s \sin\phi_s\right), \;\text{ and} \\ &b_s = \exp\!\left(\frac{j2\pi}{\lambda}\Delta_x \sin\theta_s \cos\phi_s\right) \end{aligned} \tag{17.14}$$

or equivalently, we can write (17.14) as

$$[\vartheta] = [Y]\,[W]\,[Z] \tag{17.15}$$

where $[\vartheta] = \begin{bmatrix} \vartheta(0;0) & \vartheta(0;1) & \cdots & \vartheta(0;N) \\ \vartheta(1;0) & \vartheta(1;1) & \cdots & \vartheta(1;N) \\ \vdots & \vdots & \ddots & \vdots \\ \vartheta(M;0) & \vartheta(M;1) & \cdots & \vartheta(M;N) \end{bmatrix}$; $[Y] = \begin{bmatrix} 1 & 1 & \cdots & 1 \\ a_1 & a_2 & \cdots & a_S \\ \vdots & \vdots & \ddots & \vdots \\ a_1^P & a_2^P & \cdots & a_S^P \end{bmatrix}$;

$$[W] = \text{diag}[W_1, W_2, \cdots, W_S]\,;\text{ and } [Z] = \begin{bmatrix} 1 & b_1 & \cdots & b_1^Q \\ 1 & b_2 & \cdots & b_2^Q \\ \vdots & \vdots & \ddots & \vdots \\ 1 & b_S & \cdots & b_S^Q \end{bmatrix} \tag{17.16}$$

We now form the various matrix pencils in order to extract the poles associated with the two different dimensions. Let us consider the first matrix pencil, defined by

$$[A_1] - \lambda_1 [B_1] = 0 \tag{17.17}$$

where $[A_1]$ and $[B_1]$ are defined as follows:

$$[A_1] = \begin{bmatrix} \vartheta(1;0) & \vartheta(1;1) & \cdots & \vartheta(1;N) \\ \vartheta(2;0) & \vartheta(2;1) & \cdots & \vartheta(2;N) \\ \vdots & \vdots & \ddots & \vdots \\ \vartheta(M;0) & \vartheta(M;1) & \cdots & \vartheta(M;N) \end{bmatrix} \tag{17.18}$$

$$[B_1] = \begin{bmatrix} \vartheta(0;0) & \vartheta(0;1) & \cdots & \vartheta(0;N) \\ \vartheta(1;0) & \vartheta(1;1) & \cdots & \vartheta(1;N) \\ \vdots & \vdots & \ddots & \vdots \\ \vartheta(M-1;0) & \vartheta(M-1;1) & \cdots & \vartheta(M-1;N) \end{bmatrix} \tag{17.19}$$

The eigenvalues of $[B_1]^{\dagger}[A_1]$ are the solution for the first set of exponents associated with one dimension. Let us define them to be $\{a_i\text{: } i = 1, \ldots, I\}$. Here $[B_1]^{\dagger}$ is the Moore–Penrose pseudo-inverse of $[B_1]$ and is defined by

$$[B_1]^{\dagger} = \left([B_1]^H [B_1]\right)^{-1} [B_1]^H \tag{17.20}$$

Consider the second matrix pencil to be of the form

$$[A_2] - \lambda_2 [B_2] = 0 \tag{17.21}$$

where $[A_2]$ and $[B_2]$ are defined as follows:

$$[A_2] = \begin{bmatrix} \vartheta(0;1) & \vartheta(0;2) & \cdots & \vartheta(0;N) \\ \vartheta(1;1) & \vartheta(1;2) & \cdots & \vartheta(1;N) \\ \vdots & \vdots & \ddots & \vdots \\ \vartheta(M;1) & \vartheta(M;2) & \cdots & \vartheta(M;N) \end{bmatrix} \tag{17.22}$$

$$[B_2] = \begin{bmatrix} \vartheta(0;0) & \vartheta(0;1) & \cdots & \vartheta(0;N-1) \\ \vartheta(1;0) & \vartheta(1;1) & \cdots & \vartheta(1;N-1) \\ \vdots & \vdots & \ddots & \vdots \\ \vartheta(M;0) & \vartheta(M;1) & \cdots & \vartheta(M;N-1) \end{bmatrix} \tag{17.23}$$

The eigenvalues of $[B_2]^{\dagger}[A_2]$ are the solution for the second set of exponents along the other dimension, which are defined to be $\{b_j\!: j = 1, \ldots, J\}$, where $[B_2]^{\dagger}$ is the Moore–Penrose pseudo-inverse of $[B_2]$ defined in (17.23).

We now assume that the complete solution for all the exponents is given by the tensor product of these two single exponents found before (i.e., the exponents are a possible product combination of the different pairs $\{(a_i, b_j), i = 1, \ldots, I;\ j = 1, \ldots, J\}$ that define the direction of arrival of each signal). Of course, some of them will not be related to the actual signal. Those components are eliminated when we look at the final residue.

Based on the tensorial product of the two sets of the one-dimensional solution, we form all possible pairs or combinations. If the number of one-dimensional solutions for the first matrix pencil are I and for the second are J, the total number of combinations or possible pairs will be $I \times J = T$ and it will be greater than S. Here we estimate the amplitude for all possible combination pairs by solving the following matrix equation for the residues or complex amplitudes $[W]$, which is defined by the column vector $[W_1, W_2, \ldots, W_T]$:

$$\begin{bmatrix} 1 & 1 & \cdots & 1 \\ a_1 & a_2 & \cdots & a_T \\ \vdots & \vdots & \ddots & \vdots \\ a_1^P & a_2^P & \cdots & a_T^P \\ b_1 & b_2 & \cdots & b_T \\ a_1 b_1 & a_2 b_2 & \cdots & a_T b_T \\ \vdots & \vdots & \ddots & \vdots \\ a_1 b_1^Q & a_2 b_2^Q & \cdots & a_T b_T^Q \\ \vdots & \vdots & \vdots & \vdots \\ a_1^P & a_2^P & \cdots & a_T^P \\ a_1^P b_1 & a_2^P b_2 & \cdots & a_T^P b_T \\ \vdots & \vdots & \ddots & \vdots \\ a_1^P b_1^Q & a_2^P b_2^Q & \cdots & a_T^P b_T^Q \end{bmatrix} \begin{bmatrix} W_1 \\ W_2 \\ \vdots \\ W_T \end{bmatrix} = \begin{bmatrix} \vartheta(0;0) \\ \vartheta(0;1) \\ \vdots \\ \vartheta(0;P) \\ \vartheta(1;0) \\ \vartheta(1;1) \\ \vdots \\ \vartheta(1;Q) \\ \vdots \\ \vartheta(P;0) \\ \vartheta(P;1) \\ \vdots \\ \vartheta(P;Q) \end{bmatrix} \tag{17.24}$$

This matrix is solved in a least squares fashion for all the complex amplitudes. Once we have the amplitudes for all possible pairs, we fix a threshold to eliminate the undesired pairs and take only those signals as possible solutions whose amplitudes are greater than this threshold. The number of signals T over this threshold must be greater than S, and that is equal to the total number of signals that impinge on the array.

The azimuth and elevation angle (ϕ and θ) can be obtained using the following equations obtained from the ordered set of T pairs $\{(a_i, b_j), i = 1, \ldots, I;\ j = 1, \ldots, J\}$:

$$\phi_s = \arctan\left[\left(\frac{\ln a_i}{\ln b_j}\right)\frac{\Delta_x}{\Delta_y}\right] \tag{17.25}$$

$$\theta_s = \arcsin\left[\left(\frac{\lambda}{j2\pi\,\Delta_x}\right)\sqrt{\left(\frac{\Delta_x}{\Delta_y}\ln a_i\right)^2 + \left(\ln b_j\right)^2}\right] \tag{17.26}$$

The complex amplitudes associated with them are given by W_k, $k = 1, 2, \ldots, T$. This completes the solution process.

However, the method outlined in [7] is more accurate when the signal-to-noise ratio of the data is low. The method in [7] takes more time than the method outlined here.

17.3 PROBLEM FORMULATION FOR SIMULTANEOUS ESTIMATON OF DOA AND THE FREQUENCY OF THE SIGNAL

We now extend the formulation presented in the previous section to deal with the DOA estimation of signals with different frequencies using a three-dimensional antenna array using a single snapshot. First, we present the methodology using point sources and then we illustrate how to extend it to deal with realistic antenna elements located in a conformal array. Let us assume that there are a total of P antenna elements along the x-direction, Q antenna elements along the y-direction, and R antenna elements along the z-direction. They are all uniformly spaced resulting in a three-dimensional antenna array. Therefore, in terms of the problem, if we have the nth antenna element located in space at (x_n, y_n, z_n), the voltage ϑ_n induced at that element will be given at a particular instance of time by the sum of S signals impinging on the array. Each of the sth signal is arriving from an azimuth angle ϕ_s, and an elevation angle of θ_s and has an operating frequency of f_s. Each of the sth signal arriving at the array has an amplitude of A_s and a phase angle of ψ_s. Therefore,

$$\begin{aligned}\vartheta_{\phi,\theta,f}(p,q,r) &= \sum_{s=1}^{S} A_s \exp(j\psi_s)\exp\left[\frac{j2\pi}{\lambda_s}\left\{\begin{matrix}x_p\cos\phi_s\sin\theta_s + \\ y_q\sin\phi_s\sin\theta_s + \\ z_r\cos\theta_s\end{matrix}\right\}\right] \\ &= \sum_{s=1}^{S} A_s \exp(j\psi_s)\exp\left[\frac{j2\pi f_s}{\upsilon}\left\{\begin{matrix}x_p\cos\phi_s\sin\theta_s + \\ y_q\sin\phi_s\sin\theta_s + \\ z_r\cos\theta_s\end{matrix}\right\}\right]\end{aligned} \tag{17.27}$$

where the frequency f_s is related to the wavelength λ_s through the velocity of light υ.

We now extract the necessary information (azimuth angle ϕ_s , elevation angle θ_s, and the operating frequency f_s) related to the S signals from the voltages ϑ_n received in antennas arranged as a three-dimensional array resulting in a data cube or data collected on two planar orthogonal arrays situated along three dimensions. These three dimensions can be the three spatial coordinates or data from two spatial coordinates at different instants of time. So, by the terms of the problem the voltages $\vartheta\,(p,q,r)$ at the spatial locations x_p, y_q, z_r are given for $p = 1, \ldots, P$; $q = 1, \ldots, Q$; $r = 1, \ldots, R$; and the goal is to find A_s, ψ_s, θ_s , ϕ_s , f_s for $s = 1, \ldots, S$. If we further assume that the antenna elements are uniformly spaced, then we have

$$\begin{aligned} x_p &= (p-1)\Delta_x \\ y_q &= (q-1)\Delta_y \\ z_r &= (r-1)\Delta_z \end{aligned} \tag{17.28}$$

where Δ_x, Δ_y, and Δ_z are the antenna element spacing along the x, y, and z-direction , respectively. In summary, (17.27) can be written in a compact form as

$$\vartheta(p,q,r) \;=\; \sum_{s=1}^{S} W_s\, a_s^{p-1} b_s^{q-1}\, c_s^{r-1} \tag{17.29}$$

where, $W_s = A_s \exp(j\psi_s)$; $a_s = \exp\left(j\dfrac{2\pi}{\lambda_s}\Delta_x \cos\phi_s \sin\theta_s \right)$; $b_s = \exp\left(j\dfrac{2\pi}{\lambda_s}\Delta_y \sin\phi_s \sin\theta_s \right)$ and $c_s = \exp\left(j\dfrac{2\pi}{\lambda_s}\Delta_z \cos\theta_s \right)$. Graphically, we can then represent the set of voltages ϑ by a data cube as shown in Figure 17.1a. Inside the data cube the data may be characterized by a set of planes along each dimension as seen in Figure 17.1b.

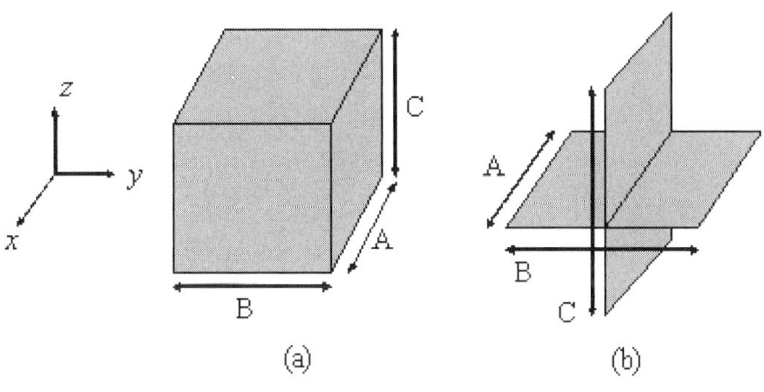

Figure 17.1. Graphical representations of 3-D data.

At the first stage of this procedure, we need to build three pencil of matrices to extract the three sets of poles a_s, b_s, c_s, given by (17.29) independently. In the second stage we need to associate the three sets (pair them) in the correct fashion as to what a_s goes with the correct b_s and with the appropriate c_s and relate them appropriately. The third stage is to find their complex amplitudes. Therefore, at the first stage we extract the three sets of poles: Let us assume that we have a data cube, which has a total of $P \cdot Q \cdot R$ data samples. We can also say that we have R planes (or slices) of $P \cdot Q$ data samples situated in a two-dimensional space (see Fig. 17.2a). Using any of these planes we can extract two sets of poles, e.g., a_s and b_s as given by (17.29). If we choose another perpendicular plane to the last one, that is, a plane of dimension $P \cdot R$, we are able to extract the sets of poles a_s and c_s (see Fig. 17.2b) using the procedure outlined in the previous section. This can be shown graphically in the Figure 17.2. Therefore, using the data samples from the plane shown in the Figure 17.2a, we can extract the set of poles $\{(a_s, b_s)\ ;\ s = 1, ..., S\}$ by using the following Pencil of Matrices. The first set of poles a_s, are extracted from the first Matrix Pencil, defined by

$$[A_1] - \lambda_1 [B_1] \quad ;\text{with}$$

$$[A_1] = \begin{bmatrix} V(2;1;h) & V(2;2;h) & \cdots & V(2;Q;h) \\ V(3;1;h) & V(3;2;h) & \cdots & V(3;Q;h) \\ \vdots & \vdots & \ddots & \vdots \\ V(P;1;h) & V(P;2;h) & \cdots & V(P;Q;h) \end{bmatrix}_{\text{for different values of } h} \text{, and}$$

$$[B_1] = \begin{bmatrix} V(1;1;h) & V(1;2;h) & \cdots & V(1;Q;h) \\ V(2;1;h) & V(2;2;h) & \cdots & V(2;Q;h) \\ \vdots & \vdots & \ddots & \vdots \\ V(P-1;1;h) & V(P-1;2;h) & \cdots & V(P-1;Q;h) \end{bmatrix}_{\text{for different values of } h} \tag{17.30}$$

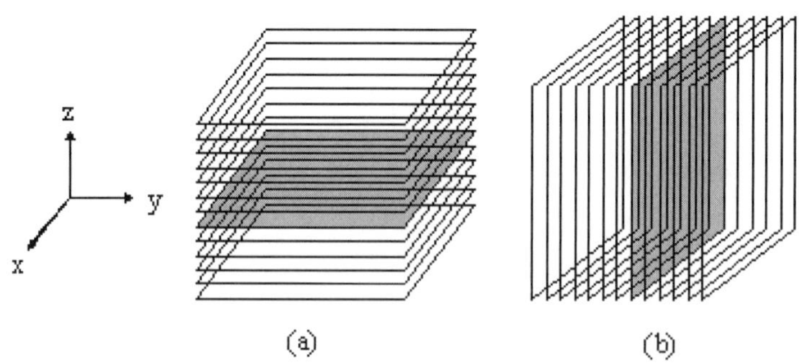

Figure 17.2. Graphical representations of Matrix Pencil formation.

The different matrices for different values of h can either be averaged together, or better still, concatenated one followed by the other as illustrated in [7, 8], to obtain a more accurate value for the frequency parameter particularly in the presence of noise. The eigenvalues of $[B_1]^{\dagger}[A_1]$ are the solution for the first set of exponents associated with a one-dimensional search. These eigenvalues correspond to $\{a_s;\ s = 1, \ldots, S\}$, which are the first set of poles that we are solving for. Here $[B_1]^{\dagger}$ is the Moore-Penrose pseudo-inverse of $[B_1]$ and is defined by

$$[B_1]^{\dagger} = \left([B_1]^{\mathrm{H}}[B_1]\right)^{-1}[B_1]^{\mathrm{H}} \tag{17.31}$$

The first set of Pencil of Matrices in (17.30) is shown graphically in Figure 17.3a where $[A_1]$ and $[B_1]$ are represented by square boxes and the voltages induced at the different antenna elements are denoted by black circles.

The second set of poles are extracted next by using the following Pencil of Matrices $[A_2]-\lambda_2[B_2]$; with

$$[A_2] = \begin{bmatrix} V(1;2;h) & V(1;3;h) & \cdots & V(1;Q;h) \\ V(2;2;h) & V(2;3;h) & \cdots & V(2;Q;h) \\ \vdots & \vdots & \ddots & \vdots \\ V(P;2;h) & V(P;3;h) & \cdots & V(P;Q;h) \end{bmatrix}_{\text{for different values of } h} \quad \text{and}$$

$$[B_2] = \begin{bmatrix} V(1;1;h) & V(1;2;h) & \cdots & V(1;Q-1;h) \\ V(2;1;h) & V(2;2;h) & \cdots & V(2;Q-1;h) \\ \vdots & \vdots & \ddots & \vdots \\ V(P;1;h) & V(P;2;h) & \cdots & V(P;Q-1;h) \end{bmatrix}_{\text{for different values of } h} \tag{17.32}$$

The Pencil of Matrices of (17.32) is formed using the methodology of Figure 17.3b.

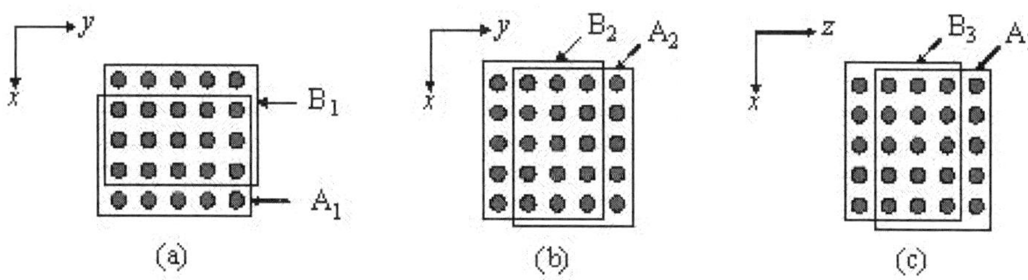

Figure 17.3. Graphical representation of Matrix Pencil Formation

The different matrices for different values of h can either be averaged together, or better still, concatenated one followed by the other as illustrated in [8], to obtain a more accurate value for the frequency parameter particularly in the presence of noise. Now, the eigenvalues of $[B_2]^{\dagger}[A_2]$ provide the second set of poles given by $\{b_s; s = 1, \ldots, S\}$.

Finally, the last set of poles can be extracted by using the data from another perpendicular plane. For example, if we use the samples shown in the Figure 17.2b we can obtain the third set of poles $\{c_s; s = 1, \ldots, S\}$ related to the following matrix pencil: $[A_3] - \lambda_3[B_3]$ with

$$[A_3] = \begin{bmatrix} V(1;h;2) & V(1;h;3) & \cdots & V(1;h;R) \\ V(2;h;2) & V(2;h;3) & \cdots & V(2;h;R) \\ \vdots & \vdots & \ddots & \vdots \\ V(P;h;2) & V(P;h;3) & \cdots & V(P;h;R) \end{bmatrix}_{\text{for different values of } h} \quad \text{and}$$

$$[B_3] = \begin{bmatrix} V(1;h;1) & V(1;h;2) & \cdots & V(1;h;R-1) \\ V(2;h;1) & V(2;h;2) & \cdots & V(2;h;R-1) \\ \vdots & \vdots & \ddots & \vdots \\ V(P;h;1) & V(P;h;2) & \cdots & V(P;h;R-1) \end{bmatrix}_{\text{for different values of } h} \tag{17.33}$$

The Pencil of Matrices of (17.33) is formed as illustrated in Figure 17.3c.

From the above analysis it is seen that we have used data on just two planes. So it is necessary to use only two planes of data to obtain a solution. However, the remainder of the data can be used efficiently in the presence of noise to reduce its effects on the parameters of the SOI.

The next step is how to group the frequencies, i.e., to pair which a_s is associated with the proper b_s.

We have to group the frequencies. To group the frequencies in a correct fashion we apply the following procedure. Based on the tensorial product of the two sets of the one-dimensional solutions, we form all the possible pairs or combinations. If the number of one-dimensional solutions for the first matrix pencil are M_1, for the second are M_2 and for the third are M_3, then the total number of combinations or possible pairs will be $M_1 \times M_2 \times M_3 = T$. We now estimate the amplitude for all the possible combinations or the *triples* by solving the following matrix equation for the residues or complex amplitudes $[W]$, which is defined by the column vector $[W_1, W_2, \ldots, W_T]$.

$$[ABC][W] = [V]$$

$$
\begin{bmatrix}
c_1 b_1 a_1 & c_2 b_2 a_2 & \cdots & c_T b_T a_T \\
c_1 b_1 a_1^2 & c_2 b_2 a_2^2 & \cdots & c_T b_T a_T^2 \\
\vdots & \vdots & \ddots & \vdots \\
c_1 b_1 a_1^P & c_2 b_2 a_2^P & \cdots & c_T b_T a_T^P \\
c_1 b_1^2 a_1 & c_2 b_2^2 a_2 & \cdots & c_T b_T^2 a_T \\
c_1 b_1^2 a_1^2 & c_2 b_2^2 a_2^2 & \cdots & c_T b_T^2 a_T^2 \\
\vdots & \vdots & \ddots & \vdots \\
c_1 b_1^2 a_1^P & c_2 b_2^2 a_2^P & \cdots & c_T b_T^2 a_T^P \\
\vdots & \vdots & \vdots & \vdots \\
c_1 b_1^Q a_1 & c_2 b_2^Q a_2 & \cdots & c_T b_T^Q a_T \\
c_1 b_1^Q a_1^2 & c_2 b_2^Q a_2^2 & \cdots & c_T b_T^Q a_T^2 \\
\vdots & \vdots & \ddots & \vdots \\
c_1 b_1^Q a_1^P & c_2 b_2^Q a_2^P & \cdots & c_S b_S^Q a_S^P \\
c_1^2 b_1 a_1 & c_2^2 b_2 a_2 & \cdots & c_S^2 b_S a_S \\
c_1^2 b_1 a_1^2 & c_2^2 b_2 a_2^2 & \cdots & c_T^2 b_T a_T^2 \\
\vdots & \vdots & \ddots & \vdots \\
c_1^2 b_1 a_1^P & c_2^2 b_2 a_2^P & \cdots & c_T^2 b_T a_T^P \\
\vdots & \vdots & \vdots & \vdots \\
c_1^2 b_1^Q a_1 & c_2^2 b_2^Q a_2 & \cdots & c_T^2 b_T^Q a_T \\
c_1^2 b_1^Q a_1^2 & c_2^2 b_2^Q a_2^2 & \cdots & c_T^2 b_T^Q a_T^2 \\
\vdots & \vdots & \ddots & \vdots \\
c_1^2 b_1^Q a_1^P & c_2^2 b_2^Q a_2^P & \cdots & c_T^2 b_T^Q a_T^P \\
\vdots & \vdots & \vdots & \vdots \\
c_1^R b_1^Q a_1 & c_2^R b_2^Q a_2 & \cdots & c_T^R b_T^Q a_T \\
c_1^R b_1^Q a_1^2 & c_2^R b_2^Q a_2^2 & \cdots & c_T^R b_T^Q a_T^2 \\
\vdots & \vdots & \ddots & \vdots \\
c_1^R b_1^Q a_1^P & c_2^R b_2^Q a_2^P & \cdots & c_T^R b_T^Q a_T^P
\end{bmatrix}
\begin{bmatrix}
W_1 \\ W_2 \\ \vdots \\ W_T
\end{bmatrix}
=
\begin{bmatrix}
V(1;1;1) \\ V(2;1;1) \\ \vdots \\ V(P;1;1) \\ V(1;2;1) \\ V(2;2;1) \\ \vdots \\ V(P;1;1) \\ \vdots \\ V(1;Q;1) \\ V(1;Q;1) \\ \vdots \\ V(P;Q;1) \\ V(1;1;2) \\ V(2;1;2) \\ \vdots \\ V(A;1;2) \\ \vdots \\ V(1;Q;2) \\ V(2;Q;2) \\ \vdots \\ V(P;Q;2) \\ \vdots \\ V(1;Q;R) \\ V(2;Q;R) \\ \vdots \\ V(P;Q;R)
\end{bmatrix}
\tag{17.34}
$$

This matrix is solved in a least squares fashion for all the complex amplitudes. Once we have the amplitudes for all the possible pairs we fix a threshold to eliminate the undesired pairs and take only those signals as possible solutions whose amplitudes are greater than this threshold. The number of signals over this threshold must be $S \leq T$ and is equal to the total number of 3-D sinusoids to be solved for our problem.

From (17.29) using the expressions for each set of poles one can find the spatial frequencies and the frequency of the signal by using $D = \text{Imaginary part of}\left[\dfrac{ln\ a_s}{2\pi\ \Delta_x}\right]$; $E = \text{Imaginary part of}\left[\dfrac{ln\ b_s}{2\pi\ \Delta_y}\right]$; and $F = \text{Imaginary part of}\left[\dfrac{ln\ c_s}{2\pi\ \Delta_z}\right]$. From which we obtain

$$\phi_s = \operatorname{arc\,cot}\left(\frac{D}{E}\right);\ \theta_s = \arctan\left(\frac{\sqrt{D^2+E^2}}{F}\right);\ \lambda_s = \sqrt{\frac{1}{D^2+E^2+F^2}} \tag{17.35}$$

Next we look at the quality of the solution in the presence of noise.

17.4 CRAMER-RAO BOUND FOR THE DIRECTION OF ARRIVAL AND FREQUENCY OF THE SIGNAL

In this section, the Cramer-Rao bound (CRB) for the various parameters are derived to study the accuracy in estimating the parameters of interest in signals contaminated by noise. The Cramér-Rao lower bound expresses a lower bound on the variance of estimators of a deterministic parameter. In its simplest form, the bound states that the variance of any unbiased estimator is at least as high as the inverse of the Fisher Information Matrix. An unbiased estimator which achieves this lower bound is said to be efficient. Such a solution achieves the lowest possible mean squared error among all unbiased methods, and is therefore the minimum variance unbiased (MVU) estimator. The noise contaminated signal $\tilde{v}$ is given by

$$\tilde{v}(a;b;c) = \sum_{i=1}^{I} M_i\, \mathrm{e}^{[j(\gamma_i + \frac{2\pi a}{\lambda_i}\Delta_x \cos\phi_i \sin\theta_i + \frac{2\pi b}{\lambda_i}\Delta_y \sin\phi_i \sin\theta_i + \frac{2\pi c}{\lambda_i}\Delta_z \cos\theta_i]} + w(a,b,c) \tag{17.36}$$

where $w(a,b,c)$ is a zero mean Gaussian white noise with variance σ^2. A 3-D array of omni-directional, isotropic, point sensors are considered in this section. We define the probability density function of $\tilde{v}$ as

$$P(\tilde{v}\,|\,\varphi) = \frac{1}{\left(\sqrt{2\pi\sigma^2}\right)^{PQR}} e^{\left(-\frac{1}{2\sigma^2}|\tilde{v}-v|^2\right)} \tag{17.37}$$

where $|\bullet|$ denotes the 2-norm. σ^2 is the variance of the noise. φ is the $I \times 1$ column vector of the unknown parameters defined by

$$\varphi = [\varphi_1 \quad \varphi_2 \quad \cdots \quad \varphi_I\ ]^T \tag{17.38}$$

and

$$\varphi_i = \begin{bmatrix} M_i \\ \gamma_i \\ f_i \\ \phi_i \\ \theta_i \end{bmatrix} \tag{17.39}$$

The $I \times I$ Fisher Information Matrix $[\mathbf{F}]$ is defined by

$$[\mathbf{F}(\varphi)]_{ij} = -E\left(\frac{\partial^2}{\partial\varphi_i\partial\varphi_j}\log\left(p(\tilde{v}\,|\,\varphi)\right)\right) \tag{17.40}$$

where $\mathbf{F}_{ij}$ is the (i,j) the element of the matrix $\mathbf{F}$. In addition the matrix $\mathbf{F}$ can be partitioned as $\mathbf{F} = \{\mathbf{F}_{ij} \quad ,i=1,\ldots,I \quad ,j=1,\ldots,I\}$ where $\mathbf{F}_{ij}$ is a 5×5, $(i,j)^{\text{th}}$ block matrix of $\mathbf{F}$. $E(\bullet)$ is the expectation operator, and $\partial/\partial\varphi_i$ is the partial derivative with respect to the ith element φ_i of φ, and log is the natural logarithm. Therefore,

$$\mathbf{F}_{ij} = \frac{1}{\sigma^2}2\,\mathrm{Re}\left(\frac{\partial v^H}{\partial\varphi_i}\frac{\partial v}{\partial\varphi_j}\right) \tag{17.41}$$

where Re($\bullet$) is the real part of. In addition,

$$\frac{\partial v}{\partial M_i} = \left\{ e^{j\left[\gamma_i + \frac{2\pi f_i}{\upsilon}\left(\cos\phi_i \sin\theta_i\, a\,\Delta_x + \sin\phi_i \sin\theta_i\, b\,\Delta_y + \cos\theta_i\, c\,\Delta_z\right)\right]} \right\} \tag{17.42}$$

where υ is the speed of light, where $\lambda = \upsilon/f$.

$$\frac{\partial v}{\partial \gamma_i} = \left\{ j\,M_i\, e^{j\left[\gamma_i + \frac{2\pi f_i}{\upsilon}\left(\cos\phi_i \sin\theta_i\, a\,\Delta_x + \sin\phi_i \sin\theta_i\, b\,\Delta_y + \cos\theta_i\, c\,\Delta_z\right)\right]} \right\} \tag{17.43}$$

$$\frac{\partial v}{\partial f_i} = \left\{ \begin{array}{l} j\,M_i \dfrac{2\pi}{\upsilon}\left(\cos\phi_i \sin\theta_i\, a\,\Delta_x + \sin\phi_i \sin\theta_i\, b\ \Delta_y + \cos\phi_i\, c\,\Delta_z\right) \\ \times\, e^{j[\gamma_i + \frac{2\pi f_i}{\upsilon}\cos\phi_i \sin\theta_i\, a\,\Delta_x + \frac{2\pi f_i}{\upsilon}\sin\phi_i \sin\theta_i\, b\,\Delta_y + \frac{2\pi f_i}{\upsilon}\cos\theta_i\, c\,\Delta_z]} \end{array} \right\}$$

(17.44)

$$\frac{\partial v}{\partial \phi_i} = \begin{Bmatrix} j\, M_i \dfrac{2\pi f_i}{\upsilon}\left(-\sin\phi_i \sin\theta_i \; a\; \Delta_x + \cos\phi_i \sin\theta_i \; b\; \Delta_y\right) \\ \times\, e^{j[\gamma_i + \frac{2\pi f_i}{\upsilon}\cos\phi_i \sin\theta_i \; a\; \Delta_x + \frac{2\pi f_i}{\upsilon}\sin\phi_i \sin\theta_i \; b\; \Delta_y + \frac{2\pi f_i}{\upsilon}\cos\theta_i \; c\; \Delta_z]} \end{Bmatrix} \tag{17.45}$$

$$\frac{\partial v}{\partial \theta_i} = \begin{Bmatrix} j\, M_i \dfrac{2\pi f_i}{\upsilon}\left(\cos\phi_i \cos\theta_i \; a\; \Delta_x + \sin\phi_i \cos\theta_i \; b\; \Delta_y - \sin\theta_i \; c\; \Delta_z\right) \\ \times\, e^{j[\gamma_i + \frac{2\pi f_i}{\upsilon}\cos\phi_i \sin\theta_i \; a\; \Delta_x + \frac{2\pi f_i}{\upsilon}\sin\phi_i \sin\theta_i \; b\; \Delta_y + \frac{2\pi f_i}{\upsilon}\cos\theta_i \; c\; \Delta_z]} \end{Bmatrix} \tag{17.46}$$

$$2\,\mathrm{Re}\left[\frac{\partial v^H}{\partial M_i}\frac{\partial v}{\partial M_j}\right] = $$
$$2\sum_{a=0}^{A-1}\sum_{b=0}^{B-1}\sum_{c=0}^{C-1}\cos\left(\gamma_i - \gamma_j + \frac{2\pi}{\upsilon}\begin{pmatrix} \left(f_i \cos\phi_i \sin\theta_i - f_j \cos\phi_j \sin\theta_j\right)\; a\; \Delta_x \\ +\left(f_i \sin\phi_i \sin\theta_i - f_j \sin\phi_j \sin\theta_j\right)\; b\; \Delta_y \\ +\left(f_i \cos\theta_i - f_j \cos\theta_j\right)\; c\; \Delta_z \end{pmatrix}\right) \tag{17.47}$$

$$2\,\mathrm{Re}\left[\frac{\partial v^H}{\partial \gamma_i}\frac{\partial v}{\partial M_j}\right] = $$
$$2\sum_{a=0}^{A-1}\sum_{b=0}^{B-1}\sum_{c=0}^{C-1} M_j \sin\left(\gamma_i - \gamma_j + \frac{2\pi}{\upsilon}\begin{pmatrix} \left(f_i \cos\phi_i \sin\theta_i - f_j \cos\phi_j \sin\theta_j\right)\; a\; \Delta_x \\ +\left(f_i \sin\phi_i \sin\theta_i - f_j \sin\phi_j \sin\theta_j\right)\; b\; \Delta_y \\ +\left(f_i \cos\theta_i - f_j \cos\theta_j\right)\; c\; \Delta_z \end{pmatrix}\right) \tag{17.48}$$

$$2\,\mathrm{Re}\left[\frac{\partial v^H}{\partial \gamma_i}\frac{\partial v}{\partial M_j}\right]$$
$$= -2\sum_{a=0}^{A-1}\sum_{b=0}^{B-1}\sum_{c=0}^{C-1} M_i \sin\left(\gamma_i - \gamma_j + \frac{2\pi}{\upsilon}\begin{pmatrix} \left(f_i \cos\phi_i \sin\theta_i - f_j \cos\phi_j \sin\theta_j\right)\; a\; \Delta_x \\ +\left(f_i \sin\phi_i \sin\theta_i - f_j \sin\phi_j \sin\theta_j\right)\; b\; \Delta_y \\ +\left(f_i \cos\theta_i - f_j \cos\theta_j\right)\; c\; \Delta_z \end{pmatrix}\right) \tag{17.49}$$

$$
2\,\mathrm{Re}\left[\frac{\partial v^H}{\partial M_i}\frac{\partial v}{\partial f_j}\right]
= -2\sum_{a=0}^{A-1}\sum_{b=0}^{B-1}\sum_{c=0}^{C-1}\left(\begin{array}{l}
M_j\dfrac{2\pi}{\upsilon}\left(\cos\phi_j\sin\theta_j\,a\Delta_x+\sin\phi_j\sin\theta_j\,b\Delta_y+\cos\theta_j\,c\Delta_z\right)\\
\times\ \sin\left(\gamma_i-\gamma_j+\dfrac{2\pi}{\upsilon}\left(\begin{array}{l}
\left(f_i\cos\phi_i\sin\theta_i-f_j\cos\phi_j\sin\theta_j\right)a\Delta_x\\
+\left(f_i\sin\phi_i\sin\theta_i-f_j\sin\phi_j\sin\theta_j\right)b\Delta_y\\
+\left(f_i\cos\theta_i-f_j\cos\theta_j\right)c\Delta_z
\end{array}\right)\right)
\end{array}\right)
\tag{17.50}
$$

$$
2\,\mathrm{Re}\left[\frac{\partial v^H}{\partial f_i}\frac{\partial v}{\partial M_j}\right]
= -2\sum_{a=0}^{A-1}\sum_{b=0}^{B-1}\sum_{c=0}^{C-1}\left(\begin{array}{l}
M_i\dfrac{2\pi}{\upsilon}\left(\cos\phi_i\sin\theta_i\,a\Delta_x+\sin\phi_i\sin\theta_i\,b\Delta_y+\cos\theta_i\,c\Delta_z\right)\\
\times\ \sin\left(\gamma_i-\gamma_j+\dfrac{2\pi}{\upsilon}\left(\begin{array}{l}
\left(f_i\cos\phi_i\sin\theta_i-f_j\cos\phi_j\sin\theta_j\right)a\Delta_x\\
+\left(f_i\sin\phi_i\sin\theta_i-f_j\sin\phi_j\sin\theta_j\right)b\Delta_y\\
+\left(f_i\cos\theta_i-f_j\cos\theta_j\right)c\Delta_z
\end{array}\right)\right)
\end{array}\right)
\tag{17.51}
$$

$$
2\,\mathrm{Re}\left[\frac{\partial v^H}{\partial M_i}\frac{\partial v}{\partial \phi_j}\right]
= 2\sum_{a=0}^{A-1}\sum_{b=0}^{B-1}\sum_{c=0}^{C-1}\left(\begin{array}{l}
M_j\dfrac{2\pi f_j}{\upsilon}\left(-\sin\phi_j\sin\theta_j\,a\Delta_x+\cos\phi_j\sin\theta_j\,b\Delta_y\right)\\
\times\ \sin\left(\gamma_i-\gamma_j+\dfrac{2\pi}{\upsilon}\left(\begin{array}{l}
\left(f_i\cos\phi_i\sin\theta_i-f_j\cos\phi_j\sin\theta_j\right)a\,\Delta_x\\
+\left(f_i\sin\phi_i\sin\theta_i-f_j\sin\phi_j\sin\theta_j\right)b\Delta_y\\
+\left(f_i\cos\theta_i-f_j\cos\theta_j\right)c\,\Delta_z
\end{array}\right)\right)
\end{array}\right)
\tag{17.52}
$$

$$2\,\mathrm{Re}\left[\frac{\partial v^H}{\partial \phi_i}\,\frac{\partial v}{\partial M_j}\right]$$
$$= -2\sum_{a=0}^{A-1}\sum_{b=0}^{B-1}\sum_{c=0}^{C-1}\left(\begin{array}{l} M_i\,\dfrac{2\pi f_i}{\upsilon}\left(-\sin\phi_i \sin\theta_i\; a\,\Delta_x + \cos\phi_i \sin\theta_i\; b\,\Delta_y\right) \\ \times\ \sin\left(\gamma_i - \gamma_j + \dfrac{2\pi}{\upsilon}\left(\begin{array}{l}\left(f_i\cos\phi_i\sin\theta_i - f_j\cos\phi_j\sin\theta_j\right)a\,\Delta_x \\ +\left(f_i\sin\phi_i\sin\theta_i - f_j\sin\phi_j\sin\theta_j\right)b\,\Delta_y \\ +\left(f_i\cos\theta_i - f_j\cos\theta_j\right)c\,\Delta_z\end{array}\right)\right)\end{array}\right) \tag{17.53}$$

$$2\,\mathrm{Re}\left[\frac{\partial v^H}{\partial M_i}\,\frac{\partial v}{\partial \theta_j}\right]$$
$$= 2\sum_{a=0}^{A-1}\sum_{b=0}^{B-1}\sum_{c=0}^{C-1}\left(\begin{array}{l} M_j\,\dfrac{2\pi f_j}{\upsilon}\left(\cos\phi_j \cos\theta_j\, a\,\Delta_x + \sin\phi_j \cos\theta_j\, b\,\Delta_y - \sin\theta_j\, c\,\Delta_z\right) \\ \times\ \sin\left(\gamma_i - \gamma_j + \dfrac{2\pi}{\upsilon}\left(\begin{array}{l}\left(f_i\cos\phi_i\sin\theta_i - f_j\cos\phi_j\sin\theta_j\right)a\,\Delta_x \\ +\left(f_i\sin\phi_i\sin\theta_i - f_j\sin\phi_j\sin\theta_j\right)b\,\Delta_y \\ +\left(f_i\cos\theta_i - f_j\cos\theta_j\right)c\,\Delta_z\end{array}\right)\right)\end{array}\right) \tag{17.54}$$

$$2\,\mathrm{Re}\left[\frac{\partial v^H}{\partial \theta_i}\,\frac{\partial v}{\partial M_j}\right].$$
$$= -2\sum_{a=0}^{A-1}\sum_{b=0}^{B-1}\sum_{c=0}^{C-1}\left(\begin{array}{l} M_i\,\dfrac{2\pi f_i}{\upsilon}\left(\cos\phi_i \cos\theta_i\, a\,\Delta_x + \sin\phi_i \cos\theta_i\, b\,\Delta_y - \sin\theta_i\, c\,\Delta_z\right) \\ \times\ \sin\left(\gamma_i - \gamma_j + \dfrac{2\pi}{\upsilon}\left(\begin{array}{l}\left(f_i\cos\phi_i\sin\theta_i - f_j\cos\phi_j\sin\theta_j\right)a\,\Delta_x \\ +\left(f_i\sin\phi_i\sin\theta_i - f_j\sin\phi_j\sin\theta_j\right)b\,\Delta_y \\ +\left(f_i\cos\theta_i - f_j\cos\theta_j\right)c\,\Delta_z\end{array}\right)\right)\end{array}\right) \tag{17.55}$$

$$2\,\mathrm{Re}\left[\frac{\partial v^H}{\partial \gamma_i}\frac{\partial v}{\partial \gamma_j}\right]$$

$$= 2\sum_{a=0}^{A-1}\sum_{b=0}^{B-1}\sum_{c=0}^{C-1} M_i M_j \cos\left(\gamma_i - \gamma_j + \frac{2\pi}{\upsilon}\left(\begin{array}{l}\left(f_i \cos\phi_i \sin\theta_i - f_j \cos\phi_j \sin\theta_j\right) a\,\Delta_x \\ +\left(f_i \sin\phi_i \sin\theta_i - f_j \sin\phi_j \sin\theta_j\right) b\,\Delta_y \\ +\left(f_i \cos\theta_i - f_j \cos\theta_j\right) c\,\Delta_z\end{array}\right)\right)$$

(17.56)

$$2\,\mathrm{Re}\left[\frac{\partial v^H}{\partial \gamma_i}\frac{\partial v}{\partial f_j}\right]$$

$$= 2\sum_{a=0}^{A-1}\sum_{b=0}^{B-1}\sum_{c=0}^{C-1}\left(\begin{array}{l} M_i M_j \dfrac{2\pi}{\upsilon}\left(\cos\phi_j \sin\theta_j\, a\,\Delta_x + \sin\phi_j \sin\theta_j\, b\,\Delta_y + \cos\theta_j\, c\,\Delta_z\right) \\ \times\cos\left(\gamma_i - \gamma_j + \dfrac{2\pi}{\upsilon}\left(\begin{array}{l}\left(f_i \cos\phi_i \sin\theta_i - f_j \cos\phi_j \sin\theta_j\right) a\,\Delta_x \\ +\left(f_i \sin\phi_i \sin\theta_i - f_j \sin\phi_j \sin\theta_j\right) b\,\Delta_y \\ +\left(f_i \cos\theta_i - f_j \cos\theta_j\right) c\,\Delta_z\end{array}\right)\right)\end{array}\right)$$

(17.57)

$$2\,\mathrm{Re}\left[\frac{\partial v^H}{\partial f_i}\frac{\partial v}{\partial \gamma_j}\right]$$

$$= 2\sum_{a=0}^{A-1}\sum_{b=0}^{B-1}\sum_{c=0}^{C-1}\left(\begin{array}{l} M_i M_j \dfrac{2\pi}{\upsilon}\left(\cos\phi_i \sin\theta_i\, a\,\Delta_x + \sin\phi_i \sin\theta_i\, b\,\Delta_y + \cos\theta_i\, c\,\Delta_z\right) \\ \times\cos\left(\gamma_i - \gamma_j + \dfrac{2\pi}{\upsilon}\left(\begin{array}{l}\left(f_i \cos\phi_i \sin\theta_i - f_j \cos\phi_j \sin\theta_j\right) a\,\Delta_x \\ +\left(f_i \sin\phi_i \sin\theta_i - f_j \sin\phi_j \sin\theta_j\right) b\,\Delta_y \\ +\left(f_i \cos\theta_i - f_j \cos\theta_j\right) c\,\Delta_z\end{array}\right)\right)\end{array}\right)$$

(17.58)

$$
\begin{aligned}
&2\,\mathrm{Re}\left[\frac{\partial v^H}{\partial \gamma_i}\frac{\partial v}{\partial \phi_j}\right] \\
&= 2\sum_{a=0}^{A-1}\sum_{b=0}^{B-1}\sum_{c=0}^{C-1}\left(\begin{array}{l}
M_i\,M_j\,\dfrac{2\pi f_j}{\upsilon}\left(-\sin\phi_j\sin\theta_j\,a\,\Delta_x+\cos\phi_j\sin\theta_j\,b\,\Delta_y\right)\\
\times\;\;\cos\left(\gamma_i-\gamma_j+\dfrac{2\pi}{\upsilon}\left(\begin{array}{l}
\left(f_i\cos\phi_i\sin\theta_i-f_j\cos\phi_j\sin\theta_j\right)a\,\Delta_x\\
+\left(f_i\sin\phi_i\sin\theta_i-f_j\sin\phi_j\sin\theta_j\right)b\,\Delta_y\\
+\left(f_i\cos\theta_i-f_j\cos\theta_j\right)c\,\Delta_z
\end{array}\right)\right)
\end{array}\right)
\end{aligned}
\tag{17.59}
$$

$$
\begin{aligned}
&2\,\mathrm{Re}\left[\frac{\partial v^H}{\partial \phi_i}\frac{\partial v}{\partial \gamma_j}\right] \\
&= 2\sum_{a=0}^{A-1}\sum_{b=0}^{B-1}\sum_{c=0}^{C-1}\left(\begin{array}{l}
M_i M_j\,\dfrac{2\pi f_i}{\upsilon}\left(-\sin\phi_i\sin\theta_i\,a\,\Delta_x+\cos\phi_i\sin\theta_i\,b\,\Delta_y\right)\\
\times\;\;\cos\left(\gamma_i-\gamma_j+\dfrac{2\pi}{\upsilon}\left(\begin{array}{l}
\left(f_i\cos\phi_i\sin\theta_i-f_j\cos\phi_j\sin\theta_j\right)a\,\Delta_x\\
+\left(f_i\sin\phi_i\sin\theta_i-f_j\sin\phi_j\sin\theta_j\right)b\,\Delta_y\\
+\left(f_i\cos\theta_i-f_j\cos\theta_j\right)c\,\Delta_z
\end{array}\right)\right)
\end{array}\right)
\end{aligned}
\tag{17.60}
$$

$$
\begin{aligned}
&2\,\mathrm{Re}\left[\frac{\partial v^H}{\partial \gamma_i}\frac{\partial v}{\partial \theta_j}\right] \\
&= 2\sum_{a=0}^{A-1}\sum_{b=0}^{B-1}\sum_{c=0}^{C-1}\left(\begin{array}{l}
M_i\,M_j\,\dfrac{2\pi f_j}{\upsilon}\left(\cos\phi_j\cos\theta_j\,a\,\Delta_x+\sin\phi_j\cos\theta_j\,b\,\Delta_y-\sin\theta_j\,c\,\Delta_z\right)\\
\times\;\;\cos\left(\gamma_i-\gamma_j+\dfrac{2\pi}{\upsilon}\left(\begin{array}{l}
\left(f_i\cos\phi_i\sin\theta_i-f_j\cos\phi_j\sin\theta_j\right)a\,\Delta_x\\
+\left(f_i\sin\phi_i\sin\theta_i-f_j\sin\phi_j\sin\theta_j\right)b\,\Delta_y\\
+\left(f_i\cos\theta_i-f_j\cos\theta_j\right)c\,\Delta_z
\end{array}\right)\right)
\end{array}\right)
\end{aligned}
\tag{17.61}
$$

$$
2\,\mathrm{Re}\left[\frac{\partial v^H}{\partial \theta_i}\frac{\partial v}{\partial \gamma_j}\right]
= 2\sum_{a-0}^{A-1}\sum_{b-0}^{B-1}\sum_{c-0}^{C-1}
\left(\begin{array}{l}
M_i\,M_j\,\dfrac{2\pi f_i}{\upsilon}\left(\cos\phi_i\cos\theta_i\,a\,\Delta_x+\sin\phi_i\cos\theta_i\,b\,\Delta_y-\sin\theta_i\,c\,\Delta_z\right)\\
\times\;\cos\left(\gamma_i-\gamma_j+\dfrac{2\pi}{\upsilon}\left(\begin{array}{l}\left(f_i\cos\phi_i\sin\theta_i-f_j\cos\phi_j\sin\theta_j\right)a\,\Delta_x\\+\left(f_i\sin\phi_i\sin\theta_i-f_j\sin\phi_j\sin\theta_j\right)b\,\Delta_y\\+\left(f_i\cos\theta_i-f_j\cos\theta_j\right)c\,\Delta_z\end{array}\right)\right)
\end{array}\right)
\tag{17.62}
$$

$$
2\,\mathrm{Re}\left[\frac{\partial v^H}{\partial f_i}\frac{\partial v}{\partial f_j}\right]
= 2\sum_{a=0}^{A-1}\sum_{b=0}^{B-1}\sum_{c=0}^{C-1}
\left\{\begin{array}{l}
M_i\,M_j\left(\dfrac{2\pi}{\upsilon}\right)^2\\
\times\;\left(\cos\phi_i\sin\theta_i\,a\,\Delta_x+\sin\phi_i\sin\theta_i\,b\,\Delta_y+\cos\theta_i\,c\,\Delta_z\right)\\
\times\;\left(\cos\phi_j\sin\theta_j\,a\,\Delta_x+\sin\phi_j\sin\theta_j\,b\,\Delta_y+\cos\theta_i\,c\,\Delta_z\right)\\
\times\;\cos\left(\gamma_i-\gamma_j+\dfrac{2\pi}{\upsilon}\left(\begin{array}{l}\left(f_i\cos\phi_i\sin\theta_i-f_j\cos\phi_j\sin\theta_j\right)a\,\Delta_x\\+\left(f_i\sin\phi_i\sin\theta_i-f_j\sin\phi_j\sin\theta_j\right)b\,\Delta_y\\+\left(f_i\cos\theta_i-f_j\cos\theta_j\right)c\,\Delta_z\end{array}\right)\right)
\end{array}\right\}
\tag{17.63}
$$

$$
2\,\mathrm{Re}\left[\frac{\partial v^H}{\partial f_i}\frac{\partial v}{\partial \phi_j}\right]
= 2\sum_{a=0}^{A-1}\sum_{b=0}^{B-1}\sum_{c=0}^{C-1}
\left\{\begin{array}{l}
M_i\,M_j\left(\dfrac{2\pi}{\upsilon}\right)\left(\dfrac{2\pi f_j}{\upsilon}\right)\\
\times\;\left(\cos\phi_i\sin\theta_i\,a\,\Delta_x+\sin\phi_i\sin\theta_i\,b\,\Delta_y+\cos\theta_i\,c\,\Delta_z\right)\\
\times\;\left(-\sin\phi_j\sin\theta_j\,a\,\Delta_x+\cos\phi_j\sin\theta_j\,b\,\Delta_y\right)\\
\times\;\cos\left(\gamma_i-\gamma_j+\dfrac{2\pi}{\upsilon}\left(\begin{array}{l}\left(f_i\cos\phi_i\sin\theta_i-f_j\cos\phi_j\sin\theta_j\right)a\,\Delta_x\\+\left(f_i\sin\phi_i\sin\theta_i-f_j\sin\phi_j\sin\theta_j\right)b\,\Delta_y\\+\left(f_i\cos\theta_i-f_j\cos\theta_j\right)c\,\Delta_z\end{array}\right)\right)
\end{array}\right\}
\tag{17.64}
$$

$$2\,\mathrm{Re}\left[\frac{\partial v^H}{\partial \phi_i}\frac{\partial v}{\partial f_j}\right]$$

$$= 2\sum_{a=0}^{A-1}\sum_{b=0}^{B-1}\sum_{c=0}^{C-1}\left(\begin{array}{l} M_i\, M_j\left(\frac{2\pi}{\upsilon}\right)\left(\frac{2\pi f_i}{\upsilon}\right)\left(-\sin\phi_i \sin\theta_i\, a\,\Delta_x + \cos\phi_i \sin\theta_i\, b\,\Delta_y\right) \\ \times \quad \cos\left(\gamma_i - \gamma_j + \frac{2\pi}{\upsilon}\left(\begin{array}{l}\left(f_i \cos\phi_i \sin\theta_i - f_j \cos\phi_j \sin\theta_j\right) a\,\Delta_x \\ +\left(f_i \sin\phi_i \sin\theta_i - f_j \sin\phi_j \sin\theta_j\right) b\,\Delta_y \\ +\left(f_i \cos\theta_i - f_j \cos\theta_j\right) c\,\Delta_z\end{array}\right)\right) \\ \times \quad \left(\cos\phi_j \sin\theta_j\, a\,\Delta_x + \sin\phi_j \sin\theta_j\, b\,\Delta_y + \cos\theta_j\, c\,\Delta_z\right)\end{array}\right) \tag{17.65}$$

$$2\,\mathrm{Re}\left[\frac{\partial v^H}{\partial f_i}\frac{\partial v}{\partial \theta_j}\right] = 2\sum_{a=0}^{A-1}\sum_{b=0}^{B-1}\sum_{c=0}^{C-1} M_i\, M_j\left(\frac{2\pi}{\upsilon}\right)\left(\frac{2\pi f_j}{\upsilon}\right)$$

$$\left(\begin{array}{l} \times \quad \left(\cos\phi_j \cos\theta_j a\Delta_x + \sin\phi_j \cos\theta_j b\Delta_y - \sin\theta_j\, c\,\Delta_z\right) \\ \times \cos\left(\gamma_i - \gamma_j + \frac{2\pi}{\upsilon}\left(\begin{array}{l}\left(f_i \cos\phi_i \sin\theta_i - f_j \cos\phi_j \sin\theta_j\right) a\,\Delta_x \\ +\left(f_i \sin\phi_i \sin\theta_i - f_j \sin\phi_j \sin\theta_j\right) b\,\Delta_y \\ +\left(f_i \cos\theta_i - f_j \cos\theta_j\right) c\,\Delta_z\end{array}\right)\right) \\ \times \left(\cos\phi_i \sin\theta_i\, a\,\Delta_x + \sin\phi_i \sin\theta_i\, b\,\Delta_y + \cos\theta_i\, c\,\Delta_z\right)\end{array}\right) \tag{17.66}$$

$$2\,\mathrm{Re}\left[\frac{\partial v^H}{\partial \theta_i}\frac{\partial v}{\partial f_j}\right]$$

$$= 2\sum_{a=0}^{A-1}\sum_{b=0}^{B-1}\sum_{c=0}^{C-1}\left(\begin{array}{l} M_i\,M_j\left(\frac{2\pi}{\upsilon}\right)\left(\frac{2\pi f_i}{\upsilon}\right)\left(\cos\phi_i\cos\theta_i\,a\Delta_x+\sin\phi_i\cos\theta_i\,b\Delta_y-\sin\theta_i\,c\Delta_z\right) \\ \times\quad\cos\left(\gamma_i-\gamma_j+\frac{2\pi}{\upsilon}\left(\begin{array}{l}\left(f_i\cos\phi_i\sin\theta_i-f_j\cos\phi_j\sin\theta_j\right)a\Delta_x \\ +\left(f_i\sin\phi_i\sin\theta_i-f_j\sin\phi_j\sin\theta_j\right)b\Delta_y \\ +\left(f_i\cos\theta_i-f_j\cos\theta_j\right)c\Delta_z\end{array}\right)\right) \\ \times\quad\left(\cos\phi_j\sin\theta_j a\Delta_x+\sin\phi_j\sin\theta_j b\Delta_y+\cos\theta_j c\Delta_z\right)\end{array}\right)$$

(17.67)

$$2\,\mathrm{Re}\left[\frac{\partial v^H}{\partial \phi_i}\frac{\partial v}{\partial \phi_j}\right]$$

$$= 2\sum_{a=0}^{A-1}\sum_{b=0}^{B-1}\sum_{c=0}^{C-1}\left(\begin{array}{l} M_i\,M_j\left(\frac{2\pi f_i}{\upsilon}\right)\left(\frac{2\pi f_j}{\upsilon}\right)\left(-\sin\phi_i\sin\theta_i\,a\Delta_x+\cos\phi_i\sin\theta_i\,b\Delta_y\right) \\ \times\quad\cos\left(\gamma_i-\gamma_j+\frac{2\pi}{\upsilon}\left(\begin{array}{l}\left(f_i\cos\phi_i\sin\theta_i-f_j\cos\phi_j\sin\theta_j\right)a\Delta_x \\ +\left(f_i\sin\phi_i\sin\theta_i-f_j\sin\phi_j\sin\theta_j\right)b\Delta_y \\ +\left(f_i\cos\theta_i-f_j\cos\theta_j\right)c\Delta_z\end{array}\right)\right) \\ \times\quad\left(-\sin\phi_j\sin\theta_j\,a\Delta_x+\cos\phi_j\sin\theta_j\,b\Delta_y\right)\end{array}\right)$$

(17.68)

$$2\,\mathrm{Re}\left[\frac{\partial v^H}{\partial \phi_i}\frac{\partial v}{\partial \theta_j}\right]$$

$$= 2\sum_{a=0}^{A-1}\sum_{b=0}^{B-1}\sum_{c=0}^{C-1}\left(\begin{array}{l} M_i\,M_j\left(\frac{2\pi f_i}{\upsilon}\right)\left(\frac{2\pi f_j}{\upsilon}\right)\left(-\sin\phi_i\sin\theta_i\,a\Delta_x+\cos\phi_i\sin\theta_i\,b\Delta_y\right) \\ \times\quad\cos\left(\gamma_i-\gamma_j+\frac{2\pi}{\upsilon}\left(\begin{array}{l}\left(f_i\cos\phi_i\sin\theta_i-f_j\cos\phi_j\sin\theta_j\right)a\Delta_x \\ +\left(f_i\sin\phi_i\sin\theta_i-f_j\sin\phi_j\sin\theta_j\right)b\Delta_y \\ +\left(f_i\cos\theta_i-f_j\cos\theta_j\right)c\Delta_z\end{array}\right)\right) \\ \times\quad\left(\cos\phi_j\cos\theta_j\,a\Delta_x+\sin\phi_j\cos\theta_j\,b\Delta_y-\sin\theta_j\,c\Delta_z\right)\end{array}\right)$$

(17.69)

$$
2\,\mathrm{Re}\left[\frac{\partial v^H}{\partial \theta_i}\frac{\partial v}{\partial \phi_j}\right]
= 2\sum_{a=0}^{A-1}\sum_{b=0}^{B-1}\sum_{c=0}^{C-1}
\left(
\begin{array}{l}
M_i\, M_j\left(\dfrac{2\pi f_i}{\upsilon}\right)\left(\dfrac{2\pi f_j}{\upsilon}\right)\left(-\sin\phi_j \sin\theta_j\, a\,\Delta_x + \cos\phi_j \sin\theta_j\, b\,\Delta_y\right) \\
\times \quad \cos\left(\gamma_i - \gamma_j + \dfrac{2\pi}{\upsilon}\left(\begin{array}{l}\left(f_i\cos\phi_i\sin\theta_i - f_j\cos\phi_j\sin\theta_j\right) a\,\Delta_x \\ +\left(f_i\sin\phi_i\sin\theta_i - f_j\sin\phi_j\sin\theta_j\right) b\,\Delta_y \\ +\left(f_i\cos\theta_i - f_j\cos\theta_j\right) c\,\Delta_z\end{array}\right)\right) \\
\times \quad \left(\cos\phi_i\cos\theta_i\, a\,\Delta_x + \sin\phi_i\cos\theta_i\, b\,\Delta_y - \sin\theta_i\, c\,\Delta_z\right)
\end{array}
\right)
\tag{17.70}
$$

$$
2\,\mathrm{Re}\left[\frac{\partial v^H}{\partial \theta_i}\frac{\partial v}{\partial \theta_j}\right]
= 2\sum_{a=0}^{A-1}\sum_{b=0}^{B-1}\sum_{c=0}^{C-1}
\left(
\begin{array}{l}
M_i\, M_j\left(\dfrac{2\pi f_i}{\upsilon}\right)\left(\dfrac{2\pi f_j}{\upsilon}\right) \\
\times \quad \left(\cos\phi_i\cos\theta_i\, a\,\Delta_x + \sin\phi_i\cos\theta_i\, b\,\Delta_y - \sin\theta_i\, c\,\Delta_z\right) \\
\times \quad \left(\cos\phi_j\cos\theta_j\, a\,\Delta_x + \sin\phi_j\cos\theta_j\, b\,\Delta_y - \sin\theta_j\, c\,\Delta_z\right) \\
\times \quad \cos\left(\gamma_i - \gamma_j + \dfrac{2\pi}{\upsilon}\left(\begin{array}{l}\left(f_i\cos\phi_i\sin\theta_i - f_j\cos\phi_j\sin\theta_j\right) a\,\Delta_x \\ +\left(f_i\sin\phi_i\sin\theta_i - f_j\sin\phi_j\sin\theta_j\right) b\,\Delta_y \\ +\left(f_i\cos\theta_i - f_j\cos\theta_j\right) c\,\Delta_z\end{array}\right)\right)
\end{array}
\right)
\tag{17.71}
$$

$$
\mathbf{F}_{ij} = \frac{1}{\sigma^2}\,2\,\mathrm{Re}\left(\frac{\partial v^H}{\partial \varphi_i}\frac{\partial v}{\partial \varphi_j}\right)
\tag{17.72}
$$

$$\mathbf{F}_{ij}=\frac{1}{\sigma^2}2\,\mathrm{Re}\left\{\begin{bmatrix} \frac{\partial v^H}{\partial M_i}\frac{\partial v}{\partial M_j} & \frac{\partial v^H}{\partial M_i}\frac{\partial v}{\partial \gamma_j} & \frac{\partial v^H}{\partial M_i}\frac{\partial v}{\partial f_j} & \frac{\partial v^H}{\partial M_i}\frac{\partial v}{\partial \phi_j} & \frac{\partial v^H}{\partial M_i}\frac{\partial v}{\partial \theta_j} \\ \frac{\partial v^H}{\partial \gamma_i}\frac{\partial v}{\partial M_j} & \frac{\partial v^H}{\partial \gamma_i}\frac{\partial v}{\partial \gamma_j} & \frac{\partial v^H}{\partial \gamma_i}\frac{\partial v}{\partial f_j} & \frac{\partial v^H}{\partial \gamma_i}\frac{\partial v}{\partial \phi_j} & \frac{\partial v^H}{\partial \gamma_i}\frac{\partial v}{\partial \theta_j} \\ \frac{\partial v^H}{\partial f_i}\frac{\partial v}{\partial M_j} & \frac{\partial v^H}{\partial f_i}\frac{\partial v}{\partial \gamma_j} & \frac{\partial v^H}{\partial f_i}\frac{\partial v}{\partial f_j} & \frac{\partial v^H}{\partial f_i}\frac{\partial v}{\partial \phi_j} & \frac{\partial v^H}{\partial f_i}\frac{\partial v}{\partial \theta_j} \\ \frac{\partial v^H}{\partial \phi_i}\frac{\partial v}{\partial M_j} & \frac{\partial v^H}{\partial \phi_i}\frac{\partial v}{\partial \gamma_j} & \frac{\partial v^H}{\partial \phi_i}\frac{\partial v}{\partial f_j} & \frac{\partial v^H}{\partial \phi_i}\frac{\partial v}{\partial \phi_j} & \frac{\partial v^H}{\partial \phi_i}\frac{\partial v}{\partial \theta_j} \\ \frac{\partial v^H}{\partial \theta_i}\frac{\partial v}{\partial M_j} & \frac{\partial v^H}{\partial \theta_i}\frac{\partial v}{\partial \gamma_j} & \frac{\partial v^H}{\partial \theta_i}\frac{\partial v}{\partial f_j} & \frac{\partial v^H}{\partial \theta_i}\frac{\partial v}{\partial \phi_j} & \frac{\partial v^H}{\partial \theta_i}\frac{\partial v}{\partial \theta_j} \end{bmatrix}\right\} \tag{17.73}$$

By using the equations as outlined, the Fisher Information Matrix is formed. It is a 5×5 matrix. The Cramer-Rao bound (CRB) is defined as

$$\mathrm{var}(\hat{\varphi}_i)\ge\left[\mathbf{F}^{-1}(\varphi)\right]_{ii} \tag{17.74}$$

and

$$\mathrm{var}(\hat{\varphi}_i)\ge\mathbf{F}^{ii} \tag{17.75}$$

where $\mathbf{F}^{ii}$ is the ith diagonal element of $\mathbf{F}^{-1}$. The CRB on the variance of the unbiased estimate of the ith parameter φ_i is the ith diagonal element of the inverse of the matrix $\mathbf{F}^{-1}(\varphi)$. So, $\mathrm{var}(\hat{M}_i)$, $\mathrm{var}(\hat{\gamma}_i)$, $\mathrm{var}(\hat{f}_i)$, $\mathrm{var}(\hat{\phi}_i)$, and $\mathrm{var}(\hat{\theta}_i)$ will be the diagonal elements of the inverse of the matrix $\mathbf{F}^{-1}(\varphi)$, respectively.

17.5 EXAMPLE USING ISOTROPIC POINT SOURCES

In this section, illustrative computer simulation results are provided to illustrate the performance of this novel technique. The noise contaminated signal is modeled by (17.36). A three-dimensional array of omni-directional isotropic point sensors are considered in this study. The separation distance between the antenna elements are along the *x*-direction $\Delta_x=0.5m$, along the *y*-direction $\Delta_y=0.5m$, and along the *z*-direction $\Delta_z=0.5m$. The size of the antenna array along the three respective coordinate axes are $a=1,\ldots,A$, $b=1,\ldots,B$ and $c=1,\ldots,C$. It is assumed that there are three signals that are impinging on the array with amplitudes $M_1=M_2=M_3=1$. Numerical examples illustrate the performance of the MP estimator in the presence of white Gaussian noise. The attributes of the signals are given in Table 17.1. The three signals are assumed to have a phase of $\gamma_i=0$ degrees.

Table 17.1. Summary of the Signal Features Incident on the 3-D Antenna Array.

	Signal 1	*Signal 2*	*Signal 3*
Frequency	300 MHz	290 MHz	280 MHz
ϕ	30°	40°	50°
θ	45°	35°	25°

The number of the antenna elements along each of the three axes are, $A = B = C = 10$. The voltages measured at the antenna elements are noisy. The estimated DOA and their associated wavelength will have a bias and a variance due to noise. In the case of noisy data, the estimated values will also be random variables. The stability/accuracy of the results needs to be expressed in terms of its statistical properties, which in this case are the estimated values such as the mean, variance, and so on of the estimate. These results can be obtained with Monte Carlo simulations.

The Cramer-Rao bound (CRB) measures the goodness of an estimator. This bound is the smallest limit for the variance of the estimated values under noisy measurements with white Gaussian noise. The bound is found from using the Fisher Information Matrix, whose diagonal elements are the corresponding CRB of that element. The Fisher Information Matrix and how it relates to the CRB has been shown in section 17.4. The current simulation results show that the variance of the estimators approaches the CRB. The inverse of the sample variance of the estimates of ϕ_1 (azimuth angle), θ_1 (elevation angle), and λ_1 (wavelength) is compared against the corresponding CRB versus signal-to-noise ratio (SNR) of the incoming signals and are plotted in Figures 17.2–17.4.

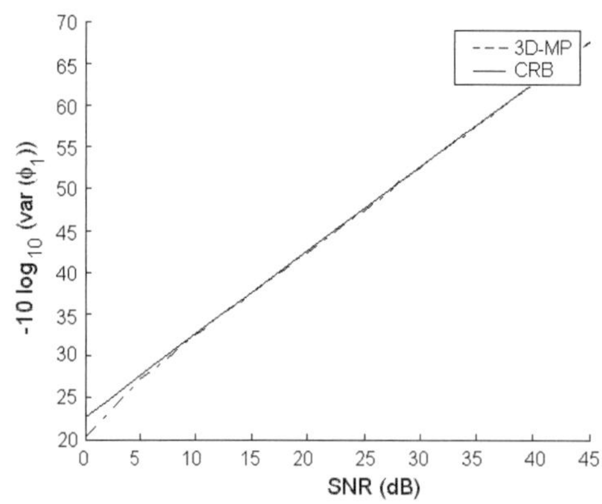

Figure 17.2. The variance $-10\log_{10}\left(\text{var}\left(\phi_1\right)\right)$, 3-D MP and the CRB are plotted against the SNR.

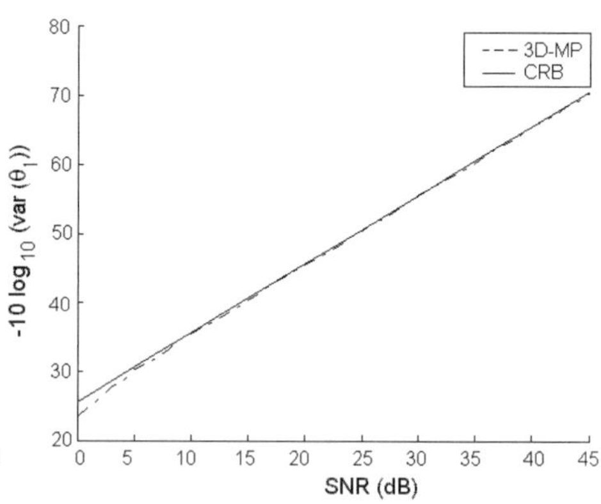

Figure 17.3. The variance $-10\log_{10}\left(\text{var}\left(\theta_1\right)\right)$, 3-D MP and the CRB are plotted against the SNR.

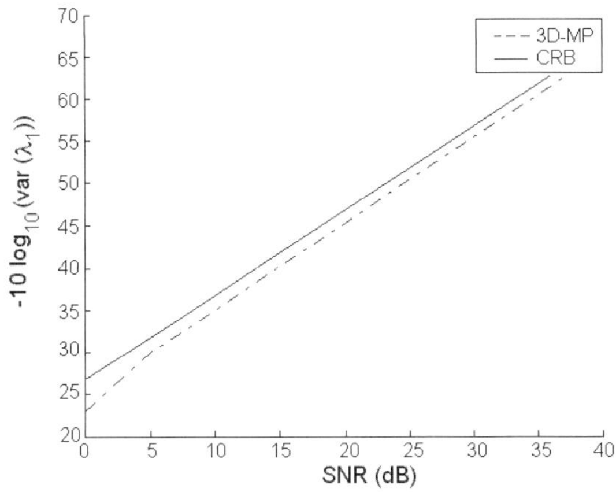

Figure 17.4. The variance $-10\log_{10}\left(\mathrm{var}\left(\lambda_1\right)\right)$, 3-D MP and thc CRB are plotted against the SNR.

Different values of SNR are plotted along the x- axis and the inverse of the variance of the estimated azimuth, elevation angles and wavelength are in logarithmic domain, $-10\log_{10}\left(var(\ldots)\right)$ is shown along the y-axis. The variance of the estimated values of elevation and azimuth angles and the wavelength of the sources plotted against SNR are shown below. The results are based on 1000 Monte Carlo simulations.

The scatter plot of the estimated elevation and azimuth angles are shown in Figures 17.5a–17.5d for different signal-to-noise (SNR) ratios of SNR = 5 dB, SNR = 10 dB, SNR = 15 dB, and SNR = 25 dB. The results are based on 200 Monte Carlo simulations. As it is expected, when the SNR increases, the estimated values approach to its true values in the scatter plot.

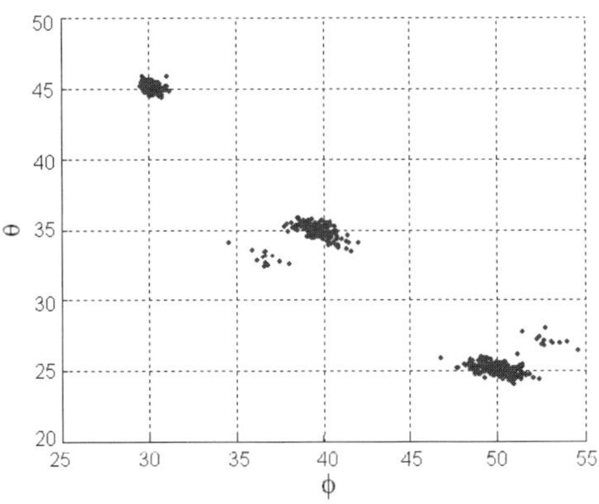

Figure 17.5a. The scatter plot of elevation of azimuth angles, SNR = 5 dB.

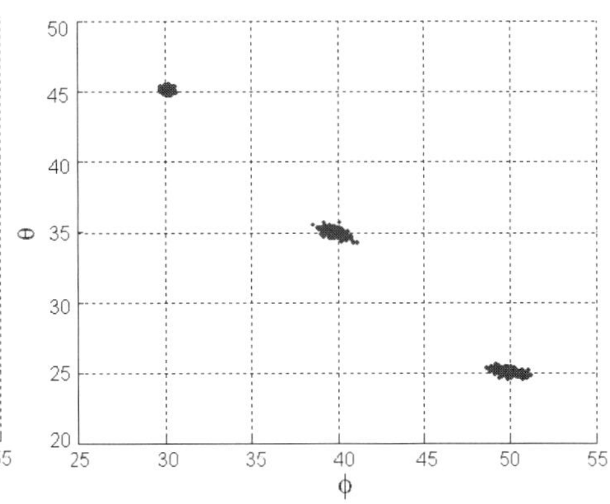

Figure 17.5b. The scatter plot of elevation and azimuth angles, SNR = 10 dB.

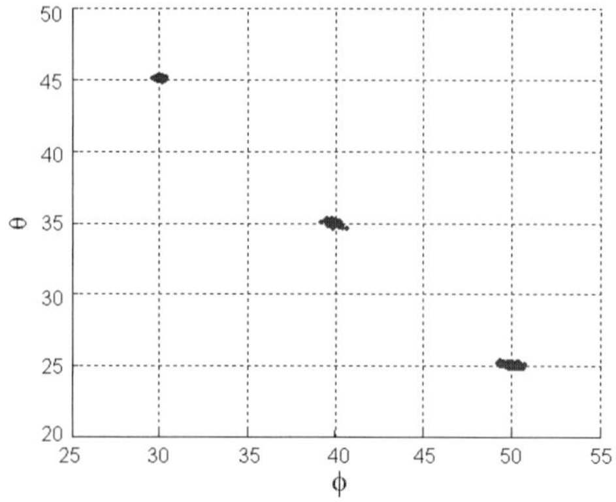

Figure 17.5c. The scatter plot of elevation and azimuth angles, SNR = 15 dB.

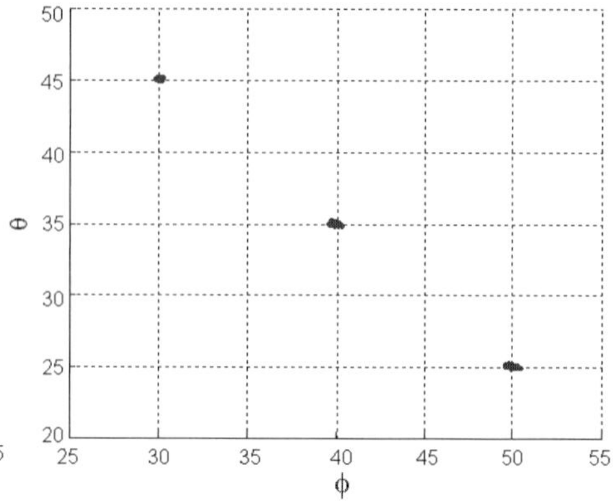

Figure 17.5d. The scatter plot of elevation and azimuth angles, SNR = 25 dB.

The histogram of the estimated elevation and azimuth angles and the wavelength of the sources are shown in Figures 17.6–17.8 for different signal-to-noise ratios of SNR = 5dB, SNR = 10dB, SNR = 15dB, and SNR = 25dB. The results are based on 1000 Monte Carlo simulations. As can be seen, for the increased SNR values, the estimated values approach to its true values in the histogram plot.

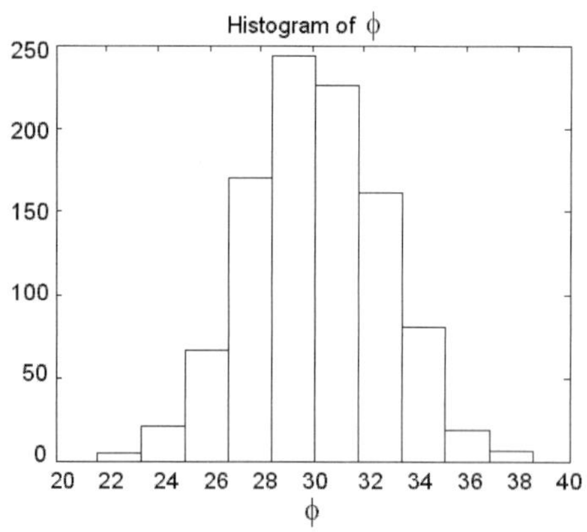

Figure 17.6a. Histogram of azimuth angle, SNR = 5 dB.

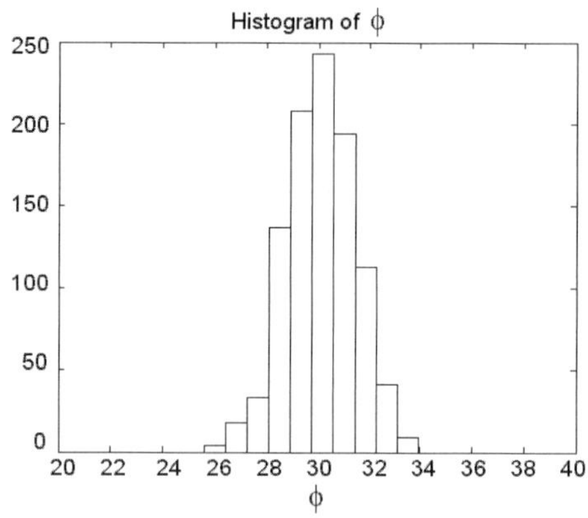

Figure 17.6b. Histogram of azimuth angle, SNR = 10 dB

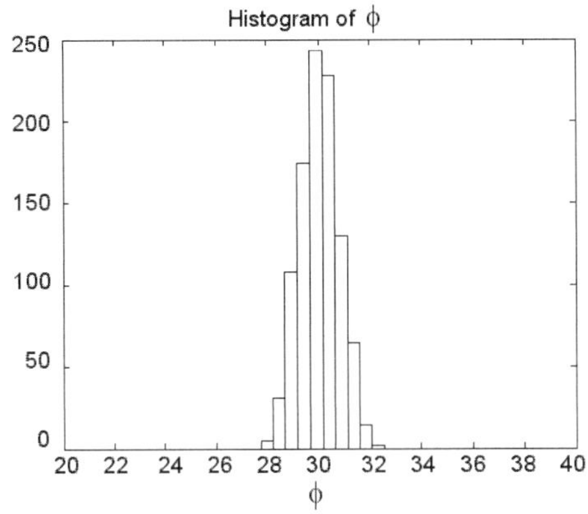

Figure 17.6c. Histogram of azimuth angle, SNR = 15 dB.

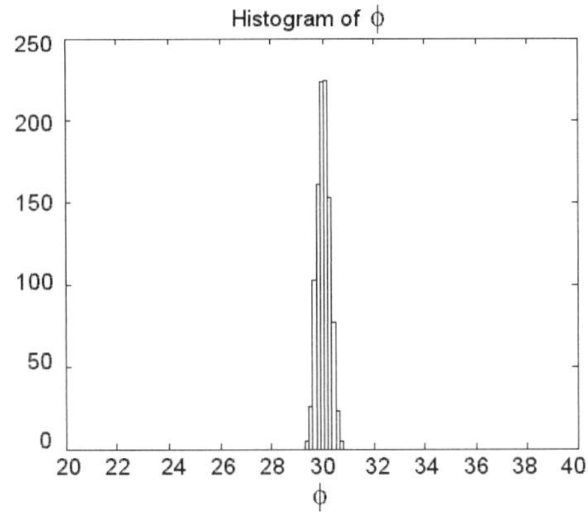

Figure 17.6d. Histogram of azimuth angle, SNR = 25 dB.

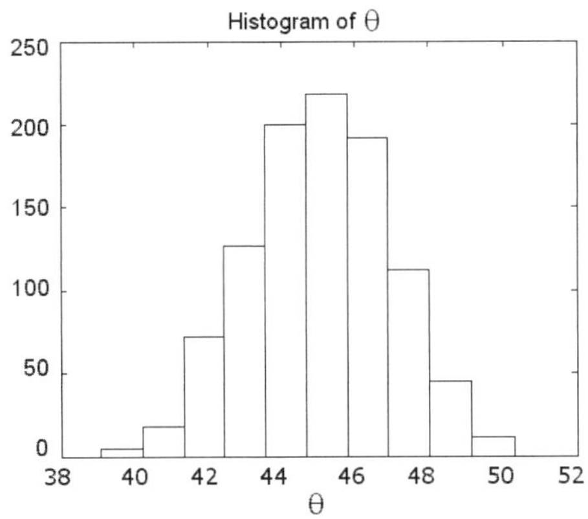

Figure 17.7a. Histogram of elevation angle, SNR = 5 dB.

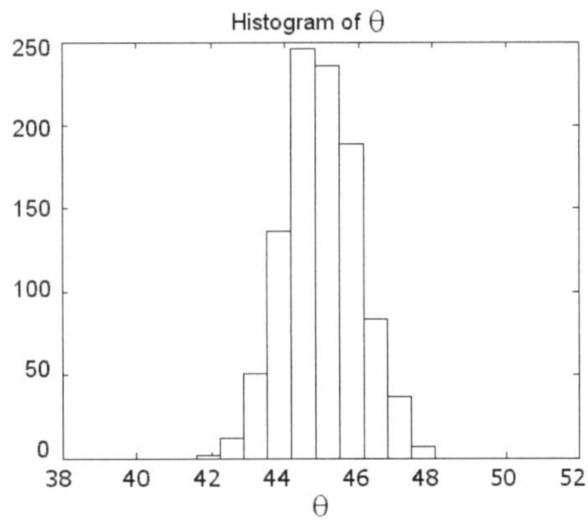

Figure 17.7b. Histogram of elevation angle, SNR = 10 dB.

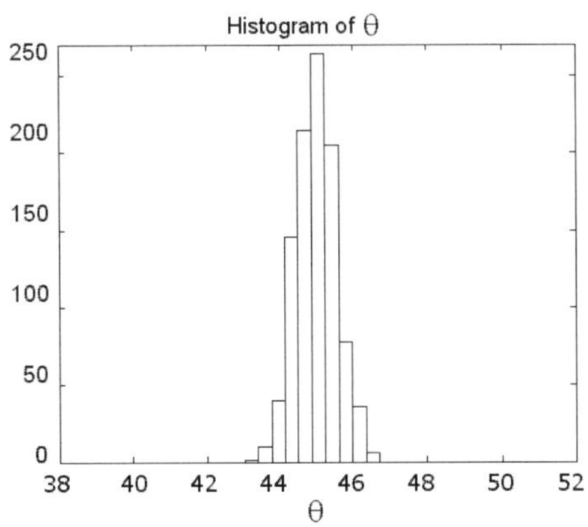

Figure 17.7c. Histogram of elevation angle, SNR = 15 dB.

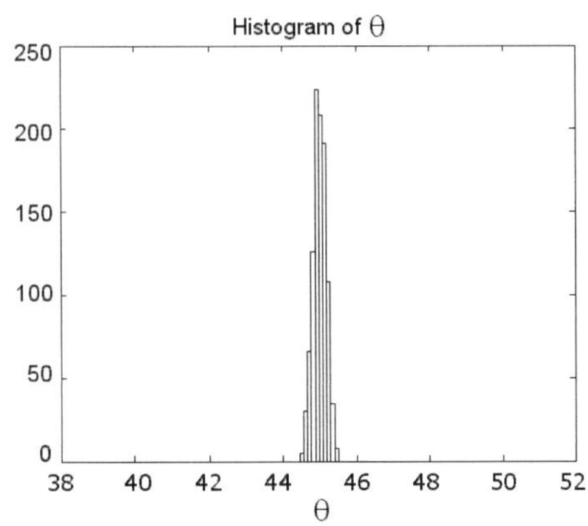

Figure 17.7d. Histogram of elevation angle, SNR = 25 dB.

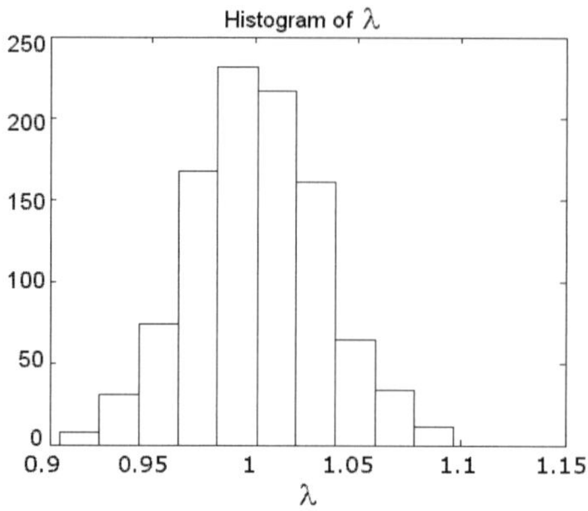

Figure 17.8a. Histogram of wavelength, SNR = 5 dB.

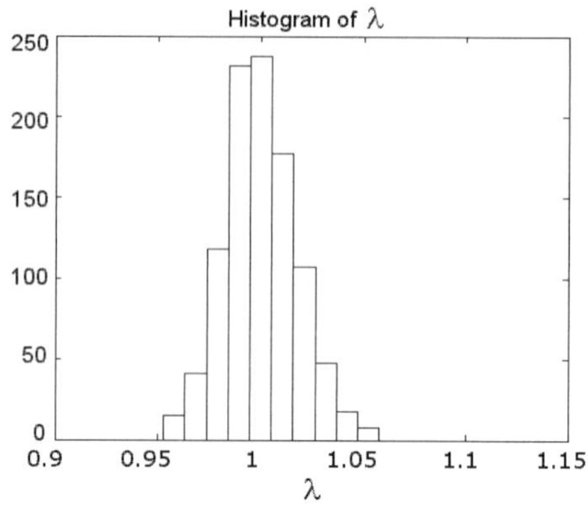

Figure 17.8b. Histogram of wavelength, SNR = 10 dB.

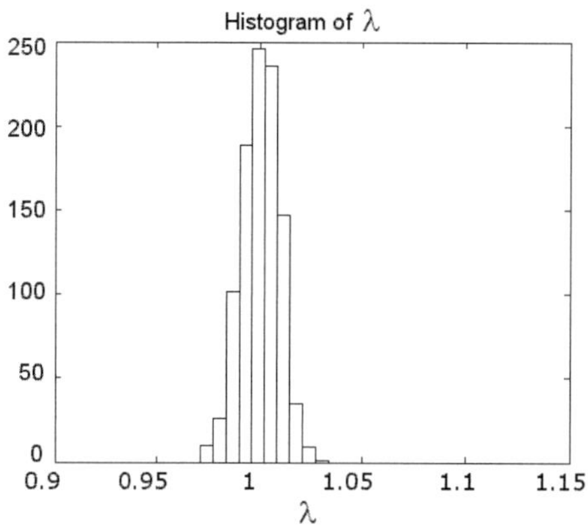

Figure 17.8c. Histogram of wavelength, SNR = 15 dB.

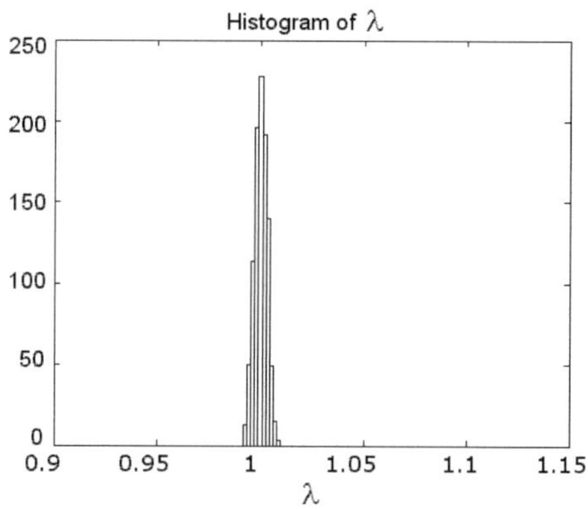

Figure 17.8d. Histogram of wavelength, SNR = 25 dB.

The bias of the estimator has also been studied to observe the efficiency of this new method. For the elevation and azimuth angles and the wavelength, the bias of the estimator has also been computed. The bias is calculated as

$$bias(\phi) = \mathcal{E}\left(\hat{\phi}\right) - \phi \tag{17.76}$$

$$bias(\theta) = \mathcal{E}\left(\hat{\theta}\right) - \theta \tag{17.77}$$

$$bias(\lambda) = \mathcal{E}\left(\hat{\lambda}\right) - \lambda \tag{17.78}$$

where $\mathcal{E}(\bullet)$ denotes the expected value, $\hat{\phi}$, $\hat{\theta}$, and $\hat{\lambda}$ are the estimates of ϕ, θ, and λ, respectively. The bias (in dB) of the estimator for azimuth and elevation angles, and the wavelength versus SNR is shown in Figures 17.9–17.11. For higher values of the SNR, the bias of the estimator decreases, as expected.

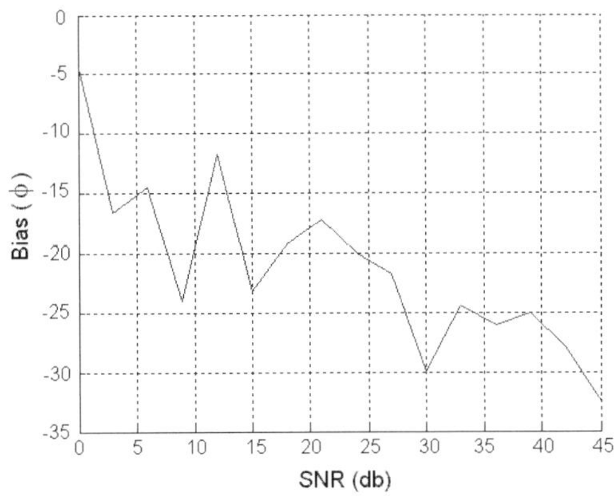

Figure 17.9. Bias (in dB) of the estimator for azimuth angle versus SNR.

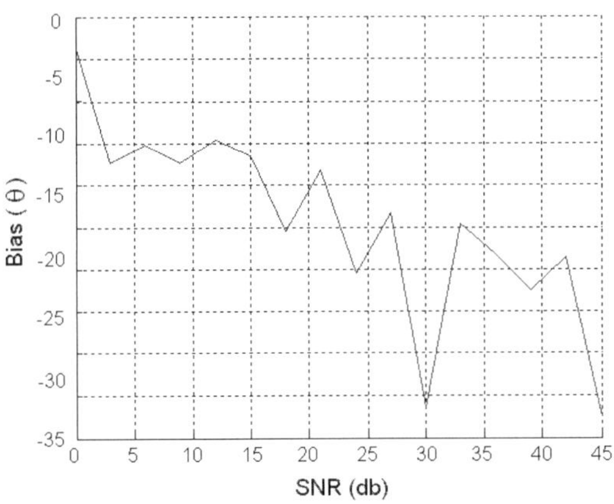

Figure 17.10. Bias (in dB) of the estimator for elevation angle versus SNR.

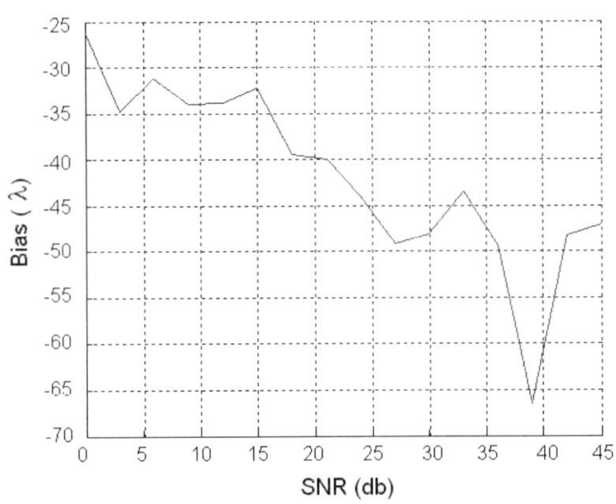

Figure 17.11. Bias (in dB) of the estimator for wavelength versus SNR.

17.6 EXAMPLE USING REALISTIC ANTENNA ELEMENTS

To deal with realistic antenna elements located in an array, we need to transform the voltages that are measured at the terminals of the actual antenna elements in the array into a uniformly spaced virtual array, called ULVA in order to compensate for the effect of mutual coupling between the elements of the array [5,9,10]. In addition this mutual coupling may be frequency dependent and hence this transformation matrix needs to be frequency independent as well. Therefore, we map the actual voltages received at the feed points of the elements due to signals arriving from different directions with different frequency components in an array (not necessarily uniformly spaced but in the presence of mutual coupling between the elements and near-field coupling effects of near field scatterers) into the voltages induced in an omnidirectional array of isotropic elements in the absence of mutual coupling and near field scatterers. This interpolation matrix is built offline, that is, in the preprocessing mode, we only have to multiply the vector that contains the output of the signals from our M sensors by the interpolation matrix to map it in a vector that contains the voltages that would be induced in an array of isotropic elements as we have said. Next, we have to apply a method like the Matrix Pencil (MP) to the preprocessed voltages to estimate the desired parameters of the signals such as the DOA's. In order to build the interpolation matrix we have to do the following steps: we store the voltages induced in the antenna elements of the real array due to incident waves arriving from a particular direction to a matrix called $[A]$. This matrix $[A]$ is generated in this case using numerical simulations using an electromagnetic simulator. On the other hand, we build a matrix $[B]$ that consists of the voltages induced in a uniformly spaced virtual array consisting of omni-directional point radiators placed in a region spanned by the original real array. We generate a transformation matrix $[\Im]$, which will convert the induced voltages in the real array $[A]$ into the voltages of the virtual array $[B]$. This interpolation matrix, valid on a predefined sector, is independent of the angle of arrival of the incident wave and will be used to correct for all the undesired electromagnetic effects. In the outlined interpolation technique we obtain by measurements or by simulations using a numerical electromagnetics code, the voltages induced in the array due to incident waves coming from different directions (matrix $[A]$) at a given frequency. To deal with multiple frequencies within a given band, the induced voltages from the different directions of the sector at different frequencies is used to build a super matrix $[A]$. In the same way, we obtain the induced voltages induced in the omni-directional virtual array at different frequencies and we store them in a super matrix $[B]$. The new super matrices $[A]$ and $[B]$ are obtained by stacking the matrices at different frequencies. It is assumed that the incoming signals can have any frequency within the assumed band and not necessarily coinciding with one of the chosen frequencies to build the super matrices. The method is mathematically expressed as:

$$[B_1|B_2|......|B_N]_{V\times(D.N)} = [\Im]_{V\times R}\,[A_1|A_2|......|A_N]_{R\times(D.N)} \tag{17.79}$$

where $[A_i]$ is the matrix of voltages induced at the real antenna elements at the i-th frequency and $[B_i]$ is the voltage induced in the uniformly spaced virtual array consisting of isotropic omnidirectional point sources radiating in free space. V represents the number of virtual antenna elements, R stands for the number of real antenna elements, D implies the number of directions that we consider in a given sector in which the calibration has been made, and N is the number of different frequencies used. Hence, objective in this methodology is to estimate any signal arriving at the array in this specified frequency range. Equation (17.79) is now solved using the total least squares method by using the singular value decomposition:

$$[\,\Im\,] = [B][A]^H \left\{[B][A]^H\right\}^{-1} \tag{17.80}$$

We note that we use only one snapshot of the data. The block diagram of the method is depicted in the Figure 17.12 where we can see how the interpolation matrix $[\Im]$ is built. For a given snapshot of the voltages at a particular instance of time t, given by the vector a (see Fig. 17.12), and after multiplying a by the interpolation matrix $[\Im]$, one obtains the voltages that will be induced in an ideal array consisting of isotropic elements radiating in free space through the vector b. We apply the three dimensional MP technique to estimate the frequency and the DOA.

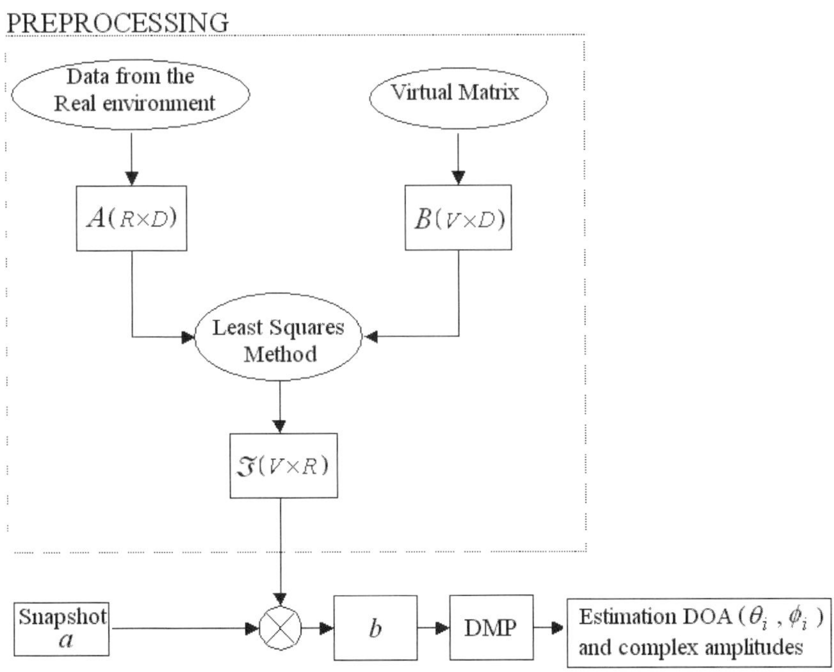

Figure 17.12. Block diagram for preprocessing to deal with mutual coupling.

When using real antenna elements, there will be some electromagnetic (EM) effects as opposed to using the isotropic point sensors. These EM effects in the antenna arrays are simulated by using very efficient, and accurate EM solvers. In this example, an array of 5 × 5 × 5 dipoles oriented along the *x*, *y*, and *z* direction is considered. The geometry of the array is depicted in the Figure 17.13. Each element of the array is a dipole of length 0.475 λ designed at the central frequency f_c = 300 MHz (λ = 1 m). There are 5 elements along the *X*-axis, 5 elements along the *Y*-axis, and 5 elements along the *Z*-axis. Each dipole antenna has a radius of 0.001λ.

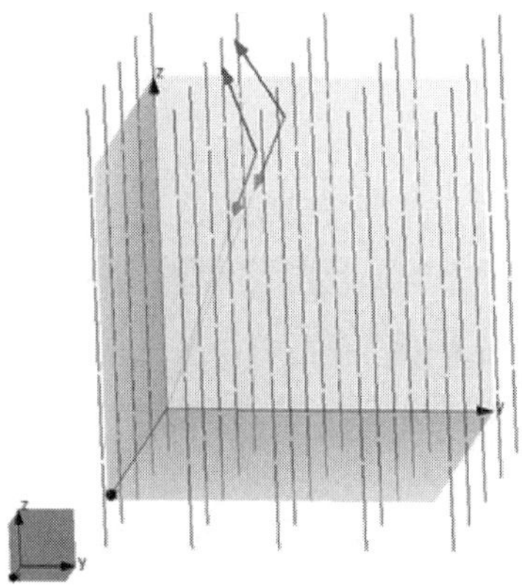

Figure 17.13. The geometry of the 5 × 5 × 5 dipole antenna array.

We consider two different signals arriving at the array from different directions and with different operating frequencies. In the Table 17.2 the parameters of these signals are shown.

Table 17.2. Signal Features Coming to the 3-D Antenna Array.

	φ	θ	Frequency (MHz)
Signal 1	33^0	50^0	298
Signal 2	44^0	55^0	303

In this array configuration, we are going to show the response of the proposed technique for the two signals that arrive at (ϕ,θ) = (33°, 50°), and (ϕ,θ) = (44°, 55°), the frequency of the signals are from 298 MHz and 303 MHz. The geometry of the antenna arrays are shown from the different planes and angles are shown in detail in Figures 17.15 and 17.16.

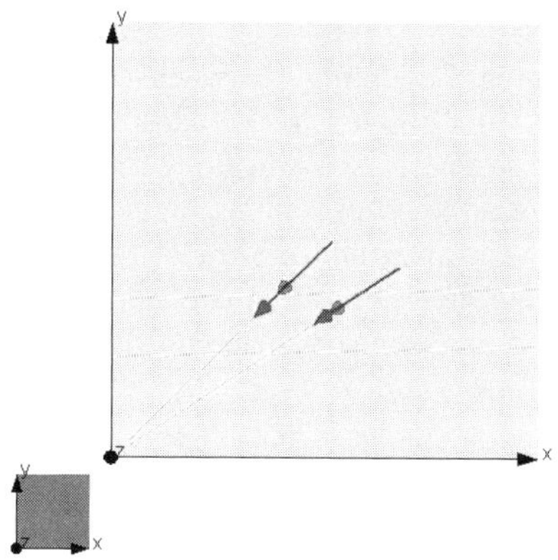

Figure 17.14. XY view of the geometry of the 5 × 5 × 5 dipole antenna array.

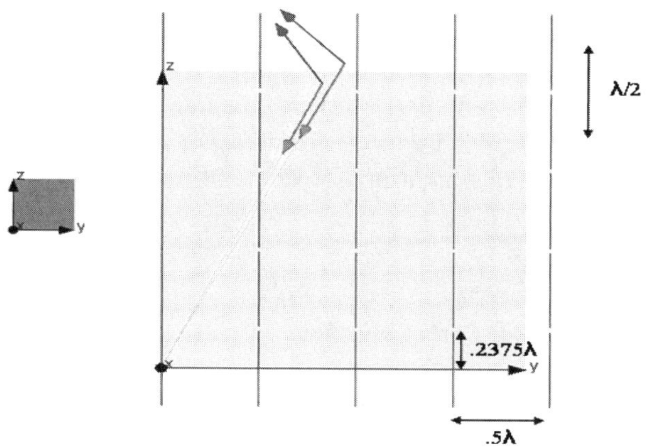

Figure 17.15. ZY view of the geometry of the 5 × 5 × 5 dipole antenna array.

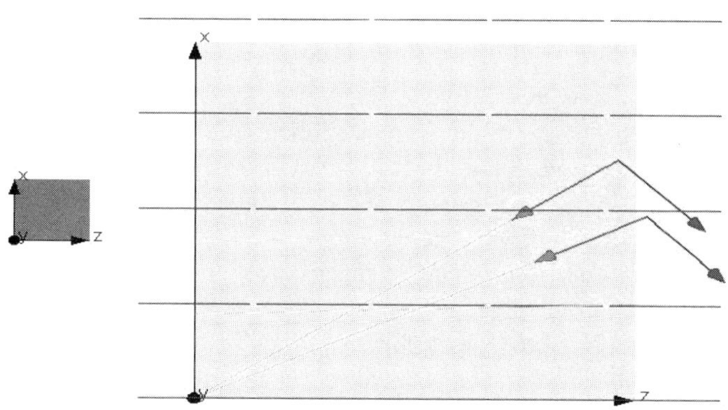

Figure 17.16. XZ view of the geometry of the 5 × 5 × 5 dipole antenna array.

In order to compensate for the undesired electromagnetic effects displayed through mutual coupling between the antenna elements, the transformation matrix is first applied to the data before processing it using the Matrix Pencil method. The matrices A and B are created as defined earlier in the section. For different azimuth and elevation angles and for each frequency [10], the supermatrix matrix is formed and the transformation matrix is obtained by solving (17.79). The transformation matrix is generated as follows: first the span of elevation and azimuth angles is defined. In this simulation the sector of the angles are $(\phi_q, \phi_{qq}) = (30°, 60°)$ and $(\theta_q, \theta_{qq}) = (30°, 60°)$ for azimuth and elevation, respectively, with an angular step size of $\Delta_\phi = \Delta_\theta = 1°$. The incoming signals may arrive from any direction within this sector. The polarization of the incoming waves is E_θ. The compensation is done for each frequency located within the band. That is how we form the supermatrix. In the simulation, the band of interest is from 295 MHz to 305 MHz. Reducing the step size and frequency range will increase the accuracy, but will also increase the computational complexity which implies that if more frequency samples are available to calibrate the array then the error will be less.

We now apply the broadband technique over the sector $(\phi_q, \phi_{qq}) = (30°, 60°)$ and $(\theta_q, \theta_{qq}) = (30°, 60°)$, in constructing the transformation matrix $[\Im]$. The estimated values of the DOA and the frequency of these signals are shown in Table 17.3. The interpolation error is 86.54, and the condition number of the transformation matrix is 2.23×10^{11}.

Table 17.3. Estimated Features of Signal Impinging on the 3-D Antenna Array, SNR = 20 dB, 2000 Monte Carlo Simulation.

	φ	θ	Frequency (MHz)
Signal 1	33.45°	50.24°	298.20
Signal 2	44.42°	54.88°	302.74

For the next example, consider an array of 21 by 2 dipoles oriented along the x- and z-axis, respectively. The geometry of the array is depicted in Figure 17.17. Each element of the array is a dipole of length 0.475λ. There are 21 elements along x-axis separated by 0.5λ. In addition, there are 2 elements along z-axis separated also by 0.5λ. The dipoles are designed at the central frequency f_c = 300 MHz (λ_c = 1m).

To apply the broadband compensation technique for the mutual coupling between the antennas, we consider the sector $(\phi_q, \phi_{qq}) = (60°, 120°)$ and between $(\theta_q, \theta_{qq}) = (70°, 110°)$ and sample it with an angular step size of $\Delta_\phi = \Delta_\theta = 1°$. The polarization of the incoming waves is E_θ. We consider four different signals arriving at the array from different directions and with different frequencies. In Table 17.4 the parameters of the signal are numerated.

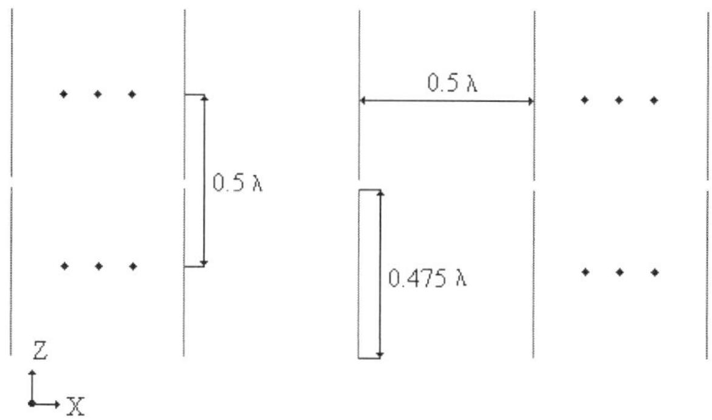

Figure 17.17. Geometry of the dipole antenna array.

Table 17.4. Parameters of the Four Signals Arriving at the 21 × 2 Dipole Array.

	φ	θ	Frequency (MHz)
Signal 1	70°	85°	280
Signal 2	85°	105°	290
Signal 3	100°	90°	305
Signal 4	110°	80°	310

Let us define N as the number of frequencies used to build the matrices $[A]$ and $[B]$ to form the composite matrix of (17.79). Let us also define the bandwidth of the interpolation as:

$$BW(\%) = \frac{\left(f_{final} - f_{initial}\right)}{f_c} \times 100 \qquad (17.81)$$

where $f_{initial}$ and f_{final} define the frequency of the first and the last point in the simulation, respectively, and f_c is the center frequency. We are going to show how the accuracy of the estimation of the signals varies with N and with the bandwidth $BW(\%)$. For each case we fix a bandwidth and we vary N from 4 to 9. In Table 17.5 we present the different cases considered in this simulation. As we can see in the Figures 17.18 and 17.19, the interpolation error and the error in estimating the DOA decrease when N is increased. This implies that it is much easier to interpolate if we have more frequency points. Following this line of thought, we then conclude that for a fixed N, the wider the bandwidth of the interpolation it is more difficult to interpolate and so there will be larger interpolation error and still bigger error in the estimation of the DOA.

Table 17.5. Parameters of the Simulations Over the 21 × 2 Dipole Array.

Case	$f_{initial}$ (MHz)	f_{final} (MHz)	BW (%)	N
1	280	320	13.33	4, 5, 6, 7, 8, 9
2	275	325	16.67	4, 5, 6, 7, 8, 9
3	270	330	20.00	4, 5, 6, 7, 8, 9
4	265	335	23.33	4, 5, 6, 7, 8, 9
5	260	340	26.67	4, 5, 6, 7, 8, 9

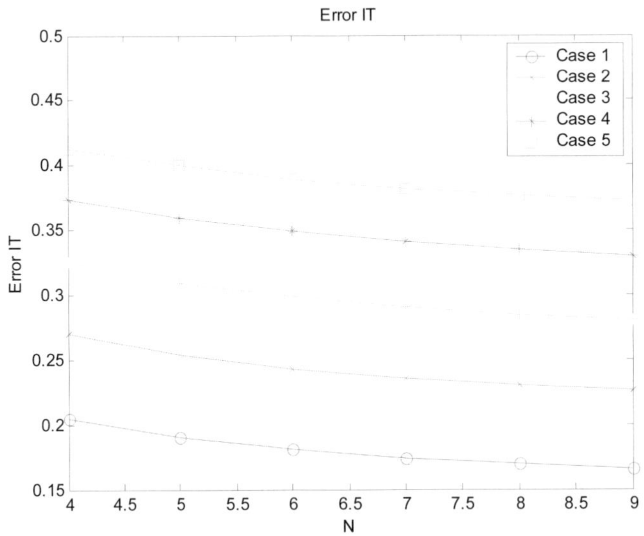

Figure 17.18. Interpolation error for the 21 × 2 dipole array with the four signals of Table 17.4.

For the final example, consider now an array of 6 by 3 horn antennas located conformal to a cylindrical surface oriented along the z-axis (6 antennas along the azimuth and 3 antennas along the z-axis as shown in Figures 17.20 and 17.21). The radius of the cylinder is 800 mm. The geometry of this array is depicted in the Figures 17.20 and 17.21. The center frequency of the array is f_c = 2.3 GHz. The sector under consideration for constructing the transformation matrix is defined by (ϕ_q, ϕ_{qq}) = (70°, 110°) and (θ_q, θ_{qq}) = (75°, 105°) and this sector is sampled with an angular step size of $\Delta_\phi = \Delta_\theta = 1°$. The polarization of the incoming waves is E_θ.

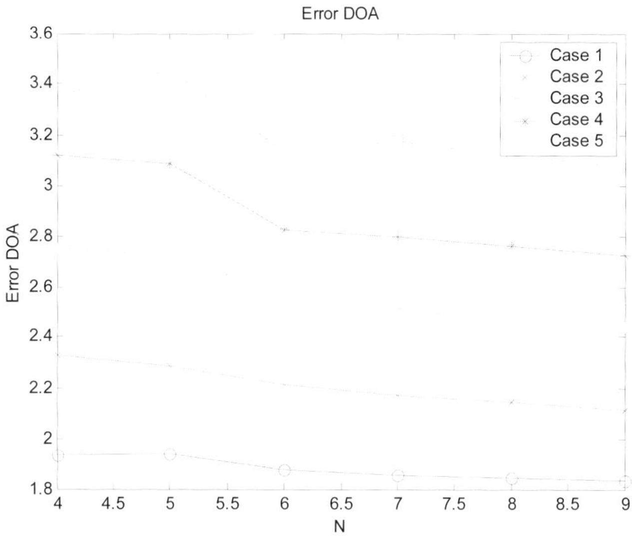

Figure 17.19. DOA error for the 21 × 2 dipole array with the four signals of Table 17.5.

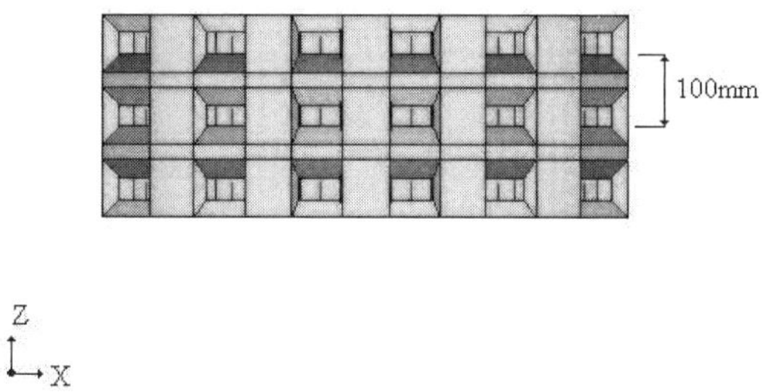

Figure 17.20. XZ view of the geometry of the 6 × 3 horn antenna array.

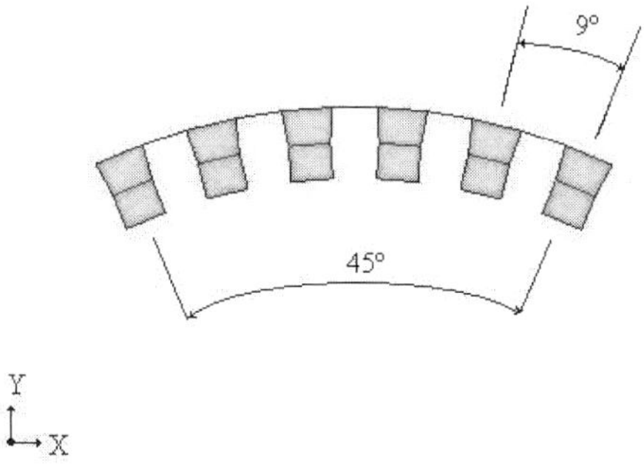

Figure 17.21. XY view of the geometry of the 6 × 3 horn antenna array.

Let us consider two signals arriving at the array with following parameters defined in Table 17.6.

Table 17.6. Parameters of the Two Signals Arriving at the 6 × 3 Conformal Horn Array.

	φ	θ	Frequency (GHz)
Signal 1	80°	85°	2.4
Signal 2	95°	100°	2.2

Using the broadband interpolation technique described earlier in the section, we choose 7 different values for N as defined in Table 17.7, along with the other relevant parameters. The interpolation error is shown in Figure 17.22 for this compensation and the mean squared error in the estimation of the DOA is plotted in Figure 17.23.

Table 17.7. Parameters of the Simulations Over the 6 × 3 Horn Antenna Array.

Case	$f_{initial}$ (GHz)	f_{final} (GHz)	BW (%)	N
1	2.15	2.45	13.04	3, 4, 5, 6, 7, 8, 9
2	2.10	2.50	17.39	3, 4, 5, 6, 7, 8, 9
3	2.07	2.53	20.00	3, 4, 5, 6, 7, 8, 9

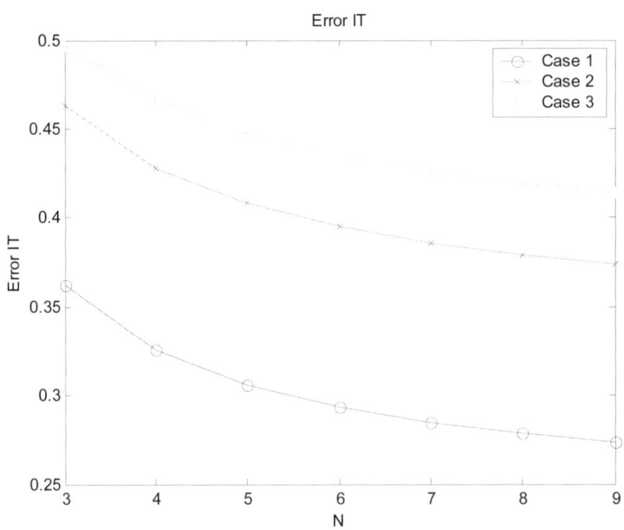

Figure 17.22. Interpolation error for the 6 × 3 horn array with the two signals of Table 17.6.

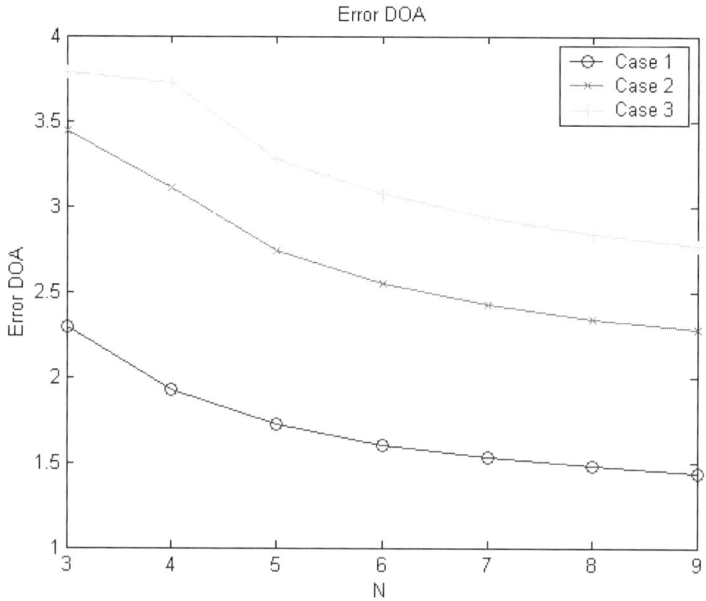

Figure 17.23. DOA error for the 6 × 3 horn array with the two signals of Table 17.6.

17.7 CONCLUSION

A new technique to estimate DOA when signals have different frequencies has been presented. In this presentation it has been shown that using a three-dimensional array, or equivalently three orthogonal planar arrays, it is possible to extract all the unknown parameters of the unknown signal in the presence of noise. In this study, a new methodology has been presented to extract the 3-D frequencies from the sampled values of the voltages induced in an array. This 3-D estimation method can easily be applied to a generalized DOA estimation problem, where one can estimate not only the azimuth and elevation angles of the various signals impinging on the array but also their wavelengths of operation. It has been shown how to pair the respective poles to obtain the corresponding angle of arrivals and the wavelengths. Numerical examples are provided to illustrate the validity of the new method. The statistical performance parameters such as variance of the estimator have also been compared with the Cramer-Rao bound to observe the accuracy of the new method.

REFERENCES

[1] Y. Hua and T. K. Sarkar, "Matrix Pencil Method for Estimating Parameters of Exponentially Damped/Undamped Sinusoids in Noise", *IEEE Transactions on Acoustics, Speech and Signal Processing*, Vol. 38, No. 5, May 1990.

[2] Y. Li, J. Razavilar, and K. J. Liu, "A High Resolution Technique for Multidimensional NMR Spectroscopy", *IEEE Transactions on Biomedical Engineering*, Vol. 45, No. 1, pp. 78-86, January 1998.

[3] A. J. van der Veen, P. Ober, and E. F. Deprettere, "Azimuth and Elevation Computation in High Resolution DOA Estimation", *IEEE Transactions on Signal Processing*, Vol. 40, No. 7, pp. 1828-1832, July 1992.

[4] K. Takao and N. Jijuma, "An Adaptive Array Utilizing an Adaptive Spatial Averaging Technique for Multi-path Environments", *IEEE Transactions on Antennas and Propagation*, Vol. 35, No. 12, pp. 1389-1396, 1987.

[5] T. K. Sarkar, M. C. Wicks, M. Salazar-Palma and R. J. Bonneau, *Smart Antennas*, John Wiley & Sons, Hoboken, NJ, 2003.

[6] N. Yilmazer, J. Koh and T. K. Sarkar, "Utilization of a Unitary Transform for Efficient Computation in the Matrix Pencil Method to Find the Direction of Arrival", *IEEE Transactions on Antennas and Propagation*, Vol. 54, No. 1, pp.175 –181, Jan. 2006.

[7] Y. Hua, "Estimating Two-Dimensional Frequencies by Matrix Enhancement and Matrix Pencil", *IEEE Transactions on Signal Processing*, Vol. 40, No. 9, Sep. 1992.

[8] S. Burintramart and T. K. Sarkar "Estimation of the Two-Dimensional Direction of Arrival by the Diagonal Matrix Pencil Method", *The Applied Computational Electromagnetics Society Conference* (ACES Conference), Syracuse, NY, April 2004.

[9] K. Kim, T. K., Sarkar, and M. Salazar-Palma, "Adaptive Processing Using a Single Snapshot for a Nonuniformly Spaced Array in the Presence of Mutual Coupling and Near-Field Scatterers", *IEEE Transactions on Antennas and Propagation*, Vol. 50, No. 5, pp. 582-590, May 2002.

[10] R. S. Adve and T. K. Sarkar, "Compensation for the Effects of Mutual Coupling on Direct Data Domain Adaptive Algorithms", *IEEE Transactions on Antennas and Propagation*, Vol. 48, No. 1, pp. 86-94, Jan. 2000.

18

ADAPTIVE PROCESSING OF BROADBAND SIGNALS

18.0 SUMMARY

In this chapter we present a novel algorithm for adaptive processing of signals with different operating frequencies and bandwidths. This methodology is a direct extension of the direct data domain least squares method described earlier in chapter 6, where it was applied to deal with narrowband signals operating at the same frequency. Narrowband adaptive algorithms cannot handle signals operating at different frequencies. This chapter presents an application of the direct data domain least squares method (D3LS) to deal with signals of finite bandwidths operating at different frequencies. The goal is to extract the signal of interest in the presence of broadband interferers. Numerical examples are presented to illustrate the application of this novel methodology limited to the extraction of narrowband signals in the presence of broadband interferers.

18.1 INTRODUCTION

In conventional adaptive algorithms one cannot process signals very easily which has a finite bandwidth. So, a new adaptive technique is needed to adaptively enhance signals in the presence of jammers and interferers which may be operating at different unknown frequencies over a finite bandwidth. The proposed algorithm in this section can estimate the complex amplitude of the Signal of Interest (SOI) in the presence of interferers operating at different frequencies with a finite bandwidth which can be quite large. Some earlier work has been done on similar problems [1–3]. However, the procedure is quite cumbersome as the antenna array pattern has to be generated at different frequencies. The new methodology presented in this chapter is based on the space-time adaptive processing using the D3LS approach [4–7]. The technique is simple to describe, quite accurate and even easy to implement on a signal processing chip.

Conventional adaptive processing or space-time adaptive algorithms are based on statistical approaches which estimate the covariance matrix of the various interferences. The covariance matrix is estimated by utilizing secondary data obtained from multiple range cells assumed to be "signal free," and considered to be independent identically distributed, and thus representative of the interference in the range cell under test for extracting the SOI. The formation of the covariance matrix is quite time consuming and so is the evaluation of its inverse. Also, as one requires multiple snapshots of the data to generate an estimate of the covariance matrix, it is generally assumed that the environment remains stationary during the data collection process. In a highly transient or dynamic environment, this stationarity assumption of the signals is no longer valid, and this leads to poor estimation for the covariance matrix.

Recently, a D3LS algorithm and an associated Space-Time Adaptive Processing (STAP) algorithm have been proposed to overcome these problems. A D3LS approach has certain advantages related to the computational issues associated with the adaptive array processing problem. Specifically, it adaptively analyzes the data at each snapshot as opposed to forming a covariance matrix of the data from multiple snapshots, and then solving for the weights utilizing that information. However, the D3LS approaches were developed using a narrowband assumption of the signals. In those methodologies, the SOI and interference has to have the same frequency of operation. So, the usual D3LS approach presented in section 6.3 cannot handle the case that the SOI and interference that has different frequencies and operate over a finite bandwidth.

In this chapter, a new algorithm is described based on the D3LS algorithms to handle signals operating at different frequencies having a finite operating bandwidth. In section 18.2 we describe this novel methodology. Section 18.3 presents some simulated numerical examples followed by conclusions in section 18.4

18.2 FORMULATION OF A DIRECT DATA DOMAIN LEAST SQUARES METHOD FOR ADAPTIVE PROCESSING OF FINITE BANDWIDTH SIGNALS HAVING DIFFERENT FREQUENCIES

In this section we describe the novel D3LS for extracting a SOI in the presence of finite bandwidth interferers operating at different frequencies. We develop three different techniques of achieving this goal and follow along the methodology described earlier in chapter 6. Here, we address the radar problem, where we know the direction of arrival of the signal and its operating frequency.

18.2.1 Forward Method for Adaptive Processing of Broadband Signals

Consider an array composed of N antenna elements consisting of isotropic omni-directional point radiators separated by a distance d as shown in Figure 9.1. In narrowband adaptive processing, the operating frequency of the SOI and the interferers are the same. Here, that assumption is not valid. Let us define $S(m)$ to be the complex voltage received at the mth antenna element at a particular instance of time, representing a snapshot of the voltages across the entire array. We further stipulate that the voltage $S(m)$ is due to a signal of unity amplitude that is incident on the array from a known azimuth angle θ_s. Hence, the signal-induced voltage under the assumed array geometry and a narrow band signal assumption will result in a complex sinusoidal signal given by

$$S(m) = \exp\left\{ j 2\pi f_s \frac{d}{\upsilon}(m-1)\cos\theta_s \right\} \tag{18.1}$$

where θ_s and f_s is the angle of arrival and frequency of operation of the SOI. Here, υ is the velocity of light.

Let $X(m)$ be the actual measured complex voltages in the antenna elements at a particular instance of time, constituting a snapshot of the data voltages. The actual sampled voltages $X(m)$ will contain the SOI of amplitude α (α is a complex quantity), which is unknown and needs to be determined along with all the other coherent or noncoherent undesired interferences. There is also a contribution to the measured voltages from the thermal noise generated at the receiver. Hence the actual measured voltages $X(m)$ will be composed of

$$X(m) = \alpha \ \exp\left\{ j 2\pi f_s \frac{d}{\upsilon}(m-1)\cos\theta_s \right\} + \text{ Interference} + \text{Thermal noise} \tag{18.2}$$

Here, we have assumed that all the signals are operating at the same frequency. However if the frequency of the SOI and the interferers are different, the above equations cannot be used in an adaptive algorithm to estimate the complex amplitude of the SOI. So, Eq. (18.1) has to be reformulated to handle the multiple narrowband signals with different frequencies. To solve this problem we need to consider temporal sampling. Then the data at the mth antenna element and for the nth time sample can be represented as

$$S(m,n) = \exp\left\{ j 2\pi f_s \frac{d}{\upsilon}(m-1)\cos\theta_s \right\} \times \exp\left\{ j 2\pi f_s (n-1) T \right\} \tag{18.3}$$

where T is the temporal sampling period. Now (18.3) can be rewritten in the following form as

$$S(m,n) = \exp\left\{ j 2\pi f_s \ \frac{d}{\upsilon} \ (m-1) \ \cos\theta_s + j 2\pi (n-1) \ \frac{f_s}{f_r} \right\} \tag{18.4}$$

where $f_r = 1/T$ and is the sampling frequency. Using this line of reasoning, the actual complex voltages measured at the mth antenna element and for the nth time instance can be represented as

$$X(m,n) = \alpha \exp\left\{ j2\pi f_s \frac{d}{\upsilon}(m-1)\cos\theta_s + j2\pi(n-1)\frac{f_s}{f_r} \right\} + \text{Interference} + \text{Noise} \tag{18.5}$$

Using this two-dimensional data set consisting of the space and time samples, one can employ a two-dimensional filtering in space based on the angle of arrival and on time related to the operating frequency of the SOI. Therefore, one can employ a two-dimensional D3LS method for estimating the complex amplitude of the SOI for a given angle of arrival and a specified operating frequency out of the out of band interferers and clutter.

This can be accomplished by focusing our attention to the following difference equation given by

$$X(m,n) - \alpha S(m,n) \tag{18.6}$$

In (18.6), the differencing eliminates the SOI from the measured voltages at the mth antenna element and for the nth time sample, leaving all the undesired signals including noise plus interference. It is important to note that the complex amplitude of the SOI given by α is still an unknown quantity and has to be determined. Based on (18.4) and (18.6), a two-dimensional matrix pencil can be generated whose solution will result in a weight vector which will null out the interferers and extract the SOI as described in chapter 6. The appropriate matrices of this matrix pencil can be constructed by sliding a window (box) over the 2-D data, as shown by the shaded plane in Fig. 6.5. By creating a vector using the elements in the window, each window position generates a row in the matrix S and X as shown in chapter 6.

The window size along the element dimension is N_a, and N_t along the pulse dimension. Selection of N_a determines the number of spatial degrees of freedom, while N_t determines the temporal degrees of freedom. Typically for adaptive processing, N_a and N_t must satisfy the following equations

$$N_a \leq \frac{N+1}{2} \tag{18.7}$$

$$N_t \leq \frac{M+1}{2} \tag{18.8}$$

Therefore, since the terms $X(m,n)-\alpha\, S(m,n)$ eliminate the SOI, then these elements represent the contribution due to the unwanted signal multipaths, jammers, unwanted signals, and receiver thermal noise. In the D3LS adaptive processing, the goal is to take a weighted sum of these matrix elements defined in (18.6) and extract the SOI which is going to be α. So, the total number of degrees of freedom, Q, for the D3LS method is

$$Q = N_a \times N_t \tag{18.9}$$

Next, it is illustrated how a D3LS approach is taken for the extraction of SOI. The least squares procedure for the 2-D case is available in [7] and has been described in chapter 6. In real-time applications, it is difficult to numerically solve for the generalized eigenvalue problem in an efficient and stable way, particularly if the value $N_a N_t$ representing the total number of weights is large and the matrix is highly rank deficient. For this reason, we convert the solution of a generalized eigenvalue problem to the solution of a linear matrix equation.

Let the element to element off set of the SOI in space and time, respectively, as

$$Z_1 = \exp\left\{ j2\pi f_s \frac{d}{\upsilon} \cos(\theta_s) \right\} \tag{18.10}$$

$$Z_2 = \exp\left(j2\pi \frac{f_s}{f_r} \right) \tag{18.11}$$

It is important to note that the angle of arrival of θ_s and the operating frequency f_s for the SOI is known. As has been presented earlier, three types of difference equations can then be generated as given by

$$X(m,n) - X(m+1,n)\, Z_1^{-1} \tag{18.12}$$

$$X(m,n) - X(m,n+1)\, Z_2^{-1} \tag{18.13}$$

$$X(m,n) - X(m+1,n+1)\, Z_1^{-1} Z_2^{-1} \tag{18.14}$$

Note that in (18.12), the signal component (SOI) is canceled from the samples taken from different antenna elements at the same time. Similarly (18.13) represents signal cancellation from samples taken at the same antenna elements at different time instances. Finally, (18.14) represent signal cancellation from neighboring samples in both space and time. Therefore, we are performing a filtering operation simultaneously using $N_a N_t$ samples of the space time data. The cancellation rows of the matrix $[F]$ can now be formed using (18.12)–

(18.14). In this case, the dots in Fig. 6.5 represent the induced voltages, $X(m,n)$ as defined in (18.5). The rows are formed by performing an element-by-element subtraction between the elements of the windows and then arranging the resulting data into a row vector. The window is then slid one space to the right and 3 more rows are generated, and so on. After this window has reached the second column to the far right and 3 rows are generated, the window is lowered a row and shifted back to the left side of the data array, and the generation of rows continues. The elements of this row can be obtained by placing a $N_a \times N_t$ window, such as window #1 as shown in Fig 6.5, over data. In order to restore the signal component in the adaptive processing, we fix the gain of the subarray (in both space and time) formed by fixing the first row of the matrix $[F]$. The elements of the first row are given by

$$\left[1 \quad Z_1 \quad Z_1^2 \cdots Z_1^{N_a-1} \quad Z_2 \quad Z_1 Z_2 \quad Z_1^2 Z_2 \cdots Z_1^{N_a-1} Z_2 \quad Z_2^2 \quad Z_1 Z_2^2 \cdots Z_1^{N_a-1} Z_2^{N_t-1}\right] \tag{18.15}$$

The first row of the system matrix is used to set the gain of the system in the look direction, and at the appropriate given frequency of operation. For the 2-D case the look direction is specified by the angle of arrival, θ_s, and the frequency, f_s, of the SOI.

By setting the product of $[F]$ and $[W]$ equal to a column vector $[Y]$ then the matrix equation is completed and it becomes a square system. The first element of $[Y]$ consists of the constraint gain C, and the remaining elements are set to zero in order to complete the cancellation equations. The resulting matrix equation is then given by

$$[F][W] = [Y] = \begin{bmatrix} C \\ 0 \\ \vdots \\ 0 \end{bmatrix} \tag{18.16}$$

where C is a complex constant and the matrices $[F]$, $[W]$, and $[Y]$ are of dimensions $Q \times Q$, $Q \times 1$, and $Q \times 1$ matrices, respectively. In solving this equation one obtains the weight vector $[W]$ which places space time nulls along the direction of the interferers while maintaining a gain along the direction of the SOI and at the given frequency. The amplitude of the SOI can be estimated using

$$\alpha \approx \frac{1}{C} \sum_{e=1}^{N_a} \sum_{h=1}^{N_t} W\{N_t(e-1)+h\}\ X(h;e) \tag{18.17}$$

18.2.2 Backward Method

Next we reformulate the problem using the same data to obtain a second independent estimate for the solution. This is achieved by reversing the data sequence and then complex conjugating each term of that sequence.

It is well known in the parametric spectral estimation literature that a sampled sequence which can be represented by a sum of exponentials with purely imaginary arguments can be used either in thc forward or in the reverse direction resulting in the same value for the magnitude of the exponent. From physical considerations we know that if we solve a polynomial equation with the weights as the coefficients then its roots provide the direction of arrival for all the unwanted signals including the interferers. Therefore, whether we look at the snapshot as a forward sequence, as presented in the last section, or by a reverse conjugate of the same sequence the final results for W_i must be the same. Hence for these classes of problems we can observe the data either in the forward direction or in the reverse direction. This is equivalent to creating a virtual array of the same size but located along a mirror symmetry line. Therefore, if we now conjugate the data and form the reverse sequence, then one gets an independent set of equations similar to (18.12)–(18.14). This is represented by

$$X^*(m,n) - X^*(m-1,n)\, Z_1^{-1} \tag{18.18}$$

$$X^*(m,n) - X^*(m,n-1)\, Z_2^{-1} \tag{18.19}$$

$$X^*(m,n) - X^*(m-1,n-1)\, Z_1^{-1} Z_2^{-1} \tag{18.20}$$

The form of this linear matrix equation is similar to that of the forward algorithm resulting in

$$[B][W] = [Y] \tag{18.21}$$

where the matrices $[B]$, $[W]$ and $[Y]$ are of dimensions $Q \times Q$, $Q \times 1$, and $Q \times 1$ matrices, respectively.

Using equations (18.18)–(18.20) the SOI is removed from the windowed data. Once the weights are solved for from the given system of equations similar to (18.16), the complex amplitude of the desired signal is estimated from

$$\alpha \approx \frac{1}{C} \sum_{e=1}^{N_a} \sum_{h=1}^{N_t} W\{N_t(e-1)+h\}\; X^*\left(M-h+1; N-e+1\right) \tag{18.22}$$

18.2.3 Forward-Backward Method

Finally, in this section we combine both the forward and the backward method to double the given data and thereby increase the number of weights or the degrees of freedom over that of either the forward or the backward method. In the forward-backward model we double the amount of data by not only considering the data in the forward direction but also conjugating it and reversing the direction of increment of the independent variable. This is always true for the adaptive array case. For the forward-backward method, the number of degrees of freedom can be increased without increasing the number of antenna elements as explained in chapter 6. This is accomplished by considering the forward and backward versions of the array data. For this case, the maximum number of degrees of freedom can reach $[(N+0.5)/1.5+1] \times [(M+0.5)/1.5+1]$. By using the samples from N antenna elements and M time sample we formulate the following matrix equation:

$$[FB][W]=[Y] \tag{18.23}$$

Here again the system matrix $[FB]$ consists of constraint rows and cancellation rows. The constraint rows preserve the SOI along the given direction and at the same operating frequency during the adaptive process. The remaining rows in $[FB]$ consist of cancellation equations that are formed in both along the forward and backward directions. Once the weights are solved for, the complex amplitude of the SOI can be obtained as explained in chapter 6.

18.3 NUMERICAL SIMULATION RESULTS

For the first example, we consider the SOI to be arriving from $\theta_s = 65°$ operating at the frequency of 1300 MHz. Five interferers with different operating frequencies are also used in the simulation. The signal-to-noise (SNR) ratio at each antenna element is set at 30 dB. The signal-to-interference ratio (SIR) for each interferer is kept at –9.5 dB, which implies that all the interferers are much stronger than the SOI. In this simulation, the antenna element spacing d is set to $0.5\,\lambda$ at the given operating frequency of the SOI. N is set to 11 and M is set to 17. All the numerical values of the signals intensities, their directions of arrival along with their operating frequencies are summarized in Table 18.1. Figure 18.2 shows all the signals including the SOI along with the interferers. The SOI is marked with '+' and the interferers are marked as 'o'.

Table 18.1. Parameters for the SOI and Interference.

	AOA (Degree)	Frequency (MHz)	
SOI	65°	1300	SNR = 30 dB
Discrete Interferers	110°	900	SIR = –9.5 dB
	80°	1500	
	120°	1300	
	150°	1200	
	65°	700	

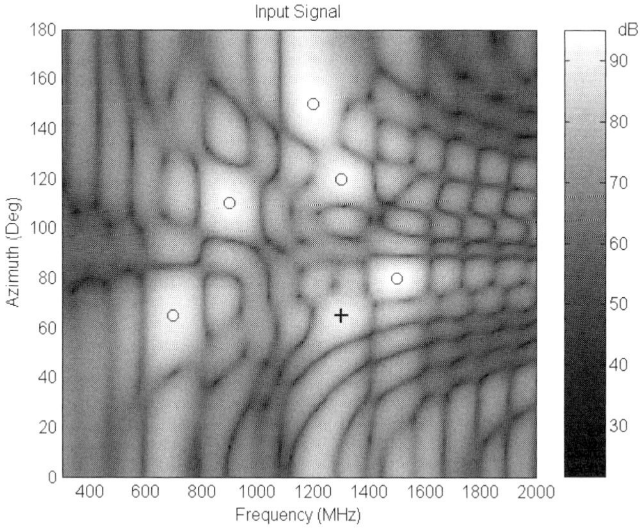

Figure 18.2. The parameters of all the signals.

Figure 18.3 shows the estimated output using the forward method for each angle and frequency after adaptive processing with the appropriate weights. It is clear that the SOI is preserved while all the interferers along the correct angles and frequencies have been nulled. Some interferers share the same angle of arrival or the frequency with the SOI, but these interferers have also been properly eliminated.

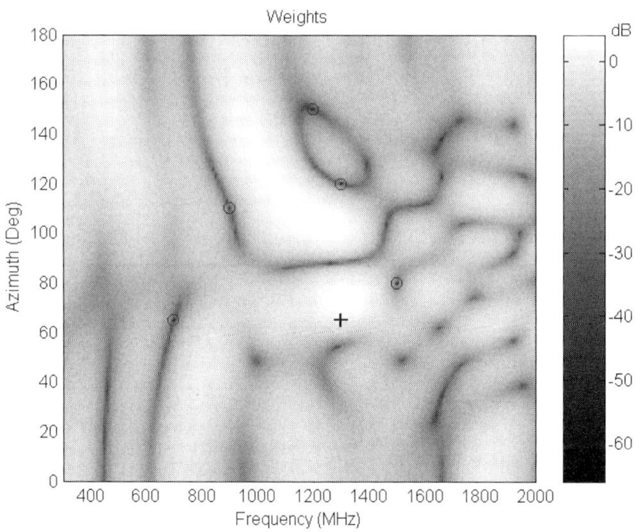

Figure 18.3. The estimated output using the forward method.

Figures 18.4 and 18.5 provide a second and a third independent estimate for the output using the backward and the forward-backward method. It is quite clear that the interferers not sharing exactly the same angle of arrival and operating frequency as the SOI have all been nulled out.

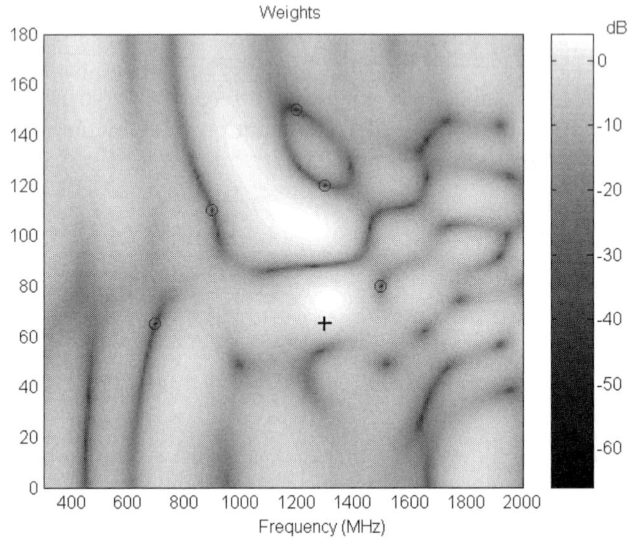

Figure 18.4. The estimated output using the backward method.

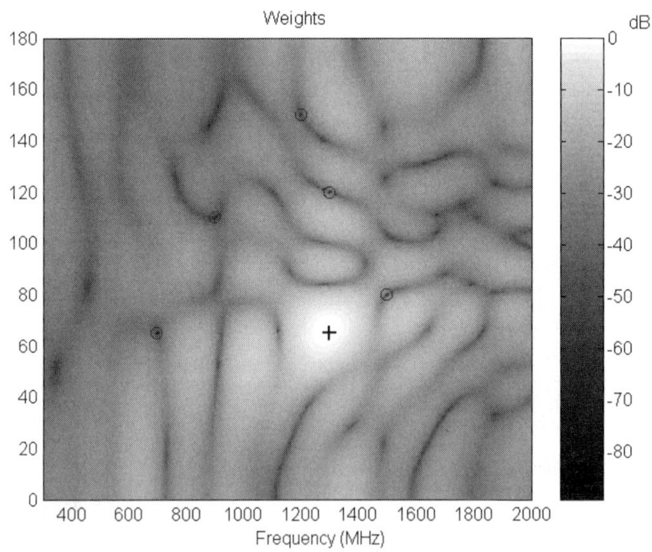

Figure 18.5. The estimated output using the forward-backward method.

For the next simulation, we consider the SOI to be arriving from θ_s = 65° operating at a frequency of 1300 MHz. Five interferers with different operating frequencies were simulated. The SNR at each antenna element is set to 30 dB. Each of the interferers is much stronger than the signal and the SIR for each one is −14.3 dB. In this simulation, the antenna element spacing d is set to 0.5 λ at the frequency of the SOI. Also N is set to 11 and M is set to 17. But in this example, one of thc interferers has a broad operating frequency ranging from 1100 MHz to 1500 MHz. All the parameters of the SOI and the interferers have been summarized in Table 18.2. Figure 18.6 plots the parameters of the input signal. Figure 18.7 shows the estimated output obtained by the forward method of the D3LS adaptive methodology. In this figure, one can see the interferer, which has a wide bandwidth has been properly eliminated corresponding to the appropriate angle of arrival and the appropriate operating frequency bandwidth.

Table 18.2. Parameters for the SOI and Interference.

	AOA (Degree)	Frequency (MHz)	
SOI	65°	1300	SNR = 30 dB
Discrete Interferers	60°	1000	SIR = −14.3 dB
	80°	1100	
	50°	1500	
	120°	1700	
	120°	1100 ~ 1500	

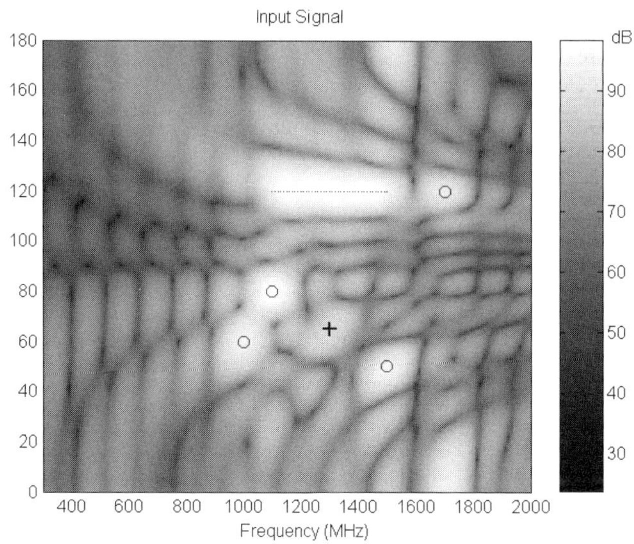

Figure 18.6. Parameters of the various input signals.

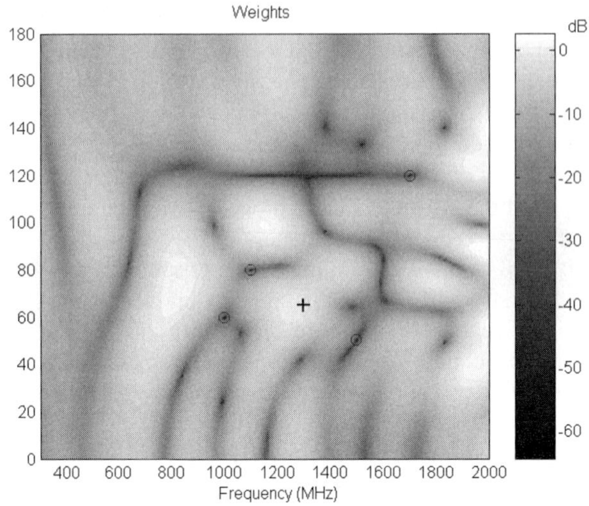

Figure 18.7. The estimated output using the forward method.

Figures 18.8 and 18.9 show the estimated output obtained by the backward method and the forward-backward method, respectively, using the D3LS adaptive methodology. In these figures, one can see that the interferer, which has a wide bandwidth has been properly eliminated corresponding to the appropriate angle of arrival and the appropriate operating frequency bandwidth. In addition, the SOI has been appropriately identified.

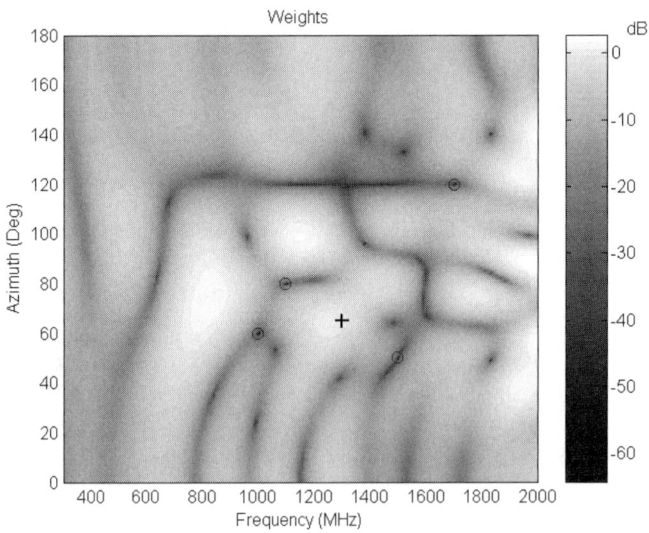

Figure 18.8. The estimated output using the backward method.

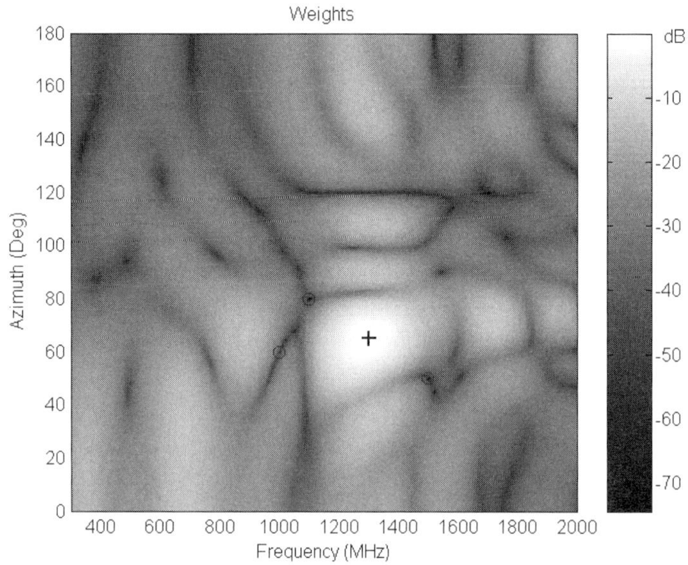

Figure 18.9. The estimated output using the forward-backward method.

18.4 CONCLUSION

In this chapter, an adaptive algorithm based on the D3LS methods has been presented to deal with signals operating at different frequencies. In addition, it is shown that this method can also null broadband interferers. However, the signals of interest are narrowband. Numerical results have been presented to illustrate the application of this novel methodology. Work is in progress to deal with the case when the SOI has a finite operating bandwidth.

REFERENCES

[1] W. Liu, S. Weiss, and L. Hanzo, "A Generalized Sidelobe Canceller Employing Two-Dimensional Frequency Invariant Filters", *IEEE Trans. Antennas and Propagation*, Vol. 53, pp. 2339-2343, July 2005.

[2] W. Liu and S. Weiss, "A New Class of Broad Arrays with Frequency Invariant Beam Patterns", in *Proc. IEEE International Conference on Acoustics, Speech, and Signal Processing*, Montreal, Canada, Vol. 2, pp. 185-188, May 2004.

[3] P. P. Moghaddam and H. Amindavar, "Direction of Arrival Estimation: A New Approach", *Signal Processing Conference (NORSIG2000)*, Nordic, Sweden, June 2000.

[4] T. K. Sarkar, S. Park, J. Koh, and R. A. Schneible, "A Deterministic Least Squares Approach to Adaptive Antennas", *Digital Signal Processing - A Review Journal*, Vol. 6, pp. 185-194, 1996.

[5] T. K. Sarkar, H. Wang, S. Park, J. Koh, R. Adve, K. Kim, Y. Zhang, M. C. Wicks and R. D. Brown, "A Deterministic Least Squares Approach to Space Time Adaptive Processing (STAP)", *IEEE Trans. Antennas and Propagation*, Vol. 49, pp. 91-103, January 2001.

[6] T. K. Sarkar, S. Nagaraja, and M. C. Wicks, "A Deterministic Direct Data Domain Approach to Signal Estimation Utilizing Nonuniform and Uniform 2-D Arrays", *Digital Signal Processing – A Review Journal*, Vol. 8, pp. 114-125, 1998.

[7] T. K. Sarkar, M. C. Wicks, M. Salazar-Palma, and R. J. Bonneau, *Smart Antennas*, Hoboken, NJ, John Wiley and Sons & IEEE Press, 2003.

19

EFFECT OF RANDOM ANTENNA POSITION ERRORS ON A DIRECT DATA DOMAIN LEAST SQUARES APPROACH FOR SPACE-TIME ADAPTIVE PROCESSING

19.0 SUMMARY

The equivalent isotropically radiated power (EIRP) degradation due to random position errors for Space-Time Adaptive Processing (STAP) is presented based on the relationship between the EIRP degradation and the standard deviation or the uniform bound of the probability distribution of location of the antenna elements. We simulate a direct data domain least squares (D3LS) approach for STAP. In this simulation two situations are considered. In the first case, the antenna elements at every time instance are assumed to have different spatial positions from the previous time instance. For the other case, the antennas are assumed to be randomly located, but are assumed to be fixed in a coherent processing interval (CPI). It is demonstrated that whether the locations of the antenna are fixed or they vary within a CPI with identical random errors, the output signal to interference and noise ratio (SINR) are almost the same. When the antenna elements are moved randomly, the output SINR is less than the unperturbed case. It is possible to develop a bound on the EIRP degradation due to the D3LS approach to STAP based on the random position errors.

19.1 INTRODUCTION

The D3LS approaches are based on the work of Sarkar and Sangruji [1]. Then Sarkar and his coworkers [2–4] presented two more variations to this approach, the backward processor and the forward-backward processor. These algorithms were then extended to the space-time domain, where Park [5] described a generalized eigenvalue based processor and Sarkar et. al. [6] implemented the D3LS processors. A review of these techniques is available in chapter 6.

The D3LS STAP algorithm has been used for suppressing highly dynamic clutter and interference. This enables the system to detect potentially weak target returns. In this analysis, we assume that the system consists of a linear array of N equally spaced antenna elements. We assume that the system processes a number of pulses (M) within a coherent processing interval (CPI). Each pulse repetition interval (PRI) consists of the transmission of a pulsed waveform of finite bandwidth and the reception of the reflected energy by the radar aperture and receivers, with a bandwidth matched to that of the radar pulse. Data can be arranged into a three-dimensional matrix commonly referred to as a data cube [7] as explained in section 6.3. We assume that the signals entering the array are narrowband, with a planar phase front, and consist of a SOI along with interference plus noise. The noise (thermal noise) originates in the receiver and is understood to be independent across elements and pulses. The interference is external to the receiver and consists of clutter, jammers, and signal multipath, which may be coherent or incoherent. For a uniformly spaced linear array (ULA), the complex envelope of the received SOI with unity amplitude, for the mth pulse and n^{th} antenna element, can be described by

$$S_{m,n} = \exp\left[j2\pi \left\{ \frac{(n-1)d}{\lambda}\cos(\phi_s) + \frac{(m-1)f_s}{f_r} \right\} \right], \quad \text{for } m = 1, 2, \cdots, M \quad n = 1, 2, \cdots, N \tag{19.1}$$

where d is the inter-element spacing between the antenna elements as shown in Fig. 9.1. λ is the wavelength, f_r is the pulse repetition frequency, ϕ_s is the angle of arrival and f_s is the Doppler frequency of the SOI. For the nth antenna element and m^{th} pulse, the complex envelope of the received signal is given by

$$X_{m,n} = \alpha_s \times S_{m,n} + \text{Interference} + \text{Noise} \tag{19.2}$$

where α_s is the complex amplitude of the SOI entering the array. For a particular row of (19.2) the column-to-column phase difference, due to the SOI, is

$$Z_1 = \exp[\, j2\pi \frac{d}{\lambda}\cos(\phi_s)] \tag{19.3}$$

The row-to-row phase difference in a given column, due to the Doppler of the SOI, is given by

$$Z_2 = \exp(\, j2\pi \frac{f_s}{f_r}) \tag{19.4}$$

For the two-dimensional case these differences are performed with elements offset in space, time, and jointly in space and time. The three types of difference equations as explained in section 6.3, are then given by

$$X_{m,n} - Z_1^{-1} \cdot X_{m,n+1} \tag{19.5}$$

$$X_{m,n} - Z_2^{-1} \cdot X_{m+1,n} \tag{19.6}$$

$$X_{m,n} - Z_1^{-1} \cdot Z_2^{-1} \cdot X_{m+1,n+1} \tag{19.7}$$

Now, one can form the cancellation rows in matrix $[F]$ using equations (19.5)–(19.7), which will deal with only the undesired signals. And the elements of the first row of matrix $[F]$ is implemented as follows

$$[1, Z_1, Z_1^2, \cdots, Z_1^{N_a-1}, Z_2, Z_1Z_2, Z_1^2Z_2, \cdots Z_1^{N_a-1}Z_2, Z_2^2, Z_1Z_2^2, \cdots, Z_1^{N_a-1}Z_2^{N_t-1}] \tag{19.8}$$

where N_a is the spatial dimension and N_t is the temporal dimension of the window. N_a and N_t must satisfy the following equations

$$N_a \leq \frac{N+1}{2} \tag{19.9}$$

$$N_t \leq \frac{M+1}{2} \tag{19.10}$$

The resulting matrix equation is then given by

$$[F][W] = [Y] = [C \;\; 0 \;\; \dots \;\; 0]^T. \tag{19.11}$$

The second process is to use the backward processor, which can be implemented by conjugating the element-pulse data and processing this data in reverse. We also use forward-backward process that is formed by combining the forward and backward solution procedure [3, 4] as described in section 6.3.

When a D3LS STAP is used, we assume that each antenna element is designated to be placed at a specific location [8]. In a real environment, however, it is possible for the location of the antenna elements to fluctuate due to an installation error or environmental effects. Research on random perturbations in the excitation coefficients of the antenna arrays has been the subject of several papers [9,10]. These papers consider random perturbations of the amplitude and phase of the amplifiers. As a follow-on to the paper by Zaghloul [9] on the deterioration of the EIRP due to random perturbations in the excitation coefficients of the antenna array, Hwang and Sarkar [11] presented allowable tolerances for acceptable performance in adaptive arrays for the position of antenna elements. In this section, we illustrate how random position errors can affect a D3LS approach when applied to STAP. Our goal is to find how the system will perform when the antenna elements are randomly moved.

In section 19.2, we discuss EIRP degradation of the array antennas due to random position errors in terms of the normalized EIRP degradation. In section 19.3, to explore the utility of the EIRP degradation, a probabilistic analysis of a simple phased array antenna is studied. In section 19.4, the D3LS approach for STAP is simulated using the relationship between the EIRP degradation and position errors due to random locations of the elements in an array.

19.2 EIRP DEGRADATION OF ARRAY ANTENNAS DUE TO RANDOM POSITION ERRORS

Consider an antenna array consisting of N antenna elements fed by N separate amplifiers. Except for a scaling factor, the far field $E(\theta,\phi)$ is given by

$$E(\theta,\phi) = \sum_{i=1}^{N} a_i \, e^{j(u x_i + v y_i + w z_i)} \, E_{el}(\theta, \phi) \tag{19.12}$$

where

a_i = $|a_i| e^{j\psi_i}$ complex excitation coefficient for the ith antenna element

x_i = the random perturbation of the x-coordinate of the ith antenna element

y_i = the random perturbation of the y-coordinates of the ith antenna element

z_i = the random perturbation of the z-coordinates of the ith antenna element

u = $k \sin\theta \cos\phi$, $v = k \sin\theta \sin\phi$, $w = k \cos\theta$, $k = 2\pi / \lambda$

$E_{el}(\theta,\phi)$ = the antenna element pattern along the direction (θ,ϕ).

The probability density functions of the random perturbations x_i, y_i, and z_i may be uniform or have Gaussian distributions [12]. For uniform distributions, the probability density of x_i, y_i, and z_i are given by

$$\begin{aligned} p(x_i) &= \frac{1}{2\Delta_x}, \quad \bar{x}_i - \Delta_x < x_i < \bar{x}_i + \Delta_x \\ &= 0, \qquad \text{otherwise} \end{aligned} \tag{19.13}$$

$$\begin{aligned} p(y_i) &= \frac{1}{2\Delta_y}, \quad \bar{y}_i - \Delta_y < y_i < \bar{y}_i + \Delta_x \\ &= 0, \qquad \text{otherwise} \end{aligned} \tag{19.14}$$

$$\begin{aligned} p(z_i) &= \frac{1}{2\Delta_z}, \quad \bar{z}_i - \Delta_z < z_i < \bar{z}_i + \Delta_x \\ &= 0, \qquad \text{otherwise} \end{aligned} \tag{19.15}$$

For Gaussian distributions, the probability density functions for x_i, y_i, and z_i are given by [12]

$$p(x_i) = \frac{1}{\sqrt{2\pi}\sigma_x} e^{-(x_i - \bar{x}_i)^2 / 2\sigma_x^2} \tag{19.16}$$

$$p(y_i) = \frac{1}{\sqrt{2\pi}\sigma_y} e^{-(y_i - \bar{y}_i)^2 / 2\sigma_y^2} \tag{19.17}$$

$$p(z_i) = \frac{1}{\sqrt{2\pi}\sigma_z} e^{-(z_i - \bar{z}_i)^2 / 2\sigma_z^2} \tag{19.18}$$

where $\bar{x}_i, \bar{y}_i$, and $\bar{z}_i$ are the mean values of the x_i, y_i, and z_i-positions respectively, and σ_x^2, σ_y^2 and σ_z^2 are the variance of the x, y, and z-coordinates respectively.

According to Zaghloul [9], EIRP along the direction (θ,ϕ) is proportional to the power radiated in the same direction. Therefore EIRP, is given by

$$P = \sum_{i=1}^{N} |G_i|^2 + \sum_{i=1}^{N} \sum_{j=1}^{N} e^{j[u(x_i - x_j) + v(y_i - y_j) + w(z_i - z_j)]} G_i G_j^* \tag{19.19}$$

in which * denotes the complex conjugate, and

$$G_i = a_i \cdot E_{el}(\theta,\phi). \tag{19.20}$$

The probability distribution of the EIRP will be Gaussian with the mean of $\bar{P}$ and a standard deviation of $\sigma(P)$ using the principles of the central limit theorem [12]. The probability distribution of the EIRP will be within the confidence interval of $\left[\bar{P} - n\sigma(P)]/P_0,\ [\bar{P} + n\sigma(P)]/P_0\right]$. The left side of the confidence interval in the probability distribution of the EIRP, $[\bar{P} - n\sigma(P)]/P_0$, represents the worst case scenario for all possible cases. Normalizing $\bar{P}$ and $\sigma(P)$ with respect to the unperturbed power P_0 presents the degradation in the EIRP of the array. This quantity, $[\bar{P} - n\sigma(P)]/P_0$, is the normalized EIRP degradation. Along the same line of the amplitude and phase perturbation in [9], we choose $n = 3$, or three standard deviations, so that the worst-case degradation of the EIRP is within 99.7 percent of all possible cases for the given random position distributions.

The expected value $\bar{P}$ and the standard deviation $\sigma(P)$ of P can be expressed in separable terms containing the statistical parameters for x_i, y_i and z_i. Each amplifier is located at each x, y, and z-position and is considered to be independent. They have the same distribution. An expression for the influence of random locations of the antenna elements can be calculated by integrating the product of the probability distribution and the power P. The expected value $\bar{P}$ is obtained as

$$\bar{P} = \sum_{i=1}^{N} |G_i|^2 + F_x F_y F_z \sum_{i=1}^{N} \sum_{j=1}^{N} e^{j[u(\bar{x}_i - \bar{x}_j) + v(\bar{y}_i - \bar{y}_j) + w(\bar{z}_i - \bar{z}_j)]} G_i G_j^* \tag{19.21}$$

where F_x is the x-position variation dependent factor, F_y is the y-position variation dependent factor and F_z is the z-position variation dependent factor. These expressions are given in Table 19.1.

Table 19.1. *x*, *y*, and *z*-position Factors in the Expected Value Expression.

Factor	Uniform Distribution	Gaussian Distribution
F_x	$(\sin u\Delta_x / u\Delta_x)^2$	$e^{-u^2\sigma_x^2}$
F_y	$(\sin v\Delta_y / v\Delta y)^2$	$e^{-v^2\sigma_y^2}$
F_z	$(\sin w\Delta_z / w\Delta_z)^2$	$e^{-w^2\sigma_z^2}$

The next step in calculating the standard deviation $\sigma(P)$ indicates the extent of the array EIRP degradation. The standard deviation is given by

$$\sigma(P) = \sqrt{\text{var}(P)} \tag{19.22}$$

and the variance is defined as

$$\text{var}(P) = \mathcal{E}[P^2] - \bar{P}^2 \tag{19.23}$$

in which $\mathcal{E}[\cdot]$ denotes the expectation value. We obtain $\mathcal{E}[P^2]$ as

$$\begin{aligned}
\mathcal{E}[P^2] &= \sum_{i=1}^{N} |G_i|^4 + 2\times\sum_{i=1}^{N}\sum_{j=1}^{N}|G_i|^2|G_j|^2 \\
&+ 4\times F_{1,x}F_{1,y}F_{1,z}\sum_{i=1}^{N}\sum_{j=1}^{N} e^{j[u(\bar{x}_i-\bar{x}_j)+v(\bar{y}_i-\bar{y}_j)+w(\bar{z}_i-\bar{z}_j)]}|G_i|^2 G_i G_j^* \\
&+ 4\times F_{1,x}F_{1,y}F_{1,z}\sum_{k=1}^{N}\sum_{i=1}^{N}\sum_{j=1}^{N} e^{j[u(\bar{x}_i-\bar{x}_j)+v(\bar{y}_i-\bar{y}_j)+w(\bar{z}_i-\bar{z}_j)]}|G_k|^2 G_i G_j^* \\
&+ F_{2,x}F_{2,y}F_{2,z}\sum_{i=1}^{N}\sum_{j=1}^{N} e^{j2[u(\bar{x}_i-\bar{x}_j)+v(\bar{y}_i-\bar{y}_j)+w(\bar{z}_i-\bar{z}_j)]}G_i^2 G_j^{*2} \\
&+ F_{3,x}F_{3,y}F_{3,z}\sum_{k=1}^{N}\sum_{i=1}^{N}\sum_{j=1}^{N}[e^{j[u(\bar{x}_k+\bar{x}_i-2\bar{x}_j)+v(\bar{y}_k+\bar{y}_i-2\bar{y}_j)+w(\bar{z}_k+\bar{z}_i-2\bar{z}_j)]}G_k G_i G_j^{*2} \\
&+ e^{j[u(-\bar{x}_k+2\bar{x}_i-\bar{x}_j)+v(-\bar{y}_k+2\bar{y}_i-\bar{y}_j)+w(-\bar{z}_k+2\bar{z}_i-\bar{z}_j)]}G_k G_i^2 G_j^*] \\
&+ F_{4,x}F_{4,y}F_{4,z}\sum_{k=1}^{N}\sum_{l=1}^{N}\sum_{i=1}^{N}\sum_{j=1}^{N} e^{j[u(\bar{x}_k-\bar{x}_l+\bar{x}_i-\bar{x}_j)+v(\bar{y}_k-\bar{y}_l+\bar{y}_i-\bar{y}_j)+w(\bar{z}_k-\bar{z}_l+\bar{z}_i-\bar{z}_j)]}G_k G_l^* G_i G_j^*
\end{aligned} \tag{19.24}$$

where the factors $F_{*,x}$, $F_{*,y}$, and $F_{*,z}$ result from estimating constituents of the second moment of the power *P*, and are given in Table 19.2 for the Uniform and Gaussian distributions.

Table 19.2. *x*, *y*, and *z*-position Factors in the Variance Expression.

Factor	Uniform Distribution	Gaussian Distribution
$F_{1,x}$	$\left(\sin u\Delta_x / u\Delta_x\right)^2$	$e^{-u^2\sigma_x^2}$
$F_{1,y}$	$\left(\sin v\Delta_y / v\Delta_y\right)^2$	$e^{-v^2\sigma_y^2}$
$F_{1,z}$	$\left(\sin w\Delta_z / w\Delta_z\right)^2$	$e^{-w^2\sigma_z^2}$
$F_{2,x}$	$\left(\sin 2u\Delta_x / 2u\Delta_x\right)^2$	$e^{-4u^2\sigma_x^2}$
$F_{2,y}$	$\left(\sin 2v\Delta_y / 2v\Delta_y\right)^2$	$e^{-4v^2\sigma_y^2}$
$F_{2,z}$	$\left(\sin 2w\Delta_z / 2w\Delta_z\right)^2$	$e^{-4w^2\sigma_z^2}$
$F_{3,x}$	$\left(\sin u\Delta_x / u\Delta_x\right)^2 \left(\sin 2u\Delta_x / 2u\Delta_x\right)$	$e^{-3u^2\sigma_x^2}$
$F_{3,y}$	$\left(\sin v\Delta_y / v\Delta_y\right)^2 \left(\sin 2v\Delta_y / 2v\Delta_y\right)$	$e^{-3v^2\sigma_y^2}$
$F_{3,z}$	$\left(\sin w\Delta_z / w\Delta_z\right)^2 \left(\sin 2w\Delta_z / 2w\Delta_z\right)$	$e^{-3w^2\sigma_z^2}$
$F_{4,x}$	$\left(\sin u\Delta_x / u\Delta_x\right)^4$	$e^{-2u^2\sigma_x^2}$
$F_{4,y}$	$\left(\sin v\Delta_y / v\Delta_y\right)^4$	$e^{-2v^2\sigma_y^2}$
$F_{4,z}$	$\left(\sin w\Delta_z / w\Delta_z\right)^4$	$e^{-2w^2\sigma_z^2}$

19.3 EXAMPLE OF EIRP DEGRADATION IN ANTENNA ARRAYS

To explore the utility of the EIRP degradation, a probabilistic analysis on a simple linear phased array antenna as shown in Figure 19.1 is performed. Using this example, we investigate how the EIRP degradation behaves as a function of the various parameters.

Consider a linear array consisting of 10 elements as shown in Figure 19.1. The array consists of ten isotropic omnidirectional point radiators and is fed by 10 amplifiers. The elements of the linear array are equally spaced and their inter-element spacing along the x-direction is given by 0.4775λ (λ = 1m). To simplify the example, we assume that all the elements are located at $z = 0$ so that the z-positions of all the elements are fixed. Therefore the EIRP degradation of the linear array is caused by random perturbations in the x and y position which can have two types of probability distribution, either uniform or Gaussian. F_z is 1 in the Expected value of Eq. (19.21) and $F_{*,z}$ is also 1 in the second moment of the power of Eq. (19.24). When the positions of all the elements are unperturbed, F_x and F_y equal unity, and also $\bar{P}$ becomes the unperturbed power P_0. As we expect, the variance is identically zero, as given by Eq. (19.23).

The expected value and the standard deviation now are applied to the linear array case to get the normalized EIRP degradation with n = 3 and $[\bar{P} - 3\sigma(P)]/P_0$. Of special interest is the EIRP degradation within the half-power beam width between $\phi = 90°$ and $\phi = 95.33°$. Figures 19.2 and 19.3 show the worst case degradation with $3\sigma(P)$. Figure 19.2 represents the case for the Gaussian perturbation in position with $\sigma_{x \text{ and } y}$ = 0.0, 0.03, 0.05, and 0.07 m, and Figure 19.3 deals with the uniform perturbation in position with $\Delta_{x \text{ and } y}$ = 0.0, 0.03, 0.05, and 0.07 m. As expected, the unperturbed case shows a 0 dB value for all the angles. The EIRP degrades with the larger standard deviation for the Gaussian perturbation and for the uniform bounds in a uniform perturbation. The EIRP due to the random positions of the antennas is significantly decreased.

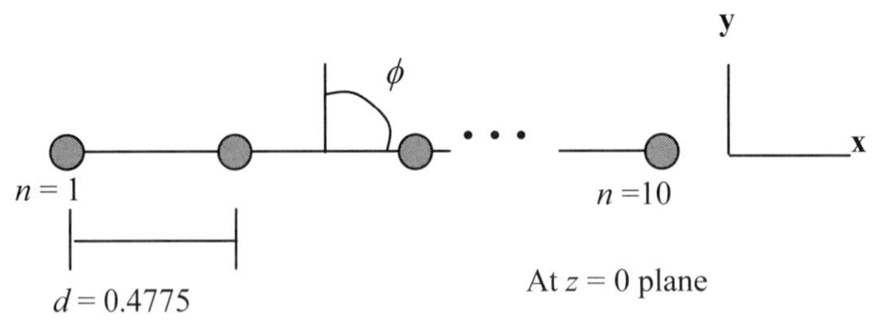

Figure 19.1. The uniformly spaced linear array configuration.

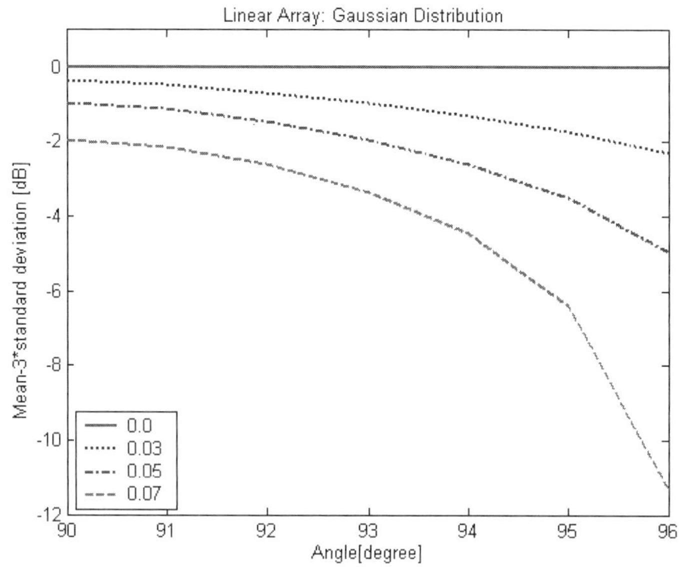

Figure 19.2. Normalized EIRP degradation with n = 3 for Gaussian perturbation with standard deviation of $\sigma_{x \text{ and } y}$ = 0.0, 0.03, 0.05, and 0.07 m.

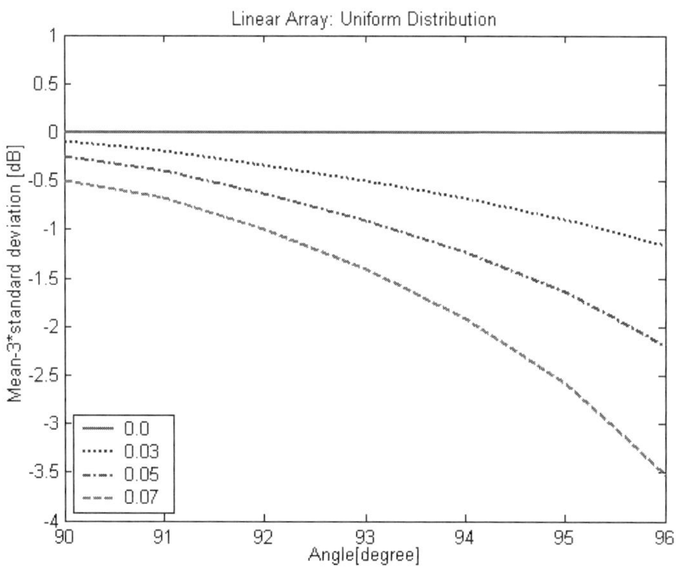

Figure 19.3. Normalized EIRP degradation with n = 3 for uniform perturbation with different bounds of $\Delta_{x \text{ and } y}$ = 0.0, 0.03, 0.05, and 0.07 m.

The normalized EIRP degradation, $[\bar{P} - 3\sigma(P)]/P_0$, represents the worst case for 99.7 percent of all possible cases as mentioned before. This value, however, depends on different values of σ for a Gaussian distribution or has a different Δ for a uniform distribution, as we see from Figures 19.2 and 19.3. The question is what is the allowable tolerance for σ and Δ and for the elements to operate appropriately in the antenna system for adaptive signal processing. We focus on the main lobe, especially inside the half-power beam width. To choose the appropriate $\sigma_{x \text{ and } y}$ and $\Delta_{x \text{ and } y}$, we select the point that makes the 3 dB loss below the optimum for the normalized EIRP degradation graph, at the angle corresponding to the half-power beam width. At this –3 dB point, the antenna performance using conventional beam forming will be significantly decreased. However, the 3 dB loss of the normalized EIRP degradation at the angle of the half-power beam is the allowable worst case for an antenna system to operate properly compared to the unperturbed case. From these σ and Δ obtained from the 3 dB loss point of the normalized EIRP degradation, we investigate how the system behaves if the antenna elements are randomly perturbed by those σ and Δ.

As illustrated in Figure 19.4, the –3 dB point of the normalized EIRP degradation with n = 3 of a uniform linear array occurs at 95.33° with $\sigma_{x \text{ and } y}$ = 0.042 m and $\Delta_{x \text{ and } y}$ = 0.072 m. Figure 19.4 also illustrates that the relationship between the normalized EIRP degradation with n = 3 and $\sigma_{x \text{ and } y}$, and between the normalized EIRP degradation and $\Delta_{x \text{ and } y}$ at a specific angle, 95.33°. The larger the $\sigma_{x \text{ and } y}$ and $\Delta_{x \text{ and } y}$, the lower the normalized EIRP degradation. We apply this value, $\sigma_{x \text{ and } y}$ = 0.042 m and $\Delta_{x \text{ and } y}$ = 0.072 m, and then investigate how the random position errors affect the D3LS approach for STAP.

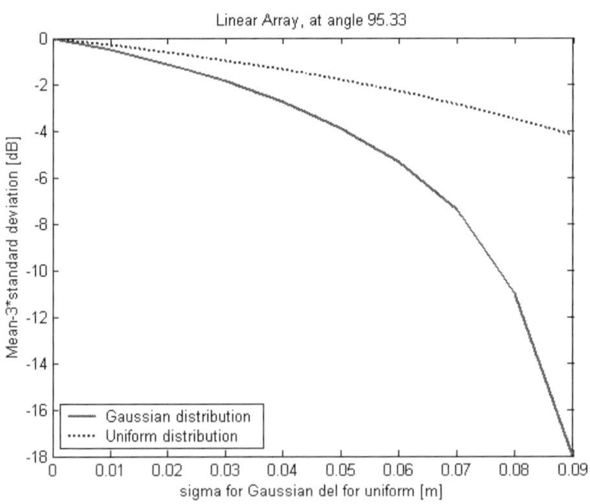

Figure 19.4. The relationship between the normalized EIRP degradation with n = 3, and $\sigma_{x \text{ and } y}$ and $\Delta_{x \text{ and } y}$ at the angle, 95.33° for a linear Array.

19.4 SIMULATION RESULTS

The received signals modeled in this simulation consist of the SOI, main beam clutter, discrete interferers, jammers, and thermal noise. The clutter is modeled as point scatterers placed approximately every 0.1 degrees apart. The amplitude of the clutter signals are modeled with a normal distribution about a mean that result in a signal to clutter ratio, SCR, of −12.7 dB. Ten strong point scatterers are modeled in this simulation, based on the angle of arrival (AOA) and Doppler parameters as defined in Table 19.3. Thermal noise generated in each receive channel is independent from one another. The resulting SNR is approximately 30 dB. The jammer is modeled as a broadband noise signal that arrives from 100° in azimuth, and covers all Doppler frequencies of interest. Summing the power of the interfering sources received by the first channel and comparing it to the power of the SOI, the average input signal to interference plus noise ratio (SINR) is evaluated as −18.6 dB. And random position errors of the antenna element are $\sigma_{x \text{ and } y} = 0.042$ m and $\Delta_{x \text{ and } y} = 0.072$ m. While processing the signal, the antenna elements are randomly moved. In this simulation we consider two situations. One in which the antenna elements at every time instance have different spatial positions from the previous time instance. Another situation is that the antenna random locations are fixed within a CPI.

Table 19.3. Parameters Related to the Simulation.

<table>
<tr><td colspan="2">Wavelength</td><td>1 m</td><td rowspan="3">Signal</td><td>AOA</td><td>60°</td></tr>
<tr><td colspan="2">Pulse Rep. Freq</td><td>4 kHz</td><td>Doppler</td><td>1169 Hz</td></tr>
<tr><td colspan="2">Number of Antennas</td><td>$N = 10$</td><td>SNR</td><td>30 dB</td></tr>
<tr><td colspan="2">Number of Pulses</td><td>$M = 16$</td><td rowspan="2">Jammer</td><td>AOA</td><td>100°</td></tr>
<tr><td colspan="2">Half-Power Beam Width</td><td>10.66°</td><td>Doppler</td><td>Covers all Doppler frequencies of interest</td></tr>
<tr><td rowspan="2">Processor</td><td>Forward</td><td>$N_a = 7$, $N_t = 9$</td><td rowspan="4">Discrete interferes</td><td rowspan="2">AOA</td><td rowspan="2">[85° 120° 40° 35° 30° 140° 100° 70° 50° 125°]</td></tr>
<tr><td>Forward-Backward</td><td>$N_a = 8$, $N_t = 9$</td></tr>
<tr><td rowspan="2">Clutter</td><td>Extent</td><td>Main Beam</td><td rowspan="2">Doppler</td><td rowspan="2">[400 −800 1700 −1400 325 −1650 950 −1200 −125 1450] Hz</td></tr>
<tr><td>Point Scatterers</td><td>0.1° Apart in Angle</td></tr>
</table>

With $N = 10$ channels and $M = 16$ pulses, the performance of the three D3LS algorithms, namely the forward, backward, and the forward and backward method is evaluated based on the weights and the output SINR. Forward and backward methods utilize 7 spatial (N_a) and 9 temporal (N_t) degrees of freedom (DOF) resulting in a total of 63 DOF, while the forward-backward method employs 8 spatial and 9 temporal DOF, for a total of 72 DOF. The spectrum of the input signal is shown in Figure 19.5. Here the circles indicate the locations of the discrete interferers and the triangles define the location of the main beam clutter and the + mark indicates the location of the SOI.

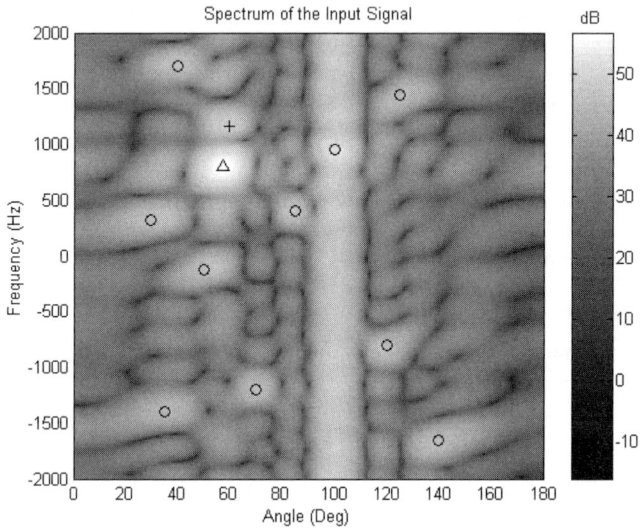

Figure 19.5. The spectrum of the input signal.

Table 19.4 shows the calculated output SINR for the unperturbed case, and perturbed case with a Gaussian profile, and a uniform profile. These values are averaged over 100 runs. The resulting weights for the unperturbed case and for the perturbed case with Gaussian and uniform density functions using the forward method are shown in Figures 19.6–19.8, when the antenna elements at every time instance have different spatial positions from the previous time instance. Antenna positions are randomly located with $\sigma_{x \text{ and } y} = 0.042$ m or $\Delta_{x \text{ and } y} = 0.072$ m. The Output SINR for the Gaussian and uniform perturbed cases are lower than the Output SINR of the unperturbed case. Even though we choose different σ and Δ, we get approximately an output SINR of ±11 dB. This shows that we can obtain EIRP degradation and this can be calculated analytically. From Figures 19.7 and 19.8, the system with random position locations generates nulls slightly moved from the angle of arrival (AOA) of discrete interferers

Table 19.4. The Output Signal to Interference Plus Noise Ratio.

		Case A (When antenna random locations are changed every time instance)	Case B (When antenna random locations are fixed in a CPI)
Forward	Unperturbed	18.21 dB	
	Gaussian	11.46 dB	11.03 dB
	Uniform	11.48 dB	10.52 dB
Backward	Unperturbed	18.21 dB	
	Gaussian	11.46 dB	11.03 dB
	Uniform	11.48 dB	10.52 dB
Forward-Backward	Unperturbed	19.21 dB	
	Gaussian	11.48 dB	11.15 dB
	Uniform	11.84 dB	11.18 dB

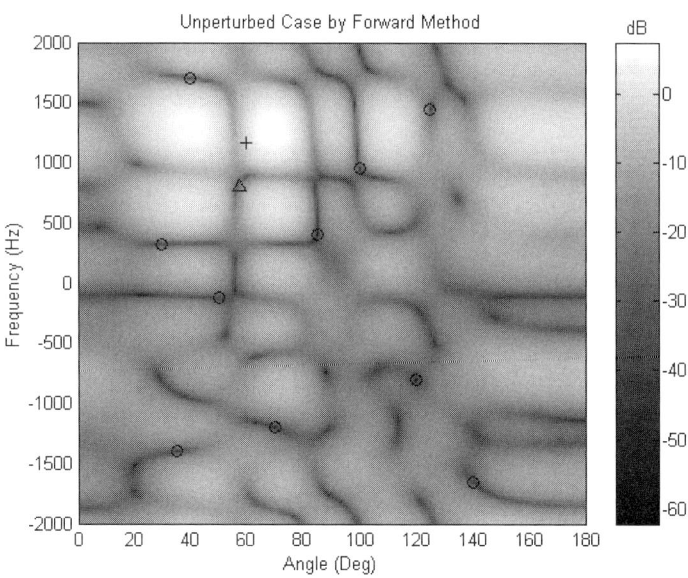

Figure 19.6. Forward method weight spectrum for the unperturbed case.

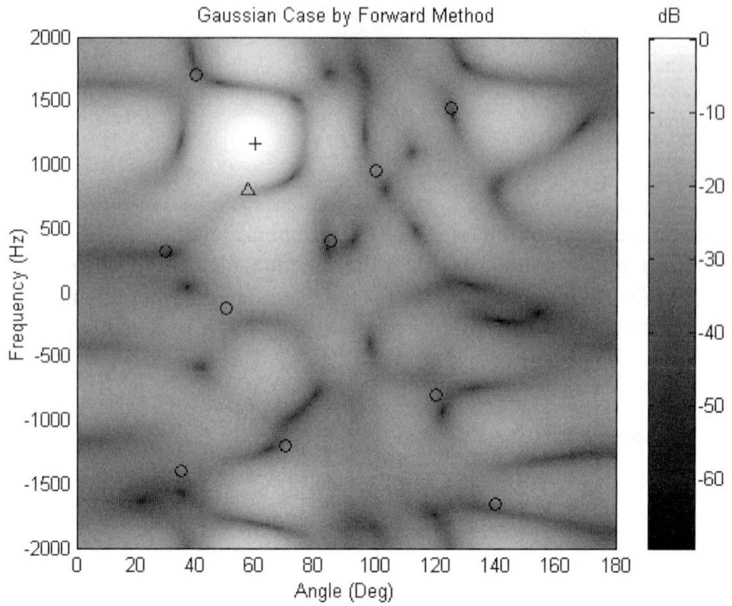

Figure 19.7. Forward method weight spectrum for the perturbed Gaussian profile with $\sigma_{x \text{ and } y} = 0.042$ m.

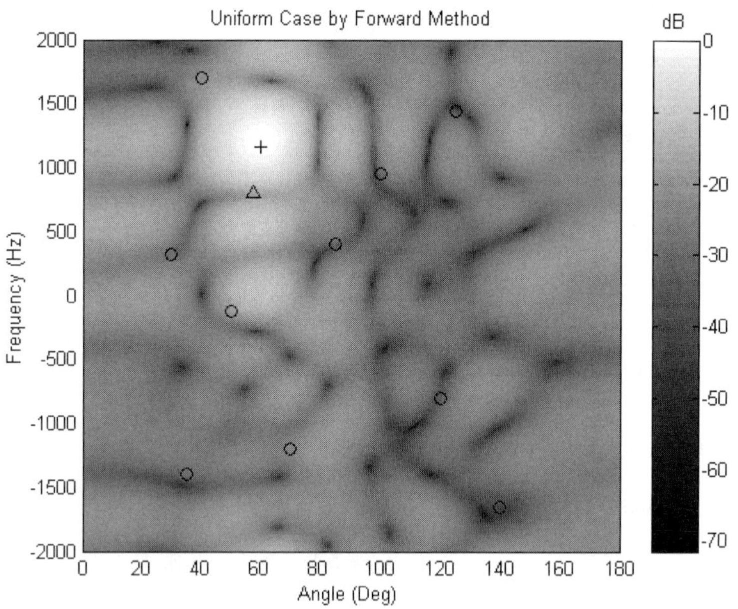

Figure 19.8. Forward method weight spectrum for the perturbed uniform profile with $\Delta_{x \text{ and } y} = 0.072$ m.

as compared to the unperturbed case as of Fig. 19.6. The output SINR of the unperturbed case is about 18-19 dB and the output SINRs for the Gaussian and uniform density distributions, which are about 11 dB when the antenna elements at every time instance have different spatial positions from the previous time instance. This perturbed system still removes the jammer, clutters, discrete interferers, and noise. When the random locations of the antenna are fixed in a CPI, one also obtains almost the same output SINR. The value of the output is dependent on the value of the tolerance of the antenna elements.

From these results, as long the antenna element position errors are within the limits set by $\sigma = 0.042$ m for the Gaussian distribution and $\Delta = 0.072$ m for the uniform distribution, the adaptive array has a reasonable performance for STAP. If we want to provide stricter tolerances in the distribution of the antenna elements, i.e., $\sigma < 0.042$ m for the Gaussian distribution and $\Delta < 0.072$ m for the uniform distribution, then one can get higher output SINR which may be close to the output SINR for the unperturbed case. As a result, one can predict how much the system would be degraded by analytically solving the problem using the normalized EIRP degradation.

19.5 CONCLUSION

In this chapter, expressions are derived for the expected value, standard deviation, and the normalized EIRP degradation with $n = 3$, standard deviations due to the random locations of the array elements. The normalized EIRP degradation represents the worst case for 99.7 percent of all possible cases. We have investigated the effects of random antenna positions on a D3LS approach for STAP using these EIRP degradations according to the random position of the array elements. Even though $\sigma = 0.042$ m for the Gaussian distribution and $\Delta = 0.072$ m for the uniform distribution are given for the antenna elements, we can still get an acceptable output SINR, where the wavelength is 1 m. When the antenna elements at every time instance have different spatial positions from the previous time instance, the output SINR is almost the same as in the case where the antenna elements are located at fixed random locations within a CPI. In this chapter, the results for the EIRP degradation have been predicted for the Gaussian and for the uniform perturbed cases. Random antenna position errors degrade the output SINR of a D3LS approach for STAP.

REFERENCES

[1] T. K. Sarkar and N. Sangruji, "An Adaptive Nulling System for a Narrowband Signal with a Look Direction Constraint Utilizing the Conjugate Gradient Method", *IEEE Transactions on Antenna and Propagation*, Vol. 37, pp. 940-944, July 1989.

[2] R. Schneible, *A Least Square Approach for Radar Array Adaptive Nulling*, Doctoral Dissertation, Syracuse University, May 1996.

[3] T. K. Sarkar, J. Koh, R. Adve, R. Schneible, M. Wicks, S. Choi and M. Salazar-

Palma, "A Pragmatic Approach to Adaptive Antennas", *IEEE Antennas and Propagation Magazine*, Vol. 42, No. 2, pp. 39-55, April 2000.

[4] T. K. Sarkar, S. Park, J. Koh and R. A. Schneible, "A Deterministic Least Squares Approach to Adaptive Antennas", *Digital Signal Processing – A Review Journal*, Vol. 6, pp.185-194, 1996.

[5] S. Park, *Estimation of Space-Time Parameters in Non-homogeneous Environment,* Doctoral Dissertation, Syracuse University, May 1996.

[6] T. K. Sarkar, H. Wang, S. Park, J. Koh, R. Adve, K. Kim, Y. Zhang, M. C. Wicks and R. D. Brown, "A Deterministic Least Square Approach to Space Time Adaptive Processing (STAP)", *IEEE Trans. on Antenna and Propagation*, Vol. 49, pp. 91-103, January 2001.

[7] J. T. Carlo, T. K. Sarkar and M. C. Wicks, "A Least Squares Multiple Constraint Direct Data Domain Approach for STAP", *IEEE Radar Conference Proceedings*, pp. 431-438, 2003.

[8] T. K. Sarkar, M. C. Wicks, M. Salazar-Palma, and R. J. Bonneau, *Smart Antennas*, Hoboken, NJ, John Wiley & Sons-IEEE Press, 2003.

[9] A. I. Zaghloul, "Statistical Analysis of EIRP Degradation in Antenna Arrays", *IEEE Trans. Antennas Propagat.*, Vol. AP-33, pp. 217-221, Feb. 1985.

[10] P. Snoeij and A. R. Vellekoop, "A statistical model for the Error bounds of an active phased array antenna for SAR applications", *IEEE Trans. Geosci. Remote Sensing*, Vol. 30, pp. 736-742 July 1992.

[11] S. Hwang and T. K. Sarkar, "Allowable Tolerances in the Position of Antenna Elements in an Array Amenable to Adaptive Processing", *Microwave and Optical Technology Letters*, Vol. 45, Nos. 5, pp.388-393, June 2005.

[12] A. Papoulis, *Probability, Random Variables, and Stochastic Processes*, 3rd ed, Kogakusha, Tokyo, McGraw-Hill, 1991.

INDEX

A

acoustics 1, 47, 48, 49, 64, 105
adapted weights 250
adaptive algorithms 249
adaptive array 247
adaptive estimation 215
adaptive procedure, real time 224-6
adaptive processing 217-8, 246
  amplitude-only 279
  D3LS amplitude-only 283-9
  D3LS phase-only 311-21
  D3LS STAP 292-8
    of broadband signals 521-33
adaptive signal processing algorithms 275-99
adaptive system 304-11
adaptive weighting 209
adaptive weights 229, 249, 275, 303
  minimum norm of 255, 258-73
adaptivity on transmit 391-95
additive white Gaussian noise (AWGN) 128, 174, 176, 181
Air Force Research Laboratory (AFRL) 240
Airborne Early Warning (AEW) systems 238
Airborne Early Warning and Control System (AWACS) 226
Alamouti coding scheme 180-82
Alamouti, Siavash M. 180
Ampère, André M. 2
Ampère's law 7-8, 12, 26-7
amplitude-only weights 276
Andrews, Dr. James R. 59, 102, 105
angle-Doppler 233
angle-of-arrival (AOA) 350
antenna 1-22
  bi-blade 93-94
  bicone 71-4, 458-9
  bow-tie 100
  circular disc dipole 99
  circular loop 30-2
  cone-blade 94-6
  conjugately matched 114, 117, 130-8, 148
  conical 103, 105-6
  conical spiral 88-90
  D*dot 106-7
  diamond dipole 85-6
  dipole 131-3, 458-9
    finite-sized 32-4, 35-6, 49
    half-wave 37-40, 45-6
    in channel capacity simulations 132-46
    resistive loading of a 65-71
  electrically small matched 437-8
  elements, dummy 383-4
  helical 396-410
  horn 458-9, 516-9
    in channel capacity simulations 134-43
  impulse radiating (IRA) 97-8
  linear 34-6, 49
  log-periodic dipole-array (LDPA) 78-80
  monofilar helix 86-8
  monoloop 90-1
  monopole 107-8
  near and far field properties of an 36-46
  planar slot 101
  quad-ridged circular horn 91-2
  spiral 80-3
  TEM horn 103
    resistive loading of 74-7
  traveling-wave 59
  ultrawideband (UWB) 59-60
  Vivaldi 96-7
  volcano smoke 83-5
antenna array 36, 226, 227, 258, 356
  arbitrary shaped 461-77
  adaptive 275
  airborne 242
  aircraft mounted 474-6

DOA estimation w/ 3 different electrically small 446-59
EIRP degradation of 538-43
in the presence of near-field scatterers 470
LOS case studies 199-206
mixed elements 469-70
of non-uniformly spaced dipoles 468
of uniformly spaced dipoles 466-8
semi-circular (SCA) 385
superdirectivity 120-1
three dimensional 480, 486-92, 503
uniform linear (ULA) 378
uniform linear virtual (ULVA) 375, 380
antenna loading 451-53
Aristotle 116
array factor 43, 44
AT&T 224
AWAS software 39, 51

B

BAC1-11 aircraft 238-43
backward method 221-22, 237, 261, 268-88, 283-9
amplitude-only D3LS STAP 293-8
D3LS broadband signals 532
D3LS in mutual coupling 376-7
D3LS multiple beamforming 328-32
D3LS phase-only 310, 312-7
D3LS STAP 291-2
phase-only 318-9, 338-45
D3LS using real weights 281-2
D3LS, in multiple beamforming 327
STAP 336-7
for adaptive processing of broadband signals 526-7, 530
for mutual coupling 379-82
backward processor, least squares 236-7
beamforming 33, 36-49, 209,
Bekenstein, Jacob 162-3
Bell Laboratories Layered Space-Time (BLAST) 168
Bell Telephone Laboratories 158
Bennett, Charles H. 162
Bernoulli, Daniel 149
black holes 162-3
blocking matrix 353, 355
Bluetooth 127-8
Boltzmann, Ludwig 150, 151, 152, 153, 154, 158, 162
Boltzmann's constant 151, 153, 154, 161
Brownian motion 160
Brukner, Caslav 161

C

calibration angles 462-4
caloric theory 148, 149
Capon's mimimum variance technique 412
Carnot,
cycle 149
engine 154
Lazare Nicholas Marguerite 148-9
Nicholas Leonard Sadi 149
Cauchy-Schwarz inequality 211
central limit theorem 44
Centro-Hermitian matrices 412, 413-4
century bandwidth 93-4
channel capacity 38, 43, 113-164, 168, 169, 185, 188-9
Gabor 114, 117, 121, 126, 132, 157
Hartley 117, 122-4, 125-6, 132-9
Hartley-Nyquist-Tuller 123-4, 132-3
historical formulations of 118-24
IEEE standard for 119

of MIMO systems 176-9
Shannon 132-9, 141, 142, 145, 146, 147-8, 185-6
various simulations illustrating 131-46
channel mismatch 356-7
channel modelingof free space 104-5
Cicero, Marcus T. 116
Clausius, Rudolf 149-50, 161
Clavier, A. G. 126, 159
closed-loop system 178
clutter 47, 209, 215, 216, 217, 218, 219, 226, 228, 229, 240-2, 253, 263-4, 278, 348
mathematical model of 349
suppressing highly dynamic 536
coherent bandwidth 170
coherent distance 171
coherent processing interval (CPI) 226, 227, 350, 535, 549,
coherent time 169
complementary property 257
complex amplitude 251
computational bottleneck 225
computational electromagnetics 130
Conceptual Inadequacy of the Shannon Information in Quantum Measurement 161
conjugate gradient 216, 224-5, 257, 280, 309, 376
nonlinear 322
constellation diagram 129
constellation points 129
covariance matrix 250, 253, 299, 303-4, 322, 323, 348, 352, 363-7, 372, 386, 411, 412, 480, 521-2
Cramer-Rao bound (CRB) 437-60, 492-3, 503, 504-5, 519
cross-spectral (GSC) method 347
cross-spectral generalized sidelobe canceller (CS-GSC) method 355, 361, 363-7
Cugnot, Nicholas Joseph 148
current distribution 32-3

D

data cube 536
data snapshots 348
data transfer rate 174
deep space communication 124, 157
degrees of freedom (DOF) 209, 218, 222, 223, 226, 231, 235, 237, 242, 282, 283, 292, 310-1, 348, 352, 368, 377
delay diversity 173
Delmarva Peninsula 240, 242
diffraction 36
digital signal processing (DSP) chip 220, 224, 225, 226, 411
digital signal processor 299, 386
direct data domain method 480
direct data domain least squares (D3LS) 250-1, 276
amplitude-only 277-99
approach to STAP 226-38
approaches to adaptive processing 215-38
comparison of, with statistical methods for STAP 347-68
for 1D adaptive problems 251-3
for adaptive processing of broadband signals 521-8
for STAP 253-5
formulation of, solution for phase-only adaptive system 303-21
in approximate compensation for mutual coupling 371-86
in multiple adaptive beamforming 323-45
STAP 535-49
STAP alogorithms 356-7
direction of arrival (DOA) 47, 48, 61, 243-8, 219, 222, 228-9, 236, 246, 249, 250-3, 255-8, 259-60, 272, 279, 281, 303, 310, 372, 373, 376

direction of arrival (DOA)
estimation, 411-33
broadband 479-519
non-conventional least squares optimization for 461-77
w/electrically small antenna and Cramer-Rao bound 437-60
directivity, 41-3, 119-21
discrete Fourier transform (DFT) 47, 170, 461, 462, 463, 467, 476, 477
discrete interferers 348
mathematical model 350
diversity,
frequency 170
gain 169, 182-3, 206
Alamouti Scheme 180-2
polarization 395, 396-410
space 170-1, 172
spatial 43-6, 199, 390, 391
time 169
diversity-multiplexing tradeoff 182-3, 194-6
Doppler 210, 235, 244, 245
Doppler filtering 348
Doppler frequency 226, 229, 232, 233, 237, 242, 245, 249, 254, 263, 265, 291, 324, 332-3, 345
Doppler radar 227
DSP32C 224

E
Earth 45-6, 52, 161, 172
channel capacity of matched antennas as related to height from the 133-40
eigen decomposition 178
eigenvalue algorithm, generalized 233
eigenvalue method 218-9
eigenvalue processor, generalized 230-2
Einstein, Albert 3
electric field intensity 27, 33
electrically small antenna 437-60
electromagnetic analysis code 372 378, 396, 398
electromagnetic software modeling code 465
electromagnetic software simulator 446
electrostatics, Gauss's law of 8-9
entropy 113, 115-8, 119, 125, 185
evolution of information concept of 156-64
evolution of thermodynamic concept of 148-54
equivalent isotropically radiated power (EIRP) degradation 535, 537-44, 549
estimation of signal parameters using rotational in-variance technique (ESPRIT) 253, 411, 412-3, 461, 480
estimation of signal parameters using rotational in-variance technique (ESPRIT)
comparison with MPM 428-33
eye pattern 128-9

F
fading channel 170
far field 29-30, 35, 36
far field differences between near
Faraday, Michael 4
Faraday's law 4-6
fast Fourier transform (FFT) 47, 60, 216, 224-5
Federal Communications Commission (FCC) 59
field concepts, introductory 28-36
Fisher information matrix 442-3, 492, 503, 504
Fitzgerald, George Francis 4
five constraint algorithm 259-73
Fletcher-Reeves 309
forward method 220-1, 222, 260, 265-7, 283-9
D3LS adaptive methodology
amplitude only 279-80
for broadband signals 531-2

for multiple beamforming 324-7, 328-32
formulation of using real weights 277-81
in mutual coupling 373-6
phase-only 304-10, 312-7
D3LS space-time adaptive processing (STAP) 289-91
amplitude-only 277-81, 293-8
for broadband signals 522-6, 529
for multiple beamforming 332-6
in mutual coupling 379-82
phase-only 318-9, 338-45
forward processor, least squares 232-6
forward-backward method 222-4
D3LS adaptive methodology 262, 270-3, 282, 283-9
for broadband signals 527-8, 530, 532-3
in multiple beamforming 328-32
in mutual coupling 377-8, 379-82
phase-only 310-1, 312-7
D3LS space-time adaptive processing (STAP) 292, 363-7
amplitude-only 293-9
in multiple beamforming 328-32, 337-8
phase-only 318-9, 338-45
forward-backward processor, least squares 237-8
Foschini, Gerard J. 168
frequency domain, propagation modeling of 49-57
Fresnel number 36
Frii's transmission formula 105
full-rank optimum STAP 351, 358-9
full-rank statistical method 347, 363-7
Fundamental Principles of Equilibrium and Movement 148

G

Gabor, D. 114, 117, 121, 126, 132, 157
gain 41-3, 119-21
gain combining 171, 172-3
Gauss, Karl F. 2
Gauss's law
of electrostatics 8-9
of magnetostatics 9-10
generalized sidelobe canceller (GSC) method 347, 353
Gibbs, J. Willard 150 , 152-3, 160, 162
global positioning system (GPS) 124, 128
gravity 162
Guerci, J. R. 348
Gupta, I. J. 372

H

Haardt, M. 413
Hankel matrix 256 , 415, 418-9, 433, 481
Hankel structure 216, 225
Hansen, R. C. 43
Hartley, R. V. L. 117, 123-4, 125, 126, 127, 128, 129, 147-8, 156-7, 158, 159, 160, 164
Hartley's law 133, 136, 137, 147, 148, 156-7, 159
Hata, M. 50-1
Hawking, Stephen 162-3
Heaviside, Oliver 3, 4
Helmholtz, Hermann Ludwig Ferdinand von 150
Hertz, Heinrich 3-4
Hertzian dipole 25-30, 48-9
Hua, Y. 412
Huang, K. C. 413
Huygen's principles 211

I

IBM 162
impulse radiating antenna (IRA) 97-8
impulse response 1, 49
information content 114-6, 118, 124-30
information entropy 116, 125
information symbols 169
information theory 116, 157
  history of 156-64
in-situ antenna element patterns 372, 386
interference 115, 119, 122, 123, 127, 129, 131, 137-40, 146, 157, 158, 170, 172, 190, 197, 214, 250, 255
interferer 47
interleaving 169
inverse Fourier transform (IFFT) 64

J

Jammer 215, 218, 228, 229, 232, 348, 372
  Blinking 209
  mathematical model of 350
Jet Propulsion Laboratory 161
Jupiter 157

K

King, R. W. P. 60
Kronecker product 349, 424
Ksienski, A. A. 372
Kumeresan, R. 257
Küpfmuller, Karl 126, 155, 164

L

Lagrange multiplier 213
Landauer, Rolf 162
Lavoisier, Antoine 148
least squares 232-8
least squares optimization, non-conventional 464-77
Lewis, Gilbert Newton 118
L-H polarization 408-10
light meter 65
line of sight (LOS) MIMO systems 199-203
Lodge, Oliver 4
Lord Kelvin 149, 155
Lorentz's Reciprocity Theorem 64-5
Lorenz gauge condition 13
Lovelock, James 161-2
Lundheim, L. 159

M

magnetic field intensity 26, 33
magnetic vector potential 26, 30-1, 33, 34-5,
magnetostatics 9-10
mainlobe 323, 371
Massachusetts Institute of Technology 157
matched filter 209, 210-2, 214-5
matrix pencil 230, 306, 524
matrix pencil method (MPM) 414-7, 430, 437, 441-2, 462
  broadband DOA estimation using the 479-519
  comparison with ESPRIT 428-33
  two dimensional (2-D) 411, 412-3
maximal ratio combiner (MRC) 169, 173
Maxwell-Poynting theory 114, 132, 134, 141
Maxwell, James Clerk 2, 150, 155
Maxwell-Boltzmann statistics 153-4
Maxwell-Heaviside-Hertz equations 3, 4-10
Maxwellians, the 4
Maxwell's demon 118, 155, 162, 164
Maxwell's electromagnetic theory 2-4
Maxwell's equations 10-5, 32, 39, 43, 48-9, 51, 64, 130, 134, 139, 143, 145, 147, 184, 199, 201

history of development and acceptance of 2-10
microwave 61
minimum norm 256-8
minimum variance unbiased (MVU) estimator 492
modulation alphabet 129
modulation symbols 129
monocycle input pulse 61
Moore-Penrose pseudo-inverse 484-5, 489
Multichannel Ariborne Radar Measurement (MCARM) 238-46
multipath fading 168, 171, 191-2, 206, 389-90, 323-45
multiple signal classification (MUSIC) 412
multi-input-multi-output (MIMO) 41, 167-206
case studies of 189-99
beamforming 174
electromagnetic nature of 183-9
-ODFM 171
polarization adaptivity in a near field,environment 389-410
multiple-input-single-output (MISO) 172, 197-9, 205
multiplexing gain 173, 176, 182-3, 191-4, 206
mutual coupling 183, 201, 246, 305, 437, 440, 446-7, 448-9, 451, 454-6, 459, 461, 474, 510-1, 514
approximate compensation for 371-86

N

NASA 161
National Institute of Standards and Technology (NIST) 103
near field 29-30, 35-6, 40-1,
negantropy 118
Newcomen, Thomas 148
noise 119, 122, 124, 251, 256, 257, 258, 259, 260, 274, 304, 373
historical importance of, in information tranmisison capacity 157-60
nonlinear conjugate gradient 322
nonlinear transmission line (NLTL) 102
Nossek, J. A. 413
nuclear magnetic resonance imaging 412, 417
Nyquist sampling 475
Nyquist, Harry 121, 123, 125, 126, 127, 128, 129, 132, 156-7, 158, 159, 160, 164

O

Oersted, Hans C. 2
Okamura, T., electromagnetic field measurements of 49-52
On Kinetic Theory of Gases 150
On the Average Distribution of Energy 150
On the Equilibrium of Heterogeneous Substances 152
optimum filters 210-5
orthogonal frequency division multiplexing (OFDM) 170, 171
orthogonal mode 187-8
output energy filter 209-11
output energy filter 213-4, 215

P

Palmer, Tim 161
parallel decomposition 178, 191
parallel processing 324
parametric spectral estimation 221, 236, 281, 291, 310, 327, 337, 376
Park, S. 535
pencil of matrices 488-90
pencil parameter 481
Penrose, Roger 162
Pentium PC 244
perfect electric conductor (PEC) 196-7
phase sweeping 173

phased array 185, 201, 240, 371, 437
EIRP degradation of 542-4
of electrically small antenna 438-9
theory,
acoustic 46-7
electromagnetic 47-9
phase-only weight control 275
picket fence effect 461, 467
Picosecond Pulse Laboratory 102, 105
Pisarenko 256
Planck, Max 153-4
point source 30
Polak-Ribière formula 309
polarization 389-410
pole-paring, 2-D UMPM 423-4
port admittance matrix 386
power density 32
power flow density 27-8
power spectral density 121, 126, 127
Poynting vector 27-8, 29, 126, 130, 390
principle component generalized sidelobe canceller (PC-GSC) technique 355, 360, 363-7
probability theory 44
Prony 256-7
propagation modeling 49-57
pseudo-inverse 421
pulse repetition interval (PRI) 227, 536

Q

QPSK constellation mapping 192-3
QR decomposition 353
quadrature carriers 129
quantum theory 154

R

Radar 211, 216, 226-7, 250, 259, 274, 347-8
airborne 227, 231, 276
imaging 412, 417
radiation 113, 114
random position errors 538-41, 545-9
ray tracing 52-7
real amplitude-only nulling algorithms (RAMONA) 276
real time 250, 386, 411, 461,
real weights 279, 289-98
received signal level 127
receiver sensitivity 126-7, 129
reciprocity 389-90, 391, 392-4, 395, 406
reciprocity theorem 15, 194, 372, 378
reduced-rank STAP 352-6, 368
relative importanceof the eigenbeam (RIE) method 347, 352-3, 358, 360, 363-7
Rényi, Alfréd 162
reradiated fields 372
resistive loading 59, 60, 103
resistive loading profile 60, 62-3, 65-83
R-H polarization 407-10
rich scattering 176
Root-MUSIC 412

S

Sangruji, N. 535
Sarkar, T. K. 412, 535
Saturn 157
Savery, Thomas 148
scalar power density 28
Schrödinger, Erwin 151, 161
selection combining 171
Shannon Channel Capacity Theorem 43, 118-25, 133, 135, 136-8, 139, 141, 142, 145, 146, 147-8, 185-6
Shannon, Claude E. 115, 117, 118-9, 122, 123, 124, 125, 126-7, 128, 129, 156, 157, 158-61, 164, 176
Shuffling matrix 420
Sidelobe 226, 239, 283

signal cancellation 279
signal enhancement , adaptivity on transmit 391-5
signal of interest (SOI) 47, 216-8, 219, 221, 226, 228-9, 230-2, 233-5, 249, 250, 251-3,254, 255-9, 274, 278-9, 305-8, 311-8, 323-4, 345, 348, 349, 372, 373-4, 378
D3LS extraction of 522-8
signal processing 257
adaptive algorithms 303, 305
adaptive array 323
signal to interference plus noise ratio (SINR) 219, 226, 348, 351, 353-4, 368, 379-85, 535, 549
signal to noise ratio (SNR) 115, 121, 126, 128, 129, 157, 160, 169-72, 173, 181-2, 206, 209, 211, 259, 263, 265, 268, 411
signaling alphabet 156
single-input-multiple-output (SIMO) 172, 189, 191
single-input-single-output (SISO) 167-8, 174, 175, 176, 177, 178, 189, 190, 197, 200, 202, 204
comparison with MIMO 184-9
singular value decomposition (SVD) 174, 353, 405, 416-7, 420, 424, 430, 477, 481
snapshot 209, 216, 217-8, 222, 223-4, 226, 228-9, 238, 246, 251-2, 323, 372, 411, 412, 460, 461-2, 480, 486, 511
Sommerfeld, Arnold 39, 51-3
sonar 183
space diversity 170-1, 172

space-time adaptive processing (STAP) 210, 249, 255, 263, 273, 276, 318-21, 324, 521
comparison of D3LS and statistical methods for 347-68
D3LS 226-8, 535-49
D3LS phase-only 332-44, 345
statistical methods 347-8, 351-6
space-time weights 233
spatial diversity 43-6, 199, 390, 391
spatial mode 201, 202, 203, 204, 205
spatial selectivity 1
spatial smoothing 412
spectral estimation 258
speed of light 493
steam engine 148, 149, 163
steering vector 249, 250, 251, 279, 290, 304, 349, 350, 356-7, 373, 461, 462, 463, 464, 467
stochastic method , JDL 245-6
stochastic processes 160
Stoke's theorem 5
string theory 163
Strominger, Andres 163
subarray 235, 253, 526
superdirectivity 41-3
supermatrix 510 , 514
superposition 394-5
symmetric and amplitude only control (SAOC) 276
system capacity 114-5, 122
Szilárd, Leo 118, 155-6

T

Tai, C. T. 41
thermal noise power 185
thermodynamics 117-8
history of 147-56
MaxEnt school of 160
Thomson, William 149, *see also Lord Kelvin*
time diversity 169
time division duplex (TDD) 171
time domain reflectometry (TDR) 104, 107
Tokyo 49
training data 347, 348, 351, 355, 356, 358, 359, 363, 364, 367, 368
transformation matrix 437, 439-42, 449, 454, 458-9, 510, 514
traveling-wave 66

traveling-wave antenna 59, 62-3
Tsallis, Constantino 162
Tufts, D. W. 257
Tuller, W. G. 123, 125, 126, 127, 128, 129,132, 157-8, 159, 164

U

ultrawideband (UWB) antenna 59-60
  experiments of Dr. James R. Andrews 102-9
  performance of various 83-101
uncertainty 125, 126-7, 153, 157, 158-9, 161
uniform linear array (ULA) 414, 454, 463
uniform linear virtual array (ULVA) 375, 440, 510
unitary matrix 413-4, 481
unitary matrix pencil method (UMPM) 417-22
  2-D 422-8
unitary transform 411, 412, 422, 433
unmanned aerial vehicles (UAV) 477

V

Vafa, Cumrun 163
velocity of light 26, 486

W

water-filling algorithm 179-80
wave number estimation 412, 417
waveguide 176
weight vector 232, 279, 303, 336, 348, 351, 353, 375, 524
  phase-only 308-9
Weiner filter 209, 210, 212-3, 214-5
Weiner solution 354
Weiner, Norbert 158
What is Life? 161
Wikipedia 148
wireless communications 37, 49, 113, 114
Wu, T. T. 60

X

Y

Yeh,C.C. 413

Z

Zaghloul, A. I. 539
Zeilinger, Anton 161